EXPERT SYSTEMS

A Software Methodology for Modern Applications

EXPERT SYSTEMS

A Software Methodology for Modern Applications

Peter G. Raeth

IEEE Computer Society Press Reprint Collection

Expert Systems
A Software Methodology for Modern Applications

Peter G. Raeth
Wright Research and Development Center
Dayton, Ohio

IEEE Computer Society Press
Los Alamitos, California

Washington • Brussels • Tokyo

Published by

IEEE Computer Society Press
10662 Los Vaqueros Circle
P.O. Box 3014
Los Alamitos, CA 90720-1264

Cover designed by Jack I. Ballestero

Printed in United States of America

IEEE Computer Society Order Number 1904
Library of Congress Number 89-46387
IEEE Catalog Number EH0303-8
ISBN 0-8186-8904-8 (casebound)
ISBN 0-8186-5904-1 (microfiche)
SAN 264-620X

Additional copies may be ordered from:

IEEE Computer Society
10662 Los Vaqueros Circle
P.O. Box 3014
Los Alamitos, CA 90728-2578

IEEE Service Center
445 Hoes Lane
P.O. Box 1331
Piscataway, NJ 08855-1331

IEEE Computer Society
13, Avenue de l'Aquilon
B-1200 Brussels
BELGIUM

IEEE Computer Society
Ooshima Building
2-19-1 Minami Aoyama
Minato-ku, Tokyo 107, JAPAN

THE INSTITUTE OF ELECTRICAL AND ELECTRONICS ENGINEERS, INC.

Dedication

This book is dedicated to my wife, Marilyn, and to the rest of my family. They are the visible sign on earth of my ultimate goal.

Preface

The physical realization of the more general field of artificial intelligence is most often found in expert systems, those systems that have the knowledge and expertise of humans built into software and data. Compared with conventional software systems, which are very procedurally oriented and code intensive, expert systems are knowledge oriented and use relatively little code. The power of an expert system is found in the knowledge base (the database that contains the facts making up a domain of human knowledge) and not in extensive procedural code.

Due to the radical difference in the way expert systems are constructed, compared with conventional systems, expert systems are used easily in application domains that do not lend themselves to implementation or maintenance via conventional software techniques. This factor makes expert systems an important technology. Expert systems are a subfield of artificial intelligence that is real, understandable, and manageable.

The automation practitioner who is well grounded in conventional computing techniques, and who may have taken a course or two in artificial intelligence, may still find it difficult to obtain material that effectively teaches this relatively new field of expert systems. Random sampling of the literature can leave a person confused as can some books that gloss over important basic subjects. Faced with this situation and automation problems not lending themselves to conventional computing methods, the editor spent considerable time acquiring theoretical and applications knowledge of expert systems development.

Contained in this reprint collection is the fruit of that labor. Here are the papers the editor found to be most useful. The papers are arranged in the following seven chapters: Introducing Expert Systems, Evaluating Expert System Shells, Application-Specific Techniques, Knowledge Representation, Knowledge Acquisition, Reasoning Methods, and Issues and Commentary. These chapters guide the reader from fundamental ideas, through details, and on to issues affecting the use of expert systems technology. Those who want to add the computing methods of artificial intelligence to their expertise will find ready support in these papers.

For those needing a broad introduction before buying an expert systems development tool, Chapters 1 and 2 provide the fundamentals. Chapter 3 presents a wide range of successful applications. Chapters 4, 5, and 6 cover incorporating expert systems techniques in existing or new software. Chapter 7 ties all the previous chapters together by discussing issues affecting the use of software in expert systems techniques.

Finally, an appendix lists those books that provide more in-depth discussions and guidance during the early experiments.

Acknowledgments

A book such as this never sees the light of day unless a large team of people give their eager support. My thanks go to the following groups and individuals:

The staff and reviewers of the IEEE Computer Society Press for their comments and overall guidance. This especially applies to Editor-in-Chief Ez Nahouraii and Editorial Board Member Frederick Petry.

Mary Murphy and Darlene Goldsby of the Eglin Technical Library, Doris Mullins and Carole Steele of the Eglin Base Library, and Ron Lundquist and Dan Sell of the Wright-Patterson Technical Library for their help during the extensive literature search.

Doug Nation and Vicky Deiter of the 3246th Test Wing for giving this book its first review.

H. W. Wilson Company for the excellent support given to users of the Wilsondisc product.

Scott Cline and Jan McKoy of the Wright Research and Development Center for their assistance in automating the entry of hard copy to disk files.

Debbie Ables of the Wright Research and Development Center for her administrative support.

Table of Contents

Dedication v

Preface vii

Acknowledgments viii

Chapter 1: Introducing Expert Systems 1

Expert Systems, Knowledge Engineering, and AI Tools—An Overview 2
C. Williams (IEEE Expert, Winter 1986, pages 66-70)

Introduction to Expert Systems 7
W. Myers (IEEE Expert, Spring 1986, pages 100-108)

The Basic Principles of Expert Systems 17
W.B. Gevarter (Extracted from National Bureau of Standards Report No. NBSIR 82-2505, May 1982)

Chapter 2: Evaluating Expert System Shells 33

The Nature and Evaluation of Commercial Expert System Building Tools 34
W.B. Gevarter (Computer, May 1987, pages 24-41)

Selecting a Shell 52
R. Citrenbaum, J.R. Geissman, and Roger Schultz (AI Expert, September 1987, pages 30-39)

Expert System Benchmarks 62
L. Press (IEEE Expert, Spring 1989, pages 37-44)

27-Product Wrap-Up: Evaluating Shells 69
R. Freedman (AI Expert, September 1987, pages 69-74)

Exploring Expert Systems 76
V. Barr (Computer, November 1988, pages 68-73)

Two PC-Based Expert System Shells for the First-Time Developer 82
P.G. Raeth (Computer, November 1988, pages 73-81)

Chapter 3: Application-Specific Techniques 91

Intelligent Tutoring Systems: An Overview 92
M. Yazdani (Expert Systems, July 1986, pages 154-162)

Automating the VLSI Design Process Using Expert Systems and Silicon Compilation 101
A.C. Parker and S. Hayati (Proceedings of the IEEE, June 1987, pages 777-785)

Knowledge-Based Decision Support in Business: Issues and a Solution 110
V. Dhar and A. Croker (IEEE Expert, Spring 1988, pages 53-62)

PlanPower: A Comprehensive Financial Planner 120
J.L. Stansfield and N.R. Greenfeld (IEEE Expert, Fall 1987, pages 51-60)

Expert System Technology for the Military: Selected Samples 130
J.E. Franklin, C.L. Carmody, K. Keller, T.S. Levitt, and B.L. Buteau (Proceedings of the IEEE, October 1988, pages 1327-1366)

An Expert Systems Approach to Decision Support in a Time-Dependent, Data Sampling Environment 170
P.G. Raeth (Proceedings: IEEE National Aerospace and Electronics Conference [NAECON], 1988, pages 1200-1207)

Chapter 4: Knowledge Representation 179

Important Issues in Knowledge Representation 180
W.A. Woods (Proceedings of the IEEE, October 1986, pages 1322-1334)
Constructs and Phenomena Common to the Semantically-Rich Domains 193
B. Adelson (International Journal of Intelligent Systems, 1986, Volume 1, pages 1-14)
Representation and Use of Metaknowledge 205
L. Aiello, C. Cecchi, and D. Sartini (Proceedings of the IEEE, October 1986, pages 1304-1321)
Procedural Knowledge 223
M.P. Georgeff and A.L. Lansky (Proceedings of the IEEE, October 1986, pages 1383-1398)
Nonmonotonic Inference Rules for Multiple Inheritance with Exceptions 239
E. Sandewall (Proceedings of the IEEE, October 1986, pages 1345-1353)
Maintaining Knowledge about Temporal Intervals 248
J.F. Allen (Communications of the ACM, November 1983, pages 832-843)
A Software Engineering Tool for Expert System Design 260
J.M. Francioni and A. Kandel (IEEE Expert, Spring 1988, pages 33-41)

Chapter 5: Knowledge Acquisition 269

A Formal Methodology for Acquiring and Representing Expert Knowledge 270
N.M. Cooke and J.E. McDonald (Proceedings of the IEEE, October 1986, pages 1422-1430)
Natural Language Interactions with Artificial Experts 279
T.W. Finin, A.K. Joshi, and B.L. Webber (Proceedings of the IEEE, July 1986, pages 921-938
Interactive Classification: A Technique for Acquiring and Maintaining Knowledge Bases 297
T.W. Finin (Proceedings of the IEEE, October 1986, pages 1414-1421)
A Case Study: Acquiring Strategic Knowledge for Expert System Development 305
D. Sharman and E.J.M. Kendall (IEEE Expert, Fall 1988, pages 32-40)
Choosing Knowledge Acquisition Strategies for Application Tasks 314
C.M. Kitto and J.H. Boose (Proceedings: WESTEX-87—Western Conference on Expert Systems, 1987, pages 96-103)

Chapter 6: Reasoning Methods 323

Rete: A Fast Algorithm for the Many Pattern/Many Object Pattern Match Problem 324
C.L. Forgy (Artificial Intelligence, Volume 19, 1982, pages 17-37)
A Formal Logic of Plans in Temporally Rich Domains 343
R. Pelavin and J.F. Allen (Proceedings of the IEEE, October 1986, pages 1364-1382)
Analogical Problem-Solving and Expert Systems 362
L.B. Eliot (IEEE Expert, Summer 1986, pages 17-28)
Processing of Semantic Nets on Dataflow Architectures 373
L. Bic (Artificial Intelligence, Volume 27, 1985, pages 219-227)
Depth-First Iterative Deepening: An Optimal Admissible Tree Search 380
R.E. Korf (Artificial Intelligence, Volume 27, 1985, pages 97-109)
Inexact Reasoning in Expert Systems—An Integrating Overview 391
R.J.P. Groothuizen (National Aerospace Laboratory NLR, Report No. NLR TR 86009 U, January 1986)
Fuzzy Logic 407
L.A. Zadeh (Computer, April 1988, pages 83-93)

Chapter 7: Issues and Commentary 419

Towards a Science of Expert Systems 420
P.J. Denning (IEEE Expert, Summer 1986, pages 80-83)
Knowledge Engineering in Practice 424
W.G. Rolandi (AI Expert, December 1986, pages 58-62)
Maintenance and Language Choice 430
M.H. Matthews *(AI Expert,* September 1987, pages 42-48)
Testing Your Knowledge Base 438
B. Marcot *(AI Expert,* August 1987, pages 42-47)
Verification and Validation of Expert Systems 444
C.J.R. Green and M.M. Keyes (Proceedings: WESTEX-87—Western Conference on Expert Systems, 1987, pages 38-43)
Using Expert Systems: The Legal Perspective 450
J.S. Zeide and J. Liebowitz (IEEE Expert, Spring 1987, pages 19-22)
Teaching an Introductory Course in Expert Systems 454
A.T. Bahill and W.R. Ferrell (IEEE Expert, Winter 1986, pages 59-63)

Appendix: Further Reading 459

About the Author 460

Chapter 1: Introducing Expert Systems

Before using expert systems technology it is important to become well grounded in the basic principles. The three articles reprinted here provide a detailed explanation of the major topics. Together they offer the reader all the important concepts and a foundation for the rest of the book. Williams leads with an excellent overview. Myers continues by going beyond basic concepts to include a discussion of languages, hardware, and systems. The last article is by one of the deans of expert systems. Gevarter offers a detailed discussion of inferencing techniques, knowledge representation, and knowledge acquisition. He also covers architectures and applications.

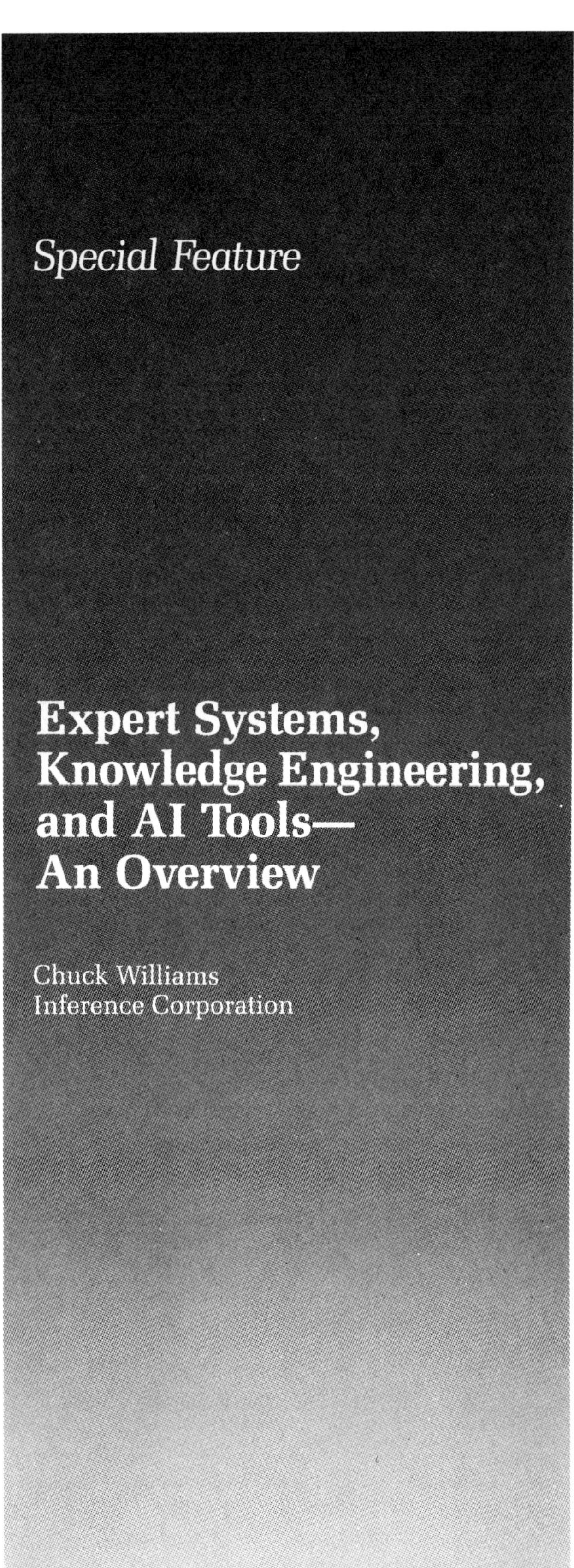

Expert Systems, Knowledge Engineering, and AI Tools—An Overview

Chuck Williams
Inference Corporation

Two decades in research labs and universities have guided AI technology to commercial thresholds. New applications demonstrate AI's tremendous potential to bring inexpensive processing power to tasks once thought beyond computer capability. Effects on both business and industry will be widespread, for AI technology can become a major competitive force in virtually every marketplace.

Successful commercial applications such as Digital Equipment Corporation's XCON configuration advisor for the VAX line have caused major corporations to recognize AI technology's potential; many have either initiated AI research and development groups or are actively integrating AI into their business and products. Many applications—financial management support and manufacturing planning systems, computer-aided engineering workstations, defense systems and resource scheduling tools—will soon incorporate AI technology. And AI technology materializes most often in expert systems—computer programs capturing and applying a human expert's knowledge to diverse tasks ranging from medical diagnoses to financial planning.

Like human experts, expert systems attempt to reason within specific knowledge domains. Since they are not subject to human frailties such as boredom, forgetfulness, or tunnel vision, fully developed expert systems frequently outperform their human models. Companies incorporating such systems into their products, services, or internal operations will realize distinct competitive advantages.

As products become smarter, they become easier to use and more helpful. For example, newspaper layout and press production tools are incorporating expert systems to automate design. Credit authorization decisions for many types of purchases and loans will soon be made by expert system analysis including credit history stored in databases. Some companies investing in expert systems have already realized significant paybacks in terms of streamlined internal operations. DEC's XCON expert system, mentioned above, saves $18 million a year.

Expert systems—a new software technology

Some key differences exist between expert systems and conventional software programs. For one, they require fundamentally different developmental approaches. Because building an expert system differs from building conventional software, potential developers must familiarize themselves with knowledge engineering.

Conventional software programs deal with problems that have been solved beforehand; that is, conventional software technology aims at turning known procedures or algorithms into code. Situations to be considered, decision points, and responses are all identified during the design process. Ideally, conventional software development's top-down approach minimizes changes needed later in the coding stage. The top-down approach is critical to conventional software development because it is so difficult to change system design once coding has begun. Due to the many interdependencies among control paths, data values, state sequences, and variable instances, too much change makes programmers quickly lose control of the change process—causing delays and escalating development costs.

Unfortunately, top-down methodology fails when significant design errors occur or (as is more often the case) when application specifications change during design and implementation. This results in cost and budget overruns, and unstructured design fixes reducing software quality and reliability. Moreover, unforeseen changes or enhancements usually surface after applications are in the field. Using conventional technology, difficulties in accommodating changes after products are released accounts for high software maintenance costs.

Creating expert systems, on the other hand, differs from conventional software's top-down approach. Far exceeding hard facts and rules, expert systems must encode subtle, heuristic forms of knowledge including symbols, strategies, and relationships. Representing such concepts is extremely difficult with conventional programming languages like Pascal or C. But even more important, no one understands beforehand exactly how human experts form conclusions—not even the experts themselves. Therefore, developing expert systems is never a linear process. Based upon repeated interviews with human experts, systems attain increasing expertise in successive iterations. Design and coding are integrated into a single iterative process.

Human experts have two important qualities in common: (1) considerable knowledge in specific domains, and (2) effective strategies for quickly sorting through that knowledge when faced with problems—strategies consisting of deductive rules, formal heuristics, and intuitions (or gut feelings). In all cases, human experts solve problems by efficiently applying a storehouse of domain-specific knowledge and experience.

Computer-based expert systems approximate expert reasoning through two major components: (1) a knowledge base containing the expert's broad range of information, and (2) an inference engine to apply that knowledge efficiently. Knowledge bases contain facts, rules, and relationships derived from expert experience relating directly to specific applications. Implicit in the knowledge base concept is a knowledge representation language enabling programmers to map human expertise into computer-usable form.

In traditional programming, we expend considerable effort guiding the computer's application of domain-specific knowledge (that is, control). Indeed, this effort substantially increases the complexity of modifying traditional programs—because you must ensure that change conforms to procedural conventions or restrictions implemented around the programs changed. Inference engines apply rule-based knowledge opportunistically; that is, they consider an inference whenever an inference can be made. Once the knowledge engineer has entered relevant knowledge into the knowledge base, the inference engine automatically identifies and considers the knowledge applicable to each given situation. Uncoupling knowledge from its application makes a data-driven system much easier to modify as the knowledge-engineering process evolves.

A major challenge when building expert systems is determining and encoding expert knowledge—difficult processes because relevant knowledge is not always immediately apparent. Overall knowledge structure and key concepts may not be obvious at first. Furthermore, experts may relate general rules but forget critical special cases until actually faced with exceptional situations. For these reasons, expert system development necessarily becomes an iterative process of successive refinement. Expert system development tools facilitate evolutionary programming much better than traditional software-engineering tools do.

A key player—the knowledge engineer

So far we have discussed only the expert—the knowledge source. But who turns that knowledge into a working expert system? That challenge requires a knowledge engineer. Knowledge engineers (1) define the expert system domain, (2) elicit desired information from human experts, (3) incorporate that knowledge into acceptable form for the knowledge base, then (4) test the system to evaluate its robustness and accuracy.

Because expert system development contrasts so sharply with traditional software engineering, experienced programmers do not necessarily make the best knowledge engineers. In fact, rather than emphasizing specialized programming expertise, a knowledge engineer's most desired skills are the abilities to think rationally and communicate well. The traditional system analyst more closely resembles our knowledge-engineering ideal, although scientists and engineers in specialties other than computer science often make excellent knowledge engineers.

Experienced knowledge engineers today are few, numbering perhaps in the hundreds. However, demand is growing quickly as corporations become convinced of expert system benefits. We will soon face a tremendous shortfall of qualified personnel—a bottleneck we can overcome by introducing powerful, productivity-enhancing development tools.

Our need for industrial-strength tools

Designers created traditional software languages to implement solutions based on clearly defined algorithms; their control structures, computational facilities, and procedural orientation reflect this. To map expert knowledge into computer-usable form, however, knowledge engineers need different programming elements—new elements forming a knowledge representation language that can express knowl-

Expert systems are research tools— and as products become smarter, they become easier to use.

edge comprised of numerous facts and heuristics rather than rigid algorithms.

Essentially, knowledge engineers must express in the formal knowledge representation language provided by the expert system development tool what they discover by interviewing experts. Depending on the tool used, a knowledge engineer can develop the operational skill required to perform this job in as little as six to eight weeks. Of course, much depends on the language's expressiveness; the more naturally concepts can be mapped into knowledge representation language, the easier the task. Rich representation constructs in the development tool improve productivity since they free knowledge engineers from having to create artificial constructions when symbolizing concepts.

Originally, researchers developed expert systems in AI languages like Lisp. Nonetheless, their efforts generally led to special languages built on top of Lisp to help them deal more directly with knowledge constructs and reasoning mechanisms. These special metalanguages, called expert system frameworks, have everything required for a working expert system except domain-specific knowledge. Expert system frameworks make knowledge engineering much easier by allowing high-level information coding that more closely resembles the way human experts express knowledge. Also, sophisticated AI techniques such as search strategies and knowledge base maintenance are conveniently packaged; in effect, knowledge engineers do not have to reinvent the wheel.

Today, many companies are investing cautiously in AI. Unfortunately, those acquiring overly simple expert system development tools may find their caution counterproductive. Most new knowledge engineers will quickly outpace today's less sophisticated expert system tools. A preferable strategy provides robust tools that continue to serve well as novice knowledge engineers expand their interest and capabilities for solving more significant problems. This approach preserves the knowledge base investment developed during earlier prototyping, and also eliminates the retraining needed when development tools must be upgraded—a situation that could occur in as little as six months.

Major programming elements needed for efficient knowledge engineering include

- Symbolic representation,
- Knowledge frames,
- Rule and logic programming,
- Forward and backward search strategies,
- Viewpoints, and
- Truth maintenance.

Symbolic representation. The heart of any expert system is its symbolic model of the domain in which it reasons and of the particular problem instance it currently faces. A symbolic model contains the entire state of a data-directed expert system. Unlike conventional procedural programming, no notion of controlling a "program counter" exists in the executing program. Therefore, the symbolic model's richness determines how deeply a program understands both the current problem situation and its own current processing context (its objective, for example, or where it is in its analysis). Good symbolic representation supports arbitrary *n*-ary relationships among objects noted by symbols, and provides for taxonomic organization of the object and relations. Symbolic representation is complemented by a pattern-matching language allowing reasoning rules to identify substructures within the symbolic model that suggest problem-solving actions such as drawing an inference, asking a question, or updating a display.

Knowledge frames. The knowledge frame (or schema) represents a major feature of expert system frameworks. Schemas enable engineers to organize knowledge as object collections sharing common properties. Because schemas are hierarchical, objects can inherit characteristics from a general object class and then add their own specific attributes. By allowing knowledge abstraction, the schema mechanism effectively manages complexity. Knowledge engineers can represent abstract concepts directly as a description of properties common to each abstract class instance. Underlying detail is available to the inference engine during expert system execution through successively elaborate descriptions of more concrete concepts.

Rule and logic programming. Expert systems are data- rather than procedure-oriented. Instead of prescribing action sequences, knowledge engineers represent knowledge by stating facts and using rules to define relationships. A rule's first part specifies a condition or situation. Its second part declares a result that may be an action or a change to the knowledge base. When the rule fires, it is dependent only on the data at execution time—knowledge engineers neither know nor care exactly when a rule will be activated. Data selects appropriate procedure (rule) instead of procedure selecting appropriate data (exactly opposite to traditional procedural programming). In each situation, the inference engine automatically accesses all relevant knowledge independent of processing state.

Logic programming can be viewed as a restriction of rule-based programming, where each rule application corresponds to a logically valid inference; in logic programming, data modifications or retractions typically do not occur. We call this monotonic logic. Viewpoints, as described below, allow nonmonotonic reasoning within these restrictions.

Forward and backward chaining. For several millennia, scholars have debated the relative merits of Socratic dialectic versus Aristotelian rhetoric—of inductive versus deductive logic. For some years, an ongoing controversy has also been waged over the best of two methods for locating specific information in the knowledge base—forward or backward chaining. Like inductive logic, forward chaining reasons forward from existing facts and rules to derive additional facts that must hold, while following all possibilities suggested by the data. Like deductive logic, backward chaining reasons backwards from a given goal, searching the knowledge base for facts or rules supporting that goal and declaring them true.

Both methods work well. Many expert systems incorporate one or the other. But a synergistic effect occurs when we integrate both into one inference engine. Backward chaining acts to focus or constrain forward chaining to a knowledge base subset determined by known goals. Forward chaining then swiftly reasons to conclusions or generates new goals continuing the process. This combined forward-backward action enables expert systems to converge on answers much more quickly and efficiently than if either method were used alone.

Viewpoints. Expert systems typically must reason under assumptions or about situations involving uncertainty. Viewpoints free these systems to compare and contrast multiple hypothetical alternatives until one alternative can be selected with a degree of certainty. Thus, when a situation suggests two possible outcomes or interpretations, two viewpoints are automatically generated to pursue subsequent reasoning about both possibilities.

Besides allowing expert systems to trace multiple paths of reasoning simultaneously, the viewpoint mechanism keeps track of assumptions and successes along the way. When a viewpoint leads to an obvious contradiction, it is eliminated from consideration. When one viewpoint is clearly superior, it is selected and the competing alternatives are abandoned. Rules coded into the knowledge base determine what viewpoint, among competing viewpoints, should be selected.

Truth maintenance. When earlier assumptions are proven false, conclusions that followed from those assumptions are automatically marked as erroneous and excluded from believed facts or conditions. Expert systems can deal with changes over time as a result, altering conclusions in response to new data. This is another instance where expert systems are truly data-driven since they respond to data changes rather than blindly following a program counter within a procedural sequence.

Truth maintenance, an important characteristic of sophisticated expert systems, also applies to multiple viewpoints. When generated by conclusions that subsequently become false due to changing conditions, as described above, viewpoints will be deleted automatically.

Knowledge engineering in practice

First, knowledge engineers must consider whether proposed applications are suitable expert system candidates. One affirmative indicator is when traditional programming techniques prove insufficient because either the problem is poorly understood or the concepts are too abstract or complex. A second affirmative indicator arises when the problem to be addressed changes substantially as the business environment evolves. Expert system applications should satisfy these additional criteria:

- Applications must have well-defined domains;
- One or more experts must have the knowledge required (experts are people widely recognized as performing the desired task in a superior manner); and
- Those experts must be able to verbalize desired task performance.

Knowledge engineers begin by (1) interviewing experts and isolating key concepts and tasks involved in the target domain, thereby creating appropriate schemas (or abstractions) within which to represent expert knowledge, and (2) identifying an appropriate reasoning architecture within which to solve problems. Knowledge engineers then enter true statements about the general environment and the special area of interest that must be understood to solve the problem. Schemas provide a terminology to express these statements. Reasoning architecture defines the required type and form of statements. Pertinent statements include facts, deductive and inductive rules, heuristics, and strategies normally employed by the expert. Since each statement is logically independent (with no ordering of the statements and no statement explicitly referencing any other statement), great freedom exists in system development methodology and in the order of knowledge elicitation and encoding. Modularization, in terms of problem-solving tasks or knowledge specialization areas, frequently evolves during system implementation. Each statement can be thought of as a declarative assertion; the inference engine takes care of proceduralization automatically by automatically noticing the conditions making a statement relevant when these conditions occur.

Once knowledge engineers have built credible prototypes, they exercise them—presenting test situations and observing the results. Because expert system execution approximates normal human expert activity, we often refer to this execution as emulation. During emulation, we ask human experts to critique an expert system's reasoning and conclusions. In this way, we can detect exceptions, misunderstandings, logical errors, and other faults. Critical for effective evaluation, we need a good tracing mechanism to monitor progress as our prototype applies its knowledge.

A powerful development environment aids expert system refinement by simplifying modification via special user interface features. Such features include

- Access to several types of program information through multiple windows,
- Menus and a pointing device for easy system control,
- Simultaneous viewing of evolving knowledge base contents and execution sequences,
- Graphical knowledge schema representation with hierarchical inheritance relationships, and
- Facilities to incrementally add, delete, or modify any knowledge-base statement at any time.

Refinement continues until we achieve the required breadth and accuracy—a judgment often made by human experts and by comparing system performance to past decisions and their outcomes.

Software maintenance is not a critical issue with expert systems because we consider evolution an integral part of development. Since we can modify systems simply by changing or adding facts and rules, change has minimal impact. Due to the data-driven nature of expert systems, knowledge engineers need not consider procedural effects such as rule order (although they must with conventional software).

Knowledge delivery systems

Transporting expert systems from development lab to end user introduces additional concerns such as performance, user interface, and integration with existing computing resources. Expert systems, because they do so much real-time processing to apply their large knowledge bases, generally require significant computer power. Currently, two factors make expert systems practical: the falling cost of computing power, and the compiled knowledge base. The first provides more raw horsepower per dollar, while the second makes expert systems efficient enough to run on smaller computers. Expert system compilers process knowledge bases into more efficient form, doing considerable work beforehand, so that applications can proceed more quickly.

For end users to maximize expert system advantages, a convenient user interface must be provided. Expert system tools with interface features such as support for mixed-initiative interaction, interactive graphics, and window management relieve much of the effort required to develop easy-to-use expert systems. We can enhance graphics abilities when, using hierarchical schemas, we define a descriptive lexicon of primitives (such as line, polygon, and area fills), attributes (line type, area type, and color), and transformations (rotate, scale, and translate). This allows complex graphical entities to be expressed as abstract graphical-icon schemas. The rule-based inference engine maps graphic displays to and from conclusions and steps in the reasoning process—thereby yielding a powerful mechanism giving users direct access to and control over expert system reasoning.

A tool's usefulness also depends on its ability to access other data and programs. Effective integration of expert systems into large computing environments requires methods to access programs written in conventional languages. Ideal expert system tools permit multilevel access to programming tools—starting with the expert system framework, dropping to a general-purpose AI language like Lisp, and finally accessing traditional, procedure-oriented languages like Fortran or C. These should be completely integrated so that any combination can be used simultaneously within a program.

While most expert systems will distribute knowledge already verified by human experts, we are increasing expert system modeling of proposed hypotheses to test for validity. In effect, expert systems are research tools. Just as engineers use modeling languages such as GPSS or Dynamo to simulate complex physical or social systems in terms of algorithmic relationships, they are using expert systems to model theories based on more abstract relations and heuristics. Once hypothetical knowledge bases have been modeled, they can be executed using real-world facts to see if they generate expected outcomes. Correlating expert system conclusions with actual observations indicates that the proposed knowledge is, in fact, accurate. Such applications could prove extremely valuable in amplifying the human researcher's reasoning power. ■

Chuck Williams is the executive vice president and chief technical officer at Inference Corporation, a Los Angeles-based commercial expert systems tool supplier. He has over 13 years experience in software systems design, concentrating on AI applications for 10 of those years. While at Information Sciences Institute in Marina Del Rey, California, Williams developed new paradigms for implementing AI tools. As a professor at USC, he taught the theory and design of compilers and graphics systems. Since the founding of Inference Corporation in 1979, Williams has led AI research and product development, participated in the company's management, and originated and enhanced the Automated Reasoning Tool product.

His address is Inference Corp., 5300 W. Century Blvd., Los Angeles, CA 90045.

Introduction to Expert Systems

Ware Myers
Contributing Editor

The line dividing intelligent from nonintelligent systems keeps shifting as computers and their software become more capable. When computers were new 35 years ago and could do little except arithmetic functions, some called them giant brains, implying intelligence. Today, we do not regard the giant brain's successor, the pocket calculator, as intelligent. In fact, we do not regard a payroll program of hundreds of thousands of lines of code, running on a mainframe now orders of magnitude more powerful than its ancestors, as intelligent.

There are current systems, however, that we do regard as intelligent. Expert systems are an example; they are a subset of intelligent systems.[1] Still, as the research community explores the problems that these programs address, the resulting systems will no doubt come to seem commonplace and hence something less than intelligent. The mystery of intelligence will continue to be a moving target.

What we mean by intelligence is broadly characterized by what human beings are able to do. As human beings we have an intuitive feel for what that is. Looking at the human brain from the outside, we have considerable knowledge of what it can do. From the perspective of inside-the-brain, however, we have taken only a few steps toward understanding how the neurons operate. We do not have enough knowledge at this level to create intelligent systems by simulating the detailed structures of the brain artificially. Still, in the 30 years since John McCarthy coined the term artificial intelligence, outside-the-brain approaches to intelligent systems, based on studying the nature of particular problems, have achieved some success. In the last few years, practical applications of expert systems have appeared.[2]

A working concept of what artificial intelligence embraces is found in the topics being studied by the artificial-intelligence community, summarized in Table 1. Many of these terms cannot be simply defined. For example, one book takes an entire chapter to define "expert system" and to set the boundaries of the concept.[3]

Fundamentals of expert systems

In Figure 1, we examine the relationships of one class of intelligent system, the expert system, to the environment in which it functions. On the right side of the figure, the expert system acquires knowledge through knowledge-acquisition software tools from a knowledge engineer who, in turn, acquires the knowledge from a domain expert. Domain refers to a specific field of knowledge. Once the expert system has been built, this relationship could be ended. On the other hand, since the domain normally continues to develop, the relationship usually continues.

On the left side, the expert system interfaces to the user. The user puts into the system facts and suppositions of varying degrees of probable truth and receives answers, recommendations, or diagnoses of some degree of reliability. The user usually works through a keyboard interface in a formal language. As natural-language capability becomes available, it can make this interchange more friendly. Eventually, speech recognition and generation may replace the keyboard.

At the bottom of the figure, a general database is shown interfacing to the expert system. At present, most expert systems depend upon a specialized knowledge base. They will become more usable with existing data processing systems when they become able to draw upon general databases, either in the immediate computer system or on a network.

Turning this relationship around, large, existing databases will become more useful when intelligence can be added to them. Query languages are beginning to add intelligence to the task of searching a database. At present, it may take an unreasonable amount of time and computer power to search an existing database. An intelligent front end capable of guided search could home in on information more quickly.

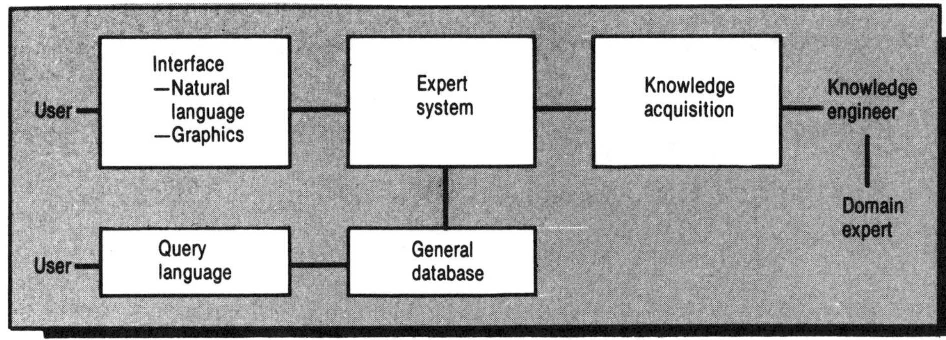

Figure 1. An expert system is related to the domain from which its knowledge is drawn, to the user for whom it solves problems, and eventually to a general database.

Expert system concept. In Figure 2, we divide the expert-system concept into a knowledge base and an inference engine. The knowledge base is unique to a particular domain, but the inference engine may be common to a number of domains that have similar characteristics. Thus, a number of inference engines were initially developed for particular applications, but were later separated from the knowledge specific to that application and used with other knowledge bases.

At the next level of detail, we identify four blocks: assertions, knowledge relationships, search strategy, and explanation tracing.

Assertions. This block, sometimes called the working memory or temporary data store, contains "declarative knowledge about the particular problem being solved and the current state of affairs in the attempt to solve the problem."[4] There are several ways to represent this data: first-order predicate logic, frames, and semantic networks.

Predicate logic represents the declaration Richard gave Jean a rose in the form *Give (Richard, Jean, rose).* A frame for this same information might be

name of frame: F1
type of frame: transfer of possession
source: Richard
destination: Jean
agent: Richard
object: rose

In the semantic network, each node contains an object and the lines between the nodes represent the relationships.

Knowledge relationships. This block contains formulas showing the relationship among several pieces of information. The most common formula is the production rule, such as

If it is clear and hot and muggy, then it is summer.

Here we have three antecedents connected by logic AND, which, when satisfied, lead to the consequence that it is

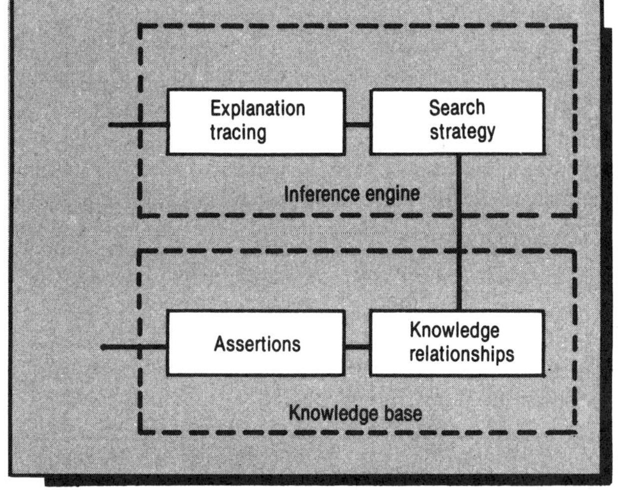

Figure 2. An expert system may be thought of as a knowledge base that an inference engine searches for answers to a particular problem, such as a diagnosis fitting a set of symptoms. The symptoms rest in the assertion block and are matched against symptom-disease relationships in the knowledge-relationships block under the control of the search strategy. The reasoning process by which the result was reached may be traced by the explanation capability.

Table 1. Session topics at recent artificial-intelligence conferences suggest the range of the field.

Automated reasoning	Logic programming, Prolog
Automatic programming	Machine learning
Cognitive modeling	Natural language, including speech
Expert systems	Perception, including visual, auditory, tactile
Intelligent databases	Problem solving and search
Knowledge acquisition	Robotics
Knowledge bases	Theorem proving
Knowledge representation	Vision
Lisp programming	

summer. In addition to antecedent-consequence, the two parts of an if-then production rule may be called a situation-action pair or a premise-conclusion pair. In the Lisp language, the relationship is called a clause and the clause has two parts: a test and a result. In Prolog, relationships are shown in the first-order predicate logic.

When all three of the antecedents, *clear, hot,* and *muggy,* are present as assertions, they match the left-hand side of the production rule, leading to the consequence, *it is summer.* In addition, there are two further steps to the making of matches: (1) The production rules may be listed in some order believed to facilitate the search for a match; and (2) the search-strategy block may be able to invoke a sequence that finds the match more quickly than random search does.

Search strategy. The simplest arrangement of the production rules is to list them in no particular order. With this arrangement, new rules may be tacked on, making it easy to grow the system as more is learned about the problem. Each asserted fact is then run through the production rules until the match is found. With a small number of rules, given the high speed of computer operation, it is practical to search a random list.

If the number of rules is large, they may be partitioned into sublists, or contexts, on some logical basis. The search strategy then uses a higher-level rule, or a metarule, based on the logic of the partition, to determine which sublist to run through first, thus reducing the length of the search.

Another arrangement is to chain the production rules to one another, so that the consequent of one becomes the antecedent of another. For example, *if it is clear, then it is hot; if it is hot, then it is muggy; if it is muggy, then it is summer.* A number of these chains may be arranged in a tree structure or a graph of some sort. The search strategy then becomes a matter of searching this structure.

If nothing is known about which is the better path to follow, then the search may be either breadth-first or depth-first. In breadth-first, the search passes from node to node across the breadth of the tree structure. In depth-first, the search passes vertically down one branch of the inverted tree, then returns to the top and searches down the next branch, etc. Moreover, the domain expert may have some rules of thumb, or heuristics, that enable him or her to indicate that one part of the tree is more promising than the other parts. These heuristics may be incorporated in the search strategy.

The sample chain given above—*if it is clear, then it is hot; if it is hot, then it is muggy; if it is muggy, then it is summer*—leads to another complication. As phrased, the sequence indicates 100 percent certainty from one rule to the next. In reality, the probability that one rule follows the preceding one is much less than 100 percent in this case, so expert-system builders have devised various methods of incorporating uncertainty into their rules. For instance, the truth value of each production rule in this sequence may be only 0.5 on a scale of 0 to 1. The probability of all three rules being valid in a particular instance is still less.

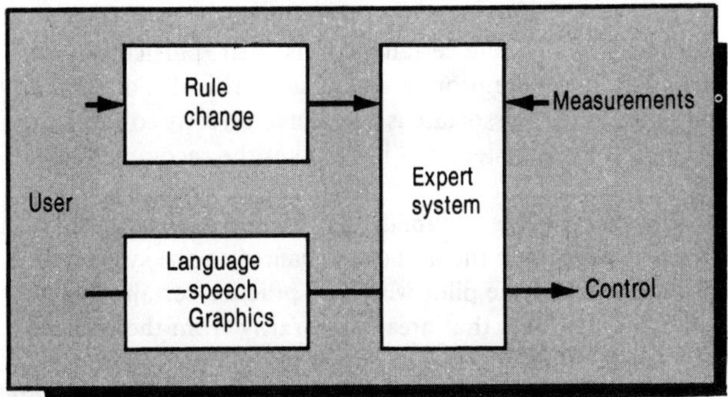

Figure 3. The expert system of the future will help a human operator control some machine, such as an aircraft.

Explanation trace. When the expert system comes up with the answer, *it is summer,* we may have some doubts about the correctness of this conclusion. The credibility of the system is greatly enhanced if it can explain to the user the line of reasoning that led to this finding. It can do this, essentially, by retracing the chain of production rules that led to this finding and translating them into some form readily intelligible to the user, such as English. Then with the user's common sense knowledge about, in this case, the local climate, he may check the expert system's chain of reasoning, noting the uncertainties associated with each step.

Expert assistant. A typical expert system supplies solutions in the form of data or information, and the user is responsible for making use of it. The concept of the expert system may be extended to the expert assistant, diagrammed in Figure 3. The expert assistant receives data directly from the environment in which it operates in the form of real-time measurements from sensors. With some direction from the user, the expert assistant generates signals in real time to control some aspect of the environment. The user interfaces to the expert assistant through a natural-language interface, a graphics interface, and ultimately speech. The user may change the rules under which the expert assistant is operating to help the assistant adapt better to the environment as it changes.

The expert assistant takes on more meaning if we think of it as a pilot's associate. An aircraft's instrument system now supplies the pilot with hundreds of readings, far too many to comprehend fully in an emergency. The pilot's associate could monitor this data, note interrelationships, figure out the significance of the information, and present a conclusion to the pilot in a form ready for implementation.

One step further, the pilot could preauthorize the pilot's associate to institute certain actions itself, particularly control actions where prompt response is critical. For this purpose the pilot's associate is connected directly to the aircraft's control system and itself takes the recommended action.

The rule-change capability enables the pilot to specify from time to time the authority granted to the system. For example, when the pilot wishes to practice certain flying skills, he removes that area temporarily from the jurisdiction of the pilot's associate, while leaving in force safety precautions. On the other hand, when he decides to devote his attention to some nonflying task, such as checking his position, he assigns more responsibility for operating the plane to the pilot's associate.

Instead of an aircraft, any kind of complex machinery or process could be directed by a human operator with an expert assistant. This complex machinery could be a robot.

AI languages and their hardware

The computer age got under way computing numbers, but artificial intelligence called for processing symbolic information, such as words and phrases. Aware of this need, John McCarthy invented Lisp in 1959. In 1972, Alain Colmerauer at the University of Marseilles began the development of Prolog, based on a different principle.

Lisp. Four commands are central to Lisp's symbol-manipulation capability. (The names of three of these commands are nonmnemonic as the result of "a regrettable historical convention," in Winston's phrase,[5] so we won't bother with them here.) One returns the first element in a list of elements, for example, the first word in a list of words. The second removes the first element from a list of elements and returns the remainder of the list. By using these two commands alternately, the second, third, or *n*th word in a list may be picked out. To put it another way, these commands take a list apart.

The next two commands put things together. One strings together the elements of two or more lists supplied as arguments. The other inserts a new first element in a list.

In addition, there are other commands that when applied to a predicate determine whether it is true or false. By definition a predicate is a function whose value is limited to true or false.

The last command that we will mention (there are many more) cruises down a list of "clauses," each composed of "a test and a result." When its argument matches the test side of a clause, it returns the corresponding result. In effect, the test and the result could be the two sides of an if-then rule. This capability may also be used to direct the flow of program control.

Thus, these basic commands can sort out symbols, can build up lists, can determine the truth or falsity of a function, and can match the if-side of a production rule. In general, Lisp's goal is to evaluate something and return a value.

Lisp has many advantages, and we will mention two: storage space is allocated dynamically, enabling programs to be larger than they would otherwise be; and there is a macro capability, permitting the language to be extended. In fact, Micro-planner, Conniver, Scheme, and other AI languages were written as extensions in Lisp.

Partly because of this extension capability, partly because of the need to adapt the language to different computers, and partly because of the tendency of researchers to implement new ideas, many versions of Lisp have been developed. These versions seem to have flowed primarily in two streams, one originating in the MIT Artificial Intelligence Laboratory, called Maclisp, and the other from Bolt, Beranek, and Newman, and later from Xerox Palo Alto Research Center, called Interlisp. Many of the Maclisp descendants are now in the process of being standardized as Common Lisp.

Prolog. Prolog, for programming in logic, was developed in Europe between 1970 and 1980. Alain Colmerauer, Robert Kowalski, and Phillip Roussel are the names associated with the early development. A compiler/interpreter was developed by David Warren, Fernando Pereira, and Lawrence Byrd at the University of Edinburgh in 1975. A modular version, MProlog, was produced in Hungary between 1979 and 1982.

The language received impetus from its selection in 1981 as the basis for the Japanese Fifth Generation Computer project. It is now gaining acceptance in the United States as well. For example, it is the language basis for the Aquarius project at the University of California at Berkeley.[6] This project is developing a high-performance computing system for combined symbolic and numeric applications. Also, Prolog is being used by organizations such as Applicon, Argonne National Laboratory, Caltech Jet Propulsion Laboratory, Fairchild, Gould SEL, Lockheed Missiles and Space Company, and SRI International.

The language works with English-like statements of three types: facts, rules, and questions.

(1) A fact is something such as *Richard works for Jean.* This fact indicates there is a relationship between Richard and Jean and that the relationship, as expressed, goes from Richard to Jean. This fact translates into Prolog like this:

```
works(richard,jean)
```

(2) A rule follows the form: Someone is the manager of an employee if the employee works for that someone. A rule is expressed as follows:

manager(X,Y) :- works(Y,X), where the expression ":-" is read as "if."

(3) A question might be: "Who is the manager of Richard?" It is expressed in this way:

?- manager(X,richard)

The Prolog language returns the answer:

X = jean

As we saw, a Lisp program consists of a series of commands that manipulate symbols. A Prolog program, on the other hand, consists of statements of facts and rules. This information is used by asking questions about it.

Prolog's adherents believe that it is easier to learn and use than Lisp. They say that it uses less memory and is more easily moved from one computer to another. In the past, it has run with reasonable speed only on mainframes, but recent modifications are running satisfactorily on smaller machines.

Software development tools for large computers

Advisor is from the ICL Knowledge Engineering Business Centre in Manchester, England. It runs on ICL VME mainframes and will be available on smaller systems. Written in Pascal, it costs about 15,000 pounds.

ART, Automated Reasoning Tool, is provided by Inference Corporation, Los Angeles. It runs on the machines from Lisp Machine and Symbolics and on the DEC Vax series. The price range is $60,000 to $80,000. It is a second-generation tool, the company says, distinguished from first-generation tools such as Emycin and OPS5 because it provides a sound base for commercial development, rather than exploring a paradigm for research purposes.

HPRL, Heuristic Programming and Representation Language, is being developed in the Computer Research Laboratory, Hewlett Packard, Palo Alto, California. It is considered to be an extension of FRL, Frame Representation Language, and to have goals in common with RLL and Age. It runs on the Vax 11/780 and the HP-9836.

IKE, Integrated Knowledge Environment, was announced by Lisp Machine at the 1985 International Joint Conference on Artificial Intelligence. It was introduced at $15,000 and runs on the company's Lisp machines. The package uses menu-based natural-language commands to customize the program's vocabulary to the application, to build the knowledge base, and to run interactive sessions to test the developing expert system. A graphic interface displays parse trees of rules as they are entered, thus verifying that the system has "understood" them.

KEE, Knowledge Engineering Environment, is available from IntelliCorp, Menlo Park, California. It runs on the Xerox 1100 series, Symbolics 3600, and the Lisp Machine, Inc., Lambda and is being made available on additional machines. The price is $60,000 for the first unit with substantial discounts for additional units. Release 2.0 in August 1984 extends the tool to support larger expert systems and adds a logic-based query language for KEE knowledge bases.

KES, Knowledge Engineering System, is supplied by Software A & E, Inc., Arlington, Virginia. It runs under various versions of Lisp, including Franz Lisp, Zetalisp, and Spice Lisp, a dialect of Common Lisp that runs on the Perq computer. In 1985, the tool was being reengineered in C to achieve further transportability, better performance, and the support of embedded applications.

Knowledge Craft is a product of the Carnegie Group, Pittsburgh, Pennsylvania. Priced at $50,000, including training and support, it runs on Texas Instruments' Explorer, the Symbolics 3600, and the DEC Vax series. It was derived from research at Carnegie-Mellon University, and the company considers it to be a "next generation" tool.

Picon, Process Intelligent Control, is the first of a series of tools being developed by Lisp Machine to make the technology more readily usable in particular industries. It helps operators deal with emergency situations in complex processes, such as refineries and chemical plants. The basic system may be adapted to different plants. It runs on the company's Lambda/Plus, meaning one Lisp machine plus one 68010 microprocessor. The purpose is to combine the real-time data-acquisition and numeric-processing capabilities of the microprocessor with the expert-system capabilities of the Lisp machine.

Reveal is from the ICL Knowledge Engineering Business Centre in Manchester, England. It runs on ICL VME mainframes and will be available on smaller systems. Written in Fortran, it costs about 30,000 pounds.

S.1 from Teknowledge, Inc., Palo Alto, California, is in the $50,000 class (less for additional copies). Written in Lisp, it runs on the Xerox 1108 workstation under Interlisp-D, Symbolics 3600 and 3670 machines in Zetalisp, and DEC Vax 11/750 and 11/780 under VMS in Franz Lisp. It is intended for large-scale systems, in contrast to the company's microcomputer tool, M.1.

S.1 Version 2 has been completely rewritten in the C language for use on computers that run the Unix operating system, including the AT&T PC7300, DEC Microvax, Xerox 1109, Symbolics 3670, NCR Tower 32, Gould Power Node 6031, Tektronix 4404, Sun-2/120, and Apollo DN30 and DN550. The development version, available in early 1986, will sell in the $25,000 to $40,000 range.

TIMM, The Intelligent Machine Model, from General Research Corporation, McLean, Virginia, is available in mainframe and personal-computer versions. Written in Fortran, the mainframe version runs on 4MB virtual machines with a full ANSI Fortran-77 compiler, including IBM, Amdahl, DEC Vax, and Prime. The first license is $39,500, less in quantity.

Lisp hardware. The Lisp language, of course, was developed on traditional, numerically oriented mainframes and later came to be widely used on minicomputers. Artificial-intelligence programs tend to be large and to use vast amounts of memory. Moreover, the structure of the language makes it execute slowly and inefficiently on this kind of computer. In addition, when several researchers had to timeshare a large computer, response time was still worse.

By the early 1970s, researchers realized that these problems could be ameliorated by developing an individual workstation optimized to run Lisp. The Lisp Machine project was launched at the MIT Artificial Intelligence Laboratory in 1974. A first-generation machine was completed in 1976 and a second-generation in 1978. Aware of a potential commercial market for the machine, two groups split off from the laboratory in 1980, one becoming Lisp Machine, Inc., with offices both in Andover, Massachusetts, and Los Angeles, and the other, Symbolics, Inc., in Cambridge, Massachusetts.

Also in 1974, researchers at Xerox PARC began an implementation of Interlisp for the Alto, a small personal computer internal to Xerox. The performance of this implementation, called AltoLisp, was not satisfactory for such reasons as the limited amount of main memory. By 1980, PARC had developed two more suitable personal computers, the Dorado and the Dolphin.

In general, the various Lisp machines took advantage of advancing technology to provide more main memory, more cache memory, stack architecture, faster processing speeds, pipelining and parallelism, and high-resolution interactive display screens. In addition, these machines implemented hardware specific features that directly supported the needs of Lisp.

In the descendants of the MIT development, the machine instruction set corresponds closely to Zetalisp, one of the successors to Maclisp. In the Symbolics 3600, for instance, a machine instruction takes the first element of a list and pushes it onto the stack.[7]

The 3600 features a 36-bit word and two of these bits are used to code the relationship between the first element and the rest of the list. This code avoids the need for using an additional word to hold a pointer to the next element, reducing memory needed for this purpose by one half.

In the operation of Lisp programs, some objects become orphans—they are no longer referenced by anything else and so they are no longer needed. Since memory space is valuable, a process called garbage collection identifies these objects and reclaims their space. Originally, this process had to read through the entire storage space to establish that there were no references to the stored objects being considered for disposal.

In the 3600, three hardware features speed up garbage collection: a two-bit field indicates whether the word contains a pointer, that is, a reference to a word in virtual memory (main memory plus disk); a page-tag bit indicates whether the garbage-collecting process need even scan the

Figure 4. Lisp Machine, Inc., began delivering the Lambda 3x3 last summer (1985). The model designation refers to the fact that the system has three processors and accommodates three programmers. When an application requires great power, the $135,000 system's three Lisp processors may be pipelined together over a 37.5 MB/s multiprocessor bus.

Figure 5. Introduced in 1983, the Symbolics 3600 supports the Zetalisp software environment and includes object-oriented programming or message passing. The entry-level configuration is about $80,000. Available as an option is the Interlisp Compatibility Package which accommodates Interlisp-10, Interlisp-VAX, and Interlisp-D.

Figure 6. In August 1984, Tektronix introduced the 4404 AI System and, in August 1985, the 4405 and 4406 AI Systems at about $12,000, $15,000, and $24,000. The three products feature the Smalltalk-80 programming environment. Tek Common Lisp, Franz Lisp, or MProlog are available as options.

Figure 7. Texas Instruments' Explorer Symbolic Processing computer is derived from technology licensed from MIT and Lisp Machine. Introduced in late 1984, it is available in six configurations ranging from $52,500 to $66,500. The principal language is Common Lisp, but a Prolog interpreter may be called from the Lisp environment.

page; and multiword read-operations fetch several words at a time to the processor. In addition, garbage collection is incremental, running in the background.

Lisp Machine, Symbolics, (see Figures 4 and 5) and Xerox (in Pasadena, California) continue to produce workstations for artificial-intelligence applications. In addition, there are two recent entrants, Tektronix and Texas Instruments, illustrated in Figures 6 and 7.

At the low end of the machine spectrum are the personal computers. Versions of Lisp are beginning to appear for them. One is ExperLisp for the Macintosh from ExperTelligence, Santa Barbara, California, for $495. It is "the first complete implementation of Lisp on a microcomputer," the company says. According to an analysis by the president of the company, Denison Bollay, the relative performance of an 8088-based machine (IBM PC) is two, an 80286-based machine (IBM PC AT) or a 68010-based machine (Macintosh or Tektronix 4404) is 10, a future 68020-based machine would be 60, and the Symbolics 3600 is 100.[8]

As a foretaste of hardware to come, Texas Instruments is developing a custom VLSI Lisp processor chip based on sub two-micron CMOS technology for DARPA's Strategic Computing Program. The chip will be software-compatible with TI's Explorer Symbolic Processing workstation. "The chip is being designed to provide up to 10 times the processing power of today's commercial symbolic processors at substantially lower cost and physical size," the company says.

Prolog hardware. Thus far Prolog in different versions is running on general-purpose computers, including the DEC Systems 10 and 20, the DEC Vax series, and IBM's large mainframes, as well as the IBM PC and the Apple II. MProlog from Expert Systems, Ltd., Oxford, England, for example, is available on the IBM/370 under VM/370-CMS, the Siemens 7,000 under BS2000, and the Vax-11 under VMS. Quintus Computer Systems of Palo Alto, California, has tried to increase the availability of Prolog by developing portable, hardware- independent systems on standard general-purpose hardware.[9]

Machines aimed at running Prolog are under development. The Japanese Institute for New Generation Computer Technology, ICOT, is developing the PSI machine to maximize Prolog performance. NEC is working with ICOT on a higher performance machine, and ICOT projects a very high performance unit by 1992.

The Aquarius project at the University of California at Berkeley, under way since 1983, plans to radically improve performance by (1) using a variant of logic programming (initially Prolog) as the primary control mechanism for the problem-solving process, (2) tailoring the MIMD processing elements to the requirements of the intended set of applications, and (3) exploiting parallelism at several levels of concurrency.[6]

Again, as in the case of Lisp implementations, performance varies over a wide range. Kahn and Warren give the following figures:

Quintus Prolog on Sun-2 workstation under Unix: 20K LIPS (logical inferences per second);
Quintus Prolog on Vax 11 under VMS and Unix 4.2: 23K LIPS;
The ICOT PSI machine (still under development): 33K LIPS;
The fastest Prolog on the market: 43K LIPS; and
Japanese goal for 1992: 100M LIPS[9]

Despain and Patt compare a dozen or more machines on several benchmarks.[6] Performance on general benchmark programs ranges from 10 LIPS (on the Apple II) to 205K LIPS. The latter figure is derived from a simulation of their Berkeley Programmed Logic Machine, the first experimental processor under the Aquarius project.

Software development tools

In the course of building the early expert systems, researchers noticed that the facts and if-then rules were specific to the problem domain, but the inference engines were very much the same from one domain to another similar one. Of course, in the first expert systems the knowledge bases and the inference engines were not neatly divided, but by the end of the 1970s researchers had sorted out a number of domain-independent versions of expert systems.

Emycin (for Essential Mycin) was derived from Mycin, a production rule-oriented expert system for diagnosing infectious blood diseases; KAS from Prospector, a minerals prospecting system; and Expert from Casnet, used in the diagnosis and treatment of glaucoma. These three are classified as skeletal systems.[3]

Representation languages are a class of general-purpose programming languages developed for knowledge engineering. Less constrained than skeletal systems, they may be applied to a broader range of tasks. This category includes Rosie, OPS5, RLL, and Hearsay-III.

A third category, computer-aided design tools for expert systems, is represented by the tool, Age. With it, the user may choose from several kinds of knowledge representation and processing methods.

With the domain-specific knowledge separated from the code in the inference engine, it would seem to be merely a matter of knowledge engineering to put together an expert system. Little programming should be necessary. Unfortunately, there may still be problems. The greatest one is that the inference tool fails to match the new problem area closely enough to be applied unchanged. A lesser one is that all task-specific knowledge has not been rooted out of the tool.

After investigating the use of these tools, Donald A. Waterman and Frederick Hayes-Roth offered some suggestions for choosing a tool appropriate to the problem:

"Do not pick a tool with more generality than you need.
Test your tool early by building a small prototype system.
Choose a tool that is maintained by the developer.
Choose a tool with explanation/interaction facilities when development speed is critical.
Use the problem characteristics to determine the tool features needed."[3]

The first large, comprehensive expert-system development tool to be marketed commercially, according to the company that developed it, was KEE, the Knowledge Engineering Environment, from IntelliCorp. It was introduced in August 1983 at the conference of the American Association for Artificial Intelligence. Since then, a dozen or more tools have appeared for use on mainframes and minicomputers. Still more recently, expert-system development tools to run on personal computers have become available.[10]

Shortcomings of tools. One tool vendor observes that most present-day tools do not support the integration of the expert system with existing software such as database management systems.[11] The ability to transport Lisp- or Prolog-based systems from one machine environment to another is limited. Their performance is often inadequate for production applications.

Silogic Incorporated of Los Angeles points out that current expert systems can operate only on expensive or special-purpose hardware and are too slow on small or conventional machines. Furthermore, they cannot operate on conventional machines in an industrial environment. They have problems going back and forth to a large database. They cannot use databases constructed on other systems.

In the process of interacting with its Fortune 100 clients, Teknowledge discovered that they were strongly wedded to their installed hardware and software base. "American and international business show no sign of abandoning their huge investment in currently installed hardware, software, databases, operating systems, and personnel training," according to Lee M. Hecht, the company's chief executive officer.

It was possible to persuade them to use Lisp-based or Prolog-based expert systems and the corresponding special-purpose computers at a research or development level. In general, they would not install hundreds of such systems for corporate-wide applications. There were logistic problems to introducing another type of hardware. Existing programming personnel were not comfortable with Lisp or Prolog.

Recognizing this reality, Teknowledge during 1985 reprogrammed its expert-system tools in C and ported them to conventional computers supporting the Unix operating environment. Moreover, in doing this, the company achieved "much higher performance," according to Earl D. Sacerdoti, vice president and general manager of Knowledge Engineering Products and Training.

End of the beginning

Every one has seen glowing accounts of expert systems performing remarkable functions. It is true that some of the university-developed systems are now in practical application, that the university spin-off companies are developing useful systems, and that some large corporations have established artificial-intelligence groups.

"Today, there are fewer than 25 knowledge systems in the field," write Hayes-Roth and London, "but several hundred more are under development."[12]

As computer professionals, however, we should remember that, although the origins of artificial intelligence go back 30 years, the arrival of practical expert systems goes back little more than three years. As a new art, we should expect problems:

Tools for personal computers

ESP Advisor from Expert Systems International, King of Prussia, Pennsylvania, runs on MS-DOS 2.0 with 256K of memory. It is priced at $895.

ExperOPS5 was implemented by Science Applications International Corporation for ExperTelligence of Santa Barbara, California. The original OPS5 was created at Carnegie Mellon University. The present version, said by the company to be "a complete implementation," operates on a Macintosh 512K with ExperLisp and an add-on floppy or hard-disk drive. It was introduced in the summer of 1985 at a list price of $195.

Expert-Ease was developed by Jeffrey Perrone & Associates, Inc, San Francisco, and lists for about $700. It is written in Pascal and runs on IBM PCs and compatibles, DEC, and Victor. Perrone says that the tool enables users to produce "models" of processes that require specialized knowledge or skills. Models have been developed for property purchasing decisions, trouble shooting, analyzing logic designs, preliminary diagnoses of dental pain, assistance in building code compliance, analyzing expense claims, and recommending disciplinary actions. Perrone notes that some might not consider these to be "true" expert systems.

Exsys, from Exsys, Inc., Albuquerque, New Mexico, lists at $200 and accommodates up to 400 rules with 128K of memory or 3000 rules with 640K. The large version runs on MS DOS 2.00.

Insight 2 is a $495 product from Level 5 Research, Melbourne, Florida, that runs on IBM, DEC, and Victor personal computers with 192K of memory.

KDS, from KDS Corporation, Wilmette, Illinois, allows for up to 16,000 rules per knowledge module. The development system requires 256K, MS DOS 2.00, and is priced at $795.

Knowledge Workbench, based on Silog, a proprietary extension of Prolog, comes from Silogic Inc. of Los Angeles. The system incorporates not only a knowledge-base manager and inference engine, but a natural-language interface. It is being designed to run on MC68000-based personal computers.

M.1 and **M.1a** are Prolog versions of Teknowledge's microcomputer tool, the first at $10,000 and the second, $2000. Both operate under MS-DOS 2.0. M.1a is an evaluation subset of M.1 containing the basic features and sufficient power to develop a proof-of-concept system.

Last October Teknowledge announced the conversion of M.1 to C. The Version 2 development system is priced at $5000 with the delivery version at $500 per copy. Rule capacity has been expanded from 200 to 1000. Performance has improved four or five times over the original Prolog product.

MicroExpert at only $49.95 is a product of McGraw-Hill Book Co., New York. It requires 128K and MS DOS 2.00.

Micro KES, listing at $4000, is the personal-computer version of KES. It is implemented in IQLisp for the IBM PC XT with 640K. Software A & E says that it is functionally equivalent to KES running on a Vax, but its performance is much less—comparable to KES on a heavily loaded Vax. Thus, development on a personal computer is difficult, but execution is believed to be adequate.

NaturalLink Technology Package is a 1985 release from Texas Instruments at $1500 for TI's Professional Computer. It enables a programmer to construct English sentences that the underlying application program recognizes as commands. The ultimate user constructs an input sentence, not by typing, but by selecting words and phrases from lists of choices displayed in menus.

Personal Consultant is a Texas Instruments product that runs on TI's Professional Computer and is compatible with PC-DOS software products. Listing at $3000, it requires at least 512K. Using the Professional Computer, a user may develop an expert system of up to 400 rules. On a larger, higher capacity computer, such as TI's Explorer, he can go up to more than 1000 production rules and then run the developed system on the Professional Computer.

RuleMaster is available from Radian Corporation, Austin, Texas, in the $5000 to $15,000 range. It runs under MS-DOS 3.0 on the IBM PC XT or under Xenix on the IBM PC AT, and requires 600K.

TIMM-PC, The Intelligent Machine Model, is supplied by General Research Corporation, McLean, Virginia, at $9500 for the first license, less in quantity. It runs on MS-DOS 2.0 with 640K.

(1) There are two major languages, Lisp and Prolog, each supported with almost religious intensity by passionate advocates. Moreover, each exists in many versions. It is not clear which one or which version is best in some absolute sense, but it is evident that developers are successfully building systems in both languages and many versions.

(2) Others believe that conventional languages capable of running on conventional computers better meet the desires of the marketplace.

(3) There are more than a score of software development tools. They are often aimed at different applications; they may be based on different principles; they are implemented in different languages; and they operate on different computers. It takes considerable effort to find out which one is best for a particular application.

(4) There are mainframes and minicomputers, special-purpose Lisp machines and forthcoming Prolog machines, and workstations and personal computers. Which sizes and types are suitable for which applications?

(5) Moreover, there may be a great gap between a prototype and a successful product, experienced developers know. Some of the difficulties include: making the product more rugged, reliable, and user-friendly; arranging to run the software on hardware suitable for the application—hardware that is accessible and cost effective; defining a specific range of problems over which the system is to work, so that the user can easily recognize whether his particular problem is within the range served by the expert system; and meeting the users' need for an application they can afford.

(6) Some knowledgeable people believe that knowledge engineering and expert-system building are too complicated to be entrusted to engineers, programmers, or domain experts who are relatively inexperienced in the art. Others believe that the software development tools are sufficiently comprehensive to be used by people with little experience. Which belief is true now? How much further development of tools is necessary before inexperienced people will be able to build systems? Alternatively, what training and experience must people have to build expert systems? E

References

1. Robert M. Glorioso and Fernando C. Colon Osorio, *Engineering Intelligent Systems: Concepts, Theory, and Applications,* Digital Press, Bedford, Mass., 1980, 472 pp.
2. Pamela McCorduck, *Machines Who Think,* W. H. Freeman and Company, San Francisco, 1979, p. 96.
3. Frederick Hayes-Roth, Donald A. Waterman, and Douglas B. Lenat, editors, *Building Expert Systems,* Addison-Wesley Publishing Co., Reading, Mass., 1983, 444 pp.
4. Dana S. Nau, "Expert Computer Systems," *Computer,* Vol. 16, No. 2, Feb. 1983, pp. 63-85.
5. Patrick H. Winston, *Artificial Intelligence,* Addison-Wesley Publishing Co., Reading, Mass., 1977, 444 pp.
6. Alvin M. Despain and Yale N. Patt, "Aquarius—A High Performance Computing System for Symbolic/Numeric Applications," *Digest of Papers,* Compcon Spring 85, pp. 376-382.
7. Curtis B. Roads, Symbolics 3600 Technical Summary, Symbolics, Inc., Cambridge, Mass., 1983, 140 pp.
8. Tom Manuel, "The Pell-Mell Rush Into Expert Systems Forces Integration Issue," *Electronics,* July 1985, pp. 54-59.
9. Patti Kahn and David Warren, "Application Development in Prolog," *Proc. Artificial Intelligence & Advanced Computer Technology Conf. 1985,* Tower Conference Management Co., Wheaton, Ill., pp. 207-212.
10. Tom Schwartz, "AI Development on the PC: A Review of Expert System Tools," *The Spang Robinson Report,* Nov. 1985, pp. 7-14.
11. Andrew B. Ferrentino, "Expert System Development Environments: Current Limitations and Future Expectations," *Proc. Artificial Intelligence & Advanced Computer Technology Conf. 1985,* Tower Conference Management Co., Wheaton, Ill., p. 195.
12. Frederick Hayes-Roth and Phillip London, "Software Tool Speeds Expert Systems," *Systems & Software,* Aug. 1985, pp. 71-75.

Ware Myers is a freelance writer specializing in computer subject matter and a contributing editor for *IEEE Expert*. From 1965 until Xerox withdrew from the mainframe business in 1975, he was a member of the Systems Development Group, Computer Systems Divison, in El Segundo, California, where he worked on the development of analog instruments, color display stations, a microprogrammed controller, and several MOS memories. His principal contribution to these developments was the preparation of design specifications, technical descriptions, operating instructions, and reference manuals. From 1956 to 1965 he was with the Consolidated Electrodynamics Corporation and its subsidiary, Consolidated Systems Corporation. He has also worked as an instructor and lecturer in engineering at the University of California, Los Angeles.

Myers received his BS from the Case Institute of Technology and his MS from the University of Southern California. He is a member of Tau Beta Pi and the IEEE Computer Society.

Myers' address is c/o *IEEE Expert*, 10662 Los Vaqueros Circle, Los Alamitos, CA 90720.

THE BASIC PRINCIPLES OF EXPERT SYSTEMS

Excerpted From:

An Overview of Expert Systems

Author: William B. Gevarter

National Bureau of Standards Report # NBSIR 82-2505
National Technical Information Center Document # PB83-217562
(Originally published May 1982)

Preface

Expert systems is probably the hottest topic in Artificial Intelligence (AI) today. Prior to the last decade, in trying to find solutions to problems, AI researchers tended to rely or non-knowledge-guided search techniques or computational logic. These techniques were successfully used to solve elementary problems or very well structured problems such as games. However, the search spaces for real and complex problems tend to expand exponentially with the number of parameters involved. For such problems, these older techniques have generally proved to be inadequate and a new approach was needed. This new approach emphasized knowledge rather than search and has led to the fields of Knowledge Engineering and Expert Systems.

This report reviews the basic principles of Expert Systems -- what an expert system is and the techniques employed.

Reprinted from "An Overview of Expert Systems" by W.B. Gevarter, *National Bureau of Standards Report No. NBSIR 82-2505,* May 1982.

I. Introduction

In the 70's, it became apparent to the AI community that search strategies alone, even augmented by heuristic(1) evaluation functions, were often inadequate to solve real world problems. The complexity of these problems was usually such that (without incorporating substantially more problem knowledge than had heretofone been brought to bear) either 1) a combinatorial explosion occurred that defied reasonable search times, or 2) the ability to generate a suitable search space did not exist. In fact, it became apparent that, for many problems, expert domain knowledge was even more important than the search strategy (or inference procedure). This realization led to the field of "Knowledge Engineering," which focuses on ways to bring expert knowledge to bear in problem solving.(2) The resultant expert systems technology, limited to academic laboratories in the 70's, is now becoming cost-effective and it has entered into commercial applications.

Because of the acknowledged value of expert systems, funding for research and implementation has become more readily available. To date, the government has been the principal source of funds for work in expert systems. The funding sources in the government for expert systems are:

DARPA (Defense Advanced Projects Agency)
NIH (National Institutes of Health)
NSF (National Science Foundation)
ONR (Office of Naval Research)
NLM (National Library of Medicine)
AFOSR (Air Force Office of Scientific Research)
USGS (U.S. Geological Survey)
NASA (National Aeronautics and Space Administration)

II. What is an Expert System?

Feigenbaum, a pioneer in expert systems, (1982, p. 1) states: An "expert system" is an intelligent computer program that uses knowledge and inference procedures to solve problems that are difficult enough to require significant human expertise for their solution. The knowledge necessary to perform at such a level, plus the inference procedures used, can be thought of as a model of the expertise of the best practitioners of the field.

The knowledge of an expert system consists of facts and heuristics. The "facts" constitute a body of information that is widely shared, publicly available, and generally agreed upon by experts in a field. The "heuristics" are mostly private, little-discussed rules of good judgment (rules of plausible reasoning, rules of good guessing) that characterize expert-level decision making in the field. The performance level of an expert system is primarily a function of the size and quality of the knowledge base that it possesses.

III. The Basic Structure of an Expert System

An expert system consists of:

1) a knowledge base (or knowledge source) of domain facts and heuristics associated with the problem:
2) an inference procedure (or control structure) for utilizing the knowledge base in the solution of the problem;
3) a working memory - "global data base" - for keeping track of the problem status, the input data for the particular problem, and the relevant history of what has thus far been done.

A human "domain expert" usually collaborates to help develop the knowledge base. Once the system has been developed, in addition to solving problems, it can also be used to help instruct others in developing their own expertise. Thus, Michie (1980, pp. 3-5) observes: ...that [ideally] there are three different user-modes for an expert system in contrast to the single mode (getting answers to problems) characteristic of the more familiar type of computing:

(1) getting answers to problems -- user as client;

(2) improving or increasing the system's knowledge user as tutor;

(3) harvesting the knowledge base for human use -- user as pupil.

Users of an expert system in mode (2) are known as "domain specialists." It is not possible to build an expert system without one.

An expert systems acts as a systematizing repository over time of the knowledge accumulated by many specialists of diverse experience. Hence, it can and does ultimately attain a level of consultant expertise exceeding(3) that of any single one of its "tutors."

It is desirable, though not yet common, to have a natural language interface to facilitate the use of the system in all three modes. In some sophisticated systems, an explanation module is also included, allowing the user to challenge and examine the reasoning process underlying the system's answers. Figure 1 is a diagram of a idealized expert system. When the domain knowledge is stoned as production rules , the knowledge base is often referred to as the "rule base" and the inference engine the "rule interpreter."

FIGURE 1: BASIC STRUCTURE OF AN EXPERT SYSTEM

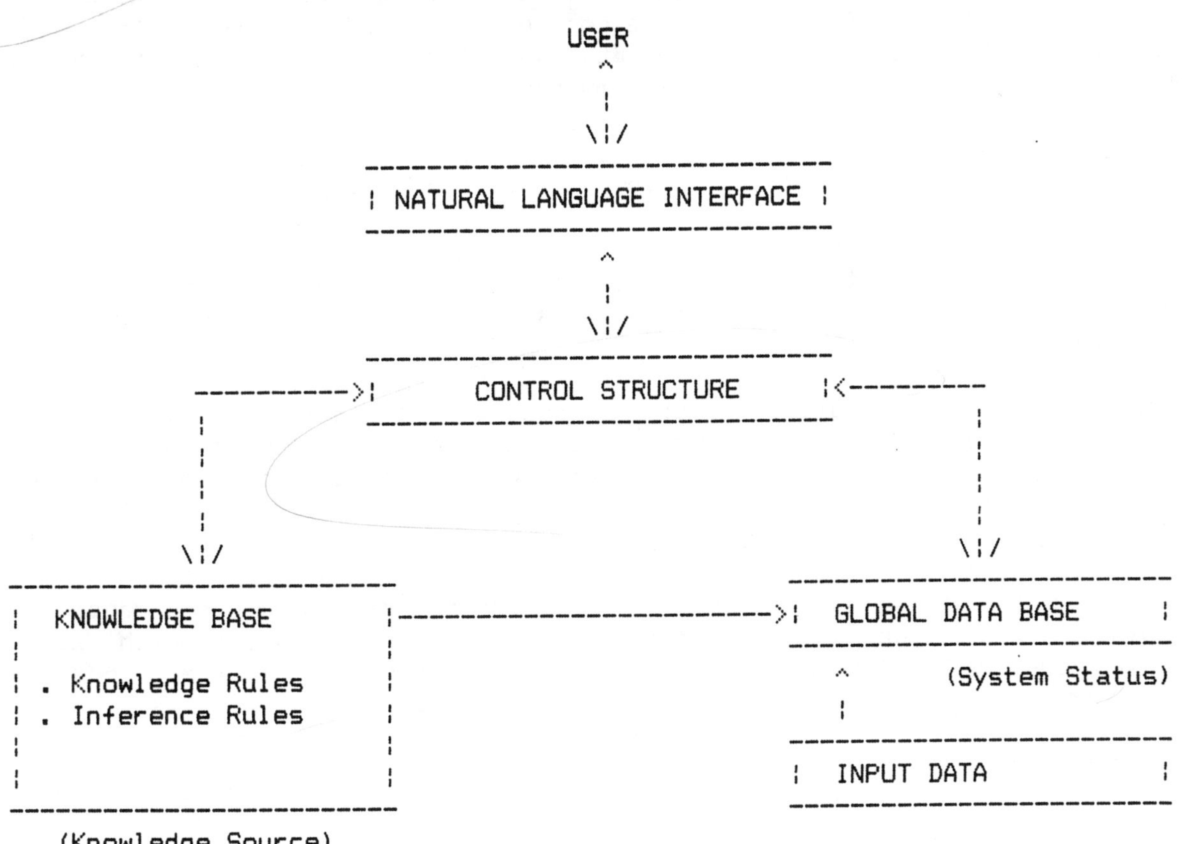

An expert system differs from more conventional computer programs in several important respects. Duda (1981, p. 242) observes that, in an expert system, "...there is a clear separation of general knowledge about the problem (the rules forming a knowledge base) from information about the current problem (the input data) and methods for applying the general knowledge to the problem (the rule interpreter)." In a conventional computer program, knowledge pertinent to the problem and methods for utilizing this knowledge are all intermixed, so that it is difficult to change the program. In an expert system, "...the program itself is only an interpreter (or general reasoning mechanism) and [ideally] the system can be changed by simply adding or subtracting rules in the knowledge base."

IV. The Knowledge Base

The most popular approach to representing the domain knowledge needed for an expert system is by production rules (also referred to as "SITUATION-ACTION rules'' or "IF-THEN rules").(4) Thus, often a knowledge base is made up mostly of rules which are invoked by pattern matching with features of the task environment as they currently appear in the global data base.

The rules in a knowledge base represent the domain facts and heuristics - rules of good judgment of actions to take when specific situations arise. The power of the expert system lies in the specific knowledge of the problem domain, with potentially the most powerful systems being the ones containing the most knowledge.

Duda (1981, p. 242) states: Most existing rule-based systems contain hundreds of rules, usually obtained by interviewing experts for weeks or months. In any system, the rules become connected to each other [by association linkages] to form rule networks. Once assembled, such networks can represent a substantial body of knowledge.

An expert usually has many judgmental or empirical rules, for which there is incomplete support from the available evidence. In such cases, one approach is to attach numerical values (certainty factors) to each rule to indicate the degree of certainty associated with that rule. (In expert system operation, these certainty values are combined with each other and the certainty of the problem data, to arrive at a certainty value for the final solution.)

Michie (1980, p. 6) indicates that the cognitive strategies of human experts in more complex domains are based "...not on elaborate calculations, but on the mental storage and use of large incremental catalogs of pattern-based rules." Thus, human chess masters may be able to acquire, organize and utilize as much as 50,000 pattern-based rules in achieving their remarkable performance. Michie (1980, p. 20-21) also indicates that such rules are so powerful that only some 30 rules are needed for expert system performance for a chess subdomain such as King and Knight against King and Rook, which has a problem space size of roughly 2,000,000 configurations. He further observed for chess that the number of rules required grows slowly relative to the increase in domain complexity. Thus, in chess and other complex domains (such as industrial routing and scheduling) it appears that well chosen pattern sets may maintain control over otherwise intractable explosions of combinatorial complexity.

V. The Inference Engine

The problem-solving paradigm, and its methods, organizes and controls the steps taken to solve the problem. One commonplace but powerful paradigm involves the chaining of IF-THEN rules to form a line of reasoning. If the chaining starts from a set of conditions and moves toward some (possibly remote) conclusion, the method is called forward chaining. If the conclusion is known (e.g., it is a goal to be achieved), but the path to that conclusion is not known, then working backwards is called for, and the method is backward chaining. (Heuristic Programming Project, 1980, p. 6)

The problem with forward chaining, without appropriate heuristics for pruning , is that you would derive everything possible whether you needed it or not. Backward chaining works from goals to subgoals (by using the action side of rules to deduce the condition side of the rules). The problem here, again without appropriate heuristics for guidance, is the handling of conjunctive subgoals. In general to attack a conjunction, one must find a case where all interacting subgoals are satisfied, a search for which can often result in a combinatorial explosion of possibilities. Thus, appropriate domain heuristics and suitable inference schemes and architectures must be found for each type of problem to achieve an efficient and effective expert system.

The knowledge of a task domain guides the problem solving steps taken. Sometimes the knowledge is quite abstract--for example, a symbolic model of "how things work" in the domain. Inference that proceeds from the model's abstractions to more detailed (less abstract) statements is called model-driven inference. Always, when one is moving from more abstract symbolic statements to less abstract statements, one is generating expectations, and the problem-solving behavior is termed "expectation driven". Often in problem solving, however, one is working "upwards" from the details or the specific problem data to the higher levels of abstraction (i.e., in the direction of "what it all means"). Steps in this direction are call "data driven". If you choose your next step either on the basis of some new data or on the basis of the last problem-solving step taken, you are responding to events, and the activity is called "event driven". (Heuristic Programming Project 1980, p. 6).

As indicated earlier, an expert system consists of three major components, a set of rules, a global data base and a rule interpreter. The rules are actuated by patterns (which, depending on the direction of search, match either the IF or THEN sides of the rules) in the global data base. The application of the rule changes the system status and therefone the data base, enabling some rules and disabling others. The rule interpreter uses a control strategy for finding the enabled rules and deciding which rule to apply. The basic control strategies used may be top down (goal driven), bottom up (data driven), or a combination of the two that uses a relaxation-like convergence process to join these opposite lines of reasoning together at some intermediate point to yield a problem solution.

Befone proceeding with the architecture of expert systems, it is appropriate to briefly pause to motivate the further pursuit of expert systems, by outlining some potential uses for such systems.

VI. Uses of Expert Systems

The potential uses of expert systems appear to be virtually limitless. They can be used to:

- diagnose
- monitor
- analyze
- interpret
- consult
- plan
- design
- instruct
- explain
- learn
- conceptualize

Thus they are applicable to:

- Mission planning, monitoring, tracking and control
- Communication
- Signal analysis
- Command and control
- Intelligence analysis
- Targeting
- Construction and manufacturing
  - design, planning, scheduling, control . Education
  - instruction, testing, diagnosis
- Equipment
  - design, monitoring, diagnosis, maintenance, repair, operation, instruction
- Image Analysis and. Interpretation
- Professions (law, medicine, engineering, accounting, law enforcement)
  - Consulting, instruction, interpretation, analysis
- Software
  - Specification, design, verification, maintenance, instruction
- Weapon Systems
  - Target identification, electronic warfare, adaptive control.

VII. Architecture of Expert Systems

A. Introduction

One way to classify expert systems is by function (e.g. diagnosis, planning, etc.). However, examination of existing expert systems indicates that there is little commonalty in detailed system architecture that can be detected from this classification.

A more fruitful approach appears to be to look at problem complexity and problem structure and deduce what data and control structures might be appropriate to handle these factors.

The Knowledge Engineering community has evolved a number of techniques which can be utilized in devising suitable expert system architectures. These techniques(5) are described in the following portions of this section. The use of these techniques in existing expert systems is summarized in Table 1.

The approaches used in the various expert systems are different implementations of two basic ideas for overcoming the combinatorial explosion associated with search in real complex problems. These two ideas are:

(1) Find ways to efficiently search a space,

(2) Find ways to transform a large search space into smaller manageable chunks that can be searched efficiently.

It will be observed from Table 1 that there is little architectural commonalty based either on function or domain of expertise. Instead, expert system design may best be considered as an art form, like custom home architecture, in which the chosen design can be implemented using the collection of techniques discussed below.

B. Choice of Solution Direction

1. Forward Chaining

When data or basic ideas are a starting point, forward chaining is a natural direction for problem solving. It has been used in expert systems for data analysis, design, diagnosis, and concept formation.

2. Backward Chaining

This approach is applicable when a goal or a hypotheses is a starting point. Expert system examples include those used for diagnosis and planning.

3. Forward and Backward Processing Combined

When the search space is large, one approach is to search both from the initial state and from the goal or hypothesis state and utilize a relaxation type approach to match the solutions at an intermediate point. This approach is also useful when the search space can be divided hierarchically, so both a bottom up and top down search can be appropriately combined. Such a combined search is particularly applicable to complex problems incorporating uncertainties, such as speech understanding as exemplified in HEARSAY II.

TABLE 1: CHARACTERISTICS OF EXAMPLE EXPERT SYSTEMS

System: MYCIN Function: Diagnosis Domain: Medicine
Search Direction: Backward
Control Mechanizm: Exhaustive Search
Search Space Transformations: None

System: DENDRAL Function: Data Interp. Domain: Chemistry
Search Direction: Forward
Control Mechanizm: Generate and Test
Search Space Transformations: None

System: EL Function: Analysis Domain: Electric Circ.
Search Direction: Forward
Control Mechanizm: Guessing & Relevent Backtracking
Search Space Transformations: None

TABLE 1 (continued): CHARACTERISTICS OF EXAMPLE EXPERT SYSTEMS

System	Function	Domain	Search Direction	Control Mechanizm	Search Space Transformations
GUIDON	Instruction	Medicine	Event Driven	Per User	None
META-DENDRAL	Learning	Chemistry	Forward	Generate and Test	None
AM	Concept Formation	Mathematics	Forward	Generate and Test	None
VM	Monitoring	Medicine	Event Driven	Exhaustive Search	None
GA1	Data Interp.	Chemistry	Forward	Generate and Test	None
R1	Systems Design	Computers	Forward	Exhaustive Search	Break Into Sub-Problems
NOAH	Planning	Robotics	Backward	Least Committment	Break Into Sub-Problems
SYN	Design Circuits	Electric	Forward	Exhaustive Search	Multiple Models
MOLGEN	Design	Genetics	Backward and Forward	Guessing, Relevent Backtracking, & Least Committment	Break Into Sub-Problems, Meta Rules, & Hierarchical Refinement
HERARSAY II	Signal Interp.	Speach Under.	Backward and Forward	Multilines of Reasoning & Network Editor	Hierarchical Resolution
CRYSALIS	Data Interpretation	Cystallography	Event Driven	Generate and Test	Meta Rules

4. Event Driven

This problem solving direction is similar to forward chaining except that the data or situation is evolving over time. In this case the next step is chosen either on the basis of new data or in response to a changed situation resulting from the last problem solving step taken. This event driven approach is appropriate for real-time operations, such as monitoring or control, and is also applicable to many planning problems.

C. Reasoning in the Presence of Uncertainty

In many cases, we must deal with uncertainty in data or in knowledge. Diagnosis and data analysis are typical examples. As discussed below, numeric techniques can be utilized or the uncertainties can be handled by incorporating a form of backtracking.

1. Numeric Procedures

Numeric procedures have been devised to handle approximations by combining evidence. MYCIN utilizes "certainty factors" (related to probabilities) which use the range of 0 to 1 to indicate the strength of the evidence. Fuzzy set theory, based on possibilities, can also be utilized.

2. Belief Revision or "Truth Maintenance"

Often, beliefs are formed or lines of reasoning are developed based on partial or errorful information. When contradictions occur, the incorrect beliefs or lines of reasoning causing the contradictions, and all wrong conclusions resulting from them, must be retracted. To enable this, a data-base record of beliefs and their justifications must be maintained. Using this approach, truth maintenance techniques can exploit redundancies in experimental data to increase system reliability.

D. Searching A Small Search Space

Many straightforward problems in areas such as design, diagnosis and analysis have small search spaces, either because 1) the problem is small or 2) the problem can be broken up into small independent subproblems. In many cases, a single line of reasoning is sufficient and so backtracking is not required. In such cases, the direct approach of exhaustive search can he appropriate, as was used in MYCIN and R1.

E. Techniques for Searching A Large Search Space

1. Hierarchical Generate and Test

State space search is frequently formulated as "generate and test" - reasoning by elimination. In this approach, the system generates possible solutions and a tester prunes those solutions that fail to meet appropriate criteria. Such exhaustive reasoning by elimination can be appropriate for small search spaces, but for large search spaces more powerful technique are needed. A "hierarchical generate and test" approach can be very effective if means are available for evaluating candidate solutions that are only partially specified. In this case, early pruning of whole branches (representing entire classes of solutions associated with these partial specifications) is possible, massively reducing the search required. "Hierarchical generate and test" is appropriate for many large data interpretation and diagnosis problems, for which all solutions are desired, providing a generator can be devised that can partition the solution space in ways that allow for early pruning.

2. Dependency-Directed Backtracking

In the "generate and test" approach, when a line of reasoning fails and must be retracted, one approach is to backtrack to the most recent choice point (chronological backtracking). However, it is often much more efficient to trace errors and inconsistencies back to the inferential steps that created them, using dependency records as is done in EL and MOLGEN. Backtracking that is based on dependencies and determines what to invalidate is called dependency-directed (or relevant) backtracking. Note the similarity to truth maintenance. When dealing with errorful information, backtracking is referred to as truth maintenance. Backtracking to recover from incorrect lines of reasoning, is referred to as dependency-directed backtracking.

3. Multiple Lines of Reasoning

This approach can be used to broaden the coverage of an incomplete search. In this case, search programs that have fallible evaluators can decrease the chances of discarding a good solution from weak evidence by carrying a limited number of solutions in parallel, until which of the solutions is best is clarified.

F. Methods for Handling A Large Search Space by Transforming the Space

1. Breaking the Problem Down Into Subproblems

a. Non-Interacting Subproblems

This approach (yielding smaller search spaces) is applicable for problems in which a number of non-interacting tasks have to be done to achieve a goal. Unfortunately, few real world problems of any magnitude fall into this class.

b. Interacting Subproblems

For most complex problems that can be broken up into subproblems, it has been found that the subproblems interact so that valid solutions cannot be found independently. However, to take advantage of the smaller search spaces associated with this approach, a number of techniques have been devised to deal with these interactions. Among these are:

(1) Find A Fixed Sequence of Subproblems So That No Interactions Occur

Sometimes it is possible to find an ordered partitioning so that no interactions occur. The R1 system for configuring VAX computers successfully takes this approach.

(2) Least Commitment

This technique coordinates decision-making with the availability of information and moves the focus of problem-solving activity among the available subproblems. Decisions are not made arbitrarily or prematurely, but are postponed until there is enough information. In planning problems this is exemplified by methods that assign a partial ordering of operators in each subproblem and only complete the ordering when sufficient information or the interactions of the subproblems is developed.

(3) Constraint Propagation

Another approach is to represent the interaction between the subproblems as constraints. Constraints can be viewed as partial descriptions of entities, or as relationships (subgoals) that must be satisfied. Constraint propagation is a mechanism for moving information between subproblems. By introducing constraints instead of choosing particular values, a problem solver is able to pursue a least commitment style of problem solving.

(4) Guessing or Plausible Reasoning

Guessing is an inherent part of heuristic search, but is particularly important in working with interacting subproblems. For instance, in the least commitment approach the solution process must come to a halt when it has insufficient information for deciding between competing choices. In such cases, heuristic guessing is needed to carry the solution process along. If the guesses are wrong, then dependency-directed backtracking can be used to efficiently recover from them. EL and MOLGEN take this approach.

2. Hierarchical Refinement into Increasingly Elaborate Spaces -- Top Down Refinement

Often, the most important aspects of a problem can be abstracted and a high level solution developed. This solution can then be iteratively refined, successively including more details. An example is to initially plan a trip using a reduced scale map to locate the main highway, and then use more detailed maps to refine the plan. This technique has many applications as the top level search space is suitably small. The resulting high level solution constrains the search to a small portion of the search space at the next lower level, so that at each level the solution can readily be found. This procedure is an important technique for preventing combinatorial explosions in searching for a solution.

3. Hierarchical Resolution into Contributing Sub-Spaces(6)

Certain problems can have their solution space hierarchically resolved into contributing subspaces in which the elements of the higher level spaces are composed of elements from the lower spaces. Thus, in speech understanding, words would be composed of syllables, phrases of words, and sentences of phrases. The resulting heterogeneous subspaces are fundamentally different from the top level solution state. However the solution candidates at each level are useful for restricting the range of search at the adjacent levels, again acting as an important restraint on combinatorial explosion. Another example of a possible hierarchical resolution is in electrical equipment design where subcomponents contribute to the black box level, which in turn contribute to the system level. Similarly, examples car be found in architecture, and in spacecraft and aircraft design.

G. Methods for Handling A Large Search Space by Developing Alternative or Additional Spaces

1. Employing Multiple Models

Sometimes the search for a solution utilizing a single model is very difficult. The use of alternative models for either the whole or part of the problem may greatly simplify the search. The SYN program is a good example of combining the strengths of multiple models by employing equivalent forms of electrical circuits.

2. Meta Reasoning

It is possible to add additional layers of spaces to a search space to help decide what to do next. These can be thought of as strategic and tactical layers in which meta problem solvers choose among several potential methods for deciding what to do next at the problem level. The strategy, focusing, and scheduling meta rules used in CRYSTALIS and the use of a strategy space in MOLGEN fall into this category.

H. Dealing with Time

The following are two approaches to dealing with time in terms of time intervals.

1. Situational Calculus

Situational calculus was an early approach by McCarthy and Hayes (1969) for representing sequences of actions and their effects. It uses the concept of "situations" which change when sufficient actions have taken place, or when new data indicates a situational shift is appropriate. Situations determine the context for actions and, through the use of "frames,"(7) can indicate what changes and what remains the same when an action takes place. VM uses the situation approach for monitoring patient breathing.

2. Planning with Time Constraints

NOAH was an early parallel planner which dealt with interacting subgoals. The method of least commitment and backward chaining initially produced a partial ordering of operators for each plan. When interference between subgoal plans was observed, the planner adjusted the ordering of the operators to resolve the interference to produce a final parallel plan with time ordered operators. DEVISER (Vere, 1981) is a recent derivative of NOAH which extends this parallel planning approach to treat goals with time constraints and durations. The principal output of DEVISER is a partially ordered network of parallel activities for use in planning a spacecraft's actions during a planetary flyby.

VIII. Constructing An Expert System

Duda (1981, p. 262) states that to construct a successful expert system, the following prerequisites must be met: There must be at least one human expert acknowledged to perform the task well. The primary source of the expert's exceptional performance must be special knowledge, judgment, and experience. The expert must be able to explain the special knowledge and experience and the methods used to apply them to particular problems. Finally, the task must have a well-bounded domain of application. Randy Davis (1982) noted that a good expert system application: doesn't require common sense, takes an expert a few minutes to a few hours, and has an expert available and willing to be committed. Hayes-Roth (1981, p. 2) adds that "...the problem should be nontrivial but tractable, with promising avenues for incremental expansion."

Having found an appropriate problem and an accessible expert, it is then necessary to have available an appropriate system-building tool. Realistic and incremental objectives should then be set. Major pitfalls to be avoided in developing an expert system are choosing an inappropriate problem, excessive aspirations, and inadequate resources.

Lenat et al. (1982) report that usually that first body of knowledge extracted from the expert are terms, facts, standard procedures, etc.; as one might find in textbooks or journals. However this information is in sufficient to build a high performance system. Thus a cycle procedure is followed to improve the program--a sample case is run, with the expert disagreeing with its reasoning at some point. This forces the expert to introspect on what additional knowledge as needed. This often elicits a judgmental rule (heuristic) from the expert. As more and more heuristics are added to the program, the system incrementally approaches the competence of the expert at the task.

The time for construction of early expert systems was in the range of 20-50 man-years. Recently, breadboard versions of simple expert systems have been reported to have been built in as little as 3 man-months, but a complex system is still apt to take as long as 10 man-years to complete. Using present techniques, the time for development appears to be converging towards 5 man-years per system. Most systems take 2-5 people to construct, but not more. (It takes one to two years to develop an engineer or computer scientist into a knowledge engineer.)

Randy Davis (1982) indicates that the stages of development of an expert system can be considered to be:

1. System design
2. System development (conference paper level)
3. Formal evaluation of performance
4. Formal evaluation of acceptance
5. Extended use in prototype environment
6. Development of maintenance plans
7. System release

Buchanan and Duda (1982) summarize some of the actual considerations and experiences involved in expert system construction. A good report of the difficulties actually experienced in the development of R1 is provided by McDermott (1981).

IX. Knowledge Acquisition and Learning

A. Knowledge Acquisition

The key bottleneck in developing an expert system is building the knowledge base by having a knowledge engineer interact with the expert(s). Expert systems can be used to facilitate the process.

The most ambitious of these systems is TEIRESIAS (Davis and Lenat, 1982) which supervises interaction with an expert in building or augmenting a MYCIN rule set. TEIRESIAS uses a model of MYCIN's knowledge base to tell whether some new piece of information "fits in" to what is already known, and uses this information to make suggestions to the expert. An appropriate expert may not always be continuously available during the construction of the expert system, and in many cases may not have all the expertise desired. In these cases other approaches to acquiring the needed expertise are desirable.

B. Self-learning and Discovery

Michie (1980, p. 11) observes that "the rule-based structure of expert systems facilitates acquisition by the system of new rules and modification of existing rules, not only by tutorial interaction with a human domain specialist but also by autonomous "learning." A typical functional application is "classification," for which rules are discovered by induction for large collections of samples (Quinlin, 1979). Michie (1980, p. 12) provides a list of examples of various "learning" expert systems.

DENDRAL, for obtaining structural representations of organic molecules, is the most widely used expert system. As the knowledge acquisition bottleneck is a critical problem, a META-DENDRAL expert system was written to attempt to model the processes of theory formation to generate a set of general fragmentation rules of the form used by DENDRAL. The method used by META-DENDRAL is to generate, test and refine a set of candidate rules from data of known molecule structure-spectrum pairs. For META-DENDRAL and several of the other learning expert systems, the generated rules were found to be of high quality (Feigenbaum, 1980 and Michie, 1980).

Another attempt at modeling self-learning and discovery is the AM Program (Davis and Lenat, 1982) for discovery of mathematical concepts, beginning with elementary ideas in set theory. AM also uses a "generate and test" control structure. The program searches a space of possible conjectures that can be generated from the elementary ideas in set theory, chooses the most interesting, and pursues that line of reasoning. The program was successful in rediscovering many of the fundamental notions of mathematics, but eventually began exploring a bigger search space than the original heuristic knowledge given to it could cope with. A more recent project - EURISKO - is exploring how a program can devise new heuristics to associate with new concepts as it discovers them.

Footnotes

(1) Heuristics are "rules of thumb" (empirical rules), common-sense knowledge or other techniques that can be used to help guide search.

(2) One important aspect of the knowledge-based approach is that the combinatorial complexity associated with real-world problems is mitigated by the more powerful focusing of the search that can be obtained with rule-based heuristics usually used in expert systems as opposed to the numerical heuristics (evaluation functions) used in classical search techniques. In other words, the rule-based system is able to reason about its own search effort, in addition to reasoning about the problem domain. (Of course, this also implies that the search strategy is incomplete. Solutions may be missed, and an entire search may fail even when here is a solution "within each" in the problem state defined by the domain.)

(3) There are not yet many examples of expert systems whole performance consistently surpasses that of an expert. And currently, there are even fewer examples of expert systems that use knowledge from a group of experts and integrate it effectively. However the promise is there.

(4) Not all expert systems are rule-based. The network-based expert systems MACSYMA, INTERNIST/CADUCEDUS, Digitalis Therapy Advisor, and PROSPECTOR are examples which are not. Factors involved in choosing an appropriate knowledge representation are discussed in Barr and Feigenbaum (1981, Chapter III). Buchanan and Duda (1982) state that the basic requirements in the choice of an expert system knowledge representation scheme are extendibility, simplicity and explicitness. Thus, rule-based systems are particularly attractive.

(5) This chapter is largely derived from information contained in the excellent tutorial by Stefik et al. (1982).

(6) This approach has certain similarities to breaking the problem down into subproblems.

(7) A frame is a data structure for describing a stereotyped situation.

References

1. Michie, Donald, "Knowledge-based Systems," University of IL at Urbana-Champaign, Report 80-1001, Jan. 1980.

2. Feigenbaum, E. A., "Knowledge Engineering: The Applied Side of Artificial Intelligence," Computer Science Dept., Memo HPP-80-21, Stanford University, July, 1980.

3. Feigenbaum, E. A., "Knowledge Engineering for the 1980's," Computer Science Dept., Stanford University, 1982.

4. Nii, H. P., and Aiello, N., "AGE (Attempt to Generalize): A Knowledge-Based Program for Building Knowledge-Based Programs," Proceedings of the Sixth International Conf. on Artificial Intelligence, (IJCAI-79), Tokyo, Aug., 20-23, 1979, pp. 645-655.

5. Buchanan, B. G., "Research on Expert Systems," Stanford University Computer Science Department, Report No. STAN-CS-81-837, 1981.

6. Hayes-Roth, F., "AI The New Wave - A Technical Tutorial for R & D Management," (AIAA-81-0827), Santa Monica, CA: Rand Corp., 1981.

7. Duda, R. O. "Knowledge-Based Expert Systems Come of Age," Byte, Vol. 6, No. 9, Sept. 81, pp. 238-281.

8. Heuristic Programming Project 1980, Computer Science Dept., Stanford University, 1980.

9. Stefik, M., et al., "The Organization of Expert Systems: A Prescriptive Tutorial", XEROX, Palo Alto Research Centers, VLSI-82-1, January 1982 (also in Artificial Intelligence 18(2), Mar. 1982, pp. 135-173).

10. Quinlin, J. R., "Discovering Rules by Induction from Large Collections of Examples," in Expert Systems in the Micro-Electronic Age, D. Michie (Ed.), Edinburgh: Edinburgh University Press, 1979, pp. 168-201.

11. Davis, R, and Lenat, D. B. Knowledge-Based Systems in Artificial Intelligence, New York: McGraw-Hill, 1982.

12. Lindsay, R. K., et al., Applications of Artificial Intelligence for Organic Chemistry: The DENDRAL Project, New York: McGraw-Hill, 1980.

13. McCarthy, J., and Hayes, P. J., "Some Philosophical Problems from the Standpoint of Artificial Intelligence," in Machine Intelligence 4, Meltzer, B., and Michie, D. (Eds.)., Halsted Press, 1969, pp. 463-502.

14. Vere, S., Planning in Time: Windows and Durations for Activities and Goals, Pasadena, CA: JPL, Nov. 1981.

15. The Seeds of Artificial Intelligence: Sumex-AIM, NIH Publ. No. 80-2071, Bethesda, MD: NIH Div. of Research Resources, March 1980.

16. Buchanan, B. G. and Duda, R. O., "Principles of Rule-Based Expert Systems," Heuristic Programming Project Report No. HPP 82-14, Dept. of Computer Science, Stanford Univ., Aug. 1982. (To appear in Advances in Computers, Vol. 22, Yorit, (ed.), New York: Academic Press).

17. Davis, R., "Expert Systems: Where Are We? and Where Do We Go From Here?" AI Magazine 3(2), 1982, pp. 3-22.

18. Brooks, A., "A Comparison Among Four Packages for Knowledge-Based Systems," Proc. of Int. Conf. on Cybernetics and Society, 1981, pp. 279-283.

19. McDermott, J., "R1, The Formative Years," AI Magazine Summer 1981, pp. 21-30.

20. Ennis, S. P., "Expert Systems: A User's Perspective of Some Current Tools," AAAI-82, pp. 319-321.

21. Barr, A., and Feigenbaum, E. A., The Handbook of Artificial Intelligence, Vol. 1, Los Altos, CA: W. Kaufman, 1981.

22. Barr, A., and Feigenbaum, E. A., The Handbook of Artificial Intelligence,
Vol. 2, Los Altos, CA: W. Kaufman, 1982.

23. Lenat, D. B., Sutherland, M. R., and Gibbons, J., "Heuristic Search for New Microcircuit Structures: An Application of Artificial Intelligence," AI Magazine, Summer 1982, pp. 17-33.

24. AAAI-82 - Proc. of the National Conference on AI, CMU & U. of Pittsburg, Pittsburg, PA, Aug. 18-22, 1982.

25. IJCAI-81 - The International Joint Conference on AI, Vancouver, Aug. 1981.

Chapter 2: Evaluating Expert System Shells

Upon learning the basics, the fastest way to put expert systems technology to work is to use a shell. Expert system shells permit a human expert to capture and use knowledge without having to write a great deal of supporting code. Shells have already incorporated in them various inferencing techniques (methods of using knowledge) and an environment that enhances the knowledge acquisition process. What is difficult, especially for someone new to expert systems, is finding the right shell for the given application. It is essential to know what shells are available and the capabilities of those shells. Then the shell's capabilities must be compared against the requirements of the application and the needs of the end-user. The six articles in this chapter assist the developer in making those determinations. The article by Gevarter discusses the features shells should have and charts the specific features of several shells. Citrenbaum and Geissman discuss shells from the point of view of various types of users and applications. Press details some benchmark performance metrics that can be applied to expert systems in general. The next three papers illustrate a number of shells for small computers. Freedman compares the features of 27 different shells. Barr and Raeth, in separate papers, each review in detail expert systems shells that they found well suited for the developer embarking on that first project. The shells ESIE (Barr) and CLIPS (Raeth) are available quite inexpensively through distributors of "user supported" software.

The Nature and Evaluation of Commercial Expert System Building Tools

Reprinted from *Computer*, May 1987, pages 24-41.

William B. Gevarter
NASA Ames Research Center

ESBTs make it possible to build an expert system in an order of magnitude less time than is possible with Lisp alone. This article reviews such tools.

The development of new expert systems is changing rapidly—in terms of both ease of construction and time required—because of improved expert system building tools (ESBTs.) These tools are the commercialized derivatives of artificial intelligence systems developed by AI researchers at universities and research organizations. It has been reported that these tools make it possible to develop an expert system in an order of magnitude less time than would be required with the use of traditional development languages such as Lisp. In this article, I review the capabilities that make an ESBT such an asset and discuss current tools in terms of their incorporation of these capabilities.

The structure of an expert system building tool

The core of an expert system consists of a knowledge base and an accompanying inference engine that operates on the knowledge base to develop a desired solution or response. If one is to use such a system, an end-user interface or an interface to an array of sensors and effectors is required for communication with the *relevant world*. (A "relevant world" is a system or situation operated on by or in contact with the expert system.) In addition, to facilitate the development of an expert system, an ESBT must also include an interface to the developer

- so that the requisite knowledge base can be built for the particular application domain for which the system is intended,
- so that the appropriate end-user interface can be developed, and
- to incorporate any special instructions to the inference engine (reasoning system) that are required for the particular domain.

The character and quality of these interfaces are two of the main differentiations between commercial tools and ESBTs developed at universities and used in research. Also important in the structure of ESBTs are

- interfaces to other software and databases, and
- the computers on which the ESBTs will run—not only the computers used for development of expert systems, but also those used for their delivery to an end user.

Figure 1 summarizes the structure of an ESBT.

Knowledge representation

The knowledge that can be easily represented by the tool is a key consideration in choosing an ESBT. As indicated by Figure 2, there are three aspects of knowledge representation that are fundamental to these tools—*object descriptions* (declarative knowledge such as facts), *certainties*, and *actions*. One method of representing objects is by frames with or without *inheritance*. (Inheritance allows knowledge bases to be organized as hierarchical collections of frames that inherit information from frames above them. Thus, an inheritance mechanism provides a form of inference.) *Frames* are tabular data structures for organizing representations of prototypical objects or situations. A frame has slots that are filled with data on objects and relations appropriate to the situation. One version of programming referred to as *object-oriented programming* utilizes objects that incorporate provisions for message passing between objects; attached to these objects are procedures that can be activated by the receipt of messages. Declarative knowledge can also be represented by parameter-value pairs, by use of logic notation, and, to some extent, by rules.

Actions change a situation and/or modify the relevant database. Actions are most commonly represented by rules. These rules may be grouped together in modules (usually as subparts of the problem) for easy maintenance and rapid access. Actions may also be represented in terms of *examples*, which indicate the conclusions or decisions reached. Examples are a particularly desirable form of representation for facilitating knowledge acquisition, and inductive systems capitalize on them. Examples are much easier to elicit from experts than rules, and may often be a natural form of domain knowledge. Actions can also be expressed in logic notation, which is a form of rule representation. Finally, actions can be expressed as procedures elicited by either

- messages (in object-oriented programming) or
- changes in a global database that are observed by *demons*. ("Demons" are procedures that monitor a situation and respond by performing an action when their activating conditions appear.)

In addition to the representation of objects and actions, one must consider the degree to which the knowledge or data is known to be correct. Thus, most ESBTs have provisions for representing certainty. The most common approach is to incorporate "confidence factors"; this approach is a derivative of the approach used in the Mycin expert system.[1] Fuzzy logic and probability are also used.

An alternative way of handling uncertainties or tentative hypotheses is to consider multiple worlds in which different items are true or not true in these alternative worlds. Another consideration is whether or not a *deep model* (which is a structural or causal model) of the system can readily be built with the tool in question as an aid in model-based reasoning. (The same underlying model can often be employed for other uses, such as preservation of knowledge and training.) Finally, system size (for example, as measured by the number of rules needed) can be of critical importance, as it can have an important effect on memory requirements, memory management, and runtimes.

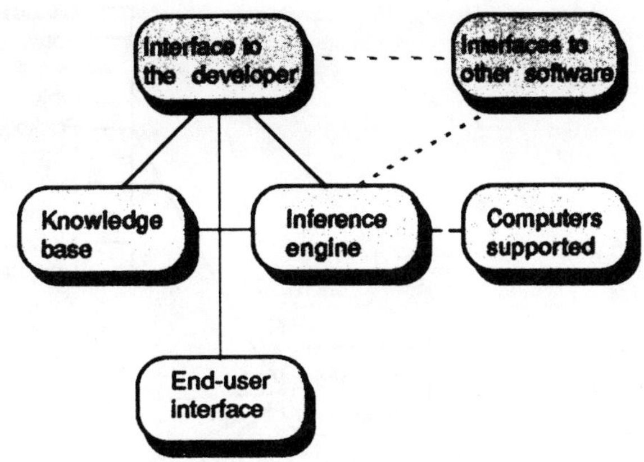

Figure 1. The structure of an expert system building tool. (Solid lines represent basic relationships, and broken lines represent related aspects.)

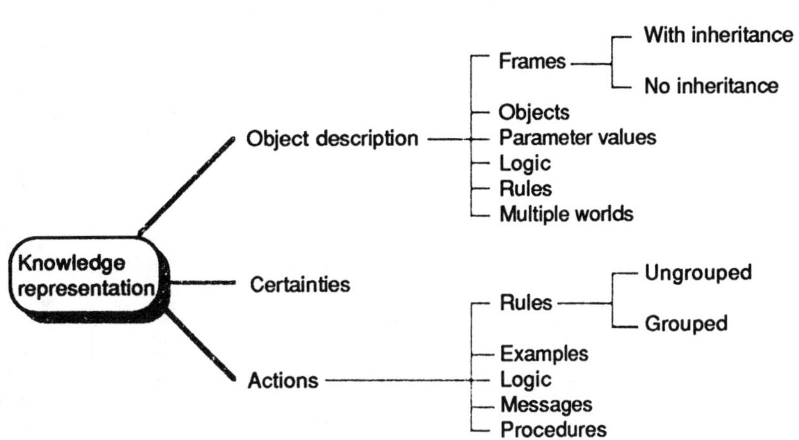

Figure 2. Different methods of knowledge representation.

Inference engine

Figure 3 indicates the major alternative means by which an ESBT performs inferencing. The most usual approach is *classification*, which is appropriate for situations in which there is a fixed number of possible solutions. Hypothesized conclusions from this set are evaluated as to whether they are supported by the evidence. This evaluation is usually done by *backward chaining* through *if-then* (that is, antecedent-consequent) rules, starting with

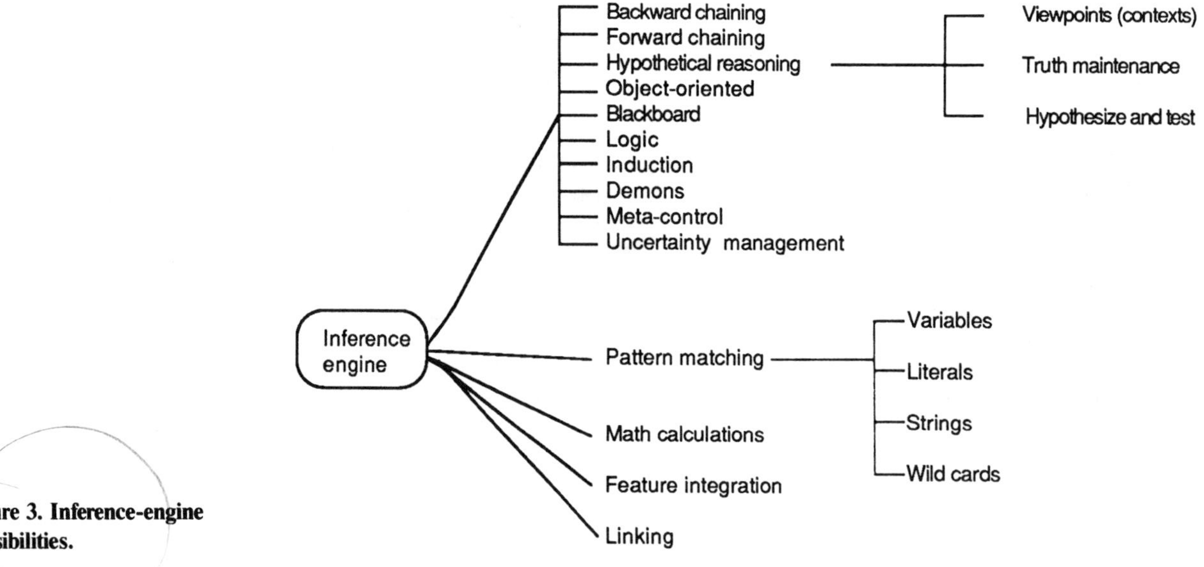

Figure 3. Inference-engine possibilities.

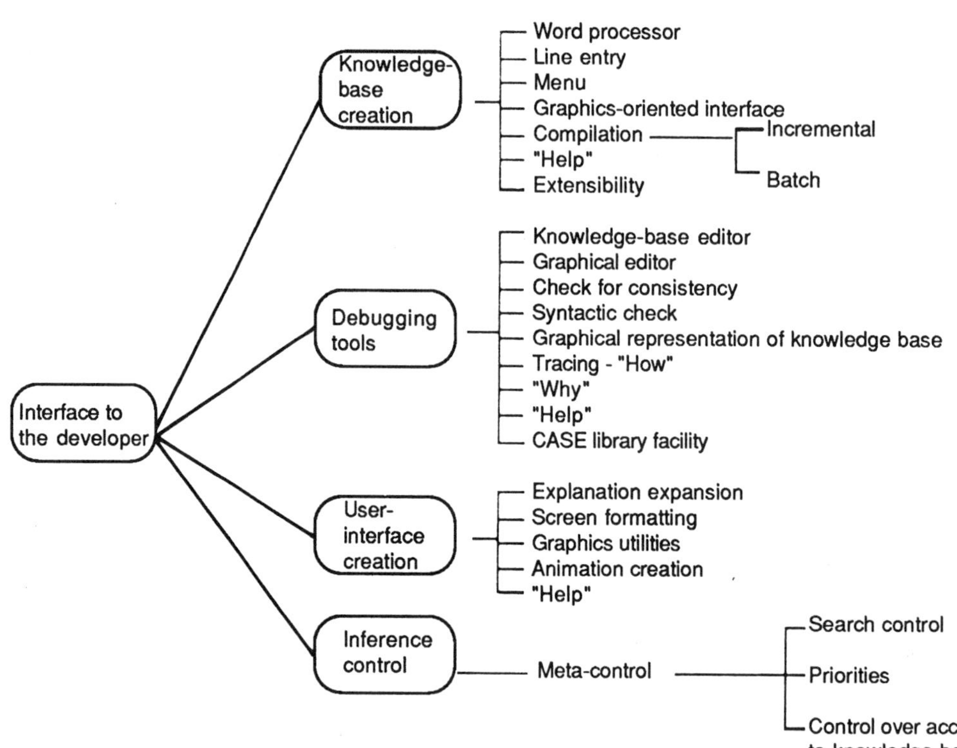

Figure 4. Possibilities for the interface to the developer.

rules that have the hypothesized conclusions as their *consequents*. Rules are then searched for those that have as their consequent the conditions that support the *antecedents* (input conditions) in the hypothesized conclusion rule. This process continues recursively until the hypothesis is fully supported or until either a negation or a dead-end is reached. If either of the latter two events happens, additional hypotheses may be tried until some conclusion is reached or the process is terminated. This depth-first, backward-chaining approach was popularized by the Mycin expert system. The corresponding Emycin ESBT shell[2] is the prototype of virtually all the hypothesis-driven (that is, *goal-driven*) commercial ESBTs currently available.

Forward chaining starts with data to be input or with the situation currently present in a global database. The data or the situation is then matched with the antecedent conditions in each of the relevant rules to determine the applicability of the rule to the current situation. (The current situation is usually represented in the global database by a set of attributes and their associated values.) One of the matching rules is then selected (for example, by the use of *meta-rules*, which help determine the order in which the rules are tried, or by *priorities*), and the rule's consequents are used to add information to the database or to actuate some procedure that changes the global situation. Forward chaining proceeds recursively (in a manner similar to that of backward chaining), terminating either when a desired result or conclusion is reached or when all relevant rules are exhausted. Combinations of forward and backward chaining have also been found useful in certain situations.

Forward reasoning (a more general form of forward chaining) can be done with data-driven rules or with data-driven procedures (demons).

Hypothetical reasoning refers to solution approaches in which assumptions may have to be made to enable the search procedure to proceed. However, later along the search path, it may be found that certain assumptions are invalid and therefore have to be retracted. This *nonmonotonic reasoning* (that is, reasoning in which facts or conclusions must be retracted in light of new information) can be handled in a variety of ways. One approach that reduces the difficulty of the computation is to carry along multiple solutions (these solutions represent different hypotheses) in parallel and to discard inappropriate ones as evidence that contradicts them is gathered. This approach is referred to as *viewpoints*, *contexts*, and *worlds* in different tools. Another approach is to keep track of the assumptions that support the current search path and to backtrack to the appropriate branch point when the current path is invalidated. This latter approach has been referred to by names like *nonchronological backtracking*. A related capability is *truth maintenance*, which removes derived beliefs when their conditions are no longer valid.

Object-oriented programming is an approach in which both information about an object and the procedures appropriate to that object are grouped together into a data structure such as a frame. These procedures are actuated by messages that are sent to the object from a central controller or another object. This approach is particularly useful for simulations involving a group of distinct objects and for real-time signal processing.

The *blackboard inference approach* is associated with a group of cooperating expert systems that communicate by sharing information on a common data structure that is referred to as a "blackboard." An agenda mechanism can be used to facilitate the control of solution development on the blackboard.

In the case of ESBTs, *logic* commonly refers to a theorem-proving approach involving *unification*. "Unification" refers to substitutions of variables performed in such a way as to make two items match identically. The common logic implementations are versions of a logic-programming language, Prolog, that utilize a relatively exhaustive depth-first search approach.

An important inference approach found in some tools is the ability to generate rules or decision trees inductively from examples. Human experts are often able to articulate their expertise in the form of examples better than they are able to express it in the form of rules. Thus, inductive learning techniques (which are currently limited in their expressiveness) are frequently ideal methods of knowledge acquisition for rapid prototyping when examples can be simply expressed in the form of a conclusion associated with a simple collection of attributes. The human builders of the resultant expert system can then refine it iteratively by critiquing and modifying the results inductively produced. *Inductive inference* usually proceeds by starting with one of the input parameters and searching for a tree featuring the minimum number of decisions needed to reach a conclusion. This *minimum-depth tree* is found by cycling through all parameters as possible initial nodes and using an *information theoretic approach* to select the order of the parameters to be used for the remaining nodes and to determine which parameters are superfluous. An "information theoretic approach" is one that chooses the solution that requires the minimum amount of information to represent it. The depth of the tree is usually relatively shallow (often less than five decisions deep), so large numbers of examples usually result in broad, shallow trees.

Some tools incorporate demons that monitor local values and execute procedures when the actuation conditions of the demons appear. These tools are particularly appropriate for monitoring applications.

A number of tools offer a choice of several possible inference or search procedures. In systems built with such tools, means are usually made available to the system builder to control the choice of the inference strategy, which the builder causes to be dependent on the system state. Such control is referred to as *meta-control*. One form of meta-control is the use of *control blocks*, which are generic procedures that tell the system the next steps to take in a given situation so that the search will be reduced, enabling a large number of rules to be accommodated without the search space becoming combinatorially explosive.

As the certainty of data, rules, and procedures is usually less than 100 percent, most systems incorporate facilities for certainty management. Thus, they have various approaches for combining uncertain rules and information to determine a certainty value for the result.

Pattern matching is often required for mechanizing inference techniques, particularly for matching rule antecedents to the current system state. The sophistication of the pattern-matching approach affects the capabilities of the system. Types of pattern matching vary—from matched identical strings to variables, literals, and wildcards, and can even include partial and/or approximate matching that can serve as analogical reasoning.

Other ESBT capabilities vary from tool to tool. Some inference engines offer rapid and sophisticated math-calculation capabilities. One of the more valuable capabilities is supplied by inference engines that can manage modularized knowledge bases or modularized solution subproblems by accessing and linking these modules as needed.

Another important consideration in a tool is the degree of integration of its various features. Full integration is desirable so that all the tool features can be brought to bear, if needed, on the solution of a single problem. For example, in the case of ESBTs incorporating both object representations and forward and backward chaining rules, it is desirable that expert-system developers be able to mix forward and backward chaining rules freely and be able to reason about information stored in objects when these actions are appropriate.

The interface to the developer

Various tools offer different levels of capabilities for the expert-system builder to use to mold the system. The simpler tools are shells into which knowledge is inserted in a specific, structured fashion. The more sophisticated tools are generally more difficult to learn, but allow the system developer a much wider choice of knowledge base representations, inference strategies, and the form of the end-user interface. Various levels of debugging assistance are also provided. Figure 4 provides an indication of the possible options

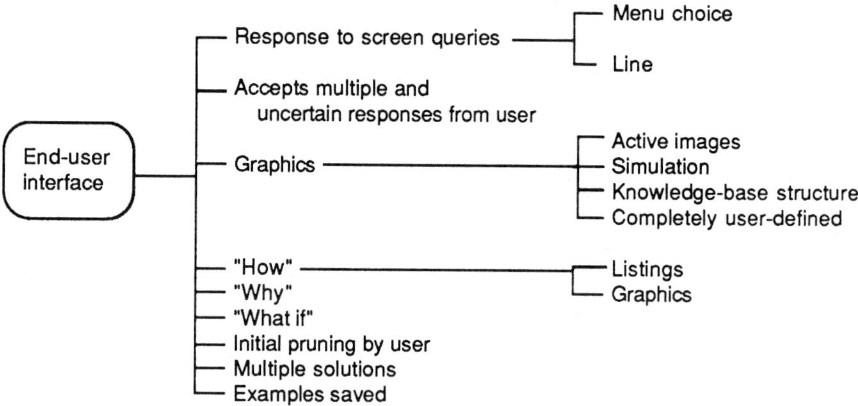

Figure 5. Possibilities for the end-user interface.

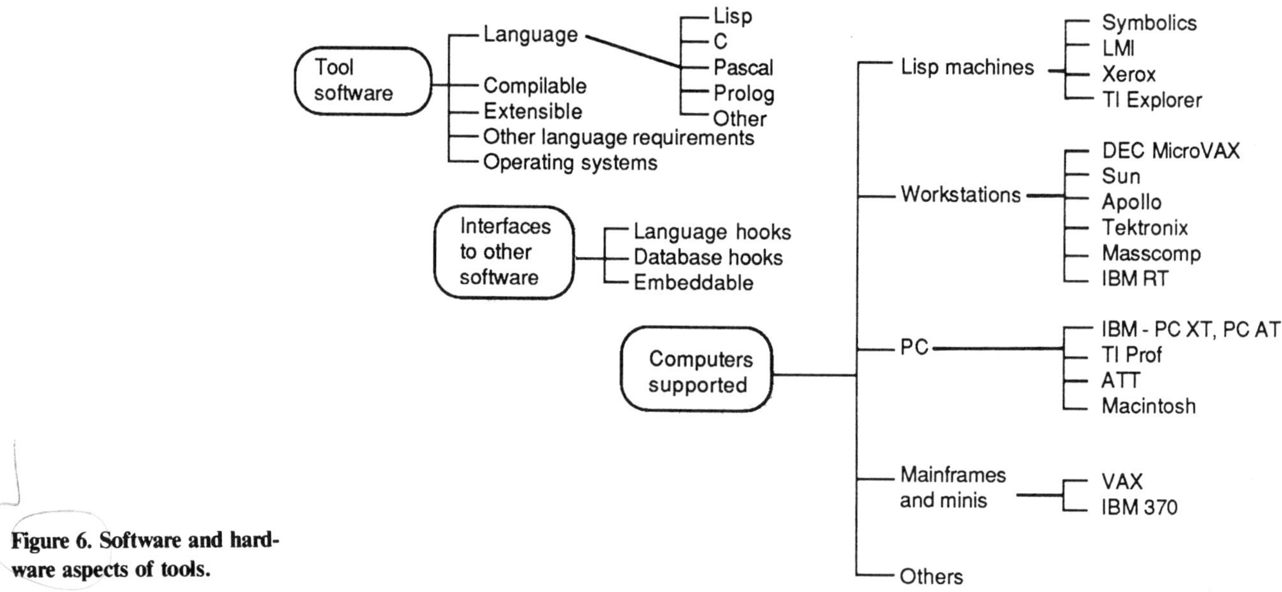

Figure 6. Software and hardware aspects of tools.

(options are tool dependent) that are available for each aspect of the interface to the developer.

End-user interface

Once the expert system has been built, its usability depends in large part on the end-user interface. Figure 5 provides an indication of the range of end-user facilities found in ESBTs. Since most expert systems are really intelligent assistants, the end-user interface is often designed to allow interactive dialogue. This dialogue and/or the initial input most often appear to the user as structured data-input arrangements incorporating menu choices that allow the user to answer requests by the system for information. In some cases, to increase system flexibility, systems will accept multiple and uncertain user responses and still arrive at conclusions (though the certainty of the resultant conclusions is reduced). In sophisticated systems, graphics are often used to show the line of reasoning when the system responds to users' "how" questions; in simpler systems, a listing of the rules supporting the system's conclusions may be employed. ESBTs often answer a user's "Why do you need this information?" question by quoting the rule for which the information is required. The ability of the system to answer the user's "why" and "how" questions is important, for it increases the end user's confidence in the system's decision-making ability.

Other capabilities often found in ESBTs are facilities that allow the end user to select alternative parameter values and observe the effect on the outcome (these facilities support "what if" queries), facilities to allow the user to perform an initial pruning of the line of questioning so that the system need not pursue areas that the user feels are irrelevant or unnecessary, and the capability to save examples for future consideration or use.

Very sophisticated tools often include interactive graphics and simulation facilities that increase the end user's understanding and control of the system being represented. Above all, the end-user interface needs to be user friendly if the system is to be accepted.

Programming-language considerations

In addition to the structure and the paradigms supported by a tool, the programming language in which the tool is written is of major importance. The language determines whether the expert system is compilable and, if it is, whether incrementally or in a batch mode. Compilability reduces the memory requirements and increases the speed of the expert system; incremental compilability speeds development. Figure 6 is illustrative of the aspects related to the tool-language choice.

In general, the more sophisticated tools have been written in Lisp. However, even these tools are now being rewritten in languages such as C to increase speed, reduce memory requirements, and to promote availability on a larger variety of computers. However, some new approaches to mechanizing Lisp may reduce the speed and memory advantages associated with C.

The user can usually extend tools written in Lisp by writing additional Lisp functions. This is also true of some of the other languages, for example, Prolog and Pascal. Similar extensibility is usually found in tools having language hooks for accessing other programs or database hooks for accessing other information. In some cases, the expert system generated by the tool is fully embeddable in other systems, which produces increased autonomy. Whether or not a system is fully embeddable in other systems and is therefore capable of autonomous operations is becoming increasingly important, now that expert systems are moving from prototypes to being fielded. Reliability and memory management (in Lisp, the latter takes the form of *garbage collection*) are often important considerations for fielded systems. "Garbage collection" is the collection of no-longer-used memory allocations; these allocations can slow the system operation.

The computers supported by the various tools are primarily a function of the language and operating system in which the tools are written, and the computer's memory, processing, and graphics-display capabilities. The trend toward making expert system shells available on personal computers (such as those made by IBM) results in part from the increasing capabilities of these computers. However, this trend is also partly owing to the writing of tools in faster languages, such as C, and to taking advantage of modularization in building the knowledge base. As mentioned earlier, such modularization involves decomposing the problem into subproblem modules and providing appropriate linking between these modules as required during operation.

1. Classification
   a. Interpretation of measurements
      Hypothesis selection based on evidence
   b. Diagnosis
      Measurement selection and interpretation
      (often involves models of system organization and behavior)
2. Design and synthesis
   Provide constraints as well as guidance
3. Prediction
   Forecasting
4. Use advisor
   "How to" advice
5. Intelligent assistant
   Provide decision aids
6. Scheduling
   Time-ordering of tasks, given resource constraints
7. Planning
   Many complex choices affect each other
8. Monitoring
   Provide real-time, reliable operation
9. Control
   Process control
10. Information digest
    Situation assessment
11. Discovery
    Generate new relations or concepts
12. Debugging
    Provide corrective action
13. Example-based reasoning
    The source of most rules

Figure 7. AI function capabilities.

Function capabilities

Of primary consideration are the function applications that can readily be built with a particular ESBT. A review of the major function applications follows (see also Figure 7).

Classification. By far the most common function addressed by expert systems is *classification.* "Classification" refers to selecting an answer from a fixed set of alternatives on the basis of information that has been input.

Below are some subcategories of classification.

• *Interpretation of measurements.* This refers to hypothesis selection performed on the basis of measurement data and corollary information.

• *Diagnosis.* In *diagnosis,* the system not only interprets data to determine the difficulty, but also seeks additional data when such data is required to aid its line of reasoning.

• *Debugging, treatment, or repair.* These functions refer to taking actions or recommending measures to correct an adverse situation that has been diagnosed.

• *Use advisor.* An expert system as a front end to a computer program or to a piece of machinery can be very helpful to the inexperienced user. Such systems depend both on the goals of the user and the current situation in suggesting what to do next. Thus, the advice evolves as the state of the world changes. Use advisors can also be helpful in guiding users through procedures in other domains (for example, auto repair and piloting aircraft).

Classification and other function applications can be considered to be of two types: *surface reasoning* and *deep reasoning.* In surface reasoning, no model of the system is employed; the approach taken is to write a collection of rules, each rule asserting that a certain situation warrants a certain response or conclusion. (These situation-response relationships are usually written as heuristic rules garnered from experience.) In deep reasoning, the system draws upon causal or structural models of the domain of interest to help arrive at the conclusion. Thus, systems employing deep models are potentially more capable and may degrade more gracefully than those relying on surface reasoning.

Table 1. A subjective view of the importance of various expert system tool attributes for particular function applications.

● VERY ○ SOMEWHAT LITTLE	DIAGNOSIS AND CLASSIFICATION	DATA ANALYSIS AND INTERPRETATION	DESIGN AND SYNTHESIS	PREDICTION AND SIMULATION	MONITORING	USE ADVISOR	INTELLIGENT ASSIST.	PLANNING AND SCHEDULING	CONTROL
INFERENCE APPROACH									
BC	●	○	○		○	○	○	○	○
FC & FORWARD REASONING	○	●	●	●	●	●	○	●	●
BC, FC, & FORWARD REASON.	●	●	●	○	○	○	●	●	○
HYPOTHETICAL REASONING	●	●	●	○	○	○	○	●	
OBJECT-ORIENTED	○	○	○	●	○	○	○	○	●
BLACKBOARD	●	●	●	○	○	○	○	●	
INDUCTION	●	○	○			○	○		○
OBJECT DESCRIPTION									
FRAMES	●	○	●	○	○	○	○	●	○
FRAMES W/ INHERITANCE	●	○	●	○	○	○	○	●	○
OBJECTS	○	○	○	●	○	○	○	○	●
PARAMETER VALUES PAIRS	○	○	○		○	○			○
LOGIC	○	○	○	○	○	○	○	○	○
RULES	○	○	○	○	○	○	○	○	○
CERTAINTIES	●	●	○	○	○	○	○	○	○
ACTIONS									
RULES	●	●	●	●	●	●	●	●	●
EXAMPLES	●	○	○	○		○	○		○
LOGIC	●	●	●	●	○	○	○	○	○
MESSAGES	○	○	●	●	○	○	○	●	●
PROCEDURES	○	○	●	●	●	○	○	●	●

Design and synthesis. "Design and synthesis" refers to configuring a system on the basis of a set of alternative possibilities. The expert system incorporates constraints that the system must meet as well as guidance for steps the system must take to meet the user's objectives.

Intelligent assistant. Here the emphasis is on having a system that, depending on user needs, can give advice, furnish information, or perform various subtasks.

Prediction. "Prediction" refers to forecasting what will happen in the future on the basis of current information. This forecasting may depend upon experience alone, or it may involve the use of models and formulas. The more dynamic systems may use simulation to aid in the forecasting.

Scheduling. "Scheduling" refers to time-ordering a given set of tasks so that they can be done with the resources available and without interfering with each other.

Planning. "Planning" is the selection of a series of actions from a complex set of alternatives to meet a user's goals. It is more complex than scheduling in that tasks are chosen, not given. In many cases, time and resource constraints do not permit all goals to be met. In these cases, the most desirable outcome is sought.

Monitoring. "Monitoring" refers to observing an ongoing situation for its predicted or intended progress and alerting the user or system if there is a departure from the expected or usual. Typical applications are space flights, industrial processes, patients' conditions, and enemy actions.

Control. Control is a combination of monitoring a system and taking appropriate actions in response to the monitoring to achieve goals. In many cases, such as the operation of vehicles or machines, the tolerable response delay may be as small as milliseconds. In such a case, the system may be referred to as a *real-time system.* "Real-time" is defined as "responding within the permissible delay time" to the end that the system being controlled stays within its operating boundaries.

Digest of information. A system performing this function may take in information and return a new organization or synthesis. One application may be the inductive determination of a decision tree from examples. Others may be the assessment of military or stock market situations on the basis of input data and corollary information.

Discovery. Discovery is similar to digest of information except that the emphasis is on finding new relations, order, or concepts. This is still a research area. Examples include finding new mathematical concepts and elementary laws of physics.

Others. There are other functions, such as learning, that are directly subsumable under the ones I have enumerated thus far. In many cases, these functions (and some of those already mentioned) can be ingeniously decomposed into functions discussed previously. Thus, for example, design and some other functions can often be separated into subtasks that can be solved by classification.

Importance of various ESBT attributes for particular function applications

Table 1* is an attempt to relate the various attributes that are found in different ESBTs to their importance in facilitating the building of expert systems that perform different functions. A solid circle indicates an attribute that is very worthwhile in helping to build that function. An open circle indicates that it is a lesser contributor. A empty cell indicates an attribute that does not provide a significant contribution. As indicated earlier, the evaluation is subjective because, depending on the insight and ingenuity of the system developer, some of the functions can be decomposed into other functions. Thus, Table 1 reflects what I see as obvious and perhaps necessary attributes for straightforward construction of expert systems that perform the indicated functions.

*In the future, various ESBT approaches may be shown to be Turing Machine equivalents, which would mean that any computation could be performed by them. Therefore, it usually cannot be said definitively that ESBT *x* cannot perform function *y*. Thus, Table 1 in the sidebar is really an attempt to reflect my perception of which ESBT attributes simplify the programming of various expert-system functions.

Brief descriptions of commercial ESBTs

The following are descriptions of some of the current commercial expert system building tools in common use. The attributes of these tools are summarized in Tables 1 and 2 for easy comparison. This sidebar is not intended to be an exhaustive survey. For example, VP-Expert, an inexpensive (under $100) but capable rule-based ESBT for PCs, has recently been introduced by Paperback Software in Berkeley, Calif. GEST, an evolving university-supported ESBT from Georgia Tech, provides high-order capabilities (such as multiple knowledge representations) at a fraction of the cost of commercial, more polished tools offering similar capabilities. GURU, from mdbs in Lafayette, Ind., a composite ESBT integrated with a database spreadsheet and natural-language front end, is also available.

ART

ART is a versatile tool that incorporates a sophisticated programming workbench. It runs on advanced computers and workstations such as those produced by Symbolics, LMI, TI, Apollo, and VAX. ART's strong point is *viewpoints,* a technique that allows hypothetical *nonmonotonic reasoning*; in nonmonotonic reasoning, multiple solutions are carried along in parallel until constraints are violated or better solutions are found. At such points, inappropriate solutions are discarded. ART provides graphics-based interfaces for browsing both its viewpoint and *schema* (frame) networks. ART is primarily a forward-chaining system with sophisticated user-defined pattern matching; the pattern matching is based on an enhanced version of an indexing scheme derived from OPS5. (OPS5 is discussed below.) Object-oriented programming is made available by attaching *procedures* (active values) to objects (the objects are called *schemata*). ART has a flexible graphics workbench with which to create graphical interfaces and graphical simulations. ART was designed for near-real-time performance. To achieve this performance, ART compiles its frame-based as well as its relational knowledge into logic-like assertions (the latter are called *discrimination networks*). Applications particularly suited for ART are planning/scheduling, simulation, configuration generation, and design. Currently written in Lisp, ART employs a very efficient, unique memory management system that virtually eliminates garbage collection. A C-language version is now available. (Further information on ART is available from Inference Corp., 5300 W. Century Blvd., Los Angeles, CA 90045; (213) 417-7997.)

KEE

Kee, which runs on advanced AI computers, is the most widely used programming environment for building sophisticated expert systems. Important aspects of KEE are its multifeature development environment and end-user interfaces, which incorporate windows, menus, and graphics. KEE contains a sophisticated frame system that allows the hierarchical modeling of objects and permits multiple forms of inheritance. KEE also offers a variety of reasoning and analysis methods, including object-oriented programming, forward and backward chaining of rules, hypothetical reasoning (which is incorporated as KEE Worlds), a predicate-logic language, and demons. It has an open architecture that supports user-defined inference methods, inheritance roles, logic operators, functions, and graphics. KEE has a large array of graphics-based interfaces that are developer/user controlled, including facilities for graphics-based simulation (the graphics-based simulation facility, Sim Kit, is available at extra cost). KEE has been used for applications in diagnosis, monitoring, real-time process control, planning, design, and simulation. (Further information on KEE is available from IntelliCorp, 1975 El Camino Real West, Mountain View, CA 94040; (415) 965-5500.)

Knowledge Craft

Knowledge Craft (KC) is a hybrid tool based on frames that have user-defined inheritance. It is an integration of the Carnegie Mellon version of OPS5 and of Prolog and the SRL frame-representation language. It is a high-productivity tool kit for experienced knowledge engineers and AI system builders. Frames are used for declarative knowledge; procedural knowledge is implemented by the attaching of demons. KC is capable of hypothetical (nonmonotonic) reasoning when Contexts (a facility offering alternative worlds) is employed. Search is user defined. A graphics-based simulation package (Simulation Craft) is available. Designed to be a real-time system, KC is particularly appropriate for planning/scheduling and, to an extent, is appropriate for process control, but it is something of an overkill where simple classification problems are concerned. (Further information on Knowledge Craft is available from Carnegie Group, Inc., 650 Commerce Court, Station Square, Pittsburgh, PA 15219; (412) 642-6900.)

Picon

Picon is designed as an object-oriented expert system shell for developing real-time, on-line expert systems for industrial automation and other processes that are monitored with sensors during real time; such processes are found in some aerospace and financial applications. Picon operates on the LMI Lambda/Plus Lisp machine and the TI Explorer, which combine the intelligent processing power of a Lisp processor with the high-speed numeric processing and data-acquisition capabilities of an MC68010 processor. The two processors operate simultaneously, enabling Picon to monitor the system in real time, detect events of possible significance in process, diagnose problems, and decide on an appropriate course of action. Picon's icon editor and graphics-oriented display enable a developer with minimal AI training to construct and represent a deep model of the process being automated. Rules about the process are entered by means of a menu-based natural-language interface. Picon supports both forward and backward chaining. (Further information on Picon is available from Lisp Machine, Inc., 6 Tech Dr., Andover, MA 01810; (617) 669-3554.)

S.1

S.1 is a powerful commercial ESBT aimed at structured classification problems. Facts are expressed in a frame representation; judgment-type knowledge is expressed as rules. Though ostensibly a backward-chaining system, S.1 performs forward reasoning by means of a patented *procedural control block technique.* Control blocks can be viewed as implementations of flow diagrams; they guide the system procedure by telling the system the next step to take in the current situation. Control blocks can invoke other control blocks or rules, or they can initiate interactive dialogue. Control blocks are a powerful, knowledge-based means of controlling the search,

Table 1. Attributes of some commercial ESBTs. (This table is continued on page 34.)

NAME	ART 3.0	KEE 3.0	KNOWLEDGE CRAFT	PICON	S.1	ES ENVIRON/ VM OR MVS	ENVISAGE	KES
FUNCTIONAL USES								
Classification	X	X	X	X	X	X	X	X
Design	X	X	X		X			
Planning/Scheduling	X	X	X					
Process Control	X	X	X	X				
KB REPRESENTATION								
Rules								X
Structured Rules	X	X	X	X	X	X	X	
Certainty Factors					X	X	X	X
Rule Limit								
Frames w/Inherit.	X	X	X	X	No inherit.			
Object-Oriented	X	X	X	X				
Logic	X	X	X		X			
Examples								
Structured Examples								
Procedures	X	X	X		X	X	X	X
INFERENCE ENGINE								
Forward Chaining	X	X	X	X	FR	X	X	
Backward Chaining	(goal rules)	X	X	X	X	X	X	X
Demons	X	X	X	X	X		X	
Blackboard-Like	X							
Time-Modeling	X	X		X				
Truth Maintenance	X	X						
Meta Control	X	X	X	X	X	X	X	
Logic	X	X	X		X			
Induction								
Math Calculations	X	X	X	X	X		X	X
Context (Viewpoints, Worlds)	X	X	X					
PATTERN MATCHING								
Variables	X	X	X	X				X
Literals	X	X	X	X	X			
Sequences	X		X	X			X	
Segments	X		X	X				X
Wildcards	X		X					
DEVELOPER INTERFACE								
KB Creation								
Word Processor	X		X		X	X	X	X
Line Entry		X	X	X	X			X
KB Editor	X	X	X	X	X	X		
Menu	X	X		X	X			
Check for Consistency		X		X	X			
Graphics Rep of KB	X	X	X	X	X			
Inference Tracing	X	X	X	X	X	X	X	X
Graphics Utilities to build end-user interface	X	X	X	X	e			
Screen Format Utilities	X	X	X	X	X	X		
Graphics Simulation Capabilities	X	Sim Kit (Extra cost)	Simulation Craft (Extra cost)	X	e			
Why	X	X	X		X	X	X	X
How	X	X	X	X	X	X	X	X
Explanation Expansion	X	X	X	X	X	X	X	X
Online Help	X	X	X		X	X	X	
Syntax Help	X	X	X	X	X		X	
Tool Extensible	X	X	X	X			X	

	M.1	NEXPERT OBJECT	PERSONAL CONSUL.+	EXSYS. 3.0	EXPERT EDGE	ESP FRAME ENGINE	INSIGHT 2+	TIMM	RULE-MASTER	KDS 3	1st CLASS
	X	X X	X	X	X	X X	X	X	X X	X X	X
		X X						X	X	X X	
	X	Cataloged	X	X	X	X	X		X		X
	X 1000	2000	X 800	X 5000	X		X 2000	X	X	X 16000 from 4000 examples	X
		X	X			X				X	
						X					X
								X	X	X	X
		X	X			X					
	FR X	X X	X X	X X	X	X X	X X	X	X X	X X	X X
	X	X	X			X				X	X
		X			X					X	
	X	X	X			X X					
	X	X	X	X	X	X	X	X	X	X X	X
	X	X X		Numeric	Numeric	Numeric & String	Numeric		X		Numeric & Logical
				X							
					X						X
	X	X	X	X X	X	X	X	X	X	X	X X
	X X	X X	X X	X X	X X		X	X	X	X X	X X
	X	X X	X	X	X	X X		X	X	X	X X
		X X	X	X	X	X	X	X		X X	X
	X	X X	X	X	X		X	X		X	X e
	X X	X X	X X	X X	X X	X X	X X		X X	X X	TREE
	X X	X X	X X	X X	X X	X X	X X	X	X X	X X	X X
	X	X X	X X		X X	X			X X	X X	X

Table 1. Attributes of some commercial ESBTs. (This table begins on page 32.)

NAME	ART 3.0	KEE 3.0	KNOW-LEDGE CRAFT	PICON	S.1	ES ENVIRON/ VM OR MVS	ENVISAGE	KES
SOFTWARE								
Language of Tool	COMMON-LISP	COMMON-LISP	COMMON-LISP	ZETA-LISP.C	C	PASCAL	PASCAL	C
Compilable	Incremental	Incremental	Incremental	X	X	Incremental	X	X
Other Lang. Req.	COMMON-LISP	COMMON-LISP	COMMON-LISP	LISP		CMS, GDDM, PASCAL/VS		
Operating Sys.	VMS, UNIX				UNIX VM/CMS VMS	VM/CMS or MVS	VMS	MSDOS on PC
COMPUTERS SUPPORTED								
Symbolics	X	X	X					
LMI	X	X	X	X				
TI Explorer	X	X	X	X				
VAX	X		X		VMS ULTRIX		X	X
XEROX		X						
IBM PC								X
Macintosh								
TI Prof.								
APOLLO		X			X			X
SUN	X	X			X			X
IBM 370					X	X		
Others		X	X		X		X	X
SYSTEMS INTERFACE								
Lang. Hooks	Via Host Computer	Via Host Computer	Via Host Mach	Via Host Mach	C, Others	X	PASCAL, etc.	C
DB Hooks	Via Host Computer	Via Host Computer	Via Host Mach	Via Host Mach	Via "	X	X	X
END USER INTERFACE								
Screen Capabilities								
Line		X	X	X	X	X	X	X
Menu	X	X	X	X	X	X	X	X
Graphics	X	X	X	X	e			e
Simulation	X	Sim Kit	Simulation-Craft	X	e			
Why	X	X	X	X	X	X	X	X
How	X	X	X	X	X	X	X	X
Help	X	X	X	X	X	X	X	X
Initial Pruning	X	X						
Multiple Solution	X	X	X					X
Examples Saved			X	X	X	X	X	X
Mult. & Uncertain User Resp. OK	X	X			X		X	X
What If	X	X	X	X	X	X	X	X
COMPANY	Inference	Intellicorp	Carnegie Group	LMI	Teknow-ledge	IBM	Sys. Designers Software	S/W Arch. & Engr.
COST (1st unit)	$65K	$52K includes training	$50K includes training	$60K	$25K 1 user mach. $45K mult. user	$35K	$25K on VAX $15K on MicVAX	$4K PC $7K WkStns $25K VAX
Run Time Sys. Cost	$1-8K	PC Host + VAX $20K + 0.5K/PC	None Yet		$9.5K	$25K		10% of Develop Sys.

M.1	NEXPERT OBJECT	PERSONAL CONSUL.+	EXSYS. 3.0	EXPERT EDGE	ESP FRAME ENGINE	INSIGHT 2+	TIMM	RULE-MASTER	KDS 3	1st CLASS
C X	C Incremental	LISP X	C Incremental	C X	Prolog 2 Incremental	Turbo-PASCAL X	FORTRAN X	C X	8086 Assembly X	PASCAL X
MSDOS	MSDOS VMS	MSDOS	MSDOS VMS	MSDOS	MSDOS	MSDOS		C Compiler DOS 3.0 UNIX, VMS	MSDOS	MSDOS
			VMS UNIX				X	X		
X	AT	X	X	X	X	X	X	AT	X	X
	X	X					X			X
								X X		
	X					X	X	X		
X X	C. PASCAL DBASE III	LISP DBASE III	X X	X X	Prolog 2 Via "	Y DBASE II	X	Most X	PASCAL, BASIC Reads Assem. Lang.	ANY X
X	X	X X	X	X X	X X	X	X	X	X	X
	X X	e	e						X X	e
X X	X X	X X	X X	X X	X X	X X	X	X X	X X	X
X	X	X X	X	X X	X	X X		X	X X	X
	X X	X X	X X	X X	X		X		X X	X
X	X	X X	X X	X X		X X	X X		X X	X
Teknow-ledge	Neuron Data	TI	Exsys, CA Intell	Human Edge	Exp Sys Inter	Level-5-R	GRC	Radian	KDS Dev Sys	Progr. in Motion
$5K	$5K PC-AT $3K MAC	$3K	$395 PC $5K VAX	$2500 Adv $5000 Pro	$895	$485	$1.9K PC $19K others	$995 PC $5K Wkst $17.5K VAX	$1495	$495
$50	$1K	$75 First Unit	One Time Fee	$50	By Agreement	Negoti.	No Fee	$100 PC $500 UNIX/VMS	Based on Quality	No Fee

and thus they have made it possible for one to write programs containing thousands of rules without being overwhelmed by a combinatorial explosion (runtimes tend to be linear with the number of rules). S.1 is written in C and executes very rapidly. A major advantage of S.1 is that it can readily be integrated into existing software. A delivery version is available without the system-development portion of S.1; the delivery version can be completely embedded in applications. S.1 is not aimed at exploratory programming; it is aimed at commercial applications in which iterative development of solutions to solvable problems is desired. S.1 has an excellent user interface that features mouse-driven, graphical representations of both the knowledge bases and the inference traces. Problems can be solved in terms of subproblems, which can be linked to handle the complete problem (consistency checking is performed as part of linkage). All S.1 features are expressed in an integrated, strongly typed, block-structured language that facilitates system development and long-term maintenance. (Further information on S.1 is available from Teknowledge, Inc., 1850 Embarcadero Rd., PO Box 10119, Palo Alto, CA 94303; (415) 424-0500.)

Table 2. A composite view of some commercial ESBTs.

● STRONG
○ FAIR
□ USER PROG.

SYSTEM	KNOW. REP.				INFERENCE ENGINE							FUNCTIONAL CAPABIL.					DEV. INTERFACE					END USER INTER.			
	OBJ. REP.	CERTAINTIES	ACTIONS	EXAMPLES	PATTERN MATCH.	BC	FC	OBJ.-ORIENTED	LOGIC	INDUCTION	HYPO. REASON.	CLASS. INTERPRETA. ADVICE (SHALLOW)	CLASS. INTERPRETA. ADVICE (DEEP)	DESIGN	PLANNING	MONITOR/CONTROL	KB CREATION	DEBUGGING	GRAPHICS	USER INT. CREA.	INFER. CONTROL	WHY/HOW	SCREEN INTER.	GRAPHICS	WHAT IF
ART	●	□	●		●	●	●	●	●		●	●	●	●	●	○	●	●	●	●	○	●	●	●	●
KEE	●	□	●		●	●	○	●	●		●	●	●	●	●	○	●	●	●	●	●	●	●	●	●
KNOWLEDGE CRAFT	●	□	●		●	●	●	●	●		●	●	●	●	●	○	●	●	●	●	□	●	●	●	○
PICON	●		●		●	●	●	●				●	●			●	●	●	●	●	○	●	●	●	●
S. 1	○	C	●		○	●	FR		○			●	●	○			●	●	●	●	●	●	●	○	●
ES ENVIRON/VM OR VMS		C, F	●		○	●	○					●				○	●	○		●	○	●	○		●
ENVISAGE (ADVISOR)	○	F, B	●		○	●	○					●	○				○	●		○	○	●	○		○
KES		C, B	●		○	●	○				●	●	○				○	○		○		●	○		●
M. 1		C	●		○	●	FR		○			●	○	○	○	○	●	●		●	○	●	○		
NEXPERT OBJECT	●	□	●		○	●	●					●	○	○	○	○	●	●	●	●	●	●	●	●	●
PERSONAL CONSULTANT +	○	C	●			●	○					●	○				●	●	○	●	●	●	●	○	●
EXSYS 3.0		C	○		○	●	○					○					●	●		○		●	●	○	●
EXPERT EDGE		B	○		○	●						●					●	●		●		●	●		○
ESP FRAME ENGINE	●		●		○	●	○		●			●	●				●	●	○	○	○	●	●		
INSIGHT 2 +		C	○		○	●	○					●					●	●		○		●	○		●
TIMM		C, I	○	●	●		●			●		●	○			○	●	●		○		○	●	●	○
RULEMASTER 3.0		F	○	●	○	●	○			●		●		○		○	●	●		○		○	○		
KDS 3	○	C		●		●	●	○		●		●	○	○	○	○	●	○	○	●	○	○	●	●	○
1st CLASS		C	○	●	○	○	○			●		●					●	●	○	○		○	○		○
OPS5+			●		●		●					●	○	○	○	○	●	●			○		○		

FR -- FORWARD REASONING

*
C — CONFID. FACTOR
F — FUZZY LOGIC
B — BAYSIAN LOGIC
I — INCOMPLETE INFO

**
L — LISP
PA — PASCAL
A — ASSEMBLY
F — FORTRAN
PR — PROLOG
O — OTHERS
C — C

@
D — DISCRIM. NET
R — RULES
IN — INCREMENTAL
Y — YES
DT — DECIS. TREE

ES Environment/VM or MVS (ESE/VM or ESE/MVS)

ESE is an improved version of Emyçin; it is designed for classification problems, but does allow for forward chaining. It consists of two components: a development interface and a consultation interface. A Focus Control Block mechanism has been added to allow the developer to modify and control the flow of inference and, thus, to increase the system speed. ESE/VM and ESE/MVS have good utilities for enabling the developer to fashion the user interface and to incorporate graphics in the user interface when appropriate. ESE is particularly suitable for IBM mainframe users who must interface with existing software and databases. (Further information on ESE is available from IBM, Dept. M52, 2800 Sand Hill Rd., Menlo Park, CA 94025; (415) 858-3000.)

Envisage

Envisage is a Prolog-derived tool. Thus, instead of entering rules, one enters logical assertions. Non-Prolog features include demons, fuzzy logic, and Bayesian probabilities. Envisage is primarily aimed at classification problems. (Further information on Envisage is available from System Designers Software, Inc., 444 Washington St., Suite 407, Woburn, MA 01801; (617) 935-8009.)

KES

KES is a three-paradigm system that supports production rules, hypothesize-and-test rules (hypothesize-and-test rules use the criterion of minimum set coverage to account for data), and Bayesian-type rules for domains in which knowledge can be represented probabilistically. KES is primarily geared to classification-type problems. KES can be embedded in other systems. The hypothesize-and-test approach starts with a knowledge base of diagnostic *conclusions* (that is, classifications) with their accompanying *symptoms* (also called "characteristics"). The session begins with the selection by the system of the set of all diagnoses that match the first symptom of the given problem; the system then reduces this set as the remaining problem symptoms are considered. If the initial set of diagnoses does not include all the remaining symptoms, new diagnoses are added to the set to cover these cases. (Further information on KES is available from Software Architecture and Engineering, Inc., 1600 Wilson Blvd., Suite 500, Arlington, VA 22209-2403; (703) 276-7910.)

M.1

M.1 is a PC-based ESBT targeted for solvable problems rather than for exploratory programming. It is a basically a backward-chaining system designed for classification. It includes the capability for meta-level commands that direct forward reasoning. Written in C, it can readily be integrated into existing conventional software. Its main drawback is that it has no true object-description capability and therefore cannot readily support deep systems. However, M.1 does have a good set of development tools and developer- and user-friendly interfaces. (Further information on M.1 is available from Teknowledge, Inc., 1850 Embarcadero Rd., PO Box 10119, Palo Alto, CA 94303; (415) 424-0500.)

Nexpert Object

Nexpert Object is a powerful, rule-based tool coded in C to run on a Macintosh with 512K of RAM, the Mac Plus, or the IBM PC AT. It has editing facilities comparable with those found on a large tool designed to run on the more sophisticated AI machines. The system allows the developer to group rules into categories so that the rules need be called up only when they are appropriate. Nexpert Object supports variable rules and combinations of forward and backward chaining. The system can automatically generate graphical representations of networks of rules; these representations of networks indicate how the rules relate to one another. Similar networks can be generated to show rule firings that take place in

S/W H/W CHARAC.				
TOOL LANG. **	COMPILABLE @	LANG/DB HOOKS	COMPUTERS @@	1st UNIT COST $K
L	D, IN	○	V, Su, S, L, T	65
L	IN	○	A, S, L, T, X, O	52† + SIM KIT
L	IN	○	L, S, T, V, O	50† + SIM
L, C	Y	●	L, T	60
C	Y	●	370, V, Su, H, O	25, 45
PA	IN	●	370	60
PA	Y	●	V, μV	25, 15
C	Y	○	I, A, V, O, Su, μV	4, 7†, 25†
C	Y	●	I	5
C	IN	●	I, M, O	3, 5
L	Y	●	I, TP, T	3
C	IN	○	I, V	0.4, 5
C	Y	○	I	2.5, 5
PR	IN	○	I	0.9
PA	Y	●	I, O	0.5
F	R	●	TP, I, V, O	1.9, 19
C	DT	●	I, A, Su, V	1, 5, 17.5
A	R	●	I, TP	1.5
PA	DT	●	I, TP	0.5
C	Y	○	A, Su, I, M, O	1.8, 3

@@

370 — IBM 370
M — MACINTOSH
TP — TI PC
I — IBM PCs
S — SYMBOLICS
L — LMI

T — TI EXPLORER
X — XEROX
V — VAX
μV — MICROVAX
Su — SUN
A — APOLLO
O — OTHERS

† — INCLUDES TRAINING

response to a particular consultation. Nexpert Object includes the capabilities of both frame representations that have multiple inheritance and of pattern-matching rules, so deep reasoning is facilitated. Nexpert Object is a sophisticated system with a focus on the graphical representation of both the knowledge bases and the reasoning process, which makes possible natural and comprehensible interfaces for both the developer and end user. (Further information on Nexpert Object is available from Neuron Data Corp., 444 High St., Palo Alto, CA 94301; (415) 321-4488.)

Personal Consultant+ (PC+)

PC+ is an attempt to provide on a personal computer many of the advanced features found in more sophisticated tools; such tools include KEE. Thus, PC+ utilizes frames with attribute inheritance, and rules. PC+ supports the backward-chaining approach derived from Emycin. It also includes forward-chaining capabilities without variable bindings. PC+ has an extensive set of tools for both development and execution that incorporate user-friendly interfaces. The new 2.0 version supports up to 2M bytes of expanded or extended memory for increased knowledge-base capacity. It also supports the IBM Enhanced Graphics Adapter and access to the popular dBase II and III database packages on the IBM PC. A version of PC+ is also available for the TI Explorer Lisp Machine. PC Easy, a simplified version of PC+ without frames, is also offered. (Further information on PC+ is available from Texas Instruments, Inc., PO Box 209, MS 2151, Austin, TX 78769; (800) 527-3500.)

Exsys 3.0

Exsys 3.0 is written in C for PCs as an inexpensive, rule-based, backward-chaining system oriented toward classification-type problems. Rules are of the if-then-else type. Exsys includes a runtime module and a report generator. Exsys can interface to the California Intelligence company's after-market products: Frame (to provide frame-based knowledge representation) and Tablet (to provide a blackboard knowledge-sharing facility that incorporates tables). (Further information on Exsys 3.0 is available from Exsys, Inc., PO Box 75158, Contr. Sta. 14, Albuquerque, NM 87194; (505) 836-6676.)

Expert Edge

Expert Edge is basically a rule-based, backward-chaining system aimed at rapidly prototyping and delivering classification applications in the 50-to-500 rule range. It uses probabilities and Bayesian statistics to handle uncertainties and lack of information. Its outstanding features are its excellent developer and end-user interfaces, which feature pop-up windowing environments. These are accompanied by a natural-language interface and very good debugging facilities. The professional version interfaces with a video disk and is also able to do extended mathematical calculations. (Further information on Expert Edge is available from Human Edge Software Corp., 1875 S. Grant St., San Mateo, CA 94402-2669; (415) 573-1593.)

ESP Advisor and ESP Frame-Engine

ESP is a Prolog-based system that is particularly appropriate for designing expert systems that guide an end user in performing a detailed operation involving technical skill and knowledge. The developer builds the system by programming in KRL (Knowledge Representation Language), a sophisticated and versatile language that supports numeric and string variables, including facts, numbers, categories, and phrases. Prolog's heritage is clearly apparent in the system's ability to support a full set of logic operators, which enables the developer to write efficient, complex rules. The ESP consultation shell offers a well-designed, multipanel display that makes good use of color. A *text-animation* feature allows the developer to insert text at any point in a consultation. Though ESP Advisor was designed as an introductory prototype tool, its extensibility makes expert systems of greater complexity possible. ESP Frame-Engine supports frames with inheritance, forward and backward chaining rules, and demons. (Further information on the ESP products is available from Expert Systems International, 1700 Walnut St., Philadelphia, PA 19103; (215) 735-8510.)

Insight 2+

Insight 2+ is primarily a rule-based, backward-chaining (that is, goal-driven) system, but it can support forward chaining as well. Facts are represented as elementary objects with single-value or multivalue attributes. Rules are entered in PRL (Production Rule Language). The knowledge base is compiled prior to runtime. Uncertainty is handled by means of confidence factors and thresholds. Because Insight 2+ lacks methods for representing deep models, it is best used for heuristic problems, for which it is a useful tool. Its ability to access external programs and databases is a major enhancement. (Further information on Insight 2+ is available from Level Five Research, Inc., 503 Fifth Ave., Indialantic, FL 32903; (305) 729-9046.)

TIMM

TIMM is an inductive system that builds rules from examples. Examples are first translated into rules, which are then used to build more powerful generalized rules. TIMM handles contradictory examples by arriving at a certainty that is based on averaging these examples' conclusions. Partial-match analogical inferencing is used to deal with incomplete or non-matching data. TIMM indicates the reliability of its results. The expert systems that result from it can be embedded in other software programs. (Further information on TIMM is available from General Research Corp., 7655 Old Springhouse Rd., McLean, VA 22102; (703) 893-5900.)

Rulemaster 3.0

Though Rulemaster is capable of independent forward and backward chaining, its major distinguishing feature is its capability for inductively generating rules from examples. It also offers fuzzy logic. Interaction with the knowledge base is accomplished by means of a text editor. If they prefer, knowledge engineers can develop Rulemaster applications by writing code directly in the high-level Radial language of Rulemaster instead of using examples. However, a strong programming background is required for easy usage. Rulemaster can generate C or Fortran source code for fast execution, compactness, and for creation of portable expert systems that can interface to other computer programs. (Further information on Rulemaster is available from Radian Corp., 8501 Mo-Pac Blvd., PO Box 9948, Austin, TX 78766; (512) 454-4797.)

KDS3

KDS3 inductively generates rules from examples. Examples can be grouped to develop knowledge modules, which KDS calls *frames* and which can be chained together to form very large systems. Both forward and backward chaining are supported. KDS3 can take input from external programs and sensors and can drive external programs. Expert systems built with KDS3 can be made either (a) interactive or (b) fully automatic for intelligent process control. The entire system is written in assembly language for very rapid execution on PCs. Graphics can be incorporated automatically from picture files or, if one makes use of built-in KDS3 color graphics primitives, they can be drawn in real time. KDS3 incorporates a blackboard by means of which knowledge modules can communicate. KDS2 without the blackboard facility is also available. (Further information on KDS3 is available from KDS Corp., 934 Hunter Rd., Wilmette, IL 60091; (312) 251-2621.)

1st-Class

This is an induction system that generates decision trees, which are elaborate rules, from examples given in spreadsheet form. Problems can be broken down into modules derived from sets of examples; the modules can be chained together with forward or backward chaining. Rules can also be individually built or edited in graphical form on the screen. Several algorithms are available for inferencing: The system can match queries to examples that exist in the database, or the system can utilize the rule trees either as generated or in the *preferred mode,* which employs optimized rule trees that ask questions in the best order. Because all the rules are compiled, the system is very fast. The 1st-Class induction system is designed to interface readily with other software. (Further information on 1st-Class is available from Programs in Motion, Inc., 10 Sycamore Rd., Wayland, MA 01778; (617) 653-5093.)

OPS5

Various versions of the OPS5 expert-system-development language, developed at Carnegie Mellon University, are available. OPS5 is a forward-chaining, production-rule tool with which many famous expert systems used at DEC, such as R1/XCON, have been built. OPS5 pattern-matching language permits variable bindings. However, OPS5 does not have facilities for sophisticated object representations. In general, the development environment is unsophisticated, although some debugging-and-tracing capability is usually provided. The use of a sophisticated indexing scheme (the Rete algorithm) for finding rules that match the current database makes OPS5 one of the tools that executes the fastest. Unfortunately, it is not an easy tool for the nonprogrammer to use. Variations of a representative version, OPS5+, can be obtained for the IBM PC, Macintosh, and the Apollo Workstation. (Further information on OPS5 is available from Computer*Thought Corp., 1721 West Plano Pkwy., Suite 125, Plano, TX 75075; (214) 424-3511.)

Attributes of particular commercial ESBTs

The sidebar entitled "Brief descriptions of commercial ESBTs" presents some of the better-known commercial ESBTs. Attributes of these ESBTs are listed in Table 1 of that sidebar. Inclusion of an ESBT in the sidebar in no way represents an endorsement of that product. The descriptions and listings have been constructed from company and noncompany literature, discussions with company representatives, demonstrations, exploratory use of the tools, and so on. However, some incompleteness, errors, and oversights are inevitable in such an endeavor, so it behooves the interested person to use this material as a guide and to examine the systems directly. Direct examination is particularly important because increasing competition is forcing ESBT developers to make rapid improvements and changes in both their systems and their prices.

A comparative, composite view of the various tools

Table 2 of the sidebar provides a composite view of the various ESBTs. Many of the attributes have been integrated to provide a more easily understandable picture of the capability of the tools in each subcategory (for example, the rule and procedure attributes have been combined into "representation of actions"). A solid circle indicates that the tool appears to be strong in a subcategory, an open circle indicates that it appears to be fair, and an empty cell indicates little or no capability in that area. Note that by relating each tool's attributes to its functional importance, I have attempted to indicate each tool's suitability for developing various function applications. Also, note that the more expensive (and correspondingly more sophisticated) tools have the widest applicability. This is often because they are a collection of different paradigms incorporated into a single tool. As a result, they may often be regarded as higher order programming languages and environments, instead of as simple shells into which information is inserted to create an expert system directly. The shell model is more nearly true of the simpler induction systems; such systems can be considered as knowledge-acquisition and rapid-prototyping tools from which more com-

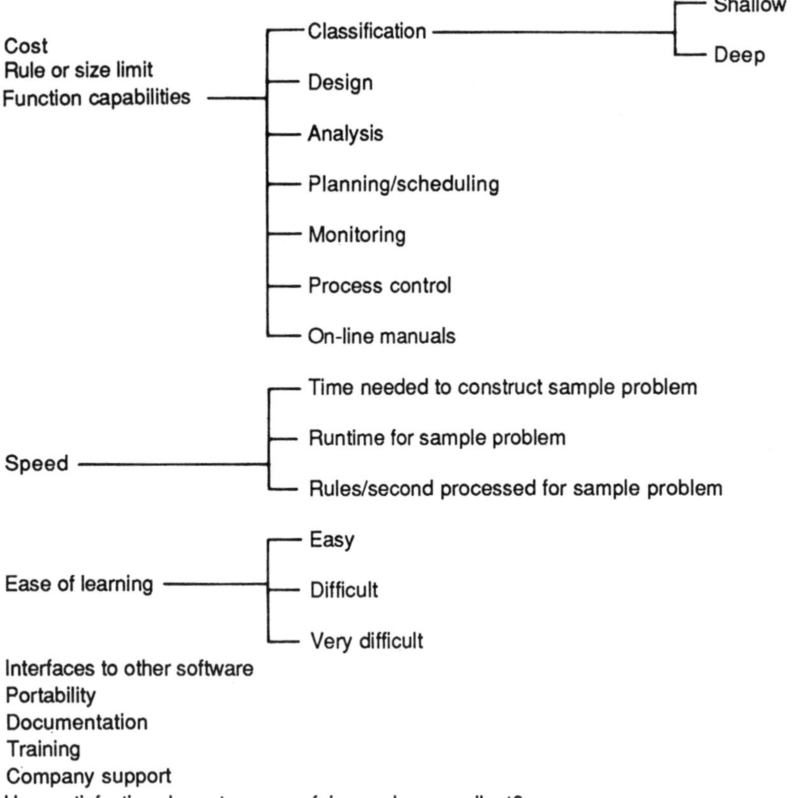

Figure 8. Considerations for assessing the overall usability of a tool.

plex systems can be built by means of other tools by enlarging upon the simple rules inductively generated.

Overall usability of a tool

Figure 8 summarizes some of the aspects that enter into the critical ESBT attribute "overall usability of a particular tool." In addition to obvious factors such as costs and function applicability (function applicability is a measure of which functions are easily accomplished with a tool and which are difficult to accomplish with it), tool choices should be guided by the size of the system to be built, how rapidly a system of the given size and complexity can be built with the tool, and the speed of operation of the tool both during development and, particularly, during end use. (During end use, sub-elements of the tool act as a software delivery vehicle for the developed expert system.) Perhaps the most important factor, however, is the degree of satisfaction of both the developer and the end user. This is related to how obvious the uses of the tool features are, how direct the lines of action to the user's or developer's goals are, the control the developer and end user sense that they have over the system, the nature of the interaction or display (for example, whether they take place by means of menu or graphics), how easy it is to recover from errors, the on-line help that is furnished, and the perceived esthetics, reasonableness, and transparency of the system. Also of major importance is how easy it is to learn the system. This often depends on many of the factors already discussed, but is also closely related to how apparent the choice is at each step (for example, the apparent choice when menus are used is different from the apparent choice when programming is required), the quality of the documentation and on-line help, and the ESBT's structure. Manufacturer-sponsored courses help; however, these are often expensive and inconvenient. A related factor is manufacturer support of the tool, particularly the availability of help over the telephone when it is required.

Finally, such factors as the system's portability, the computers it will run on, the delivery environment, the system's capability of interfacing with other programs and databases, and whether the developed system can be readily embedded in a larger system are all important in an evaluation of a tool. A more difficult factor to evaluate is the ease of prototyping versus life cycle cost. As prototypes are expanded into fielded systems and as they are iteratively further expanded and updated, difficulties are often encountered in system stability, runtime, and memory management.

Though many of these factors can be deduced from the tool's specifications and from system demonstrations, in many cases one can properly differentiate between two tools intended for the same application only if he or she learns both systems and attempts to build the same set of applications with each one. Nevertheless, the factors described in this article and the initial evaluation furnished in the sidebar should prove useful as initial guides to potential users.

To date, ESBTs have made possible productivity improvements of an order of magnitude or more in constructing expert systems. Current tools are only forerunners of ESBTs yet to come. The trend is toward less expensive, faster, more versatile, and more portable tools that will readily make possible development of expert systems that can directly communicate with existing conventional software such as databases and spreadsheets, and can also be embedded into larger systems, with resulting autonomous operations. Higher-end ESBTs are now moving from Lisp machines to more conventional workstations that are less expensive. Lower-end systems are becoming more capable and now appear on IBM PCs and Macintoshes. Delivery systems, which utilize a subset of the complete ESBTs (ESBTs with the development portion removed) are now allowing the completed expert system to be delivered on personal computers or workstations. In addition, versatility will be enhanced with increased choices of inference engines such as blackboard systems. Also in the works are modular ESBTs that will allow the developer to choose various knowledge representations and inference techniques as he or she desires and still be able to build an integrated system. Already appearing are ESBTs coupled to other software systems such as databases and spreadsheets. Also beginning to appear are expert systems that are specialized to specific functions such as scheduling, process control, and diagnosis.

Finally, the developer and end-user interfaces are getting friendlier and more capable. One of the things providing greater capability is the increased use of graphics and graphical simulations. It is expected that as these friendlier systems emerge, there will be increased development of expert systems directly by the experts themselves.

The rich and growing variety of ESBTs may make it more difficult to choose a tool, but if it is properly selected, the tool will be more closely matched to developer and end-user needs. □

References

1. E.H. Shortliffe, *Computer-Based Medical Consultations: Mycin,* Elsevier-North Holland, New York, 1976.
2. W. van Melle, "A Domain-Independent System that Aids in Constructing Knowledge-Based Consultation Programs," tech. report No. 820, 1980, Computer Science Dept., Stanford University, Stanford, Calif.

Suggested reading

Gevarter, W.B., *Intelligent Machines,* Prentice-Hall, Englewood Cliffs, N.J., 1985.

Gilmore, J.F., and K. Pulaski, "A Survey of Expert System Tools," *Proc. Second Conf. on Artificial Intelligence Applications,* Dec. 11-13, 1985, Computer Society Press, Los Alamitos, Calif., pp. 498-502.

Gilmore, J.F., K. Pulaski, and C. Howard, "A Comprehensive Evaluation of Expert System Tools," *Proc. SPIE Applications of Artificial Intelligence,* Apr. 1986, SPIE, Bellingham, Wash.

Harmon, P., and C.D. King, *Artificial Intelligence in Business,* John Wiley & Sons, New York, 1985.

Hayes-Roth, F., D.A. Waterman, and D.B. Lenat, *Building Expert Systems,* Addison-Wesley, Reading, Mass., 1983.

Karna, A., and A. Karna, "Evaluating Existing Tools for Developing Expert Systems in PC Environment," *Proc. Expert Systems in Government,* K.N. Karna, ed., Oct. 24-25, 1985, Computer Society Press, Los Alamitos, Calif., pp. 295-300.

Riley, G.D., "Timing Tests of Expert System Building Tools," NASA JSC Memorandum FM 7 (86-51), Apr. 3, 1986, FM7/Artificial Intelligence Section, Johnson Space Center, Houston, Tex.

Richer, M.H., "An Evaluation of Expert System Development Tools," *Expert Systems,* Vol. 3, No. 3, July 1986, pp. 166-183.

Waterman, D.A., *A Guide to Expert Systems,* 1986, Addison-Wesley, Reading, Mass., pp. 336-379.

"AI Development on the PC: A Review of Expert System Tools," *The Spang Robinson Report,* Vol. 1, No. 1, Nov. 1985, pp. 7-14.

"Expert Systems-Building Tools," *Expert Systems Strategies,* P. Harmon, ed., Vol. 2, No. 8, Aug. 1986, Cutter Information, Arlington, Mass., pp. 17-24.

"Small Expert Systems Building Tools," *Expert Systems Strategies,* P. Harmon, ed., Vol. 1, No. 1, Sept. 1985, Cahners Publishing, Newton, Mass., pp. 1-10.

Catalogue of Artificial Intelligence Tools, A. Bundy, ed., Springer-Verlag, New York, 1985.

PC (issue on expert systems), Vol. 4, No. 8, Apr. 16, 1985, pp. 108-189.

William B. Gevarter is a computer scientist with the Artificial Intelligence Research Branch at NASA Ames Research Center. Before being transferred to NASA Ames in 1984, he spent several years at the National Bureau of Standards on special assignment from NASA writing a series of NASA/NBS overview reports on artificial intelligence and robotics. Prior to that assignment, he was manager of automation research at NASA headquarters. His current interests are expert systems and human and machine intelligence.

Gevarter received his BS (1951) in aeronautical engineering from the University of Michigan, his MS (1955) in electrical engineering from the University of California at Los Angeles, and his PhD (1966) in aeronautics and astronautics with a specialization in optimal control from Stanford University.

Readers may write to William Gevarter at NASA Ames Research Center, MS 244-17, Moffett Field, CA 94035.

BY RONALD CITRENBAUM, JAMES R. GEISSMAN, AND ROGER SCHULTZ

Reprinted with permission from *AI Expert*, September 1987, pages 30-39. Copyright © 1987 by Miller Freeman Publications, 500 Howard Street, San Francisco, CA 94105

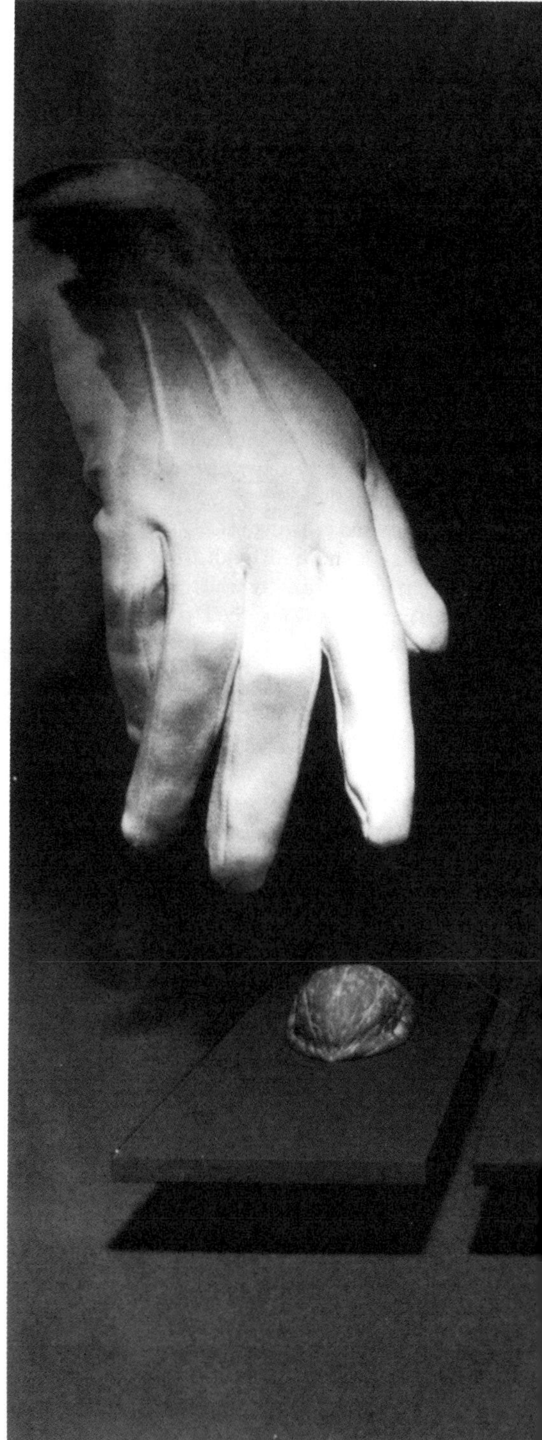

SELECTING A SHELL

So many different purposes and hardware environments exist that a single shell is unlikely to satisfy them all

What are the most desirable features to look for in an expert system development environment? Some of the features discussed in this article are widely available in commercial products; others will undoubtedly appear in the future. So many different purposes and hardware environments exist that a single shell is unlikely to satisfy all of them.

Our observations are based on our experience designing and developing more than 20 expert systems over the past four years. We have utilized a wide range of software and hardware (such as symbolic computers, minicomputers, and PCs), AI languages (like LISP and PROLOG), specialized knowledge processing paradigm languages (OPSn), multifunction shells (such as Inference Corp.'s ART), and simpler shells/rapid prototypers (for example, Insight 2+ from Level Five Research Inc.).

Our target audience consists of three groups: first-time buyers of expert system shells who need to know which features are most significant for their specific needs; persons with a major expert system responsibility (for example, at a corporate level) who are trying to establish high-level standards to help control the multiplicity of tools, languages, and even basic concepts to improve productivity; and future developers of expert system shells since some features we mention are not yet available in the marketplace.

WHAT IS A SHELL?

In the technical literature and common usage, expert system shells can lie anywhere on a continuum from interpreters of relatively simple languages to very elaborate development environments. Each has its own purposes and strengths and can complement other shells by being used at different times in a project's life cycle.

What shells have in common is the minimum feature set of a knowledge representation scheme, an inference or search

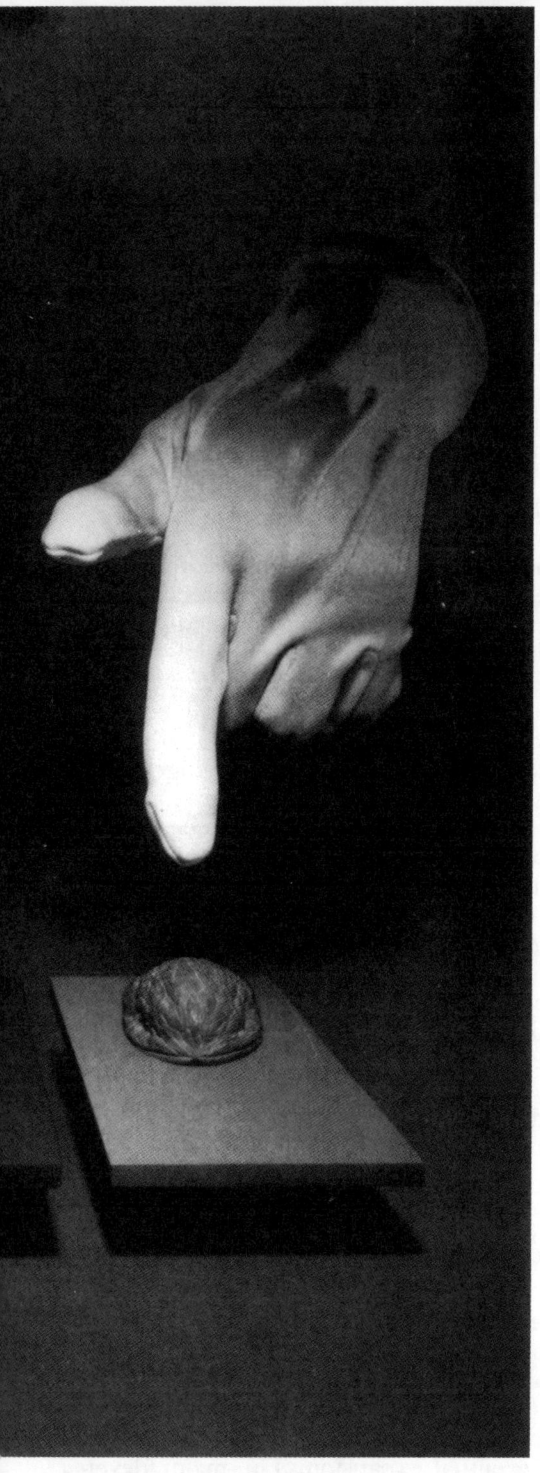

mechanism, a means of describing a problem, and a way to determine the status of a problem while it is being solved (see W.B. Gevarter's "The Nature and Evaluation of Expert System Building Tools").

Although not everyone agrees on what the word "shell" should mean, products like LISP and PROLOG are frequently used to develop expert systems. This kind of product supports an expert system language specialized for the expression of knowledge processing statements and can result in programming productivity increases of approximately one order of magnitude; a program that is 1,000 lines long in FORTRAN can typically be written in less than 100 PROLOG statements.

Expert system language environments can be specialized to a single knowledge representation for maximum efficiency—such as OPS5 and OPS83 (Production Systems Technologies Inc.), which implement the forward-chaining, rule-based production system model—or more general (like LISP and PROLOG). The shell can be extended by tools that run the gamut from interpreters and compilers to symbolic debuggers.

A much different approach is taken by inductive products like ExpertEase (Expert Software International Ltd.) and its descendants, which build a system from a statement of knowledge and its relationships. With an inductive shell, the user states the knowledge used to arrive at conclusions. These shells take a set of discriminating examples, in tabular or other form, and the resultant conclusions and produce an optimized query tree, implicitly determining the intermediate nodes.

Other shells take an explicit set of rules and goals and interact with the user to determine the facts required to satisfy the goals. With Insight 2+, for example, the knowledge takes the form of *IF-THEN* rules. Some shells perform analysis to optimize the dialog and allow (or require) the user to exert control over operations.

The more elaborate systems available on symbolic computing hardware and powerful minicomputers, such as ART and IntelliCorp's KEE, are also called expert system shells, although some observers would say that this is too limited a name for these tools and that a term like "knowledge programming environment" should be used instead.

In this article, what we mean by "expert system shell" covers all of the previously mentioned concepts, although probably no single product can serve all possible purposes. The general notion is an environment for the development of an expert system containing the four basic functions of knowledge representation, inference mechanism, problem description, and status determination.

However, the specific requirements for selecting a shell as discussed here do not point to a single product because the activities performed during the expert system development cycle are carried out by people with varied interests and experience, and are targeted at different products.

FOUR STAGES OF DEVELOPMENT

We identify four stages in expert system development: problem selection, initial prototype, expanded prototype, and delivery system. Each stage benefits from a shell, and each stage stresses different features. An overview of the four-stage expert system development methodology is presented in Figure 1.

The problem selection stage, involving system specification and problem determination, corresponds to the requirements analysis stage in a conventional software development project. The major objective is to insure that the project will both satisfy a real need and be technically feasible. The main steps are to determine whether an expert system approach is most suitable for the problem, carefully select an initial prototype subset problem so a successful demonstration can occur relatively quickly, and discover the problem's underlying knowledge requirements so appropriate knowledge representations and tools (such as a shell) can be brought to bear.

The major objective of the initial prototype stage is to quickly demonstrate the technical and economic feasibility of the desired expert system. An early demonstration has several advantages, especially from a management perspective, where there may be reluctance to fund a major development in a risky or unknown technology—especially when there is no history of past success.

The initial prototype is typically concerned with only a central subset of the problem and does not provide the full range of ultimate functions. Functions such as data base interface, real-time performance, and superintelligent user interface may be missing, but an explanation facility should be present to validate the reasoning and promote user acceptance.

Development of the initial prototype consists of devising a suitable expert system architecture and knowledge representation. The strategy taken will depend on the depth and complexity of the problem, whether it is forward chaining or data driven, the anticipated strength of inferences possible (how certain the conclusions must be), and the extent to which subproblems are likely to interact. Obviously, the more flexible the tool(s) used, the more responsive the design can be to subtle details of the problem.

Two major problems usually encountered in this phase are the requirement of prompt project completion within a limited budget and the need for sufficient knowledge engineering to insure that all essential parameters are included. Depending on the tool, the initial prototoype may be completed by the domain experts themselves, although the services of specialized knowledge engineers (expert system programmers) are often recommended to avoid becoming trapped in unsuitable representations.

Following an initial prototype demonstration of the concept and project approval, the major objective of the expanded prototype stage is to develop the full set of expert system functions required to deal with the complexity of the complete problem. The subset problem selected for the initial prototype is expanded to the full complexity of the domain area, and interactions with related systems such as data bases, measuring equipment, video, voice I/O, and so forth are included.

It may be reasonable to enhance the initial prototype iteratively or discard it (keeping the knowledge) and move to a different model; this often depends on the capabilities of the shell selected for the initial prototype. A quick-implementation shell with limited power often makes sense for the initial prototype, even though it cannot support eventual expansion. The major development problems encountered in this stage tend to be technical in nature, resulting from the complexity and sophistication of the features built into the system.

The expanded prototype may be suitable for deployment as a delivery system if only a few copies are needed and the prototype performance is sufficient for the target environment. However, in many cases, an operational environment based on different hardware (for example, a 68000 workstation or PC instead of a Symbolics machine) may be required, necessitating redeployment of the system.

The major objective at this stage is to

FIGURE 1. Four-stage expert system development methodology.

PROBLEM DETERMINATION AND SPECIFICATION
- Identify candidate opportunities
- Build on analogous successes
- Determine knowledge requirements
- Specify system functions

INITIAL PROTOTYPE
- Select inference mechanism
- Select knowledge representation
- Use existing advanced tools
- Limit initial scope
- Minimize initial use of experts
- Determine feasibility

EXPANDED PROTOTYPE
- Expand use of experts
- Utilize rapid prototyping
- Expand scope of system
- Provide I/O interfaces
- Add bells and whistles

DELIVERY SYSTEM
- Optimize speed
- Target to appropriate hardware
- Customize user interface
- Maintain system

port the expanded prototype system to the target environment. Typically, a delivery system differs from the expanded prototype in that it is widely deployed geographically and thus must run on inexpensive hardware such as a PC. Usually it must also satisfy more stringent performance and robustness requirements. The major development problems encountered in this phase typically result from design and function trade-offs required to make the system faster, smaller, and portable.

USERS AND USES OF SHELLS

One group of requirements on an expert system shell results from its application in the stages of the expert system development cycle already discussed. Other requirements result from the users of expert system shells and the activities they perform. The thumbnail sketches that follow are profiles of personnel on an expert system development team. The end user who develops an expert system for his or her own use will adopt several of these roles (for example, domain expert, knowledge engineer, prototyper, and developer) during the course of system development.

Students: Students use a shell to learn about expert systems and knowledge-based programming. By definition, they cannot be expected to have any background in the subject and hence have to be supported by the shell, especially the user interface. Students will use a shell for knowledge engineering and initial prototyping but will probably not pursue development to enhanced prototypes or delivery systems.

Domain experts: Domain experts are involved in the activities of knowledge engineering and prototyping, both initial and expanded. They may be assisted or led by knowledge engineers or, with suitable tools or an agreeable shell, function on their own. One of the most important aspects of their contact with the expert system is in testing and extending prototypes to deal with progressively larger and more complete problems.

Knowledge engineers: Knowledge engineers are concerned with defining the initial problem, knowledge engineering, and developing prototypes. They can be presumed to have some familiarity and skill with the internals of the knowledge representation methodology and the expert system shell.

Builders and programmers: Builders and programmers are concerned with implementation, especially the enhanced prototype and delivery system phases, and are likely to be involved earlier. Their special skills relate to the computer and operating system being used, as well as the details of the expert system shell. The correctness of the knowledge in the system remains the responsibility of the knowledge engineer and domain expert.

Because needed features differ at each stage of expert system development, a single expert system shell may not be capable of completing an entire project (unless it is a big, expensive one). For development and deployment on PCs, an integrated set of tools may be a superior option, with each tool specializing in certain functions.

The following features are some of the most important, from the user's perspective. They are not an exclusive set of attributes that define the ideal shell; instead the list is organized in a way that is convenient for attaching our observations on how such a shell might be used.

KNOWLEDGE REPRESENTATION AID

An expert system shell should not only accept one or more knowledge representations, it should help the user select the appropriate representation and develop it for the specific problem. A shell that does this will have the following attributes:

- Utilize clear knowledge representations. Representations that are relatively intuitive (or graphic) are advantageous in the initial stages when it may not be clear to the user whether or not the representation is correct. After the user better understands and elaborates the representation, this requirement may cease to be important.
- Provide brainstorming aids for poorly structured or poorly understood problems.
- Compare alternative representations of a given set of knowledge in some relatively obvious format.
- Help users select the representation scheme. This could be a tutorial function in which the shell takes the lead and actively participates in the knowledge representation selection process. Alternatively, the shell might come with worked examples, which the user could take and transform into an expert system to deal with his or her own problem area. A shell that required no programming background would be especially attractive to inexperienced users.
- Provide translation to a standard knowledge representation for portability, at least within a specific integrated set of tools. Ideally, there would be an ANSI standard for knowledge import and export (like the Initial Graphics Exchange Specification (IGES) graphics standard), allowing a knowledge base to be ported widely between shells. This standard would have one form for rules, another for schemata, and so forth. For example, Abacus Programming Corp. developed the EXCABL Space Shuttle cabling system in OPS5, enhanced it in ART, and then automatically translated it to OPS83. This was possible because a strict subset of ART dealing only with production

The expanded prototype may be suitable for deployment as a delivery system if performance is sufficient and only a few copies are needed

system models was used.

In the short term, one approach is to use a tool that has been implemented on a variety of systems, such as ART (on LISP machines, Digital Equipment Corp.'s VAX, and C-based workstations), OPS83 (on C-based workstations, VAX, and PCs), or Common LISP.

- Provide a variety of interlinked representations that utilize ordinary knowledge forms as well as more sophisticated AI forms (for instance, a rule cites a cell in a spreadsheet that inherits properties from a frame in which a slot value comes from a rule or data base table). The transfer of knowledge from one representation to another should be relatively transparent. This implies support for a variety of low-level data types (symbols, integers, real numbers, record types, etc.) and the means to build up representations from primitive elements.

KNOWLEDGE ENGINEERING

An expert system shell should provide knowledge engineering tools to assist users who do not have the assistance of expert knowledge engineers. Such tools would also be useful to knowledge engineers reviewing their own work. Relevant tools should:

- Allow the user to first gather and assert knowledge and then shape it while providing feedback regarding its structure and interrelationships.
- Provide knowledge consistency and completeness checking with a high-quality explanation facility to indicate the cause of any inconsistencies found.
- Develop rules and/or decision trees from a set of examples (inductive knowledge engineering).
- Optimize query sequences in rule-based systems. The user would only be required to enter rules; the shell would work out the needed dialog and optimum rule sequence.
- Allow users to change knowledge representations easily, with transformations between compatible forms.
- Provide reasonable defaults for all options, slots, etc. Ideally, these defaults would be provided both at the system level (global defaults) and separately for different classes of frequently occurring problems. Problem classes with their own defaults could include the areas of diagnostics, classification, and configuration.
- Provide built-in, high-level domain expertise in certain basic areas to guide the user.
- Interpret a reasoning audit trail maintained by the inference engine and suggest ways to reach desired conclusions more quickly and/or surely.

INFERENCE ENGINE FEATURES

The inference engine is the processor that uses the knowledge in the selected representations to solve problems. An effective inference engine must deal efficiently with the knowledge representations the shell can provide and offer a variety of problem-solving strategies.

Ideally, the inference engine should:

- Support multiple paradigms and search strategies. A flexible inference engine should be able to deal with a given knowledge representation in more than one way. For example, with a set of production rules and some asserted facts, forward chaining creates new facts; with the addition of a goal, backward chaining can operate on the same rules and facts.
- Allow user modification to modify or tune the basic control mechanisms.
- Allow dynamic user influence of hypothesis generation and search strategy; for example, by assigning priorities to rules or intermediate states.
- Allow variables as well as literals within rules.
- Provide belief maintenance, updating all related knowledge, conclusions, and reasoning strategies as individual facts become known or change.
- Support uncertain reasoning, dealing with less-than-certain conclusions and a multiplicity of possible reasoning paths. This should be done while maintaining knowledge of how certain each conclusion is, without irrevocably pruning off alternatives if their probabilities might later be improved.
- Provide explanation and audit trail. This involves giving a clear account of the path followed to arrive at conclusions and hypotheses in terms the user can understand. Exactly what the user wants to know in this regard and how much he or she can understand may vary widely.

COMPATIBILITY/PORTABILITY

Compatibility and portability requirements mean that an expert system developed under one shell should not be left high and dry in one representation, running on only a small set of possible processors. If a problem grows or rehosting to new hardware is necessary for reasons not connected to the knowledge processing, it should be possible to transport the knowledge base to other environments and add additional standard features to the shell.

In this regard, shells should:

- Provide compatibility with other shells or tools that specialize in different phases of the expert system development cycle. This means that where different tools are used for different tasks (for example, knowledge engineering vs. delivery system), interfaces or hooks should go from one to the next. (Another approach is to have a single tool provide a full set of functions and operate on a wide range of hardware.)

- Fit into a logical tool migration path. For example, the shell could be both an interpreter and compiler for a knowledge representation.
- Provide links to widely used productivity products, including data bases and spreadsheets (for example, Microrim Inc.'s R:base, Ashton-Tate Inc.'s dBASE, and Lotus 1-2-3 on IBM PC; Digital Equipment Corp.'s RDB on VAX; SAS Institute's SAS on mainframes).
- Follow a standard interface between individual expert systems to facilitate distributed processing systems.
- Interface with algorithmic or procedural languages for specific performance-critical functions within an expert system.

USER INTERFACE SPECIALIZATION

A shell's user interface is very important. If knowledge engineering is provided by a special tool, it may have its own interface that users should be able to customize with application-specific templates. A relatively fixed, structured, and easy-to-learn interface with a large number of standard or default features would be suitable for the initial prototype, while a highly programmable one should be available for delivery systems that are likely to require customization (such as company colors, type fonts, logos, and slogans).

An ideal user interface should:

- Provide standard default features enabling plain vanilla expert systems to be constructed with minimum effort.
- Provide guidance. A very simple user interface may be suitable for new users or domain experts with no expert system development experience (or interest). The interface should provide guidance, either explicitly or implicitly, through the model of the world and default values it presents. Essentially, the shell should put together an interesting expert system without requiring the user to make decisions that require knowledge of the internals of expert system operations.
- Deal with graphics. Some user interfaces deal easily with graphics, especially image capture and display (for example, dialog such as "which picture looks most like the one you're thinking of?").
- Aid in prototyping. The user interface may allow the prototype builder to dummy up unimplemented user interface features for demonstration purposes.

PERFORMANCE/PRODUCTIVITY

A shell should include development tools and an environment to maximize programmer productivity and system performance. It should:

- Provide access to system functions, data bases, etc.
- Support modular design and independent development.
- Provide strong debugging features.
- Accommodate a large knowledge base work space.
- Produce a high-performance, small-size delivery product by dropping unneeded (but available) features.
- Produce versions that are portable to various delivery systems, including low-cost hardware like PCs.
- Provide high-performance features for the delivery system (compiler, parallel rule evaluation).

ADVANCED FEATURES

A number of improvements in expert system processing are being introduced in the laboratory and will probably be available soon in commercial tools. These include improvments in expert system performance through using optimization algorithms and taking advantage of improved hardware. Advanced features are frequently said to be required if expert systems are to function in critical real-time environments.

In this article we have not paid much attention to this dimension because hardware and software advances resulting in several orders of magnitude improvements are sure to occur in a few years without our requesting them. Also, we are principally concerned with how to make expert system technology easier to use and apply, however fast the programs work. Advanced features, nevertheless, should include taking advantage of parallel processing abilities in hardware and supporting distributed expert systems composed of a number of autonomous processes by incorporating some problem subdivision mechanisms and communication or blackboarding protocol.

Expert systems should be portable to a number of different environments, ranging from corporate mainframes to home-based PCs. If a given product is available both on the mainframe for a mainframe price and on the PC for a brown-bag price, cost-conscious users will of course purchase the PC version.

Incentive to develop shells for major computers can be maintained by providing modular shells. Different models of the same shell might be provided at different prices. The shell could be modular with plug-in functions or a number of shells could be provided, each offering a different subset of features. In this way, users could avoid purchasing unneeded features.

Another way to maintain incentive is to allow portability to low-cost hardware. It will generally be to the advantage of a shell if it can run on inexpensive hardware (PCs), although a trade-off of features is likely. A

An ideal user interface provides standard default features and guidance, deals with graphics, and aids in prototyping

shell that is portable to various hardware is desirable, especially with extra features available on more powerful hardware.

USER OBJECTIVES

A shell shouldn't be judged solely by whether it has a particular feature or representation scheme but whether it allows its users to effectively do what they need and want to do. An effective shell might have all the features but conceal them from most users. Gevarter's article (see references) indicates which shells are most useful for different functional uses (classification, design, planning/scheduling, and process control).

We view shells from a different perspective, deriving user objectives from the intersection of four user categories (student, domain expert, knowledge engineer, developer) and five possible uses (learning, knowledge engineering, initial prototype, expanded prototype, and delivery system). This cross-tabulation could result in 20 classes of user objectives, but in practice not all of them occur (for example, student users will by definition not be developing a delivery system).

FIGURE 2. Significant features for learning about expert systems.

STUDENT USER
- **Most important features**
  - Intuitive knowledge representation(s)
  - Low-cost hardware/software platform
  - Casual user interface
  - Useful default values
- **Other features**
  - Simple explanation facility
  - Examples

DOMAIN EXPERT
- **Most important features**
  - Intuitive knowledge representation(s)
  - Domain expert interface
  - Knowledge base browsing
  - Examples
- **Other features**
  - Useful default values
  - Inductive knowledge acquisition
  - Casual user interface
  - Low-cost hardware/software platform
  - Training and support

KNOWLEDGE ENGINEER
- **Most important features**
  - Useful default values
  - Multiple representation schemes
  - Casual user interface
  - Integration of representation schemes
- **Other features**
  - Good documentation
  - Appropriate examples
  - Training and support
  - Low-cost hardware/software platform

EXPERT SYSTEM SOFTWARE DEVELOPER
- **Most important features**
  - Good documentation
  - Useful defaults
  - Casual user interface
  - Examples
- **Other features**
  - Low-cost hardware/software platform
  - Access to source code
  - Multiple knowledge representations
  - Multiple reasoning paradigms
  - Training and support

LEARNING EXPERT SYSTEMS

One important use for expert system shells is in education. Although undertaken principally by student users, these observations are relevant to all user categories. The student may be a software manager or a college student who wants to broaden his or her understanding by seeing how an expert system works.

What a shell should do to assist in learning depends on the student's technical background: a programmer may learn best from a close-up look at underlying inference mechanisms while someone less technical may prefer to see a number of carefully selected examples from an application area he or she understands. In either case, learning is likely to involve performing some knowledge engineering and building a small system similar to an initial prototype.

Figure 2 highlights the most signficant features a shell can provide each user type in learning about expert systems. Depending on the amount the users want to learn about expert system concepts and knowledge engineering, the most important feature a shell can have is intuitiveness or transparency of knowledge representation. The expert system paradigm(s) and knowledge representation(s) supported by the shell should closely parallel the world of everyday knowledge and the user interface should provide an easy-to-grasp view of the structure of the knowledge the system uses (probably with graphics). Without a transparent representation and an interface that displays it clearly, users with little or no expert system background may not penetrate the subject far enough to grasp the basic processes.

Other important features for learning about expert systems include low software cost and PC hardware compatibility (especially for individuals), a user interface that spells out everything needed to make the system function so casual or infrequent users can use it, a useful set of default values, and an explanation facility that relates the shell's operation to both the shell's documentation and everyday knowledge. A shell with these features will simplify the student's task of building a demonstration system without requiring the problem to be oversimplified.

KNOWLEDGE ENGINEERING

Knowledge engineering includes defining the appropriate representation scheme and collecting and organizing applicable knowledge under that scheme. Knowledge engineering may be overlooked if it is thought of as merely a preliminary to the more important activity of programming the expert system.

Knowledge engineering (and developing the initial prorotype) are just as related to requirements gathering as programming. To facilitate knowledge gathering, a shell should permit new knowledge to be added without upsetting existing knowledge. Knowledge engineering also involves representation, which, as an activity, seems to involve twisting the knowledge scheme this way and that until something obviously right and operationally useful for the problem at hand is found.

The most signficant features a shell can provide each of the user types in knowledge engineering are listed in Figure 3. The most important shell features for use in knowledge engineering center on the knowledge representations supported and the knowledge-gathering and verification tools provided. For students and domain experts, these tools should use relatively transparent representations; the specialist knowledge engineer may be concerned with representations that are more sophisticated and less intuitive.

After a scheme is laid out and some knowledge has been collected, the tool can assist greatly by permitting browsing through the knowledge base and performing automatic consistency and completeness checks. If the knowledge engineering user is not an expert, the other features mentioned in Figure 3 are important, too.

INITIAL PROTOTYPE

The purpose of the initial prototype is proof of concept, not system production. The user's needs at this stage, therefore, probably do not extend to complex structures and mechanisms but to speed of implementation and the ability to simulate specific user interface features where required for verisimilitude. The goal is to take a knowledge representation and turn it into a functioning system simply and quickly.

The most important shell features for developing an initial prototype center on knowledge engineering (it should be easy to set up the knowledge), the user interface (should be accessible to casual users and easy to make a complete-looking mock-up without a lot of tedious programming), and the explanation facility (for checking and debugging). The most important aspect of user interface support appears to be a system's ability to develop dialogs and ask the user pertinent questions without the need to program the dialog explicitly. The expert system shell generates the dialog from an analysis of all of the unknowns in the knowledge base.

For student users, a lead-by-the-hand user interface, where the tool makes sure the developer provides it with all the information it needs for normal operation, is valuable (although the delivery system builder probably does not care for this). The most signficant features a shell can provide each user type in initial prototyping are listed in Figure 4.

EXPANDED PROTOTYPE

The expanded prototype is the stage most concerned with powerful inferencing and knowledge representation mechanisms. A key issue is the capability of dealing with the multitude of special cases and exceptions usually ignored in the simplified initial prototype; these may produce a large problem space. And it can be a system's ability to handle special cases that finally convinces experts and management that an expert system is not a toy.

Another important issue at this stage is the user interface, which should be easy to define and alter, particularly where special cases require fancy user interaction or access to special knowledge such as data bases or telemetry.

The most important shell features for developing an expanded prototype are the availability of sophisticated knowledge re-

STUDENT USER
- **Most important features**
  - Intuitive knowledge representation(s)
  - Low-cost hardware/software platform
  - Casual user interface
- **Other features**
  - Multiple knowledge representation schemes
  - Integration of multiple schemes
  - Useful default values

DOMAIN EXPERT
- **Most important features**
  - Intuitive knowledge representation(s)
  - Knowledge elicitation support
  - Knowledge base browsing
- **Other features**
  - Inductive knowledge acquisition
  - Useful default values
  - Domain-specific expertise

KNOWLEDGE ENGINEER
- **Most important features**
  - Multiple representation schemes
  - Integration of representation schemes
  - Representation selection aids
  - Knowledge gathering aids
  - Knowledge base browsing facility
- **Other features**
  - Knowledge completeness and consistency checks
  - Useful default values
  - Explanation facility
  - Logical tool migration path
  - Brainstorming aids
  - Inductive knowledge acquisition aids

EXPERT SYSTEM SOFTWARE DEVELOPER
- **N/A**

FIGURE 3. Significant features for knowledge engineering.

STUDENT USER
- **Most important features**
  - Intuitive knowledge representation(s)
  - Low-cost hardware/software platform
  - Casual user interface
- **Other features**
  - Useful default values
  - Examples

DOMAIN EXPERT
- **Most important features**
  - Explanation facility
  - Consistency and completeness checking
  - Inductive knowledge acquisition aids
- **Other features**
  - Built-in domain expertise
  - Knowledge base browsing
  - Useful default values

KNOWLEDGE ENGINEER
- **Most important features**
  - Useful default values
  - Built-in domain expertise
  - Consistency and completeness checking
  - Knowledge gathering aids
  - Knowledge base browsing
- **Other features**
  - Logical tool migration path
  - Explanation facility

EXPERT SYSTEM SOFTWARE DEVELOPER
- **Most important features**
  - Interface simulation capability
  - Graphics support
- **Other features**
  - Casual user interface
  - Friendly developer interface
  - Interface with popular productivity tools

FIGURE 4. Significant features for initial prototype.

STUDENT USER
- N/A

DOMAIN EXPERT
- **Most important features**
  - Explanation facility
  - Consistency and completeness checking
- **Other features**
  - Knowledge base browsing

KNOWLEDGE ENGINEER
- **Most important features**
  - Consistency and completeness checking
  - Large problem space
  - Multiple integrated representation schemes
  - Uncertainty support
  - Belief maintenance
- **Other features**
  - Logical tool migration path
  - Explanation facility
  - Modifiable inference engine

EXPERT SYSTEM SOFTWARE DEVELOPER
- **Most important features**
  - Large problem space
  - External product and language bridges
  - Modular
  - User interface development kit
  - Graphics support
- **Other features**
  - Friendly developer interface

FIGURE 5. Significant features for expanded prototype.

presentations and a full range of inferencing approaches (Figure 5). Knowledge consistency checking, hypothetical reasoning, belief maintenance, and uncertainty support are some of the more specialized functions useful at this stage.

It can be important for the builder of the expanded prototype to be able to exercise close control over the reasoning process to achieve realistic results in difficult cases. To do this, the shell should be able to link with other systems specialized in data bases, graphics, or operating system functions. The student user would not by definition be doing this, so an easy-entry user interface is not as significant.

If the expanded prototype is to be followed by a delivery system, there should be a migration path to the delivery environment so the knowledge base and special programming to deal with special cases, user interface, and so forth can be ported to the delivery environment.

DELIVERY SYSTEM

Developing a delivery system is more like conventional software engineering than the earlier steps (although most managers and budgeters would be happy if new programming could be avoided altogether by this point). Presumably, by this time the problem is well understood and the users have had opportunities to experiment with the earlier prototypes. Therefore the software engineering issues of performance, integration with other systems, portability, modularity, debugging support, and adherence to standards become more important, as opposed to "gee-whiz" features that have already been displayed in the expanded prototype. Of course, the delivery system shell must support all of the sophisticated representations (such as uncertainty and belief maintenance) used in the expanded prototype.

The most important shell features for developing a delivery system are those related to high performance and programmer productivity. Examples of these features are source code provision or other ways to rewrite the inference engine, compilability, an operating system interface, and linkability of object code. Figure 6 lists the most significant features for delivery system support.

WHAT IT ALL MEANS

Two main conclusions can be drawn from this complex set of desirable features. First, the requirements for an expert system shell do not exist a priori but are derived from the uses to which the shell will be put. These uses in turn depend on who the user is and at which stage in an expert system project he or she is working.

Second, because of the large number of attributes (some of which are contradictory), no single shell may have all the attributes needed for all purposes. Therefore, a shell should be compatible and portable in knowledge representation with other products so expert systems can be transported to the most appropriate tools for each user in each task. AI

REFERENCES

Citrenbaum, R., and Geissman, J.R. "A Practical, Cost-Conscious Expert System Development Methodology," in *Proceedings of the Second Annual AI and Advanced Computer Technology Conference.* Wheaton, Ill.: Tower Conference Management, 1986.

Gevarter, W.B. "The Nature and Evaluation of Commercial Expert System Building Tools." *IEEE Computer* 20(5): 24-41 (May 1987).

Gilmore, J.F., K. Pulaski, and C. Howard. "A Comprehensive Evaluation of Expert System Tools," in *Proceedings, SPIE Applications of Artificial Intelligence.* Bellingham, Wash.: SPIE, 1986.

Harmon, P., and C.D. King. *Artificial Intelligence in Business.* New York, N.Y.: John Wiley and Sons, 1985.

Hayes-Roth, F., D.A. Waterman, and D.B. Lenat. *Building Expert Systems.* Reading, Mass.: Addison-Wesley, 1983.

Ronald Citrenbaum, Ph.D., is chair of the board of Abacus Programming Corp., Van Nuys, Calif., where he also heads the Abacus AI Group.
James R. Geissman is AI project manager at Abacus.
Roger Schultz, Ph.D., is principal scientist in the Abacus AI Group.

STUDENT USER
N/A

DOMAIN EXPERT
Most important features
Explanation facility
Consistency and completeness checking
Knowledge maintenance interface
Other features
Knowledge base browsing

KNOWLEDGE ENGINEER
Most important features
Consistency and completeness checking
Large problem space
Multiple integrated representation schemes
Uncertainty support
Belief maintenance
Other features
Logical tool migration path
Modifiable inference engine

EXPERT SYSTEM SOFTWARE DEVELOPER
Most important features
High speed
Large problem space
External product and language bridges
Modular
Strong debugging tools
Drop unneeded features
User interface development kit
Graphics support
Other features
Friendly developer interface
Portable to various hardware including PC

FIGURE 6.
Significant features for developing delivery system.

Expert System Benchmarks

Larry Press, California State University at Dominguez Hills

While many expert system shells exist on the market, which should you use for your application? And which computer should you use to develop and deliver that application? No simple answers exist, but we can seek enlightenment through benchmark programs. Researchers have long used benchmarks to compare computers and procedure-oriented language processors. Some approaches run real applications on different computers or with different language processors; others use standard benchmarks.

One of the first and most widely used standard benchmarks is the Whetstone program.[1] Whetstone's instruction mix is derived from empirical observation of programs for numerical computation; consequently, it emphasizes floating-point calculation. The Dhrystone benchmark[2] (also used widely) derives its instruction mix from empirical observation of system programs. The Sieve of Eratosthenes prime-number benchmark[3,4] is simple and widely used, but is not based on empirical program analysis. Comprising only 24 lines in Pascal, it is representative of nonnumeric programs (featuring memory references, structured control statements, and simple input/output). Its only arithmetic is integer addition.

Rather than dealing with procedure-oriented languages, this article concerns benchmarks for use with shells. We have pursued both approaches mentioned above; that is, running a shell with a real knowledge base on different machines and running an artificial, stylized benchmark knowledge base with different shells on the same machine.

EH0303-8/90/0000/0062$01.00 © 1989 IEEE

Although we have not done so, we could also use stylized benchmarks to compare performance on different machines.

We can use benchmarks to compare shell loading and execution time, file size, and memory requirements. Published reviews of shells discuss many features, typically, but do not report these efficiency measures. For example, I have found two published reviews of Texas Instruments' PC Plus.[5,6] Neither gives quantitative or qualitative data on efficiency. After surveying 179 firms that use expert systems, Stevens found knowledge base size limitations and execution speed identified as two of five general shell weaknesses.[7] An application will fail if the knowledge base doesn't fit in memory, if it takes too long to load, or if execution is painfully slow and we waste hours of development time while shell editors "collect garbage."

Still, we should view benchmarking in its proper perspective. In addition to efficiency, we must consider various factors including

(1) The method of knowledge representation,
(2) The developer interface,
(3) The user interface,
(4) The interface with other programs and data files, and
(5) The shell vendor's stability and commitment in evaluating shells.

Other researchers have discussed these factors in shell evaluation.[8-10] We must also validate expert systems once shells are chosen and knowledge bases prepared. Our references present a comprehensive framework for expert system validation.[11]

We will summarize our experience testing a realistic knowledge base on different machines, describe our approach to stylized benchmarking, and present sample results.

A realistic knowledge base

We developed our realistic benchmark with TI's PC Plus, a Lisp-based shell and development environment with an inference engine that uses backward chaining (forward chaining is also possible). PC Plus runs on computers ranging from a 640-Kbyte PC-compatible to TI's Explorer Lisp machines.

Table 1 shows benchmark times, using a 62-rule/19-parameter knowledge base developed with PC Plus. We ran tests for consultation and for development tasks, measuring the time required to load the shell and knowledge base, run the consultation (using a playback file for user inputs), modify the knowledge base by adding one rule and changing another, print the knowledge base to disk, and save the knowledge base.

Such tests are useful when choosing hardware to run a particular shell. Table 1 shows the Compaq roughly two to four times faster (for most tasks) than a standard 6-MHz IBM PC AT, which translates into hours of wasted time during expert system development. Consequently, the clear choice for economical yet responsive knowledge base development with PC Plus is an Intel 80386-based computer — the Compaq 386, for instance. To our surprise, we found the Compaq faster than TI's Explorer.

Table 1. Consultation and development benchmark times using a Texas Instruments PC Plus shell on several computers.

Test	Machine				
	Compat 1	Compat 2	Compat 3	Compat 4	Lisp
Consultation:					
Load shell	26.1	21.7	49.4	10.8	22.2
Load KB	9.0	7.7	17.4	3.8	26.4
Run consultation	83.0	69.6	99.9	23.7	49.4
Development:					
Modify KB	51.9	48.2	92.0	28.2	32.5
Print KB to disk	134.8	111.7	264.8	58.5	230.7
Save KB (source)	64.5	51.0	161.2	25.9	43.1

Compat 1 — AT compatible at 6 MHz with 1-Mbyte memory
Compat 2 — AT compatible at 10 MHz with 1-Mbyte memory
Compat 3 — PC compatible at 4.77 MHz with 640-Kbyte memory
Compat 4 — Compaq 386 with 1-Mbyte memory
Lisp — TI Explorer with 8-Mbyte physical and 128-Mbyte virtual memory

```
A sequential knowledge base with four body rules (B = 4):
        goal = v0                         (the goal)
        v5                                (v5 is true)
        if v1 then v0                     (the rule satisfying the goal)
        if v2 then v1                     (the knowledge base body)
        if v3 then v2
        if v4 then v3
        if v5 then v4

A disjunctive knowledge base with four two-variable body rules (B = 4 and W = 2):
        goal = v0                         (the goal)
        v5                                (v5 and v6 are true)
        v6
        if v1 or v2 then v0               (the rule satisfying the goal)
        if v3 or v4 then v1               (the knowledge base body)
        if v3 or v4 then v2
        if v5 or v6 then v3
        if v5 or v6 then v4

A conjunctive knowledge base with four two-variable body rules (B = 4 and W = 2):
        goal = v0                         (the goal)
        v5                                (v5 and v6 are true)
        v6
        if v1 and v2 then v0              (the rule satisfying the goal)
        if v3 and v4 then v1              (the knowledge base body)
        if v3 and v4 then v2
        if v5 and v6 then v3
        if v5 and v6 then v4
```

Figure 1. Examples of stylized sequential, disjunctive, and conjunctive knowledge bases in the Teknowledge M1 shell language. In disjunctive and conjunctive cases, the number of body rules (*B*) must be an even multiple of the width of their conditional portion (*W*).

Unfortunately, this approach is costly. We developed the test knowledge base as part of a consulting contract. Undertaking such a development solely for the purpose of benchmarking would have been economically infeasible. Moreover, our knowledge base is specific to PC Plus and would have to be rewritten for another shell. Using relatively standardized procedure-oriented languages, we can develop benchmarks that are easily portable from one compiler to another. Since expert system shells are not standardized, only stylized benchmarks can be easily ported.

Stylized benchmark knowledge bases

We have used stylized knowledge bases with different shells to compare loading and execution time, file size, and memory requirements. Like the Sieve of Eratosthenes benchmark, we can adapt stylized knowledge bases easily to different shells and they are based on a typical operation; namely, backward chaining.

Our stylized knowledge bases are either sequential, disjunctive, or conjunctive (see Figure 1). Each has four parts: the goal (*v*0), the rules to set final variables True, a rule that satisfies the goal, and the test case body. All variables are Boolean and we assume 100-percent certainty.

Sequential knowledge bases are simplest, since there is only one variable on each body rule's left-hand side. When executing sequential knowledge bases, backward-chaining shells begin with the goal *v*0, establish a subgoal for each rule in the test case body, then "bottom out" by finding a True variable in the cache. Figure 2 traces this process. If *B* represents the number of body rules, then

The number of goals sought = *B*+1,
The number of goals not set by cache hits = *B*,
The number of goals set by cache hits = 1, and
The number of cache hits = 1.

In the sequential case, each body rule's left-hand side has a single variable. In the disjunctive case, left-hand sides have two or more variables connected with Or operators. When executing a disjunctive knowledge base, the shell creates fewer subgoals because it finds each body rule's first condition to be True (see Figure 3). If *B* represents the number of body rules, and *W* represents the number of variables on the left-hand sides of rules (their "width"), then

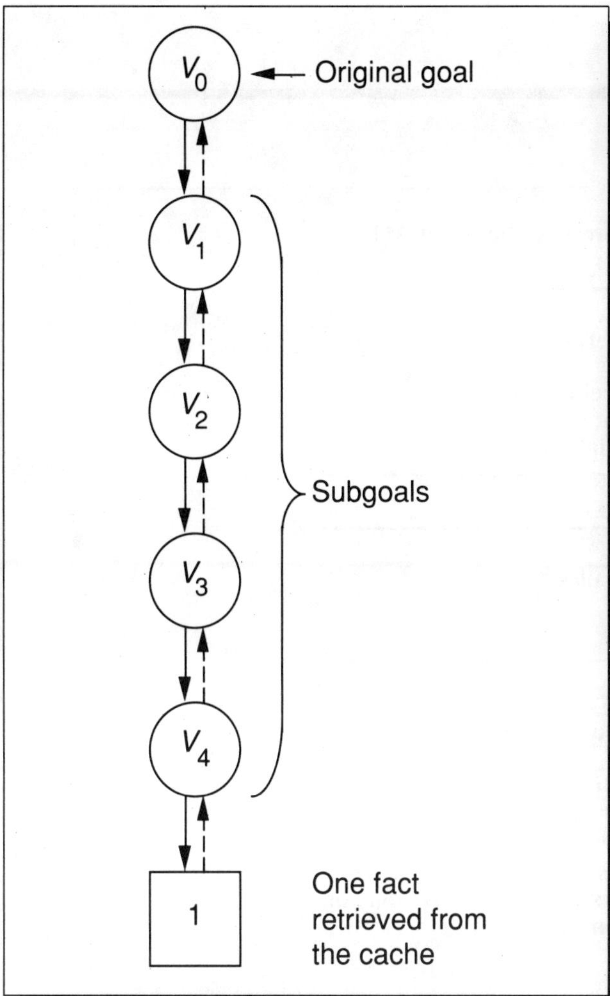

Figure 2. An execution trace for Figure 1's sequential knowledge base.

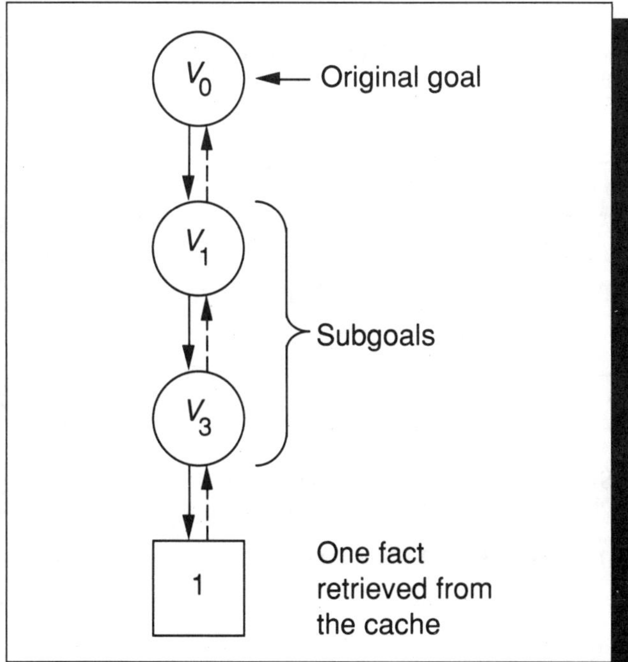

Figure 3. An execution trace for Figure 1's disjunctive knowledge base.

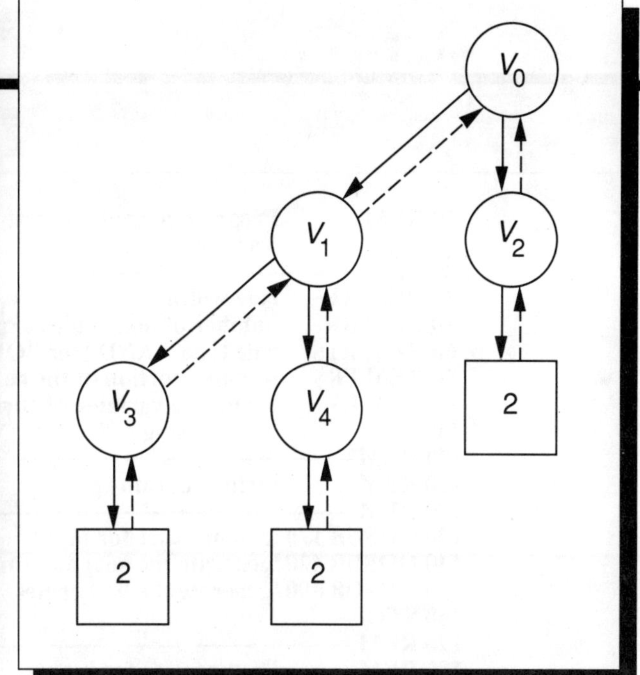

Figure 4. An execution trace for Figure 1's conjunctive knowledge base.

The number of goals sought = *B/W* + 1,
The number of goals not set by cache hits = *B/W*,
The number of goals set by cache hits = 1, and
The number of cache hits = 1.

In conjunctive knowledge bases, the left-hand sides of body rules have multiple variables joined with And operators. While execution of conjunctive knowledge bases does not go as deep as with sequential ones, the shell establishes more subgoals because all conditions in each rule must be True for the rule to be True (see Figure 4). That takes extra execution time. This is partially offset, however, since values of some intermediate variables are known (in the cache) by the time they are needed (for instance, *v*3 and *v*4 in Figure 4). Consequently, for the conjunctive case,

The number of goals sought = *B*+1,
The number of goals not set by cache hits = *B/W*,
The number of goals set by cache hits = *B* + 1 - *B/W*, and
The number of cache hits = *WB* + *W* - *B*.

Generating stylized knowledge bases

Simple programs can generate these knowledge bases because they are regular in form, which makes benchmarking economical. Figure 5 lists a program that generates stylized knowledge bases in the M1 shell syntax. We would have to modify the program slightly for a shell with

```
10 REM         Program to generate test knowledge bases for M1
20 REM         Variables used:
30 REM ------------------------------------------------------------
40 REM RW      rule width
50 REM NR      number of body rules to generate
60 REM RT$     rule type " AND " or "OR "
70 REM PR$     premise portion of the rule being generated
80 REM BV      beginning variable of current premise
90 REM                single character variables are temporary
100 REM -----------------------------------------------------------
110 REM        main program
120 REM -----------------------------------------------------------
130 GOSUB 170 'prompt user for inputs
140 GOSUB 420 'generate the goal and final rules
150 GOSUB 490 'generate the body rules
160 STOP
170 REM -----------------------------------------------------------
180 REM        Prompt user for inputs
190 REM -----------------------------------------------------------
200 INPUT      "Simple, conjunctive, or disjunctive (S/D/C): ", T$
210 RT$ = ""
220 IF T$ = "s" OR T$ = "S" THEN RT$ = ""
230 IF T$ = "c" OR T$ = "C" THEN RT$ = " AND "
240 IF T$ = "d" OR T$ = "D" THEN RT$ = " OR "
250 IF RT$ = "" THEN PRINT: PRINT "Answer S, D or C only" :PRINT: GOTO 200
260 IF RT$= "" THEN RW = 1: GOTO 280 ' simple rules have only one variable
270 INPUT "How many variables in each premise: ", RW
280 INPUT "How many body rules: ", NR
290 IF NR MOD RW = 0 THEN RETURN
300 PRINT: PRINT "The number of body rules must be an even"
310 PRINT "multiple of the number of variables.": PRINT: GOTO 270
320 REM -----------------------------------------------------------
330 REM generate a premise of width RW, type RT$, beginning with variable BV
340 REM -----------------------------------------------------------
350 PR$ = "if "
360 FOR I = 0 TO RW-1
370            PR$ = PR$ + "v" + RIGHT$ (STR$(BV+I), LEN(STR$(BV+I)) - 1)
380            IF I<>RW-1 THEN PR$ = PR$ + RT$
390 NEXT I
400 RETURN
410 REM -----------------------------------------------------------
420 REM generate the goal and final-variable rules
430 REM -----------------------------------------------------------
440 PRINT "goal = v0."
450 FOR I = 1 TO RW
460            PRINT "v" + RIGHT$( STR$ (NR+I), LEN (STR$(NR+I))-1) + "."
470 NEXT I
480 RETURN
490 REM -----------------------------------------------------------
500 REM generate the body of the knowledge base
510 REM -----------------------------------------------------------
520 FOR R = 0 TO NR
530            IF R = 0 THEN BV = 1: GOSUB 320: GOTO 550 ' the goal rule
540            IF (R-1) MOD RW = 0 THEN BV = R + RW: GOSUB 320
550            PRINT PR$
560            PRINT "then " + "v" + RIGHT$ (STR$(R), LEN(STR$(R))-1) + "."
570 NEXT R
580 RETURN
590 END
```

Figure 5. A program to generate stylized knowledge bases in the M1 syntax.

Width (*W*)	2		4		6		8	
Body rules (*B*)	Load	Exec	Load	Exec	Load	Exec	Load	Exec
100	5.5	2.2	6.3	1.4	7.2	1.0	8.1	0.8
200	11.8	4.7	13.3	2.3	15.6	1.8	18.2	1.4
300	19.2							

Table 2. Load and execution times for disjunctive knowledge bases using M1.

Width (*W*)	2		4		6		8	
Body rules (*B*)	Load	Exec	Load	Exec	Load	Exec	Load	Exec
100	5.6	4.0	6.2	5.4	7.5	7.2	8.4	8.5
200	11.6	8.4	13.7	10.8	16.3	14.0	17.8	18.2
300	19.5	13.8	22.4	16.9	25.8	22.2	28.6	26.2

Table 3. Load and execution times for conjunctive knowledge bases using M1.

	Source load time			Fast-load time		
Body rules (*B*)	PC-P	M1	Ratio	PC-P	M1	Ratio
50	6.7	2.8	2.4	2.9	0.8	3.6
100	12.1	4.9	2.5	5.5	1.2	4.6
150	19.6	7.4	2.6	6.5	1.6	4.1
200	25.5	10.3	2.5	8.6	1.9	4.5
250	33.0	14.0	2.4	11.7	2.1	5.6

Table 4. Comparative load times for sequential knowledge bases using PC Plus and M1.

	Source file size			Fast-load file size		
Body rules (*B*)	PC-P	M1	Ratio	PC-P	M1	Ratio
50	11.2	0.9	12.5	17.8	4.1	4.3
100	21.6	1.8	12.1	34.1	5.9	5.8
150	32.1	2.8	11.7	50.5	7.7	6.5
200	42.7	3.7	11.4	67.0	9.6	7.0
250	53.3	4.7	11.3	83.4	11.3	7.4

Table 5. Comparative file sizes for sequential knowledge bases using PC Plus and M1.

Body rules (*B*)	PC-P	M1	Ratio
50	4.4	1.9	2.3
100	11.1	3.7	3.0
150	25.0	5.8	4.3
200	56.9	8.5	6.7
250	NA	11.9	NA

Table 6. Comparative execution times for sequential knowledge bases using PC Plus and M1.

different syntax; for example, Paperback Software's VP Expert shell syntax.

When executed, the program prompts users for

(1) The test case type (sequential, disjunctive or conjunctive);

(2) The file name to which we write the benchmark knowledge base;

(3) The number of variables in each rule's left-hand side (*W*) (for the sequential case, this is automatically set to 1); and

(4) The number of rules in the knowledge base body (*B*). *B* must be an even multiple of *W*.

In a few seconds, this program can generate large test knowledge bases (say, $B = 1000$ and $W = 8$).

Some stylized-benchmark results

Tests were run using Teknowledge's M1 and Texas Instruments' PC Plus shells. Written in C, M1 has excellent developer and user interfaces, represents knowledge with rules resembling those of PC Plus, and runs on PC-compatible computers.

We measured the time required to load a knowledge base, the time to execute it, and the amount of disk space required in source and fast-load formats. We tested M1 for sequential, disjunctive, and conjunctive knowledge bases. We tested PC Plus only for sequential knowledge bases because it stores knowledge bases as complex Lisp lists (rules have to be entered manually). While tedious, that task required only two hours of clerical time due to its simple, repetitive rule format. We have since written a program that automatically generates stylized PC Plus benchmarks.

Our first test determined the maximum knowledge base size that could be executed without running out of memory (stack overflow) on a 640-Kbyte PC compatible. We found the largest feasible sequential knowledge bases for M1 and PC Plus to be $B = 265$ and 205, respectively. Since conjunctive and disjunctive cases generate fewer subgoals, larger values of *B* could be tested depending upon *W*.

Table 2 shows load and execution times for disjunctive knowledge bases using M1. Load time grows somewhat faster than linearly, but not much over the limited range of feasible problem sizes. Execution time falls as width grows because fewer subgoals must be established. We ran this and all the following tests on an 8-MHz AT compatible.

Table 3 shows load and execution times for conjunctive knowledge bases using M1. Increases in time are roughly linear over the range of feasible *B* and *W* values. While a 400-rule knowledge base with $W = 2$ runs out of memory, we ran one larger test with $B = 1000$ and $W = 8$; load and execution times were 150 and 106.9 seconds, respectively.

Table 4 compares load times for sequential knowledge bases using PC Plus and M1. We measured times with the knowledge bases stored in source and fast-load formats. Typically, source format would be used during development and fast-load for distribution to users. As the normalized ratio times indicate, M1 runs from 2.4 to 5.6 times faster than PC Plus.

Table 5 compares file sizes for sequential knowledge bases using PC Plus and M1. M1 files are from 4.3 to 12.5 times smaller than those of PC Plus.

Table 6 compares execution times for PC Plus and M1. M1 runs from 2.3 to 6.7 times faster than PC Plus. Since $B = 250$ exceeds maximum size for PC Plus in 640-Kbyte memory, times were not available.

While we conducted these tests to illustrate our approach, we also discovered that M1 ran considerably faster and used less memory and storage than PC Plus. However, since other factors must also be considered when comparing shells, one should not conclude that M1 is superior overall.

Using a single shell on various machines, we can test realistic knowledge bases; however, such testing consumes time and yields no comparative information on the shells. We can use stylized knowledge bases to compare shell speed plus memory and storage requirements. If shells accept ASCII input from an external editor, test knowledge bases can be generated automatically using a simple program. Even if we must enter knowledge bases manually, an hour or two of clerical time suffices to generate sequential knowledge bases.

Interested readers can obtain a copy of the source and executable versions of an M1 knowledge base generator (including documentation) by sending a self-addressed, stamped mailer and a formatted 5.25-inch floppy disk to the author. The M1 generator can be modified easily for other shells.

References

1. H.J. Curnow and B.A. Wichman, "A Synthetic Benchmark," *Computer Journal,* Feb. 1976, pp. 43-49.

2. R.P. Weicher, "Dhrystone: A Synthetic Systems Programming Benchmark," *Comm. ACM,* Oct. 1984, pp. 1013-1030.

3. J. Gilbreath, "A High-Level Language Benchmark," *Byte,* Sept. 1981, pp. 180-198.

4. J. Gilbreath, "Eratosthenes Revisited," *Byte,* Jan. 1983, pp. 283-326.

5. E.R. Tello, "Personal Consultant Plus," *Byte*, Oct. 1987, pp. 242-244.

6. S.J. Shepard, "TI's Personal Consultant," *AI Expert*, Mar. 1987, pp. 73-79.

7. B. Stevens, "Expert System Shell Survey," tech. report, Applied AI Systems, Inc., PO Box 2747, Del Mar, CA 92014, 1987.

8. S.J. Shepard, "Simplifying the Shell Game," *Computer Language*, Jan. 1987, pp. 23-28.

9. B. Olsen, B. Pumplin, and M. Williamson, "The Getting of Wisdom: PC Expert System Shells," *Computer Language*, Mar. 1987, pp. 117-150.

10. H. Firdman, "The Importance of Being Earnest in Selecting an Expert System Shell," *AI Expert*, Oct. 1986, pp. 75-77.

11. D.E. O'Leary, "Validation of Expert Systems with Applications to Auditing and Accounting Systems," *Decision Sciences*, Summer 1987.

Larry Press is a professor of computer information systems at California State University (Dominguez Hills). A member of the IEEE Computer Society, he received his PhD from UCLA. He has implemented several expert systems using personal computer shells. In addition to expert systems, he is researching object-oriented user-interface tools and the history of computing.

The author can be reached at the Computer Information Systems Dept., School of Management, California State University—Dominguez Hills, 1000 Victoria St., Carson, CA, 90747.

SOFTWARE REVIEW

BY ROY FREEDMAN

27-PRODUCT WRAP-UP: Evaluating Shells

New expert system shells seem to be announced every week. As each company seeks to distinguish its product from the pack, prospective buyers are presented with a confusing welter of features. Because expert systems development is still a relatively young field, finding out which packages come closest to what you really need can be a difficult and frustrating assignment.

The accompanying evaluation matrices (Tables 1-3) present a compact summary of the most relevant features contained in a representative sample of expert system shell products. This format is an experiment to try to provide a great deal of useful information in a small amount of space. It is meant to be a representative sample of available low- to mid-range shell products, not a comprehensive list. Let us know whether or not this format works for you and how we might improve upon it for upcoming (and more comprehensive) surveys.

MATRIX FEATURES

Most expert system shell product matrices represent knowledge about expert system shells in a binary fashion: if a feature is present (or absent) in the shell, a "yes" (or "no") is indicated in the appropriate row and column. From an AI perspective, this expert knowledge is too shallow.

Users typically require a set of features that can be customized into evaluation criteria for their own particular circumstances to indicate how well a feature satisfies their particular objectives. Such features can also be weighted by potential users so matrix entries for each shell can be summed and shells scored according to user relevance.

The accompanying tables represent "how" knowledge by providing features quantified by numbers, selections, and rankings. The features are defined in terms of concepts developed in books and courses on AI. In general, deciding how a shell incorporates these features is very subjective. Consequently, to insure a fair evaluation, shell vendors were asked to evaluate their own products with respect to specific features. Vendor responses were then compared with my evaluations to check consistency in the final feature entries.

Shell features are grouped into three categories: syntax, semantics, and pragmatics. Syntactic features describe the form or structure of a shell's language, semantic features describe the intended or denoted meaning of different components, and pragmatics features describe guidelines and criteria associated with practical use.

The description of expert system shells in terms of syntax, semantics, and pragmatics is borrowed from linguistics and is used to describe aspects of human and computer languages (or other formal systems of notation). For shells, syntax and semantics are chiefly concerned with knowledge representation and problem solving. An example of matrix entries for a sample shell looks like this:

Number of shell language keywords: 25
Select the numerical facilities supported (Integer, fiXed point, fLoating point, Trigonometric, Matrix) : IXL
Rank in order of availability and applicability the knowledge representations used (Frame, Production, Logic, Message-passing-object) : PF

Number features are used for counting or accumulation. In the preceding example, the sample shell has 25 keywords. Select features indicate which features are supported by the shell; in the list of features capital letters are used to abbreviate features selected in the matrix entry. In the preceding example, the sample shell supports integer, fixed point, and floating point operations.

Rank features are similar to select features—the significant difference is that these features are ranked in a priority order. In the preceding example, the sample shell best incorporates the knowledge repre-

Finding out which packages come closest to what you really need can be difficult and frustrating

sentations of productions and then frames. Rank features are important discriminants for shells based on multiple paradigms because they can help you decide which feature is most appropriate for the shell by vendor preference. A dash (—) indicates no response or "not applicable."

SYNTAX

Syntactic knowledge representations attempt to model the general concept of the rule. Rules are human-derived constructs that encompass relations, properties, actions, and experience.

Modern syntactic forms of a rule are based on knowledge representations such as production systems, frames (including schemas, scripts, message-passing objects, and semantic networks), and logic. A shell may include either one or several knowledge representations.

Shell knowledge representation facilities form a programming language whose grammar (or syntax) may be specified in terms of a variation of Backus-Naur Form (BNF) productions. For example, a shell vendor may describe the appearance of rules in the following way:

```
<rule> ::= (DEFRULE <rule-number>
  IF [ ( <left-hand-side-conditions> ) ]
  THEN [ ( <right-hand-side-actions> ) ] )
```

This BNF production says that a rule is specified by a *rule-number,* followed by one or more conditions and one or more actions. *DEFRULE, IF,* and *THEN* are special keywords in the shell. The symbols *<rule>*, *<rule-number>*, *<left-hand-side-conditions>*, and *<right-hand-side-actions>* denote syntactic categories. Different shells provide different shell language keywords that are used to create, access, and modify the syntactic representations.

Some shells provide special-purpose editors based on the shell syntax to aid users during rule development. In general, the number of BNF productions and the number of keywords can indicate the complexity of the shell.

Shells may also allow other conventional languages (or operating system primitives) to manipulate the representations. In this case, shell syntax also depends on the syntax of these languages and, indirectly, on certain language data types. One example is the way the shell represents numbers.

TABLE 1. Syntax features.

PRODUCT	NO. OF KEYWORDS	NO. OF BNF	REPRESENTATION[1]	LANGUAGES[2]	EDITORS[3]	NUMERICAL[4]	CONDITIONS[5]	QUERIES[5]	QUERYING[6]	RESPONSES[7]
ADS	110	100	PF	OSCF	RS	IFT	ANO	A	HWFLO	YDT
Arity Expert	79	21	PFLM	FCFSO	S	IFTX	AINO	AINO	HWL	YDTN
Envisage	140	100	PL	SCFLPAO	S	IXFMO	AINO	AINO	HWLFO	YDTNM
ESP ADVISOR	24	—	PL	PC	S	IXF	ANO	ANO	HWL	YDT
Expert-2	50	40	PFL	O	S	I	AO	ANO	—	Y
EXSYS	71	—	PF	CBASPLO	R	IXFT	AINO	AN	HWLFO	T
FLOPS	50	—	PF	C	S	IFT	AN	A	—	T
GURU	336	106	P	OC	SR	IF	AINO	AINO	HWL	DTM
HUMBLE	35	23	PM	O	SR	XFT	ANO	A	HFO	YDT
IN-ATE	—	—	FPL	L	SR	IF	—	—	HWLFO	DM
INSIGHT 2+	—	—	PL	—	S	IFXT	AINO	AINO	HWLFO	YO
Intelligence/Compiler	115	77	FLPM	CS	RSO	IFXT	ANOI	ANOI	HWL	YDTM
KDS3	20	27	FLPM	CSOFLP	RO	IXFT	AN	ANO	HWL	D
KES	88	200	PFM	CO	S	IFT	AOIN	AOIN	HWLO	YDT
Keystone	200	12	FMP	LCO	S	IFT	ANO	AN	HFWO	YDTM
KnowledgeMaker	5	5	P	SP	S	FI	A	A	H	YTM
KnowledgePro	114	34	FP	OS	SO	IF	ANO	ANO	H	YDTM
MacSMARTS	—	5	LM	—	RS	—	ANIO	OAN	HWL	DTM
NEXPERT/Object	100	—	PFM	CSF	R	FIXT	ANO	AN	HWLFO	DTM
PC-Easy	65	45	P	—	R	IFX	AIN	AINO	HWL	YDT
PC-PLUS	73	53	PFM	LCSPFAB	R	IXFT	AINO	AINO	HWL	YDTM
Rbest	12	18	PFL	CO	R	F	AINO	AINO	HWLFO	M
Superexpert	15	—	P	—	S	I	I	I	WL	T
VP-Expert	67	—	P	CSOFBLA	SO	IFT	AO	AO	HWLO	DT
WizdomMice	28	5	FMP	CSBFO	S	I	ANOI	AINO	HWL	YDTNM
Xi Plus	59	25	LP	CO	SR	IF	ANO	NO	HWLFO	YDT
1st-CLASS	—	—	PL	SCABPLO	SRO	IXF	ANOI	AIN	WHFO	YDMT

1. F—Frame, P—Production, L—Logic, M—Message-passing object.
2. F—FORTRAN, L—LISP, P—PROLOG, B—BASIC, C—C, A—Ada, S—PaScal, O—Other.
3. S—Standard text, R—Representation-specific, O—Other.
4. I—Integer, X—fiXed point, F—Floating point, T—Trigonometric, O—Other.
5. A—And, I—Implies, N—Not, O—Or.
6. H—How, W—Why, L—heLp, F—Find all, O—find One.
7. Y—Yes/no, D—yes/no/Don't know, T—arbitrary Text, N—Natural language, M—Mouse click on icon.

Shell representations may include variables for quantification (expressing concepts such as "for all") and binding. These variables are used to form conditions, goals, and queries. The exact syntactic structure of a user response to queries (ranging from arbitrary text to mouse-click input) also differs from shell to shell.

Vendors were asked the following questions about the syntax features of their products:

1. Number of shell language keywords.
2. Number of shell language BNF productions in shell language definition.
3. Rank the knowledge representations used in order of availability and applicability (Frame, Production, Logic, Message-passing object).
4. Rank the languages in order of preference for building user-defined functions (FORTRAN, LISP, PROLOG, BASIC, C, Ada, Pascal, Other).
5. Select the type of editors available for the shell (Standard text, Representation-specific rule editor, Other).
6. Select the numerical facilities supported (Integer, fiXed point, Floating point, Trigonometric, Matrix).
7. Select the shell support for building knowledge-based conditions (And, Implies, Not [negation], Or [disjunction]).
8. Select the shell support for building knowledge-based queries (goals) (And, Implies, Not [negation], Or [disjunction]).
9. Select the shell support for user-querying facilities (How, Why, heLp, Find all, find One).
10. Select the shell-supported mechanisms for user responses to queries (Yes/no, yes/no/Don't know, arbitrary Text, Natural language sentence, Mouse click on icon).

SEMANTICS

The semantics of shell representations denote the meanings of particular constructs for particular expert system task domains. Some shells are more appropriate for certain domains. In some sense, the semantics of a shell representation are based on mechanisms of problem solving, inference, and control.

Problem-solving methods based on classical AI paradigms distinquish shells from conventional programming languages. Shells are distinquished from each other by their support for different methods such as

TABLE 2. Semantic features.

PRODUCT	DOMAIN[1]	STRATEGIES[2]	INFERENCE[3]	CONFLICT[4]	ACTIONS[5]	DEFINITIONS[6]	SIDE EFFECTS[7]	RECURSION[8]	UNCERTAINTY[9]	ACQUISITION[10]
ADS	CDMQEPS	M	TBSO	PSO	ADGP	L	HMIC	PL	P	—
Arity Expert	DCM	UBMG	T	RS	G	P	CIM	B	PAR	F
Envisage	DCQE	MBG	TBS	TP	PDAG	L	—	BPL	PAR	TM
ESP ADVISOR	CD	MB	TB	T	AP	P	—	B	—	—
Expert-2	QCDEP	GM	T	T	A	P	IM	P	—	—
EXSYS	CDMQSEPA	GBM	TBS	T	PMG	P	CMI	BPL	P	FT
FLOPS	PCDE	MGB	BO	RSPO	AP	P	CMI	P	AP	M
GURU	DCPEM	B	T	TPO	AP	PL	MI	L	RA	M
HUMBLE	DCS	GM	TB	T	MAG	P	—	—	—	—
IN-ATE	DMQE	G	S	RP	A	P	HMIC	BLR	P	—
INSIGHT 2+	DCQPM	MB	BFDS	PT	PAG	D	IM	LRP	PR	—
Intelligence/Compiler	DPMSQAE	UBGMR	TBDSO	PTR	PGADM	PL	MHI	BLRP	PAR	MFT
KDS3	DCMESP	MB	BTS	R	PMAG	P	MI	PL	A	MF
KES	CDMP	MGB	TB	TP	PD	P	—	PL	P	T
Keystone	CDMQSEP	MB	BTDSO	TSPO	PMDAG	L	—	PLR	—	F
KnowledgeMaker	CD	B	T	T	AP	—	M	B	P	T
KnowledgePro	CDMQSEP	GMB	TBDS	TO	PMDAG	P	MH	LBR	AR	TM
MacSMARTS	DCE	MB	TB	RP	AGP	P	M	B	P	T
NEXPERT/Object	DMEPC	RBU	TBS	PR	APMD	D	CI	BL	AR	FM
PC-Easy	DC	BM	TBS	PTS	PA	L	I	P	P	—
PC-PLUS	PCMD	BM	TBS	PTRS	PDAG	LD	IC	PL	P	M
Rbest	DCQMPS	MURB	BT	T	AG	D	IMH	PBL	PR	F
Superexpert	QDCP	M	TB	S	A	—	—	—	A	T
VP-Expert	QCDMSP	BMG	TB	T	PAG	P	MI	LB	PA	—
WizdomMice	CDPQE	BMRUG	TBSDO	STR	G	DPL	IC	B	R	—
Xi Plus	DCQME	UBRM	TBO	TP	PM	PL	MI	PB	A	—
1st-CLASS	CDMQ	MRU	STBO	TPO	AG	P	MI	L	PA	T

1. C—Classification, D—Diagnosis, M—Monitoring, Q—Querying, S—Simulation, A—Animation, E—dEsign, P—Planning.
2. G—Generate and test, M—Match, R—Resolution, U—Unification, B—Backtracking.
3. T—Top down, B—Bottom up, D—breaDth first, S—beSt first, O—Other.
4. T—Top down, R—Recency, S—Specificity, P—rule Priority, O—Other.
5. P—Procedural attachment, M—Message passing, D—Demons, A—production Actions, G—subGoal instantiations.
6. P—declarative Program sections, L—Lexical scoping, D—Dynamic scoping.
7. I—Input/output, C—control of Conflict resolution, M—working Memory assignments, H—inHeritance.
8. B—Backtracking, P—Production, L—explicit Loops, R—explicit Recursive functions.
9. P—Predefined set, A—Arbitrarily (user) defined, R—Run-time modifiable.
10. T—Table, M—Metarule, F—Frame instantiation.

matching (for production systems), unification (for logic programming systems), or generate-and-test.

Expert system shells use two classical representations of inference. Top-down (also referred to as "backward chaining") inference reasons backward from proposed conclusions to hypotheses until no contradictions are established. Bottom-up (also referred to as "forward chaining") inference derives new facts from old by using established facts and rules to create new facts.

These two extremes are seen by considering problem solving as tree traversal. Variations of these representations include breadth-first and best-first representation. Some shells have other methods of controlling inference.

Shell control semantics reduces to the behavior of the interpreter or program executor. Some of the specific issues associated with shell control include determining how working memory elements are used to collect a set of rules that are eligible to fire (matching), determining which rules should fire from a set of eligible rules (conflict resolution), and ascertaining what types of actions rules are allowed to perform.

The semantics of shell rule actions may also be based on classical AI syntactic paradigms. These actions may result in some input or output operation to a file or display or a change in the state of working memory.

Some state changes are side effects and may indirectly change the order of rule firings, explicitly (or implicitly) causing a loop. Other side effects may involve the propagation of certain properties through an inheritance network. Such actions may further involve certain shell-defined structures or global variables: the way such structures are defined impacts the maintainability of the expert system.

Finally, shell semantics for representing uncertain and incomplete knowledge are also important for many applications. Uncertain information is usually represented by incorporating a statistical model into the knowledge representations and inference, which may be predefined by the shell or arbitrarily defined by the user.

Incomplete information may be represented if the form of the missing knowledge is known. In this case, missing knowledge can be automatically acquired by instantiating a frame, firing a rule that creates another rule (a metarule), or matching similar entries in a table.

Vendors were asked the following questions about semantic features:

TABLE 3. Pragmatic features.

PRODUCT	DOCUMENTATION	ERRORS	TRAINING DAYS	COST ($K)	TRAINING COST ($K)	RUN-TIME COST ($K)	EXAMPLES	DEVELOPMENT[1]	CONFIGURATION MANAGEMENT[2]	MAINTENANCE[3]	INTERFACE[4]	REHOSTABILITY[5]	CHOICE[6]	USER CHOICE[6]
ADS	720	100	3	7.0–60.0	1.5	0.75–25.0	6	DHE	FM	CL	FDC	DO	PWTK	PWTK
Arity Expert	270	30	3	0.3	0.75	—	1	ED	FIM	CL	CFD	D	PTW	PWT
Envisage	190	50	5	5.0–40.0	—	1.0	2	HED	FI	CL	FC	VUDO	PWTMK	PWTMK
ESP ADVISOR	200	15	3	0.9	—	0.045–0.135	5	HD	F	—	CF	DVU	WT	T
Expert-2	150	0	10	0.15	—	—	4	TE	I	L	C	DO	T	T
EXSYS	272	4	3	0.4–5.0	—	0.60	21	HTED	FM	CL	CF	DUVO	WT	T
FLOPS	300	0	30	0.5	0.5	—	30	D	MF	—	FC	DUV	T	T
GURU	900	300	3–6	6.5–60.0	0.75	0.40	—	TH	FIG	—	FCD	DVU	PWK	PWTKM
HUMBLE	124	0	5	0.4–1.5	1.5–2.0	—	2	ED	FM	UL	C	DUO	PTM	PWTMK
IN-ATE	300	—	2	10.0	1.5	5.0	5	HTE	FIG	CUL	FCD	UV	PWTMK	PWTMK
INSIGHT 2+	320	—	5	0.5	0.75	0.095	18	HTED	FMG	LC	FCD	V	PW	—
Intelligence/Compiler	275	36	5	0.99	—	—	50	THED	FIMG	CU	CFD	DU	PWTMK	PWTMK
KDS3	265	—	5	1.5	—	0.15–0.495	—	HED	FMG	CUL	CD	D	PKW	PWK
KES	300	—	4	4.0–60.0	1.2	0.4–6.0	—	ED	FM	CL	F	DUVO	T	PWTMK
Keystone	200	—	3	9.75	0	0.5	2	ED	FIM	L	CF	DUVO	PWTMK	PWTMK
KnowledgeMaker	60	20	1	0.099	—	—	3	HTED	F	—	FC	D	PWTMK	PWTMK
KnowledgePro	150	20	2	0.495	—	—	15	HTED	F	U	CF	D	PWTMK	PWTMK
MacSMARTS	85	12	1	0.15	—	0.05	5	H	MF	U	DF	UV	PWTM	TPMW
NEXPERT/Object	500	—	5	5.0–8.0	0.65–2.0	0.5–1.2	5	HTED	FIMC	CUL	CDF	DUVO	PWTMK	PWTMK
PC-Easy	900	—	5	0.5	1.0	0.095–2.0	2	HTED	FMG	CL	FCD	DVU	PWT	PWT
PC-PLUS	1200	—	10	3.0	1.0	0.095–2.0	4	HTED	FMG	CL	FCD	DVU	PWTM	PWTM
Rbest	80	0	3	negotiable	—	—	4	HED	FIM	L	CF	UDOV	MT	—
Superexpert	230	20	5	0.2	—	0.30–1.0	20	HE	F	CU	DF	D	PWT	T
VP-Expert	380	26	2	0.1	—	0.50	22	HTED	F	CU	FCD	D	WTKP	TKP
WizdomMice	300	—	3	0.75–15.0	0.75	0.30	3	ETD	FMG	LCU	CF	DUO	T	TWM
Xi Plus	463	140	3	1.25	0.75	0.275	6	HTED	FI	CLU	CFD	DO	PTK	PT
1st-CLASS	440	53	1	0.5–1.3	—	—	25	HTED	FM	CU	CFD	D	PTMK	WTMK

1. H—on-line Help, E—on-line Examples, D—Debugging aids.
2. F—saving Files, M—saving Multiple knowledge bases, G—saving Graphics.
3. C—Consistency checking, U—run-time User modifiability, L—Loop detection.
4. F—Files, C—Call facility, D—Data base format conventions.
5. D—DOS, U—UNIX, V—VMS, O—Other.
6. P—Pop-up menus, W—Windows, T—Text dialog, M—Mouse-sensitive regions, K—Key bindings.

1. Rank in order of applicability the most appropriate task domains for the shell (Classification, Diagnosis, Monitoring, Querying, Simulation, Animation, dEsign, Planning).

2. Rank in order of applicability the most appropriate problem-solving strategies supported (Generate and test, Match, Resolution, Unification, Backtracking).

3. Rank in order of preference the most appropriate (default) control of inference (Top down [backward chaining], Bottom up [forward chaining], breaDth first, beSt first, Other).

4. Rank in order of preference the most appropriate (default) methods of conflict resolution (Top down and left to right, Recency, Specificity, rule Priority, Other).

5. Select how the shell represents actions (state changes) (Procedural attachment, Message passing, Demons, production right-hand-side Actions, subGoal instantiations).

6. Select how the shell represents definitions (declarative Program sections, Lexical scoping, Dynamic scoping).

7. Rank in order of preference the most appropriate reliance on side effects and global variables (Input/output, control of Conflict resolution, production working Memory assignments, inHeritance).

8. Select how the shell supports recursion and iteration (Backtracking, Production contexts, explicit Loops, explicit Recursive functions).

9. Select how the shell specifies uncertainty representations (Predefined set, Arbitrarily defined by user, Run-time modifiable).

10. Rank in order of preference the most appropriate shell support for automated knowledge acquisition (inference from a Table, Metarule, Frame instantiations).

PRAGMATICS

Pragmatic aspects of shells concern issues relevant to getting the job done. This includes factors such as cost, documentation, interfaces to other tools, the ability to save and restore intermediate results, input and output devices, maintenance and debugging support, delivery (run-time) support, and shell rehostability. I/O operations (involving numerics, text, or graphics) and operations involving the user interface may incorporate menus that pop up and then vanish; others may only involve windows, programmable mouse-sensitive areas, and the ability to bind terminal keys to signal an appropriate action.

Probably the most basic aspect of a shell is usability. Shell usability is indicated by the amount of documentation provided, the types of on-line development facilities supported, and the number of days required to become proficient in shell use. Shell complexity (indicated by the syntax features) also should be used to judge usability.

Vendors were asked the following questions about the pragmatics features of their packages:

1. Number of pages of documentation.

2. Number of shell-defined errors documented in user manual.

3. Number of training days to become proficient in shell use.

4. Number (dollars) cost of shell for development environment.

5. Number (dollars) cost of shell (vendor) training.

6. Number (dollars) cost of shell run time for delivery, if available.

7. Number of expert system examples provided in documentation.

8. Select the shell-supported mechanisms for development (on-line Help, on-line Tutorial, on-line Examples, Debugging aids).

9. Select the shell support for configuration management (saving Files, saving screen Images, saving Multiple knowledge bases [blackboards], saving Graphics).

10. Select the shell support for quality assurance and maintenance (Consistency or completeness checking, run-time User modifiability [belief maintenance], or Loop detection).

11. Rank in order of preference the most appropriate shell support for the interface to other tools (Files, Call facility [operating system], Data base format conventions).

12. Rank the operating systems in order of ease of shell rehostability (DOS, UNIX, VMS, Other).

13. Select the shell choice facilities available during expert system development (Pop-up menus, Windows, Text dialogue, Mouse-sensitive regions, Key bindings).

14. Select the shell choice facilities available for the deliverable user interface that are programmable by the shell user (Pop-up menus, Windows, Text dialog, Mouse-sensitive regions, Key bindings).

USING THE MATRIX

Tables 1-3 are meant to help shell users select suitable shell candidates for expert system development and delivery. Table 4 lists manufacturer names and addresses.

The goals of development and delivery correspond to two measures of the suitability of a shell:

- Application-specific knowledge is easily expressible in terms of shell representations by an expert system builder or shell user.
- The demand on computational resources during the actual execution of the expert system (built with the shell) is within workable limits.

The first measure is referred to as developmental efficiency, the second as run-time efficiency. Developmental efficiency is crucial for an initial demonstration of conceptual and operational feasibility. Run-time ef-

Roy Freedman, Ph.D., is an associate professor of computer science at Polytechnic University, New York, N.Y. He is also president of Roy S. Freedman Consulting Ltd., a firm that specializes in applying advanced software technology in the financial services and electronics industries.

ficiency is crucial for system delivery.

Overemphasis on developmental efficiency has led many people to conclude that knowledge-based systems are run-time inefficient. On the other hand, overemphasis on run-time efficiency can lead to the conclusion that many applications are not possible (even though they might be if the most suitable shell or knowledge representation techniques are utilized). It is important to note that the shell or representation used for development need not be the same as the shell or representation used for delivery.

Once a set of candidate shells is selected with the matrix, shell users should develop a small sample problem based on their application to further distinquish the candidates. The criteria used in the final selection should be based on the previously stated goals for expert system development and delivery. Of course, this final selection also depends on the availability of trained knowledge engineers. **AI**

I want to thank Debora Riccio, Fran Tischler, Isidore Sobkowski, and Robert Frail.

TABLE 4. Expert system shells and manufacturers.

ADS AION Corp. 101 University Ave. Palo Alto, Calif. 94303 (415) 328-9595	**HUMBLE** Xerox Special Information Systems 250 N. Halstead St., Box 5608 Pasadena, Calif. 91107-0608 (818) 351-2351	**NEXPERT/Object** Neuron Data 444 High St. Palo Alto, Calif. 94301 (415) 321-4488
Arity Expert Arity Corp. 30 Domino Dr. Concord, Mass. 01742 (617) 371-1243	**IN-ATE** Automated Reasoning Corp. 290 W. 12th St., Ste. 1D New York, N.Y. 11014 (212) 206-6331	**PC-Easy, PC-PLUS** Texas Instruments Corp. P.O. Box 809063 Dallas, Texas 75380
Envisage Systems Designers International 5203 Leesburg Pike Falls Church, Va. 22041 (703) 820-2700	**INSIGHT 2+** Level Five Research Inc. 503 Fifth Ave. Indiatlantic, Fla. 32903 (305) 729-9046	**Rbest** Titan Systems Inc. 20151 Nordhoff St. P.O. Box 2123 Chatsworth, Calif. 91313 (818) 709-9685
ESP ADVISOR Expert Systems International 1700 Walnut St. Philadelphia, Pa. 19103 (215) 735-8510	**Intelligence Compiler** IntelligenceWare Inc. 9800 S. Sepulveda Blvd. Los Angeles, Calif. 90045 (213) 417-8897	**Superexpert** Softsync Inc. 162 Madison Ave. New York, N.Y. 10016 (212) 685-2080
Expert-2 Mountain View Press P.O. Box 4656 Mountain View, Calif. 94040 (415) 961-4103	**KDS3** KDS Corp. 934 Hunter Rd. Wilmette, Ill. 60091 (312) 251-2621	**VP-Expert** Paperback Software Inc. 2830 Ninth St. Berkeley, Calif. 94710 (415) 644-2116
EXSYS EXSYS Inc. P.O. Box 75158, Contr. Stn. 14 Albuquerque, N.M. 87194 (505) 836-6676	**KES** Software A&E Inc. 1600 Wilson Blvd., Ste. 500 Arlington, Va. 22209-2403 (703) 276-7910	**WizdomMice** Machine Intelligence Corp. 1593 Locust Ave. Bohemia, N.Y. 11716 (516) 589-1676
FLOPS Kemp-Carraway Heart Institute 1600 N. 26th St. Birmingham, Ala. 35234 (205) 226-6697	**Keystone** Technology Applications Inc. 6621 Southpoint Dr. N. #310 Jacksonville, Fla. 32216 (904) 737-1685	**Xi Plus** Portable Software Inc. 650 Bair Island Rd. Redwood City, Calif. 94063 (415) 367-6264
GURU Micro Data Base Systems Inc. P.O. Box 248 Lafayette, Ind. 47902 (317) 463-2581	**KnowledgeMaker/KnowledgePro** Knowledge Garden Inc. 473A Malden Bridge Rd., RD2 Nassau, N.Y. 12123 (518) 766-3000	**1st-CLASS** Programs in Motion 10 Sycamore Rd. Wayland, Mass. 01788 (617) 653-5093
	MacSMARTS Cognition Technology 55 Wheeler St. Cambridge, Mass. 02138 (617) 492-0246	

NEW PRODUCT REVIEWS

Editor: Richard Eckhouse, MOCO, Inc., PO Box A, 91 Surfside Rd. Scituate, MA 02055; Compmail+, r.eckhouse

Exploring expert systems

Valerie Barr, Pratt Institute

Expert systems—also, and perhaps more properly, known by the term knowledge-based systems—seem to be everywhere. Various applications have appeared in corporate settings, including the well-known Xcon system designed by Digital Equipment and the Cooker system designed by Campbell Soup. Digital developed the former to automate the organization of computer system components and to ensure that order shipments included all necessary parts and cables. Campbell developed the latter when a 40-year employee who serviced the cooking machines approached retirement. The company incorporated his knowledge about the machines into the Cooker program to provide expert help to new service employees.

Various applications developed for home use largely serve as an introduction to knowledge-based systems. You can find these programs as samples in the various expert system shells, as shareware, and in ads in the back pages of popular computer magazines. The applications are often food-oriented—picking the proper cheese or wine to accompany a particular meal, for example.

Looking only at these two extremes can daunt computer users who believe they have an application suited for an expert system, but who are unclear on what expert systems actually are or how to build one. Expert system shells are programs designed to aid a user in building an expert system. The shell essentially provides a framework into which the user can plug the details of the application.

In this review, I take the perspective of a user developing an expert system for the first time. The shells reviewed fall in the under-$500 price range, so the initial outlay is not prohibitive. For the most part, you do not need a knowledge of expert systems to use the shells, since many of them provide some introductory material as well as examples.

Introduction to expert systems

In expert systems, unlike other types of programs, we tell the computer what to know, not what to do. When constructing the expert system, we do not provide a set of instructions; rather, we provide knowledge and advice. If we have a step-by-step algorithm for solving the problem at hand, then a "traditional" program is in order. However, if we have no such step-by-step method, then an expert system is called for. Problems of this sort, for which we cannot write down a step-by-step solution, are considered ill-structured problems.

Ill-structured problems are rarely numeric, and you might find it impossible to write a set of specifications for them. Their solutions rely on a large body of knowledge or on things that cannot be quantified, such as 40 years of experience repairing cooking machines.

The AI approach to ill-structured problems looks at the ways people normally solve them. Following this approach, we want to accumulate knowledge and store it in such a way that we can then use it to solve future problems of the same type. In effect, we collect, formalize, and represent within a knowledge-based system some large body of knowledge specific to a task, then use that knowledge as if we had it in our own heads.

Knowledge-based systems have two basic parts: the inference engine and the knowledge base. The knowledge base contains the task-specific information. The inference engine is the mechanism for using the information in the knowledge base.

In a rule-based system, the knowledge base breaks down further into the rule base and the working memory (or database). The system usually (I would hope) has some kind of explanatory interface and a user interface. Also, some systems will have a knowledge acquisition module.

A rule base is a set of rules specific to the problem at hand. For example, a knowledge base system for auto repair would have rules such as

IF	the engine does not start
AND	the starter motor does not turn over
THEN	the problem is in the electrical system
IF	the engine does not start
AND	the starter motor does turn over
THEN	the problem is in the fuel system

Each rule has two parts: the antecedent or premise (IF and AND parts) and the consequent or conclusion (THEN part). A rule is triggered if information in the database matches the antecedent of the rule. When the action specified by the rule is carried out, the rule fires and the conclusion of the rule joins the working memory.

Working memory, or the database, consists of a set of facts that describe the current situation. Generally, you will start out with very few facts. These will expand as you learn more about the situation at hand based on the rules executed.

The inference engine, also called the rule interpreter, has two tasks. First, it examines facts in working memory and rules in the rule base, and adds new facts to the database (memory) when possible. That is, it fires rules. Second, it determines in what order rules are scanned and fired. If the system is designed to ask the user for more information, then the inference engine does the asking, through the user interface. The inference engine is also responsible for telling the user what conclusions have been reached.

You use the knowledge acquisition module to add rules to the rule base and to modify existing rules. In a very simple system, you will have to use an editor or word processor outside of the expert system. More sophisticated shells include an editor so that you have a mechanism for entering rule information.

The user interface is usually a natural language interface, which gives the user some degree of flexibility and a feeling of familiarity when interacting with the system. The user interface will provide some introductory message as well as ask questions of the user in a seemingly natural way.

The explanatory interface is that component of the system that lets the user see how the system reached a particular conclusion. Various degrees of sophistication are possible in the explanatory interface. For example, at any point the interface might allow the user to ask how the system reached an intermediate conclusion. Again, when the system asks the user for additional information, the interface might permit the user to ask the system to explain why it needs the particular information requested.

An expert system shell contains all or most of these components, except the knowledge base and the actual statements issued by the user interface. The person developing the system adds the rules, thus providing the knowledge base. The end user of the resulting expert system provides the information that goes into the database or allows information from the knowledge base to be placed into the database.

Knowledge-based systems employ two basic control strategies: backward chaining and forward chaining. To illustrate, assume a knowledge base containing the rules

R1: IF A AND B THEN C
R2: IF C THEN D
R3: IF E THEN B

and a database containing A and E.

Forward chaining involves searching the rules in the order listed to see which one can be fired based on the appearance of its antecedent in the database. R3 is the only one, since E is in the database. Fire R3 and add B to the database. Search the rules again. R1 can be fired because A and B are in the database. Add C to the database. Now R2 can be fired, and D is added to the database.

The forward-chaining inference-engine algorithm summarizes this, as shown in Figure 1.

Backward chaining starts by trying to prove C, which is the consequent of the first rule. To prove C, we must have A and B in the database. A is already there. To get B, we look for a rule that has B as its consequent and try to establish the antecedent of that rule. A search finds that R3 has B as its consequent, with E as the antecedent. Since E is in the database, we can add B to the database. Now R1 can be fired, and C is proven or can be added to the database, or returned to the user as the result of the system. If the system was set up to allow multiple conclusions, it would proceed to R2. Since C was proven by R1, D could be proven by R2.

The backward-chaining inference-engine algorithm summarizes this, as shown in Figure 2.

Applications that use forward chaining, such as process control, we call data-driven. Applications that use backward chaining we call goal-driven.

You should use forward chaining if you have a small set of relevant facts, where many facts lead to few conclusions. A forward chaining system must have all its data at the start, rather than asking the user for information as it goes.

You should use backward chaining for applications having a large set of facts, where one fact can lead to many conclusions. A backward-chaining system will ask for more information if needed to establish a goal.

The expert system shells reviewed below all implement a backward-chaining strategy, allowing you to design interactive knowledge-based systems, where the end user supplies the data.

In the interest of fair testing, I took an existing set of rules written with no particular shell in mind and set up systems with each shell using those rules. The following reviews describe the expert system shells ESIE, Exsys, and VP-Expert, and discuss my creation of expert systems using this given set of rules.

When a new fact is added to the database,

(1) Scan the knowledge base to find all rules using that fact.
(2) Discard those not ready to fire.
(3) Add those ready to fire to the active list.
(4) Fire the rule at the front of the active list.

Figure 1. The forward-chaining algorithm.

(1) Find the list of rules that conclude about the current attribute A.
(2) If there are no such rules, ask the user for the value of attribute A.
(3) Otherwise, for each such rule R, use the following to determine the truth of the premise of the rule:
For each clause C in the premise of R:
Look in the database for the value of the attribute mentioned in C.
If there is no information in the database about that attribute
Start from the top with that attribute as the current one.
Evaluate the truth of C using the information in the database.
If the premise of the rule evaluates to "true," make the conlusion shown (given in the rule).

Figure 2. The backward-chaining algorithm.

ESIE

ESIE, from Lightwave, stands for Expert System Inference Engine. This shareware package comes on one disk containing 19 files. It includes on-disk documentation consisting of a user's manual, tutor, knowledge engineer's manual, history (a brief history of AI), and a novice guide (for new users also unfamiliar with AI). ESIE is available directly from Lightwave or from shareware groups.

For the $145 registration fee, you get a copy of the latest version of ESIE, printed versions of the manual, PC-Write, and access to the ESIE help line. If you register with Lightwave as a purchaser of ESIE, you will get the next version of ESIE when it is released.

System requirements are an MS-DOS computer with at least 128 Kbytes of RAM (256 Kbytes recommended), DOS 2.0, and a word processor that produces ASCII files.

Only three files are essential to run ESIE: Esie.com, Esie.cfg (created by the Config program), and the file containing the knowledge base. This last is built by the expert system developer in the word processor.

Once in ESIE, you have only four top-level commands: Go, which starts the consultation (meaning you run the expert system); Exit, which returns to DOS; Trace On, which helps you debug the knowledge base; and Trace Off, which turns the trace off.

The trace is off by default when you enter ESIE, and you must change its status before the consultation begins. If you are in the midst of the consultation and suspect a bug in the knowledge base, you cannot turn on the trace at that point. Instead, you have to finish the consultation, turn the trace on, and try again. A consultation will continue until it completes or an error is found in the knowledge base logic.

When you first turn the trace on, it tells you the number of rule lines (ESIE counts each IF, AND, and THEN separately, so it really does measure rule lines, not rules), the number of questions, and the number of legal answers in the knowledge base. Then, as the consultation proceeds, you see the subgoals or subproblems, that is, the information the inference engine is trying to find so that it can fire rules. The program displays new answers as they are determined, either directly from information you provide or from the rules.

The system has seven types of statements that you can put into a knowledge base. "Introtext is <text>" defines the message printed when the consultation starts. "Goal is <variable>" defines the variable for which the consultation is seeking a value. The rules are handled in a standard IF-AND-THEN form. In addition, the knowledge-base designer can specify legal answers, the text of questions asking the end user for information, text preceding the solution found by the system, and termination text signifying the end of the consultation.

With ESIE, it took some maneuvering to get my set of rules into the proper format. I have two suggestions that might help novice expert-system designers:

(1) Even if you have your application in mind, don't start writing rules until you know what shell you will use, because it can be harder to rework existing rules than to think them up in the proper form.

(2) I found it helpful to set up two windows in the word processor. In one, I put one of the sample knowledge bases. In the other, I wrote my own rules, following the general format given in the sample.

A knowledge base set up with ESIE can accept 1,000 rule lines (between 300 and 500 actual rules), 300 questions, 50 legal answers, 400 variables and values, 12,000 bytes of text, one goal, and one answer text. ESIE will look for up to 50 different rules to satisfy at a time (in attempting to prove a THEN), and it can learn up to 200 facts before it runs out of space in memory.

In setting up the rule base, you can put the rules and questions in any order. Since each question includes the variable about which it asks, you can have all the rules first, followed by all questions, or each rule followed by any related questions.

ESIE is a good system for someone who wants to try out expert systems or who has an application without complex rules.

I did encounter a few frustrating points in using ESIE.

For one thing, you cannot edit rules from inside ESIE. When an error crops up, you have to exit ESIE, go to your word processor, and then return to ESIE after (you hope) fixing the error.

Another frustration, particularly when learning ESIE, is that you cannot load a new knowledge base from within ESIE. You have to return to DOS and reenter ESIE, specifying the new knowledge base.

In general, ESIE is a good system for someone who wants to try out expert systems or who has an application that, while it may involve a lot of rules, does not involve complex rules. It is certainly possible to develop very useful knowledge-based systems using ESIE.

Exsys (Version 3)

In comparison, Exsys from Exsys Inc. is a much more powerful system than ESIE. The first hint of this shows up in the system requirement of 320 Kbytes of RAM. The Exsys demo allows execution of knowledge bases with up to 80 rules, but you can save only 25 rules to disk. In the full system on a PC, you can run up to 5,000 rules. Exsys will use all available memory and requires about 64 Kbytes for every 700 rules (assuming an average of six or seven conditions per rule). On a VAX computer, you will have essentially no limit on the number of rules.

Exsys costs $395 for a single-user PC version, $600 for a noncommercial runtime license for PCs (allowing unlimited copies), $600 for a commercial runtime license for PCs (allowing 30 installations), and $7,500 for a VAX version (allowing up to 16 users; $12,500 for more than 16 users). The recently released Exsys Professional costs $795 for a single-user PC version and $12,000 for a 16-user VAX version.

The many features offered in Exsys include "why" and "what-if" capabilities. For the "why," Exsys shows the chain of rules used to arrive at a result. For the "what-if," the end user can change some of the answers given during a consultation and rerun the consultation. It is then possible to look at the old and new results together and compare them. You can also save input and results to disk.

Exsys allows the system designer to specify the probability that a certain result is true. You can set up an Exsys system in three different ways. First, a true/false (or 1/0) method gives all answers as yes or no. Second, the solutions use probability values between 0 and 10. Third, solutions give probability values from −100 to 100. When the program displays the results of the consultation, the end user can view all results or only those above a certain probability. Once you have chosen the probability mode for a system, you cannot change it.

The rules in Exsys can get more complicated. For example, you can carry out numeric comparisons in the IF portion and computations in the THEN part. Furthermore, the THEN part can have ANDs, so you can get multiple results based on the condition(s) satisfied in the IF portion of the rule. You can also have an ELSE section of a rule, so you can specify what results hold if the IF portion is not satisfied. Each part of a rule (IF, THEN, ELSE) can have up to 126 conditions. You can build OR conditions into the IF portion, but only with regard to the value of a single variable ("IF the color is blue or green" is okay, but "IF the color is blue or the time is night" is not okay).

Basically, the rules can ask only two types of questions: multiple choice or numeric value. In the first case, the user is asked a question and presented with choices, each preceded by a number. The user responds by entering the number(s) of the choice(s) selected. The response must fall within the range of numbers displayed or Exsys will not allow the user to proceed with the consultation. In the second case, the user is asked a question (like "The length of the room is __?") and gives a numeric answer.

When executing an Exsys-based system, you can call external programs for data acquisition, calculation, or result display, for example. You can also pass data back to Exsys for use in analysis during a consultation. This capability allows the designer to build systems with a great deal of power and flexibility.

Exsys' internal rule editor provides prompting, so you can often select material for a rule based on information supplied for prior rules. The editor takes a bit of getting used to, but after you enter about five rules, things start to go very quickly. Qualifiers receive numbers within the system when first introduced, so you can recall them for use in subsequent rules.

As designer, you also enter introductory and concluding text into the system, as well as your name as author and a name for the expert system. You also select the data derivation options (apply all possible rules or stop after the first successful rule) and the default for rule display. If rule display is on, then each rule used will be shown when it completes firing, with either the IF and THEN sections or just the ELSE section highlighted, depending on whether or not the condition in the IF section was satisfied. Regardless of the default you choose, a person using the expert system can override it.

During system design, you can add additional information to the rules in the form of notes and references. A note is for the user's information and is displayed with the rule if rule display is on. A reference can explain the source for the rule. This information is displayed only if the user asks for it.

I encountered a few bothersome things about the demo disk. First of all, the file allocation table was faulty, causing the demo of the report generator to bomb. Furthermore, while the demo is very good and can certainly teach you enough to start building a system, there is no way to move through it quickly. So, if you want to repeat parts of the demo to refresh your memory, you have restart from the beginning. The demo also does not tell you how to quit in the middle (Ctrl-Break works).

Peculiarly, the Exsys manual does not refer to the antecedent portion of a rule (the IF part). Instead, it refers to a "condition" with two distinct parts: the qualifier, which contains a "verb" (for example, "the color is"), and "values," which are possible completions of the condition. While it doesn't take long to get used to this, it can confuse you at the start.

The demo's major flaw is that it does not clearly explain that, if you have a rule in which the THEN section sets a qualifier, then you must have an ELSE section that sets the qualifier to some other value. Otherwise, the system will ask the user for it.

In executing a system, Exsys tries to be efficient where space allows. When you create a knowledge base, Exsys builds two files for it, a .rul and a .txt file. When a consultation begins, Exsys reads in the .rul file, which contains an encoding of the rules. Then, if there is space, Exsys also reads into memory the .txt file, which contains the actual text of the rules. This allows the consultation to proceed more quickly. If the .txt file does not fit in memory, the program will access it from disk when needed, slowing down the process.

Overall, Exsys has a myriad of features that allow you to design very powerful expert systems in a fairly efficient manner. In addition to the features discussed here, it has a report generator that lets you design systems that output to files (or the printer) various information generated by the expert system. It can also do a mix of forward chaining and backward chaining or pure forward chaining. Use of these and other advanced capabilites involves some study of the manual, particularly the section on command-line options.

Clearly, Exsys is designed for the development of sophisticated and complex knowledge-based systems, but it also works well for small-scale systems. I think it is a good choice for a user who plans to ultimately design large systems, but would like to start out with smaller scale systems.

VP-Expert (Version 1.2)

Of the three packages, VP-Expert by Paperback Software is the one most representative of the current generation of PC software, both in look and feel as well as capabilities. It is on a par with Exsys in terms of system configuration, requiring 384 Kbytes of RAM and DOS 2.0 or higher. While you can probably build knowledge bases as large as the ones built with Exsys, what sets this shell apart is its use of the PC and its compatibility with other PC software packages. It costs $100.

The program source for VP-Expert is in Microsoft C, allowing you to build knowledge bases that use all of primary memory (not extended memory). The program has a limit of 20 conditions in a rule and 20 levels of backward chaining. VP-Expert also has a knowledge base chaining feature, allowing you as designer to build knowledge bases which otherwise would not fit into memory. At the end of one knowledge base, the program saves the current database values. A Chain command then opens a new knowledge base. Finally, the database values are restored.

VP-Expert does not have an on-disk demo or tutor. However, a number of sample knowledge bases come as part of the package. These are referred to in the first chapter of the manual, which serves as a tutorial. The screen displays shown in the tutorial chapter are very accurate. The only discrepancy is that the book consistently states that the light bar will come up on option 1 (Help), but it consistently comes up on option 2 instead (which varies from screen to screen). After the first chapter, I noticed slight differences between the screen displays pictured in the book and the ones that came up on the computer, probably due to changes in the sample knowledge bases.

Each screen in VP-Expert comes up with command choices clearly displayed along the bottom of the screen. You can select one of these choices by typing the first letter, typing the command number, hitting the function key corresponding to the command number (F1 for command 1, etc.), or using the cursor control keys to move the light bar to the desired command and pressing Enter. However, when you use the built-in editor, only the latter two methods work. The others cause unwanted characters in your file. I suggest choosing a single method early, one you can use for everything. Otherwise, the editor will take you by surprise when you start using it.

The VP-Expert main menu presents six options (besides Help and Quit). FileName and Path let you call a new knowledge base or change the default drive and directory. A novice will see the options Consult and Edit most frequently. Selecting Consult takes you to a consultation window, where an expert system executes. During development, the consultation window takes up half the screen; a rules window and a results window occupy the lower half of the screen. The rules window shows the activity of the inference engine, while the results window notes conclusions derived during the consultation. In a finished system, ready for the end user, you can remove the extra windows by entering a single statement into the knowledge base to get a consultation window the size of the entire screen.

While in the consultation window, when the consultation has ended or paused (waiting for user input), you can obtain a subsidiary menu by typing a slash (/). This menu gives choices for How (how the value of a variable was found), Why (why the current question is being asked), and Slow and Fast, which slow down or reset the speed of knowledge base execution (so that you can read the display in the rules and results windows).

Other options available from the consultation window are

- WhatIf—lets the user change a single response and rerun the previous consultation. All other variables keep their prior value unless changed by rules in the rule base. The user is asked for new information if needed to satisfy some rule which had not fired during the initial consultation.
- Variable—lets the user select any variable to see the value it had at the end of the previous consultation.
- Rule—lets the user specify a rule number to see that rule displayed in the rule window. If no rule window is open, nothing happens.
- Set—displays a submenu that has commands for Trace, Slow, and Fast. Selecting Trace causes the path of the inference engine during the next consultation to be stored in a file with the same prefix as the knowledge base and a trc extension. The user can view this later. Slow and Fast are the same as above. The Slow command has a cumulative effect if issued several times in succession.

Selecting Edit from the main menu puts you into VP-Expert's built-in editor. An excellent editor for this purpose, it has relatively few quirks. When you enter the editor, the function key assignments appear at the bottom of the screen. If you press Ctrl, Alt, or Shift, the display changes to show the assignments for the function keys in combination with the depressed key (Ctrl-F1, Ctrl-F2, etc.). One apparently undocumented feature in the editor can give you quite a surprise—Ctrl-Backspace deletes the word before the cursor.

A very nice feature of this package is the connection between the Edit and Consult modes. If a problem with the knowledge base materializes during a consultation, then VP-Expert returns to Edit mode at the point in the knowledge base where the problem was detected.

Knowledge bases designed with VP-Expert have a basic layout of actions, rules, and statements. The actions block provides overall direction for the consultation, displaying an introductory message, defining the goal variable for which the system is to find a value or values, and displaying a final message with the result. The actions block can also include clauses specifying more complex tasks involving databases and spreadsheets.

The rules used are if-then rules, broken down into name, premise, and conclusion. Each rule is introduced by the keyword "rule" followed by a unique rule label, a string up to 40 characters long. The premise, or IF portion, can include up to 10 conditions. Each condition compares a variable to a value, using standard relational operators. You can also write conditions that compare a variable to the current value of another variable.

Conditions are combined using AND and OR but, unlike most other uses, OR takes precedence over AND. One problem with the introductory material is that the first mention of AND and OR does not explain the precedence.

The conclusion, or THEN portion, must contain at least one equation assigning a value to a variable. Conclusions can have confidence factors with them, using an integer between 0 and 100 to indicate the degree of certainty that the conclusion is valid. (Confidence factors can also be entered by the user at question prompts, reflecting the user's degree of confidence in the response. The system assumes a factor of 100 if you give no factor; a factor of less than 50 is treated as a negative response.) Multiple conclusions can be listed and do not use an AND for connection.

Comparison of sample rules from ESIE and VP-Expert

The expert systems set up for this example are designed to tell you what kind of drink to serve, based on information that you—as the end user—provide about your dinner guests and the entree you have chosen to serve. The rules are shown in the format used for each shell.

Sample rules from ESIE

```
If  expensive_wine is yes
    And New-Year's is yes
    Then type.drink is Bond's-Champagne

If  expensive_wine is yes
    And steak is yes
    Then type.drink is Chateau-Earl-of-Bartonville-Red

If  cheap_wine is yes
    And chicken is yes
    And well-liked is no
    Then type.drink is Honest-Henry's-Apple-Wine
```

Sample rules from VP-Expert

```
RULE 1
IF      Expensive_Wine = yes AND
        New_Year's_Eve = yes
THEN    Drink=Bonds_Champagne;

RULE 2
IF      Expensive_Wine = yes AND
        Entree=steak
THEN    Drink=Chateau_Earl_ofBartonville_Red;

RULE 3
IF      Cheap_wine = yes AND
        Entree=chicken AND
        Well_liked = no
THEN    Drink=Honest_Henrys_Apple_Wine;
```

The statements section contains information about the consultation itself. In this section you enter the Ask command for those variables whose values the end user will supply. The Plural command indicates that a variable can have more than one value at a time during a consultation.

The Runtime command suppresses the rules and results windows during consultation. You can use other commands to change the background color of the consultation screen or to provide the end user with a menu of choices when asked the value of a variable.

VP-Expert contains on the order of 50 keywords, predominantly statements or clauses. The clauses are used in the actions block or in the conclusion of a rule. In the actions block, they are executed sequentially. However, clauses in a rule conclusion are executed only if the premise of the rule is true. The clauses let you display text on the screen, eject the printer, clear the screen, specify that the value of a variable should be found, and link VP-Expert to database and spreadsheet files.

VP-Expert's implementation of backward chaining is as follows: When the inference engine needs the value of a variable, it looks for a rule with that variable in the conclusion and attempts to prove the premise of the rule. If it finds no such rule, then it looks for an Ask statement for the variable to ask the user for information. If it finds an Ask, it will also look for a Choices statement for the same variable, which causes display of a menu with the options possible for the variable. If it finds no Choices statement, then it asks the user for the value and the user types a response.

Other important main-menu choices are Induce and Tree. Induce generates a knowledge base from an induction table. You can build this table in any ASCII text-editor, or you can use database or spreadsheet files (from VP-Info, dBase, VP-Planner, Lotus 1-2-3, and dBase and Lotus work-alikes). Each line of the table becomes a rule of the knowledge base.

Even when setting up your first simple system, it can help to build a table and use Induce. VP-Expert has strict rules about punctuation, and using Induce results in a model that you can then follow when adding additional rules to the knowledge base. Although the induction process is not very complex, it would be helpful to have more than one example of induction tables in the manual.

The Tree menu lets you view the path previously saved in a trc file. You get options for text or graphics, the latter requiring a graphics card. The text display has three types of lines: <variable name>, which indicates that the value of the variable is currently being sought by the expert system; "Testing #," which indicates that the rule with the specified number is being tried; and (text), which means that the value in parentheses has been assigned to the variable being sought. Levels are indicated by indentation marked with ! to show the tabs.

The graphics display shows a tree construction with the root at the left-hand side of the screen. You can then zoom in to areas of the tree to see the detailed information, which is the same as that displayed in the text version.

The one complaint I had about the Tree menu is that it is the only menu that does not tell you how to exit and return to the main menu.

Overall, VP-Expert is an excellent package, particularly well suited to people familiar with light-bar menus and other features common in PC packages. The accompanying text is excellent as well, including, in addition to the beginner's guide (Chapters 1-4), chapters on database access and worksheet access, a topical reference, a menu command reference, an alphabetical keyword reference, a math function reference, and several helpful appendixes. The index is also fairly good. One weakness is that it does not often reference the first four chapters, even though many topics apppear there first. So, if you want to go back and review that material, often you must flip through those sections until you find the appropriate part.

A final note: Paperback Software is now shipping Version 2, for $249. They have made a number of substantial changes, including: forward chaining and "whenever" rules that fire whenever the condition in the premise is satisfied, even if backward chaining is being used (to have pure backward chaining, do not use this kind of rule in the rule base); hypertext with mouse support; dynamic images that allow display of graphical images (like a thermometer) that are tied to a variable so, when the variable changes, the image changes, and vice versa; automated access to dBase files; more powerful access to external programs; and smart forms ability.

Even at the new price, VP-Expert is an extremely powerful product for the money and at least as powerful as packages in higher price ranges.

Two PC-based expert system shells for the first-time developer

Peter G. Raeth, Air Force Systems Command

Have you ever wanted to try out a new software technology but run into the classical Catch-22? You know your company needs new technology, but the company wants proof of its cost effectiveness before spending much money. Yet you can't generate the proof because the seed money is just not available.

By using a few "nickels and dimes" scraped up here and there, software developers can begin exploring new software approaches using technologies already in place, such as microcomputers. This personal effort can generate the proof required for larger expenditures.

An increasingly popular software technology well worth exploring is expert systems. A fundamental difference distinguishes an expert system shell from a traditional programming language: shells are object oriented and knowledge intensive, while traditional programming languages are procedure oriented and code intensive. This fundamental difference affects even the way you have to think about an automation problem.

Shells can greatly simplify certain automation problems, particularly those requiring great flexibility and incremental additions of capability. Using a natural mechanism, shells permit the capture of knowledge in the expert's own language. The key is that the shell doesn't change just because the knowledge base changes. Traditional programming techniques require that all the expertise be embedded in the code. The code has to change every time the knowledge represented by the code changes. Shells permit the developer to concentrate on the expert's knowledge instead of computer code.

Expert systems offer many implementation techniques. A frequently used technique that focuses on object-action pairs requires specific actions to be performed when specific objects exist. (However, objects do not necessarily exist in any predetermined order.) Usually, we think of these object-action pairs as if-then rules. If-then rules state, "If a given object exists, then perform a given action."

Choosing that first shell can be difficult because of all the various features. The accompanying recommended reading list contains articles that offer a good discussion on features for new users. Besides features, the new user should look for a shell at a reasonable price that he or she can use to develop meaningful applications and later expand to accommodate even larger applications.

Two PC-based expert system shells on the market support the idea that the cost of a tool should not exceed its value to the user and that the tool should grow with the user's needs. Clips, for C Language Production System, was written by NASA and is available on the user-supported market. Personal Consultant from Texas Instruments is a commercial product available as a series of additive modules. Both Clips and Personal Consultant come with a tutorial guide, a reference manual, example knowledge bases, and the possibility of on- or off-site training. In their most basic form, both shells work on the IBM PC and PC clones. For larger applications, they both require PC-AT class machines. However, even their most basic forms permit the implementation of useful expert systems.

I discuss these two shells from the viewpoint of the developer exploring expert systems technology for the first time. The basic question I'll address is, "Given a new developer and a detailed specification, did the shell prove effective?" To answer this question, I'll begin with an overview of an application's specifications. Next, I'll look at each shell's performance in meeting the specifications. As you'll learn, each shell stands on its own as a useful tool that meets the goals of expandability and value.

The application

For this review I implemented the Statistical Strategist, which can—given the assumptions attendant on a given data set and the purpose of a given analysis—recommend appropriate methodologies. This application was originally written in Turbo Pascal for the IBM PC, using expert systems techniques. It is supported by PC-File, a relational database management system, and PC-Write, a word processor. Figure 1 shows a diagram of the application's logical structure.

The following paragraphs list major features of the Statistical Strategist application.

• User help facility. The user can ask for an explanation of words, phrases, and suggested methodologies without interrupting the flow of interaction with the system.

• Easy knowledge base entry. The domain expert, a statistician in this case, can easily enter and change methodologies and their related assumptions and purposes. A simple frame-based entry method captures this information in a relational database. You enter explanations into a word processor. You can enter both types of information in a random manner. The statistician does not have to worry about rules, information ordering, or special syntax in the development of the knowledge base.

• Interactive conversation that makes no initial assumptions. The Statistical Strategist makes no assumptions about which methodology will be appropriate. It queries the user to find out what it needs to know and makes recommendations as it acquires sufficient information. Its conversation is very direct, guiding the user quickly to an applicable methodology. It also explores methodologies that require more detailed assumptions.

• Machine translation of the knowledge base. There is no human interaction in the process of translating the knowledge base into something the inference engine can easily deal with. Once this translation is complete, the computer can initiate an interactive conversation with the user to determine the purpose of the analysis and the assumptions made on the data. As this conversation progresses, the computer recommends appropriate statistical methodologies.

• Logic trace facility. At any point in the conversation, you can ask the computer to show the reasoning so far. In this way, you can see the facts used by the computer to make its recommendations. At the end of the conversation, you can see which assumptions and purposes resulted in which methodologies.

• Restart capability. You can tell the computer to ignore the previous conversation and start the session again. This permits the user to change the answers to some questions, resulting in new questions being asked.

• No time delay between questions and recommendations. No delay occurs between the time the user answers a question and the time the inference engine delivers the next question or the

EH0303-8/90/0000/0082$01.00 © 1988 IEEE

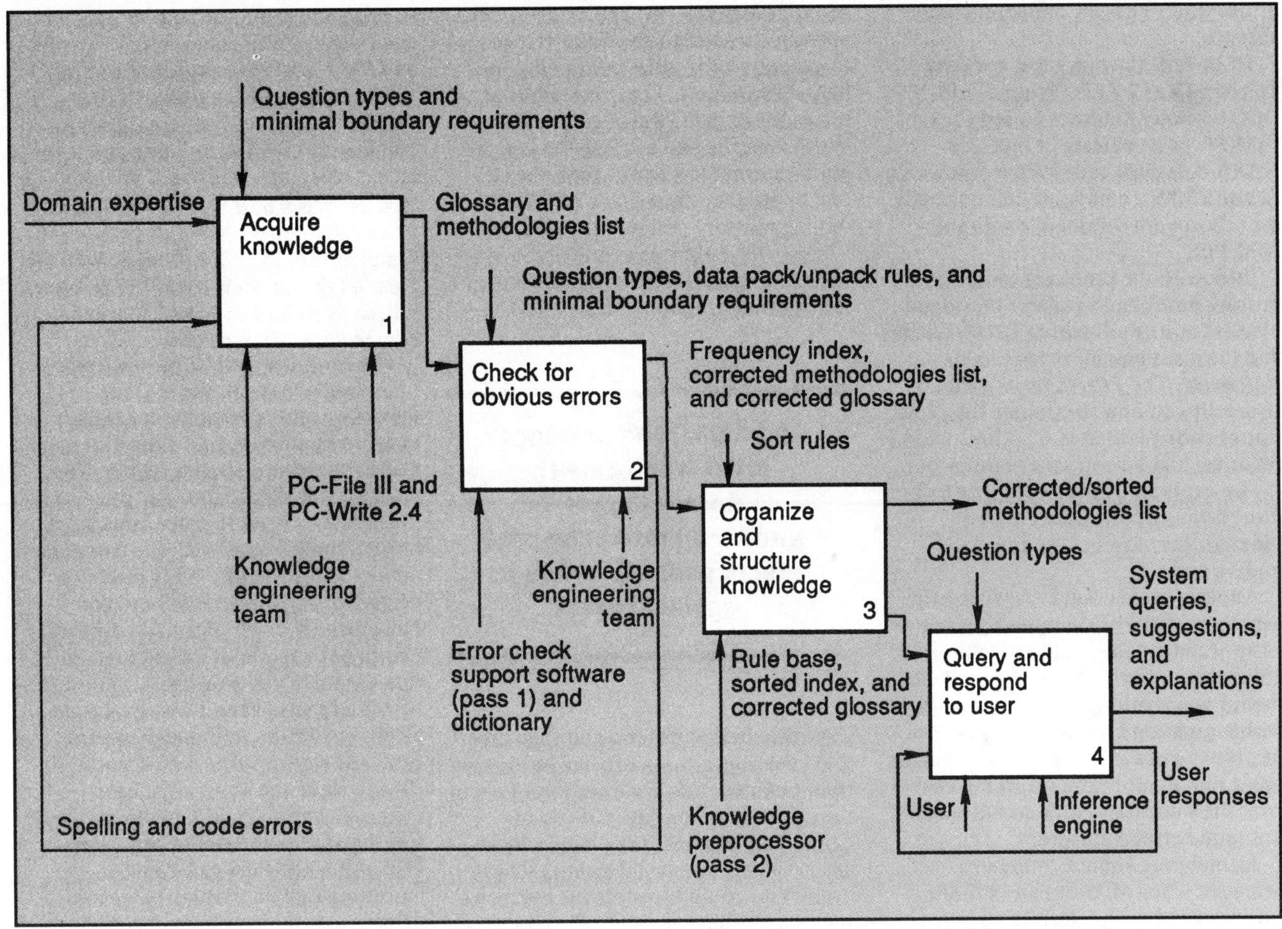

Figure 1. Logical structure of the Statistical Strategist.

next appropriate recommendation.

• Debugging capability. A knowledge base preprocessor helps the user catch entry errors, misspellings, and nonfactual entries.

Now let's see how Clips and Personal Consultant performed during implementation.

Clips (Version 4.11)

According to Clips' documentation, the Artificial Intelligence Section at NASA's Johnson Space Center has encountered a number of problems delivering Lisp-based expert systems to NASA users. Three problems in particular have hindered the use of expert systems within NASA: low availability of Lisp on conventional computers, high cost of current technology Lisp tools and hardware, and poor integration of Lisp with other languages, making embedded applications difficult. Clips was developed to solve these problems.

Clips is command driven. Its primary representation methodology is a forward-chaining rule language with an inferencing technique based on the Rete algorithm. The basic elements of Clips are a fact list that forms a global memory for data, a knowledge base that contains all the rules and initial conditions, an inference engine that controls execution and decides the rules to be executed based on the available facts, and external software written by the user in C or some other language. C language code can be linked directly into the tool.

Getting started. I recommend starting with the *User's Guide*. Written in a conversational tutorial style, this guide is essential to the effective use of Clips' capabilities. If you are new to expert system shells, expect to spend at least two hours per day for two weeks reading the guide and working through two meaningful exercises per chapter. The exercises are very important because they teach not only how to use Clips, but also good rule-programming style. Indeed, without good style, you cannot successfully complete the later exercises. Moreover, good style will keep you out of many traps as you go on to develop that first real application.

One thing missing from the tutorial is an installation guide. You need this supplement to specify the hardware and operating system expected by the given version of the shell. The guide should also explain how to set up the software.

Without an installation guide, I went down many a wrong path. My version of Clips runs on an IBM-PC. It nominally needs 230 Kbytes of RAM and PC-DOS version 3.x. It also turns out that Command.com must be on the boot-drive disk's root directory. Not knowing that last fact caused me no end of frustration. Clips will report that certain commands executed correctly, even if they don't due to Command.com's absence.

To be fair, the disk label does say that MS-DOS 2.11 is required. MS-DOS typically indicates a need for an IBM PC or its clones. In fact, the NASA-compiled code even worked on a Zenith Z100, a computer that normally is not software compatible with the IBM PC.

Even with the problems caused by a missing installation guide, I must laud NASA for its well-written *User's Guide* and the companion *Programmer's Reference*. The *Programmer's Reference* tells you how to compile the Clips source code (written in C), how to add features, and how to take features out. It also explains Clips in complete technical detail. Both guides are well indexed, easy to use, and easy to understand.

Another great point is the user help desk provided by Computer Sciences Corp. (under contract to NASA, for government users only). Before I told them I was writing an article, I called with a question I knew they would not be able to answer readily. They took my name and number and called me back later the same day with a well-thought-out (and correct) answer.

A final note about Computer Sciences: They offer on- and off-site training. They can also fill you in on Space Operations Automation and Robotics (SOAR) conferences where Clips tutorials are given. I have not attended one of those tutorials, but I have a copy of the handouts. If those who teach the tutorial are as knowledgeable as the ones who run the help desk, the SOAR tutorial would be worth going to—once you have worked through the *User's Guide*.

Clips' syntax. Clips' syntax is very similar to Inference Corp.'s ART (Automated Reasoning Tool) and takes several lessons from Lisp. Even those with some experience in Lisp will find Clips' syntax and case sensitivity unnatural at first. However, with a little experience, you will see that the syntax does make a lot of sense and that it permits a more powerful and flexible system, especially when you consider that even Clips' user commands can be activated by rules. So, have patience with the syntax.

Implementing an application using Clips. The following paragraphs detail the major features of the Statistical Strategist from the Clips point of view.

User help facility. A user help facility was simplicity itself to implement. I could have taken two approaches. One approach uses autotranslator techniques to generate Clips rules containing the help information. The other approach leaves the help information in a PC-Write data file and accesses it via a program external to Clips. I took the second approach, since that saved Clips' on-line memory. When the user types in "help," the help rule executes the System command. This command activates the external program, which asks the user what help is desired and then displays the appropriate off-line information. Granted, this is slower than having the information on-line directly in Clips. This is one of the places where, as developer, you could modify Clips itself. You could translate the code used in the external program to C and integrate it into the Clips software. That would speed things up considerably while keeping all the text information on disk.

In Clips, the developer must translate the domain expert's knowledge into rules and facts while following a special syntax.

The method I chose to add the help capability to Clips took about five minutes to implement, thus showing one of the real powers of Clips. Like the original Statistical Strategist, written in Pascal, the Clips-Rules version was built incrementally. I found it much easier to add capability to the Clips version than to the Pascal version because of Clips' object orientation versus Pascal's procedural orientation.

The type of application I used seems more natural in the object-oriented setting. Of course, there are times when, even in this setting, you must write some procedural code. Clips makes it possible to have the best of both worlds. I found that Clips let me concentrate on what I wanted to do rather than on the details of how to do it.

Easy knowledge base entry. There are two ways to develop a Clips knowledge base. You can enter rules and facts directly into the system, test them, and then save them to disk. Alternatively, you could use a text editor (Clips has one built in), type in Clips rules and facts, and then use the Load command to bring the result into the system. You can then perform tests, enter new rules and facts, and save the result to disk.

The commands Defrule, Deffacts, Assert, Reset, and Run are the major commands used for building and testing an expert system in Clips. However, the system has a host of other commands and capabilities. All of these made the implementation of the Statistical Strategist a fairly straightforward task, once I discovered how to install Clips and gained a bit of experience.

The problem with Clips' method of knowledge-base entry is that the developer must translate the domain expert's knowledge into rules and facts while following a special syntax. The original Statistical Strategist permitted the domain expert to enter knowledge into a simple frame, with the computer doing the translation. So, I used the original frame-based entry method implemented in PC-File and wrote a two-page program in Pascal that reads that database and translates it into Clips rules and facts. Then I wrote a header of about 10 rules to implement some standard features that would not change when the knowledge base changed. The original knowledge base translated into about 30 Clips rules. When I loaded the header and autogenerated knowledge bases into Clips, the resulting system worked much like the original.

Interactive conversation that makes no initial assumptions. This feature followed naturally from the autogenerated knowledge base and the handwritten header. It was not difficult to add more and more conversational power to the system. I just had to extend the autogenerator's rule outlines to offer more information and feedback.

Machine translation of the knowledge base. You must enter Clips knowledge bases directly in a special rule/fact syntax. No higher level entry method is provided. Of course, you could always extend Clips to do this, since it comes with the source code.

Logic trace facility. Again, I found this an easy task to implement. As the rules execute, the system asserts facts that Clips ignores but still stores in the order created. These facts contain the assumptions and purposes indicated by the user and the resulting methodologies suggested by Clips. When the user asks "why," the Why rule executes the Facts command. At that command, Clips lists the facts previously asserted. Due to continuous facts cleanup, the list contains no extraneous facts. Only the trace of the conversation appears in the list. Once the previously asserted facts are listed, Clips picks up the conversation where it left off.

Restart capability. Clips easily empties its store of facts and begins executing anew. I only had to have a restart rule to execute the Reset command. Restart can be initiated at any time by the user or automatically by Clips, once there is nothing else for Clips to ask or recommend.

No time delay between questions and recommendations. On a minimally configured IBM PC, Clips does exhibit a slight delay between user responses and the next question or recommendation. These delays result from the Rete algorithm's attempts to find the next set of appropriate rules based on any new facts. This delay is less than one second.

I didn't notice a significant increase in the delay even with 342 rules and five initial facts loaded. The delay did increase somewhat, to about five seconds between the first response and the second question, in a test with 250 rules and 150 initial facts. This time decreased as the conversation progressed.

Clips took about 1 minute 23 seconds to load the Statistical Strategist knowledge base and begin the conversation.

Debugging capability. Clips' Watch command permits the developer to observe facts being asserted and retracted. You can also see rules executed.

Clips provides complete syntax checking. It clearly shows the error and its location. The *Programmers Reference* provides complete details on each error message.

The system makes no attempt to verify entries to the knowledge base. Clips assumes that, once the syntax is correct, the rule or fact is correct.

Clips in summary. Clips could not directly implement all of the original Statistical Strategist without additional programming in C or some other language. However, you must realize that Clips is a general purpose, rule-based shell. As such, it works well and is worth the price. Since it comes with the source code, Clips can be expanded to deliver additional capability. It can also execute external programs written in any language.

Be aware that Clips is not a commercial product. Thus, it does not have all the support some people—particularly nongovernment users—are used to. So far, NASA appears to have a solid commitment to its tool. This should continue as long as there is a strong government audience. Given that ART's syntax is so similar to Clips (according to NASA), it should be possible to upgrade to this commercial tool if necessary.

As a government-produced, user-supported program, Clips is not copyrighted. You can freely distribute both the compiled and source code. Thus, you only need to buy Clips if you don't know someone who already has it.

TI's Personal Consultant modules support the beginner while giving the experienced knowledge engineer extensive power.

Clips runs on the DEC VAX 11/780 under VMS, Sun workstation under Unix, Apple Macintosh, and IBM PC and AT or clones under PC-DOS 3.x and MS-DOS 2.11.

Since C is supposedly a very portable language, you should be able to compile and run Clips if you can get the source code into your machine.

Contact information. Government users send six diskettes or a tape of the type used by the intended computer. Contact Steven Baudenistel, Computer Sciences Corp., Clips Users Help Desk/M30, 16511 Space Center Blvd., Houston, TX 77058, phone (713) 280-2233 or (713) 280-2430.

Nongovernment users pay $250. Specify MSC-21208 and be sure to say what media you want. Contact Director, Software Distribution, COSMIC, University of Georgia, 382 E. Broad St., Athens, GA 30602, phone (404) 542-3265.

Personal Consultant Plus (Version 3.02)

Texas Instruments has developed a choice set of integrated tools for the knowledge engineer, presented as a group of upwardly compatible modules called the Personal Consultant series. These modules permit a new developer to get started without a lot of pain and to gradually move on to the development of very sophisticated expert systems. They support the beginner while giving the experienced knowledge engineer extensive power.

TI's modules include PC Easy, the foundation module, for $495; PC Plus, an extended form of PC Easy, for $2,950; PC Scheme, an integrated Lisp, for $95; PC Images, a graphics-oriented user interface, for $495; and PC Online, a real-time inferencing package, for $995. If you purchase PC Plus within six months of PC Easy, the price drops by $495. PC Scheme comes with PC Plus, but also as a stand-alone package. PC Images and PC Online are expander packages for PC Plus. Figure 2 shows how these modules are integrated.

Personal Consultant is menu driven. A rule language forms the basis for this expert system shell. While it can do forward chaining, its most natural inferencing style is backward chaining. For each goal conclusion, it goes sequentially through the goal's list of premises and thereby seeks to prove the conclusion. Each goal premise that cannot be proven directly becomes a subgoal. The rules that affect the subgoal are executed in an attempt to prove the subgoal. Premises of subgoal rules that cannot be proven directly result in other subgoals. This process continues until the original goals are either proven, disproven, or become undeterminable.

Personal Consultant's inferencing method is supported by several major categories of information:

- rules: IF-THEN premise-conclusion pairs
- meta rules: rules affecting the execution of rules
- frames: groups of rules that can inherit attributes
- premises: what must be proven during inferencing
- goals: the results of the inferencing process
- parameters: objects that make up premises and goals
- variables: names of values accessible to the application
- functions: developer-defined Scheme modules
- text tags: blocks of text accessible to the application
- confidence factors: certainty of premise and goal values
- global properties: information defining the system's environment
- external software: user-written in Scheme or another language
- external data: sequential and dBase files

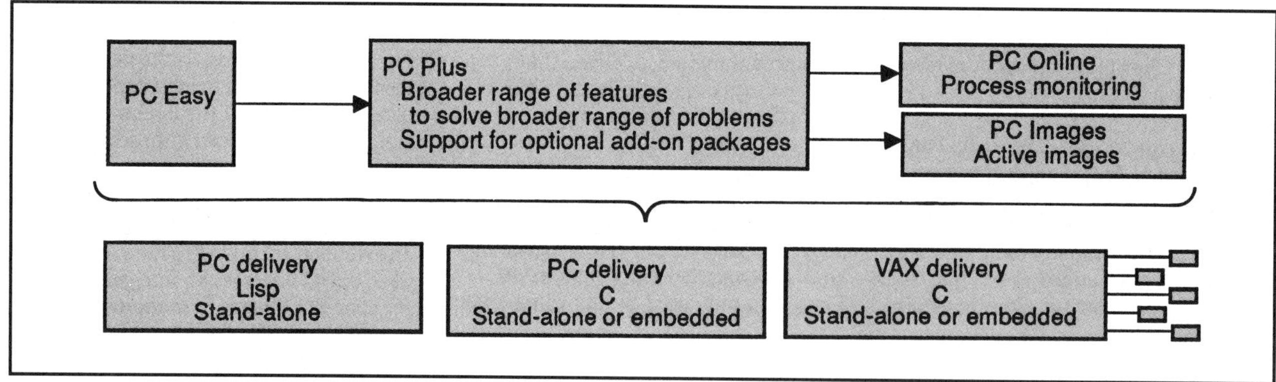

Figure 2. The integrated modules of the Personal Consultant series.

- still graphics: explanatory screen frames stored on disk
- moving graphics: screen images that move to assist user input

Getting started. Normally, I would advise you to start with the user's guide. However, the Personal Consultant series comes with more than nine manuals.

You should not just dive into this package without some prior thought. Start with the flyer "The TI Personal Consultant Series," published by the Texas Instruments marketing department. It gives a good technical and functional overview of the modules and how they are integrated. This short pamphlet will help you decide on the module(s) you need and the minimal hardware necessary to run them.

Pay special attention to the hardware requirements. For instance, do not purchase something that will restrict you to PC Easy or CGA graphics. On the memory requirements, I recommend 640 Kbytes of main memory even for PC Easy. (The Statistical Strategist is about all PC Easy can handle on a minimally configured IBM PC.) You will need at least 2 Mbytes of extended memory to achieve useful speed with the other modules. TI's software will work in less memory, but garbage collection and disk accesses for overlays slow things down considerably.

Table 1 summarizes the software options and the minimum hardware requirements of the Personal Consultant series of modules. I carried out this review of Personal Consultant on a fully configured Zenith Z248 using Personal Consultant Plus.

Once you have gotten an overview of the modules, you are ready to look at the manuals. Basically, each module comes with two manuals, a user's guide written to teach you how to use the module and a technical reference to answer detailed questions. Each manual has two tracks though it, one track for the person new to expert systems and a second track for experienced knowledge engineers learning Personal Consultant's syntax and conventions.

If you are new to expert system shells, expect to devote two to three hours per day for five days with each manual. Be sure to work through the examples. Start with PC Easy's user's guide and explore your first application with that module before you purchase any of the other modules.

The sample expert systems sent with the Personal Consultant Series are discussed in the user's guide. They do a good job of showing the power of the shell and specialized tools like graphics screens. My complaint about the still graphics frames is that TI does not give you a lot of help to put them together. The manuals make a vague reference to a third-party graphics development tool called Dr. Halo, but give no ordering information. (I discovered elsewhere that Dr. Halo, and an associated scanner, are available from Diamond Flower Electric Co. of Sacramento, California.)

The technical references are quite good. Liberal examples show you how to use each element discussed. The indexes, page headings, and tables of contents make the manuals easy to navigate. However, I did not like TI's use of spiral bindings for some of the manuals and the overstuffing of some of the three-ring binders. In my experience, spiral bindings make a manual impossible to maintain, and overstuffed binders make it easy to tear the pages.

Single commands allow you to install and start each module. You will have no trouble here. TI has done a good job of documenting and simplifying both installation and initiation. Be careful to follow the instructions to the letter.

On- and off-site training is readily available to all TI customers. Be sure to work through the user's guide before going to any of these workshops. A help desk is available for three months after your purchase of PC Plus. You can extend this period by paying an additional fee. I used this help desk and I must confess that I was a bit disappointed. TI does not provide a toll-free number during the initial three-month period. When I called, the answering machine kept me on hold for one and a half minutes. The call was then picked up by a data-entry person who asked me a host of questions, all but two of which had nothing to do with my problem. Then he gave me a "sequence number" and said a technician would return my call. This first call lasted 10 minutes and produced no results. The technician did call back that day. The question I had was a real curve ball. The technician handled it well and gave a correct answer. This person really knew what he was talking about.

Syntax. The syntax for PC Plus is simplicity itself. For the most part, unless you write directly in Scheme, you need not learn a special syntax. You must be conscious of only a few things, such as the use of the equal sign. On the IF side of a rule, " = " is a logical operator. On the THEN side of a rule, " = " is an "is-replaced-by" operator.

You must also pay attention to the type of parameter (or variable) used. For instance, some are Boolean (can be true or false), some are multivalued (can have several values at once), some are single-valued (can have only one of several values at once), and some are numeric values of limited range. How special functions operate depends on the type of parameter they are given. The differences are important, but can be subtle.

AND and OR logical relationships are interpreted through a hierarchy of

Table 1. Personal Consultant's modules and their minimum hardware requirements.

	PC Easy	PC Plus	PC Online	PC Images
Hardware	TI Business-Pro, IBM PC, PC-XT, PC-AT, PS/2, or true compatible	TI Business-Pro, IBM PC-AT, PS/2, or true compatible	TI Business-Pro, IBM PC-AT, PS/2, or true compatible	TI Business-Pro, IBM PC-AT, PS/2, or true compatible
Base Memory	512 Kbytes	640 Kbytes	640 Kbytes	640 Kbytes
Extended/Expanded Memory	—	—	512 Kbytes	512 Kbytes
Graphics Adapter/Monitor	CGA or EGA	CGA or EGA	CGA or EGA	EGA only
Recommended Configuration	640-Kbyte memory Hard disk EGA	1-Mbyte expanded memory EGA	1-Mbyte expanded memory EGA	1-Mbyte expanded memory EGA

operations or through standard parenthetical statements. If you have written in a common procedural language such as Basic, Fortran, or Pascal, you will have no trouble constructing AND/OR statements that reflect your meaning.

Implementing with PC Plus. Implementing the Statistical Strategist in TI's Personal Consultant Plus was quite an interesting experience that resulted in a 27-rule application. The original implementation was based on a non-rule, forward-chaining concept. PC Plus uses rules and a backward-chaining concept that does not easily lend itself to the original's directed binary inferencing. It also requires a strict definition of parameters and goals not originally required. However, the enforcement of this discipline resulted in some added power to the application.

User help facility. Without writing in Scheme, you cannot implement the generic help facility of the original application. It is also not a simple task to capture a generic user input such as "help," call out to an external procedure that will perform the appropriate operations, and return to the conversation where you left off.

However, TI designed PC Plus to offer specific help. In this regard, they have really done a fantastic job. The knowledge engineer can associate specific help information with each parameter and rule. The user easily accesses this information during a conversation by pressing the F1 and F2 keys and selecting the Why and How options. (Why tells why a given question is being asked; How tells how a given conclusion was reached.) In this way you can offer the user literally thousands of explanations.

Easy knowledge base entry. By providing a sophisticated environment, TI has made it easy for a knowledge engineer to enter, maintain, and improve knowledge bases. PC Plus prompts you for the IF and THEN parts of each rule. It also prompts you for the information required to initiate a new application. It keeps track of a number of details to ease the workload and permit you to concentrate on the application. For instance, when you make a change anywhere in the knowledge base, all implied structural changes are made automatically.

I could not bypass the PC Plus environment to simply enter domain knowledge directly. You must provide the interface between the domain expert and PC Plus.

Three quirks about text entry are worth mentioning here.

(1) The Home key toggles the cursor between the beginning and end of the line. It would be more natural for the Home and End keys to move the cursor to the appropriate line positions.

(2) The cursor moves from beginning to end very slowly.

(3) No command will let you automatically center a line of text.

These three quirks make text-screen development and maintenance a little tedious.

Interactive conversation that makes no initial assumptions. Personal Consultant generally presents a good interactive interface to the end user. It assumes that all goals must be proven, disproven, or found to be undeterminable. Thus, it can spend a lot of time asking questions about goals that will never be proven. The knowledge engineer can keep this from becoming a problem by placing the parameter that tends to be the biggest differentiator at the head of the list of parameters in the goal rules.

For instance, I have a rule group that determines the purpose of the statistical calculation to be performed. Once that purpose is determined, the system ignores all questions concerning methodologies that do not meet that purpose. You can do this by having a trigger parameter that is set when a purpose parameter is determined. Then, no other questions about such parameters are asked. The remaining purposes are undetermined, leading directly to undetermined goals (methodologies, in this case).

Machine translation of the knowledge base. PC Plus is a tool for knowledge engineers, not for domain experts. It provides no environment for the domain expert to enter information for later translation. Since TI provides an interface to Scheme, you could write a domain expert interface to support knowledge capture at that level. You could then translate the domain expert's input into Scheme or the disk-storage version of ARL (TI's Abbreviated Rule Language). However, the manuals don't explain the disk-storage version of ARL. While they do explain Scheme in detail, it is a low-level language requiring a large programming effort for automatic code generation.

Logic trace facility. You will find that PC Plus provides a good toolkit for inference tracing. All details of the inferencing process are shown during a trace. Especially useful is the display of the entire effort to backtrack to a parameter value that can be directly determined. This is followed by the trace of that parameter's use during the search for secondary parameter values.

You can put the trace in a file, on the screen, or on the printer. You can have it start or stop at any user-input point in the conversation.

Using the trace facility, I had no trouble observing the inference process caused by my knowledge base. In fact, I learned a lot about backward chaining from watching it.

Restart capability. Again, I found this an easy task. PC Plus makes a command menu available at all user-input points in the conversation. At any time, the end user can ask for a restart, trace, explanation, review, or end to the session.

PC Plus permits you to change answers within the conversation after the conversation has taken place or even in the middle of the conversation. The original application does not permit this. Restarting a conversation with this kind of "what if" game playing gives you quite a powerful feature.

No time delay between questions and recommendations. Since Personal Consultant backward chains from the goals list, you can expect some variable delay as the conversation continues. PC Plus uses the facts it has gathered from the end user to determine other facts and to explore goals that may or may not be applicable. The more facts that are determined, the less it has to converse with the user. I found it disquieting that no "Inferencing" message appeared on the screen during times when PC Plus went off on its own. As a developer, you can employ meta rules to force the program to execute the most useful rules first. This can speed up the inferencing process.

Another reason behind the delays between questions is that PC Plus does a lot of software- and knowledge-base overlaying and garbage cleanup (if less than full extended memory is available). It has to go to the disk numerous times, even if main memory is fully populated. The initial loading of the Statistical Strategist and the starting of the conversation took 23 seconds.

PC Plus also takes an undue amount of time to come to even trivial conclusions. It wants all available information before displaying even conclusions that use only one piece of that information. It should display conclusions as it comes to them and let the user decide if it is worth going on with the conversation. Of course, the final listing of all conclusions makes a useful report. By using the Print and Show functions and appropriate rules, the developer could force the display of conclusions as the conversation progresses. However, this violates TI's apparent philosophy that the shell should handle such details.

One last point about the conclusions report: PC Plus does not generate the report if it has made no conclusions and if the developer set the property that eliminates reports on goals not met. The developer can get around this with a rule that becomes true if all goals are false or not determined. However, you must remember to update that rule every time a goal is added.

PC Plus wants all available information before displaying even conclusions that use only one piece of that information.

A further source of delays arises from the function that rules can use to execute an external program. If the developer does not specify how many paragraphs of memory PC Plus should transfer to disk in order to make room for the program, PC Plus transfers the entire executing memory to disk and then back in again when execution of the shell is to resume. This takes far too long. In the case of a missing memory transfer value, PC Plus should figure out for itself the size of the external program. Only in the case where the external program makes dynamic use of memory should the user have to figure out how much memory to transfer. As an application changes, it can be difficult to keep track of all the changes that you must make to calls for external programs. The program could still make total memory transfers if it allowed the developer to specify a value of "All."

Debugging capability. As I mentioned earlier, PC Plus's trace, playback, and review capabilities make logic debugging a snap. The syntax debugging is not so friendly, however. For example: The program prompts you for the IF and THEN parts of each rule. Should you actually type in "if" or "then," PC Plus generates the cryptic error message "ERROR: Possible invalid use of $AND." I spent quite some time trying to deduce what this error message meant, especially the reference to $AND. Finally, I found $AND in the appendix to one of the manuals. It did not appear in any index. The information did not help. Intuition eventually said that if the system prompted for IF and THEN, you should not actually type in those words. To be fair, the user's and reference manuals' examples always have "IF" and "THEN" in a color other than that indicating user input.

More about Personal Consultant. I did not use the Images or Scheme modules to implement the Statistical Strategist. However, I did look at both.

PC Images. Currently in version 1.0, PC Images offers the powerful concept of using graphics instead of text for user input. For instance, instead of getting the user to type in the position of a machine's switch settings, the program presents a picture of the machine. The user moves a cursor from switch to switch and setting to setting. This is potentially much easier than trying to get the user to understand a text description of the machine and its switches.

You can also use PC Images to create an on-screen data entry form that the user fills in instead of answering a lot of questions. From running the example included with PC Images, I gather that you cannot allow the user to move a picture's cursor "backward." If the user makes an error, he or she must retrace the conversation from the beginning to change the appropriate answer. This amounts to a lot of work just to correct a typing error.

The text data-entry forms have a wraparound feature, but the demo did not allow access to all the elements on the form.

Another problem is that, if the user selects "Continue" after a consultation (in which case a "New Start" is supposed to occur), the initial data-entry screen is only partially displayed.

One last point: You will find it difficult to set very small gauges because of the resolution limitations of EGA graphics. So, you will need to watch out for some image size limitations.

All in all, PC Images is a good idea, but version 1.0 needs some work.

PC Scheme. PC Scheme, TI's Lisp implementation, is now in version 3.0. PC Plus is written in Scheme, so it's easy to integrate Scheme modules into an application. The ability to link developer-defined modules directly into a generic shell is an important improvement in the field. Previously, developers would start with a generic shell and then have to implement their applications again in a custom shell due to the limitations of the generic shell.

Summing up Personal Consultant Plus. PC Plus could not directly implement every detail of the Statistical Strategist without some programming in Scheme (to do the generic Help facility and domain expert knowledge input and quality control). However, the power of the application increased because of the parameter discipline enforced by the development environment. The original Statistical Strategist has no knowledge of "purpose," "population size," or the meaning of other parameters. It simply has data tokens that can be either true or false. Therefore, in a backward chaining, rule-oriented environment, many duplicate questions can be asked. For instance, a data token that involves a population size greater than 30 is not necessarily related to one that involves a size greater than 20. Personal Consultant wants to know the type of data token so that a population size greater than 30 is automatically greater than one of 20. The definition of data token types permitted PC Plus to identify methodologies that would not be identified by the original Statistical Strategist.

Personal Consultant takes considerable commitment for you to learn to apply it effectively. Once this learning process reaches its critical mass, however, Personal Consultant is a powerful expert systems development and delivery tool.

Recommendations

Both Clips and PC Plus proved themselves worth the effort needed to learn them. They have their limitations, but they are solid knowledge-engineering tools that permit you to start with a small learning effort and go on to develop major applications involving heavy customization. I would recommend either depending on the developer's specific needs.

As a closing note, I would like to thank the following people for their help: Ed Alexander and Jim Krug of the Air Force Armament Laboratory; Richard Eckhouse of MOCO, Inc.; and Marilyn Raeth of Humana, Inc. I also appreciate the able support of Steve Baudenistel of Computer Science Corp. and Debbie Adams and Isam Khouri of Texas Instruments.

Recommended reading

Barr, Avron, Edward A. Feigenbaum, and Paul R. Cohen, *The Handbook of Artificial Intelligence*, Vols. I, II, and III, Addison-Wesley, Reading, Mass., 1981 (Vol. I) and 1982 (Vols. II and III).

Citrenbaum, Ronald, James R. Geissman, and Ronald Schultz, "Selecting a Shell," *AI Expert*, Sept. 1987, pp. 30-39.

Forgy, Charles L., "Rete: A Fast Algorithm for the Many-Pattern/Many-Object Pattern Match Problem," *Artificial Intelligence*, Jan. 1982, pp. 17-37.

Freedman, Roy, "27-Product Wrap-Up: Evaluating Shells," *AI Expert*, Sept. 1987, pp. 64-74.

Gevarter, William B., "An Overview of Expert Systems," PB83-217562, Nat'l Tech. Info. Service, Franconia, Va., May 1982.

Gevarter, William B., "The Nature and Evaluation of Commercial Expert System Building Tools," *Computer*, Vol. 20, No. 5, May 1987, pp. 24-41.

Jackson, Peter, *Introduction to Expert Systems*, Addison-Wesley, Reading, Mass., 1986.

Naylor, Chris, *Build Your Own Expert System*, Halsted Press, New York, N.Y., 1985.

Krutch, John, *Experiments in Artificial Intelligence for Small Computers*, Howard W. Sams, Indianapolis, Ind., 1981.

Raeth, Peter G., and James M. Hardin, "Using Expert Systems Technology to Develop a Statistical Strategist," Paper presented at the 51st Meeting of the Mississippi Academy of Science, 1987. Available from Raeth at AFWAL/AAWP-1, WPAFB, OH 45433.

Chapter 3: Application-Specific Techniques

Getting that first project going can be well aided by examples of the successful work of others. This chapter presents a wide range of applications in areas of common interest. Each of these six articles discusses an application and presents specific implementation techniques. Teaching a machine to teach is an area of great importance given the amount of new knowledge the world generates daily. Yazandi overviews tutoring systems, a large subfield. Computer-aided chip design is covered next by Parker and Hayati. Dhar and Croker talk about decision-support in business. Stansfield and Greenfield follow with an article on financial planning. Franklin *et al.* then discuss a number of applications, including image recognition, diagnostics, and planning. They also offer an extensive bibliography grouped by topic area. Raeth closes with a presentation on multi-channel data correlation in support of data reduction.

EH0303-8/90/0000/0091$01.00 © 1990 IEEE

Intelligent tutoring systems: An overview

Abstract: *In this paper we look at the evolutionary development of Computer Assisted Instruction from the early days of 'linear programs' up to the use of 'expert systems' in education and training. We present the basic principles of Intelligent Tutoring Systems (ITS) which are capable of rich interaction with the student, which know how to teach, and who and what they are teaching. We point out the need for knowledge representation formalisms which can support ITS and present one such formalism (production systems). In the framework presented we describe systems developed for the teaching of modern languages, electronic trouble shooting and computer programming. Finally we point out the shortcomings of ITS and identify areas where a consensus of opinion does not exist.*

MASOUD YAZDANI

University of Exeter
Department of Computer Science
Prince of Wales Road
Exeter, EX4 4PT
England

1. Computer assisted instruction

Computer assisted instruction (CAI for short) has followed an evolutionary path since it was started in the 1950s with simple 'linear programs'. The development of such programs was influenced by the prevailing behaviourist psychological theories [1] and the programmed learning machines of the previous century. It was believed that if the occurrence of an operant is followed by the presentation of a reinforcing stimulus, the strength is increased. To this end a computer program will output a frame of text which will take the student one small step towards the desired behaviour. The student then makes some kind of response based on what he already knows, or by trial and error! Finally the program informs the student whether he is correct. A stream of such steps forms what is known as a 'linear program'. The student may work through the material at his own pace and his correct replies are rewarded immediately.

In the 1960s it was felt that one could use the student's response to control the material that the student would be shown next. In this way students learn more thoroughly as they attempt problems of an appropriate difficulty, rather than wading their way through some systematic exploration. The 'branching programs' therefore offered corrective feedback as well as adapted their teaching to students' responses. However the task of the design of the teaching materials for such systems was impossibly large, this led to the birth of 'author languages', specific languages suitable for the development of CAI material.

In the 1970s a new level of sophistication was discovered in the design of CAI systems where, in some domains such as arithmetic, it was possible to generate the teaching material itself by the computer. A random number generator could produce two numbers to be added together by the student, and then the result of the computer's solution of the addition would be compared with that of the student's, in order to generate a response. Such systems need only therefore to be given general teaching strategies and they will produce a tree of possible interaction with infinitely large numbers of branches. Such 'generative' systems could answer some of the questions from the students, as well as incorporate some sort of measure of difficulty of the task.

By looking at the development of algorithms for CAI, developed over the last thirty years, we see that they have improved on the richness of feedback and the degree of individualisation they offer the students. CAI systems seem to have improved beyond expectation in computational sophistication from their humble beginnings of replacing the programmed learning machines. However, they fall far short of being any match for human teachers. The main problem is the impoverishment of knowledge which they contain. In generative systems there is a mismatch between the program's internal processes (Boolean arithmetic) and those of the student's cognitive processes (rules and tables). None of these systems have human-like knowledge of the domain they are teaching, nor can they answer serious questions of the students as to 'why' and 'how' the task is performed.

2. The basics of intelligent tutoring systems

Intelligent Tutoring Systems (ITS for short) started as an enterprise attempting to deal with shortcomings of generative systems and can be seen as Intelligent CAI of the 1980s. This enterprise has benefitted from the work of researchers in the field of Artificial Intelligence (AI for short) who have had a long standing preoccupation with the problem of how best to represent knowledge within an intelligent system. Various techniques have been tried within AI with varying degrees of success. One of these methods seems to be well suited to the domain of tutoring [2] where it can be claimed that people seem to show some indications in the structure of their behaviour to support the feasibility of using 'production systems' as a way of modelling their behaviour. 'Production systems' have been used extensively, within the psychological experimentation for the analysis of people's behaviour, in addition to being used as a method of representing knowledge within computer based 'expert systems'. A production system is a particular method of organising knowledge into three different categories:

Reprinted with permission from *Expert Systems,* July 1986, Volume 3, Number 3, pages 154-162.

1. Facts: Factual (declarative) knowledge about a particular case (This animal has feathers).
2. es: Procedural knowledge on how to reason in a domain and expertise (If it has feathers then it is a bird).
3. Inference: Control knowledge of how to carry out reasoning from a set of given facts and rules to come up with a conclusion.

Most tutoring systems focus on communicating expertise at the second level above. Such production rules can be written as:

IF precondition1 .... precondition2
THEN conclusion.

Such a simple structure can then encode a variety of forms of knowledge. (See [2] for a detailed exposition of this). At one end it encodes general notions such as the use of analogy.

IF the goal is to write a solution to a problem and there is an example of a solution to a similar problem
THEN set a goal to map the [already known] template to the current case

At the other end it could be used to represent specific knowledge of the domain at hand. For example, in Anderson's [3] Lisp tutor, the above rule is complemented with another which is a representation of some part of Lisp:

IF the goal is to get the first element of LIST1
THEN write (CAR LIST1)

The student is expected to be able to use knowledge structures such as those above, plus example Lisp code given to him in books, such as one to define a function which translates Fahrenheit degrees to Centigrade: –

```
(DEFUN F-T-C(TEMP)
(QUOTIENT (DIFFERENCE TEMP 32)1.8))
```

as well as the text book explanation of function definition in Lisp

```
(DEFUN <function name>
   (<parameter1><parameter2> . . .)
   <process description>)
```

in order to write the answer to the exercise 'Write a function which returns the first element of a list.'

The correct solution is:

```
(DEFUN FIRST)(LIST1)(CAR LIST1))
```

If the student comes up with a wrong answer, the tutoring program would attempt to simulate the student's behaviour and see how he could have gone wrong and offer advice, which is to some extent the result of a reasoning process similar to that of the student's.

The notion that the tutoring program itself can solve the problem which it is setting for the student and in a way similar to that of the student, is the basis of a large number of ITSs [4].

3. Expert systems for tutoring

"Expert systems are such knowledge-based systems which use inference to apply knowledge to perform a task. Such systems are currently being built to handle knowledge in many areas of human thought and activity from medical diagnosis to complex engineering design, from oil technology to agriculture, military strategy to citizens' advice" [5]. It has been claimed that these systems can not only be useful for automating some part of human decision making, but also they could be used for imparting knowledge from an expert in a field to a large number of trainees [6]. After an expert system has been built by extracting the knowledge of the domain expert and embodying it into the computer system, trainees can then observe the knowledge and the line of reasoning of the program. Trainee doctors could be asked to look over the shoulder of a program such as Mycin [7], devised for the diagnosis of infectious blood diseases. Mycin's representation of medical knowledge is, in fact, a form of production system. Furthermore when all the necessary support levels are added to it to make it useful as a training tool, Mycin looks very similar to any other intelligent tutoring system developed from scratch specifically with tutoring in mind.

Guidon [8] is a system which uses Mycin for tutoring. It includes an independent module containing the teaching expertise as well as a limited amount of competence to carry out a coherent dialogue with the student. Guidon attempts to transfer expertise to the students exclusively through case dialogues where a sick patient (the 'case') is described to the student in general terms. The student is asked to play the role of a physician and ask for information which he thinks might be relevant to this case. Guidon compares the student's questions to those which Mycin would have asked and critiques him on this basis. Guidon uses a 'closed world' model where all objects, attributes and values that are relevant to a case are determined before the tutorial session begins.

Guidon can therefore provide at any point:

1. A list of all data relevant at that point
2. Sub-tasks to be performed
3. Hints on how to go about a problem
4. A summary of all evidence already discussed
5. Exposition of a task
6. Completion of a diagnosis if the student gives up.

Clancey's work on Guidon shows a great deal of promise for the use of expert systems in the next generation of CAI systems. In frame oriented CAI systems for a similar domain, the medical and teaching expertise is 'compiled' into the branching structure of the frames while in Guidon these are available for inspection and change whilst the program is running. This also provides the possibility of quick and simple modifications to be made to the system without needing to worry about the fixed structure of the dialogue in advance.

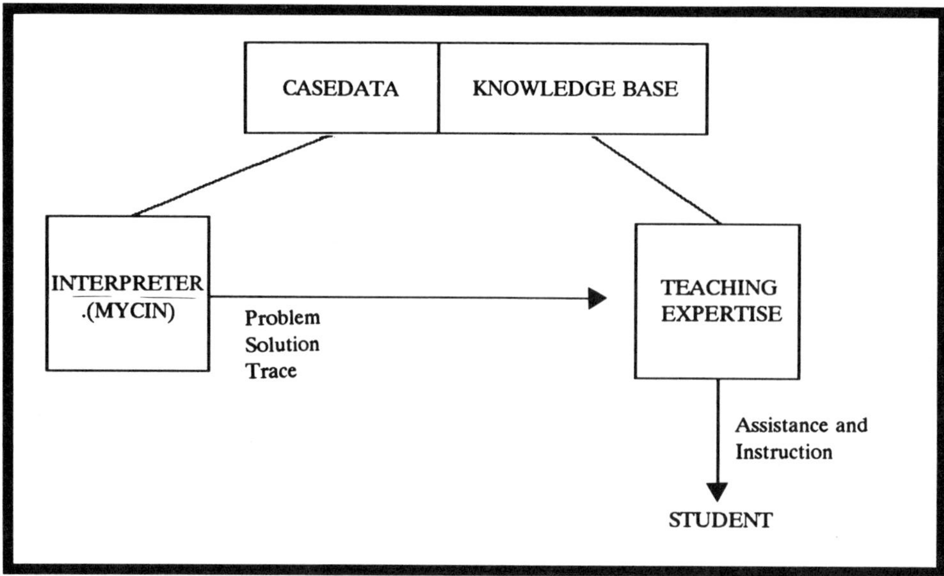

Figure 1: *The overall structure of Guidon*

4. Language learning

Most language teaching CAI systems rely on a massive store of correct sentences and derivations from them. Despite that large number of legitimate constructs with which they can deal, the programs are really nothing more than dumb, if effective, pattern matchers, linking unintelligible orders of characters to those pre-stored. As a consequence, the programs cannot recognise or comment upon errors encountered, even if the errors are frequent, unless these have been individually and specifically anticipated by the programmer. Therefore the standard of accuracy required (coupled with the time for preparation of exercises) seems very high indeed.

Instead, systems such as FROG [9] and the French Grammar Analyser (FGA for short [10]), which use a general purpose language parser, can cope with an indefinite number of possibilities without being programmed in advance to anticipate all the possibilities. These systems are also more flexible as, unlike Guidon, they are not designed with all objects and their relationships known in advance.

The FGA program began as an attempt at a rational reconstruction of Frog. However, in order to ease the task of future extensions, we decided to produce FGA in a highly modular form. The dictionary, grammar and error reporting routines are all kept as distinguishable data structures which are clearly intelligible by anybody without intimate knowledge of the system and they can be modified with ease. Like Frog, FGA anticipates grammatically incorrect input and the need to parse such input to its best degree, rather than simply giving up.

FGA, unlike other attempted natural language parsers, is based on only a small subset of the complete French grammar and dictionary. Limitations of time, speed and space, as well as the (current) impossibility of defining a full natural language grammar, dictate the need for subsets. As the result of the admission of the fact that FGA's own knowledge of the language is incomplete, there is therefore a need to handle grammatically correct but unanticipated structures, as well as those which are definitely wrong. When a seemingly erroneous lexeme, word group or structure is encountered there are, effectively, two different explanations: if we take the case of an individual word, it may appear in the dictionary but be incorrect in its context (misspelling, bad ending, incorrect agreement, wrong position/function in the sentence), or it may be potentially quite correct but not present in the subset dictionary. The same holds for structures and the grammar. We are particularly aware of this latter danger when attempting to cope with unknown words, resorting to intelligent guesses dependent upon position and spelling to identify their most probable identity and attributes and also informing the user of the assumptions made.

The fact that FGA's grammar is a restricted subset is considered by us to be one of the values of our system as we feel that systems which tend to be somewhat over ambitious in the extent of the attempted coverage of the language lose the generality, clarity and modularity which we consider so essential when the work is basically a research oriented one. We therefore have subscribed to a sort of 'equation' which reads like:

Subset + Modularity – > clear and general principles.

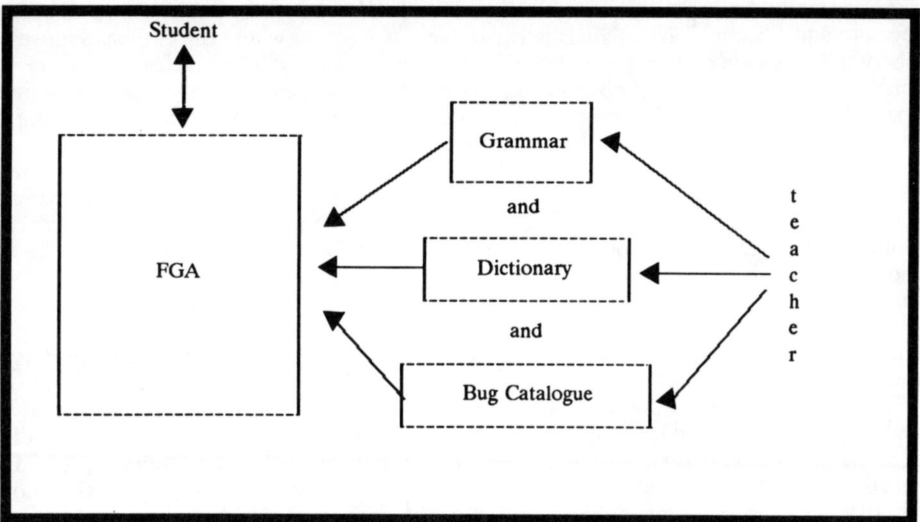

Figure 2: *The overall structure of FGA*

These features of our system have provided a variety of applications within our University's French Department. One Lecturer presents his students with English passages which students are then asked to translate into French. The student submits the French text to FGA for analysis before handing in the program's comments to the lecturer for grading. The grammar and the dictionary of our system has had to be extended continuously in order to cope with the variety of constructions used by advanced students.

We have found a new and unexpected way of using our system as the result of its current limitation of grammar. We have allowed some advanced students to play with different versions of the program, each incorporating a different grammar. The students try various sentences on these systems and try to discover the grammar used by each version. Some of us feel this rather devious use of FGA may be educationally more beneficial than the former one.

This exercise has helped to clarify our intentions. We now clearly know that we are producing an aid for the teacher and not replacing him. Hoping to build complete computational tutoring systems seems to us to be a premature exercise. Teaching is such a complex task that we should first attempt to build computational tools for practising teachers who could customise the systems to fit in with their own individual way of teaching. The situation in which teaching happens would involve students, the teacher and the computational system.

5. Electronic trouble shooting

Most ITSs (see [4]) have been designed and remain as prototypes. There are some notable exceptions; Sophie [11] is the oldest and most well known one. Sophie, sponsored by the US Department of Defense (AFHRL, ARPA, Tri Services), which, after limited use for on-site job training over the ARPA network for two years is, however, no longer maintained.

The three stages of development of Sophie (I to III) incorporate the most intensive attempt at building a complete ITS. The American Air Force's interest was in using computers in its on-board electronics trouble shooting, particularly in a laboratory setting. The intention of the researchers in development of the system has been to explore an interactive learning environment that encourages explicit development of hypotheses by the student carrying out problem solving which is then communicated to the machine so they may be critiqued.

Sophie uses a general-purpose electronic simular [12] in order to provide a simulation of the domain both for the student and itself. The key idea is to construct and 'run' an experiment in order to 'see' what happens, as opposed to the logical deduction of an answer. This is achieved by Sophie containing an articulated expert trouble shooter which it can use. It enables the student to insert arbitrary faults in the circuit and watch the expert system locate them. In order to see how Sophie works we shall look at each stage of its development separately.

I. In the first instance Sophie is created out of a simulation of the chosen electronic device (the IP-28 regulated power supply) which is capable of being faulted, a simple language interface and a database of knowledge about the electronic device. In addition, it contains the basics of an automated laboratory instructor which is itself capable of a limited amount of reasoning. Most of Sophie's intelligence resides in a collection of routines which are selected, set up and run important experiments on the circuit simulator. In this way Sophie evaluates students' hypotheses, engages in question answering, checks redundancy and accordingly provides useful hints to the student.

II. The major addition in the second stage has been to add a 'canned' articulate expert trouble shooter. This expert is not a general-purpose one but is constructed by observation of the expert's way of trouble shooting the electronic device on a defined set of problems. In return for the limited number of problems that the expert can solve, it is capable of explanation of both its tactics for choosing measurements of voltages etc. and, more importantly, explanation of its high level strategies for attacking the problem.

The expert trouble shooter creates a long explanation for each measurement it makes explaining 'why' it was made and 'what' logically follows from the value obtained. The educational aspect of this stage is that the student chooses the fault (out of a reasonably large but nevertheless limited set) and watches the tactical reasoning processes of an expert in finding the fault chosen.

At this stage Sophie still has a limited automatic laboratory instructor and coach. Most of the educational benefits of the system are the result of providing some laboratory based games as well as acting as an interactive book where written material describing a particular piece of equipment can be printed as and when it is relevant to the student's task.

III. The final stage attempts to become a complete tutor where the emphasis moves away from the use of simulation to more powerful and human-like reasoning capabilities for the system and addition of coaching strategies and student modelling. Figure 3 presents the overall structure of the full Sophie system.

The most interesting underlying philosophy of Sophie is its 'best of both worlds' approach. The old fashioned dichotomy of a 'teacher centred' *versus* 'student centred' view of education has shown its counterpart in computer based learning systems too.

AI applications in education mostly seem to fall at the two ends of a spectrum. At one end we can see open ended problem solving environments [13, 14] and, at the other end, we see directed individualistic tutoring systems [4, 15]. Both these approaches have their own shortcomings; in the former the unstructured exploration is inefficient use of time, especially when one of the purposes of computer based education is efficiency. Further, when (as occurs frequently), the student gets lost and the system cannot offer the necessary support. I have discussed some of these issues elsewhere [16]. The latter system's directed approaches do not allow the student to explore topics of interest to the student who is denied control of the interaction.

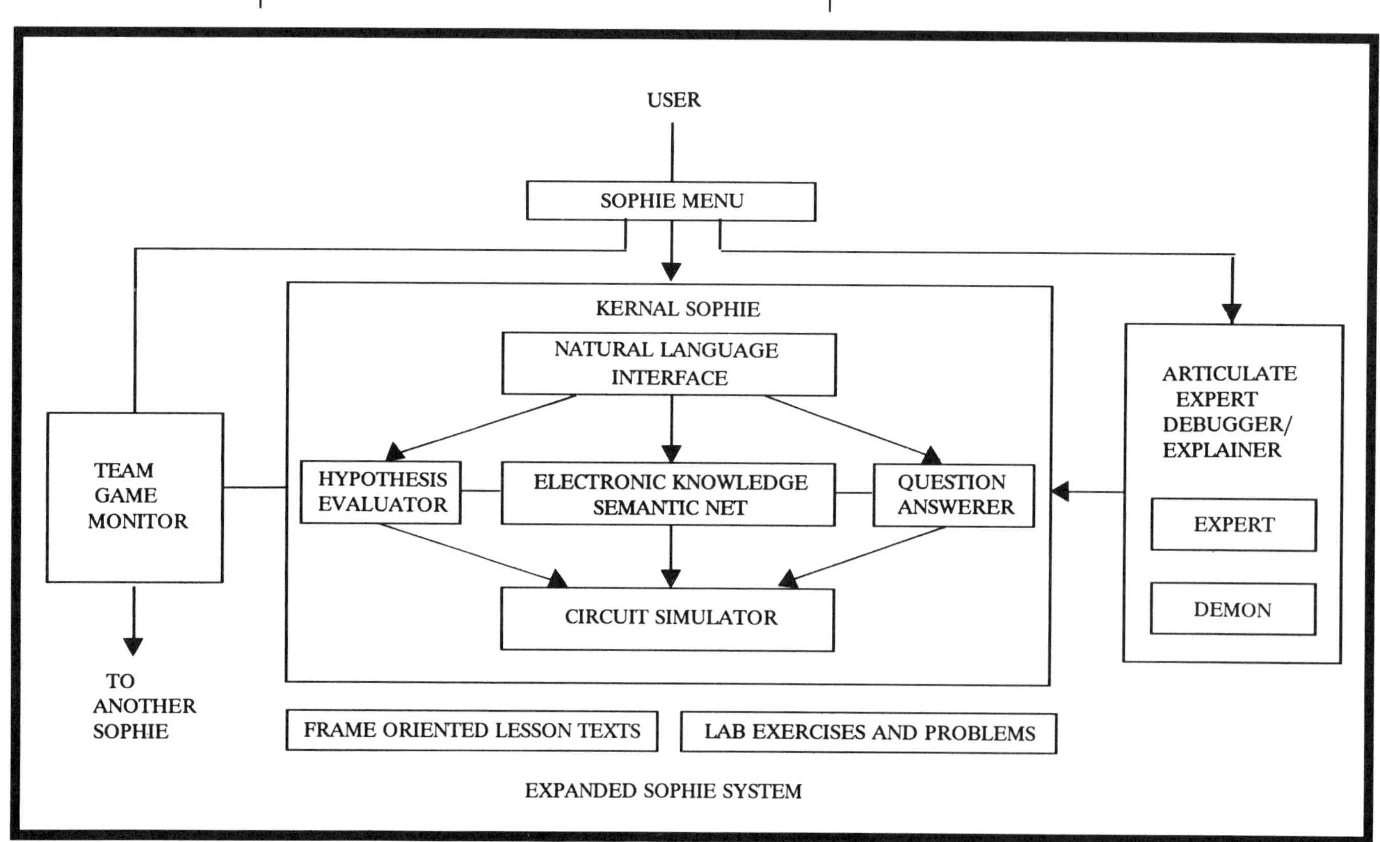

Figure 3: *Extended Sophie system*

In order to produce the 'best of both worlds' Sophie's use is part of a general training course. A typical Sophie mini course consists of four parts:

The first two hour period is used to give a brief introduction to the Sophie system itself, an intensive study of an instructional booklet on electrics and detailed explanation of the device in question (the IP-28 regulated power supply).

B. The second period covers the use of Sophie's expert trouble shooter and the student doing the trouble shooting himself in the simulated laboratory.

C. The third period consists of two activities. One is concerned with finding faults which are propagated to blow out specified components. The second takes the form of a game where two students are asked to play against one another diagnosing each other's chosen faults on the system.

D. The final part brings the two-person teams to play against each other where members of each team attempt to cooperatively diagnose the faults chosen by other teams. The course concludes with a trouble shooting test for each individual on the course.

6. Computer programming

If computer technology is so powerful, why shouldn't it be used to help people to learn computer programming? There is a growing need for training of personnel to use computers and any form of automated support system for novice programmers can speed up this process. Further, the domain of computer programming, being a very structured one, makes it the most suitable test-bed for ITS development. It is not surprising that the only two ITSs which are currently offered for sale are both in this area, one for Lisp [3] and the other for Pascal [17].

Finding syntactic errors in computer programs is a reasonably straightforward job. The reporting of these errors in a form which will make it possible for the novice to learn more about the structure of the language is not that straight-forward. Nevertheless it is clearly an accepted part of mainstream computer science. Computer scientists are primarily concerned with the analysis of what the program actually does rather than with what the user intended it to do.

Dealing with non-syntactic errors of novice programmers, especially in programming languages such as Pascal and Lisp (which are popular teaching languages at university), has attracted a good deal of attention (See [18]) in AI research in general and its applications to education in particular.

Proust (Program Understanding for Students) is an intelligent tutoring system which finds errors ('bugs') in Pascal programs written by novice programmers. It is not confined to a narrow class of error, but designed to find 'every' bug in most beginners' programs. It is claimed [17] that Proust can currently identify 70% of all the bugs in the programs that students write for moderately complex programming assignments. Having identified the bug, Proust determines how the bug can be corrected and suggests why the bug arose in the first place. Proust therefore is part of a wider system which assigns exercises to students, analyses their work and gives them helpful suggestions.

The teaching of computer programming requires a great deal of individualisation. A class of 100 students write 100 different programs, each different in design and bugs for the same programming exercise. Bugs in programming arise from a number of different sources: accidental omissions (missing variable declaration or initialisation), failure to work out how all the different procedures hang together (each piece of the program appears correct, but the program does not run properly), and misconceptions about the language (the program does not do what is expected of it).

If an automatic tutor is to cope with variations and with types of student errors, it must understand what the programmer is trying to do. Proust achieves its level of competence by being provided in advance with a description of the problems set for the students. Each of the problems which is handed out to students is also coded in a frame-based problem-description language and added to Proust's library. Furthermore, Proust has a further knowledge base of common bugs in Pascal programs. Therefore, as long as it knows what problem the student is trying to solve, and what possible mistakes are possible in the language, it can identify them in the student's various programs.

Proust synthesises each program, looking up the corresponding problem description in the library and making a hypothesis about the methods by which programmers may solve each part of the problem. If one of these hypotheses fits the student's code, then Proust concludes that the student is correct. If not, it checks its library of common bugs to see if any of them fits the code.

Proust has been tested on a large number of Yale University undergraduates. It has also been used on a bank of recordings of student programs submitted to the Pascal compiler. Proust has managed to score well over 70% in the tests in identifying all the bugs in Pascal programs. However "17% of programs are analysed partially . . . 4% of the programs, deviated from Proust's expectations so drastically, it could not analyse them at all" [17]. A major problem with Proust is that when it fails to understand a program completely, its ability to recognise bugs deteriorates dramatically. This indicates the sensitive role of a complete problem-description library.

Proust has been developed at a cost of half-a-million dollars over a four-year period at Yale University with two programming assistants. It consists of 15,000 lines of Lisp code (four megabytes of memory) running on a DEC Vax750. It takes three-five minutes but correctly identifies over seventy bugs in novices' Pascal programs. Micro-

Proust is a version of the system running on an IBM PC (512K) in Golden Common Lisp. It is claimed to have taken one programmer two months to produce and only has one-fifth of the bug catalogue of Proust, taking only ninety seconds to run but with an unknown rate of success.

7. ITS architecture

As a result of the experimental nature of work in the area of ITS, no clear general architecture for such systems can be identified as yet. However, the work of the Carnegie-Mellon University Psychologist, John Anderson and his colleagues, on the Lisp tutor [3] and the geometry tutor [19] is offering a strong hint of a breakthrough. The underlying structure of Advanced Computer Tutoring [20] principles used in systems for such diverse applications as Lisp and Geometry, seems capable of supporting other subjects too. At the same time Anderson's earlier work [21] on Adaptive Control of Thought Theory gives this approach a psychological plausibility.

There are four components to ACT's ITS architecture:

1. Domain expert: this module is capable of actually solving problems in the domain. This is sometimes also referred to as the 'ideal student' model.
2. Bug catalogue: this is an extensive library of common misconceptions and errors in a domain.
3. Tutoring knowledge: this module contains the strategies used to teach the domain knowledge.
4. User interface: this is the module which administers interaction between the tutor and student.

The 'ideal student model', or the domain expert, enables the tutor to solve for itself the problem on which the student is working. Rules of Lisp programming are represented as production rules of a Goal-Restricted Production System (Grapes). These differ from a general-purpose production system in that the conditions of a rule include a specific goal i.e.

IF the goal is to multiply
NUMBER1 by NUMBER2
THEN use the function TIMES and set as subgoal to code
NUMBER1 and NUMBER2
IF the goal is to code a recursive function, FUNCTION with integer argument INTEGER
THEN use a conditional structure and set as subgoals
1) to code the terminating case when INTEGER is 0 and
2) to code the recursive case in terms of FUNCTION (INTEGER-1)

The same formalism is used to code the 'buggy' rules representing misconceptions novice programmers often develop. The ACT theory does not claim that these productions exist in the student's head, but rather that the student's behaviour could be modelled with these productions.

In contrast to the richness of student modelling of some other ITSs, the ACT tutors seem to incorporate a very dogmatic and authoritarian approach to education. The main driving force behind these tutoring systems is the derivation from the ideal student model. Whenever the student makes a planning or coding error, the tutor guides the student back to the correct path. This obviously has some dangers, especially when the student is following a correct path but one which differs from the path that the system is following. Nevertheless, the Lisp tutor seems to be able to turn problem solving episodes into learning experiences which they would otherwise not have been.

The student's interface presents itself as a smart screen editor. As long as the student does not make an error, the tutor remains quiet and therefore is seen as no more than an editor. If the student exhibits an error in his program, the system diagnoses the error and provides feedback in the form of a hint.

The Lisp tutor contains approximately 325 production rules about planning and coding Lisp programs and 475 buggy versions of those rules. It is claimed to be "effective in diagnosing and responding to between forty-five and eighty students' errors" [3]. It can be run under VMS and Unix operating systems on DEC Vaxes. A single work station with two megabytes of memory could support one user with three-four megabytes on the Vax730, it can support two users and with six-eight megabytes it could be used as a time-sharing program. The Lisp tutor is commercially available from Advanced Computer Tutoring Inc.

8. The role of student modelling

It is possible to argue that the architecture of ITS presented in the previous section is consistent with other proposals [22, 23]. However, this is only on a superficial level as there seems to be major differences between the competing proposals in a number of issues most importantly on the role of student modelling. Hartley and Sleeman [22] have suggested that ITS should normally have four distinct knowledge bases.

1. Knowledge of the task domain
2. A model/history of the student's behaviour
3. A list of possible teaching operations
4. Mean-ends guidance rules which relate teaching decisions to conditions in the student model.

This proposal differs from Anderson's [2] inasmuch as it does not give the representation of misconceptions in the domain (the bug catalogue) primary importance, but instead introduces the student model as a primary component which is created for each individual user. Further, this proposal subsumes the user interface in a more tutoring oriented module which includes guidance rules on how to carry out an interaction with the user.

What then are the counterparts for the 'student model' and 'guidance rules' in Anderson's tutoring systems? A simple answer would be that within Anderson's overall tutoring philosophy, these two features are not necessary. The role of the student model is not as important as in its place we have two knowledge bases of ideal and buggy representations of the knowledge of the domain. Immediately a behaviour is exhibited which indicates a bug, we nudge the user to follow the correct version by presenting that ideal version. There is therefore also a simple guidance rule to discover deviations from the norm and correct them.

The five ring model presented by O'Shea *et al* [23] which bears some similarity to Hartley and Sleeman [22] shows how the difference of emphasis on student modelling and teaching strategies leads to an architecture which is radically different from Anderson's. This includes

1. Student history
2. Student model
3. Teaching strategy
4. Teaching generator
5. Teaching administrator

In this proposal the role of an explicit representation of the knowledge in the domain (ideal model) or the common misconceptions in the domain (the bug catalogue) are undermined in favour of emphasis on the importance of various teaching skills.

I believe that these three proposals can be viewed as points on a spectrum where at one end Anderson's proposal is closer to the more open-ended exploratory learning environments of Papert [13] and Yazdani [14]. At the other end of the spectrum O'Shea *et al*'s [23] proposals are closer to the traditional CAI systems which sacrificed a rich representation of the knowledge domain in favour of emphasis on general purpose teaching skills.

The choice of a position on the spectrum below is not simply a matter of conviction of the individual researchers, but is influenced by the nature of the expertise which is to be taught. Exceptionally abstract and general concepts such as model building, use of analogy, etc. can be better taught within an exploratory learning environment through the construction of an appropriate computer-based microworld [24]. The teaching of skills which are basically problem solving in a specific domain can be best achieved via problem solving monitors such as Anderson's. As the tasks become more concrete and specific the proposals of Hartley and Sleeman, O'Shea *et al* and those of traditional CAI become more appealing.

There is, however, on major drawback of this diversity of methods for the design of ITS. While the development of traditional CAI systems is greatly facilitated by the use of author languages, construction of an ITS still seems to be a one off process. O'Shea *et al*, due to the closeness to traditional CAI, have the most concrete proposal for a tool kit for ITS while any form of a counterpart for author languages in ITS seems a long way away. What is clear however is that ITS needs powerful knowledge representation formalisms and production systems but those used in Anderson's work seem to be as good as any other while offering a degree of psychological plausibility.

9. Concluding remarks

While CAI can be considered a mature technology, ITS are currently at a pre-technology phase. However, with the arrival of the first commercial fruits of research in this area (so far primarily producing prototypes), we can expect the beginning of a new phase of development where consideration of architectures and tool kits for the development of ITS will play a major role. Following in the footsteps of CAI, ITS needs a counterpart for author languages where it is estimated that an hour of CAI would require 100 hours of an experienced programmer's time to produce. With future toolkits ITS development could possibly take twice as long but, in the current state of the art, it is estimated to be in the order of magnitude higher than CAI.

In the absence of any tool kits for ITS development, any serious attempt at construction of an ITS should be based around the 'know how' of tutor construction. Some attempts are currently being made to build tutoring prototype frameworks where the system could easily be changed from application in one domain to another similar domain. Davies *et al* [25] presents one such framework which, although designed to teach the *Highway Code* in the first place, is now used for teaching flight safety regulations to air traffic controllers.

Our own work on the teaching of French grammar [10] is currently being extended [26] to other Western European languages. Our approach has been to build a Language Independent Grammatical Error Reporter (Linger) where the system acts as a general purpose shell which, when supplied with the databases specific to a language, would teach that language. The motivation behind the project lies in the duplication of effort and code involved in separate development of tutoring systems for languages which show so many common features.

The progression from the FGA to Linger can be seen as increasing movement from constructing tutoring systems for specific tasks towards

< -- >

Learning environments	Anderson's proposal	Hartley and Sleeman's	O'Shea's *et al's* proposal	Traditonal CAI

tools for even greater generality. The object is to produce flexible systems which possess knowledge of their domain and which can be put to a wide variety of potential uses according to the imagination of a human tutor or the learner himself. In this respect the focus of our work is in finding an approach even closer to exploratory microworlds than that of Anderson's on a spectrum which has CAI at its other end.

Acknowledgements

This paper is partly based upon a presentation to the Italian National Council for research, and partly on a presentation to ITT Europe Engineering Support Centre. The author gratefully acknowledges their support.

References

[1] B.F. Skinner, 'Teaching Machines,' *Science*, **128**, 1958.

[2] J.R. Anderson, *Skill Acquisition: compilation of weak method problem solutions*, Carnegie-Mellon University, 1985.

[3] J.R. Anderson and B.J. Reiser, 'The LISP Tutor,' *Byte*, **10**, 4, 1985.

[4] D. Sleeman and J.S. Brown (eds.), *Intelligent Tutoring Systems*, Academic Press, 1982.

[5] P. Alvey, *A programme for advanced information technology*, HMSO, 1982, London.

[6] T. O'Shea and J. Self, *Learning and Teaching with Computers*, Harvester Press, 1983.

[7] E.H. Shortliffe, *Computer-based Medical Consultations: MYCIN*, American Elsevier, 1976.

[8] W.J. Clancey, 'Tutoring rules for guiding a case method dialogue,' *Int. J. of Man-Machine Studies*, 1979.

[9] W. Imlah and B. du Boulay, 'Robust Natural Language Parsing in Computer Assisted Language Instruction,' *Systems*, **13**, 2, 1985.

[10] J. Barchan, B.J. Woodmansee and M. Yazdani, 'A Prolog-based Tool for French Grammar Analysers,' *Instructional Science*, **14**, 1985.

[11] J.S. Brown, R.R. Burton and J. de Kleer, 'Pedagogical, natural language and knowledge engineering techniques in SOPHIE I, II and III, in Sleeman and Brown (eds.), 1982.

[12] L.W. Nagel and D.O. Pedersen, 'Simulation program with integrated circuit emphasis,' *Proceedings of the 6th Midwest Symposium Circuit Theory*, Waterloo, Canada, 1973.

[13] S. Papert, *Mindstorms, Children, Computers and Powerful Ideas*, Harvester Press/Basic Books, 1980

[14] M. Yazdani (ed.), *New Horizons in Educational Computing*, Ellis Horwood/John Wiley, 1984.

[15] S. Ohlsson (in press), 'Some principles of Intelligent Tutoring,' *Instructional Science*, **14**.

[16] M. Yazdani (in press), 'Artificial Intelligence, powerful ideas and children's learning,' in R. Rutkowska and C. Crook (eds.), *Computers and Child Development*, John Wiley & Sons.

[17] W.L. Johnson and E. Soloway, 'PROUST,' *Byte*, **10**, 4, 1985.

[18] B. du Boulay (in press), 'Computers Teaching Programming,' in Lawler and Yazdani (eds), *Artificial Intelligence and Education*.

[19] J.R. Anderson, C.F. Boyle and G. Yost, 'The Geometry Tutor,' *Proceedings of IJCAI-85*, 1985.

[20] J.R. Anderson, C.F. Boyle and B.J. Reiser, 'Intelligent Tutoring Systems,' *Science*, 228, 1985, pp. 456–62.

[21] J.R. Anderson, 'Acquisition of Cognitive Skills,' *Psychological Review*, 89, 1982, pp. 369–406.

[22] J.R. Hartley and D.H. Sleeman, 'Towards Intelligent Teaching Systems,' *Int. J. of Man-Machine Studies*, 1973.

[23] T. O'Shea, R. Bornat, B. du Boulay, M. Eisenstad and I. Page, 'Tools for Creating Intelligent Computer Tutors,' in Elithor and Banerjii (eds), *Human and Artificial Intelligence*, North Holland, 1984.

[24] R. Lawler, 'Designing Computer-based Microworlds,' in Yazdani (ed.), 1984.

[25] N.G. Davies, S.L. Dickens and L. Ford, 'TUTOR – A Prototype ICAI system,' in Bramer M. (ed.), *Research and Development in Expert Systems*, Cambridge University Press, 1985.

[26] J. Barchan (in preparation), *Language Independent Grammatical Error Reporter*, M. Phil. Thesis, University of Exeter.

Reprinted from *Proceedings of the IEEE,* Volume 75, Number 6, June 1987, pages 777-785.

Automating the VLSI Design Process Using Expert Systems and Silicon Compilation

ALICE C. PARKER, SENIOR MEMBER, IEEE, AND SALLY HAYATI, STUDENT MEMBER, IEEE

Invited Paper

This paper describes the automatic design of custom integrated circuits from higher level specifications. The paper covers four topic areas: the problem domain and solution approach, higher level synthesis, module to layout automation systems, typically called silicon compiliers, and expert systems which control the design process.

In the first three sections, several features of the VLSI problem domain which complicate automation are listed. The VLSI design process is diagrammed and the individual steps described. The term "silicon compilation" is defined to cover the entire process, and definitions for various subcategories of silicon compilers are given.

The next section describes both algorithmic and knowledge-based techniques which perform higher level synthesis, including area estimation and module binding concurrent with synthesis. Research at Bell Labs, USC, CMU, and in Canada is described, along with other projects.

The fifth section discusses the three categories of silicon compilers: commercially available systems, experimental compilers being developed by industry, and artificial intelligence approaches from university research. A survey of several systems is provided.

The last section focuses on two systems developed at CMU and USC which plan or control design activities, Ulysses and DPE. Both systems allow the integration of various design automation tools and determine the proper tool invocation to automatically create a design. ADAM takes an autonomous approach, while ULYSSES follows user-defined scripts.

I. Introduction

VLSI design is a time-consuming, intricate task; the time for an electronic product containing VLSI components to be developed may be longer than the marketing lifetime of the product. Investments in the design process may not be recovered before the product becomes obsolete. Given this situation, cost-effective VLSI designs must be produced as quickly as possible, be functionally correct on the first pass, and meet specifications without lengthy tuning and design iterations. One way to achieve these goals is to automate the design process starting with the highest level system specification and resulting in a physical realization with masks, boards, and system configurations.

The following sections describe successfully completed projects as well as ongoing attempts to automate the design process. We describe design packages which, in some overlapping fashion, cover the design process from functional specification to mask generation. The emphasis here is on high-level synthesis, silicon compilation, and automatic control over the design process itself. Although there are numerous efforts underway to solve these problems and many commercial packages available, due to lack of space we have selected a few packages representative of the activity of the entire community.

II. The Problem Domain

Large Search Spaces

The major problem facing VLSI system designers today is the combinatoric number of design possibilities which must be examined in order to produce a design which meets input constraints and which is near optimal with respect to design goals. Example design goals include maximizing speed or minimizing cost. Constraints on the design include time delays for and between significant events, area or package count upper bounds, and maximum number of pins. Goals involve quantities which are optimized, while constraints refer to quantities which must be achieved.

Many synthesis and layout subproblems (e.g., microcode optimization, register allocation with a fixed set of registers) are known to be NP-complete, which means that the process of finding an optimal solution is believed to involve an exponential number of steps. Most synthesis tasks are thought to be at least NP-hard (harder than NP-complete problems).

Evaluation of Partial Designs is Difficult

Designs which have been carried out only to an intermediate level or which are partially complete are difficult to evaluate since timing and chip area are dependent on the physical implementation. If a critical timing path spans the chip, the wiring delays will usually impact the performance

Manuscript received July 15, 1986; revised October 7, 1986. Funding for the USC research described in this paper was provided by the Army Research Office under Contract DAAG29-83-K-0147 and the National Science Foundation through its Computer Engineering Program under Grant DMC-8310774.

The authors are with the Department of Electrical Engineering—Systems, University of Southern California, University Park, Los Angeles, CA 90089-0781, USA.

IEEE Log Number 8714501.

of the chip. Chip area consumed will be dependent on the placement of macros on the chip, since wiring area tends to dominate functional area. Thus a decision to share an adder might be made before any physical design has been done, and might result in more area consumed due to wiring than would have been consumed by duplicating the adder!

Design Strategies are Problem-Dependent

Design problems cannot all be approached using the same strategy. Pipelining works well for computation-intensive problems with low probability of data dependencies and conditional branches but not for some CPU designs; PLAs are an effective way of implementing control logic but not arithmetic.

No Fixed Order to the Steps of the Design Process

The overall design process is also problem-dependent, and varies dynamically depending on the success of each individual step. Partitioning may be performed repeatedly on a large design, but rarely on a smaller one. The critical path might be designed down to silicon before any other parts of the design are begun for a problem with tight timing constraints, but essentially ignored in a low-cost application. If a design fails to meet constraints, a few steps may have to be repeated.

Design Tasks Interact

The partitioning of the design process into discrete tasks is artificial and is performed due to efficiency considerations. Placement affects routability and routing may force placement adjustment. Sharing resources to reduce the datapath cost produces a more costly control, and the delay through the control store affects the timing requirements on the datapaths.

III. The Solution: Silicon Compilation

Silicon compilation is the process of automatically mapping a more abstract representation of digital design to the integrated circuit layout, as shown in Fig. 1.

Adapting the terminology used in [17], silicon compilers which accept a functional description and produce an implementation with predictable structural semantics are *architectural silicon compilers*. Silicon compilers which accept a functional description and produce an implemenation with structural semantics which cannot be predicted are *behavioral silicon compilers*. If a structural description is input instead of a functional description, then the compiler is referred to as a *structural silicon compiler*.

Synthesis packages form the front-end of the behavior silicon compilation process. *Register-transfer (RT) synthesis* involves producing a structural design which can carry out the operations described in the input functional specification. The *input functional specification* normally contains behavioral information: what functions the resulting implementation is to perform, but not how the functions are to be executed. It may be a data-flow graph [39], a hardware-descriptive language such as ISPS [3], or a modified programming language such as Ada. The output from RT synthesis contains the kinds and numbers of modules and registers required to adequately implement the behavior

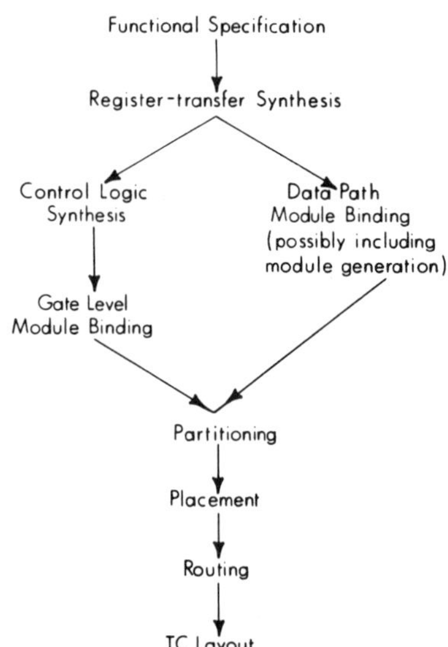

Fig. 1. The VLSI design process.

plus interconnections and scheduling of resources to execute the desired tasks. There will be many possible implementations with differing performance and cost for any one behavior. Architectural silicon compilers perform some limited register-transfer synthesis.

The next two steps of the design process differ between the control and datapath parts of the design. For the datapath, physical modules to realize the RT level modules must be determined; this is called *module binding*. The physical modules might exist in a database or might be generated during run time through a process called *macrocell compilation*. Module binding decisions have a first-order effect on the system performance and cost, since the symbolic modules may be realized with diverse physical modules (e.g., parallel or serial multipliers). Any datapath module binding must be considered before the control logic synthesis, since the required control signals cannot be determined until the actual modules have been selected.

For the control, next state equations are derived from the datapath resource schedule and the required control signals. Control *logic synthesis* is then performed by producing gate-level schematics [6], [10], selecting a microcontroller architecture, and generating microcode [30] or generating PLA personality matrices and producing the PLA layouts [11].

The last steps of the design process are the partitioning of cells onto chips or areas of a chip, their placement, and then routing.

No current system provides all the functions described above while solving all the problems we mentioned earlier. As each program or project is discussed, the problems each attempts to solve will be described. We focus on register-transfer synthesis first, in Section IV. Then, we describe current silicon compilers in Section V. Section VI presents more experimental systems which use rule-based approaches to control the design process.

IV. Register-Transfer Synthesis

Register-transfer (RT) synthesis programs must face the problem of large search spaces. Two programs we present here have evaluation capabilities for partial designs, and some variations in design strategy are supported by each program. However, none of these synthesis programs addresses the difficult issue of interacting design tasks, or provides flexibility in the ordering of design steps. All but one are algorithmic rather than knowledge-based.

The emphasis of register-transfer synthesis has been on datapath synthesis. Datapath synthesis involves two operations, *event scheduling* and *resource allocation*. Event scheduling assigns each operation to a time range corresponding to a clock cycle or part of a clock cycle. Resource allocation assigns operators to operations present in the behavior, and registers or wires to values. The mapping from operations and values to resources is not necessarily one-to-one, since resources are often shared to reduce costs. Fig. 2 shows the overall operation of a datapath synthesis program.

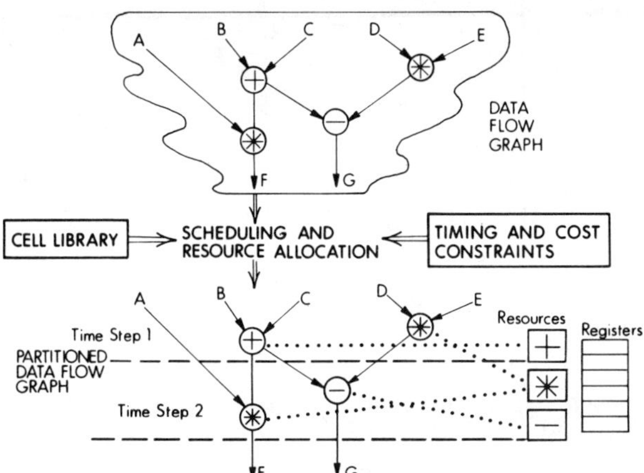

Fig. 2. The datapath synthesis process.

Searching the Design Space

The first published RT datapath synthesis program, EXPL (explore) [2], searched the design space to produce a set of designs. EXPL used a very small, well-behaved module set, so it was easy to evaluate the cost and speed of designs as they progressed. Until recently, EXPL was the only synthesis program to have such searching and evaluation capability.

Hafer's formal model of the design process [19], using a mixed integer-linear programming formulation, provided an exhaustive search capability. However, such a model was impractical for all but the smallest design problems. This model implicitly varied the style of the design produced from pipelined to parallel to serial, as the user varied the cost and speed constraints. Timing and cost constraints could be detailed and local, since they could be expressed as part of the integer-linear program constraints. Thus in theory, Hafer's model could synthesize designs with complex timing requirements, like interfaces, and could adapt its design strategy.

Limiting the Search Space

Writers of newer synthesis programs have addressed the complexity of datapath synthesis by performing only a subset of the tasks, by limiting the design styles handled by the program, by using heuristics and theory to narrow the search space, by limiting the types of constraints and goals which can be handled, and/or by allowing user interaction.

EMUCS [20], [29] and FACET [40] perform only the allocation task, assuming that event scheduling has already been done. They use heuristics to limit the search space. Both optimize only cost, since the timing is already fixed. FACET can design with shared buses and ALUs; EMUCS produces designs with shared ALUs. However, EMUCS allows user interaction, which allows structures like buses to be manually inserted into the design. Both programs have been used to design the MC6502, and FACET has also designed an IBM-370.

ELF [14], [15] is more sophisticated than EMUCS or FACET in that it performs event scheduling along with allocation. However, while EMUCS and FACET have been tested on large designs, ELF has been run on smaller test cases such as a temperature controller. ELF always attempts to reduce cost, subject to timing constraints. These timing constraints can be local, and occur throughout the design. Thus ELF handles a wider range of design strategies than other allocators, and can also produce designs like interfaces which require precise timing behavior.

MAHA [34] performs event scheduling along with allocation, and attempts to optimize either cost or speed, subject to constraints on the other. MAHA uses heuristics to limit the search space, and, like ELF, has been tested on small designs. In order to limit the search space, MAHA does not complete the allocation process, leaving the register and interconnect allocation to a later program. The decision to use shared ALUs in place of individual operators is also postponed. MAHA assumes that certain checks and transforms have already been performed on the input specification where loops occur.

The Introduction of Expert Systems

In order to avoid exhaustive search while still achieving close to optimal solutions, information about the characteristics of the problem domain can be embedded into design software in the form of heuristics. Systems employing heuristics in the form of *IF-THEN* rules are referred to as *production systems*. Expert systems are generally production systems whose knowledge has been obtained from humans expert in the problem domain. More precisely, *expert* also refers to the performance level of the program. Expert systems have three components: a *rule base*, an *inference engine*, and a *working memory*. An example rule for a synthesis program could be "*if* an increment of data is followed by storage of data, and the operation is performed repeatedly, *then* substitute a counter for the incrementer and latch."

The inference engine chooses among and interprets the rules. In a *forward chained* system, the engine matches the *if* parts of the rules with its current status and known facts. If a match is found, it uses some tie-breaking strategy to choose a single rule. The *then* part of that rule is executed, and the current status updated. This kind of expert system is *data-driven*. In a *backward-chained* system, the engine would match some desired goal with the *then* parts of the rules. It would pick a matched rule, and determine whether the *if* part of that rule is true. If that cannot be determined by direct examination of known facts, the *if* part of the chosen rule is asserted as a new goal. Such systems are *goal-driven*.

Some expert systems have multiple rule sets or rule hierarchies. Many newer systems represent domain knowledge in frames or semantic networks, and use rules just to control the reasoning process. An example of this is discussed later in this paper.

The Use of Expert Systems

In an attempt to reduce the search space, handle large designs, and complete the entire RT synthesis process in a single program, Kowalski turned towards an expert system solution. His system, DAA (Design Automation Assistant) [24], [25], designed the complete RT structure of an IBM-370. DAA designs the most parallel design which does not violate cost constraints, performing greedy scheduling during operator allocation. It uses shared ALUs and buses where possible.

Knowledge was acquired for DAA by interviewing designers. Thus DAA has rules which force certain design practices. For example, one rule states that logical operations on data should be separated from the ALU to enhance testability. DAA has about 300 rules and uses forward chaining.

DAA deals with the large search space by applying a partitioning algorithm prior to design [27]. This algorithm heavily influences the resource sharing DAA carries out. DAA attempts to measure the area of the resulting layout by estimating interconnection costs using a separate C language procedure. A possible advantage of DAA over earlier programs is the estimation of partial solutions.

The above programs, with the exception of Hafer's, produce nonpipelined designs. Girczyc has reported [16] that modifications to the ELF input allow it to perform pipelined synthesis in a limited way. This technique is general, and could be used by a number of the above synthesis packages.

The Pipelined Design Strategy

SEHWA [31], [32], a part of USC's ADAM system, performs event scheduling along with allocation for pipelined designs. SEHWA synthesizes datapaths which overlap resource sharing in a more general way than conventional pipelines, whose stages consist of mutually exclusive sets of operators. SEHWA mixes resources between stages. One resource (e.g., an adder) can be used in stages 1, 3, and 5. A second resource (e.g., a multiplier) can be used in stages 1, 4, and 7. For this reason, the designs produced are more complex than those typically created by humans. SEHWA first synthesizes the least expensive and the fastest designs, followed by a range of designs closer to a desired design goal. Finally, the user can perform exhaustive search if none of these designs meets her or his needs. Because the search space is carefully bounded, pipelines as long as ten stages can be designed in a few minutes on a workstation.

SEHWA can handle conditional branches which allow it to synthesize pipelines which execute more than one type of task. This cost-effective technique makes it possible to use a single pipeline for multiple types of tasks.

Fig. 3 shows a sample output from SEHWA. The partitioning of a data-flow graph into time steps is shown in Fig. 3(a), and the allocation of operations to operators is shown

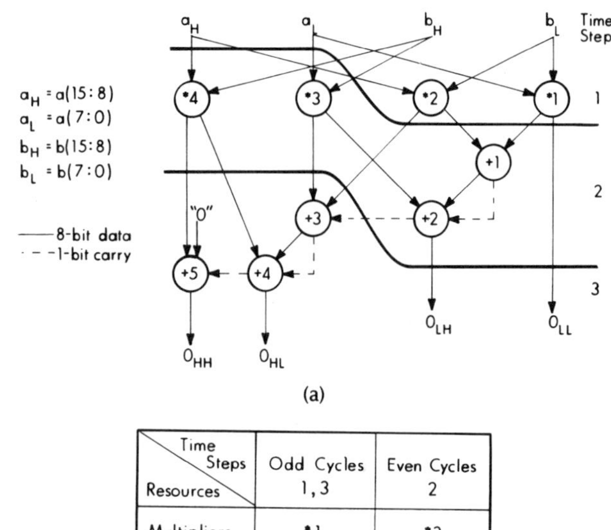

(a)

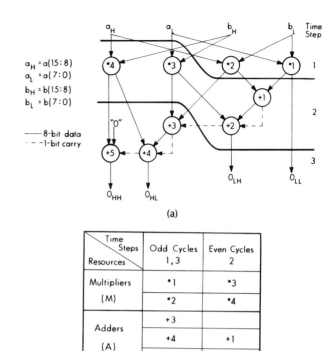

Time Steps / Resources		Odd Cycles 1, 3	Even Cycles 2
Multipliers (M)		•1	•3
		•2	•4
Adders (A)		+3	
		+4	+1
		+5	+2

(b)

Fig. 3. Output from the pipeline synthesis program SEHWA.

in the *allocation table*, an extension of a conventional reservation table, Fig. 3(b). The allocation table shows time divided into two steps. During the first step, stages 1 and 3 are being processed. During the second step, stage 2 is being processed. Two multipliers and three adders are shared for all three stages of processing.

The allocation of registers and interconnect is postponed, although their cost is estimated by SEHWA. This program narrows the search space in a unique way—by using the theory of pipelined designs to bound the search space. Those designs which are discarded by SEHWA are guaranteed to be inferior to those which are kept. SEHWA does generalize the design strategy to some extent, since it produces a nonpipelined parallel or serial design automatically if that is the best design which meets the constraints. The designer can give either cost or speed constraints, and is allowed to alternate such constraints interactively until SEHWA converges on an acceptable solution.

Evaluating Partial Solutions

The utility of the above programs would be enhanced if better data to predict the impact of design decisions on the actual area and performance of the physical implementation were available. There are two approaches to this problem: using estimation to predict cost and timing variations due to physical implementation, or constructing physical modules and a floorplan dynamically during the synthesis process.

ADAM takes the first approach. The ADAM PLEST procedure estimates area, given physical module and net lists [26]. A probabilistic model has been developed for estimating the area of standard cell blocks plus routing. The model was applied to a set of six small test chips. The estimated chip area for all six chips was within 10 percent of the measured area. Once PLEST is expanded to include RT-level modules, MAHA and SEHWA can get accurate area estimates as the design progresses.

The alternate approach of constructing the physical design in parallel with the synthesis process has been taken by McFarland in the BUD (Bottom Up Design) program [28]. BUD can accept both area and overall time constraints.

BUD will be used as a scheduling front-end for DAA, providing global direction to the local processing performed by DAA. The experience gained from the partitioner used by DAA has motivated the heuristic clusterer for BUD. During synthesis, BUD obtains actual cell information from a database. This information, along with the clustering, is used to produce a floorplan. Wire lengths are estimated from this floorplan, providing accurate area and time delays to BUD. The scheduling process also performs limited search by not considering all possible schedules. BUD produces nonpipelined designs.

V. Architectural and Structural Silicon Compilation

A major difference between conventional silicon compilers and synthesis programs has been the pressure to provide commercial silicon compilers for present applications. This section gives a brief tutorial on silicon compilation and examines three commercial compilers, GENESIL, CONCORDE, and SDL. Then we briefly describe an in-house compiler, Silc™, and turn to two research efforts incorporating AI.

The Silicon Compilation Process

Structural silicon compilers map interconnected functional units into a single, integrated layout. The functional units are arrays like RAM and ROM, registers, and operators like adders and shifters. The user specifies what each functional unit should do, but not its internal structure.

As an example, we will examine the generation of datapaths by silicon compilers. The sketch of such a datapath is shown in Fig. 4.

The compilers take as input parameters the functions, their interconnections, number of bits, and other requirements (e.g., on performance). The compiler either finds library cells which match the functions or calls routines which generate the cells to match the requirements, sizing transistors and modifying circuits as necessary. Most compilers assemble each functional unit from an array of cells by placing cells so that they precisely abut. That is, physical positions of inputs and outputs (e.g., carries through an adder) precisely match, so that cells can be placed adjacent to each other without extra wiring. Global wires, like power and clocks, can run through an array of abutted cells.

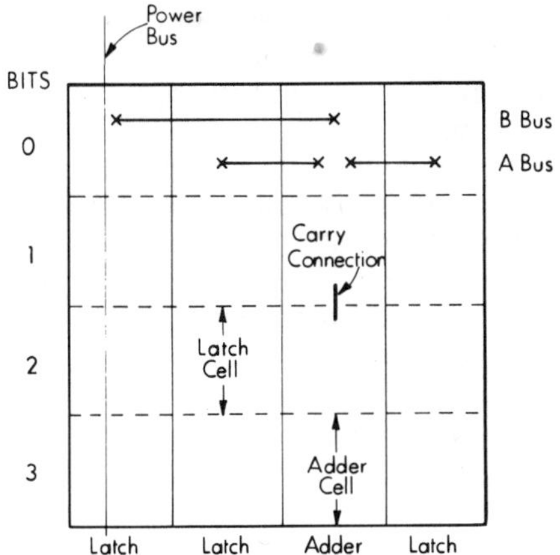

Fig. 4. Example 4-bit datapath produced by a silicon compiler.

Each functional unit is then placed adjacent to other functional units which share inputs or outputs. Again, the cells are adjusted so that the inputs and outputs precisely abut. Since most functional units have more than a single input, extra signal pathways are provided through the cells, as shown in the figure. Connections to the signal pathways are shown with an ×.

Once all major system blocks have been designed, they are placed on a floorplan by the user. The system then routes the interconnections between the major blocks, adjusting block placement if necessary to produce more wiring space.

Commercial Compilers

Because the commercial compilers surveyed must design production-quality integrated circuits, they cannot limit the search space by omitting steps of the design process. Furthermore, heuristics to bound the search space have not been fully developed. Thus these compilers must limit the design space by having the user make many decisions, and by using limited variations in design strategy.

Concorde: The Concorde VLSI compiler by Seattle Silicon Technologies [9] is primarily a structural silicon compiler. The user must determine the numbers and types of modules, yet may use Concorde to compile (create a layout of) appropriate versions of the desired modules, with the proper bit width, signals, and other parameters. A suite of module compilers are available for this task, consisting of SSI compilers (e.g., D and R-S flip-flops), MSI compilers (e.g., ripple counters), memory compilers (e.g., RAM), a datapath compiler, and a PLA compiler. A state machine (storage logic array) compiler is also available. SST also provides the SLIC™ language with which the user can create specialized module compilers; using the gate level behavior compiler it is also possible for the user to define from within Concorde any random logic function desired as a module.

To arrange these modules in a larger configuration, the user performs relative placement for the composite modules; the system determines the exact placement, thereby assuring design rule independence. Routing is performed automatically, though the user may specify net priority and edit the routing manually. Composite modules created in this way may be used in a hierarchical fashion. The system supports analog as well as digital CMOS compilation, allowing both to reside on the same chip. Simulation and timing verification are supported.

GENESIL: GENESIL, from Silicon Compilers Inc. (SCI) [13], is a structural silicon compiler for NMOS and CMOS, which grew out of Bristle Blocks, the original silicon compiler [21]. Functional and electrical parameters must be specified by the user. Forms are filled in by the user to enter data into GENESIL. The system utilizes defaults on under-specified designs. The first step of the GENESIL design process is module generation. A function is selected from the function set menu, which includes an ALU, barrel shifter, PLA, multiplexer, datapath, FIFO, RAM, ROM, and stack. As the user refines the design, the defaults are replaced with actual values and estimates of design characteristics become more accurate. GENESIL automatically creates three models: the geometric (transistor layout), the timing, and the functional models. These may be used to automatically verify and simulate the design. Functions not defined in the function library must be designed externally and incorporated into the design. Following module generation, the user performs relative placement, and then actual placement and routing are done automatically. Routing priorities may be assigned to the different signals and bus signals are routed together. A hierarchical set of objects may be created, and multiple system clocks may be defined. Both Concorde and GENESIL have been used to design production chips.

Generator Development Tools (GDT): GDT is a silicon compiler from Silicon Design Labs (SDL) [38] which enables designers to construct and verify their own structural silicon compilers. The term generator refers to the module compilers which create cells at the various levels of complexity. GDT provides the user an opportunity to use his or her own parameterized cells in design. SDL provides a library of sample generators in CMOS technology.

The core element of GDT is the L language, a procedural circuit-layout language containing primitives ranging from lines, rectangles, and polygons to wires, transistors, contacts, terminals, and hierarchical cells at the highest level. A combination of editors, simulators, rule checkers, and routers assist the user in custom designing cells, which may then be parameterized and reused. Higher level generators may call lower level generators with automatic routing (relative placement is performed by the user); in this way generators can be built for all levels of design complexity from small cells to whole systems.

Experimental Compilation

Silc™: GTE's Silc™ [5], based on the MacPitts silicon compiler [12], is representative of a number of experimental architectural silicon compilers. It accepts an algorithm expressed in a Lisp-like language as input. The input is mapped to an architecture comprised of finite-state machines functioning in parallel. Therefore, Silc™ performs some register-transfer level synthesis, namely, determination of the kinds and numbers of modules necessary to implement the algorithm, plus intermodule connections. Logic synthesis is also performed: control-logic equations are derived from internal data structures created when parsing the input specifications and a two-level logic minimization is performed which utilizes heuristics to reduce problem complexity. Silc™ is thus particularly well suited to specifications with a mixture of control and datapath elements, and is one of the few systems which can deal with both aspects of design in a unified fashion. Since the event scheduling is input to Silc™, the compiler optimizes only chip area.

Although the input is in the form of an algorithm, the user is constrained by the language to certain design methods (e.g., variables must be defined as stored or unstored and the communication protocol between finite-state machines of the design is predefined as two-line handshaking). Thus although the top level of Silc™ performs some synthesis functions, there are rigid architectural restrictions.

Partitioning of the design is performed by the user by specifying the division into FSMs. Silc™ follows a noninteractive philosophy throughout, so placement and routing are performed automatically. Because its operation is more sophisticated than most other compilers, Silc™ is still in the research and development stage.

Artificial Intelligence Approaches

Palladio (at Xerox PARC and Stanford) [7] and REDESIGN (at Rutgers) [35] support user interaction, while providing capabilities for knowledge-based design decisions.

Palladio assumes that circuit design is a process of incremental refinement. Functions are decomposed into simpler functions that are finally mapped into structures, which are then implemented in silicon. Design languages at many levels are an integral part of Palladio, and these languages are used internally to represent design information. Both behavior and structure are represented at each level of description. Palladio provides interactive editors for entering design information, an event-driven simulator, a frame-based mechanism for assigning multiple perspectives (e.g., behavior and structure) to components, and a protocol for creating new perspectives. There are also mechansims for implementing rule-based, expert-system design aids.

Palladio uses an *object-oriented* programming paradigm. The software and design data are structured as objects: software modules which encapsulate a behavior and internal data structures. Objects communicate by means of message passing. Objects can belong to classes, and every object of a class inherits all the properties of that class. Accordingly, all individual NAND gate objects would belong to the object class *NAND gate*, which itself belongs to the class *gate*. Thus these individual gates would inherit the behavior of the class NAND, but perhaps have a different circuit structure and layout. Individual NAND gate objects might belong to either the class MOS or BIPOLAR. Class inheritance is not a strict hierarchy.

An expert system within Palladio assigns IC layers to interconnect and circuit structures. Palladio has provided the inspiration for many other efforts, including the ADAM project at USC.

The REDESIGN project at Rutgers University incorporates sophisticated AI techniques in order to interactively aid designers in producing digital circuits and to learn automatically from the designers. The REDESIGN system includes the VEXED program, which interactively proposes design tasks, ranks the tasks, and allows the user to choose the task to be applied. If the user does not accept the ranking given by VEXED, the learning apprentice (LEAP) examines user design decisions and attempts to generalize and learn from them.

As VEXED is running, a history of the design is maintained. After the design is completed, if design constraints have changed, REDESIGN reruns the design process. It stops at the point where a previous design decision conflicts with the present situation and changes the plan.

The system starts with a register-transfer structure and proceeds to design the lower levels. The system at present only designs small circuits.

VI. Control Over the Design Process

In parallel with the development of specific silicon compilation tools, research has begun on the automation of the entire design process. Two projects are representative of this effort, ULYSSES at CMU and ADAM at USC.

ULYSSES

Carnegie Mellon University's ULYSSES system (Unified LaYout Specification and Simulation Environment for Silicon) [8] is a CAD tool integration environment. ULYSSES addresses the database representation problem as well as the design control problem. ULYSSES provides its own design space representation, and provides for the automatic invocation of translators (provided by the user) from other representations to the internal representation. Thus programs with arbitrary input and output formats may communicate with each other automatically.

ULYSSES has dealt with control issues in a straightforward way. Each CAD tool is treated as a knowledge source. There is a scheduler knowledge source whose task is to choose among a set of CAD tool knowledge sources whose activation conditions have been met. In practice, the ability of the scheduler to determine the proper ordering is limited; therefore, intervention by the user in this process is expected and commonly occurs. In order to decrease the user interaction, ULYSSES supports the use of scripts written by human designers to describe and automatically execute a sequence of design tools [37]. In addition to the sequencing itself, the scripts contain rules which describe actions to be taken in case of error or user intervention (for example, which tool to rerun). These rules ensure that changes made by the user are propagated and recorded properly and automatically.

Thus ULYSSES provides a way for expert designers to specify a control sequence (called a design methodology) which may subsequently be invoked by less experienced designers. New design tools may be introduced into ULYSSES with the simple addition of translators to deal with any difference in internal data structures. However, any existing scripts would need to be modified in order to take advantage of the new tools. ULYSSES, though extensible, is currently used for the integration of layout tools from floorplanning to routing.

ADAM

The ADAM (Advanced Design AutoMation) system at the University of Southern California [18] was designed specifically to solve the problems described in Section II. The long-term goals of ADAM are to support behavioral silicon compilation, while allowing varying amounts of user interaction, problem-dependent goals and constraints, and differing design styles. The approach taken is to use algorithms where appropriate, to use expert-system technology to assert control over the algorithms, and to interact with the user. The ADAM *Design Planning Engine* (*DPE*) [23] is an expert system which determines its own design control strategy.

Planning involves the definition and solution of the problem in an abstracted search space in which unimportant details have been omitted. The highest level decisions are made in this abstracted space, and subsequent restricted searches within a more detailed space fill in the solution details, following the outline of the plan. Provided that the knowledge guiding the planning process is accurate, much time-consuming search may be avoided, and yet quality solutions may be found.

Planning allows ADAM to deal with the dynamic ordering of and interactions between design tasks. The DPE efficiently searches the space of possible design strategies. The DPE can consider an almost infinite number of plans, and any design step can be considered at any point in the design process.

The DPE builds a design plan by choosing from a set of possible analysis and synthesis tasks and tools, including clocking scheme synthesis, component selection, critical path location, area estimation, and hardware allocation. DPE represents plans as sets of partially ordered operators like those described by Sacerdoti [36]. Design tasks are first selected based on characteristics of the problem specification, then added to the plan on the basis of their pre- and post-conditions and an estimate of the advisability of choosing that particular task. Once the plan is complete, the tasks are executed in the prescribed order.

Planning knowledge is contained in a collection of rule sets; domain knowledge (knowledge about digital design) is found in data structures called *frames*.[1] Frames may contain sets of rules to carry out the design processes directly, or may reference program code for a given tool. Valid inputs and outputs for each tool are specified in the frames, and preconditions which must be valid prior to tool usage are maintained in the frames and checked by the planner. New tool frames can be added without any modification of the planning rules.

Integration of the different design tools is aided by the common use of the Design Data Structure (DDS) [22]. The DDS is an integrated design representation for behavioral, structural, and physical design information. Programs invoked by the planner which do not use the DDS will require translation of data formats, and the translation task will be appended to the plan. In practice, the only outside programs which are incorporated into ADAM are the low-level design aids, such as SPICE or layout programs; higher level design tools such as MAHA and SEHWA use the DDS.

The overall structure of the design planner is shown in Fig. 5. The planner has knowledge about RT datapath design, and can build plans using such knowledge. The separation of planning and design knowledge increases the general applicability of the planner and aids in the creation of metaknowledge to guide the design process. The planner itself can be used to plan general engineering design tasks.

VII. Discussion and Conclusions

Discussion

The synthesis programs discussed above range from broad attacks on the entire RT synthesis problem to narrow approaches which solve specific parts of the problem using algorithms based on theory. DAA exemplifies an approach to solving the entire problem, while SEHWA solves a focused part of the problem, but in a more comprehensive

[1]Frames provide a way of organizing declarative and procedural information to facilitate inference about objects. The representation consists of named slots which provide a context and organization for information about the object and may establish object hierarchies (such as is-a or part-of) and allow procedural attachment.

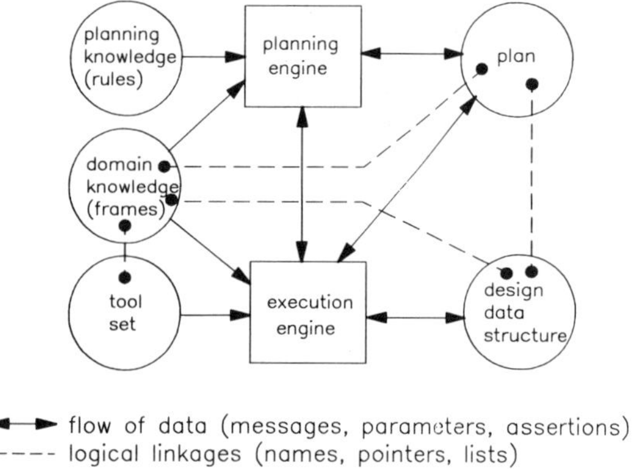

Fig. 5. Overall structure of the design planner.

fashion. ELF and BUD are a cross between the two approaches.

The commercial compilers rely on user interaction to cover a number of steps in the design process, and the user can control some style variations through module selection and architectural configuration. Silc™ attempts to incorporate a large portion of the design process, but with a narrow enough design style to be practical in the near term. Palladio and REDESIGN are long-term approaches to the problem.

ULYSSES has been designed to allow the user to provide intelligence in making choices and controlling the design process. The effort involving ADAM has concentrated on a unified design representation, a planner which could be applied to any engineering design problem, and the automatic generation of a design strategy. While it might be straightforward to add user interaction to ADAM, the system currently operates autonomously. ULYSSES has many features which support its use in an interactive design environment; ADAM has more features which allow it to make intelligent decisions on its own.

Conclusions

Three major conclusions can be drawn about the research and development efforts discussed here. While synthesis research began in the late 1960s, the efforts in the 1970s were concentrated in a few universities and research laboratories. Now there are synthesis and silicon compilation groups worldwide. The capabilities of the tools produced has also increased, and larger design problems are being used as examples. More formal understanding of the problem has occurred, so that the software can be based on theory.

Secondly, resistance to automation of the design process has decreased. Designers are accustomed to automation and aware that dealing with integrated circuit design is more difficult than designing with discrete components.

Finally, AI techniques are being adopted by CAD researchers in an attempt to solve the problems described here. AI researchers have found VLSI design to be a challenging problem domain. Constructing CAD programs utilizing AI techniques requires the expertise of both fields, since CAD problems pose new and more difficult situations than most problems solved by expert system techniques in the past.

Future Prospects

There are three directions in which silicon compilation will go in the next few years. First, major advances will come about via the use of AI for CAD applications. Furthermore, since the field of AI is advancing rapidly, newer and more powerful techniques will allow CAD researchers to solve problems more easily than current AI technology. While there is some controversy over the role of AI in CAD, a recent panel session at the 1985 International Conference on Computer-Aided Design concluded that algorithmic approaches to specific design tasks were successful, with AI providing overall control over the design process, along with problem-specific heuristics for final optimization and handling exceptions. This conclusion has been supported in [4].

Secondly, synthesis will merge with and become a part of the behavioral silicon compilation process. The attention to physical design shown by BUD will become increasingly important at the higher levels, and extend to the board, rack, and system levels of design.

Finally, attention will become focused upward on the higher levels of system design. Entire systems will be synthesized automatically. Research is underway at USC to enter system specifications into ADAM using natural language [1], and such specifications will provide the inputs to a new generation of synthesis tools.

Further Reading

For more information on synthesis and silicon compilation, the reader is urged to examine the proceedings of the 1985, 1986, and 1987 *Design Automation Conference*, the 1985 *International Conference on Computer Aided Design*, the IEEE Transactions on Computer-Aided Design, Integration: *The VLSI Journal* (North Holland), *The AI in Engineering Journal* (Computational Mechanics, Great Britain), as well as numerous recent publications on CAD from Kluwer and North Holland. A tutorial paper on synthesis can be found in [33].

References

[1] J. Granacki and A. Parker, "A natural language interface for specifying digital systems," in D. Sriram and R. Adey, Ed., *Applications of Artificial Intelligence in Engineering Problems*, pp. 215–226, Apr. 1986.

[2] M. Barbacci and D. Siewiorek, "The CMU RT-CAD system: An innovative approach to computer aided design," in *Amer. Fed. of Information Processing Societies Conf. Proc.*, vol. 45, pp. 643–655 (Amer. Fed. of Information Processing Societies, June, 1976).

[3] M. Barbacci, "Instruction set processor specifications (ISPS): The notation and its specifications," *IEEE Trans. Comput.*, vol. C-30, pp. 24–40, Jan. 1981.

[4] W. Birmingham, R. Joobbani, and J. Kim, "Knowledge-based expert systems and their application," in *Proc. 23rd Design Automation Conf.*, pp. 531–539 (ACM and IEEE), June 1986.

[5] T. Blackman, J. Fox, and C. Rosebrugh, "The Silc silicon compiler: Language and features," in *Proc. 22nd Design Automation Conf.*, pp. 232–237 (ACM and IEEE), June 1985.

[6] R. K. Brayton *et al.*, *VLSI, Computer Architecture, and Digital Signal Processing: Logic Minimization Algorithms for VLSI Synthesis.* Hingham, MA: Kluwer Academic Publ., 1984.

[7] H. Brown and M. Stefik, "Palladio: An expert assistant for integrated circuit design," Xerox Memo KB-VLSI-82-17 (working paper), Xerox PARC, 1982.

[8] M. Bushnell and S. Director, "VLSI CAD tool integration using

the ULYSSES environment," in *Proc. 23rd Design Automation Conf.*, pp. 55-61, June 1986.
[9] Seattle Silicon, *Concorde Documentation*. Bellevue, WA, 98005.
[10] J. Darringer *et al.*, "Logic synthesis through local transformation," *IBM J. Res. Develop.*, vol. 4, pp. 272-280, 1981.
[11] G. De Michelli and A. Sangiovanni-Vincentelli, "Multiple constrained folding of programmable logic arrays: Theory and applications," *IEEE Trans. Computer-Aided Design*, vol. CAD-2, no. 3, pp. 167-180, July 1983.
[12] J. Fox, "The Macpitts silicon compiler: A view from the telecommunications industry," *VLSI Des.*, vol. 4, no. 3, pp. 30-37, May-June, 1983.
[13] Silicon Compilers, Inc., *GENESIL Documentation*. San Jose, CA 95125.
[14] E. Girczyc and J. Knight, "An ADA to standard cell hardware compiler based on graph grammars and scheduling," in *Proc. 1984 Int. Conf. on Computer Design—ICCD*, pp. 726-729, Oct. 1984.
[15] E. Girczyz, R. Buhr, and J. Knight, "Applicability of a subset of Ada as an algorithmic hardware description language for graph-based hardware compilation," *IEEE Trans. Computer-Aided Design*, vol. CAD-4, no. 2, pp. 134-142, Apr. 1985.
[16] E. Girczyc, "Loop-winding—A data flow approach to functional pipelining," Tech. Rep., Univ. of Alberta, Edmonton, Alta.
[17] A. V. Goldberg *et al.*, "Approaches toward silicon compilation," *IEEE Circuits and Devices Mag.*, vol. 1, no. 3, pp. 29-39, May 1985.
[18] J. Granacki, D. Knapp, and A. Parker, "The ADAM Advanced Design AutoMation System: Overview, planner and natural language interface," in *Proc. 22nd Design Automation Conf.* (ACM and IEEE), June 1985.
[19] L. Hafer and A. Parker, "A formal method for the specification analysis, and design of register-transfer level digital logic," *IEEE Trans. Computer-Aided Design*, vol. CAD-2, no. 1, pp. 4-18, Jan. 1983.
[20] C. Y. Hitchcock and D. E. Thomas, "A method of automatic data path synthesis," in *Proc. 20th Design Automation Conf.*, pp. 484-489 (ACM and IEEE), June 1983.
[21] D. Johannsen, "Bristle BLocks: A silicon compiler," in *Proc. 16th Design Automation Conf.*, pp. 310-313, (ACM and IEEE), June 1979.
[22] D. Knapp and A. Parker, "A unified representation for design information," in *Proc. IFIP Conf. on Hardware Description Languages* (IFIP), Aug. 1985.
[23] —, "A design utility manager: The ADAM planning engine," in *Proc. 23rd Design Automation Conf.* (IEEE), 1986.
[24] T. J. Kowalski and D. E. Thomas, "The VLSI design automation assistant: Prototype system," in *Proc. 20th Design Automation Conf.* (IEEE), 1983.
[25] T. J. Kowalski, "The VLSI design automation assistant: A knowledge-based expert system," Ph.D. dissertation Carnegie-Mellon Univ., Pittsburgh, PA, Apr. 1984.
[26] F. Kurdahi and A. Parker, "PLEST: A program for area estimation of VLSI integrated circuits," in *Proc. 23rd Design Automation Conf.*, pp. 467-473 (IEEE and ACM), June 1986.
[27] M. McFarland, "Computer-aided partitioning of behavioral hardware," in *Proc. 20th Design Automation Conf.*, pp. 472-478 (ACM and IEEE), June, 1983.
[28] —, "Using bottom-up design techniques in the synthesis of digital hardware from abstract behavior descriptions, in *Proc. 23rd Design Automation Conf.*, pp. 474-480 (IEEE and ACM), June, 1986.
[29] —, "Allocating registers, processors, and connections," Carnegie-Mellon Univ., Pittsburgh, PA, Internal Rep., 1981.
[30] A. Nagle, R. Cloutier, and A. Parker, "Synthesis of hardware for the control of digital systems," *IEEE Trans. Computer-Aided Design*, vol. CAD-1, no. 4, pp. 201-212, 1982.
[31] N. Park, "Synthesis of high-speed digital systems," Ph.D. dissertation, Dept. of Electrical Engineering, University of Southern California, Los Angeles, CA, Sept. 1985.
[32] N. Park and A. Parker, "Sehwa: A program for synthesis of pipelines,"in *Proc. 23rd Design Automation Conf.*, pp. 454-460 (IEEE and ACM), June 1986.
[33] A. C. Parker, "Automated synthesis of digital systems," *IEEE Des. Test Comput.*, vol. 1, pp. 75-81, Nov. 1984.
[34] A. C. Parker, J. Pizarro, and M. Mlinar, "MAHA: A program for datapath synthesis," in *Proc. 23rd Design Automation Conf.*, pp. 461-466 (IEEE and ACM), June 1986.
[35] Rutgers Researchers, "Rutgers artificial intelligence VLSI CAD session," in *Proc. 21st Design Automation Conf.* (ACM and IEEE), June 1984.
[36] E. D. Sacerdoti, *A Structure for Plans and Behavior*. New York, NY: Elsevier Sci. Publ., 1977.
[37] R. Schank *et al.*, "SAM—A story understander," Tech. Rep. 43, The Yale AI Project, Yale Univ., New Haven, CT, Aug. 1975.
[38] Silicon Design Labs., *GDT Documentation*. Liberty Corner, NJ.
[39] E. Snow, "Automation of module set independent register transfer level design," Ph.D. dissertation, Dept. of Electrical Engineering, Carnegie-Mellon University, Pittsburgh, PA, Apr. 1978.
[40] C.-J. Tseng and D. P. Siewiorek, "Emerald: A bus designer," in *21st Design Automation Conf. Proc.*, pp. 315-321 (IEEE DATC and ACM SIGDA), June 1984.

Alice C. Parker (Senior Member, IEEE) received the M.S.E.E. degree from Stanford University, Stanford, CA, and the Ph.D. degree from North Carolina University, Raleigh, in 1971 and 1975, respectively.

She is currently an Associate Professor in the Department of Electrical Engineering—Systems at the University of Southern California, Los Angeles. Prior to that, she was with Carnegie-Mellon University, Pittsburgh, PA. Her areas of interest are in automated synthesis of digital systems, hardware verification, hardware descriptive languages, silicon compilers, and use of AI techniques in synthesis.

Sally Hayati (Student Member, IEEE) received the B.S. and M.S. degrees from the University of California, Berkeley in 1974 and 1976, respectively, and the Masters degree in computer engineering from the University of Southern California, Los Angeles, in 1985.

Presently she is a Ph.D. student in the Department of Electrical Engineering—Systems at the University of Southern California. Her research interests are in the automatic synthesis of digital systems and AI techniques for design.

Knowledge-Based Decision Support in Business: Issues and a Solution

Vasant Dhar and Albert Croker
New York University

MIS/Finance

Editor's note: Technological development involves an interplay between user needs and tools for solving user problems. Since the computer era began, business applications have exhibited needs strongly influencing computer hardware and software system development. These needs are now influencing intelligent system development as well. The MIS/Finance track represents our response: Significant business applications for intelligent systems exist—applications that will benefit from and, in turn, shape future technical developments.

Many business applications will evolve from current work in management information and decision support systems. Natural development in these areas points to an integration of database, financial-modeling, expert-system, and natural-language-processing technologies. Some applications are potentially more revolutionary, however, leading to entirely new opportunities and changing the ways business is done.

Articles in last fall's IEEE Expert *(Vol. 2, No. 3), a special issue devoted to financial applications, illustrate the wide variety of problems that researchers are currently addressing. And those articles represent but a small fraction of current work. Beginning with Dhar and Croker's article in this issue, the MIS/Finance track will provide an ongoing view of progress in this important and vigorous domain.*

—Richard O. Duda
Associate Editor

Like expert knowledge, expert systems and their uses vary significantly across problem domains. While we can identify stable theoretical or experiential models in scientific domains, managerial problem solving and decision making are often based on an evolving organization-specific context. For such problem situations, we can use intelligent systems to maintain models rather than burdening users with this error-prone and cumbersome task.

This article describes an implemented system supporting the maintenance of evolving symbolic models and the spreadsheet-like algebraic models based on them. The system's primitives are knowledge fragments that are instantiated into symbolic models; these fragments can be modified in response to changes in the task environment, and appropriate changes induced in the algebraic model to reflect changes in the symbolic model. Such functionality is particularly useful for managerial problems such as resource planning, where a model (with projections and hypothetical reasoning based on it) evolves in response to changing external conditions or internal policies.

Over the last several years, computer-based modeling systems have made it relatively easy for end users to develop powerful decision support systems in many application areas. This has been facilitated by two features of modeling systems: First, such systems are nonprocedural to some extent, thereby freeing users from the rigid discipline associated with specifying problems in procedural languages. Second, they usually contain primitives that are useful in modeling many problems but that are cumbersome to program from scratch for each new application.

Regardless of these features, there is a growing recognition that unless such systems are augmented with representational frameworks and inference mechanisms that take explicit cognizance of the intellectual component of managerial decision making, their utility as decision aids is limited. In parallel efforts in the field of AI, researchers have been concerned with similar issues, although in problem areas that would probably be regarded as more "structured" than those encountered in management. Some of the programs that have resulted from this research, commonly referred to as "expert systems," have received considerable attention because of their ability to engage in judgmental reasoning similar to that of domain experts and exhibit comparable levels of performance.

Can we build similar systems to support decision making in the management arena where problems tend to be open-ended, nonrepetitive, and not amenable to analytical solutions? To answer this question, we will first contrast the knowledge involved in business problems with problems where knowledge-based system development efforts have heretofore been directed, and the different roles for intelligent systems these problems encourage. We will then describe an architecture providing the functionality needed to support a commonly occurring problem in business organizations; namely, that of formulating and maintaining plans for resource allocation and other purposes. Abstractly, we can view the planning problem as one of generating and maintaining varying assumptions or alternatives in light of varying constraints.

Expert knowledge, expert systems and their varying uses

The type of knowledge required to solve a problem is influenced by the degree to which the task has been formalized. As domains become better understood, formal theories or normative models become established. These provide bases for understanding and solving problems within a domain. In the absence of such formalization, problem solving is likely to depend on informal, intuitive, or transient models.

In this section, we analyze model features for problem solving in domains lying at three different points of the "structuredness" spectrum: highly formalized problems where normative models used for problem solving are based on a stable theoretical body of knowledge, less structured problems where models are based on experientially derived knowledge, and managerial problems where models used in problem solving are themselves based on evolutionary knowledge.

Stable theoretical models form the basis for normative models in physical domains. Psychological researchers have frequently studied human problem solving, mostly in well-structured problem domains. Broadly stated, problems studied have either involved commonsense reasoning pertaining to everyday physical phenomena[1-2] (sometimes referred to as "naive physics"[3]) or specialized knowledge from highly formalized domains such as physics or algebra.[4]

Although humans are generally competent with naive physics, few can solve complex physics problems. Several problem-solving studies have contrasted expert and novice behavior, and have frequently found that the *nature of representation* adopted has influenced the quality and speed of solution. In turn, the quality of the representation depends on how closely it reflects general domain principles relevant to solving the problem. Expertise involves *classifying* problem descriptions into appropriate theoretical categorizations or principle-oriented schemata, and once correctly classified, the use of axiomatic knowledge to solve problems in a primarily top-down manner. For textbook problems, classification can occur rapidly—sometimes after reading a small fraction of the problem.[5] For more complex real-world problems, categorization can involve extensive problem solving.[1]

Intelligent systems in physical domains can be useful for instruction and structured problem solving. Computer-based support systems in physical domains provide a major potential use as intelligent tutoring systems that employ a normative knowledge component reflecting domain principles to be taught, and a component modeling naive student concepts about the domain. We can use these two models in concert to transfer a principled body of knowledge to novices. Several experimental systems have been built along these lines for symbolic integration, electronic troubleshooting,[6] axiomatically based mathematics,[7] probability theory,[8] and a consultative system for Macsyma.[9]

More recently, several efforts have been directed towards building knowledge problem solvers for the more structured parts of engineering design problems. Theoretically, we can view such systems as incorporating the expert component of instructional systems.

Experiential knowledge forms the basis for normative models in interpretive domains. Contrasted with the problems described above—where expertise involves mapping problem descriptions onto principle-oriented categorizations—many problems require *interpretation* of problem features comprising the problem description. This requires imposing a structure on the problem description—necessary because data comprising the description can be uncertain, ambiguous, or fragmentary. We refer to such domains as interpretive domains.

Expertise in interpretive problem domains tends to be based on experientially derived associations among data. To clarify problems when problem solving, experts must judge the reliability of facts and acquire additional evidence to discriminate among competing concepts. For example, physicians often employ heuristic problem-focusing strategies to limit decision (disease) alternatives actually considered and evaluated. In complicated cases, this strategy becomes even more important since physicians frequently must come to more than one diagnostic decision for a given case.[10] Similar observations have been made in other problem areas where "noisy" data (coupled with an inherently large search space) require the use of intelligent heuristics, typically refined through experience, to impose pragmatic constraints on complex, open-ended problems.

Intelligent systems in interpretive domains can be useful for consultation. Most expert system research has led to the development of large knowledge-based systems in problem areas that require consultative decision support for solving difficult problems. In the medical domain, for example, Pople points out that occasions often exist where physicians incorrectly interpret facts and arrive at incorrect diagnoses because of tunnel vision—a phenomenon caused by the large numbers of symptoms (data items) to be interpreted and combined into a diagnostic conclusion.[10] In effect, even the expert responsible for knowledge contained in the knowledge base may need decision support due to the sheer volume of data requiring interpretation.

In general, consultative systems can be useful in domains where experts must interpret and analyze vast quantities of data. Major efforts to date have been in geological exploration,[11] analysis of oil well logs,[12] mass spectroscopy interpretation,[13] computer layout,[14] and medicine (Casnet[15] specializes in glaucoma assessment and therapy, MYCIN[16] in antimicrobial therapy, while Caduceus[10] deals with the whole of internal medicine). Central to these systems is a relatively stable body of knowledge. Once validated, this knowledge serves the role of a theory, moving the problem closer to the structured-problems category. Apart from their pragmatic use as consultants, in fact, some researchers regard the role of intelligent systems in such domains as being an epistemological one—that is, as facilitating the development of domain theory.

Evolving symbolic knowledge often forms the basis for models in managerial domains. Much modeling in business organizations is geared toward models that support *decision making,* which can involve processing diverse data. Large-scale simulation models in the 60s and 70s were early attempts at building computer models for managerial decision making.[17] These efforts constructed detailed mathematical models of organizations for modeling complex problems in which closed-form solutions were not possible. Such systems made it possible to evaluate in concrete terms the impact of alternative policies, opportunities, and external events (all implemented as parameters of the simulation model). However, such systems were time consuming to build and extremely difficult to modify in light of changes in the task environment—major drawbacks and major reasons for their failure.

While current-day software systems provide powerful interactive modeling environments for rapid problem structuring, their functionality is similar to the simulation models of the 60s and 70s. Essentially, building a decision support model involves explicit consideration of diverse types of knowledge comprising a problem situation (typically, symbolic and quan-

Table 1. Hardware (HW).

Name	Cost (K)	MIPS	Memory (M bytes)	Annual maintenance
DEC	300	3	8	low
IBM	500	5	8	high
CDC	350	5	8	high
ATT	100	2	4	medium

Table 2. Air-conditioning units (AC).

Name	Cost (K)	Power (kW)	Space (K sqft)
Borg	20	200	2
Westinghouse	35	500	3
GE	30	300	3

Table 3. Operating systems (OS).

Name	Cost (K)	Memory required (M bytes)
UNIX	10	1.5
VMS	20	2.0
CMS	40	2.0
TOPS	5	1.0

Table 4. Database management systems (DBMS).

Name	Cost (K)	Memory required (K bytes)	Number of records (K)
INGRES	7	200	200
IMS	15	350	1000
RIM	4	100	150
DMS	10	150	200

titative types) and the construction of a mathematical model based on such knowledge. We will now describe a simple planning scenario that involves diverse knowledge types for modeling purposes. Our scenario underscores the need for tightly coupled symbolic and mathematical models that are together capable of explicitly representing these different knowledge types.

To exemplify a decision-making situation where modeling can be useful, let us consider a planning problem that involves making and keeping track of alternative sets while considering various constraints (in a planning context, since alternatives refer to possible future choices, we can regard them as assumptions). Consider a capital investment scenario involving the setup of a computer-based corporate information system. Specifically, assume that alternative hardware vendors, air-conditioning units, operating systems, and database systems have been identified. We can characterize each assumption set in terms of *attributes*. Each set of alternative assumptions, referred to as a choice set, might have associated with it a "selection function" that is assumed to represent the basis for making one selection over another; this function is defined over a set of alternatives characterized in terms of certain attributes.

Two types of constraints exist, *quantitative* and *qualitative*. Quantitative constraints are inequalities defined over choice-set attributes. Qualitative constraints establish relationships among sets of alternatives. These different knowledge fragments form a *context* within which we can formulate different symbolic models, depending on the assumptions in force. In other words, a symbolic model is an *instantiation* of this context; that is, selections (or assumptions in a planning context) have been made in all parts of the task environment in light of the applicable constraints. Once the symbolic model is in place, we can solve equations corresponding to it.

An example. Let's consider a situation involving the following knowledge fragments. For simplicity, we limit the problem to the four choice sets described above.

Choice sets. We define four choice sets: hardware (see Table 1), air-conditioning units (see Table 2), operating systems (see Table 3), and database management systems (see Table 4). Each row defines a choice-set alternative. Each alternative is defined over the set of choice-set attributes specified by the column headings. A choice set may also have an evaluation function associated with it that defines an ordering in which alternatives are to be considered. An example of an evaluation function is "maximize MIPS/Cost" defined over the hardware choice set. We use the notation $\mathbf{CS_{attr}}$ to denote the **attr** attribute of the selection made from the choice set **CS**.

Quantitative constraints. In the following insert, the first constraint states that at least 6M bytes must be available apart from memory requirements of the operating system and the database system. The second constraint states that the total cost of the various information system components should not exceed $375,000.

$$HW_{memory} - OS_{memory} - DBMS_{memory} \geq 6.0 \quad (1)$$

$$HW_{cost} + AC_{cost} + OS_{cost} + (N * DBMS_{cost}) \leq 375{,}000 \quad (2)$$

where

N = the number of database system copies to be purchased.

Qualitative constraints. In the following insert, the third qualitative constraint (5) states that if DEC hardware and the UNIX operating system are selected, then INGRES must be selected as the database management system. Similarly, constraint (8)—from which one might infer that VMS is worth running only on fast machines—states that the VMS operating system should not be installed on hardware that is less than 5 MIPS.

$$HW_{Name=IBM} \rightarrow OS_{Name=CMS} \quad (3)$$
$$HW_{Name=IBM} \rightarrow AC_{power>400} \quad (4)$$
$$HW_{Name=DEC} \text{ and } OS_{Name=UNIX} \rightarrow DBMS_{Name=INGRES} \quad (5)$$
$$OS_{Name=IMS} \rightarrow HW_{Name=IBM} \quad (6)$$
$$DBMS_{Name=INGRES} \rightarrow \sim HW_{Name=CDC} \quad (7)$$
$$HW_{Mips<5} \rightarrow \sim OS_{Name=VMS} \quad (8)$$

Equations. In the following insert, TC represents total cost, S represents space required, and N is the number of copies of the database system:

$$TC = HW_{cost} + AC_{cost} + OS_{cost} + N * DBMS_{cost} \quad (9)$$

$$S = AC_{space} + 1000 \quad (10)$$

where TC and S are exogenous variables; that is, variables that do not occur as choice-set attributes.

Similarly, we can add other choice sets and constraints to the context. For example, heavy air conditioning requirements might call for certain building and wiring alterations—all of which must be planned. For large planning problems, the interdependencies among assumptions about different task environment parts can become too complex for decision makers to handle, particularly if numerous sources contribute knowledge underlying the models.

Intelligent management systems must support formulation and maintenance of models. Apart from the complexity involved in formulating planning models, a reality of business decision making situations is that the context and hence the symbolic model can change. In the small scenario described above, for example, a previously held assumption (say that DEC would be the hardware vendor) may be discarded in favor of one that was previously passed over. In such a situation, related assumptions are also affected, with concomitant changes becoming necessary in the symbolic model. However, while most modeling systems represent some of the quantitative aspects of the problem situation, the other types of knowledge described above are seldom if at all represented explicitly in such systems.* Because of this, the user must assume full responsibility of maintaining the correspondence between the knowledge that can be expressed within the model and that which cannot. For large-scale modeling, this can be a complex task.

If we recognize that it is actually the formulation/reformulation of models based on changing problem scenarios that is problematic for a manager and his support staff, it is this problem that must be supported by an intelligent assistant. Conceptually, this type of support can be achieved by endowing modeling systems with the symbolic knowledge about the context within which models were formulated.

From a design standpoint, we need structures and mechanisms that formulate and maintain symbolic and algebraic models based on the evolving organization-specific context. This emphasis on *synthesis* of symbolic and quantitative models and their *maintenance*—as opposed to *selection* from a predefined set of models—requires a computer-based architecture somewhat different from that of most modeling systems. Specifically, we must (1) maintain

*In principle, however, if the problem can be expressed as an optimization problem, these relationships can be modeled using integer programming. However, such formulations are not always possible, and even when they are, the model can be difficult to formulate and solve.

symbolic knowledge pertinent to a problem situation and its relationship to the quantitative model, and (2) orchestrate changes in these models to reflect changing scenarios.

Architecture

Various data structures are needed to represent the symbolic and quantitative knowledge involved in formulating models that support decision making. In this section, we describe a system permitting choice-set representation plus quantitative and qualitative constraints. We will define data structures used for representing choice sets and constraints and describe the system's procedural knowledge component, which allows knowledge represented in the structures to be used in an integrated way.

Choice sets. We represent each alternative as an instance of a structured-object data type consisting of fields that correspond to the alternative's attributes. In our implementation, we represent objects using Flavors primitives.[18] Conceptually, however, we can implement choice sets using any language mechanism that allows the representation of record sets or other structured objects.

A choice set C contains a set of alternatives $\{A_1, A_2, ..., A_n\}$ where all selections are characterized by the same set of attributes (called *choice-set attributes*). These attributes correspond to instance variables of objects. In the absence of other constraints, users can express preferences for alternatives in terms of the choice-set attributes of these alternatives. Choice sets are represented as structured objects of the form

choice-set = {**id:** *string*
  attributes: *array [1:n] of string*
  alternatives: *array [1:m, 1:n] of string or numeric*
  selection: *numeric* **/* in the range 0:m */**
  eval-func: *function* **/* to order alternatives */**
  }

where **id** is the name of the choice set and is followed by the list of *n* choice-set **attributes**. Each of the *m* rows in the two-dimensional array **alternatives** represents an alternative, and consists of one value for each of the *n* attributes. The choice set slot **selection** designates which of the *m* alternatives has been selected for the choice set (the value zero indicates that no selection has been made). The **evaluation function,** if not null, determines the order in which alternatives are explored for inclusion in an evolving plan.*

Qualitative constraints. Conceptually, qualitative constraints are similar to production rules. They establish dependency relationships among alternatives in different choice sets, and are expressible in the form

$$s_{i_1, j_1}, s_{i_2, j_2}, \ldots, s_{i_{n-1}, j_{n-1}}, \rightarrow s_{i_n, j_n}$$

where each term $s_{l,m}$, which represents the selection of alternative a_m from choice set C_l, can be preceded by an optional negation sign "∼" and no two alternatives in the constraint are in the same choice set.

A term $s_{l,m}$ in a qualitative constraint is said to be *satisfied* if it is not preceded by a negation sign and alternative a_m has been selected from choice set C_l, or if it is preceded by a negation sign and another alternative has been selected from C_l. A qualitative constraint is satisfied if each of its terms is satisfied.

A qualitative constraint is represented as a structured object of the form

qual-const-term = {**id:** *string*
  sl: *array [1:n_s] of term-sequence*
  nl: *array [1:n_n] of term-sequence*
  }

where each object represents one term in the qualitative constraint. In this representation, **id** uniquely identifies an alternative over which a qualitative constraint is defined and **sl** is an array of n_s term sequences, where each term sequence corresponds to the left-hand side of a qualitative constraint that contains the non-negated right-hand side identified by **id**. Similarly, **nl** is an array of term sequences, one corresponding to each left-hand side of a qualitative constraint containing the negated right-hand side identified by **id**. The **nl** and **sl** slots serve a role similar to Doyle's *inlist* and *outlist* parts of support-list justifications in his truth maintenance system.[19]

Quantitative constraints and equations. Quantitative constraints establish relationships involving

*If the evaluation function is null, an ordering is established based on a heuristic evaluation function. The details of this function are not relevant to the discussion in this article.

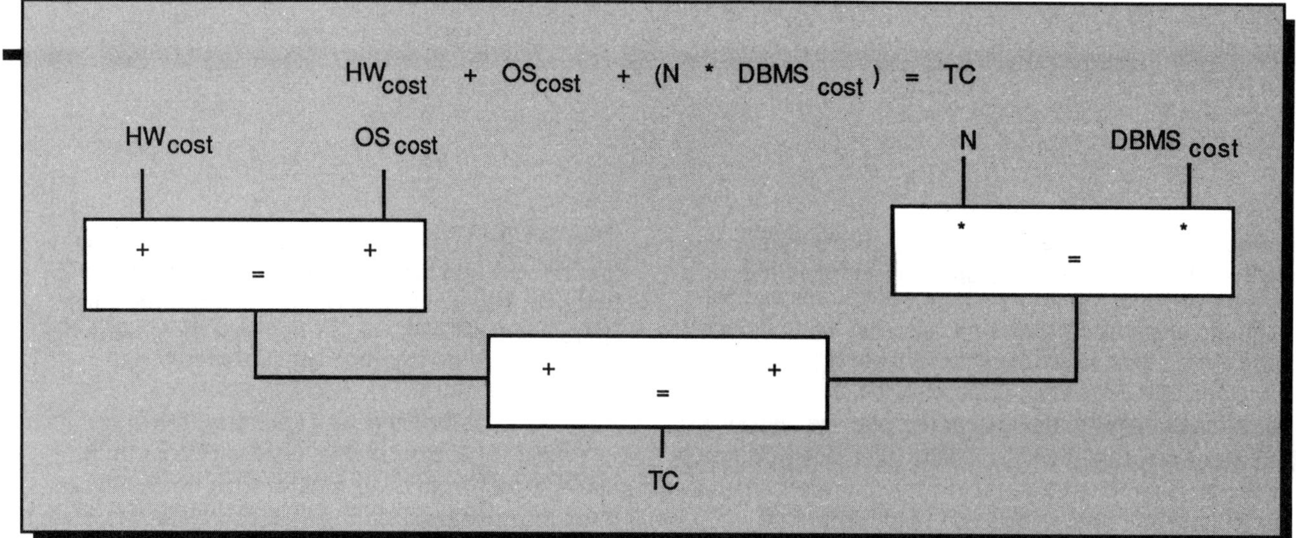

Figure 1. A numeric constraint net segment.

choice-set attributes, constants, and exogenous variables. For example,

$$HW_{cost} + OS_{cost} + AC_{cost} + DBMS_{cost} \leq 375{,}000$$

is a constraint establishing a "global" resource constraint that affects choices only indirectly. It establishes a fixed relationship among the attributes of selections that can be made from choice sets. Specifically, a constraint is a Boolean expression over choice-set attributes and constants.

In contrast, an equation necessarily involves exogenous variables. For example, the relationship

$$HW_{cost} + OS_{cost} + AC_{cost} + DBMS_{cost} = TC$$

involves the exogenous variable TC.

We represent constraints as Boolean functions that return the value *false* when the corresponding constraint is violated, and *true* otherwise. A closed constraint can be violated only after values for all of the choice-set attributes that define it are known. An attribute value is known once a selection for the corresponding choice set has been made. In our implementation, we represent closed constraints as Lisp forms that are evaluated when the needed attribute values become known.

We represent equations in the form of numeric constraint-net structures consisting of nodes, each representing a binary operator. Each node has three edges, two of which represent the operator's operands. The third is the result obtained by applying the operator. Operands can be choice-set attributes, constants, exogenous variables, or output from another node. The result can be an exogenous variable or an output to another node.

The numeric constraint net structure in Figure 1 is based on the one described by Winston.[20] Instantiating values for any two edges of a node in the network results in the third being computed. This structure allows for the propagation of values in either direction within the network. The values associated with edges can change in three ways: as a consequence of changes in the set of selections, the user changing the values of exogenous variables, or as a result of values being propagated from other parts of the network.

Search and backtracking. In addition to the structural knowledge components described above, the system has a procedural knowledge component that is used to make selections, maintain consistency among selections (that is, ensure that constraints are not violated), and propagate values through the constraint network.

The modeling system alternates between two modes of operation which, following Stefik, we term *constrained* and *heuristic* modes.[21] In the constrained mode, qualitative constraints determine selections from specified choice sets. The heuristic mode becomes operative when propagation of selections comes to a halt. In this situation, a selection must be made from some choice set in order for problem solving to continue. The selection can be specified directly by the user, or made automatically by the system based on an evaluation function defined over the choice-set attributes. The order in which choice sets are examined is based on a prioritization of the choice sets that reflects the user's view of the importance associated with the decisions corresponding to them.

If a constraint violation occurs while the system is in the constrained mode, it becomes necessary to revise a previously made selection in some choice set, and then to continue forward from that point. Specifically, all selections associated with the heuristic mode selection corresponding to the lowest priority choice set are discarded. A revised selection is then made from that choice set, and plan synthesis continues.

If a constraint violation occurs while in the heuristic mode as a consequence of a selection made from a choice set, a new selection is made from that choice set if possible. However, if all selections have been tried unsuccessfully, then, as in the case above, all selections associated with the heuristic mode selection corresponding to the lowest priority choice set are discarded and a revised selection is made from that choice set.

Once selections have been made from the choice sets (that is, values have been determined for the choice set attributes in the open constraints), the user can specify values for certain exogenous variables, while others might be computed automatically by propagation of values through the constraint net. If the user specifies values for all of the exogenous variables in an equation, it becomes a constraint. If this constraint is violated with the current set of selections, backtracking is invoked to make a revised set of selections to satisfy the new constraint.

Functionality for decision support. The system we have described has resulted from a collaborative effort with a large computer manufacturing company where managers expressed the need for a level of functionality that is not available in existing modeling systems. These systems require the manager to translate an inherently flexible situation involving choice points and various types of constraints into a set of equations, which are generally not capable of completely representing the problem situation. In these systems, what-if analyses are limited to parametric variations of an algebraic model. What a planner often needs is the ability to assess the repercussions of alternative assumptions (selections) and changes in the scenario of constraints since such changes occur frequently in business problems. To be able to provide the support needed to deal with such changes, a system must represent explicitly the types of knowledge we have described in this article.

The range of functionality provided by our system is summarized below. For purposes of illustration, we make use of the simplified problem situation described earlier.

(1) Parametric sensitivity analysis: That is, "what if N changes from 20 to 30?" This requires propagating changes through an algebraic constraint (like that in Figure 1).

(2) Change criteria for alternative evaluation: That is, "what if the criterion for making hardware choices becomes *minimize cost* instead of *maximize MIPS * Memory*?" This can change the hardware selection and, through qualitative constraints, directly affect choices in other parts of the model. Choices can also be affected indirectly through quantitative constraints.

(3) Change the value of a choice-set attribute: That is, "what if the cost of DEC hardware goes up to 400?" Apart from being a parametric variation that must be propagated through the constraint net, such a change can cause fundamental changes of choice in various model parts. First, depending on the hardware choice set's evaluation function, the selection in that set might change. Second, resulting from this change, other selections might be affected depending on the constraints. This can result in a radically altered set of choices and a differently instantiated spreadsheet model.

(4) Relax or tighten constraints: That is, "IBM hardware does not require IBM software." This "loosening up" of constraints can result in a different selection in the operating system and related parts of the model. Loosening up or removing quantitative constraints can have similar consequences. Tightening constraints (that is, "IBM hardware must use IBM software") can cause constraint violations in an existing model, requiring backtracking as described in the previous section. This can result in revised choices or an insatiable constraint scenario.

(5) User-specified choices: That is, "choose INGRES as the database system regardless of evaluation functions and constraints." This functionality essentially allows the user to specify the context within which the system must synthesize the model.

In addition to functionalities described above, the system also enables users to define new choices and constraints. In effect, it is an "open system," enabling easy modifications to its knowledge base. This functionality is important in business modeling where situations on which we base models are continually subject to change.

Discussion

Designing problem solvers based on the constraint propagation concept has its theoretical underpinnings in the works of Guzman,[22] Waltz,[23] Mackworth,[24] and Stallman and Sussman.[25] These works deal with issues relating primarily to symbolic constraint propagation.

We find a constraint propagation application in Steele and Sussman, who illustrated its use for electrical circuit analysis.[26] More recently, Stefik designed planning models for the design of molecular genetic experiments in terms of symbolic constraint propagation.[21] Dhar and Pople described a constraint-based reasoning system (Planet) designed to support planning in a computer-manufacturing organization.[27] The system we have described represents an extension and generalization of the Planet system in that it applies to any problem domain involving model synthesis and maintenance as long as domain knowledge can be expressed in terms of primitives we have designed. This general model allows for reasoning based on quantitative and qualitative constraints that prevail in many business modeling situations. Specifically, we have designed structures that maintain knowledge about alternatives generated in decision-making situations, knowledge about quantitative and qualitative constraint relationships inherent to problem situations, and a constraint network representing equality relationships. Our system's constraint network component corresponds to models typically represented in many current-day modeling systems. In addition to structural components, we have defined mechanisms that aid in making selections. This ensures consistency between selected alternatives and user-defined constraints, and propagates numeric values through constraint networks.

Although many decision-making situations are characterized by consideration of alternatives and the existence of constraint types we have described, modeling systems do not allow for explicitly encoding and integrating such knowledge into the model. Hence, users must ensure that the component of the problem modeled jibes with overall constraints pertinent to the problem situation. In contrast, our system plays a more active role in the decision-making process by assuming responsibility for making selections among the alternatives available in a decision-making situation, and enforcing quantitative and qualitative constraints. Because it can maintain this knowledge, our system also provides more sophisticated what-if reasoning. First, what-if analysis involving a change in the value of a selected alternative's choice-set attribute can result not only in a simple propagation of the new value through the constraint network, but in revised selections being made from choice sets if any constraint violation occurs. Second, the ability to specify a value or range of values for an exogenous variable can transform an equation into a constraint. If the current selections violate the resulting closed constraint, then we initiate further problem solving to revise the selections, thereby satisfying the modified set of constraints.

We have designed and implemented a powerful modeling environment for representing and reasoning about a variety of constraints. This environment is surprisingly powerful in its ability to model a range of situations that confront planning managers. In contrast to most modeling systems where decision makers are forced to maintain the correspondence between the data that can be expressed within the model and that which cannot, our system reasons about several types of symbolic and quantitative knowledge within a uniform framework. Second, for group modeling situations, it is possible that the system can enhance communication and coordination across the multiple parties involved because of its functionality to make explicit the impacts of decision changes in one part of the task environment on other parts. Similarly, it can highlight to a high-level manager or coordinator the repercussions of local changes in selections on the overall model. The ability to bring into focus previously foreclosed options provides the manager a window into the inherently underconstrained problem situation out of which specific models (instantiations) are synthesized for decision making, estimating resource requirements, or other purposes.

This view of models for decision support contrasts sharply with other domains where considerable effort may be expended by a knowledge engineer in uncovering the correct model. In contrast, we regard the model as a tentative structure that must evolve in response to changing conditions. Our assertion on model maintenance as being the core problem to support has been shaped largely by observations of modeling efforts in the financial services and manufacturing industries. This evidence, although preliminary, has led us to believe that a central concern of decision makers is one of preserving the faithfulness of models under changing conditions, particularly when the projections generated from such models involve large amounts of resources. If we recognize that much of the use of models in business organizations occurs in changing scenarios, it is necessary to have systems that synthesize models that reflect the changing reality. This problem of "getting the model right" and maintaining it in light of changing conditions is an important responsibility of a manager or his support staff that can benefit greatly from knowledge-based support. The model formulation and synthesis system described here has been designed to undertake some of this responsibility in order to effect a more balanced division of responsibility between the decision maker and the system.

References

1. K. Forbus, "Qualitative Reasoning about Space and Motion," in *Intelligent Tutoring Systems,* D. Sleeman and J.S. Brown, eds., Academic Press, New York, N.Y., 1983.
2. J. de Kleer and J.S. Brown, "Assumptions and Ambiguities in Mechanistic Mental Models," in *Intelligent Tutoring Systems,* D. Sleeman and J.S. Brown, eds., Academic Press, New York, N.Y., 1983.
3. P. Hayes, "Naive Physics I—Ontology for Liquids," working paper, Dept. of Computer Science, University of Essex, Essex, UK, 1975.
4. A. Bundy and L. Bird, "Using the Method of Fibres in Mecho to Calculate Radii of Gyration," in *Intelligent Tutoring Systems,* D. Sleeman and J.S. Brown, eds., Academic Press, New York, N.Y., 1983.
5. D.A. Hinsley, J.R. Hayes, and H.A. Simon, "From Words to Equations: Meaning and Representation in Algebra Word Problems," in *Cognitive Processes in Comprehension,* P.A. Carpenter and M.A. Just, eds., Lawrence Erlbaum Associates, Hillsdale, N.J., 1977.
6. J.S. Brown, R. Burton, and J. de Kleer, "Pedagogical, Natural Language, and Knowledge Engineering Techniques in Sophie I, II, and III," in *Intelligent Tutoring Systems,* D. Sleeman and J.S. Brown, eds., Academic Press, New York, N.Y., 1983.
7. R.L. Smith et al., "Computer-Assisted Axiomatic Mathematics: Informal Rigor," in *Computers in Education,* O. Lecareme and R. Lewis, eds., North-Holland/Elsevier, New York, N.Y.,1975.
8. A. Barzilay, *Spirit: An Intelligent Tutoring System for Probability Theory,* PhD dissertation, University of Pittsburgh, Pittsburgh, Pa. 15260, 1984.
9. M.R. Genereseth, "The Role of Plans in Intelligent Tutoring Systems," in *Intelligent Tutoring Systems,* D. Sleeman and J.S. Brown, eds., Academic Press, New York, N.Y., 1983.
10. H.E. Pople, "Heuristic Methods for Imposing Structure on Ill-Structured Problems: The Structuring of Medical Diagnostics," *Artificial Intelligence in Medicine,* P. Szolovits, ed., Westview Press, Boulder, Colorado, 1982.
11. R.O. Duda, J. Gashnig, and P.A. Hart, "Model Design in the Prospector Consultant System for Mineral Exploration," in *Expert Systems in the Micro-electronic Age,* D. Michie, ed., Edinburgh Press, Edinburgh, Scotland, 1979.
12. R. Davis et al., "The Dipmeter Advisor: Interpretation of Geological Signals," *Proc. Seventh Int'l Joint Conf. Artificial Intelligence,* Morgan Kaufmann, Los Altos, Calif., 1981.
13. R. Lindsay et al., *Applications of Artificial Intelligence for Chemical Inference: The Dendral Project,* McGraw-Hill, New York, N.Y., 1980.
14. J. McDermott, "R1: A Rule-Based Configurer of Computer Systems," *Artificial Intelligence,* Vol. 19, No. 1, 1982.
15. S. Weiss et al., "A Model-Based Method for Computer-Aided Medical Decision Making," *Artificial Intelligence,* Vol. 11, Nos. 1 and 2, 1978.
16. E. Shortliffe, *MYCIN: Computer-Based Medical Consultation,* Elsevier, New York, N.Y., 1976.
17. R. Shannon, *Systems Simulation: The Art and Science,* Prentice-Hall, Englewood Cliffs, N.J., 1975.
18. D. Moon and D. Weinreb, *Lisp Machine Manual,* MIT, Cambridge, Mass. 02139, 1981.
19. J. Doyle, "A Truth Maintenance System," Memo 521, AI Laboratory, MIT, Cambridge, Mass. 02139, 1978.
20. P. Winston, *Artificial Intelligence,* Addison-Wesley, Reading, Mass., 1984.
21. M. Stefik, *Planning With Constraints,* PhD dissertation, Dept. of Computer Science, Stanford University, Stanford, Calif. 94305, 1980.
22. A. Guzman, *Computer Recognition of Three-Dimensional Objects in a Visual Scene,* PhD dissertation, MIT, Cambridge, Mass. 02139, 1968.
23. D. Waltz, *Understanding Line Drawings of Scenes with Shadows,* PhD dissertation, MIT, Cambridge, Mass. 02139, 1972.
24. A. Mackworth, "Consistency in Networks of Relations," *Artificial Intelligence,* Vol. 8, No. 1, 1977.
25. R. Stallman and G. Sussman, "Forward Reasoning and Dependency-Directed Backtracking in a System for Computer-Aided Circuit Analysis," *Artificial Intelligence,* Vol. 9, No. 2, 1977, pp.135-196.
26. G. Steele and G.J. Sussman, "Constraints," Memo 502, AI Laboratory, MIT, Cambridge, Mass. 02139, Nov. 1978.
27. V. Dhar and H.E. Pople, "Rule-Based Versus Structure-Based Models for Explaining and Generating Expert Behavior," *Comm. ACM,* Vol. 30, No. 6, 1987, pp. 542-555.

Vasant Dhar—a full-time faculty member at New York University since 1983—is an assistant professor of information systems at the graduate school of business administration. His research interests focus on empirical and theoretical aspects of AI—including empirical investigation of problem-solving processes in domains involving design, planning, decision making, and the design of representational formalisms needed to build intelligent systems in these domains. He has written numerous articles on knowledge representation, heuristic search, and methodological issues in the development of knowledge-based systems.

Albert Croker is an assistant professor of information systems at New York University's graduate school of business. His research interests include database concurrency control theory, relational database theory, database semantics, and historical databases. He is currently involved in a research project on constraint-based reasoning systems. He received his PhD in computer science from the State University of New York at Stony Brook.

The authors can be reached at New York University, 624 Tisch Hall, Washington Square, New York, NY 10003.

PlanPower

A Comprehensive Financial Planner

James L. Stansfield and Norton R. Greenfeld

Applied Expert Systems

In the last two decades, America's financial environment has undergone tremendous change. Periods of unprecedented inflation, the decreasing regulation of the financial industry, the evolution of ever-more-complex financial products, and the marketplace's growing sophistication are all forces causing consumer and industry behavior to change. Such change has confused consumers regarding their financial affairs. Many financial products have appeared, and each seems more complicated than the last. New players in the financial services industry provide more places to buy these products. Even the 1986 "tax simplification" law is barely comprehensible to most US taxpayers.

This developing complexity has motivated the financial planning profession over the last 15 years. Financial planners—advisors who study a consumer's financial state and then guide the consumer toward best long-term advantages—should be aware of our complex financial environment, and should match consumer needs and desires with prudent action choices.

Typically, financial planners produce plans for affluent households—charging between $3,000 and $10,000, depending on the situation's complexity. Their comprehensive financial plans cover a household's current financial state and recommend positions in the following areas: cash management, tax

Photograph by Eileen Rassi Ayling

planning, investment, risk management, borrowing, retirement, special-needs funding, and estate planning. Good financial plans provide coordinated recommendations covering actions to be taken in these areas over time—all in the context of client specifics such as age, financial status, anticipated needs, and financial attitudes.

While expert systems have begun to show potential in a broad range of business applications,[1] developing a system from prototype to useful product has proven a major obstacle. APEX—Applied Expert Systems, Inc.—attempted to overcome this obstacle with its PlanPower research. PlanPower contains an operational expert system, used by financial planners, that produces high-quality and comprehensive financial plans. This article describes PlanPower's developmental history and its overall expert system architecture.

PlanPower's history and methodology

The original development team began work on the PlanPower project in 1982, defining the problem and initiating market research studies. The notion that expert systems technology might be applicable led them to build a small prototype—a simple production rule system written in Franz Lisp on a VAX 11/780 that took about four months for one programmer/knowledge engineer and one part-time expert to develop. They used this prototype to identify technical problem areas and to estimate the size and cost of building a full-scale operational system, concluding that the market was large enough and the technology appropriate. They also felt that, to support commercial applications, the technology needed to evolve. They considered tax-related issues the largest knowledge area, estimating that it would contain about 40 percent of the system's knowledge.

Applied Expert Systems (APEX) was incorporated in late 1982 and started operations at the beginning of 1983. It had a development partner—a firm that saw the potential of expert systems and wanted to be first in the market with such a financial planning tool. This customer provided further market input and system feedback throughout the development cycle, and greatly influenced decisions about what knowledge areas were included and how deeply each was covered.

From the beginning, we at APEX had a strategy about the development team's composition. We clearly separated knowledge engineers from "systems" programmers. Knowledge engineers focused on the financial domain and tended to have financial backgrounds. Some began with limited programming experience; others were well versed in traditional data processing. This mix ensured good communications between knowledge engineers, financial experts, and the systems group.

Systems programmers focused on underlying technology and had backgrounds either in AI or in traditional data processing. APEX mixed these cultures carefully, seeking a methodology for commercial expert system development. Throughout this development, tensions existed between good data-processing methodology (requirements, specifications, design, implementation, and testing) and good expert systems methodology (an evolving sequence of prototypes). Expert systems methodology dominated the early stages because the task proved complex and difficult. Knowledge engineers and systems programmers worked together during this period to develop a knowledge architecture and an appropriate development platform. As these became clearer, the two groups worked more independently.

The knowledge engineering team, with a primary expert, devoted early 1983 to learning more about financial planning. By April, they understood enough to attempt a second version—a somewhat larger prototype built in EMYCIN,[2] licensed from Stanford, and ported to the Xerox 1100. This effort required two knowledge engineers, two systems people, and about a month. Primarily, it explored production rule technology as a basis for the system. We concluded that EMYCIN's technology was inadequate, particularly in control structure and data representation areas.

We obtained an early Loops[3] version from Xerox and built our third prototype in that system—an effort taking until July and involving four knowledge engineers, four systems people, and three experts. This prototype resolved issues regarding what experts we needed and where the problem was most difficult. We discovered that tax was merely a subproblem of investment decision making. We also concluded that the system required multiple experts. Generalist financial planners could cover overall strategies and knowledge integration, but we needed specialists (an estate planner, for example) to get depth. We discovered that financial planners often work in such mixed groups.

We attempted a knowledge base specification at this point, which was relatively useful although we couldn't find a succinct way to express everything knowledge engineers knew about the depth of knowledge in each domain area the system had to cover. While it guided the next few months' work in a

PlanPower must continually examine the entire picture during planning.

general sense only, the knowledge base specification communicated to our customers what the system would generally know.

We completed the fourth prototype—a more ambitious effort aiming at both the knowledge base and the total system picture (including the user interface)—by the end of 1983. And we finally had real proof of feasibility: The problem was circumscribed, a solution structure was in place, some knowledge engineering was done for major knowledge areas, overall system architecture was known, and a first-pass user interface had been implemented. This prototype took eight knowledge engineers, six systems people, four experts, and ran on top of Loops on Xerox 1108s. While significant changes would still be made to the knowledge base architecture and the platform, the project had changed from experiment to product completion.

We started implementing the production version early in 1984. The APEX team had grown to 10 knowledge engineers, 12 systems people, and six experts. To ensure the consistency of knowledge obtained from different experts, several experts met as a panel for overall review. Management style changed from the loose prototype-and-see-where-we-get approach to a planned-milestone method. We generated and maintained specifications at both system and knowledge base levels.

By the spring of 1984, a knowledge base information-flow diagram covered one office wall. It became apparent by mid-year that we had built (1) an interface software layer, (2) our own rule-based language between Loops and the knowledge base code, and that (3) we were no longer using Loops directly. In effect, we had built a platform specific to our needs on top of Loops. New features had been added, some features had been adapted, and others were not used at all.

We felt that a great deal of overhead existed in this extra layer, so the systems team extended our own platform downward to Lisp—an implementation tailored to just what the problem demanded—and we were able to make optimizations when full generality wasn't needed. Ultimately, we obtained several-factor gains in both speed and space. We feel that—while using Loops sped development by a year—our product required a custom platform. Since that time, of course, commercial expert system shells have appeared and continuously improved.

The system went into field test in May, 1985—and testing took precedence from then on. While system-level testing was relatively straightforward, quality control of the knowledge base remains difficult for want of a methodology ensuring adequate coverage of the possibilities. We treat knowledge base testing as an art relying on carefully chosen test cases and detailed review by financial planning experts. Consequently, knowledge engineers must be closely involved in developing test cases and predicting the impact of "fixes" to the knowledge base.

One general heuristic—examining boundary conditions or extreme cases—proved useful; for example, such examination uncovered a serious bug when the client was running a deficit. The risk module "realized" that dependents would be left with insufficient income, and recommended buying far more life insurance than the client could afford. Of course, this recommendation aggravated the deficit. The correct solution addresses the deficit first.

Xerox announced the 1186 Lisp machine in August, 1985. APEX formally introduced and demonstrated PlanPower on the Lisp system at a financial planners' conference in September, 1985. Volume shipments commenced in April, 1986. Thus, PlanPower spent a full year in field test—all of it needed to achieve commercial robustness. The system is now in a maintenance-and-enhancement mode, with releases occurring every three to six months.

The resulting system

Users can run PlanPower in a fully automated manner to produce financial plans, or in an interactive mode as an expert support system. The interactive mode requires a full complement of standard workstation software (data entry forms, database and query systems, spreadsheet and word processing software, what-if modeling, and so forth). All of these were built in Lisp and integrated into a standard user interface. PlanPower uses Xerox's Tedit word processor, but we replaced Tedit's user interface to ensure system consistency.

We base client data for system input on an 80-page questionnaire; however, not all 80 pages need be completed since many values can be defaulted and inapplicable sections can be skipped. A shorter summary questionnaire can be substituted. Users enter information through a forms system that uses the workstation's graphics capabilities and provides selection

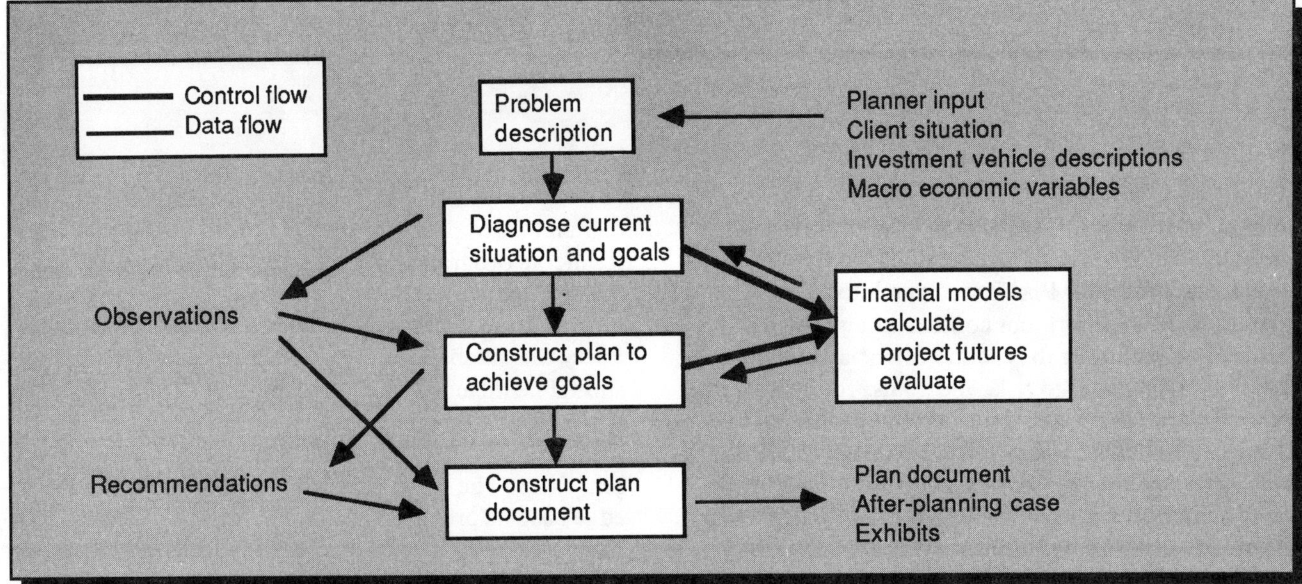

Figure 1. The expert's high-level architecture.

menus, default calculations, and error checking to expedite the process. We store the input in a data file; users can convert this into a knowledge base of frame instances by running a program called the Casemaker.

PlanPower provides clients with financial plans ranging from 20 to 120 pages, plus up to 40 tabular and graphics data exhibits. The expert system also creates an after-planning case representing the client's situation after all recommendations have been simulated. Planners can examine these by viewing exhibits and can experiment by simulating further recommendations of their own.

The system also allows planners to influence the expert's recommendations in various ways: Specific recommendations like using a certain asset as collateral for a loan may be prohibited; objectives such as target percentages for asset categories may be specified either directly or by changing properties used to determine the objective; priorities may be affected by coercing the system to consider a particular recommendation; and investment features such as yields can be altered to affect the expert system's available choices. While quite extensive, this capability has been carefully tailored to maintain reliability. For example, although we can coerce the system to consider a type of investment, it will not necessarily recommend one. Planners aren't allowed to change rules themselves, for that would raise a difficult problem—ensuring that the expert's recommendations follow a consistent planning strategy.

The following scalars provide some sense of the system's size: The expert has both a frame-based representation of static data and a procedural representation for its heuristics. The frame system contains about 500 classes, about 2500 slots, and about 1500 when-needed procedures that calculate relatively simple functions (for example, the equivalent of a taxpayer's 1040 form). Most slots (both primitive and when-needed) contain time-varying data.

The system incorporates about 1200 chunks of heuristic procedure—knowledge that tends to be grouped in pieces larger than we consider typical for production rules. We estimate these chunks roughly equivalent to 6000 simple rules (each containing about 3-5 antecedent and 1-2 consequent clauses).

InterLisp-D is about a 3M-byte memory image in a Xerox Lisp machine's compact representation. PlanPower is delivered in a memory image (including Lisp) that takes about 11M bytes. We estimate that the whole system contains about 250,000 lines of Lisp code.

Internal architecture and solution techniques

PlanPower's expert is a highly complex program embodying much detailed knowledge and orchestrating many interactions. We applied considerable effort to provide an overall architecture enabling these knowledge fragments to cooperate in producing financial plans of professional quality. Some major steps toward automatically creating financial plans follow, illustrating how expert system technology made this large-scale problem tractable. Along the way, we will discuss some trade-offs caused by the nature of the problem.

Figure 1 illustrates PlanPower's architecture at the highest level. Process input consists of a problem description including the client's financial and personal data, attitudes to risk, and goals. To control the expert, planners can add data for this specific case or the generic knowledge base. The knowledge base provides a comprehensive set of frames describing investment vehicles completely enough for the system to (1) model a financial transaction and (2) determine the complete effects of that transaction considering the applicable time frame and macroeconomic variables such as inflation rate.

Processing comprises three stages: before-planning observations, planning and recommending, and plan document production. The first stage represents a complete diagnostic expert system in itself—its major task is to analyze the client's current financial situation, projecting this into the future and providing observations summing up positive and negative aspects. The knowledge base provides generic financial observation frames from which the expert can choose to construct problem descriptions, worthwhile goals to pursue, and comparisons between the client's position and the recommended position for clients in similar stages and income levels. Observations can be quite simple (such as noting that client debt payments are high proportional to cash flow) or quite complex (such as evaluating client tax situations, allocating assets into various investment categories, and determining whether clients would achieve their goals if no recommendations were followed).

Observation. Processing in the observational stage evaluates backward-chaining if-then rules. Each rule may make an observation in its own area either by invoking other rules or by evaluating and logically combining predicates that apply to the case data. Some lower level rules could be implemented equally well in a forward-chaining manner, but higher level rules may make control decisions best handled by backward chaining. Such rules can decide not only whether a particular observation is true but what observations should be attempted.

Since general techniques for language production were not adequately developed, we provided a special language (Computed Text) enabling knowledge engineers to specify rules that build the document from facts and frames in the knowledge base that results from planning. At one extreme, the language lets knowledge engineers specify formatted boilerplate text and customize it to the data (as in traditional letter merge programs). At the other extreme, it allows them to write rules determining in top-down fashion how to arrange plans and what to express. Rules can query the database to decide whether a certain section should be included, and at what point. They can search for recommendations or explanations relevant to the point at hand, and they can decide how to present them. The language facilitates access to knowledge base information. A limited set of syntax generation primitives allows knowledge engineers to specify methods for handling constructs such as plurals and possessives.

The planning stage

Of the three major processing stages in PlanPower's financial planning expert, plan generation is the most complex. Its structure can be compared with an automobile's, which has numerous subsystems (like the fuel system and the electrical system) that interact closely. PlanPower's subsystems are modules, each of which is expert in a specific area such as retirement, insurance, asset allocation, and tax planning. These modules are carefully coordinated.

Figure 2 expands plan generation into its next architectural level, where we can more clearly see some techniques used to structure problem solving and manage the large space of constraints and possible actions. Because of space limitations, however, many details are not covered here and even entire processing modules are not described.

Planning's first step rules out as much as possible, creating a focus of attention that will drive the planning process. In effect, the expert evaluates all possible actions on all available investment vehicle types and decides whether to consider them in depth. It also examines the client's current specific assets and determines whether to modify these assets (for example, sell them). The system creates a possibilities list using backward-chaining rules that evaluate particular investment vehicles or general classes of investment vehicles. In fact, these rules are indexed into the knowledge base's frame hierarchy so that (while similar to regular if-then rules) the rules are invoked in a way that takes advantage of the object-oriented programming paradigm.[5]

Planning and recommending. The planning stage determines what actions clients should take to

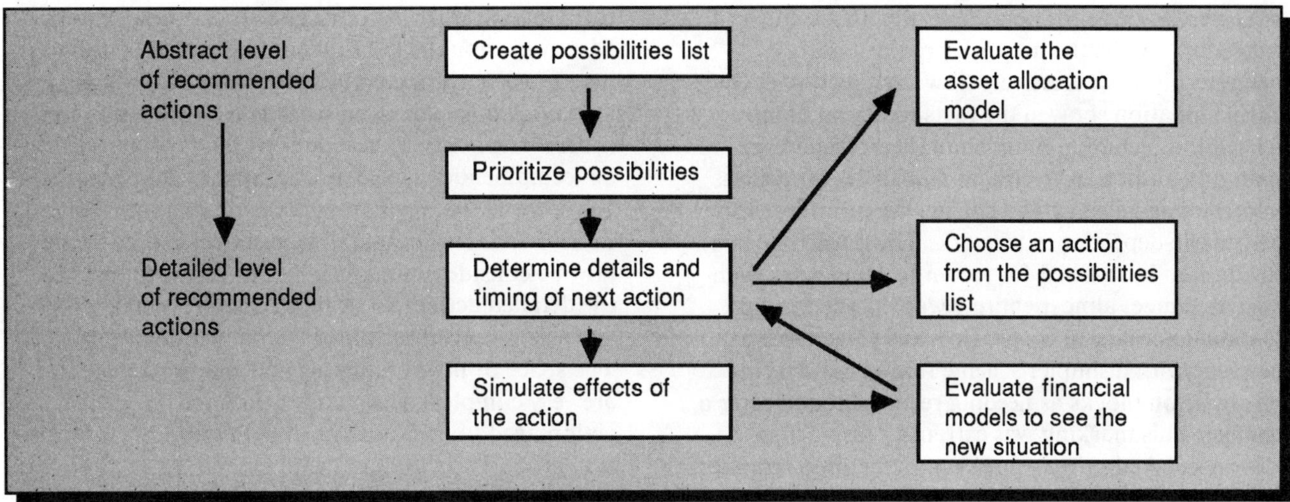

Figure 2. A partial architecture of the planning stage.

optimize their financial positions over the succeeding years. Planning is the most complex stage by far since numerous different constraints must be accounted for and the possible combination of actions is considerable. Moreover, the far-reaching effects of financial actions on client situations prevent us from independently treating goals and problems discovered in the observation stage.

During planning, the system must continually examine the entire picture. Planning produces recommendation frames, most of which detail particular actions that clients should take. These frames form a plan in the sense that the system specifies actions, their timing, and their exact parameters to address two top-level goals: increasing the client's wealth, and protecting the client's assets. Applying each action at the recommended time will (1) improve the client's position regarding these goals, (2) determine if any special funding objectives can be met, and (3) follow accepted financial planning guidelines.

While some goals may not be achievable at all (like buying a vacation home), others are not binary choices and can be satisfied to varying degrees (like diversifying assets, reducing taxes, increasing net worth, improving cash flow, and providing for retirement). Planning must evaluate goals as a whole and choose actions that optimize the total position. For example, reducing taxes is simply one possible means toward an end. It is valuable only to the extent that it contributes toward the system's top-level goals.

Besides its main task of producing action plans, the expert's planning stage also constructs explanation frames in the knowledge base. These frames justify its recommendations and record significant decisions or evaluations made during its processing. They include references to contextual data used by the expert's rules as well as pointers to general financial planning principles providing background rationale for the reasoning (while not being a part of the logical argument). The former can be viewed as the set of arguments used by rules, while the latter provides justification for using rules at all.

The knowledge base includes financial models that support planning and observation. These models consist of perhaps 1500 rules and procedures for making financial calculations regarding net worth, cash flow, tax situation, yields, investment costs, and the like. They also allow logical evaluations of resulting numbers. Although intricate, these rules are straightforward since they embody accounting procedures rather than financial planning expertise. The expert's observation and planning stages can access financial models on demand. Whenever the expert requests a fact that is part of one of these models, that model will calculate it (if necessary) and store the result until it becomes invalid.

These models, evolving over PlanPower's history, led to an interesting discovery about the financial planning domain—a discovery that forced significant redesign. Originally, the rules comprising these models were joined in a complete dependency network. Whenever data was changed, the dependency network ensured that exactly those facts depending on the change would be recalculated the next time they were needed. AI programs have used this technique for years.[4] While the network provided correct behavior, it was excessively slow.

Examination showed the facts to be so interconnected that (1) even minor changes affected large amounts of data, and (2) many facts the expert was interested in using after a change were located at the end of the dependency network. This caused most of the data to be marked for recomputation after each minor change, almost entirely wasting the network maintenance overhead. Consequently, we replaced the network with a simpler scheme that marked large information blocks as needing recomputation after a change. This marking was extremely fast. Since the system invoked rules by backward-chaining, reprocessing remained demand driven and the expert could make assertions without actually triggering recomputation until the financial models received a subsequent request for information.

In problem areas such as this, where significant numerical calculation exists, an extension to the dependency network approach looks promising. This extension has not been implemented, but can be illustrated by a procedure that finds all the assets of a person and adds up their values—a procedure requiring multiple retrievals and additions. If one asset's value changes, or if the set of assets changes, the procedure's result should be declared invalid. Rather than recalculating, as a simple dependency network would, a procedure can be invoked to modify the old result—a procedure requiring only one retrieval and one addition.

Plan document production. The third processing stage, producing the plan document from the after-planning case, presents (1) the diagnosis of the client's situation, (2) recommended actions that should be taken, (3) explanation and rationale for these actions, and (4) relevant background information about financial matters. The document had to be well organized and written in good English, which posed a problem since language generation is difficult even in limited domains and our domain was extremely large and complex.

The possibilities list forms a kind of agenda worked on throughout the planning stage. When an item is first added, criteria considered include such conditions as the state of the economy, financial rules concerning such investments, tax implications, client preferences for keeping or acquiring such investments and, of course, observations resulting from the first processing stage. The expert prioritizes possibilities by comparing them in light of the current situation and objectives. Directives input by the planner at the start of processing significantly affect this part of the reasoning.

At this point, the transactions on the possibilities list represent only abstractions of the final actions to be recommended. The specific types of investments to be used and the exact amounts and timing of the transactions are still to be determined. The problem can now be considered as a refinement process in which the detailed actions to be recommended are developed from the rough description on the possibilities list. The space of possibilities is large and the constraints are very complex. The changes produced by each action need to be evaluated in light of all the information produced by the financial models, general principles of diversification of assets among categories, and constraints provided by either the client or the planner in terms of preferences or objectives.

PlanPower addresses this problem in several ways. First, we have studied the domain thoroughly, developing heuristics to order some possibilities. For example, when transferring assets to achieve a desired target within an asset category, a particular order may be preferable. The system can prioritize specific investments or types of investment to sell or buy. In such cases, assets can be bought in that order until the goal is satisfied and then the process may stop. Other heuristics can modify this process—to prevent buying too much of one asset at one time, for example (thereby reducing risk).

When we first examined financial planning, the complexity of the process appeared overwhelming. However, ordering can sometimes govern the resolving of separate goals. For instance, the system always handles insurance needs before allocating discretionary assets. Extensive knowledge engineering revealed that many choices could be decided beforehand or could be decided more simply than expected. We should not expect the system to solve planning problems from first principles or to master financial planning. Experienced financial planners have, in effect, compiled the results of their experience. Codifying this is not easy, but is important if the resulting system is to be an effective product. It is also easier to represent when knowledge engineers have a convenient way to exert procedural control over the program.

Developing rules that estimate the effect of executing actions is a technique that allows the system to determine details, such as investment amounts, without simulating many possibilities. Estimates are then checked by simulation and adjusted if necessary. PlanPower can set up temporary contexts as hypothetical situations in which to perform these checks.

In some cases, the system must resort to a search. It may find appropriate actions for achieving goals by searching a few carefully chosen possibilities and discarding all but the best. We devoted significant effort to finding heuristics that minimize this search space. Even small changes can propagate effects over wide areas in this domain; therefore, fully evaluating single hypothetical situations is expensive.

Throughout planning, an asset allocation model modulates action recommendation—a model categorizing assets into types such as tangibles, real estate, and fixed income. This model provides objectives for investment proportions that clients should achieve in each category. Initially, the model decides these objectives based on the client's situation—on parameters such as the state of the economy and whether the client is at an early or late stage of financial development. For example, the client may be accumulating assets or may be nearing retirement. As planning proceeds, however, it may become clear that ideal asset allocations cannot be met. The model gradually relaxes its constraints while the remaining system attempts to meet its revised goals.

Producing explanations

The planning stage produces frames that provide explanations of the expert's observations and recommendations. Early in PlanPower's development, we realized that we could not rely on fully automatic explanation generation mechanisms for several reasons. First, such a mechanism would have difficulty separating out the expert's relevant decisions: Some were simply decisions about processing order; others were trivial; still others were correct but not the most relevant explanations for users. Appropriate explanations might actually have been part of the rationale for writing rules in the first place, but may not be represented in rules at all. For example, one rule states that selling treasury bills is pointless because they're short-term investments—we might as well wait until they mature. In PlanPower, this may be represented as "do not sell if x is a T-bill." The relevant rationale is not necessarily part of the operational decision. This problem has been cited before.[6]

Second, fully automatic mechanisms were unable to generate explanations fluent enough to satisfy financial planning product needs. Generating relevant explanatory information, and preparing high-quality expositions of it, are active research areas.

We addressed the first problem by allowing knowledge engineers to specify what was relevant to an explanation and what should be explained. Knowledge engineers represent these important facts and decisions as observation frames in the knowledge base. When the expert system runs, its rules create instances of these frames and fills them with context information tailoring them to the specific situation. Knowledge engineers can also specify which observations provide explanations for which others. In the worst case, particular instances can be explicitly linked. In most cases, however, a two-step approach is more convenient. First, the knowledge base is provided with generic assertions about which types of observation may support which other types. Then, when observation instances are created, they're given refinement assertions that usually refer to objects in the current environment. We find particular supporting explanations for a supported observation by filtering generic observations that could support it; by so doing, we locate those generic observations that have been given contextual tags matching the supported observation's refinement assertions. This approach allows some decoupling of supports from supported observations, enabling dependencies to be set up easily between observations created in separate rule invocations.

We created fluent English explanations by associating rules from the text production language with various observation types. These rules can access any contextual information associated with the instance of its observation frame, and knowledge engineers can have rules tailor text to that information—thereby making available the full expressiveness of the text language.

The underlying platform

While previous sections have described the PlanPower expert's features, some features of the platform used to support this expert should also be discussed. The knowledge base consists of a frames database and a rules database. Frames are organized along the lines of other frame-based systems. The system uses a class taxonomy for inheritance and each class can have instances. For example, since observations are frames in the knowledge base taxonomy, they can use the hierarchy. Observations can inherit generic dependency information and attached text production rules from more general observation types. Frames have attributes—attributes that can have values, defaults, and attached methods that either specify particular facts about an instance or generic facts about all instances of a class.

The APEX frame system has special mechanisms that handle time series of information, manage relations having inverses, and classify entities. Time series in particular were essential for this domain: We use them to represent values of facts at different points and they provide methods to (1) default values for years when they are unknown, or (2) calculate values by applying known growth rates. The knowledge base contains many time series facts—facts that proved practical for adequately representing time in the problem area. We considered that more general representations of time periods and relationships between time periods were insufficiently developed for this problem.

The APEX heuristic-knowledge language combines a typical if-then rule-based language and a set of powerful procedural constructs. It includes three types of rule: actions, predications, and designations. An action determines how to execute procedures; its results are side effects to the database. A predication determines how to test or assert statements of logical truth. A designation determines how to reference an entity or set of entities by description. Predications and designators can chain to combinations of other predications and designators or can be evaluated procedurally.

PlanPower orchestrates these three rule types in a way that allows knowledge engineers considerable control over rule invocation—an extremely important factor in PlanPower's development since (1) the domain is complex, (2) the expert combines a diagnostic system and a synthetic (plan-producing) system, and (3) the ultimate product-quality system must meet professional financial planning standards.

The frame system and rule language couple tightly. The system indexes rules closely into the frame knowledge base, taking advantage of the inheritance hierarchy and allowing some rules to be more general than others. In effect, the platform combines rule-based, procedural, and object-oriented aspects. The resulting expressiveness, combined with the graphic interface that APEX has built upon InterLisp-D, enabled a sizeable group of knowledge engineers to build a very complex system and complete it on schedule. This development environment and architecture permitted APEX to react quickly to the most significant Internal Revenue Code changes in decades.

PlanPower—generating comprehensive personal financial plans—is one of the earliest commercially available large-scale expert systems. Its development history, major processing stages, and analysis/synthesis structure provide the expert systems community with another experiential data point. The difficulty and length of time required to develop PlanPower underscore the pressing need for better methodology and development environments. Other research areas that will help enormously in the future include explanation generation, fluent and efficient natural language generation, and representation of both knowledge and control paradigms.

On the other hand, PlanPower demonstrates that today's techniques can produce operational systems at acceptable cost. ■

Acknowledgments

We acknowledge the dedication and hard work of the financial planners who shared their expertise with PlanPower, the knowledge engineers who gave that expertise to PlanPower, the systems and research groups who allowed PlanPower to receive that expertise, and those special people who had the idea in the first place and who carried it to fruition.

References

1. F.L. Luconi, T.W. Malone, and M.S. Scott Morton, "Expert Systems: The Next Challenge for Managers," *Sloan Management Rev.*, Summer 1986.
2. W. van Melle, *System Aids in Constructing Consultation Programs,* UMI Research Press, Ann Arbor, Mich., 1981.
3. D.G. Bobrow and M. Stefik, "The Loops Manual: A Data and Object Oriented Programming System for InterLisp," VLSI Design Group Memo KB-VLSI-81-13, Xerox PARC, Palo Alto, Calif., Aug. 1982.
4. R.M. Stallman and G.J. Sussman, "Forward Reasoning and Dependency-Directed Backtracking in a System for Computer-Aided Circuit Analysis," *Artificial Intelligence,* Vol. 9, No. 2, 1977.
5. M. Stefik and D.G. Bobrow, "Object-oriented Programming: Themes and Variations," *AI Magazine,* Winter 1985.
6. W.R. Swartout, "Explaining and Justifying Expert Consulting Programs," *Proc. Seventh Int'l Joint Conf. Artificial Intelligence,* Morgan Kaufmann, Los Altos, Calif., 1981.

James L. Stansfield manages the Representation and Reasoning section at Applied Expert Systems, Inc. Before joining APEX, he held a senior staff position at Computervision Corp., working on CAD tools for electronics design. He has served as a research associate at MIT's AI laboratory and as a lecturer at MIT's Division for Study and Research in Education. He received his BA and MA in mathematics from Cambridge University, England—and his PhD in artificial intelligence from Edinburgh University, Scotland.

Norton R. Greenfeld is vice president of software development at Applied Expert Systems, Inc. Before joining APEX, he was a senior scientist in the AI department of Bolt, Beranek, and Newman, Inc., where he worked on Lisp machines and did research in expert systems for output presentation. He received his BS in mathematics, and his MS and PhD in computer science, from the California Institute of Technology.

The authors can be reach at Applied Expert Systems, Five Cambridge Center, Cambridge, MA 02142.

Reprinted from *Proceedings of the IEEE*, Volume 76, Number 10, October 1988, pages 1327-1366.

Expert System Technology for the Military: Selected Samples

JUDE E. FRANKLIN, SENIOR MEMBER, IEEE, CORA LACKEY CARMODY, MEMBER, IEEE, KARL KELLER, TOD S. LEVITT, MEMBER, IEEE, AND BRANDON L. BUTEAU, ASSOCIATE, IEEE

Invited Paper

This paper is concerned with the applications of expert systems to complex military problems. A brief description of needs for expert systems in the military arena is given. A short tutorial on some of the elements of an expert system is found in Appendix I. An important aspect of expert systems concerns using uncertain information and ill-defined procedures. Many of the general techniques of dealing with uncertainty are described in Appendix II. These techniques include Bayesian certainty factors, Dempster-Shafer theory of uncertainty, and Zadeh's fuzzy set theory. The major portion of the paper addresses specific expert system examples such as resource allocation, identification of radar images, maintenance and troubleshooting of electronic equipment, and the interpretation and understanding of radar images. Extensions of expert systems to incorporate learning are examined in the context of military intelligence to determine the disposition, location, and intention of the adversary. The final application involves the use of distributed communicating cooperating expert systems for battle management. Finally, the future of expert systems and their evolving capabilities are discussed.

I. Introduction

The increasing complexity of weapon systems and the growing volume of complex information creates numerous problems for the military [1]–[5]. Commanders must make decisions faster than ever before and maintain operational readiness in spite of limitations on manpower and training. Artificial Intelligence (AI) technology can potentially solve many of these problems for the military, with some AI applications already demonstrating their utility. The military and industry have made major investments in the area of AI. This paper will concentrate on the potential of expert systems, a subordinate category of AI, to the military arena.

Within the last few years the military has witnessed an almost explosive expansion of the field of expert systems within various agencies of the Department of Defense (DOD). This has been sparked by the rapid growth in computer technology, by the development and better understanding of expert systems concepts, by the progress that has been made in sensors and control devices, and by the growing need caused by an information explosion. For example, typical military command and control centers handle thousands of messages in a single day and commanders must decide promptly on correct actions. Commercial applications in the fields of financial investment, manufacturing, and business planning have similar high information volume and time constraint problems. Although this paper concentrates on military applications of AI, there are analogies in the commercial arena.

While the field of expert systems is still a long way from solving the military's most persistent problems, this research activity for creating intelligent machines has demonstrated certain machine properties that offer great hope and promise, particularly in the area of manpower reduction. There are reasonable expectations of future computers that can learn; reason; understand text and speech; perform complex problem solving operations; recognize anomalous behaviors and warn the decision maker; understand drawings and photographs; and process signals such as speech, sonar, radar, and imagery.

Early DOD investments [3] have helped to establish the scientific foundations upon which the present U.S. capabilities and thrusts in AI and robotics are based. For example, the Office of Naval Research (ONR) and the Defense Advanced Research Project Agency (DARPA) have been supporting research in AI for over 20 years through the support of "Centers of Excellence" at several prominent universities. These centers have published extensively, hosted symposia for government and industry, and spawned technological innovations such as the DARPA-sponsored Strategic Computing Program (SCP) that has applications including the Air Force Pilot's Associate, the Army Autonomous Land Vehicle and the Naval Battle Management programs. The Navy has created a Center for Applied Research in AI at the Naval Research Laboratory (NRL) and has major programs in maintenance and troubleshooting of complex electronic equipment, target identification from radar or acoustic sensors, machine learning, and fusion of data from multiple sensors. More recently, the Army Research Office

Manuscript received January 6, 1987; revised May 19, 1988.
J. E. Franklin, C. L. Carmody, and B. L. Buteau are with Planning Research Corporation Government Information Systems, McLean, VA 22102, USA.
K. Keller is with MITRE Corporation, McLean, VA 22102, USA.
T. S. Levitt is with Advanced Decision Systems, Mountain View, CA 94043-1230, USA.
IEEE Log Number 8823835.

(ARO) has invested in a long-term AI research, development, and training effort with two major universities (the University of Texas and the University of Pennsylvania). The Army has started an AI center in the Pentagon with a concentration on information management for logistics. One of the Army's first problems is the correct distribution of critical equipment such as cellular radios for the signal corps. The Army has major programs in planning and threat analysis from multiple sensors. The Air Force has just started a consortium of universities oriented toward AI in the region around Rome Air Development Center (RADC) and has ongoing programs in software automation, indications and warning systems, and decision systems for military commanders. The Air Force Office of Scientific Research (OSR) has sponsored research in manufacturing science, spacecraft image understanding, systems automation through AI, and software automation. The Strategic Defense Initiative (SDI) is considering the use of AI techniques to accomplish accurate, time critical decisions and to provide robust adaptive communications for battle management applications.

The remainder of this paper discusses the use of expert systems in the military. Section II gives an overview of expert systems technology and what expert systems do; what the components are; and what some of the interesting problems have been. Section III presents several examples of military applications using expert systems; Section IV discusses a learning system applied to the military; Section V considers the next step of distributed problem solving; and Section VI provides a summary and conclusion with a glimpse at realistic expectations for how expert systems can help solve future military problems and what research areas need to be addressed.

II. Background on Expert Systems and Uncertainty

A. Expert System Overview

Expert systems are computer programs that attempt to duplicate results obtained by *actual* experts in a particular field or domain. Recent results have demonstrated that this technology can be used by the military. This section provides a short introduction to expert systems terminology. Concepts of an expert system are discussed more fully in Appendix I. Other sources of information on expert systems can be found in [6]-[10]. A block diagram of a typical expert system is shown in Fig. 1. The real power of an expert system is found in the knowledge base, which contains the fundamental facts and assertions necessary to solve a specific problem. The knowledge can be stored in at least three structures. The first is to represent statements about the problem domain as predicate calculus (logic). The second is to represent the knowledge as if-then rules, and the third representation is a collection of attribute-value pairs known as frames. There are other representation schemes and they are mentioned in the Appendix. The best representation schemes use a combination of several of these strategies.

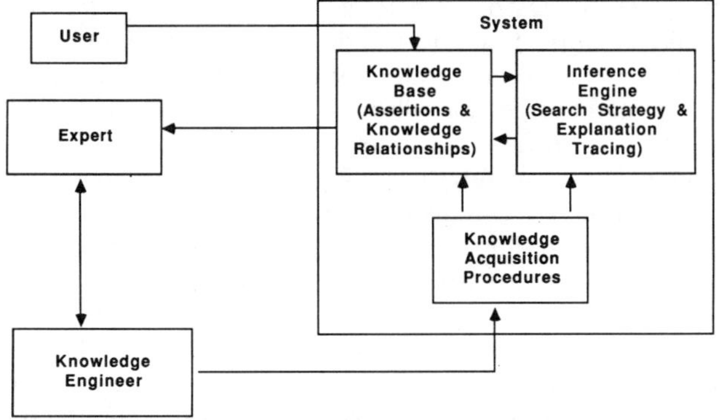

Fig. 1. Diagram of a typical expert system.

The *inference engine* accepts the input data and the information in the knowledge base to develop a means of solving the problem. The inference engine may use a goal-directed scheme that examines various potential hypotheses and determines which one is true. This technique works backwards from the goals toward the input data. Another approach is to start with the input data and infer conclusions in a forward direction. The final approach is to combine both techniques. The inference engine will usually contain various search strategies that take advantage of how best to examine alternatives and what search paths will provide the fastest solutions.

The knowledge engineer attempts to work with the actual domain expert and represent his problem-solving techniques on the computer. The knowledge engineer is the key to the total knowledge acquisition process. Frequently, the domain expert becomes the knowledge engineer, and optimum results are achieved with fewer errors and misunderstandings. Additional details about expert systems are found in Appendix I.

B. Uncertainty Representations and Approaches

A popular maxim pertinent to the study of expert systems is "In the knowledge lies the power." By design, expert systems augment human reasoning; it is a natural extension to basic reasoning approaches within expert systems to include the ability to form reasonable conclusions from uncertain and incomplete knowledge. Associating a representation of our degree of belief with the knowledge contained in a knowledge base is one of the most common approaches to reasoning with uncertainty. Unfortunately, most of the representations bear little resemblance to human cognitive processes—their primary virtue is that they can be implemented within an expert system and can help to solve the problem. Appendix II contains illustrative material. References [11]-[32] and the additional readings section also provide background for this research area.

The problem of reasoning with uncertainty takes on greater importance when the application area concerns the fusion of information from many knowledge sources, e.g., the command, control, communication and intelligence (C^3I) sensor environment. In particular, when the independent knowledge sources are sensors producing low-level perceptual outputs, it is essential to have a methodology for assigning uncertainty values to evidence from different knowledge sources, and for combining these values when evidence from these knowledge sources supports (or contradicts) the same basic conclusion.

Generally, sensor systems do not report results with 100-percent accuracy and certainty. This is a particular problem when multiple sensors must be combined. For example, when output from Sensor *A* support the conclusion **"The target is a type X relocatable weapon"** with certainty 0.42, and outputs from Sensor *B* support the same conclusion with certainty 0.87, a mechanism should exist for combining these figures in a coherent manner. Furthermore, the method must deal with the situation when a similar conclusion from Sensor *C* is incorporated at a later time, or when even later Sensor *B* retracts its conclusion or updates the certainty factor associated with its conclusion.

Approaches to dealing with uncertainty often take some variation of numeric characterization. Numeric representations usually take the form of the assignment of a point value (as the application of Bayes' Theorem or maximum entropy), intervals on a range (as in Dempster–Shafer Theory) or points within an evidence space. A variant on a numerical approach with foundations in set theory is Zadeh's Fuzzy Logic, or Fuzzy Set Theory. A discussion of these major methods is in Appendix II.

C. Methods of Dealing with Uncertainty

One of the first problems with numeric methods of uncertainty lies in allocating initial numbers. Where do the numbers come from? Are the numbers intended to capture truth (i.e., in terms of frequency data), or are they representing a consensus of belief from one or more experts? These two interpretations lead to different sets of problems. In most military applications, existing frequency data is not available, so a foundation for further calculations must be defined in another manner. There are at present no standard methods for laying this foundation.

If the numbers are to represent confidence or belief, the initial assignment of certainties must come from the expert whose assertions form the knowledge base. Different experts may well assign different certainty numbers, and may have different reasons for the assignments. The reasons for assignment are left behind, once the numbers are assigned to assertions or rules. Apart from the loss of reasons, there is some question as to the quality or precision of the initial numbers.

Another basic problem is in the interpretation of the numbers. Just what does it mean to say that **"The target under observation is an SS-25 missile"** with certainty 0.87. Does this mean that **"I'm 87% sure that it's an SS-25," "The probability is 87% that it's an SS-25," "The odds are 87 in 100 that it's an SS-25," or "87 out of 100 targets that we've tracked with these kinds of characteristics turned out to be SS-25s?"** Interpretations of certainty values vary between degrees of belief, probabilities, betting, and frequency of occurrence. A number is only a measure of how uncertain we are about a proposition and does not convey our reasons for doubt.

Within specific numeric methods, there are problems with single point implementations, such as Bayes' Theorem, since the degrees of belief and degrees of disbelief in a proposition must always sum to one; any doubt must be represented as an unknown hypothesis. Another problem with the Bayes approach is that in order to update certainty values effectively, a large amount of information is needed. Two of the applications, BATTLE and the ship identification expert system, described in Section III-A and Section III-B, use a variation and extension of the Bayesian techniques.

In contrast, the Dempster–Shafer method provides a representation which separates values for belief and disbelief (they need not always sum to one), and a propagation algorithm which will accept initial ignorance in a proposition, so that reasoning can proceed without a large amount of initial information. The Dempster–Shafer method provides a model for the gradual accrual of new knowledge into an expert system, and its inverse may be computed in order to retract evidential contributions. There are, however, significant implementation difficulties in the full scale theory as discussed in Shafer [17]. The original theory calls for creating supersets of all possible propositions and deals with independent knowledge sources. Obviously, this full approach will cause serious difficulties when the problem area exceeds a very simple problem space, or when the knowledge sources are interdependent. Ginsberg [29] and Barnett [30] have recommended ways to solve the potential difficulties, and Lowrance and Garvey [31], [32] have successfully implemented a slightly more advanced subset of Dempster–Shafer theory and applied it to the military problem of sensor fusion in support of electronic warfare. An additional flaw in the Dempster–Shafer rule of combination is discussed by Zadeh [13], and is reviewed in Appendix II, Uncertainty Methods. The Dempster–Shafer technique is used in the distributed expert system example that depicts an indications and warning problem involving terrorist activities and is discussed in Section V.

III. Applications

This section discusses several applications of expert systems that are in varying stages of development for military applications. The choice of the applications was driven by the authors' personal experiences and by no means represents an exhaustive discussion. The samples that were selected represent work at the Naval Research Laboratory (NRL), Defense Advanced Research Project Agency (DARPA), the U.S. Army, Planning Research Corporation (PRC), Mitre, and Advanced Decision Systems (ADS).

A. Combat Management for the Marine Corps

The Naval Research Laboratory, under the direction of Slagle, developed an expert consultant system for weapon allocation. The present status of the system is a working prototype tool. This system, called BATTLE [33], [34], generates weapon allocation plans for a system with requirements similar to the Marine Integrated Fire and Air Support System (MIFASS).

The BATTLE expert system evaluates the effectiveness of individual weapons to targets and then it produces complete evaluation plans that consider the possible allocation of all weapons to all targets. Normally, this would involve exhaustive search techniques. The goal of the system is to maximize the destruction (total value *D*) for all targets. In an allocation plan, the destruction value for a target is the product of the target's strategic value and the expected percentage of the target that will be destroyed in the plan. When the destruction value is maximized, the plan is considered optimal. Unfortunately, achieving this optimal plan in real conditions using exhaustive search techniques can consume too much time. For example, if we have *W* weapons and *T* targets, there are $(T + 1)^W$ different options. For an optimal plan that considered 8 weapons and 17 targets, the computer run time, using exhaustive search, was 11 minutes and 43 seconds. The BATTLE expert system took 6.75 seconds for the same problem and achieved 98 percent of optimality. Reduction in processing time was due to the heuristic search mechanism that eliminated investigating unlikely weapon-target allocations.

A few of the 55 factors used by BATTLE to arrive at its weapons allocation are shown in Table 1.

Table 1

- Range and position
- Personnel readiness
- Counterfire ability
- Resupply
- Ammunition status
- Number of tubes per group
- Experience of enemy group
- Maintenance status
- Replacement crew status
- Physical condition

The first step of the problem computes the effectiveness of a single weapon to a single target, and the second step assesses the complete matchup of all weapons to all targets. BATTLE uses rules specified by a military expert to accomplish this matchup. A portion of the single weapon target allocation network is shown in Fig. 2, and this network is a generalization of the inference networks of PROSPECTOR. PROSPECTOR was an early expert system that was developed to assist in mineral exploration.

The artillery allocation network, shown in Fig. 2, uses many different quantitative and qualitative parameters. Quantitative parameters include reliable range estimates for weapons or target counterfire capabilities. The qualitative parameters include readiness of personnel or maintenance status of equipment. The computational network shown in Fig. 2 uses several different techniques to com-

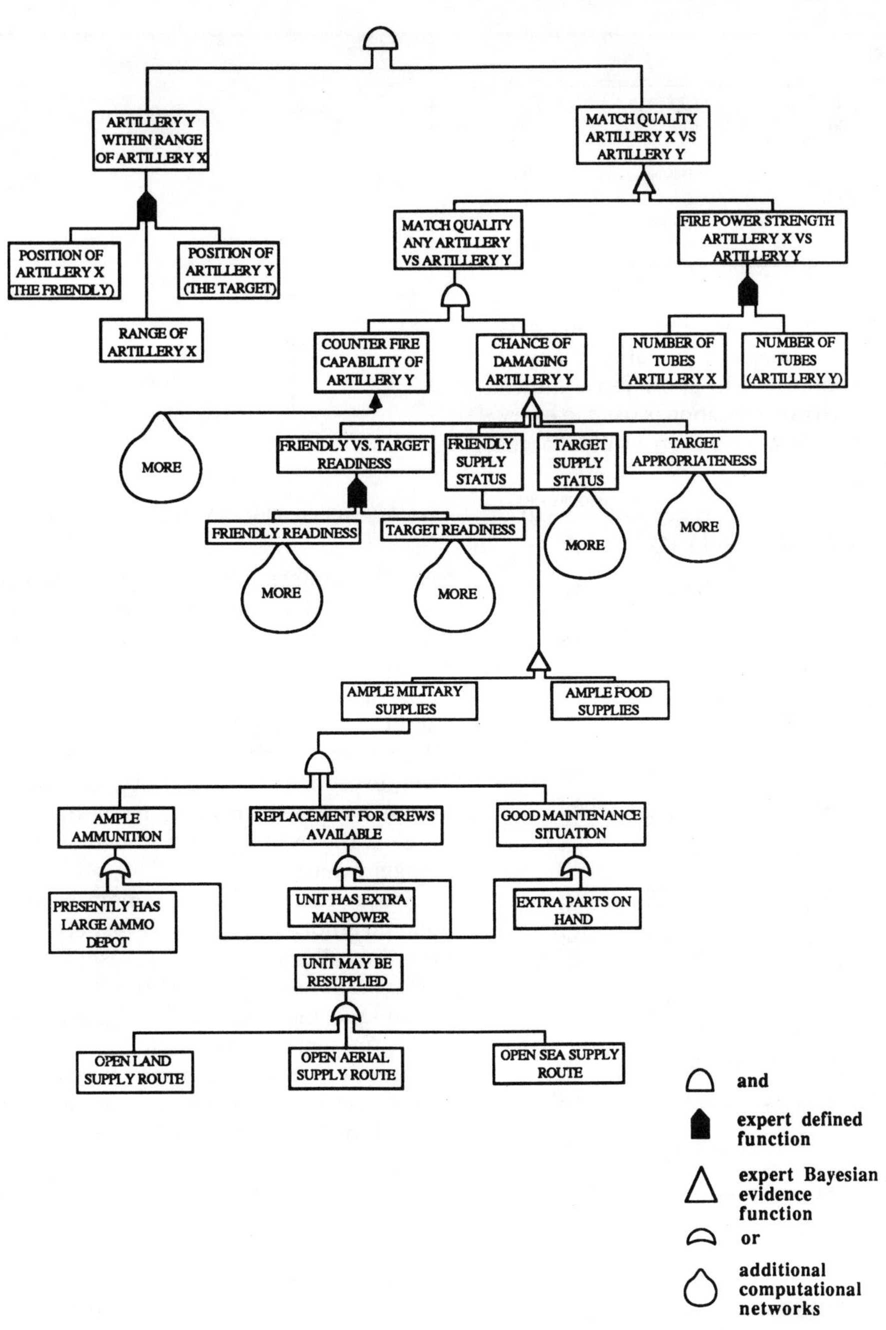

Fig. 2. Weapons allocation network [33].

bine information prior to allocating weapons to targets. Classic Boolean operators of **and** and **or** are used to combine information such as resupply alternatives and counterfire capability. The Bayesian evidence functions indicated by the triangular symbols are used to combine information such as friendly versus target readiness, friendly and target supply status, or target appropriateness. In this case, there is a degree of uncertainty in the information that is used as well as uncertainty in how this information should be combined. As discussed earlier, and in Appendix II, the Bayesian technique provides a means to combine these various forms of uncertainty. Expert defined heuristics are indicated by the filled-in symbols. These heuristics usually represent empirical results that aid in reliable weapon to target allocations.

A typical scenario for the allocation of 4 weapons to 3 targets is shown in Table 2. In this example, the target's strategic value is shown along the top, and the weapon effectiveness is the expected percentage of the target that will be destroyed by that weapon. The allocation tree is shown in Fig. 3 for the case in which every weapon is used in every allocation, massing of multiple weapons on a single target is allowed and a weapon can be fired only one time. The weapon and target assignments are shown on the left-hand side, and the destruction value is shown on the right-hand side. Fig. 3 illustrates exhaustive search with an optimal allocation of $W4$ to $T2$, $W3$ to $T3$, $W2$ to $T1$, and $W1$ to $T1$, yielding a total destruction value of 401. Fig. 4 shows the same tree but indicates that the total number of computations was reduced dramatically by the addition of a few heuristic rules to the search tree. This pruning was accomplished by some very simple considerations. One consideration was to evaluate the best weapon to target allocation to that point and then to use some look-ahead heuristics to determine if the next branch of the tree under evaluation could possibly exceed the existing best weapon-to-target destruction value. If the estimated cumulative destruction value did not exceed the existing value, this branch of the tree was no longer evaluated.

Table 2

Target	$T1$	$T2$	$T3$
Strategic value	200	100	150
Weapon effectiveness			
$W1$	60%	60%	60%
$W2$	70%	30%	0
$W3$	80%	0	90%
$W4$	90%	90%	90%

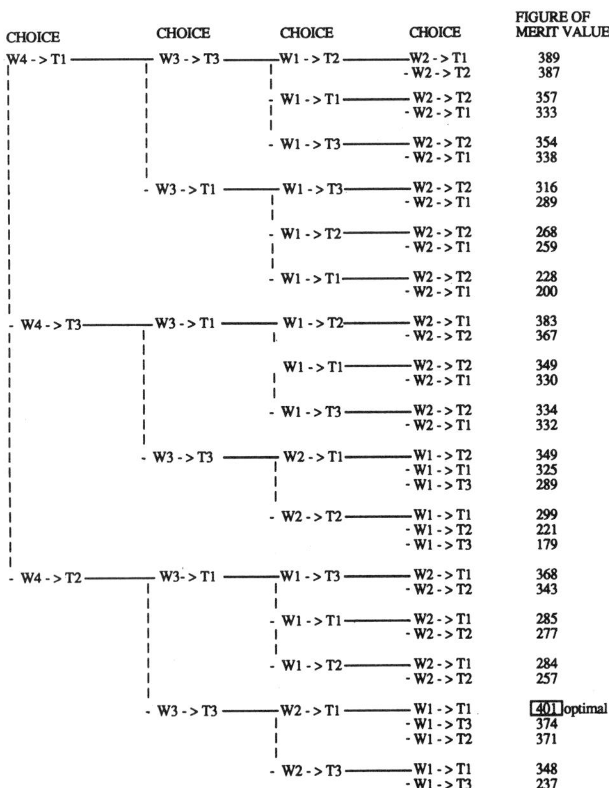

Fig. 3. Example of weapon allocation to targets and subsequent figure of merit value.

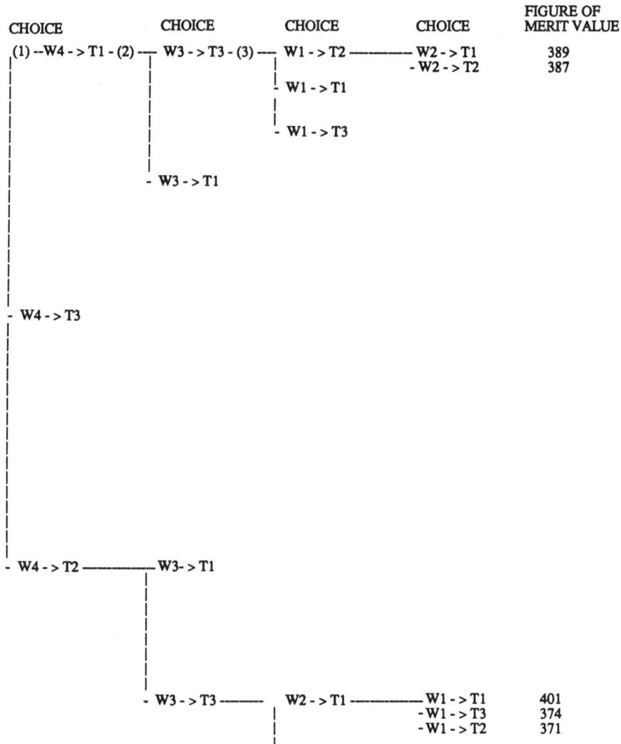

Fig. 4. Pruned allocation tree using heuristic search.

Frequently, there are situations in which all of the information is not available for the expert system to perform the required allocation. For this case, the expert system must ask questions of the user. The BATTLE system uses a new technique, merit, to ask the highest payoff questions first, and to minimize the user's time and cost to provide answers. The merit system performs this function by considering how easy it is for the user to provide the information and how useful the answer will be for the weapon-to-target allocation.

The merit system starts with the concept of the self merit. The self merit is defined as the expected variability in the answer to a question divided by the cost to acquire the answer. The potential variability in the answer to a question would be very high, for example, with matters associated with weather since it is expected to change frequently. The cost is associated with the difficulty in providing the answer. One measure of cost could be the time it takes to find an answer to a question. Therefore, the self merit of a node involving weather would be relatively high because of the large expected variability of the response divided by the relatively low cost of acquiring weather information.

Fig. 5 illustrates the merit system. The self merit associated with the node G_1 is the absolute value of the change

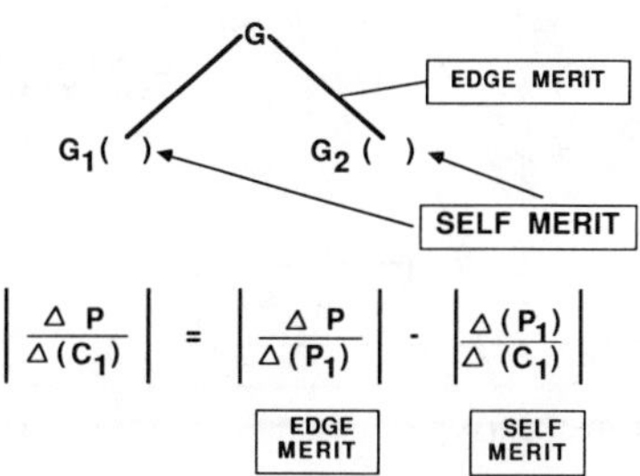

Fig. 5. The merit concept.

in the response parameter (ΔP_1), divided by the cost (ΔC_1). The next important consideration is the effect of this answer at node G_1 and the impact it will have on the next higher node G. This is referred to as the edge merit and it is illustrated in Fig. 5. It is the absolute value of the change in node $G(\Delta P)$ divided by the change at node $G_1(\Delta P_1)$. The total merit at node G associated with G_1 is the product of the self merit and the edge merit as shown in Fig. 5.

Fig. 6 shows a comparison of the merit strategy for asking

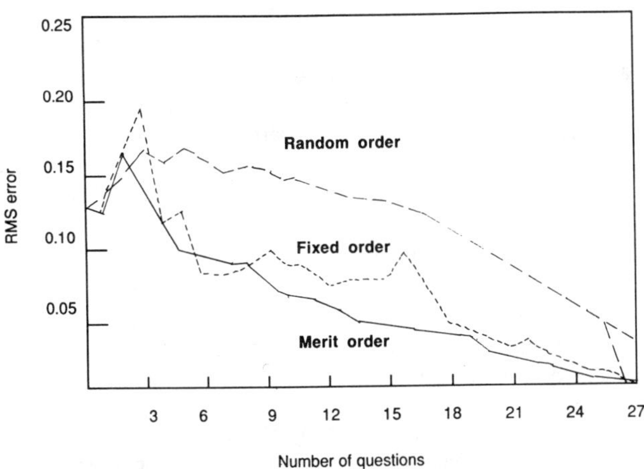

Fig. 6. Comparison of questioning strategies.

questions with two other techniques. The other two strategies are the random selection of questions and the fixed order strategy. The fixed order strategy is calculated prior to the availability of answers to any questions and it is not updated as answers become available. This reduces the need for online computation but it results in a fixed order in the set of questions and frequently takes more time. The error on the Y axis refers to the difference between the optimal plan that maximizes the destruction value and the actual destruction value for the three strategies. The merit strategy arrived at a solution that is within five percent of an optimal plan and required less than a dozen questions. The other two strategies used in the comparison would require 18 to 25 questions to achieve this same level of performance as the merit strategy. The expert system must minimize the questions asked of a Marine artillery officer when he is in a stressful combat environment, and the merit strategy helps reduce unnecessary questions.

The description of this project has been used as the initial example to show the application of AI to a real military problem, in this case one of resource allocation. Several points should be mentioned. First, the implementation of expert systems is at best an empirical problem. There is no sure tried and true technique that is universally applicable. In this case the PROSPECTOR paradigm was useful, but it needed additions and alternatives such as the merit questioning strategy and a pruning technique that was customized to the problem. This project demonstrated that the resource allocation problem was one that could be useful for the military. The next project to be discussed used many of the tools developed by BATTLE and applied them to a ship identification effort. In fact, one of the things being studied at the time was the question of whether the BATTLE expert system could be used in other domains. This is a very real problem in the extensibility of expert systems. Frequently they can work fine in one problem domain but are utterly useless in an apparently similar problem domain.

B. Applications of AI for Ship Identification

Another project at the Navy Center for Applied Research in Artificial Intelligence concerns the correct identification of ships from images. The methodologies used in this expert system are spin-offs of the BATTLE system discussed above. A prototype decision aid has been developed that demonstrates the feasibility of using AI techniques to help human interpreters recognize images of ships. The heuristic reasoning processes used by an expert analyst were modeled to perform this task. The prototype system uses heuristic rules provided by a human expert to determine the most likely classification, given the available feature data. The system has been tested on over 100 images of 10 similar classes of ships, and it provides the correct identification 84 percent of the time [35].

Accurate identification of ships in real time can be a very difficult task. It is desirable to identify these ships within minutes of detection. The images are hard to interpret for numerous reasons such as weather, turbulence, lighting, interference, and viewing angle. When these difficulties are coupled with the hundreds of classes of ships that sail the oceans and the time constraints to make accurate decisions, we can understand the need to help operators as much as possible. Experience has shown that even expert operators will have bad days or be so burdened with numerous target detections that they will become less accurate and on occasion even miss obvious clues. The average operator is frequently inconsistent and uncertain. He also has a difficult time determining what should be the next clue to be pursued to reduce the target class ambiguity and finally to narrow his decisions to a specific selection. The thrust of the NRL research headed by Booker [35] was to allow for the use of automatic equipment that performs signal processing and pattern recognition and combine this equipment with AI techniques to aid the operator. It was very important to have the expert system capture the operator's judgment, pattern recognition capabilities, the

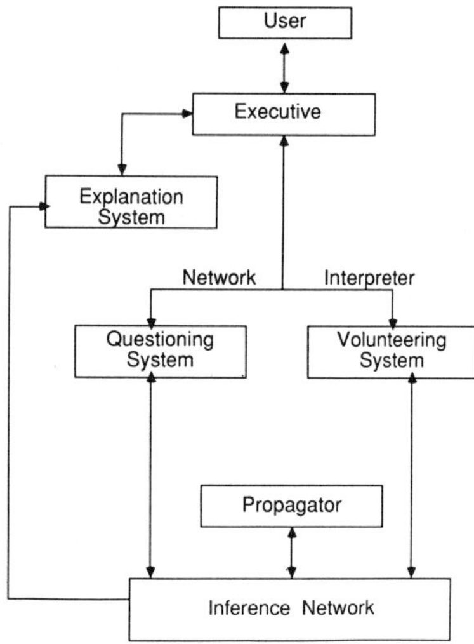

Fig. 7. Expert system design.

knowledge and the reasoning that he uses for his expert analysis.

The block diagram of the system is shown in Fig. 7. The system guides the operator to investigate specific clues or parameters and then suggests an identification for a specific unknown ship. The Explanation System uses the logic provided by the expert and can show the user why the expert system made a specific decision regarding the identification of an image as a specific ship class. The Questioning System decides what question should be asked next. It uses the merit strategy, which is also used in BATTLE, to optimize the order of the questions. The Volunteer System allows the operator to provide as much initial data as he deems fit. It also allows the operator to volunteer additional information during the questioning period. The Propagator calculates the effect of the uncertainty of the data and the operator's observations as the initial information travels throughout the computational network toward the goal of ship identification. The techniques used here are a variation of the Bayesian techniques described earlier and in Appendix II. The Inference Network combines the initial observations and clues into intermediate hypotheses that are combined further towards a final conclusion. Fig. 8 is a simple cartoon that the image expert found very useful in explaining his reasoning to the AI knowledge engineer. Concepts such as, **"do the masts have a linear taper,"** as shown in the top of Fig. 8, were important to the image expert's approach to identifying specific ships. Fig. 8 depicts other concepts such as nonlinear tapers, linear tapers with a gap, and no tapers. For example, the system might ask if there is a linear or nonlinear taper in the forward mast structure, and the answer would be an input to the expert system. After the operator is asked about the presence of a specific clue, such as "is there a linear taper in the forward section?" the operator will respond with a number between **−5** and **+5**. A **−5** indicates that the operator is sure that this characteristic does not exist, and a **+5** indicates that the operator is sure that this clue is present. A zero would indicate that the operator does not know if that feature is present or not. In this way the operator can specify his uncertainty. These answers are then used as inputs to the computation network. A typical network is shown in Fig.

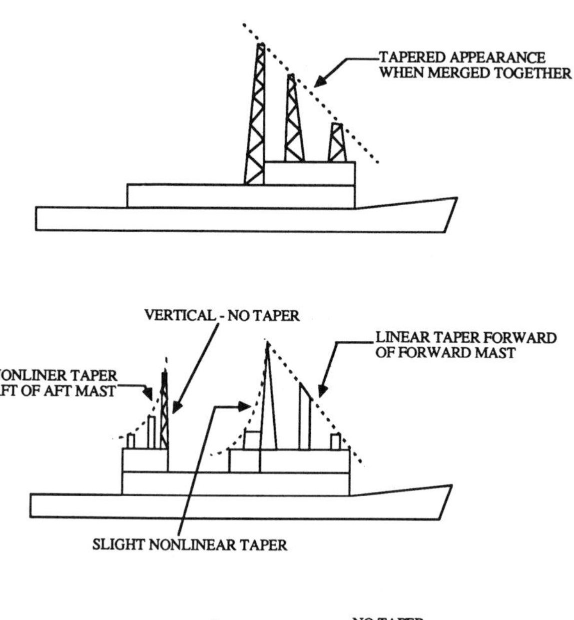

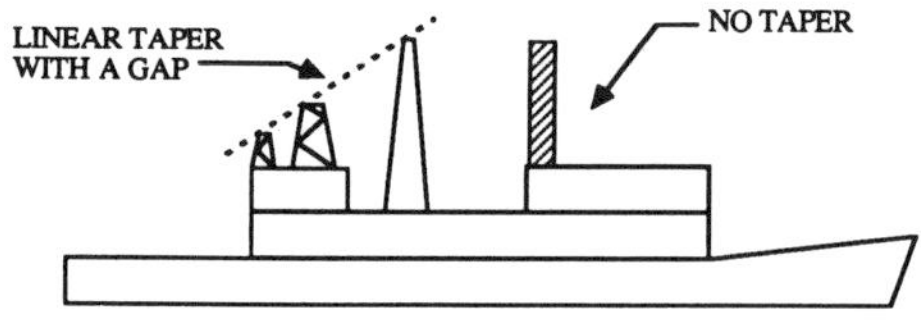

Fig. 8. Mast taper descriptions.

9. This network is based on the techniques used in PROSPECTOR [37] and in BATTLE as discussed earlier. Fig. 9 is just a small subset of the complete computation network. Here, as in BATTLE, the network combines Boolean operators and Bayesian evidence functions. The inputs to this network are related to the various clues that the image expert found useful. As can be seen from the figure, these clues are associated with locations of masts, locations of superstructure blocks, and the presence of linear and nonlinear tapers.

The order of the questions is changed dynamically by the merit strategy developed at NRL and used for the first time in BATTLE. This technique that was described previously has been shown to reduce the required number of questions needed to identify a ship. Excerpts of an actual session using this expert system are shown below. The expert system's questions and comments are shown in bold font.

Expert system question 1: **To what degree do you believe that the profile-has-isolated-directors-fwd-of-bridge-with-a-linear-taper?**

Human response to question 1: **(−3)**
Note that the operator does not believe that this condition is present.
Expert system comment: **I neither suspect nor doubt the possibility of ForestSherman, Bainbridge, Sverdlov, Coontz, Leahy, Belknap, Truxtun, California or Virginia.**
I doubt that image84 could be LongBeach

Note at this point that the machine already doubts the Long Beach ship class is likely and is neutral about the rest.

Expert system question 2: **To what degree do you believe that the profile-has-a-gap-between-aft-mast-and-aft-superstructure-block?**

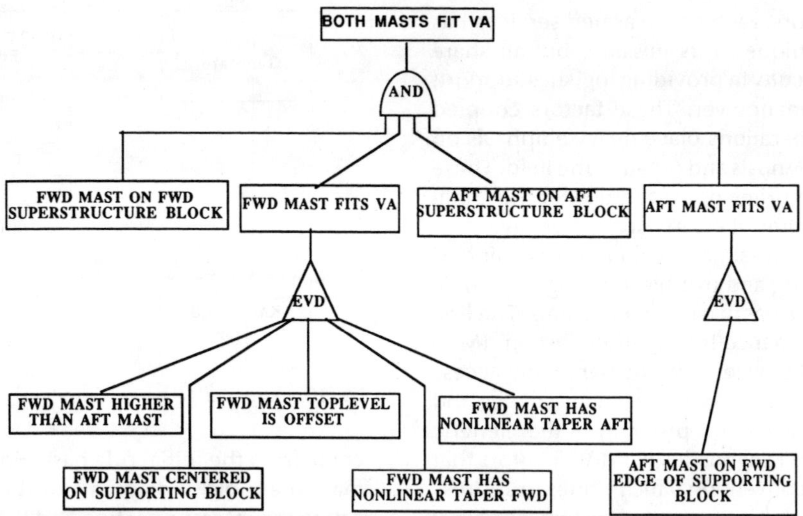

Fig. 9. Typical inference network.

Human response to question 2: **(+5)**
Note that the operator strongly believes that this clue is present.
Expert system comment: **I strongly suspect that image84 is Coontz**
I doubt that image84 could be ForestSherman, Sverdlov, Leahy, Belknap, Truxtun, California or Virginia
I strongly doubt that image84 is Bainbridge or LongBeach

Note at this point the machine has asked a question using the merit strategy to determine the most useful question to answer. Indeed, with the response from the user, it is already focusing on the Coontz class.

These questions continue, and at the end the machine responds with its list of potential classifications (identifications) in a priority order. In addition, the machine gives its confidence of these classifications as shown below.

Class	Confidence	Class	Confidence
Coontz	4.9	Virginia	−5.0
California	−4.8	Sverdlov	−5.0
Bainbridge	−5.0	Leahy	−5.0
Truxtun	−5.0	ForestSherman	−5.0
Belknap	−5.0	LongBeach	−5.0

Expert system comment: **I strongly suspect that image84 is Coontz**
I strongly doubt that the image84 is ForestSherman, Bainbridge, Sverdlov, LongBeach, Leahy, Belknap, Truxtun, California or Virginia

Note that the expert system has listed the Coontz first with a confidence factor of positive 4.9. The California class was listed second, but the expert system was not very confident of this because it has a confidence factor of −4.8. Clearly it did not believe it was the California class or any of the others that are on the list. This is an extremely useful feature since it can help the operator decide if he is satisfied with the expert system's conclusions. In this case, the expert system was very confident that it was the Coontz and none of the others.

There were 119 images investigated. This included 101 total images that were from the chosen ship classes that were represented in the expert system. The other 18 were for ships not contained in the chosen category and the expert system did not know of their existence. The success rate was 84 percent. In 85 of the 101 trials, the ship class ranked first by the expert system was the correct identification. In 94 of the 101 trials, the correct target class was listed as a plausible identification. The expert system provided consistently useful discrimination between the top ranked class and the second rated class. The summary of this discrimination is shown below.

Correct trials:
Average confidence in top ranked class was 1.7.
Average confidence in second ranked class was −2.4.
This means that the system usually comes up with only one plausible classification.
Incorrect trials:
Average confidence in top ranked class was −0.16.
Average confidence in second ranked class was −0.99.

Note that in 84 percent of the cases where the machine correctly identified the ship, the expert system average confidence was 1.7 and this indicates a relatively high positive belief that the chosen top ranked identification is correct. Further, the expert system had a confidence of −2.4 in the second ranked class and this indicates a fairly strong belief that this is not the correct class. This helps the user believe in the expert system's results because of the large separation of +1.7 and −2.4. In the cases where the machine incorrectly identified the image, its confidence in the first choice was −0.16. Clearly, the machine was not "sure" of its choice, and the user can see this and would probably want to acquire a new image and add more results prior to a final selection. More detailed results can be found in a publication by NRL's Booker [35].

C. Maintenance and Troubleshooting

Since the early 1960s military equipment has steadily increased in complexity and variety, while at the same time the pool of trained technicians has been decreasing. A major cost of operations is in fault diagnosis and repair, the procurement of maintenance equipment, and the training of technicians and operators. Each of the armed services has problems which are unique to its mission, but all share problems of space, difficulty in providing logistics support, and limited technical manpower. These factors coupled with the demands of operations place heavy emphasis on speedy and accurate diagnosis and repair in the field. These difficulties have created prime opportunities for the application of AI, and a number of efforts are underway.

All three services are investigating alternatives using AI in the area of maintaining and troubleshooting electronic equipment. The Air Force has two major programs. The first is the Integrated Maintenance Information System (IMIS) project that is designed to give flightline personnel access to all onboard diagnostic data as well as maintenance and scheduling records. The second program is the Generic Integrated Maintenance Diagnostics (GIMADS) effort that proposes to use AI and conventional techniques to address the diagnostics problem for an integrated system. The Navy has two major programs, Integrated Automatic Test Generation (IATG) and the Integrated Diagnostic Support System (IDSS). The IDSS will use expert systems to assist the equipment technician to troubleshoot by providing optimal test trees that are adaptive to the changing conditions. The following examples of AI being applied to the field of maintenance and troubleshooting will draw heavily on the efforts of DeJong, Pipitone, Shumaker, and Cantone [36], [38]–[41] at the Navy Center for Applied Research in Artificial Intelligence. Other relevant work includes Duda *et al.* [37], DeKleer [42], and Davis *et al.* [43].

There are some basic technical issues that should be addressed when AI is applied to maintenance and troubleshooting. It is not a straightforward transition from an application of medical diagnostic successes such as MYCIN to electronic troubleshooting, because too many unexpected causes of equipment malfunctions cannot be anticipated in a traditional if-then rule-based backward-chaining paradigm. For example, solder splashes that short out several components on a printed circuit board, a cable that is crushed because someone rolled a heavy piece of equipment onto it, or a piece of equipment that was doused with a pot of hot coffee are typical unfortunate circumstances that sometimes approach the routine. All of these conditions cannot be anticipated with appropriate rules, and thus a more general methodology must be used.

As an insight to the reader, the following description of an actual expert system at NRL will be used. The system design was evolutionary and several distinct systems were developed and tested. Fig. 10 shows how the knowledge base was generated in the first NRL system. The initial rule base, as shown in this figure, consists of the heuristics, ad hoc rules, and information about the specific design and the operating environment. In this way the knowledge of the designer, the operator, and the technician can all be tapped, and the results of their experience and understanding of the problem can be exploited and put into the initial knowledge base. This initial rule base will not be adequate to resolve all of the troubleshooting problems that will be encountered in the field. As an augmentation to the system, a more general source of information will be used, which is shown in the upper right-hand corner of Fig. 10. This portion of the knowledge base must use other sources of information that can be used to generate rules by the expert system. These rules should be in the same form as those that

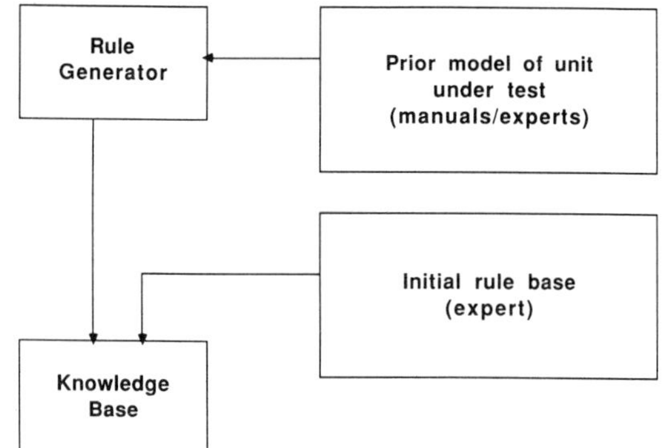

Fig. 10. Knowledge base compiler.

come from the initial rule base. Some sources of this information are block diagrams of the actual equipment indicating the interconnection and the functionality of the system components. The basic information about the circuit can be combined with generalized circuit analysis techniques and fault diagnosis techniques. In this way, new rules can be generated by the expert system from an understanding of the circuit and the problem. Reliability and maintainability analysis should also be used. This would include statistical data associated with past failures as well as failure prediction algorithms.

One of the new features of the approach by Pipitone [41] is that he has included relationships of how the module will react functionally to various faults. An example for a voltage controlled oscillator (VCO) would be that an input voltage that is outside of the desired range will cause the output to be outside of the desired frequency range, in the same direction and with the same relative magnitude. Another causal relation could specify that if the output load is reduced below a certain value the signal will decrease. Another example associated with a high pass filter would be that if the output of a high pass filter is a high dc voltage, the high pass filter is very likely to have failed. More examples can be found in recent publications [38]–[43]. One of the desirable features of an approach using a causal functional description of each module is that it can be used with slight modifications when describing a similar module in another system. Recent results indicate that the addition of this causal functional knowledge requires fewer tests than an approach that uses no functional information.

Fig. 11 shows one configuration for the block diagram for the diagnostic system. The knowledge base shown at the top is the same one that was discussed and shown in Fig. 10. The inference engine investigates and chooses what rules should be used from the rule generator in Fig. 10. If there is no appropriate rule, it will generate one that can be used to assist in the troubleshooting. An important consideration for this expert system is to decide what is the next best test to be performed.

The next best test should consider what test will cost the least and will result in the most information gain. Initial attempts at calculating the next test used game theory derived from earlier AI research. This technique would consider the machine as the opponent, similar to an opponent in a chess match, and the strategy would be to maximize your gains and minimize the anticipated losses caused by your opponent's move. This proved to be computationally infeasible [36]. Next, a Gamma Miniaverage [38] technique

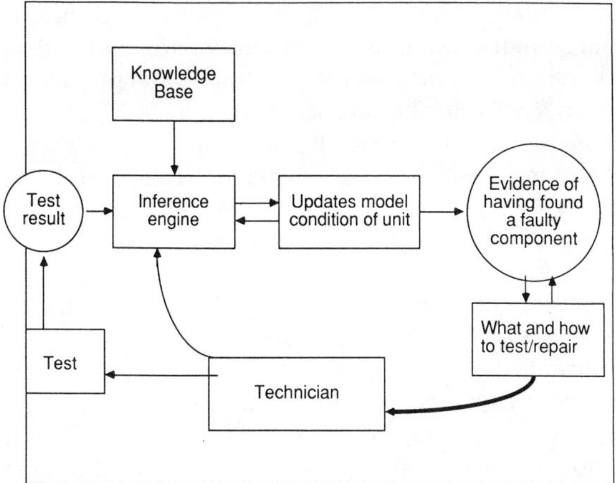

Fig. 11. Diagnostic system.

by Slagle was used. In this case, average losses were calculated, since the equipment was not actively plotted against the technician (although some technicians have argued to the contrary). The miniaverage approach was still computationally too costly. Pipitone [36], [39]–[41] introduced heuristic screening at first and then used a one level miniaverage computation. The results in the Pipitone work have shown that functional knowledge is a necessary component of the troubleshooting process and that the total number of necessary tests can be reduced. In addition, the functional knowledge can be used to show the technician how the test can be done and what results should be anticipated.

The progress to date has shown that expert systems can help in the area of diagnostics and troubleshooting. The first expert systems will be used as decision support systems to aid the technician. As progress is made, the state of the art will allow the use of expert systems to be used in an autonomous mode in applications such as generating automatic test equipment code or space-based maintenance and troubleshooting.

D. Advanced Digital Radar Imagery Exploitation System (ADRIES)

The Advanced Digital Radar Imagery Exploitation System (ADRIES) is a software prototype testbed for research on extraction of information from radar imagery. Its objectives are to provide a system for enhancing and automating various aspects of digital radar image interpretation, for applications including tactical intelligence missions, military situation assessment, and target recognition. ADRIES is capable of producing interpretations of the possible military situations with a set of radar imagery, collection parameters, *a priori* terrain data, such as maps or digital terrain databases, and other tactical data. ADRIES is currently under development. Key work on terrain, detection, and non-radar intelligence source reasoning will be presented in future publications. Here we emphasize intermediate results in knowledge and model-based military unit inference.

ADRIES is founded on a theory of model-based, Bayesian probabilistic inference. Models represent knowledge of the organization and formations of military units, and they also specify how knowledge of terrain provides evidence in support or denial of the presence of types of forces at given locations. A probabilistic certainty calculus [44] specifies how evidence extracted from SAR imagery and terrain databases is matched against the models and combined to infer the presence or absence of military forces. In particular, radar data, forces, terrain or other entities that have been modeled in the certainty calculus can be used as evidence for inferences output by the system. This is the basis for a clean split between inference and control in ADRIES.

The model database of ADRIES implicitly contains all possible chains of inference that the system can use to draw any conclusion. However, any information whatsoever can be used for control in the system. As an extreme example, if a human intelligence report was available that indicated the presence of a force in a region, ADRIES could use that to direct processing of imagery and terrain of that region without danger of circular reasoning in its conclusions. This is because any output hypotheses about forces in the region must be supported by image and terrain evidence as specified in the model database, and that inference is necessarily from the certainty calculus.

As another example, other source (other than SAR imagery) intelligence data can be used to direct search, make predictions, or act as a trigger to activate agents or other processing, but it cannot be fused to provide evidential support for system outputs unless it is modeled in the model database and in the certainty calculus. It follows that to extend ADRIES to a full fusion system, it will be necessary to do the research on the knowledge representation of other source information, and the probabilistic relationships between other source intelligence and the entities already accounted for in models and the calculus.

Fig. 12 shows the concept behind the model-based Bayesian inference in ADRIES. Evidence, such as clusters of detections, are matched against the geometry of formations that are explicitly represented in the model database. Matches lead to generation of alternative hypotheses of the presence of forces, such as batallions and companies, that are dynamically created (as imagery is processed, for example) and instantiated in a data structure called a Bayesian network. Terrain rules are applied to the terrain extracted from terrain databases of the area corresponding to the imagery, and the certainty calculus combines that evidence with the model match in computing the probabilities asso-

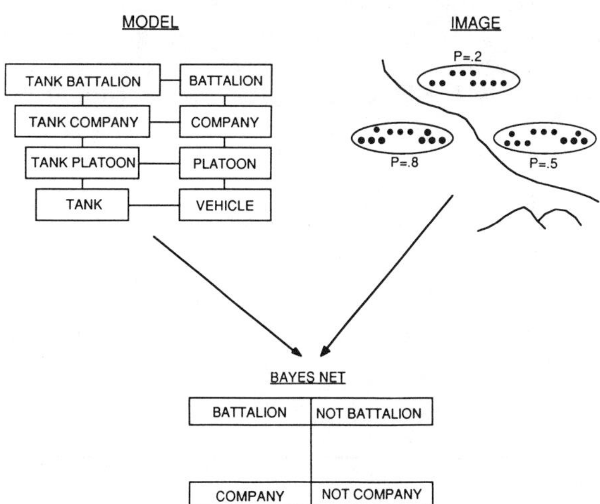

Fig. 12. ADRIES inference methodology.

ciated to the hypothesized forces in the Bayesian network. In practice it does not matter in what order the terrain evidence probabilities or goodness of formation match probabilities are calculated, as the certainty calculus will obtain the same values regardless of the processing order. The calculations used for computing prior and runtime probabilities in the calculus are presented in [45], [46], [44]. See also [47].

In the course of the ADRIES program, theoretical work has also been performed on detection algorithms [48], [49], clustering algorithms, and elicitation of terrain rules and probabilities [45], [46]. All of this work is incorporated in functionality of the relevant distributed processing agents.

ADRIES is built as a distributed set of software agents communicating by message passing. The agent decomposition for ADRIES is pictured in Fig. 13. There is also a set of databases used by multiple agents in their processing. These are pictured in ovals in Fig. 13. All agents have access, either directly or indirectly, to all databases. Individual agents may have additional databases. In the following, we briefly summarize the functionality of each agent.

Control/Inference Agent: The Control/Inference agent plans system processing to fulfill the exploitation requests (ER) received from the user interface. Basically, an ER specifies what forces to look for in which geographic locations. It posts its processing as messages sent to the Agenda agent. The Control/Inference agent maintains the Bayesian network and decides when to generate a new hypothesis in the Bayesian network based on the available evidence. It also decides when to terminate processing on an ER.

Agenda Agent: The Agenda agent receives process plan messages from the Control/Inference agent and sends them on to the Exec agent. It provides a loose coupling between the planning in the Control/Inference agent and the resource allocation in the Exec agent.

Exec Agent: The Exec agent picks up the current processing plans from the Agenda agent, and performs resource allocation to distribute processing on the distributed system. It sends back a summary of its process messages to the Control/Inference agent.

Imagery Location Agent: The Imagery Location agent is a spatially indexed database of imagery in ground coordinates. It keeps track of virtual sub-images cut from larger images, and maintains message-level records of the processing done to imagery. The imagery files are not resident with this agent; in particular, the Imagery Location agent does not send or receive actual imagery.

Registration Agent: It performs coarse registration to compute the ground coordinates of an image based on its platform parameters and flat world assumptions. The Registration agent also computes a refined registration of the image to the ground. For a given image, it computes a function that takes the elevation at a point in the terrain and outputs the corresponding point in the image.

Lo-Res Detection Agent: This agent detects potential vehicles in low resolution imagery. It also computes the likelihoods corresponding to the hypotheses vehicle versus non-vehicle.

Clustering Agent: The Clustering agent takes detections and their probabilities as inputs, and outputs clusters of detections and the probability that the cluster contains a military unit of "array" size (e.g., 8–15 vehicles). It accounts for inter-vehicle spacings, likelihood of false alarm detections, and dispersion of the cluster on the ground versus the expected extent of the array-sized formation.

Spot Mode Detection Agent: This agent performs vehicle detection on high resolution imagery. It also computes probabilities of vehicle versus non-vehicle.

Vehicle Classification Agent: This agent performs vehicle recognition on high resolution imagery, and also computes probabilities over the set of hypotheses of the vehicle type.

Other Source Agent: This agent provides signals intelli-

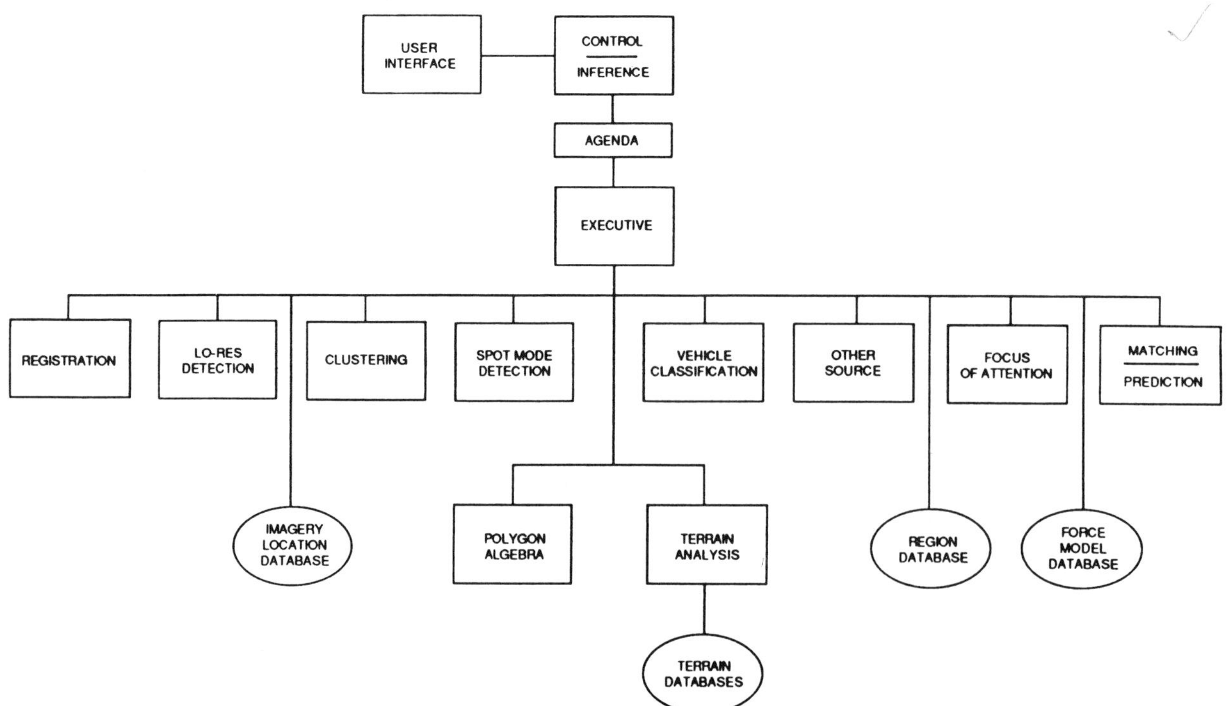

Fig. 13. Agent decomposition.

gence and other source intelligence summaries indexed by geographic location and force type.

Focus Agent: Focus takes as input a geographic region and a force type and outputs a prioritized list of sub-regions for search for forces of that type in the region. Focus uses inputs from the Terrain agents and the Other Source agent.

Matching/Prediction Agent: Matching/Prediction takes as input the locations of hypothesized forces and matches them against doctrinal formations. It also interacts with the Terrain agent to adjust military formation models to fit the underlying terrain according to military doctrine of deployment. Predictions are provided by reasoning about the forces missing from partial matches. Matching/Prediction also provides the probabilities for goodness of fit to formation used as evidence by the Control/Inference agent.

Terrain Analysis Agent: Terrain Analysis is currently performed over three different terrain rule bases. Local Terrain Analysis takes as input a force type and a geographic region and uses terrain rules to compute sub-regions of constant probability indicating the likelihood that a force of that type will be located in the sub-region. The terrain rules used are those that take as pre-conditions terrain attributes that can be gotten from direct access to the terrain database. The reason for separating these rules from "structural" terrain rules is that we have created a probability model for local terrain rules. Thus, these probabilities can be combined as evidence by the Control/Inference agent as part of the probability of output hypotheses.

The Mobility Corridors sub-agent takes as input a force type and geographic region and outputs sub-regions through which a force of a given type can move across the entire region. This sub-agent is intended to be upward compatible with a future Avenues of Approach agent that does more global reasoning over the tactical situation.

The Structural Terrain Analysis sub-agent takes as input a force type and a geographic region and uses terrain rules to compute sub-regions that are acceptable for a force of the given type to occupy. These rules are relational in nature; that is, they use as pre-conditions terrain attributes that cannot be gotten by direct access from the terrain database. No probability model currently exists for these rules. Instead, they interact with formation matching to adjust expected force formations based on terrain and military constraints.

User Interface Agent: User Interface allows the user to interactively input an exploitation request. It also displays all system results, including imagery, terrain overlays, and multi-level force hypotheses. The user can interact with the outputs to obtain explanations for the system's conclusions.

Inference in ADRIES is performed over a space of hierarchically linked hypotheses. The hypotheses typically (although not solely) represent statements of the form "There is a military force of type *F* in deployment *D* at world location *L* at time *T*." The hierarchy in the hypothesis space corresponds to the hierarchy inherent in military doctrine of force structuring. Thus, "array-level" hypotheses of military units such as companies, artillery batteries, and missile sites are linked to their component unit hypotheses of vehicles, artillery pieces, and missile launchers. Similarly, moving upward in the force hierarchy, companies are grouped to form battalion hypotheses, battalions to form regiments, etc.

The hypotheses are generated by hierarchical and partial matching of military force models to evidence available in radar imagery. Thus, ADRIES is a model-based radar vision system. Evidence of the truth (or denial) of a hypothesis is accrued numerically from probabilistic estimates about the sub-hypotheses that comprise their parent hypothesis.

Although control of processing in ADRIES can be complex, the structure of inference follows a pattern based on the models of military forces. These models consist of force types (i.e., names) and spatial-geometric deployment data relating force types. The part-of relationships for generic force models are shown in Fig. 14. Numbers next to forces indicate how many of a given force are expected as com-

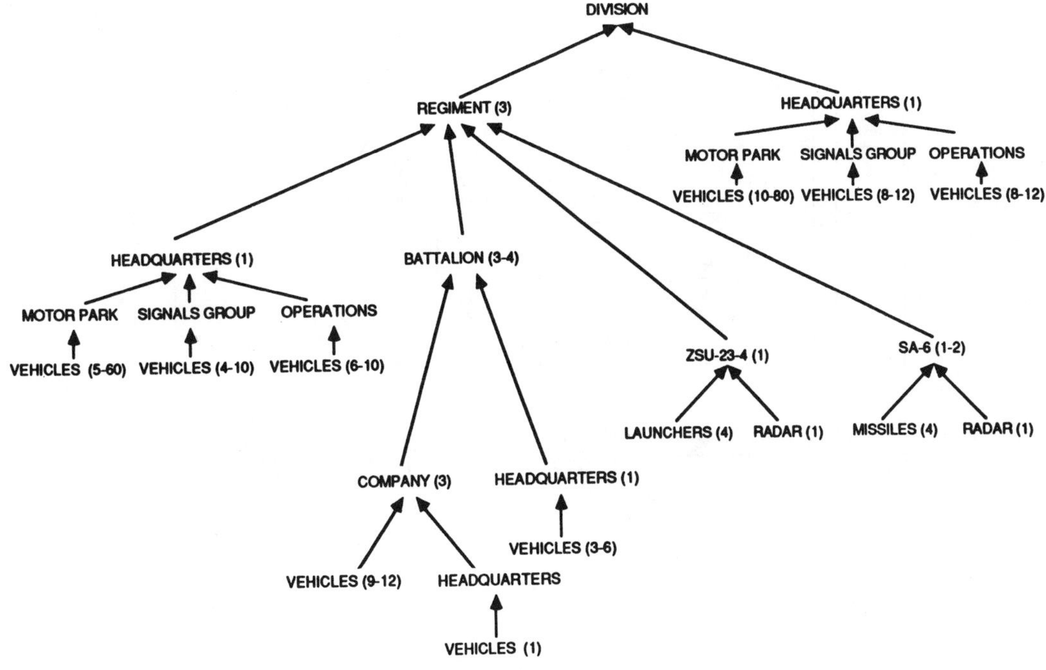

Fig. 14. Part-of hierarchy for generic forces.

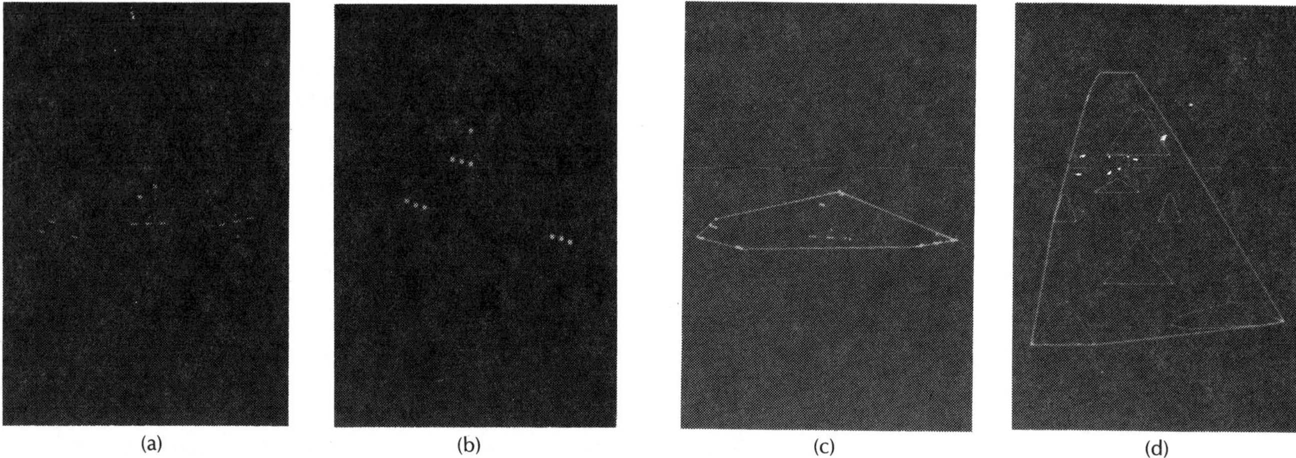

Fig. 15. Force deployment models. (a) Defensive battalion formation. (b) Defensive company formation. (c) Battalion constraints. (d) Regiment constraints.

ponents of the parent force. Fig. 15(a) shows a standard defensive deployment formation for a battalion, while Fig. 15(b) shows the formation associated to the pattern of vehicles making up a company. Models are also represented along "is-a" or type relationships by appropriately weakening model constraints. For example, a "tank company in defensive deployment" is-a "tank company" which is-a "company" which is-a "array-level force."

The models are represented as constraints in semantic nets. A constraint consists of a relation name, parameters associated with that relation, and additional model attributes. Fig. 15(c) and (d) give pictorial representations of constraints for a battalion and regiment. Fig. 16 illustrates the internal representation of the constraints in the force-deployment models as semantic nets. At the top of each constraint is a parent model item that is represented by a rectangular box. A box with rounded corners represents the relation and its associated parameters. Both parent and subordinate model items can be forces or formations.

ADRIES has a utility and model-based approach to control of force-deployment pattern matching. Algorithms are employed opportunistically, as determined by the utility of tasks in light of the currently available data, rather than necessarily being checked or applied in a fixed order. Some

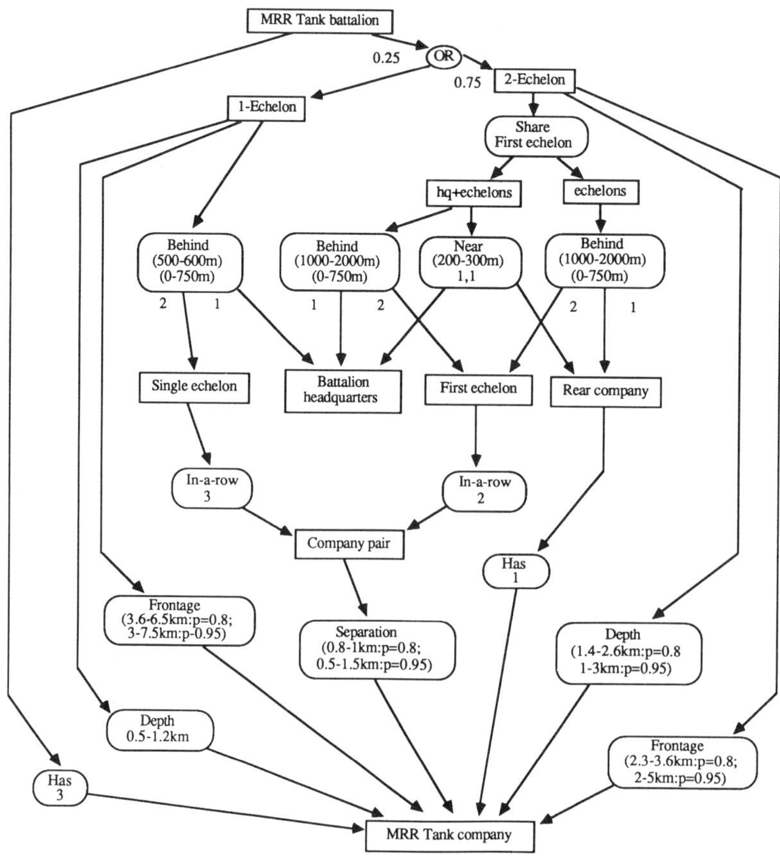

Fig. 16. Battalion semantic net.

example rules, implicit in the model-base, and inference chains are pictured in Fig. 17. (The inference chain pictured uses more models and rules than are shown in the figure.) The utility theory is derived from decision analytic methods on top of the Bayesian inference. For details of the approach, see [57], [58].

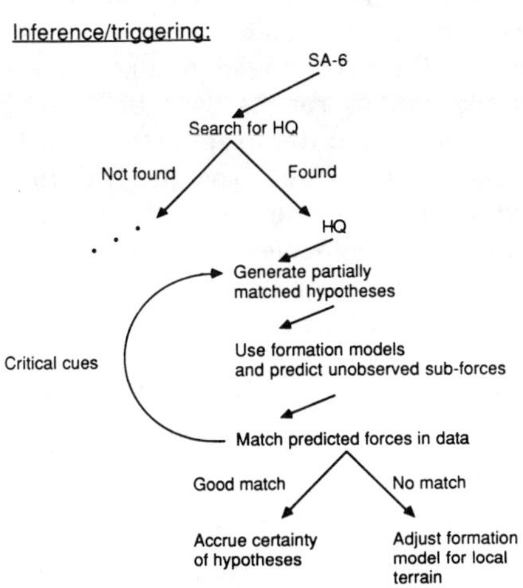

Fig. 17. ADRIES knowledge-based processing.

The concept of conflict resolution and numerical accrual of evidence is that, while it is often the case that automated matching can be locally incorrect, the weight of the global evidence will override local mismatches and result in an unambiguous interpretation of the tactical situation. The global picture is presented in ADRIES as a set of goals about what it expects to see in the SAR imagery. For example, a command to verify the presence of a division sets up a goal to find a division. The Control/Inference agent then looks up the structures according to the local terrain and military situation. The model is used to set up sub-goals recursively to find the component military units of the division. The existing force-deployment hypotheses are stored as nodes in a Bayes net. These represent the goals and subgoals that have already been pursued or have been satisfied.

Thus, the goal structures and the force-deployment hypothesis space are dual to each other. In the course of processing, control moves back and forth between them. Goal structures help predict where to look for forces in the imagery and the force deployment hypothesis space. Having localized a search area, data is retrieved from the Bayes net and matched against the models associated to the goals. Matches may trigger bottom up actions to infer other forces hypotheses, which may in turn trigger the generation of new goals.

A fundamental concept here is that, while vision system processing may be complex, with numerous feedback loops, multiple levels of resolution, recursion, etc., in the end we should be able to associate a deductive chain of evidence to a system output, along with an associated probability that supports that result.

We selected probability theory as the underlying technology for this numerical accrual of evidence. One of the major motivations for this choice is that Bayesian inference, a well-developed scientific theory, already exists for probabilistic evidential reasoning; see, for example [50]. This approach requires us to lay out, *a priori*, the links between evidence and hypotheses in the models over which the system will reason. Having laid out these links, we then need a numerical interpretation of the conditional belief, i.e., probability, in a hypothesis, given chains of evidence that support it through links. This is similar in spirit to propagation networks [51], to influence diagrams [52], and other probabilistic accrual models. Hierarchical Bayesian inference was introduced by Gettys and Willke [53], Schum and Ducharme [54], Kelly and Barclay [55] and has been carried forward by others, e.g., [44], [56], [57]. Fig. 18 shows the

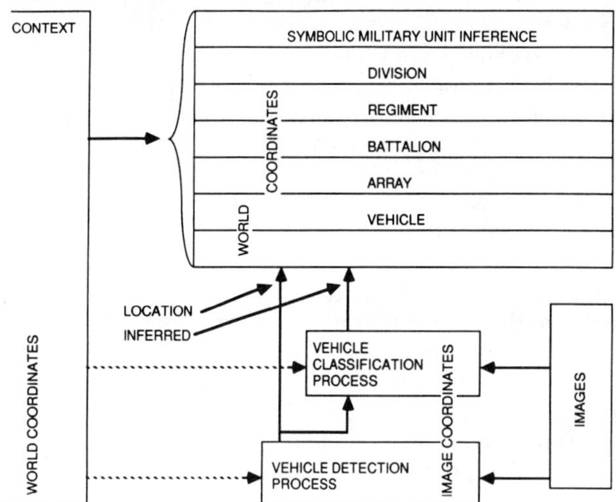

Fig. 18. Hierarchical hypothesis-space relationships.

hypothesis and evidential accrual hierarchy for Bayesian inference in ADRIES.

The evidential accrual for the generated hypothesis is performed by using a Bayesian network. This paradigm supports the propagation of evidential probabilities through a hierarchical tree of hypotheses in a coherent, stable, and efficient manner [51]. The result of the evidential accrual is a posterior belief value for each hypothesis indicating its relative likelihood. Mutually exclusive conflicting hypotheses are collected in a single Bayes net node. Each node then consists of a belief vector indicating the current relative belief that each of the conflicting hypotheses is true. For example, a node can group the beliefs for a cluster being a battalion or being a false alarm (i.e., not a battalion). The links connecting the nodes consist of conditional probability matrices indicating the relationship of the hypotheses. These probabilities are derived from terrain constraints and locality of the hypothesized forces. For example, for a Bayes link connecting a battalion node (X) with a company node (Y) these probabilities specify:

P(particular image clusters are companies | particular region is battalion)

P(particular image clusters are companies | particular region is false alarm)

P(particular image clusters are false alarms | particular region is battalion)

P(particular image clusters are false alarms | particular region is false alarm).

These conditional probabilities form the elements of a matrix, $M(Y|X)$, where the (i, j)th entry specifies $P(Yj|Xi)$.

Messages passed between Bayes nodes consist of probability vectors providing evidential support (child to parent) or model-driven support (parent to child). Evidence coming into a node will normally be in the form: *P*(image features are due to $Y|Y$).

Evidence for a Bayes company node, for example, would therefore consist of the message: [*P*(image features are due to a company | company), *P*(image features are due to false alarms | false alarm)].

A model-driven support message from a node *X* to its child *Y* consists of a vector derived by dividing (component-wise) the current belief vector by the current evidential message from *Y* to *X*.

Upon receiving a new message from a parent or child, a Bayes node, *X*, (for single parent trees) computes its overall evidential support vector, *Dx*, by multiplying (component-wise) all evidential vectors, *Dx*, from all its children *Yi*. The new resulting belief vector is the result of: $Dx * [(\text{transpose } Mwx) \text{ o } Cw]$, where $*$ indicates component-wise multiplication, o indicates matrix multiplication, *Cw* is the model-driven support message from the node's parent *W*, and *Mwx* is the conditional probability matrix from *W* to *X*. New messages are then generated for each of the node's children *Yi*, by dividing (component-wise) the belief vector by the current evidential message from *Yi* to *X*. A new message is generated for the node's parent *W*, by computing: *Mwx* o *Dx*, where *Mwx* is the conditional probability matrix from *W* to *X* and o indicates matrix multiplication (See Figs. 19 and 20).

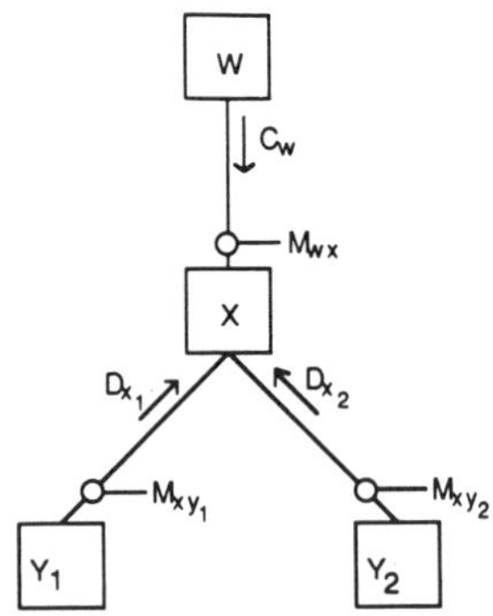

Fig. 19. Messages into Bayes node *X*.

The advantage of using Pearl's algorithm for belief propagation is that it can be implemented as an active network of primitive, identical, and autonomous processors. The primitive processors are simple and repetitive and thus can be implemented in a distributed environment. Each node in the Bayes net can be assigned a virtual processor that sends messages in the form of probabilistic belief vectors to parent or children nodes. When a virtual processor receives a new message, it updates its belief vector and propagates the change to its other relatives. Therefore,

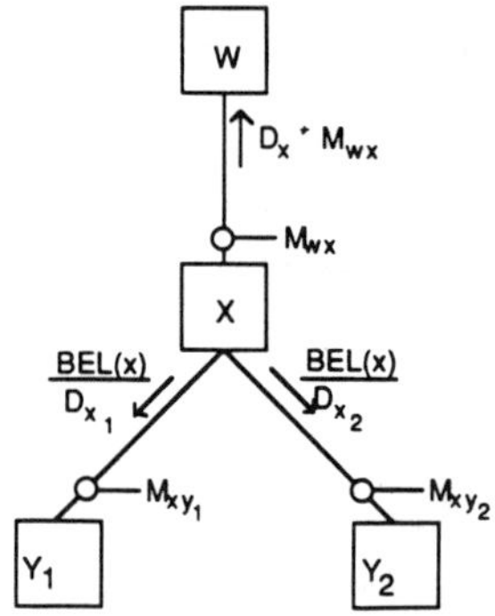

Fig. 20. Messages out of Bayes node *X*.

equilibrium is guaranteed to be reached in time proportional to the network diameter.

Pearl specified his propagation algorithm for static hypothesis spaces. His results have been extended for dynamic hypotheses in order to apply to the ADRIES application. The Bayes net evolves dynamically during processing. New hypotheses get generated for aggregation of lower level hypotheses into higher level hypotheses, while refinement stages generate more possibilities for a hypothesis. Conflicting hypotheses are then merged into single Bayes nodes. An example of the generation of a Bayes net is shown in Fig. 21. This figure shows the generation of two conflict-

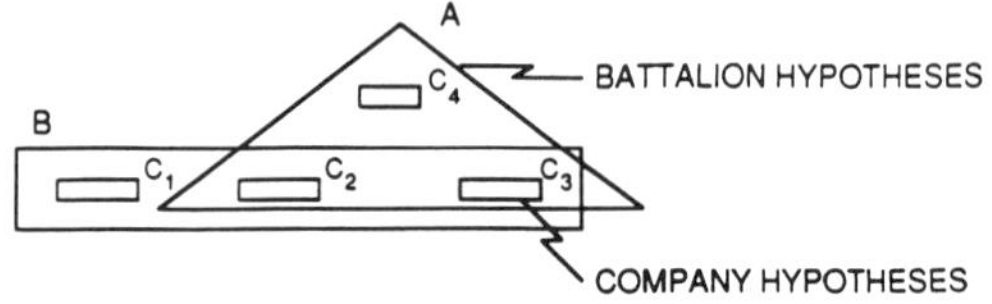

BAYES NET

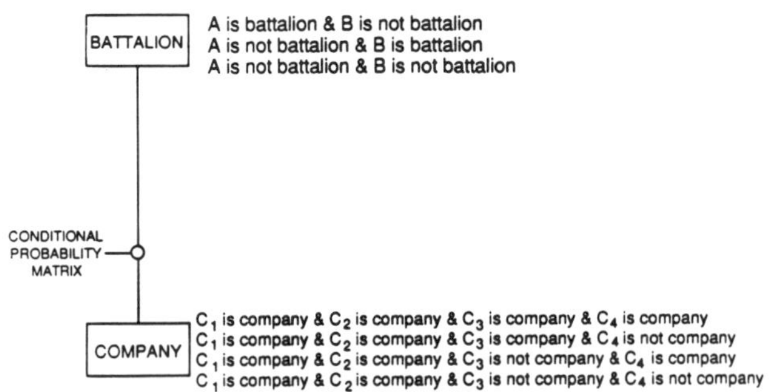

Fig. 21. Bayes net generation example.

ing battalion hypotheses based on four company hypotheses. The diagram at the top shows formation region constraints of possible battalions (the triangle and rectangle) fitted to company formation constraints (the small rectangles). The structure of the Bayes net parallels the structure of the Force Models Database. As larger force hypotheses are generated, Bayes nodes corresponding to force models further up the part-of hierarchy will be generated. As the force type and deployment type of the hypotheses are refined, Bayes nodes corresponding to force models further down the is-a hierarchy will be generated.

Here we give a simple example of Bayesian inference in ADRIES. The approach is detailed in [44]. Fig. 22(a) shows

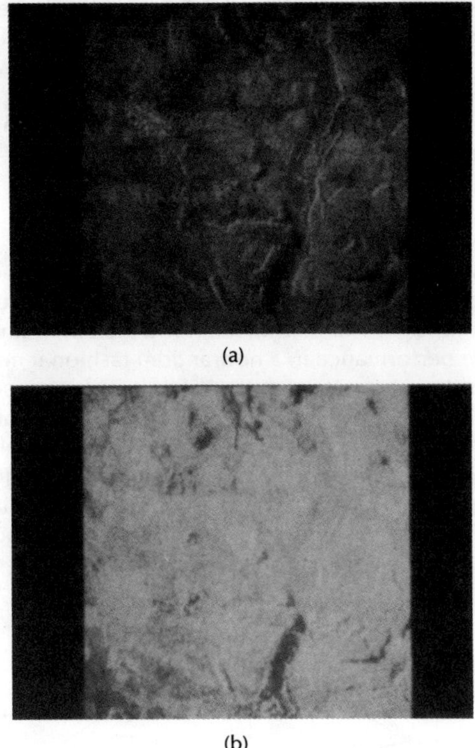

(a)

(b)

Fig. 22. Original imagery. (a) Original SAR image. (b) Ground truth for vehicles.

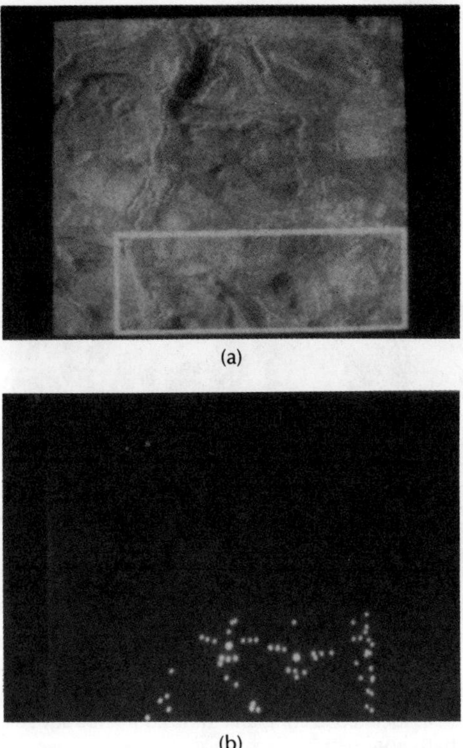

(a)

(b)

Fig. 23. Focus of attention and detection processing. (a) Focus of attention region cued from other intelligence sources. (b) Vehicle detections.

the original SAR image. Fig. 22(b) shows the original image with ground truth of tank locations indicated. Note that there are three tank companies, deployed in defensive formations. Fig. 23(a) shows a focus of attention sub-region of the original image obtained from other intelligence source cues. Fig. 23(b) shows the results of detection algorithms run on the focus of attention image region in Fig. 23(a).

Fig. 24(a) shows the results of applying the defensive formation tank company mode matching algorithm to the detections. This is the model of Fig. 15(b). Note that matching was restricted to a localized search area, rather than processing detections from the entire image. Because partial matches are allowed, many incorrect and partially correct matches are made. Because more than one company cannot claim the same vehicles or occupy the same territory, each of these tank companies is in conflict with the others. Fig. 24(b) shows the resolution of the conflicting matches.

Fig. 25(a) shows the probabilities involved in resolving conflicts between competing company hypotheses. The notation "*S*<three-digit-number>" indicates a grouping of three tanks in a line with approximate spacing corresponding to squad groupings in the defensive deployment formation being searched for by the tank company matching algorithm. The notation "*C*<two-digit-number>" is the grouping into companies formed in the matching process. Note how multiple companies attempt to claim the same tanks grouped in squads. The numbers to the side of the *C*s are the probabilities computed in support for these company hypotheses based on the calculation in [44].

Fig. 25(b) shows the match of the doctrinal battalion defensive formation to the company data. Note that the formation structure predicts the location of the (not yet found) third company and battalion headquarters.

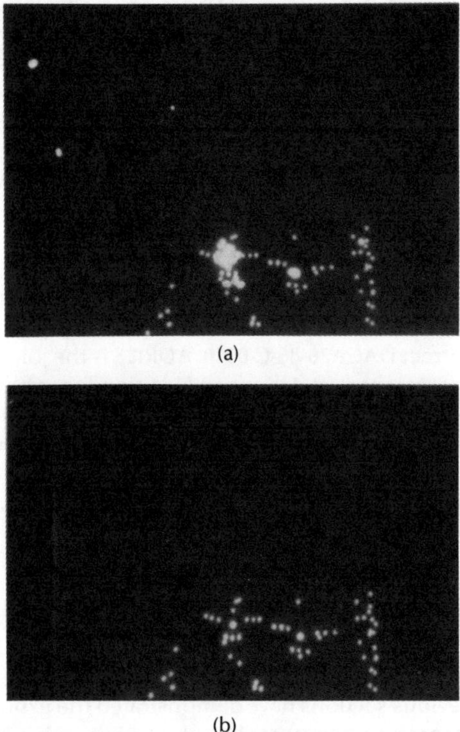

(a)

(b)

Fig. 24. Tank-company matches and resolution. (a) Conflicted company matches. (b) Resolved matches.

At the current state of system development, ADRIES successfully infers the presence and locations of military units through the regimental level in high false alarm data, demonstrating the feasibility of the approach. Knowledge is

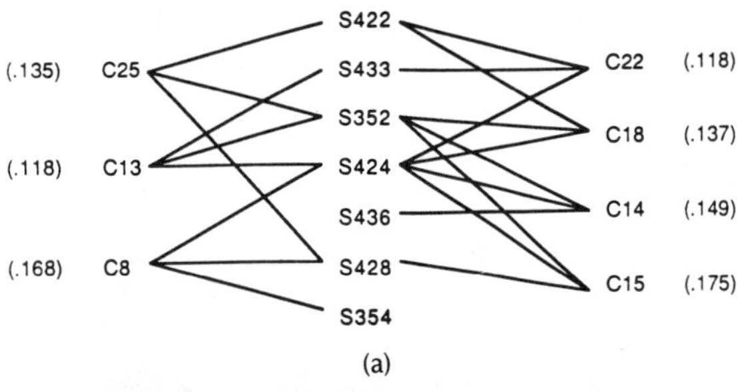

(a)

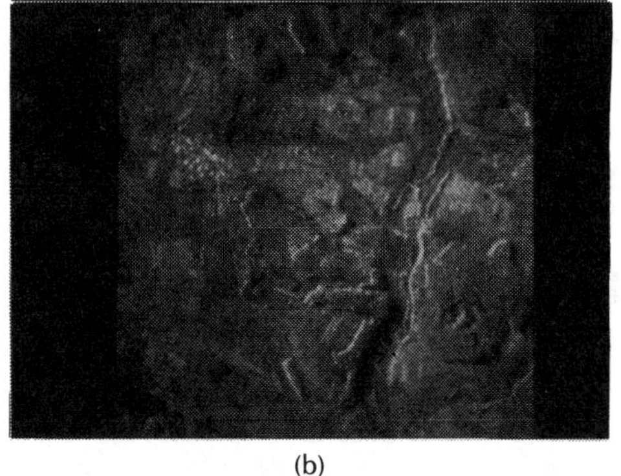
(b)

Fig. 25. Probabilistic reasoning for conflict resolution. (a) Probabilities associated to companies and conflicted squads. (b) Partial battalion match.

being added in ADRIES to accomplish reasoning about situation deployment of forces in terrain. This will enable ADRIES to predict deployment variation to accommodate terrain constraints. Capability for software simulation of input data is being developed to make statistically significant testing of ADRIES possible.

The research for ADRIES was originally supported by the Defense Advanced Research Projects Agency (DARPA), by the U.S. Army Intelligence Center and School (USAICS) under U.S. Government Contract DAEA18-83-C-0026, and by Air Force Wright Aeronautical Laboratories (AFWAL) under U.S. Government Contract F33615-83-C-1070. It is currently supported by DARPA and the U.S. Army Engineer Topographic Laboratories (USAETL) under U.S. Government Contract DACA76-86-C-0010. ADRIES is the joint work of many researchers at Advanced Decision Systems (ADS) in Mountain View, CA, Science Applications International Corporation (SAIC) in Tucson, AZ, and the MRJ Corporation in Oakton, VA. The project supervisors and principal investigators are Bob Drazovich and Tod Levitt at ADS, Dick Kruger and Larry Winter at SAIC, and Bob Ready and Chris McKee at MRJ.

IV. A Learning System for Intelligence Analysis

A. The Need For Learning

The previous sections have demonstrated the expert systems approach to computer-based problem solving. This approach is applicable to a wide variety of problems and provides a convenient mechanism for expressing knowledge. Unfortunately, there are difficulties associated with the approach that limit its intrinsic usefulness. The inability of expert systems to adapt to new situations outside of a particular sphere of knowledge (frequently termed *brittleness* or *falling off the knowledge cliff* [59], [60]) unacceptably limits performance. Machine learning is an emerging technology that can potentially solve a significant portion of this problem by making systems self-modifying, which will improve performance. In expert systems, the improvement of rules in the system can lead to an improvement in system performance.

B. The Machine Learning Approach

Given that the objective of learning is to improve performance, several design constraints can be placed on any system that claims to learn. First, for a system to improve its future performance in a nonrandom fashion it must, at some level, evaluate its current and past performance. The evaluation or critique must consist of comparing observable behavior with some desired behavior. The problem of evaluation is referred to as apportionment of credit [61], or credit assignment [62]. The apportionment of credit is to those elements of the system responsible for good performance. The diagram in Fig. 26 shows the basic feedback

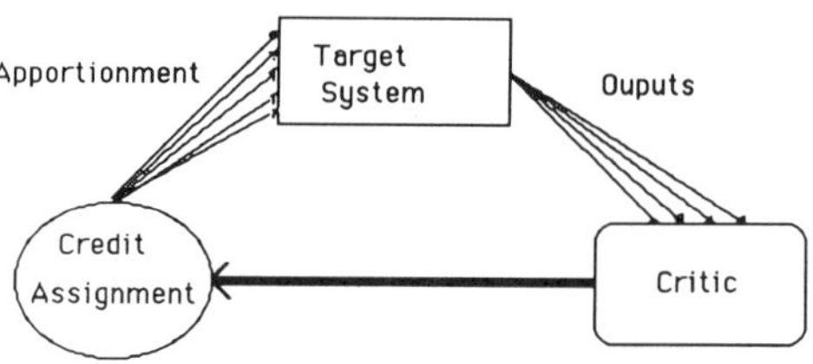

Fig. 26. General feedback mechanism.

mechanism, critic, and apportionment needed to evaluate performance. In the context of an expert system, the apportionment of credit problem boils down to rewarding good rules and (possibly) punishing bad rules. Secondly, the evaluation should lead to or support some lasting modification of the system so that the system avoids the same mistakes next time. Again, for expert systems, this requires that a system be able to alter its rules on the basis of the evaluation.

The rest of the section on learning describes research at the MITRE Corporation, an attempt to apply production system technology to problems in military intelligence. The difficulties associated with knowledge engineering of a large-scale expert system (e.g., extended knowledge acquisition, preference-biased expert knowledge, static knowledge in the face of adaptive adversaries) led to the consideration of incorporating learning capabilities.

Learning systems are usually characterized by three basic elements: domain of application, knowledge representation, and learning strategy. Sections D, E, and F discuss these elements in the context of an existing implemented system called M2. The discussion concludes with issues and further research topics. This brief overview of the basic approaches to machine learning is not intended to be a tutorial on learning; rather, it serves to place M2's approach in the context of previous research.

C. Machine Learning Paradigms

A number of approaches to incorporating learning into computer systems have been developed since the 1950s. Michalski [63] has identified three basic paradigms in the machine learning field that include the learning of concepts (with an extensive domain model and without one), and learning by self-organization.

Acquiring Concepts: Symbolic concept acquisition, SCA, attempts to formulate general concepts from specific examples. These systems have a teacher that provides explicit positive and negative instances of a target concept. In this case the feedback mechanism is simplified by the description of the example as a positive or negative instance of the concept. An example system which makes use of this approach is Quinlan's ID3 program [64]. One application of ID3 produces classifications of chess end games using King-Rook versus King-Knight positions. The approach shows significant speed improvement over more traditional search procedures like minimax and discovered winning strategies overlooked by expert human players.

Using Domain Models: A second paradigm, referred to as knowledge-intensive domain-dependent learning, KDL, uses large amounts of domain-specific knowledge to build a model of the domain from which modifications can proceed. The primary difference between this and the SCA approach is in the amount of knowledge used by the system to formulate useful concepts and their frequent use of symbolic logic and theorem-proving approaches to develop consistent concepts. A well-known system which discovered interesting mathematical concepts using this approach was Lenat's AM system [65]. In AM, the search through the space of possible concepts is guided by a utility function with multiple objectives defining the *interestingness* of the concept being explored. In general, the KDL approach has an applications orientation.

Building from the Ground Up: The last paradigm, termed self-organizing systems, often exploits sampling-based learning algorithms rather than symbolic logic approaches to improve performance [66], [67]. These systems use a series of evaluation functions and a direct feedback mechanism to the rules, nodes, or networks which determine the behavior of the system. An example of this approach is the work done in classifier systems and genetic algorithms by Goldberg [68]. This system was given a random set of rules and tasked to perform a gas pipelining problem. The system developed rules to efficiently route and control gas flow in the network. The genetic approach is discussed in more detail in Section F. As a matter of perspective, it should be noted that these paradigms are not mutually exclusive and that opportunities exist for cooperative learning efforts.

D. The Military Intelligence Problem

The problem associated with intelligence analysis is characteristic of a broad class of problems associated with substantiating hypotheses given a model and some data. Consider a set of sensors which provide the intelligence analyst with a data stream of reports of enemy activity at varying time intervals. The task of the analyst is to generate a description of the enemy units that are generating the activities detected by the sensors. This description consists of identities (e.g., UNIT 1), locations (e.g., RIVER1 at TIME *t*), and goals (e.g., to have UNIT 1 MOVE UNIT 2 to RIVER 1 at TIME *t*). The description leads to expectations of further reports and suggests experiments (sensor tasks) which the analyst should execute. An example of expectations would be:

> If I believe that the goal of UNIT 1 is to move UNIT 2 to RIVER 1 at TIME *t*, then I expect UNIT 1 to move his lead elements to RIVER 1 at TIME $t - 2$. Further, if I put a sensor there, I expect a report at TIME $t - 2$ or TIME $t - 1$.

In general, no information is received which directly corroborates the hypotheses made by the analyst. The only feedback available is in the substantiation of (hypothesis generated) expectations by the continuing stream of reports emanating from the sensors. The analyst generates the hypotheses by applying an underlying model of the enemy's behavior to the data. There are four basic sources of errors in the analyst's description that require more than the traditional expert systems approach to solve: 1) noisy and incomplete data, 2) incorrect models of enemy behavior, 3) deceptive enemy behavior, and 4) adaptive enemy behavior. Arguably, problems 2 and 3 could be solved with enough knowledge engineering, if deception is considered to be another set of rules to be captured.

These domain considerations lead to some projections about the capabilities of a learning system performing the analyst's task. First, it must generate its description without access to ground truth. This requirement is representative of the self-organizing class of learning systems; however, the strategy employed is one of inductive learning in general. See Section F for more on learning strategies. The analyst never knows absolutely whether his analysis is correct even when his hypotheses are substantiated. Second, the system will operate in noisy and possibly discontinuous search spaces in which behaviors are incompletely or improperly executed. In general, this requirement most closely reflects the capabilities demonstrated by research in the self-organizing approach. Finally, the system should take advantage of the existing organization and models of the domain, rather than learning from the ground up (an argument for the KDL approach). The existence of a model is a good reason for not starting the search (for good rules) from scratch. Any reliable search strategy will do better given a starting point closer to the objective, and it is assumed that existing models are fairly accurate to begin with. The existence of a domain model points up a need for being able to construct and manipulate high-level data structures like production rules. The structures with which M2 stores and manipulates the model are the topic of the next section.

E. Knowledge Representation

The knowledge in M2 is represented in two forms: the "declarative" knowledge and vocabulary of the system, in our case an object description and datatype language, expressed as a taxonomy of the terms of interest; and the "assertional" dynamic knowledge of the system expressed by pattern-matching production rules.

Fact Bases: In M2, objects in the domain are stored in a frame representation language. The classes of other related objects help organize the description by characterizing the object in terms of its class membership as well as its properties and their values [69]. An example of an object represented in this way appears in Fig. 27. The diagram shows the attributes of the military unit, UNIT 1, and the values of those attributes. The values of the attributes are restricted by the properties of the attributes as with the ECHELON attribute in the figure. At this point, the structure of the

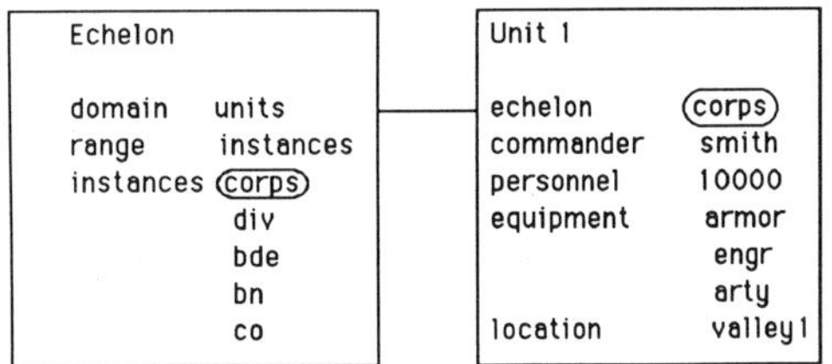

Fig. 27. Object representation.

objects of the domain are static and not subjected to the learning strategy. This was an initial design decision to focus on the learning of behaviors rather than concepts, because the objects of the domain are assumed to be well known.

Rule Bases: The behaviors associated with objects, and the relationships between object behaviors and attributes, and object goals are stored and manipulated in production rules. An example of a production rule for recognizing a river crossing goal appears in Fig. 28. The translation of the

```
IF (unit-is ?name ?equipment ?location)
   (river-is ?riv-name ?location1)
   (within-distance ?location ?location1)
   (includes ?equipment engr)
THEN
   (goal-is ?name cross-river ?location1)
```

Fig. 28. Rule representation.

rule is: if there is a unit with engineering assets near a river then its goal is to cross the river. It should be noted that the clauses of the rule in the diagram correspond to objects defined elsewhere. The clauses, frequently called relations, have the same internal structure as their object counterparts. Thus, the datatypes associated with the fields of each relation are restricted in the same way that attributes of the objects are restricted. The details of these datatype restrictions become important when the learning mechanism begins creating rules of its own. We want the learner to explore the space of *legal* rules rather than the greater set of all possible combinations of syntactic elements. The production rules that can be generated by combining relations define the space of possible rules to be searched by the learning strategy. Obviously, a simple enumeration of all the legal rules is impractical since the number of rules varies exponentially with the number of relations. The next section describes a procedure for searching the space of legal rules efficiently, focusing on areas that need improvement.

F. Strategies for Learning

A strategy is a plan or series of plans whose actions accomplish a goal. A learning strategy then, is that set of actions or the methodology that will be invoked to improve performance. The general types of strategies available are: 1) learning by rote, 2) learning by instruction, 3) learning by deduction, 4) learning by analogy, and 5) learning by induction [63]. Although a full description of these learning strategies is beyond the scope of this paper, it is useful to notice that the learning strategy defines the search mechanism employed in the space of rules. The strategy of learning by induction is used because it closely models what the human analyst must do. (A strategy is inductive when it performs the inference by generating plausible hypotheses and selecting among them.)

M2 Architecture: Previous sections have attempted to describe the constraints under which a learning system in the intelligence domain must act. The overall picture of the flow of information between the system modules is shown in Fig. 29. The system receives input from the user in the

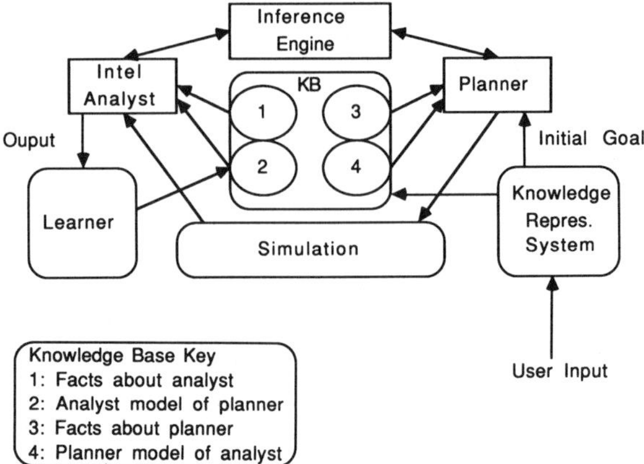

Fig. 29. Domain driven architecture.

form of a goal for the planner to achieve. The planner, without access to the analyst's knowledge bases, generates a plan to achieve the goal and sends the plan to the simulation. The simulation carries out the actions of the plan and generates activity assertions that are available to the analyst's sensor rules. Sensor rules generate reports based on their own coverage (in time and space) of the activities. The reports are used by rules to generate a description of the planner (as described above), and predictions of future reports are made. The degree of correlation between these reports and incoming reports prompts the learner to modify the analyst's knowledge base of the planner. The primary points of interest in the architecture are: 1) separation of planner and analyst knowledge bases, the ground truth restriction, 2) user input to planner in the form of some goal to be achieved, and 3) learning driven by the output of the analyst to modify the analyst model of the planner.

Learning Module: The learning module is divided into two distinct components oriented around the evaluation and modification of knowledge. Evaluation results drive the modification of knowledge. We begin by discussing M2's apportionment of credit mechanism, a variant of the bucket brigade algorithm [70], which was reimplemented in the context of our rule representation and in an OPS-like [71] inference mechanism.

Bucket Brigade and Apportionment of Credit: A complete description of the bucket brigade implementation in this system and results are presented in [72]; however, it will be useful to review the ideas presented there. As we indicated in Section E, M2's rule clauses contain either variable or literal values in their fields. Let f_i represent the total number of fields contained in the rule i's clauses, and v_i represent the number of fields which have variable values. A heuristic measure of specificity is computed, s_i, for each rule as

$$s_i = 1.0 - v_i/f_i.$$

Specificity, as computed, acts as a measure of match for a rule to a set of short-term memory items that are matched by the rule's conditions. Strength will be used as a record of a rule's performance. Strength and specificity traditionally play a part in the bucket brigade serving as objectives in a utility function known as the bid. Let $S_i(t)$ represent the current strength of rule i at time t. A constant c represents a moderator much less than 1. Let the bid, $B_i(t)$, of a rule at time t equal the product of c, s_i, and $S_i(t)$:

$$B_i(t) = c * s_i * S_i(t).$$

The bid is used to produce a linear ordering of rules in the conflict set during the conflict resolution cycle of operation. Rules which support rule i at time t will be represented by $\text{supp}_i(t)$. Finally, ENV will be used to denote the environment. It has an attribute, $\text{ENV}_{\text{payoff}}$, which represents the amount of reward the environment provides to rules (typically an order of magnitude greater than the average inter-rule payoffs described below) whose predictions of observables exactly match observables from ENV.

The somewhat lengthy definitions above help to present succinctly the operation of the bucket brigade algorithm adapted to the rule representation. Rules that match items in short term memory are entered into the conflict set. Each rule in the set submits its (closed) bid, $B_i(t)$. All rules in the set are then ordered based upon their respective bids. The highest bidder is selected for firing, is fired, and the bid made by that rule is distributed equally to $\text{supp}_i(t)$ and added to their strengths $Sj(t)$, $\{$for j an element of $\text{supp}_i(t)\}$. Conversely, this bid is subtracted from $S_i(t)$ of the winning rule. This primitive rule economy leads to reinforcement of rules on the basis of their utility to the system.

At this point, the flow of strength in a chain of rules is from last to first. The last rule in a chain has no mechanism for regaining its bid, leading to a gradual weakening of all the rules in the chain. This problem can be alleviated by ensuring that the rule base is capable of generating assertions which can be matched against observables in the system. The only observables of the analyst module are the continuing reports from the simulation. Our implementation employs a set of meta-level rules with the general pattern:

IF
  (there is a report of some kind)
  and
  (there is a prediction of that report)
THEN
  (payoff the prediction rule with $\text{ENV}_{\text{payoff}}$),

as the agents for effecting this reinforcement of the predictions. These rules take the place of the critic in Fig. 26.

The behavior of the bucket brigade is greatly influenced by the value of c. The constant acts as a moderator and restricts rules to risking only a small fraction of their wealth on any cycle. This acts as a mechanism for reducing the chances of overfitting the data. The basic procedure behind the bucket brigade is shown in Fig. 30. The form of strength revision in the bucket brigade is a form of learning because it has lasting effects on the performance of the system, i.e., rules that have done well have an increasing advantage for firing over rules that performed poorly. However, strength revision may not be adequate if the rules do not contain the right sets of relations and restrictions in the first place. This points out the need for structural modification of the rules.

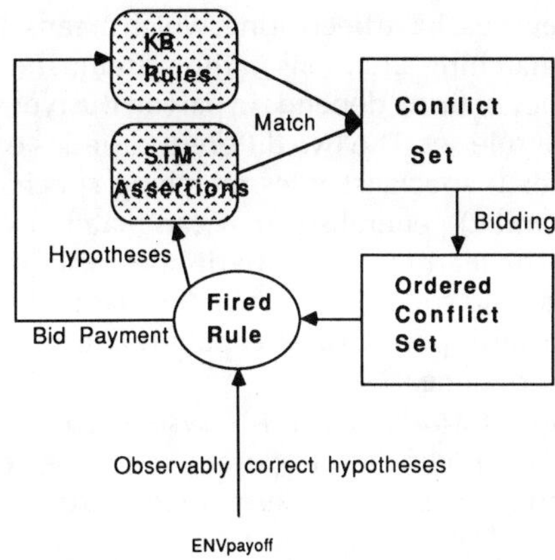

Fig. 30. Bucket brigade process.

Modification Operators: The performance of rules in the system are reflected by the rule strengths and the frequency with which the rules are used. (A rule that has never matched and never fired can hardly be evaluated beyond its general inapplicability.) The rule base in M2 is divided into rule sets whose member rules make the same class of inference. For example, rules that reason about equipment behaviors are separated from those that reason about battalion behaviors. If one considers the operation of the bucket brigade, it can be seen that a differential in strength from the beginning of the reasoning chain to the end develops. However, within a single class of inference any difference reflects more of the relative merits of the rules and less their position in a chain of inference.

This research takes advantage of the strong attributes of inductive approaches to learning developed by Holland and the genetic algorithms community. This initial approach does not preclude the use of other learning techniques nor does it provide commentary on their efficacy in the domain; rather it is a starting point. The primary effort has been in the adaptation of genetic modification operators to the M2 representation language [73]. Although genetic algorithms have been implemented in n-ary languages, they have not been implemented in systems employing rigid datatyping and hierarchically organized object and behavior spaces.

When To Modify?: The indications for modifications come in two significantly different forms. The first, cataclysmic failure on the part of the system, indicates a lack of the rule sets (or possibly the entire rule base) to lead to useful and correct predictions about the environment. Ordinarily, randomly generated rules would not exhibit such failures unless a drastic shift in the application domain of the system had just been undertaken. A second, less disheartening indicator is the gradual differentials which arise in the strengths of rules at the same reasoning level. An examination of the conditions under which this differential arises would suggest the appropriate *when* and *how* of modification. The M2 system assigns a threshold to the standard deviation of rule strengths in a rule set. Passing this threshold serves to direct the attention of the modification operators to the offending rule set. One of two things could be true in this situation: 1) some rules are winning a lot and collecting all the payoff, or 2) some rules are losing a lot and losing all of their strength. Whether the rules were winning or losing they must have been firing, which means they must have been matching situations in the system. The strength and frequency of firing depends in part on the types of rules extant in the rule set. The two different varieties of rules in rule sets are: 1) specialist rules matching specific sets of assertions, and 2) generalists, that can match a variety of assertions. The learning task is to discover the proper mix of generalists and specialists (notice the bias in the bidding functions toward specific rules) capable of producing good overall performance.

How Should I Modify?: The M2 system contains three basic kinds of modification operators: specializers, generalizers, and recombiners. In terms of the pattern-matched rules discussed above, specialization amounts to either changing variable values to literals, or adding constraints on variable values. Generalization, on the other hand, is exactly the opposite. Generalization operators change literals to variables and remove constraints [73]. Using the river-crossing rule in Fig. 28, a generalization would be to remove the constraints (WITHIN-DISTANCE ?LOCATION ?LOCATION1) and (INCLUDES ?EQUIPMENT ENGR). A specialization would be to change ?NAME to UNIT1. If the operators strictly modify the rules, then the orientation of these operators, as expressed, is completely different than that of the bucket brigade and the stated purpose of our modifications. Note that a strict replacement scheme tries to generate *the best rule* for performing in the environment, rather than generate an increasingly stratified (in terms of specificity and generality) population of rules to produce better performance. The incorporation of operator modified copies of these rules can produce the desired effect at the expense of an initially expanding rule base. The remaining requirement is that these operators have some conditions for use. The utilization of meta-rules in learning generalizations and specializations is one of M2's approaches to modification. An example of a (rabid) generalizer meta-rule in M2 is shown below:

```
IF
    (rule-is ?name ?rule-set ?strength ?frequency ?specificity)
    (rule-set-is ?rule-set ?ave-strength ?ave-specificity)
    (exceeds-threshold ?strength ?ave-strength)
    (specific-rule ?specificity ?ave-specificity)
THEN
    (associate ?new-rule (make-copy ?name))
    (remove-predicate-constraints ?new-rule)
    (variablize-literals ?new-rule).
```

This meta-rule makes a copy of the rules that satisfy its conditions, removes the constraints in the new rule, and changes its literal values to variables.

Preservation of rules and parts of rules is a natural element of the recombination operators in M2. The recombination operators make use of notions from natural genetics and were first described in the context of machine problem solving by Holland [67]. The idea behind the recombination of rules is that through fitness proportionate replication of rules and the recombination of the replicants with the original population, the average performance of the rule population will improve. The parts of the good rule are extant in the population at a greater frequency and thus appear on average in more rules. For example, consider the river-crossing rule above. The constraints of the rule, WITHIN-DISTANCE and INCLUDES, may be applicable in a broad class of rules about crossing objects, and the fitness proportionate reproduction and recombination methods would promote the use of these constraints.

The experimentation into the behavior of the modification operators in M2 is continuing. Current work is centering on strengthening the theoretical foundations of the recombinant approach in the higher-level representations M2 employs. A series of experiments to determine relative measures of performance between the two approaches to modification is planned as of this writing.

G. Issues and Future Topics

A variety of technical issues with respect to the M2 system exist. In terms of evaluation, the rate of learning is limited by the number of rules in a chain of inference before an ENV_{payoff} is received. As implemented, the system requires, as a lower boundary, a number of iterations equal to the length of the chain before the stage-setting rules at the head of the chain receive a payoff. Holland has suggested [70] (and Riolo implemented) [74] a remedy for this problem; the implementation in M2's representation remains as a future topic. A second issue related to evaluation is the use of the utility measure as a strength revision mechanism. Notice that the system as described selects for useful rules rather than strictly correct rules.

Some of the most creative work in the field of machine learning is in the development of useful modification operators. A current debate in the field is the relative merits of the logic-based approach compared to the sampling-based approaches we have described. Future research will focus on this issue particularly in the context of the required knowledge representations for each approach and the constraints those representations impose.

This section deals with expert systems as primarily notational, problem solving conventions rather than as models for the human cognitive process. The area of learning research concerned with modeling human learning [75] is not treated here. Cognitive psychologists have made significant advances toward building systems that model human learning phenomena [76]. Discoveries in these endeavors have led to formal theories of general learning mechanisms.

Machine learning is a field in its infancy. Many very difficult problems remain unsolved including issues related to the rate of learning, overfitting of data, and general problems related to efficient search and inference. A final issue related to knowledge representation is the use of the most efficient versus the most transparent representation. The performance of learning techniques in large-scale problems will determine the efficacy of approaches being developed.

V. The Next Step: Distributed Problem Solving

The potential benefit of systems such as BATTLE, ADRIES, and M2 is extensive. It is unclear, however, how the potential utility of these and other expert systems will be transferred to real applications in operational environments. Test methodology, configuration management, reliability, maintainability, performance, and system security are just some of the practical issues that must be resolved realistically for military expert systems. Failure to fully address these issues will probably cause the early (if not preemptive) retirement of many future systems.

In addition, several important military problems are not suitable for conventional expert system approaches, even though this technology has been applied to many diverse domains, including medicine, geology, mathematics, manufacturing, finance, and education. For most of the successful applications, however, at least one common characteristic can be observed: solving a typical problem requires knowledge only about a single, bounded domain. For example, a system in a hospital that makes diagnoses about patient illnesses can reach acceptable conclusions without having any knowledge of hospital staffing problems or accounting procedures.

This same characteristic is also a good reason why expert systems have not been fully exploited in certain military applications; many critical military problems require a broad perspective that spans a variety of domains. For example, the problem of determining a country's intentions based on intelligence about its local troop movements could require the effective application of knowledge about many different complex domains: the local tactical situation, the history of that country's deployment patterns, the force posture of its allies and neighbors, the political and economic situation (both within that country as well as between it and others), the public pronouncements of that country and others, the current weather, the capabilities of collection systems, and the proper interpretation of specific types of collected intelligence, including IMINT and SIGINT. Since no one human expert could master this breadth of knowledge, it is reasonable to conclude that this problem will not be solved soon by any single, monolithic expert system, either. In fact, some automated systems (such as TRICERO [77] and ALLIES [78]) have achieved modest success with this type of problem solving by distributing the work among multiple cooperating expert systems.

What is required is an open network architecture that will permit multiple expert systems to communicate and cooperate effectively and coherently [79] on the solutions to large, complex problems. Numerous advantages may be realized by distributing a large expert system problem across a set of cooperating experts, including the extension of a system across multiple physical systems, greater simplicity of each component expert, and the use of parallelism to speed up the resultant system. This technique can be applied quite naturally to some of the more intractable military analysis problems, since it mirrors the groups of human experts that work as units within large military organizations (e.g., an intelligence analysis team or a battle planning team).

A. Distributed Problem Solving Systems

The concept of a distributed problem solving system (DPSS) can be defined as a loosely coupled network of expert system nodes that communicate both information and goals to each other in the process of arriving at common solutions. A global *blackboard* (discussed later) serves as the logical communications medium. The experts to be coordinated may be both logically and physically distributed (i.e., some experts may be co-resident on one machine while others may be distributed over several machines linked via communication paths of various types). The system as a whole may be both data-driven and goal-driven, by permitting each expert node to select different categories of work dynamically.

Much of the current research into DPSS involves systems descended from the HEARSAY experiments [80]. The HEARSAY family of expert systems started with HEARSAY-I, an early experiment in speech understanding [81]. After HEARSAY-I, evolution continued with HEARSAY-II, which improved the flexibility of both knowledge representations and problem-solving control strategies [82], and HEARSAY-III, which provided problem domain independence [83]. The HEARSAY paradigm involved the processing of input data through multiple levels of detail using a blackboard data structure (blackboards are discussed in more depth in the next section).

The use of multiple levels allowed the modularization of expertise for each level, thereby permitting each module to work in a limited domain and apply contextual cues available at that level. For instance, referring to Fig. 31, the lowest

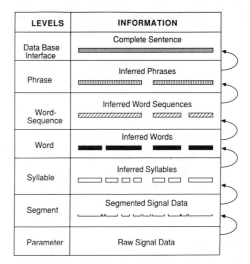

LEVELS	INFORMATION
Data Base Interface	Complete Sentence
Phrase	Inferred Phrases
Word-Sequence	Inferred Word Sequences
Word	Inferred Words
Syllable	Inferred Syllables
Segment	Segmented Signal Data
Parameter	Raw Signal Data

Fig. 31. Construction of higher-level patterns from raw data in HEARSAY processing.

level of HEARSAY divided the raw input data into *segments*; the next higher level collected segments into *syllables*, then into *words*, *word sequences*, and so on. The actual performance involved the creation of "islands" of information on the blackboard where HEARSAY could make inferences of high confidence. These islands were extended to include neighboring information and cues until enough information was correlated to create an inference at the next higher level.

This process of dividing the inference structure into multiple levels of detail can be applied directly to large, complex military problems. Consider, for example, the simplified I&W process depicted in Fig. 32. Raw sensor data is collected into messages or processed directly; messages are collected into indicators; and indicators are processed into warnings, which are sent out of the system or fed as input to other applications. The output of such a system consists of the patterns recognized in the data and message stream by the set of experts who make up the processing power of the system.

Thus the HEARSAY blackboard paradigm is appropriate for two reasons. First, it solves a problem that is structurally

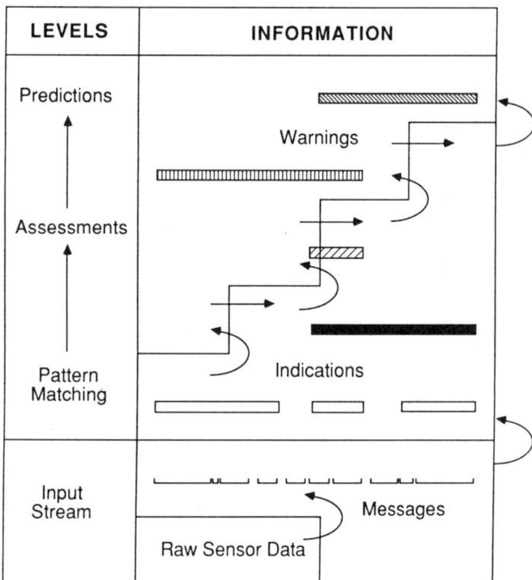

Fig. 32. Construction of higher-level patterns from raw data in I&W analysis.

similar in nature to intelligence data analysis problems, as well as to other complex military problems requiring the coherent integration of diverse expertise, such as battle management. Second, as discussed in the next section, the blackboard is an especially appropriate control structure for DPSS.

B. Blackboards in DPSS

A blackboard is a global data structure used to communicate facts, inferences, and goals between multiple expert nodes in a DPSS. Its power results from the way it provides a common framework and interface for sharing useful information both within and among expert nodes. Its flexibility supports both data-driven and goal-driven operation.

Three major types of information are represented in the full blackboard paradigm. The first type is frequently referred to as "data," but it can be further divided into *facts* and *inferences*. A fact is some datum that has been inserted into the system from an external data stream or a human user. It may have a confidence factor assigned by (or because of) its source, but it is not generated by the system itself. By contrast, an inference is produced by the application of some sort of knowledge to the facts and inferences already available to the system. The confidence factor associated with a given inference is a function of the system's confidence in the knowledge that produced the inference and in the other facts and inferences used to support that inference.

The second type of information is "control" information. Again, this is further broken down into *knowledge sources* and *goals*. Knowledge sources are the procedures that define how inferences are to be drawn and actions are to be performed by the system during its normal mode of operation. In some implementations, knowledge sources are able to pre-identify both the data required for them to be effective and the types of inferences that they can produce. Goals represent more general activities that may be undertaken by an expert, such as the generation of an inference that matches a certain pattern (if possible). In some of the literature, a separate data structure known as an agenda is established for the maintenance of goals. Expert nodes will generally have numerous goals to choose from, and will use some criteria for selecting the most valuable goal at a particular time.

The final type of information associated with a DPSS blackboard is derived from its distributed nature. This information is required to control the flow of information from one expert's blackboard to those of other experts. In a single-node expert system using a blackboard architecture, one blackboard will be the central repository of all information about the problem. With a DPSS, however, each separate expert node requires access to blackboard information. This implies that expert nodes should use local blackboards that are actually individual partitions of an abstract global blackboard. One or more strategies can be employed by each expert to select the information that should be transmitted from or accepted into the local blackboard, the other experts that should participate in an information exchange, and the conditions under which such exchanges are desirable. The alternative is a single blackboard maintained as a service for all of the experts. However, this technique has severe drawbacks when the experts are physically distributed. Since the blackboard is a vital resource to the inference process, the communications load involved in supporting a centralized blackboard would be immense. In addition, a single global blackboard would provide a single point of failure in a physically distributed DPSS.

Thus, each expert node of a networked DPSS should have a local blackboard composed of information directly available to that expert. Some of this information may be irrelevant to that expert, and some of it may be required by other experts. During the course of problem solving this "misplaced" information will be communicated from one blackboard to another according to blackboard communications criteria. For instance, when a particular confidence threshold was reached in an inference generated by one expert, it might be "leaked" to a sibling or higher-level expert for further processing. This process of leakage is detailed in Fig. 33. In the case of information needed by the receiving expert, it will be used as an inference to drive further processing. In the case of information received by an expert node via this process and substantiated, a return flow would be used to bolster the confidence of that inference in the original node. In the case of conflicting information, a reverse return flow would lower the confidence and/or spur error recovery actively in one or both experts. In the case of goals propagated between blackboards, the receiving expert node would have to choose among externally-supplied goals and its own goals when evaluating which to execute [79]. This localized control ability is discussed in more detail later in this section.

The blackboard paradigm thus provides a method of loosely coupling various expert nodes supporting conflict resolution and individual activity. The paradigm is not centralized, so that there is a graceful degradation if one or more nodes are removed from the system (assuming that the loss of one node's expertise or information does not prohibit effective problem solving altogether). The system may be data-driven as information islands are propagated upward to higher-level or sibling nodes, or it may be goal-driven by placing goals on the local blackboards attached to various nodes. Communication load may be tuned by changing the rules controlling inter-blackboard information movement.

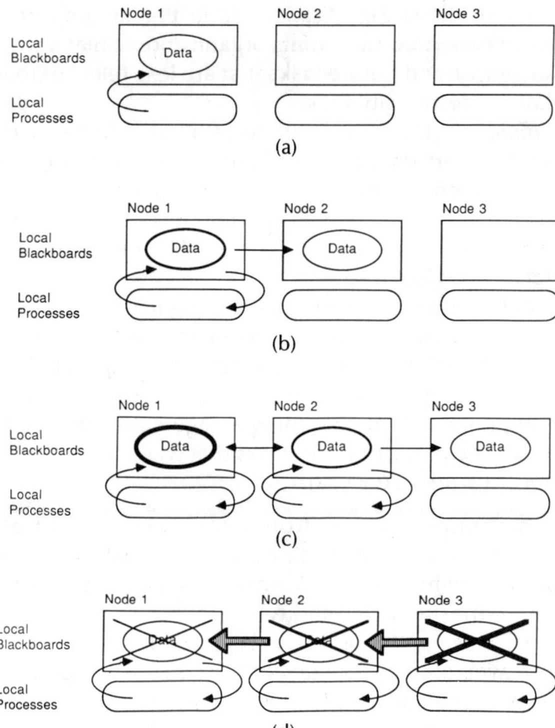

Fig. 33. Data "leakage" between local blackboards. (a) Data placed on first blackboard not certain enough to reach replication threshold. (b) Data reaches certainty threshold for replication (relative certainty is indicated by line thickness). (c) Feedback enhances certainty of data and reaches threshold for replication to next node. (d) Conflict propagates negative certainty back.

C. DPSS Attributes

A number of different attributes may be used to compare and evaluate alternative DPSS architectures. Each of these attributes can have a profound impact on the performance or even the viability of a given DPSS architecture with respect to a particular problem domain:

• *Physical Distribution:* The individual expert nodes of a DPSS can be resident in the same processor (i.e., a *logical distribution*) [84], assigned to individual processors in a tightly-coupled parallel system [85], or distributed across multiple processors communicating through a LAN. A combination of physical and logical distribution is possible within a single DPSS. The degree and type of physical distribution for a given DPSS is largely a function of the overall goals of the system architects. For example, logical distribution provides a means for flexible experimentation with different network topologies while requiring a minimum amount of processing and communication hardware.

• *Data Distribution:* During normal operation, each expert node in a DPSS will receive a different set of facts about the world and inferences from other experts. (The utility of any architecture where all experts receive the same information is unclear, due to the implicitly redundant activity required at each node.) The set of facts available to each expert node can be influenced by many factors, including the spatial distribution of data, the network topology, and the distribution of knowledge sources throughout the DPSS. The inferences available to each expert node are a function of the facts it receives, the knowledge sources it has available, the control decisions it makes, and the communication policies that regulate the transfer of inferences between experts.

• *Knowledge Source Distribution:* In some DPSS architectures, all of the knowledge sources are directly available to all of the expert nodes [86], while in others, each expert has access to only some knowledge sources [85]. The determination of which knowledge sources will be available to each expert is usually made by the system architect, based on an evaluation of the difficulty of hosting different knowledge sources on different nodes. Knowledge sources that are initially designed to be part of a given DPSS are likely to be easier to share among experts than knowledge sources that already exist when the DPSS is designed.

• *Task Distribution:* When considered as a monolithic operation, each knowledge source is capable of performing one or more units of work or *tasks.* The potential tasks that knowledge sources can perform are usually represented as goals, which define tasks in terms of the possible result produced by knowledge sources. During DPSS operation, the set of tasks under consideration at an expert node is a function of the inferences produced by its own knowledge sources, the communication policies that regulate the transfer of goals between experts, and the degree to which a given goal can be easily factored into local sub-goals. Since any one node can actually work on only a limited number of tasks (usually one) at one time, the method by which it evaluates and activates potential tasks has a critical influence on the usefulness of its problem solving activities (as discussed in the next section).

• *Communication Policy:* As already noted, each expert node in a DPSS must implement a communication policy, which need not be the same for all nodes. From the perspective of a sending node, this policy determines when the node will transmit information (either inferences or goals) to other nodes, what types of information will be transmitted, and which nodes are appropriate recipients of the information. From the perspective of a receiving node, this policy determines when the node will accept information, what types of information will be accepted, which nodes are appropriate originators of the information, and how the accepted information should be evaluated locally. The choice of a particular set of communication policies may be influenced by any predetermined organizational structure for a given DPSS. However, the possible subtle interactions of different aspects of communication policies make it difficult to establish an *optimal* set of policies without extensive experimentation in real or simulated environments [87].

• *Organizational Structure:* It is possible to bias the evaluation of tasks by each expert node so that it prefers certain activities over others, regardless of the current state of problem solving. This type of bias, in conjunction with communication policies, implements an organizational structure across the DPSS. In a hierarchical structure, for example, each node accepts tasks from at most one other node, thereby permitting higher-level nodes to control and coordinate the activities of lower-level nodes. In a flat (committee) structure, each node can accept tasks from any other node, thereby maximizing the availability of a given node's services to the whole network. Of course, many other organizational structures are possible. The combination of an organizational structure and a set of communication policies establishes a distributed problem solving strategy within a DPSS.

- *Organizational Development:* If the DPSS architect pre-establishes an organizational structure for the system, then no organizational development is required. Alternatively, the DPSS can develop an appropriate organizational structure as part of its initial problem solving activity. This can be achieved by requiring nodes to negotiate (as in the "contract net" approach of Smith and Davis [88], [89], an appropriate task distribution prior to the start of problem solving. Another option is to permit the DPSS to alter its organizational structure during the course of problem solving, by allowing nodes to detect and respond to strong mismatches between their knowledge of local task requirements and their knowledge of current organizational responsibilities [90]. The choice of an appropriate organizational development scheme for a given DPSS depends on whether an optimal organizational structure: 1) can be predetermined for every problem-solving situation, 2) can be determined once for each problem-solving situation, or 3) must change during the course of problem-solving activity.

The possible interplay among these attributes can yield complex system behavior that is not easy to anticipate. Consequently, it is important for DPSS developers to incorporate extensive internal auditing and measuring capabilities in their systems, so that they can sensibly test, understand, and modify problem-solving results and strategies. In fact, some DPSS environments [84] have been developed primarily for the purpose of measuring the effectiveness of different problem-solving strategies.

Regardless of the specific attributes of a given DPSS, its expert nodes must incorporate a method for determining when a potential task should actually be performed (i.e., for determining the *focus of attention*). Each expert may be attempting to satisfy multiple goals, and some of these goals may be generated locally, while others may be supplied by other experts. The mechanism for resolving these competing claims implements a local problem-solving strategy within each node.

D. The Focus of Attention Mechanism

The DPSS techniques that have been developed at the University of Massachusetts [90], [91] include an elaborate focus of attention mechanism. The technique functions by giving each expert node the ability to decide which of its possible goals it should pursue (through the activation of its available knowledge sources). These goals include both those generated by the expert itself and those generated by others. The focus of attention is determined by first evaluating potential knowledge source activations against various criteria, and then by selecting the most highly rated task for execution. These criteria could include, but not be limited to, the following:

- *Validity:* If the information available for a task has a high confidence level, then that task should be performed before a task for which the information has a low confidence level, as the resulting information would be more valid.
- *Internal/External Control:* Tasks that are triggered by local activity may be preferred over tasks that are received from other experts (and vice versa).
- *Organizational Significance:* Tasks that are important to an expert because of its organizational bias should be performed before tasks that are less relevant to its known responsibilities.
- *Efficiency:* If it is possible to estimate effectively the cost of performing a task before it is executed, then less expensive tasks may be preferred over more expensive ones. This criterion may be especially useful when a given goal can be achieved by more than one local knowledge source.
- *Goal Satisfaction:* Tasks that are themselves subgoals or supergoals of several other tasks may be preferred over tasks that are less connected with other activities.
- *Task Age:* If a task has been awaiting activation for a long time or is approaching an age cutoff threshold, it may be preferred over tasks that were created more recently (or vice versa).

Thus a high priority request from another expert might take precedence over a local "business as usual" task, or a short, simple task might override a long task that is based on information of questionable validity.

These focus of attention criteria can be implemented as variable, weighted parameters of the DPSS. The operation of the system can be modified by altering these parameters for a single expert or a group of experts. Changing the balance between the criteria of validity and goal satisfaction, for example, can alter system behavior to become more data-directed or more goal-directed; modifying the internal/external control parameter can determine whether individual nodes are more locally directed or more externally directed.

The use of these criteria allows the focus of attention to vary with time. During normal operation, the focus would follow a logical "train of thought" with respect to a given expert. Anomalous situations, however, could be handled by shifting the focus to different types of tasks. Requests from a system user or another expert could be handled as interrupts to the train of thought, and critical warnings from input data or other experts could be dealt with in an effective manner. Thus, this approach permits DPSS nodes to function passively, actively, or interactively, depending on dynamically changing problem solving requirements.

E. An Example

As part of its ongoing research and development program, PRC has been exploring the potential military application of distributed problem solving techniques by building a prototype DPSS. This system, called the Communicating, Cooperating Expert Systems (CCES) Framework, provides an evolving workbench for testing different DPSS concepts. A brief overview of this prototype and an example of its operation may help clarify some of these concepts.

The CCES Framework is implemented as a collection of physically and logically distributed nodes communicating via an Ethernet LAN. Each node has a similar internal structure consisting of three principal processing components (see Fig. 34): a blackboard-based expert system, a focusing tool, and a communications switcher. The expert system component exchanges information with the external world via a user interface and a message stream; its own knowledge sources use this information, in conjunction with what

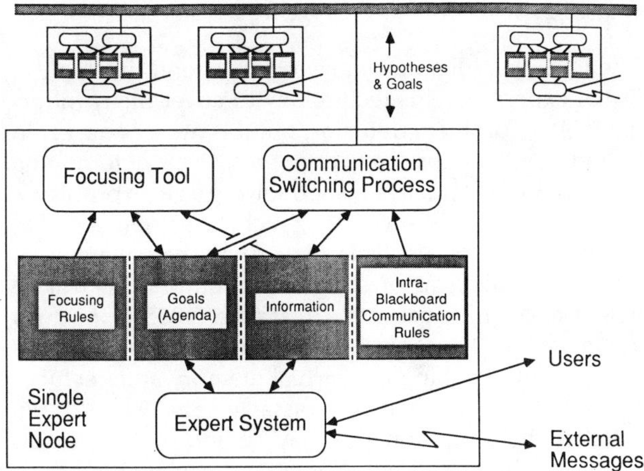

Fig. 34. PRC's CCES framework.

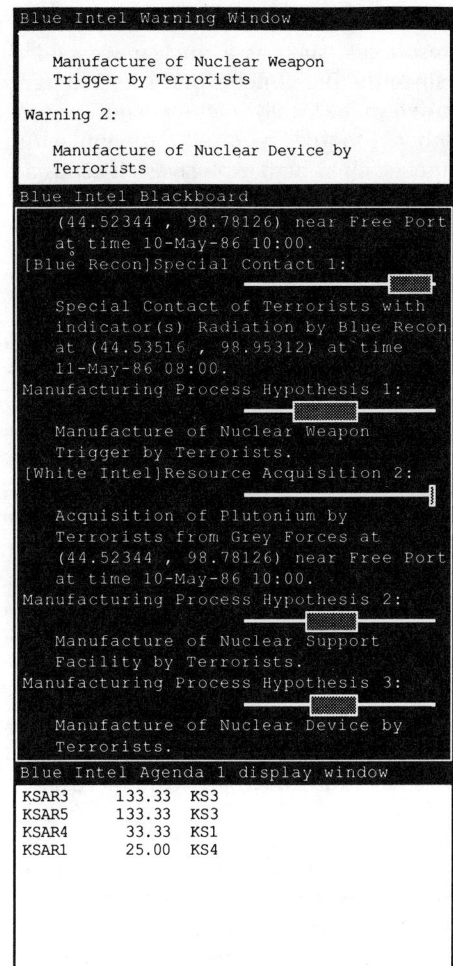

Blue Intel Warning Window
Manufacture of Nuclear Weapon Trigger by Terrorists
Warning 2:
Manufacture of Nuclear Device by Terrorists
Blue Intel Blackboard
(44.52344 , 98.78126) near Free Port at time 10-May-86 10:00.
[Blue Recon]Special Contact 1:
Special Contact of Terrorists with indicator(s) Radiation by Blue Recon at (44.53516 , 98.95312) at time 11-May-86 08:00.
Manufacturing Process Hypothesis 1:
Manufacture of Nuclear Weapon Trigger by Terrorists.
[White Intel]Resource Acquisition 2:
Acquisition of Plutonium by Terrorists from Grey Forces at (44.52344 , 98.78126) near Free Port at time 10-May-86 10:00.
Manufacturing Process Hypothesis 2:
Manufacture of Nuclear Support Facility by Terrorists.
Manufacturing Process Hypothesis 3:
Manufacture of Nuclear Device by Terrorists.
Blue Intel Agenda 1 display window
KSAR3 133.33 KS3
KSAR5 133.33 KS3
KSAR4 33.33 KS1
KSAR1 25.00 KS4

Fig. 35. Example of local blackboard.

is received from other nodes, to produce new inferences and goals. The focusing tool uses a frame-based data structure describing goal rating factors to reevaluate items on the agenda; each node may use different criteria for agenda evaluation. The communication switcher exchanges locally generated inferences and goals with other nodes, and maintains records that allow it to broadcast local changes to previously transmitted information.

These three components run as asynchronous, concurrent processes in each CCES node (although a single communication switcher can service multiple nodes that are located in the same physical processor). This type of architectural decomposition should facilitate the replacement of local node expert systems with other "foreign" expert systems or perhaps more conventional ADP components. This potential adaptiveness is important to the overall goals of PRC's research in distributed problem solving, which include the need for a capability to coordinate heterogeneous nodes such as expert systems and conventional data base systems.

In a current test scenario, the CCES Framework is configured to support four independent expert nodes, which collectively work on two problems at the same time. The four experts are: *Blue Intel*, which is assigned a current intelligence problem (monitoring the activities of *Terrorists*); *Blue Recon*, which primarily receives reports of visual contact with *Red Forces*; *Blue I&W*, which is assigned an I&W problem (monitoring the status and intentions of Red Forces); and *White Intel*, which is a source of allied intelligence. Each expert receives a different set of reports about the world, but they all have access to the same knowledge sources. Thus, each expert's name was chosen based on its data sources and assigned responsibilities, rather than on any unique local knowledge.

As reports and inferences (called hypotheses) are received or produced by an expert, they are added to the expert's local blackboard. Any knowledge source activations triggered by a new blackboard item are then added to the expert's agenda. Each expert maintains an interactive user interface to these data structures (see Fig. 35). Items are displayed on the blackboard from top to bottom in the order in which they were created. The visual representation of a blackboard item includes the name of the originating node, the type of information contained in the item, a brief textual description of the item (with times and grid locations for report items), and an indication of the expert's confidence in the item. The CCES Framework currently uses an adaptation of Dempster-Shafer technique for representing uncertainty (see Section II and Appendix II). The confidence is displayed graphically as a horizontal bar: the length of the solid left end indicates the degree of belief for the item, the length of the solid right end indicates the degree of disbelief for the item, and the length of the fuzzy middle part indicates the remaining, unaccounted-for belief (or uncertainty).

As part of each expert's blackboard display, the CCES Framework also maintains an agenda display window that describes the current status of that expert's agenda (see Fig. 35). Unsatisfied knowledge source activations are displayed in the order of their estimated value (also shown), along with the names of the knowledge sources from which they originate. Within the current test scenario, efficiency and task age are the only criteria used to evaluate the knowledge source activations.

The communication policies for this scenario generally allow one expert to transmit data to other experts only when it has first received a request for a pattern of information that matches that data. For example, the Blue Intel expert receives reports about various resources acquired by Terrorists (such as nuclear weapons expertise). These reports trigger the activation of knowledge sources capable of hypothesizing what can be manufactured from the newly acquired resources (such as a nuclear weapon trigger). However, since the Blue Intel expert's knowledge source is unable to determine locally whether or not Terrorists have acquired other resources required for manufacturing the same product (such as high explosives), it sends a pattern description of the missing information as a subgoal to the other experts. Until they receive this request, they will not transmit this type of information to Blue Intel.

A conservative communication policy of this type is appropriate when it is important to reduce the likelihood of experts flooding each other (and the scarce communication resources) with useless information. However, there are some circumstances where it may be desirable for one expert to transmit data to other experts, even when they have not expressed an interest in it. For example, when the Blue Intel expert hypothesizes that Terrorists are manufacturing a nuclear weapon trigger (and later, a nuclear device), its local knowledge source recognizes that these hypotheses are important enough to be treated as *warnings*. Warnings are displayed locally in a special window (see the top of Fig. 35) and are then transmitted to all known experts, where they are also displayed.

As additional assistance for human users and experts, each expert node maintains a situation map (see Fig. 36) that presents a symbolized, geographic representation of items on the expert's blackboard. When two or more nodes share the same physical processor, they also share the same situation map. As the blackboard is updated, the situation map is modified to reflect the changes. Icons on the map and their corresponding entries on the blackboard can be selected by the user. Additional information is available for a selected item, including the tree of knowledge sources, hypotheses, and reports that were used to generate the item. This kind of dependency tracking not only explains how a given hypothesis is derived; it also supports a network-wide truth maintenance capability that allows warnings and hypotheses to be withdrawn if their supporting data is later retracted.

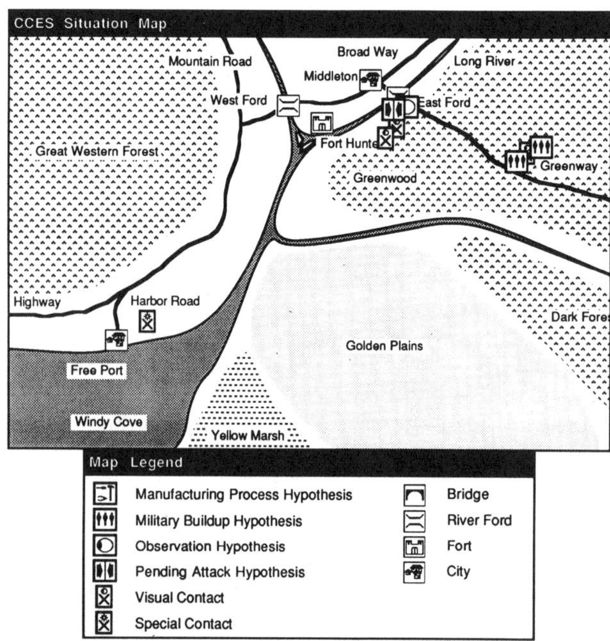

Fig. 36. Situation map.

F. Cautions

The DPSS paradigm may eventually extend the applicability of expert systems technology to cover many intractable military problem domains, but it is not a panacea. In addition to spawning new problems (e.g., determining when a DPSS has actually finished solving a given problem), it may make existing problems, such as system security, more complicated. Nor it is likely that current focus of attention mechanisms or communication policies will be efficient enough to prevent a real-time DPSS from overloading itself with inter-expert communications (the DPSS equivalent of thrashing in a virtual memory architecture). Research at the University of Massachusetts [90], [92] suggests that DPSS performance may be improved by incorporating sophisticated nodes that can reason about their own plans (potential task sequences) and the plans of other nodes. Whether these advanced techniques will be sufficient to support fielded operational systems remains to be seen.

Nevertheless, the distributed problem-solving paradigm represents a promising direction of research in expert systems technology. Without it, some of the most critical, complex military problem domains will remain without the benefit of support from expert systems.

VI. Current and Future Research Topics

This section summarizes selected expert system applications that have been used for the military. The remainder of the section addresses the key research issues that must be solved to enhance the use of expert system technology for future problems.

A. Summary Applications

The expert system process was described briefly and then several key applications were presented. The first application was a resource allocation problem that was developed to help the Marine Corps match weapons or targets. New concepts were described that reduce the amount of questions that were asked and thus reduced the total amount of time to solve the problem.

A second expert system evolved from this first example that was used to help the operator identify complex images. This system will enhance the performance of existing military sensors by guiding the operator along a near-optimal solution to the identification of unknown targets. The third expert system example examined the area of troubleshooting electronic equipment. Prototypes have been developed that can reason about varying fault conditions and can guide the operator to identify the faulty component and to fix the equipment. The fourth expert system application was a digital radar imagery exploitation system. This system produced interpretations of military situations from radar imagery.

Next, an intelligence analysis system was described that had a learning component. Machine learning is crucial to avoid the brittleness of expert systems that is commonly encountered, and this learning system indicated encouraging progress. The final system description was a departure from conventional isolated expert systems. This distributed problem solving approach used communicating, cooperating expert systems that work as a team to solve large complex problems.

B. Research Issues

The examples of expert systems in the military presented in this paper illustrate that a great deal has been learned about knowledge, its representation, acquisition, and utilization since the beginnings of AI research. We believe that there are many reasons to be optimistic about the future applications of expert systems to real-world commercial and military problems but there are critical research issues to be solved. In a recent presentation Friedland [93] made several interesting observations on the future research issues in expert systems. He points out that the initial researchers in AI felt that the power of expert systems would be in complex, general-purpose inference engines able to emulate human beings. Friedland suggests that this led to a great deal of frustration and the growing realization that the power was in the knowledge. Feigenbaum refers to this as the Knowledge Principle. A very important point is that vast amounts of knowledge must be acquired and codified in the machine. Friedland shows that there are four major bottlenecks in the utilization of this knowledge. They are listed here and then discussed:

- knowledge acquisition
- knowledge consistency and completeness
- large knowledge base manipulation
- interface technology.

Knowledge Acquisition: This involves the process of getting the knowledge into the system. The standard use of a knowledge engineer is to act as an intermediary, which can cause delays and a loss of accuracy. Major research efforts directed at eliminating this intermediate step have developed two techniques. The automatic technique would use inductive and deductive means to acquire the knowledge without the knowledge engineer. This is clearly related to certain aspects of the learning projects such as Michalski [63]. The semi-automatic technique would use the domain expert as the knowledge engineer, which has become increasingly popular in the development of expert systems. The knowledge acquisition process, which can take from 50 to 90 percent of the time and effort in the development of an expert system, is not a single event process. The realistic view is that the knowledge must change throughout the life of the system.

Knowledge Consistency and Completeness: This bottleneck involves the combination of knowledge in many different forms, from numerous, often disparate, sources with a second issue of whether there is enough knowledge for functional adequacy. As large expert systems are constructed, the sources of information will increase because no single human can possibly know more than a small percentage of the final system. Another complication is that much of the information will come from nonhuman resources. In the example in Section III on the expert system for maintenance and troubleshooting, much of the data comes from instruction manuals, design information, computer-aided design programs as well as data from the machine designers. The use of all of these different sources can lead to both apparent and real disagreements. The future research in this area involves the design of mechanisms to aid in the information entry in large systems. Knowledge source disparity can include contradictory heuristics (likely failure modes and design considerations), problems with the actual language of rules, frames, etc., and facts (tolerances, expected life-time and anticipated temperature and environmental conditions).

The second major issue in this knowledge acquisition bottleneck involves the completeness of the knowledge base. It is necessary to determine if this collection of knowledge is enough to solve the actual problem. The completeness issue can be solved for logic-based knowledge systems but no technique has been developed to solve the general problem. As systems evolve and the knowledge acquisition process continues through the life of the systems, as indicated in the first bottleneck, this problem will become even more difficult. The ultimate solution involves sufficient deep knowledge that can understand underlying principles in the realms of physics, chemistry, electronics, etc., to assess the completeness of its knowledge base.

Large Knowledge Base Manipulation: This bottleneck involves the problems of how data, procedures, and other problem-solving techniques must be manipulated in very large expert systems containing vast amounts of data. Traditional techniques of forward and backward chaining with rules, theorem proving, and object oriented programming may not be adequate as the expert systems of today scale up to the large systems of tomorrow.

The present techniques in knowledge manipulation may be adequate if there are significant increases in the power and speed of computer hardware. Undoubtedly, hardware improvements will help solve this problem, but another generation of reasoning systems is needed to solve these large problems more efficiently.

Research in this area involves the direct solution to large knowledge base manipulation, or an attempt at distributed problem solving. One distributed problem solving solution is to break the problem into many smaller communicating cooperating expert systems as discussed in Section V. Difficulties to be solved include the degree of independence in the individual expert systems, the required amount of communications among the individual problem solvers, the degree of autonomous behavior, and the use of robust learning techniques.

Interface Technology: This bottleneck involves the process of getting the information out of the machine and into a form that humans can understand and use effectively. Present systems do not use very much natural language to explain to the user what is happening and why. This problem will become increasingly difficult as the sizes of systems increase and the user requires more explicit explanations from the expert system.

Other problems, in addition to knowledge acquisition, are:

1) System Brittleness: Today's systems cannot determine if the problem they are trying to solve is within their "area of expertise." For example, there is a definite possibility that a system designed to diagnose medical problems could easily be confused between measles and rust on a car. In addition, the systems expect their inputs in a fairly tightly defined environment and changes in language or syntax can often result in disarray.

2) Common Sense Reasoning and Analogical Reasoning: A human, stumped by a particular problem, will often try to use "common sense" to help in the solution. There is not consensus on what common sense means but Friedland [93] suggests that common sense means knowing a little about a lot and recognizing from a human's vast knowledge base what is relevant and can be linked to the particular problem at hand. Lenat [94] at Microelectronics and Computer Consortium (MCC) is attempting to codify the knowledge in a desk encyclopedia, which will be used as the basis for the knowledge contained in an expert system. An example that Lenat uses to show the difficulty of common sense reasoning is the problem of distinguishing the difference between the two phrases:

- the pen is in the box
- the pen is in the corral.

Lenat hopes that once he has entered the knowledge contained in his disk encyclopedia, the computer can determine the difference in the two uses of pen. Initially Lenat felt that learning would provide a solution but now he feels that hand crafted knowledge and hard work will be the solution for the next 10 years.

The second mechanism is analogical reasoning. The process consists of two steps. The first step is to pick a potential analogous situation and the second step is to extract the common items. This sounds simple but is extremely difficult, and little progress has been made in this area of basic research. Progress made in the research on common sense reasoning may help to solve this problem too.

3) Learning: This area, discussed briefly in Section IV, needs additional development. Clearly there is a large gap between present-day expert systems and their human counterparts when it comes to learning. Friedland [93] points out that we would think a fellow human being hopelessly stupid if the same mistake were repeated endlessly. This is exactly what happens to most expert systems when they do not have sufficient knowledge and reasoning components to solve a problem. Learning becomes mandatory when one considers that large systems, such as the NASA space station, are continuously evolving. Long-term mission projects will need learning to acquire new knowledge and to re-validate the system as knowledge is added. The addition of learning to future expert systems will probably provide the biggest change from today's expert systems to those of tomorrow. Potential aid for future learning lies in the fields of neural networks and causal modeling of physical systems.

4) Synergism Among Intelligent Agents: This area has great potential. The idea, related to communicating cooperating expert systems, or problem solvers, involves using them to provide synergism. This can only be accomplished when the problem solvers are organized as teams able to work effectively and cooperatively to solve a bigger problem. The organization of these problem solvers can be in many forms including hierarchical, committee, distributed in functional or geographic terms, etc. The final synergistic configuration is ultimately related to the actual problem. One major advantage of these cooperating problem solvers is that individual ones can be made relatively small. The state of the art in small expert systems is far advanced to what can be done for very large expert systems, because it is easier to test them, there is increased confidence, and production costs are lower. The individual expert systems can then be hooked together as a team to solve a bigger problem. The existing examples of this at the University of Massachusetts, Stanford, and PRC are showing signs of promise for large synergistic systems.

5) Generic Expert Systems: It is hoped that domain independent expert systems can be used as financial advisers, medical diagnosticians, indications and warning analysts, target identifiers, and electronic troubleshooters. No single system can accomplish this. Indeed many medical diagnostic expert systems cannot be easily modified to work on apparently similar medical problems.

Another issue with generic expert systems concerns a uniform approach to handling uncertainty. The present systems are ad hoc and usually not mathematically verifiable. When certain information is concatenated, the problem usually becomes more difficult because the propagation of uncertain information is not well understood. Knowledge representation is also not handled uniformly. Further details are found in Appendixes I and II.

6) Delivery Machines for Expert Systems: There has been a big change in attitudes in the AI community. Originally everything was done on large high priced LISP machines with identical delivery machines. The user community is no longer willing to pay this price since conventional hardware has become increasingly powerful at lower cost. The new microprocessor-based workstations will be the delivery machine in many cases, and for some users the AI solutions will be provided on a mainframe. The development will still be on a mix of special AI machines and more general purpose microprocessors.

7) The Degree of Expert in an Expert System: It has become very clear that expert systems are nothing of the kind. The majority of expert systems are only slightly clever. This is not all bad. The trend in expert systems should be to improve the user from an average grade of poor to mediocre, not mediocre to great. The reason is obvious. Most of the problems that we encounter are routine, time consuming, boring and mundane, and it is these irritating little problems that take from 50 to 80 percent of the human expert's time. The solution is to use our "clever" expert systems to solve the mundane and allow the human to free up his time, by as much as 40 to 50 percent, to solve the problems that the machine cannot solve.

These are genuine problems that must be addressed and solved before AI can be used successfully in the complex world of the present. In fact, the Defense Advanced Research Project Agency, the Army, the Air Force and the Navy are addressing all of the key issues listed above. The success of this research, by some of the best researchers in the United States, is expected to hasten the routine use of expert systems by the military and in the commercial marketplace.

Appendix I Expert System Tutorial

Expert systems [6]–[10] are computer programs that attempt to imitate real experts. The desired final product is computer program outputs with the same correct results as those of human experts. This section touches upon many of the tools and techniques commonly used in the construction of expert systems. A general expert system architecture, knowledge representation schemes, and reasoning or inference mechanisms are described. Many different definitions of expert systems exist; however, the objective is a computer program that reasons at least as well as an expert in a given field. This is a lofty goal and in general is difficult to achieve. A less ambitious, and more successful, use of expert systems is not to replace the human expert but to provide him with expert or knowledgeable assistance. There are many commercial expert systems [95] but they are used mostly in nonmilitary applications. Some of these successes include PUFF, an expert system to diagnose pulmonary disease; NAVEX, an expert system to monitor controls of space shuttle flights; Cooker Advisor, used to troubleshoot electrostatic soup "cookers" for Campbell Soup; STEAMER, a system that trains steam plant operators for Navy ships; ACE, a system to troubleshoot cable; OCEAN, an expert system that checks orders and configures NCR computers and DECs; and XCON, used to configure VAX computers. This section will give an overview of expert systems, discuss some of the challenging problems, and finally review in more detail the problems associated with dealing with uncertain information.

Fig. 1 is a block diagram of an expert system including outside interfaces. The knowledge engineer works with a domain expert to acquire the critical information that is needed to clone the expert. The process of knowledge acquisition is what the knowledge engineer uses to extract data, knowledge, and techniques from the domain expert. The extracted information includes system facts and suppositions that might have varying degrees of uncertainty associated with them. Typically, the knowledge engineer will derive from the domain expert the equivalent of a data base of facts that will be put into the knowledge base. This knowledge base is custom fitted to the domain. It is convenient to think of the knowledge base as made up of two components: assertions and knowledge relationships. The assertion component is similar to a working memory or a temporary storage of data. It contains declarative knowledge about a particular problem, and the current status of the problem that is being solved. The data in the assertion component of the knowledge base can be represented as first order predicate logic, frames, semantic networks, state-space or other techniques that are convenient to that particular problem. These terms for knowledge representation will be explained in the following paragraphs.

Logic

A predicate is simply something that asserts a fact about one or more entities and has a value of true or false. An example is "Jack caught the ball." In this case we are saying that there is a catching relationship between Jack and the ball. Predicate calculus is a means of calculating the truth about propositions and it combines the notion of predicates with logical relations such as **and**, **or**, **not**, **imply** and **equivalence**. Predicate calculus alone tends to be somewhat clumsy to use, so two additional concepts were added. The first is the idea of operators or functions, and these functions are different from predicates because they are not restricted to the values of TRUE or FALSE. They can return objects, and thus a function "uncle of" when applied to Mary would return a value of John. The second additional concept is that of the predicate, **equals**, which says that two individuals X and Y are indistinguishable under all predicates and functions. With these changes, we have a variety of first order logic, and it is no longer pure predicate calculus. A simple example of first order predicate logic would look like this:

$$\forall XYZ(\text{SMALLER}(X, Y)\ \text{SMALLER}(Y, Z) \rightarrow \text{SMALLER}(X, Z).$$

In words, this example indicates "for all X, Y and Z, and the case that X is smaller than Y and Y is smaller than Z, then this implies that X is smaller than Z."

Frames, Objects and Semantic Nets

An alternate form for representing this knowledge is a frame. A frame is a means of representing a structured situation such as a typical day in school or a generic object definition. In the frame there are slots that hold different pieces of information that are important to the particular stereotyped situation or are features or attributes of the object. Some of this information concerns how the frame can be used, or what one can expect to happen or actions that should be taken if certain situations did not take place. Slots are given values to represent instances of a situation or object. For example, a target frame may have slots for target type, number, manpower requirements, and indications associated with the possible presence of the target.

Semantic networks are associative networks that link nodes together with lines or arcs. The nodes represent an object and the arcs joining the nodes represent the relationship between the nodes, such as an engine is a part of a car. In this case the nodes of the semantic net are engine and car, and the arc joining the nodes defines the relationship "is a part of." Semantic networks are frequently stored in frames.

State-Space

State-space was one of the earliest representation formalisms and it was developed for problem-solving and game-playing programs. The search space is not actually a knowledge representation but is really a structure of a problem in terms of the alternatives available at each possible state of a problem. An example would be the alternative moves available on each turn of a game, such as chess. A straightforward way of finding the winning move would be to try all of the alternative moves and then try all of the opponent's responses. Clearly, in complex situations such as chess there are too many possible combinations and this leads to a combinatorial explosion. Special search strategies must be developed to deal with this combinatorial explosion or the run time of the expert system will be unreasonable. "Knowledge up, search down" is one way of expressing an AI approach to reducing the search space. The more knowledge that the problem-solving system can apply to guide the search, the quicker the search will be.

Procedures and Algorithms

Another alternative for knowledge representation is procedural representation. One of the earliest knowledge representation implementations was on PLANNER [7]. The procedural knowledge was used to encode explicit control of a theorem-proving process within a logic-based system. In this case, the procedural representation contains knowledge of the world in small programs that know how to do specific things and how to proceed in well specified situations.

The knowledge base alternatives for the selection of the knowledge representation scheme could itself use an expert system as an advisor to the human engineer. The length of this paper limits the available discussion of this important topic of knowledge base alternatives. In the specific applications in Section III the reader will be able to see the richness and variety that is required to handle knowledge representation properly.

The second part of the knowledge base, shown in Fig. 1, is the knowledge relationships component. The most common form that is encountered is the production rule. A typical rule is in the form of **IF THESE ANTECEDENTS ARE TRUE THEN THIS CONSEQUENT IS TRUE.** Some production rules are derived from physical relationships that are defined by conventional science. These rules can often be extracted from the domain expert or codified information such as a textbook. Other production rule relationships are derived from empirical forms of knowledge and are generally referred to as heuristics. Heuristics are rules of thumb that the domain expert uses, which usually achieve the correct answer or desired results. These heuristics are not optimal in a mathematical sense, but they frequently succeed. Often, an expert will use hard physical facts combined with these rules of thumb to solve his problems successfully.

A second type of knowledge is algorithmic. The knowledge relationships portion of the knowledge base can contain algorithms. These algorithms are additional procedural knowledge that allows various facts and inputs to be combined in a calculation that will provide new information in the process of solving that specific problem. Examples of the algorithms are estimation techniques to calculate the important parameters such as speed of an object, location, relative motion, etc.

The components of the knowledge base for an expert system have been described in a very general manner. The knowledge base can contain the following:

- facts
- rules and procedures
- logical relationships
- algorithms
- heuristics.

One important thing to remember is that the knowledge base is specific to the particular problem that is being solved. The second major portion of the expert system is the inference engine, which will be discussed next. The inference engine, as opposed to the knowledge base, may be common to a number of domains with similar characteristics. The inference engine, shown in Fig. 1, is a gatekeeper between what the expert system believes and the actual expert system program [7]. This gatekeeper is responsible for adding and deleting beliefs and performing certain classes of inferencing. This inferencing can be done as facts are added or when requests for information arrive at the gatekeeper.

One way of thinking about an expert system is to lay the problem out as a graph network that is filled with branches containing **and/or** logical functions. In this case, the goal is to find a node that will solve the problem. The search space that must be investigated in the pursuit of the solution can be very large. Barr and Feigenbaum [6] point out that the number of different complete plays for an average length chess game is 10^{120}. For checkers the search space is estimated to be 10^{40}. Clearly an exhaustive search of this space would task the fastest supercomputers for an inordinate amount of time. The search time of this very large, complex space can be reduced by using smart strategies. One of these smart strategies is heuristic search, which can usually solve a problem but without a mathematical guarantee. Generally, heuristic search uses rules of thumb, tricks, simplifications or any other clever techniques that drastically limit search in a large problem space.

Space here limits a full description of these search techniques, but frequently a combination of algorithms and heuristics are used. As an example, a particular path on a decision tree could be evaluated by how much cost was consumed to a given point, and then a heuristic would be used to estimate the cost to complete the evaluation of the tree from the given point to the final desired goal. These heuristics are usually determined by interviewing the expert and determining how he estimates that one branch of a decision tree is significantly more costly or has a higher payoff to evaluate than another. These rules of thumb are then embedded in the inference engine in order that it can decide what branch of the tree should be evaluated and what branch is too costly to evaluate or does not have an expected high payoff. The inference engine is then used to determine what knowledge should be used from the knowledge base, what inputs should be used, what information should be transmitted or sent to the user, and how the problem should be addressed. As information from sensors or other inputs becomes available the inference engine will try and reason about the situation. The inference engine attempts to guide the expert system to do the required actions to solve a specific problem.

The inference engine in its role as a gatekeeper can guide the expert system toward the solution of the problem. One approach called **forward chaining** reasons forward from the inputs of the expert system toward the final solution. The object is to guide the present situation forward from its initial conditions (usually data) to the final situation that will satisfy the goal condition. An example would be a chess game in which you are given a configuration of white and black pieces and you want to use legal chess moves to achieve checkmate.

The second major technique available to the inference engine is to **backward chain**. Backward chaining involves a strategy that works from the goal of the expert system and not from the current situation. Usually the goal statement is broken into one or more subgoals that can be solved more easily than the major ultimate goal. The individual subgoals can be broken down further until each sub-subgoal is a trivial problem that can be solved. This approach is used when there are a limited number of final states that can be achieved or when there is external evidence to suspect a specific situation. For example, for a medical doctor who is diagnosing pulmonary disease, he knows that there is a relatively small set of diseases that are under consideration. In this case, one expert system called PUFF uses backward chaining from a specific diagnosis and evaluates the evidence to determine if this diagnosis is justified. As the systems become more complex, multiple initiative approaches (that combine forward and backward chaining) are frequently used by the inference engine to solve a given problem.

As the problem domain increases in complexity, there are definite tradeoffs associated with the level of reasoning. For some problems the physical system is not modeled in a very accurate way but rather is greatly simplified. In these cases, a shallow reasoning system is frequently used, and in this case the knowledge base will contain rules and procedures that do not rigorously reflect the causality of a system. This can work in an acceptable manner for some problem domains, but frequently a more sophisticated model must be invoked and a deep reasoning system must be used. An example could be an electronic diagnostic problem. If the knowledge base only contains rules such as: **"when the power amplifier levels are low then examine the power supply output voltage"** we have an example of a relatively shallow reasoning system. An example of a deeper reasoning system would use more knowledge about the "physics" of a situation. An example could be an electronic circuit diagnostic expert system in which components such as transistors are modeled and the system knows that if the base current increases the collector voltage will drop.

At this point we have just touched the tip of the iceberg with regard to the complexity and variations of the design and implementation of expert systems. The main body of this paper is concerned with giving the reader a better appreciation for the application of expert systems and a discussion of some of the interesting problems associated with these applications. We believe that expert systems will play an important role in future military applications. Expert systems research, as well as currently fielded systems, has illustrated that useful systems can be built which display a great deal of intelligence in a narrow domain. Some of the most interesting problems to be solved are briefly listed below.

1) Acquiring and Structuring a Variety of Knowledge: Some of the facets of this problem have been mentioned previously. There is no clear cut technique, and a hybrid approach has been used in the past. For example, facts can be represented as rules, logic, frames or a combination of these techniques and others.

2) Performing in Real Time: Complex military problems with thousands of inputs and hundreds of choices will literally cause the machine to grind to a halt. Heuristics search and faster computers are parts of the solution, but this represents one of the concerns for real operational expert systems.

3) Dealing with Unanticipated Events: The ability to learn from experience is a key element in intelligent behavior of living organisms, yet one that is almost totally absent in present intelligent artifacts such as expert systems. The need for learning in an expert system is exemplified by the vast amounts of knowledge needed in most expert systems for good performance. The process of imbuing these systems with knowledge is slow and error prone and hinges upon the developing art of knowledge engineering. The prolonged use of expert systems also requires an ability to learn, because as the domain knowledge itself changes, so must the knowledge in the system, if the system is to sustain good performance. The evolution of knowledge in expert systems to improve system performance must be of paramount importance if the systems are expected to operate effectively in the battlefield of the future. One example of learning is for the machine to consider alternatives with its present knowledge base and inference engine and to acquire new techniques. Encouraging projects at Stanford, University of Illinois, MIT, University of Michigan and others show that there is hope but that a great deal of work must be accomplished before we have truly autonomous expert systems that can adapt to a situation and then restructure their rules.

4) Dealing with Uncertainty: This is the situation that we face most of the time. Because we cannot count on a radar detection or a sonar detection with a high degree of confidence, we must determine how the inference engine can manipulate the facts and propagate the evidence into a satisfactory situation. Since this is such a crucial issue, the discussion in Appendix II is provided as an introduction to the problems associated with uncertainty.

Appendix II Methods of Dealing with Uncertainty

Approaches to dealing with uncertainty generally take some variation of numeric characterization. Numeric representations usually take the form of the assignment of a point value (as the application of Bayes' Theorem or maximum entropy), intervals on a range (as in Dempster–Shafer Theory) or points within an evidence space. A variant on a numerical approach with foundations in set theory is Zadeh's Fuzzy Logic, or Fuzzy Set Theory. A discussion of these major methods follows.

The sources of uncertainty in reasoning with expert systems are numerous [12]–[15]. Some of these sources include situations such as when information is deficient because it is partial or not fully reliable, or when the representation language is inherently imprecise or information from multiple sources is conflicting [14]. For example, uncertainty is necessary when one interviews an expert and receives qualitative information such as "that evidence gives credence to this diagnosis or it suggests that this circumstance could be the cause for this disease." The problem for the expert system designer is to decide the acceptable way to handle this pervasive uncertainty. Chandrasekeran [15] points out that resolution of uncertainty is something that a human is expert at doing but that a human does not use a single method for resolving uncertainties. As a result, the expert system designer needs to have a bag of tricks to handle the different situations requiring uncertainty.

For a long time, derivations of the Bayesian model have been used in expert systems. The two most well-known rule based expert system examples are MYCIN, an expert system to diagnose and recommend the therapy for infectious blood disease, and PROSPECTOR, an expert system to aid in the identification and location of high valued ore deposits from geological data. The MYCIN program defined a concept of certainty factors that were used to manage uncertainty in rule based expert systems. In general, the certainty factor concept is an empirical ad hoc technique that did not have consistent characterization. Heckerman [14] documents a clear, precise and consistent formulation of the certainty factor model, and the following discussion shows how he developed the mathematical foundation for the MYCIN certainty factor. This development was completed several years after the expert system was built and tested, and in general does not bring any clear advantages over using a probabilistic model.

MYCIN's knowledge is stored as rules in the form of "**If evidence Then hypothesis.**" Frequently in medicine there is significant uncertainty between the relationship of evidence and hypothesis. The certainty factor was used to accommodate these nondeterministic relationships. The certainty factor varies between -1 and $+1$. Positive numbers convey belief in the hypothesis from the evidence, whereas negative numbers correspond to a decrease in belief. These certainty factors do not correspond to measures of absolute belief. A convenient notation is shown below

$$E \rightarrow CF(H, E) \rightarrow H$$

where H is the hypothesis, E is the supporting evidence and $CF(H, E)$ is the certainty factor associated with the rule. It is possible for one hypothesis to be dependent on more than one piece of evidence and, further, for that hypothesis to serve as evidence for another hypothesis. Each link between evidence and hypothesis will have its own certainty factor associated with it. Fig. 37 shows a typical evidence link. In this case hypothesis H_1 is supported by evidence E_1 and E_2. As can be seen, evidence E_2 is actually a hypothesis supported by evidence E_4 and E_5.

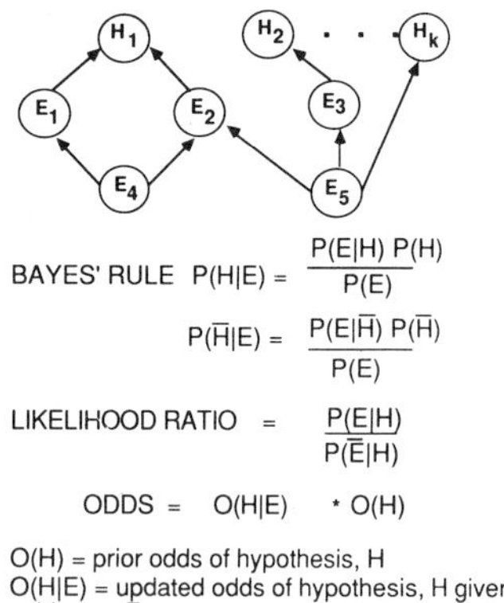

BAYES' RULE $P(H|E) = \dfrac{P(E|H)\,P(H)}{P(E)}$

$P(\bar{H}|E) = \dfrac{P(E|\bar{H})\,P(\bar{H})}{P(E)}$

LIKELIHOOD RATIO $= \dfrac{P(E|H)}{P(\bar{E}|H)}$

ODDS = O(H|E) • O(H)

O(H) = prior odds of hypothesis, H
O(H|E) = updated odds of hypothesis, H given evidence, E

Fig. 37. Evidence link.

As an example of how this system works let us concentrate on hypothesis H_1 that is supported by evidence E_1 and E_2 as shown below. One of the first questions to ask is how to combine this evidence. The simplified situation is shown below for the parallel combination of evidence:

$$\begin{matrix} E_1\ CF(H_1, E_1) \\ \searrow \\ \quad H_1 \\ \nearrow \\ E_2\ CF(H_1, E_2) \end{matrix} \quad \text{or} \quad E_1E_2 \xrightarrow{CF(H_1, E_1, E_2)} H_1$$

where $CF(H_1, E_1)$ is the certainty factor of hypothesis H_1 associated with evidence E_1 and $CF(H_1, E_2)$ is defined similarly. The equivalence of this situation is shown on the right-hand side but in this case the evidence E_1 and E_2 are shown as a single input with a composite certainty factor $CF(H_1, E_1, E_2)$. Heckerman points out that the certainty factors are combined as indicated below:

$$z = \begin{cases} x + y - xy & x, y \geq 0 \\ \dfrac{x + y}{1 - \min(|x|, |y|)} & x, y \text{ are of opposite sign} \\ x + y + xy & x, y < 0 \end{cases}$$

where $x = CF(H_1, E_1)$, $y = CF(H_1, E_2)$ and $z = CF(H_1, E_1, E_2)$.

The second situation is concerned with the combination of evidence that is configured in the sequential configuration shown below:

$$E' \xrightarrow{CF(E,E')} E \xrightarrow{CF(H,E)} H \text{ or } E' \xrightarrow{CF(H,E')} H$$

where $CF(H, E')$ denotes the combined certainty factor

$$z = \begin{cases} wx & w \geq 0 \\ -wy & w < 0 \end{cases}$$

$$w = CF(E, E')$$

$$x = CF(H, E)$$

$$y = CF(H, \bar{E})$$

$$z = CF(H, E').$$

Note that in the above case, the certainty factor, $CF(H, E')$ is dependent on $CF(H, \bar{E})$ which is the certainty factor for H given that the evidence E is not true. Heckerman [14] calls these sequential and parallel combination rules **desiderata**, and he shows that this desiderata should be used as the definitions of certainty factors. He points out that the original definitions of certainty factor as defined by Shortliffe and Buchanan [15] were inconsistent with the sequential combinations of certainty factors and caused major inconsistencies. Heckerman [14] shows that if uncertainty is to be propagated through an inference network in accordance with the desiderata for combinations, the evidence must be conditionally independent given the hypothesis and its negation. Further, the inference network must have a tree structure as opposed to the more general graph structure. The certainty factors that are used by Heckerman are the certainty factors that are defined by the axioms of the desiderata. In particular, he finds that monotonic transformations of the likelihood ratio, $\lambda(H, E)$, satisfy the desiderata where the likelihood ratio is defined as

$$\lambda(H, E) = P(E|H)/P(E|\bar{H})$$

$P(E|H)$ is the conditional probability of the evidence E given that the hypothesis H is true.

$P(E|\bar{H})$ is the conditional probability of the evidence E given that the hypothesis H is not true.

The odds-likelihood form of Bayes rule is shown below:

$$O(H|Ee) = \lambda(H, E, e)\ O(H|e)$$

where odds of an event x is $O(X) = P(x)/(1 - P(x))$ and $P(x)$ is probability of x, and $O(H|Ee)$ is the odds of hypothesis H conditioned on evidence E and prior evidence e. Notice in this expression that the likelihood ratio provides an update in the odds of a given hypothesis since λ is a multiplier factor times the prior odds and this yields updated odds. The likelihood ratio λ can vary from zero to infinity as opposed to the desired value of -1 to $+1$ for the certainty factors. This problem is easily resolved by setting $CF(H, E) = F(\lambda(H, E))$ where F is a function that maps λ into the interval of $[-1, +1]$. One function for $F(x)$ is

$$F_1(x) = (x - 1)/x, \quad x \geq 1$$
$$= x - 1, \quad x < 1.$$

One certainty factor $CF_1(H_1E)$ that satisfies the desiderata is found to be

$$CF_1(H, E) = F_1(\lambda(H, E)) = (\lambda(H, E) - 1)/\lambda(H, E), \quad \text{for } \lambda \geq 1$$
$$= \lambda(H, E) - 1, \quad \text{for } \lambda < 1.$$

This certainty factor CF_1 can also be expressed as

$$\frac{P(H|E) - P(H)}{P(H|E)(1 - P(H))}, \quad \text{for } P(H|E) > P(H)$$

$$CF_1 = \frac{P(H|E) - P(H)}{P(H)(1 - P(H|E))}, \quad \text{for } P(H) > P(H|E).$$

It is interesting to go back and compare this form with the original MYCIN work. The difference is an added factor in the denominator for each of the above expressions for the certainty factor.

There are numerous monotonic transformations of the likelihood function that allows a probabilistic interpretation for certainty factors and, in fact, Heckerman shows that every probabilistic interpretation for certainty factors is a transformation of λ. INTERNIST, an expert system for diagnosing internal medicine, uses an ad hoc scoring system that was very similar to the additive property of the log-likelihood system which states that

$$\ln [\lambda(H, E_1, E_2)] = \ln [\lambda(H, E_1)] + \ln [\lambda(H, E_2)].$$

The GLASGOW DYSPEPSIA system uses a certainty factor equal to $\ln \lambda[(h, e)]$ whereas PROSPECTOR uses $\lambda(H, E)$ and in particular a combination function of

$$\frac{\lambda(H, E) - 1}{\lambda(H, E) + 1}.$$

There are many problems concerned with the certainty factor discussed above. First, inherent in the certainty factor model, there must be conditional independence of evidence given H and given not H. Another problem is that the model does not hold for nontree networks. A third problem for systems such as MYCIN is that certainty factors were not derived from a clear operational definition. Despite all of these problems, MYCIN worked as well as an expert, and Heckerman points out that a sensitivity analysis of MYCIN's knowledge base showed that the system performance did not change significantly when large numbers of the certainty factors were changed. It is important to note that the original MYCIN certainty factors made some sense due to the application area of medical diagnosis. They tend not to make sense outside that domain—a failing of some expert system shells that use the original factors as inherent tools. Kyburg, in a recent AAAI workshop, notes that under very weak assumptions, probability is the only theory with point values that makes sense [96]. Levitt addresses the issues in "Model-Based Probabilistic Situation Inference in Hierarchial Hypothesis Spaces" [15].

Dempster-Shafer

Alternate approaches to certainty factors have been investigated since MYCIN and used on expert systems. One alternate theory was developed by Dempster [16] and an application was developed by Shafer [17]. One advantage of the Dempster-Shafer theory is to model the narrowing of the hypothesis set with the accumulation of evidence. This is a process that characterizes diagnostic reasoning in medicine, troubleshooting of electronic equipment and expert reasoning in general. Frequently an expert will use evidence that will focus his thinking on a larger subset of the total possibilities as opposed to a single hypothesis. For example, Shortliffe [15] points out that in the identification of an infecting organism, a smear showing gram negative organisms narrows the hypothesis set of all possible organisms to a specific subset. A Bayesian approach might assume equal prior probability and distribute the weight of this evidence uniformly. Shafer points out that this does not distinguish between uncertainty (or lack of knowledge) and equal certainty. In effect the Dempster-Shafer theory attributes belief to subsets as well as to individual elements of the hypothesis set. Zadeh [18] gives a simple example in explaining the Dempster-Shafer theory. Assume that Country X believes that a submarine S, belonging to Country Y, is hiding in X's territorial waters. The Ministry of Defense of X summons his experts $E_1, \cdots, E_n$, and asks each one to indicate the possible locations of S. The first m experts, $E_1, \cdots, E_m$, $m \leq n$, give the possible locations as $L_1, \cdots, L_m$. Each location L_i for $i = 1$ to m is a subset of the territorial waters. The rest of the experts $E_{m+1}, \cdots, E_n$ assert that there is no submarine in the territorial waters and the equivalent statement is that $L_{m+1} = \varnothing, \cdots, L_n = \varnothing$ where $\varnothing$ is the empty set.

If the Minister of Defense raises the question, **"Is S in a specified subset, A, of the territorial waters?"**, there are two cases that arise.

Case 1) E_i is a member of A and this implies that Expert E_i feels **certain** that S is in A.

Case 2) $E_i \cap A \neq 0$ and this implies that it is **possible** or **plausible** that S is in A.

Clearly case 1) implies case 2).

Assume the Minister of Defense aggregates his experts' opinions by averaging. If k out of n experts vote for case 1), the **average certainty (or necessity)** is k/n and if q experts vote for case 2), the **average possibility** is q/n. If the opinion of those experts who believe there is no submarine in the territorial waters is disregarded, the average certainty will be k/m and the average possibility or plausibility is q/m, respectively. The disregarding of those experts that said there is no submarine in the territorial waters is referred to as normalization. Zadeh [18] points out that normalization can lead to counterintuitive results since it suppresses an important aspect of the experts' opinions.

The Dempster-Shafer theory shows how the weighted experts' opinions can be combined. For example, if expert E_i has an opinion weighted by W_i, the **average normalized certainty $\mathbf{P_c}$** is

$$P_c = \frac{1}{K}\left(\sum_j W_j\right) \text{ for } E_j \text{ a member of area subset } A$$

and average normalized possibility P_p is

$$P_p = \frac{1}{K}\left(\sum_j W_j\right) \text{ for } E_j \text{ that could be in subset area } A$$

where $K = 1 - \sum_j W_j$ for E_j not a member of subset area A.

The average normalized certainty is the **belief** function of Dempster-Shafer, the average normalized possibility function is the **plausibility** function of Dempster-Shafer, and the weights $w_1, \cdots, w_n$ are basic probability functions of Dempster-Shafer's theory.

If the Minister of Defense wanted to know what the probability $P(A)$ was that S is in A, the normalized answer would be

$$P_c \leq P(A) \leq P_p.$$

The basic components of the Dempster–Shafer theory are the representation and the rule for combining evidence or degrees of belief in evidence. A major distinction is that certainty is not merely allocated to singleton hypotheses, but also to sets of hypotheses; this leaves some belief perhaps unallocated to a definite hypothesis, but allocated none the less to a set which may contain the correct hypothesis. The allocation of belief is in the construction of belief functions over a set of hypotheses Θ. (This set of hypotheses, or exclusive and exhaustive possibilities, is called the frame of discernment.) These belief functions are mappings from the power set of Θ to the unit interval, such that the belief in the null set is ∅, and the belief in Θ is 1.

In the framework of an I&W system, Θ could be the set of observable Soviet missiles, with a subset of strategic offensive missiles and the growing family of Soviet Mobile Missiles within the subset. The classification could proceed down to the singleton sets, containing such missiles as the SS-16, the SS-20, the SS-25, the SS-X-24, etc. The subsets of Θ form a kind of membership tree, with Θ at the top extending down to the singleton sets at the bottom. The Dempster-Shafer basic probability assignment (bpa), **m** assigns a quantity of belief to every element in the tree. This bpa corresponds to the weight *w* that was discussed before. A belief function **Bel,** (which corresponds to the average normalized certainty discussed earlier) represents the belief in a subset; **Bel** entails the belief in all subsets contained in that set by combining the values for **m(A),** for all subsets **A.** The belief function on a subset of mobile missiles, the SS-16, the SS-X-24, and the SS-25, would be represented by the sum of the basic probability assignments on all the subsets of that subset; **Bel({SS-16 SS-X-24 SS-25}) = m({SS-16 SS-X-24 SS-25}) + m({SS-16 SS-X-24}) + m({SS-16 SS-25}) + m({SS-X-24 SS-25}) + m({SS-16}) + m({SS-X-24}) + m({SS-25}).** Uncommitted belief is belief that is committed to the entire set Θ; **m(Θ).** Total ignorance is represented by the vacuous belief function; where **m(Θ) = 1,** and **m(A) = 0** for all subsets **A** of Θ. The belief interval of a subset **A** is given by **[Bel(A) 1-Bel(A^c)].** The width of this interval is the uncertainty of our belief in the hypotheses contained in **A.** For example, we might associate a belief interval **[.5, 2]** with an indicator on the above subset; this means that we are 50 percent sure that the observed target is in the above set, and 20 percent sure that it is not. We might use Dempster's rule of combination when we get a new belief interval from another knowledge source. The scheme for combining will update both our belief and our disbelief in the proposition. The Dempster rule of recombination can be a problem because when sources of evidence are combined, it is assumed that they are independent and frequently this is not the case.

Zadeh [13], [18] discusses a serious problem with the Dempster–Shafer theory; he shows that the use of the method to combine evidence from distinct sources may lead to counterintuitive results. To modify his example for our military application, suppose sensor *A* reports that the observed target is an SS-16 with certainty 0.99, or an SS-20 with certainty 0.01, but supports the assertion that the target is an SS-X-24 with certainty 0.99. Applying the orthogonal sum of Dempster–Shafer provides us with the unlikely conclusion that the belief in the assertion that the target is an SS-20 is 1.0. This is clearly a problem, and stems primarily from the fact that under Dempster–Shafer null values are not counted, but rather attributed to ignorance.

In the same paper Zadeh presents what is probably one of the more implementable views of Dempster–Shafer theory. He views it as applied to relational data base technology, as an instance of inference from second-order relations. Zadeh relates the measures of belief and plausibility to the certainty (or necessity) and possibility of a given query set *Q* for retrieval from a second-order relation in which the data entries are possibility distribution.

As an example, Zadeh shows a database called EMP2 that contains the following information:

EMP2 DATABASE

NAME	AGE
1	[22, 26]
2	[20, 22]
3	[30, 35]
4	[20, 22]
5	[28, 30].

Thus, in the case of category name, 1, the interval value for the age of 1 is known to be in the set {22, 23, 24, 25, 26}. This set contains the possible values of the variable AGE (1) or equivalently is the possibility distribution of AGE (1). A query *Q* to this database can ask a question such as "what fraction of employees satisfy the condition that AGE(i) ∈ *Q*, $i = 1, \cdots, 5$, where *Q* is the query set [20, 25]." Zadeh points out that the query set *Q* and the data entries in the column labeled AGE can be regarded as a possibility distribution. In this context, Zadeh says the database information and the queries can be described as granular with the data and the queries play the role of granules.

For the situations where the database attribute values are not known with certainty, it is proper to consider the possibility of *Q* given the possibility distribution. An example would be if the query *Q* was [20, 25] and AGE(1) is [22, 26], it is possible that AGE(1) ∈ *Q*, it is certain (or necessary) that AGE(4) ∈ *Q*, and it is not possible that AGE(5) ∈ *Q*. In general form we have:

a) AGE(i) ∈ *Q* is possible if $D_i \cap Q \neq \emptyset$ = empty set where D_i is the possibility distribution.
b) *Q* is certain (or necessary) if $Q \supset D_i$.
c) *Q* is not possible if $D_i \cap Q = \emptyset$ = empty set.

Updating the above table with test results for *Q* of [20, 25], we have:

NAME	AGE	TEST
1	[22, 26]	possible
2	[20, 22]	certain
3	[30, 35]	not possible
4	[20, 22]	certain
5	[28, 30]	not possible.

At this point, we can form an answer to the query **"What fraction of the employees are in the range of ages equal to 20, 21, 22, 23, 24, and 25?"** The response, *Resp*(*Q*), will be in two parts, one relating to certainty or necessity *N*(*Q*) and the other relating to its possibility $\pi(Q)$ and it will be written:

$$Resp(Q) = (N(Q); \pi(Q)).$$

For our example we have

$$Resp[20, 25] = (N[20, 25]) = 2/5; \pi([20, 25]) = 3/5).$$

In this case certainty also counts as possible since certainty implies possibility. The first entry for $Resp(Q)$ is $N(Q)$ and is referred to as a measure of belief in the Dempster–Shafer theory and the second entry $\pi(Q)$ is the measure of plausibility in Dempster–Shafer. If EMP2 database is a relation in which the values of age are singletons chosen from the possibility distributions in EMP2, then the response to Q of $N(Q)$ and $P(Q)$ are the lower and upper bounds.

Zadeh goes on to show that $N(Q)$ and $P(Q)$ can be computed from a summary of EMP2 which specifies the fraction of employees whose ages fall in the interval-valued entries in the AGE column. Assume EMP2 has n rows, with the entry in row $i = 1, \cdots, n$ under age D_i and that the D_i are comprised of k distinct sets $A_1, \cdots, A_k$ so that each D is one of the A_s, $s = 1, \cdots, k$. For our example, we have:

$$n = 5 \qquad k = 4$$

$$D_1 = [22, 26] \qquad A_1 = [22, 26]$$

$$D_2 = [20, 22] \qquad A_2 = [20, 22]$$

$$D_3 = [30, 35] \qquad A_3 = [30, 35]$$

$$D_4 = [20, 22] \qquad A_4 = [28, 30].$$

$$D_5 = [28, 30]$$

If EMP2 is viewed as a parent relation, its summary can be expressed as a granular distribution Δ of the form

$$\Delta = \{(A_1, p_1), (A_2, p_2), \cdots, (A_k, p_k)\}$$

in this case p_s, $s = 1, \cdots, k$, is the fraction of the D's that are A_s. For our case

$$\Delta = \{([22, 26], 1/5), ([20, 22], 2/5), ([30, 35], 1/5), ([28, 30], 1/5)\}.$$

In summary, we can express $N(Q)$ and $\pi(Q)$ defined in terms of the granular distribution D as

$$\text{Belief} = N(Q) = \sum_s p_s$$

$$\text{such that } (Q \supset A_s, s = 1, \cdots, k)$$

$$\text{Possibility} = \pi(Q) = \sum_s p_s$$

$$\text{such that } (A_s \cap Q \neq \varnothing, s = 1, \cdots, k).$$

Note that this interpretation is consistent with the earlier discussion of Dempster–Shafer and that the p_s correspond to the weights w and are the basic probability functions of Dempster–Shafer.

It can be shown that $P(Q) = 1 - N(Q^c)$ where Q^c is the complement of Q. In this explanation of Dempster–Shafer, Zadeh shows insight as to why normalization causes counterintuitive results. He also points out that in the case of definite attributes, the Dempster–Shafer rule for combination of evidence is not applicable unless the underlying granular distributions are combinable; that is, they have at least one parent relation which is conflict free. This implies that distinct probability distributions are not combinable and hence Dempster–Shafer is not applicable to such distributions.

Despite these problems, the Dempster–Shafer theory provides some advantages. There will be a great deal of work necessary before it can be used in a broad group of problems as some of the implementation problems have discussed. Even with the current restrictions, Dempster–Shafer has been used to help in the development of expert systems that deal with uncertainty.

As was stated in the introduction to this section, most of the AI methods for dealing with uncertainty do not attempt to mirror the way humans process uncertain or inexact information. Lofti Zadeh [19]–[21] has fathered a branch of AI called Fuzzy Logic, which aims at solving the lack of expressiveness that plagues classical probability theory. The problem is the fuzzy language that humans use to describe the events that we deal with and our beliefs that they will occur, and in particular with the kinds of uncertainty that one deals with in expert systems.

In probability theory, we cannot represent propositions containing fuzzy language, such as

- fuzzy predicates; *tall, old, irascible*
- fuzzy quantifiers; *most, several, few, usually*
- fuzzy events; *the boss will be in a difficult mood, this article will gain wide public acceptance, he will be in good health throughout his golden years*
- fuzzy facts; *older people look more distinguished*
- fuzzy probabilities; *likely, not too likely, pretty unlikely*
- fuzzy rules; *if you want to succeed in business, you've got to know a lot of people.*

Fuzzy language surrounds us. It is intuitive to our reasoning and thought processes. It is, quite simply, the way we think. Things are not always black or white, and not always black, white, or gray. Sometimes things are charcoal gray, pearl gray, graphite gray, or soot gray. In order to more closely approximate this cognitive representation, Zadeh developed fuzzy logic, building upon traditional set theory and aiming at the second main limitation with probability theory: the foundation of two-valued logic on which it is built [15]. In classical probability theory, the descriptive language is black or white; an event happens or it does not, an object is in a set or it is not, an item has a characteristic or it does not. In fuzzy sets, an object is not simply in or out of a given set, an object is assigned a grade or degree of membership, expressed along the unit interval (0, 1), where 0 stands for non-membership and 1 stands for (definite) membership.

Consider the following basic example for fuzzy set representation:

Let B be a bag of potatoes. What is the likelihood that the potato that you pull out to bake is a huge one? If there are n potatoes, $p_1, p_2, \cdots, p_n$, you may define the fuzzy set HUGE, and the membership function $\mu_{\text{HUGE}}(p_i)$, $i = 1, \cdots, n$ which denotes the degree of hugeness of each potato. So a 14 ounce potato p_x might have $\mu_{\text{HUGE}}(p_x) = 0.78$, whereas a 3 oz potato p_y might have $\mu_{\text{HUGE}}(p_y) = 0.2$.

Zadeh has defined a way of expressing the number of huge potatoes in the bag B, by using a concept called the *sigma-count*, which is the sum of the grades of membership, rounded to the next integer, if appropriate. Further, if the bag has a lot of little potatoes, and one does not want these small potatoes to misrepresent the HUGE count by providing enough small membership grades to raise the HUGE count by an integral number, it is possible to specify a minimum threshold, under which the potatoes will not be counted. As an example, suppose the bag of potatoes contained 10 potatoes, and we wanted to know how many are huge potatoes. Let us set the threshold at 0.3, and all potatoes with μ_{HUGE} under 0.3 will not be considered in our sigma-count. If the membership values are $\mu_{HUGE}(p_i) = (0.78, 0.2, 0.5, 0.43, 0.25, 0.64, 0.27, 0.3, 0.19, 0.8)$, with our threshold, the sigma-count would be 3.15, or simply 3 huge potatoes. Without the threshold, the sigma-count would be 5.08, or 5 huge potatoes, which seems to widen the definition of huge to include medium-sized potatoes.

Zadeh has built upon Fuzzy Logic to develop Possibility Theory, with the same kind of constructs that are found in probability theory. In the previous example, one would develop the possibility distribution, so that you could calculate the possibility of choosing a huge potato to bake. This distribution may be similar to probability distribution; the difference lies more in the interpretation of the distributed values. Let us go back to our earlier example:

Let **X** be a hostile relocatable missile in the set **A** = **{SS-20, SS-16, SS-25, SS-x-24}**. As a result of observances, we could define a probability distribution **Pr(X ∈ A)** = **{0.7, 0.1, 0, 0.2}**. That is, **X** is quite probably the SS-20, and highly unlikely that it is the SS-25, as that would be a blatant violation of the SALT treaty. In contrast, also based on observances, we could describe a possibility distribution, **P(X ∈ A)** = **{1, 1, 0.7, 1}**, meaning that it could possibly be any one of the four, but less possible that **X** is an SS-25. In possibility theory, every member of the set could have a possibility of one. An obvious difference between a possibility and a probability distribution is the requirement that the probabilities sum to one, although possibility values could be interpreted as maximum probability values.

Zadeh claims that possibility theory subsumes classical probability theory. His claim is that possibility theory, and its axioms and functions, answers well every problem stated for classical probability, and in addition, allows for the representation of fuzzy logic and inference from fuzzy sets. In spite of this apparently wide potential, fuzzy logic and possibility theory have not been widely implemented in expert systems.

Acknowledgment

The authors wish to thank Mrs. Ann Goddard for the tireless and careful preparation of this manuscript and Mrs. Bette Violette for her patient editing of this manuscript. We also thank Dr. Y. T. Chien and Ms. Michelle Younger for providing a great deal of guidance and direction. The IEEE reviewers' comments were thoughtful and well received in this revised paper, and we thank them for their useful inputs.

References

[1] A. J. Baciocco, "Artificial intelligence," *Military Science Technology*, vol. 1, no. 5, pp. 38–40, 1981.

[2] —, "Artificial intelligence and C³I," *Signal Magazine*, AFCEA, Sept. 1981.

[3] E. W. Martin, "Artificial intelligence and robotics for military systems," in *Proc. of The Army Conf. on Application of Artificial Intelligence to Battlefield Information Management*, Battelle Columbus Laboratories, (Washington, DC), Apr. 20–22, 1983.

[4] A. B. Salisbury, "Opening remarks on artificial intelligence," in *Proc. of The Army Conf. on Application of Artificial Intelligence to Battlefield Information Management*, Battelle Columbus Laboratories, (Washington, DC), Apr. 20–22, 1983.

[5] R. Shumaker and J. E. Franklin, "Artificial intelligence in military applications," *Signal Magazine*, vol. 40, p.29, June 1986.

[6] A. Barr and E. A. Feigenbaum, *The Handbook of Artificial Intelligence.* Los Altos, CA: William Kaufmann, Inc., 1981.

[7] E. Charniak and D. McDermott, "Introduction to artificial intelligence." Reading, MA: Addison-Wesley, 1984.

[8] P. Harmon and D. King, *Expert Systems.* New York, NY: Wiley, 1985.

[9] P. H. Winston, *Artificial Intelligence*, Second Edition. Reading, MA: Addison-Wesley, 1984.

[10] P. Jackson, *Introduction to Expert Systems.* Reading, MA: Addison-Wesley, 1986.

[11] D. B. Lenat, A. Clarkson, and G. Kiremidjian, "An expert system for indications and warning analysis," *IJCAI Proceedings*, pp. 259–262, 1983.

[12] R. O. Duda, P. E. Hart, and N. J. Nilsson, "Subjective Bayesian methods for rule-based inference systems," Tech. Rep. 124, Stanford Research Institute, Palo Alto, CA, Jan. 1976.

[13] L. A. Zadeh, "A simple view of the Dempster–Shafer theory of evidence and its implication for the role of combination," *AI Magazine*, vol. 7, no. 2, Summer 1986.

[14] D. Heckerman, "Probabilistic interpretation for MYCIN's certainty factors," in *Uncertainty in Artificial Intelligence*, L. N. Kanal and J. F. Lemmer, Eds. New York, NY: North Holland, 1986.

[15] B. Chandrasekaran and M. C. Tanner, "Uncertainty handling in expert systems: uniform vs. task-specific formalisms," in *Uncertainty in Artificial Intelligence*, L. N. Kanal and J. F. Lemmer, Eds. New York, NY: North Holland, 1986, pp. 35–46.

[16] A. P. Dempster, "A generalization of Bayesian inference," *J. Roy. Statis. Soc. Ser.* B 30, 1968.

[17] G. Shafer, *A Mathematical Theory of Evidence.* Princeton, NJ: Princeton University Press, 1976.

[18] L. A. Zadeh, "Review of Shafer's book, a mathematical theory of evidence," *AI Magazine*, vol. 5, no. 3, Fall 1984.

[19] —, *A Computational Theory of Dispositions.* Berkeley, CA: University of California, 1984.

[20] —, "A theory of commonsense knowledge," in *Aspects of Vagueness*, Skala, Termini, and Trillas, Eds. Dordrecht, Holland: D. Reidel Publishing Co., 1984.

[21] —, "Syllogistic reasoning as a basis for combination of evidence in expert systems," in *Proceedings of the 9th International Joint Conf. on Artificial Intelligence*, (Los Angeles, CA), Aug. 1985.

[22] R. Cohen and M. R. Grinbert, "A theory of heuristic reasoning about uncertainty," *AI Magazine*, vol. 4, no. 2, 1983.

[23] —, "A framework for heuristic reasoning about uncertainty," in *Proceedings of the 8th International Joint Conf. on Artificial Intelligence*, (Karlsruhe, West Germany), 1983.

[24] R. Cohen and M. D. Lieberman, "A report on FOLIO: An expert assistant for portfolio managers," in *Proceedings of the 8th International Joint Conf. on Artificial Intelligence*, (Karlsruhe, West Germany), 1983.

[25] Cohen *et al.*, "Representativeness and uncertainty in classification systems," *AI Magazine*, vol. 6, no. 3, 1985.

[26] P. R. Cohen, "Heuristic reasoning about uncertainty: An Artificial Intelligence approach," Great Britain: Pitman Advanced Publishing Program, 1985.

[27] J. Doyle, "Methodological simplicity in expert system construction: the case of judgements and reasoned assumptions," *AI Magazine*, vol. 4, no. 2, 1983.

[28] —, "A truth maintenance system," *Artificial Intelligence*, vol. 12, 1979.

[29] M. L. Ginsberg, "Implementing probabilistic reasoning," Working Paper HPP 84-31, Stanford University, Stanford, CA, June 1984.

[30] J. A. Barnett, "Computational methods for a mathematical theory of evidence," in *Proc. of the 7th Int. Joint Conf. on Artificial Intelligence*, (Vancouver, British Columbia, Canada), 1981.

[31] T. D. Garvey, J. D. Lowrance, and M. A. Fischer, "An inference technique for integrating knowledge from disparate sources," in *Proceedings of the 7th International Joint Conf. on Artificial Intelligence* (Vancouver, British Columbia, Canada), 1981.

[32] J. D. Lowrance and T. D. Garvey, "Evidential reasoning: and implementation for multisensor integration," SRI Technical Note 307, Dec. 1983.

[33] J. R. Slagle and H. Hamburger, "An expert system for a resource allocation problem," *Communications of the ACM*, vol. 28, no. 9, pp. 994-1004, Sept. 1985.

[34] J. R. Slagle, M. W. Gaynor, and E. J. Halpern, "An intelligent control strategy for computer consultation," *IEEE Trans. Pattern Anal. Machine Intell.*, vol. PAMI-6, Mar. 1984.

[35] L. B. Booker, "An Artificial Intelligence (AI) approach to ship classification," in *20th Annual Technical Symp.*, Washington, D.C. Chapter of ACM (Gaithersburg, MD), 1985.

[36] R. R. Cantone, F. J. Piptone, W. B. Lander, and M. P. Marrone, "Model-based probabilistic reasoning for electronics troubleshooting," in *IJCAI-83*.

[37] R. O. Duda *et al.*, "Development of a computer-based consultant for mineral exploration," Annual Report, SRI Projects 5821 and 6415, SRI International, Menlo Park, CA, Oct. 1977.

[38] R. R. Cantone, W. B. Lander, M. P. Marrone, and M. W. Gaynor, "IN-ATE™: Fault diagnosis as expert system guided search," in *Computer Expert Systems*, L. Bolc, and M. J. Coombs, Eds. New York, NY: Springer-Verlag, 1986.

[39] F. J. Pipitone, "An expert system for electronics troubleshooting based on function and connectivity," in *IEEE 1st Conf. on AI Applications*, (Denver, CO), 1984.

[40] —, "FIS, an electronics fault isolation system based on qualitative causal modeling," in *Aerospace Applications of AI Conf.*, 1985.

[41] —, "The FIS electronics troubleshooting system," in *Computer*, July 1986.

[42] J. DeKleer, "Reasoning with uncertainty in physical systems," *AI Journal*, 1985.

[43] R. Davis *et al.*, "Diagnosis based on description of structure and function," *AAAI Proceedings*, pp. 137-142, 1982.

[44] T. Levitt *et al.*, "Design of a probabilistic certainty calculus for ADRIES," Contract DACA76-86-C-0010, Advanced Decision Systems, Mountain View, CA, Apr. 1986.

[45] T. Levitt, L. Winter, T. Eppel, T. Irons, and C. Neveu, "Terrain knowledge elicitation for ADRIES: Part II," Contract DACA76-86-C-0010, Advanced Decision Systems, Mountain View, CA, Oct. 1987.

[46] T. Levitt, W. Edwards, and L. Winter, "Elicitation of a priori terrain knowledge for ADRIES," Contract DACA76-86-C-0010, Advanced Decision Systems, Mountain View, CA, Nov. 1986.

[47] DARPA, ETL, ADS, SAIC, MRJ, and TASC, "Advanced digital radar imagery exploitation system (ADRIES) program plan," Advanced Decision Systems (ADS), Mountain View, CA, Oct. 1986.

[48] R. Drazovich *et al.*, "Advanced digital radar imagery exploitation system (ADRIES) annual technical report," Contract DAEA18-83-C-0026, Advanced Decision Systems, TR-1040-02, Apr. 1985.

[49] D. Granrath, "Estimation procedure for ROC curves," Science Applications International Corporation (SAIC) Internal Memo dated April 15, 1987, Tucson, AZ, Apr. 1987.

[50] D. von Winterfeldt and W. Edwards, *Decision Analysis and Behavioral Research*. Cambridge, MA: Cambridge University Press, 1986.

[51] J. Pearl, "Fusion, propagation, and structuring in belief networks," *Artificial Intelligence*, vol. 29, pp. 241-288, 1986.

[52] R. A. Howard and J. E. Matheson, "Influence diagrams," SRI Technical Memo, Menlo Park, CA, 1980.

[53] C. F. Gettys and T. A. Willke, "The application of Bayes' Theorem when the true data state is uncertain," *Organizational Behavior and Human Performance*, pp. 125-141, 1969.

[54] D. A. Schum, "Current developments in research on cascaded inference," in *Cognitive Processes in Decision and Choice Behavior*, T. S. Wallsten, Ed. Hillsdale, NJ: Lawrence Erlbaum Press, 1980.

[55] C. W. Kelly, III, and S. Barclay, "A general Bayesian model for hierarchical inference," *Organizational Behavior and Human Performance*, vol. 10, 1973.

[56] J. Pearl, "On evidential reasoning in a hierarchy of hypotheses," *Artificial Intelligence*, vol. 28, no. 1, Feb. 1986.

[57] T. O. Binford and T. S. Levitt, "Bayesian inference in model-based machine vision," in *Proc. AAAI Uncertainty in Artificial Intelligence Workshop*, (Seattle, WA), July 1987.

[58] —, "Utility-based control for computer vision," in *Proc. AAAI Uncertainty in Artificial Intelligence Workshop*, (Minneapolis, MN), Aug. 1988.

[59] J. H. Holland *et al.*, *Induction: Processes of Inference, Learning, and Discovery*. Cambridge, MA: MIT Press, 1986, ch. 1, pp. 1-28.

[60] A. E. Feigenbaum, "Lecture at the first U.S.-China joint seminar on automation and intelligent systems," Beijing, China, May 28-June 1, 1984.

[61] A. L. Samuel, "Some studies in machine learning using the game of checkers," *IBM J. Res. Develop.*, vol. 3, pp. 211-232, 1959.

[62] M. Minsky, "Steps toward artificial intelligence," in *Computers and Thought*, A. E. Feigenbaum and J. Feldman, Eds. New York, NY: McGraw-Hill, 1963, pp. 429-435.

[63] R. S. Michalski, "Understanding the nature of learning: issues and research directions," in *Machine Learning*, vol. 2, R. S. Michalski, J. G. Carbonell, and T. G. Mitchell, Eds. Los Altos, CA: Morgan Kaufman, 1986, ch. 1, pp. 3-18.

[64] J. R. Quinlan, "Learning efficient classification procedures and their application to chess end games," in *Machine Learning*, vol. 1, R. S. Michalski, J. G. Carbonell, and T. G. Mitchell, Eds. Palo Alto, CA: Tioga Publishing, 1983, ch. 15, pp. 463-482.

[65] D. B. Lenat, "The role of heuristics in learning by discovery: three case studies," in *Machine Learning*, vol. 1, R. S. Michalski, J. G. Carbonell, and T. G. Mitchell, Eds. Palo Alto, CA: Tioga Publishing, 1983, ch. 15, pp. 249-263.

[66] G. E. Hinton *et al.*, "Boltzmann machines: Constraint satisfaction networks that learn," Tech. Rep. CMU-CS-84-119, Carnegie-Mellon Univ., Department of Computer Science, 1984.

[67] J. H. Holland, *Adaptation in Natural and Artificial Systems*. Ann Arbor, MI: Univ. of Michigan Press, 1975, ch. 6, pp. 89-120.

[68] D. E. Goldberg, "Computer aided gas pipeline operation using genetic algorithms and rule learning," Ph.D. Dissertation, Ann Arbor, MI, Univ. of Michigan Press, 1983.

[69] R. J. Brachman and J. G. Schmolze, "An overview of the KL-ONE knowledge representation system," *Cognitive Science*, vol. 9, pp. 171-216, 1985.

[70] J. H. Holland, "Properties of the bucket brigade algorithm," in *Proc. of an Int'l. Conf. on Genetic Algorithms and Their Applications*, pp. 1-7, 1985.

[71] C. L. Forgy, "Rete: A fast algorithm for the many pattern/many object pattern match problem," *Artificial Intelligence*, vol. 19, pp. 17-37, 1982.

[72] K. S. Keller and H. J. Antonisse, "Prediction-based competitive learning in the M2 system," to appear in *Proc. Expert Systems in Gov't. Conf.*, 1987.

[73] H. J. Antonisse and K. S. Keller, "Genetic operators for high-level knowledge representations," in *Proc. of 2nd Int'l. Conf. on Genetic Algorithms and Their Applications*, pp. 69-76, 1987.

[74] R. L. Riolo, "Bucket brigade performance: I. Long sequences of classifiers," in *Proc. of 2nd Int'l. Conf. on Genetic Algorithms and their Applications*, pp. 184-195, 1987.

[75] D. Klahr *et al.*, *Production System Models of Learning and Development*, D. Klahr, P. Langley, R. Neches, Eds. Cambridge, MA: MIT Press, 1987, ch. 1, pp. 1-53.

[76] J. E. Laird *et al.*, "Chunking in Soar: The anatomy of a general learning mechanism," *Machine Learning*, vol. 1, pp. 11-46, 1986.

[77] P. Nii, "Blackboard systems: Blackboard application systems," *AI Magazine*, vol. 7, no. 3, Conference, pp. 82-106, 1986.

[78] J. R. Benoit *et al.*, "ALLIES: An experiment in cooperating expert systems for command and control," in *Proc. of the Expert Systems in Gov't Symp.*, pp. 372-380, 1986.

[79] E. Durfee, V. Lesser, and D. Corkill, "Coherent cooperation among communicating problem solvers," Techn. Rep. 85-15, Department of Computer and Information Science, University of Massachusetts, Amherst, MA, pp. 2-8, 1985.

[80] P. Nii, "Blackboard systems: The blackboard model of problem solving," *AI Magazine*, vol. 7, no. 2, pp. 38-53, Summer 1986.

[81] D. R. Reddy, L. D. Erman, and R. B. Neely, "The HEARSAY speech understanding system: An example of the recognition process," in *Proc. of the 3rd IJCAI*, pp. 185-193, 1973.

[82] L. Erman, F. Hayes-Roth, V. Lesser, and D. Reddy, "The Hearsay-II speech-understanding system integrating knowledge to resolve uncertainty," *Computing Surveys*, vol. 12, pp. 213–253, 1980.

[83] R. Balzer, L. Erman, P. London, and C. Williams, "HEARSAY-III: A domain-independent framework for expert systems," in *Proc. of the 1st National Conf. on AI*, pp. 108–110, 1980.

[84] V. R. Lesser and D. D. Corkill, "The distributed vehicle monitoring testbed: A tool for investigating distributed problem solving networks," *AI Magazine*, vol. 4, no. 3, pp. 15–33, Fall 1983.

[85] L. Gasser, C. Braganza, and N. Herman, "Implementing distributed AI systems using MACE," in *Proc. 3rd Conf. on AI Applications*, pp. 315–320, 1987.

[86] D. D. Corkill, "A framework for organizational self-design in distributed problem solving networks," Tech. Rep. 82-33, Department of Computer and Information Science, University of Massachusetts, Amherst, MA, pp. 17–20, 1982.

[87] S. Cammarata, D. McArthur, and R. Steeb, "Strategies of cooperation in distributed problem solving," in *Proc. of the 8th IJCAI*, pp. 767–770, 1983.

[88] R. G. Smith and R. Davis, "Frameworks for cooperation in distributed problem solving," *IEEE Trans. Syst., Man, Cybern.*, vol. SMC-11, no. 1, pp. 61–70, Jan. 1981.

[89] R. Davis and R. G. Smith, "Negotiation as a metaphor for distributed problem solving," *Artificial Intelligence*, vol. 20, pp. 63–109, 1983.

[90] D. D. Corkill and V. R. Lesser, "The use of meta-level control for coordination in a distributed problem solving network," in *Proc. of the 8th IJCAI*, pp. 748–756, 1983.

[91] D. D. Corkill, V. R. Lesser, and E. Hudlicka, "Unifying data-directed and goal directed control," in *Proc. of the National Conf. on AI*, pp. 143–147, 1982.

[92] E. H. Durfee, V. R. Lesser, and D. D. Corkill, "Increasing coherence in a distributed problem solving network," in *Proc. of the 9th IJCAI*, pp. 1025–1030, 1985.

[93] P. Friedland, "Knowledge servers—Applications of artificial intelligence to advanced space information systems," presentation at AIAA NASA Conf., Washington, DC, June 1987.

[94] D. B. Lenat, "Overcoming the brittleness bottleneck," Keynote presentation to the IEEE Computer Science 3rd Conf. on Artificial Intelligence, KIssimmee, FL, Feb. 1987.

[95] P. Harmon, Ed., *Expert System Strategies*, vol. 2, no. 8, San Francisco, CA, August 1986.

[96] H. G. E. Kyburg, "Knowledge and certainty," in *Proceedings AAAI Uncertainty Workshop*, pp. 30–38, 1986.

Bibliography

Further Reading in Uncertainty

— B. G. Buchanan and E. H. Shortliffe, *Rule-Based Expert Systems*. Reading, MA: Addison-Wesley, 1984.

— P. Cheeseman, "A method of computing generalized Bayesian probability values for expert systems," in *Proceedings of the 8th International Joint Conf. on Artificial Intelligence*, Karlsruhe, West Germany, 1983.

— R. A. Dillard, "Computing probability masses in rule-based systems," Technical Document 545, Navel Ocean Systems Center, San Diego, CA, Sept. 1982.

— D. Dubois and H. Prade, "Combination and propagation of uncertainty with belief functions," in *Proceedings of the 9th International Joint Conf. on Artificial Intelligence*, (Los Angeles, CA), Aug. 1985.

— L. Friedman, "Extended plausible inference," in *Proc. of the 7th International Joint Conf. on Artificial Intelligence*, Vancouver, British Columbia, Canada, 1981.

— M. L. Ginsberg, "Does probability have a place in nonmonotonic reasoning?" in *Proc. of the 9th International Joint Conf. on Artificial Intelligence*, (Los Angeles, CA), Aug. 1985.

— J. Gordon and E. H. Shortliffe, "A method for managing evidential reasoning in a hierarchical hypothesis space," HPP 84-35, Stanford, CA, Sept. 1984.

— J. Y. Halpern and D. A. McAllester, *Likelihood, Probability, and Knowledge*, AAAI-84, Austin, TX, Aug. 1984.

— J. H. Kim and J. Pearl, "A computational model for causal and diagnostic reasoning in inference systems," in *Proc. of the 8th International Joint Conf. on Artificial Intelligence*, (Karlsruhe, West Germany), 1983.

— N. A. Khan and R. Jain, "Uncertainty management in a distributed knowledge based system," *Proc. of the 9th International Joint Conf. on Artificial Intelligence*, (Los Angeles, CA), Aug. 1985.

— S. Y. Lu and H. E. Stephanou, *A Set-Theoretic Framework for the Processing of Uncertain Knowledge*, AAAI-84, Austin, TX, Aug. 1984.

— W. Lukaszewicz, "General approach to nonmonotonic logic," in *Proc. of the 8th International Joint Conf. on Artificial Intelligence*, (Karlsruhe, West Germany), 1983.

— T. Niblett, "Judgemental reasoning for expert systems," in *Proc. of the 9th International Joint Conf. on Artificial Intelligence*, (Los Angeles, CA), Aug. 1985.

— H. Prade, "A synthgetic view of approximate reasoning techniques," in *Proc. of the 8th International Joint Conf. on Artificial Intelligence*, (Karlsruhe, West Germany), 1983.

— J. R. Quinlan, "INFERNO: A cautious approach to uncertain inference," Technical Note N-1898-RC, The Rand Corporation, Santa Monica, CA, Sept. 1982.

— —, "Consistency and plausible reasoning," in *Proc. of the 8th International Joint Conf. on Artificial Intelligence*, (Karlsruhe, West Germany), 1983.

— E. Rich, *Default Reasoning as Likelihood Reasoning*, AAAI-83, Washington, DC, Aug. 1983.

— C. Rollinger, "How to represent evidence—Aspects of uncertain reasoning," in *Proc. of the 8th International Joint Conf. on Artificial Intelligence*, (Karlsruhe, West Germany), 1983.

— G. Shafer, "Probability judgment in artificial intelligence and expert systems," presented at the Conference on the Calculus of Uncertainty in Artificial Intelligence, George Washington University, Dec. 1984.

— G. Shafer and A. Tversky, "Languages and designs for probability judgement," *Cognitive Science*, July-Sept., 1985.

— L. Shastri and J. A. Feldman, "Evidential reasoning in semantic networks: A formal theory," in *Proc. of the 9th International Joint Conf. on Artificial Intelligence*, (Los Angeles, CA), Aug. 1985.

— E. H. Shortliffe, *Computer-Based Medical Consultation: MYCIN*. New York, NY: American Elsevier, 1976.

— T. M. Strat, *Continuous Belief Functions for Evidential Reasoning*, AAAI-84, Austin, TX, Aug. 1984.

— M. Sullivan and P. R. Cohen, "An endorsement-based plan recognition program," in *Proc. of the 9th International Joint Conf. on Artificial Intelligence*, (Los Angeles, CA), Aug. 1985.

— T. R. Thompson, "Parallel formulation of evidential-reasoning theories," in *Proc. of the 9th International Joint Conf. on Artificial Intelligence*, (Los Angeles, CA), Aug. 1985.

— R. M. Tong, D. G. Shapiro, J. S. Dean, and B. P. McCune, "A comparison of uncertainty calculi in an expert system for information retrieval," in *Proc. of the 8th International Joint Conf. on Artificial Intelligence*, (Karlsruhe, West Germany), 1983.

— L. R. Wesley, "Reasoning about control: The investigation of an evidential approach," in *Proc. of the 8th International Joint Conf. on Artificial Intelligence*, (Karlsruhe, West Germany), 1983.

— A. P. White, "Predictor: An alternative approach to uncertain inference in expert systems," in *Proc. of the 9th International Joint Conf. on Artificial Intelligence*, (Los Angeles, CA), Aug. 1985.

— R. R. Yager, "Reasoning with uncertainty for expert systems," in *Proc. of the 9th International Joint Conf. on Artificial Intelligence*, (Los Angeles, CA), Aug. 1985.

— L. A. Zadeh, "A computational approach to fuzzy quantifiers in natural languages," *Computers and Mathematics*, vol. 9, 1983.

Further Reading for Expert Systems

— B. Buchanan, G. Sutherland, and E. Feigenbaum, "Heuristic DENDRAL: a program for generating explanatory hypotheses in organic chemistry," in *Machine Intelligence 4*, B. Meltzer and D. Michie, Eds. Edinburgh University Press, 1969.

— R. O. Duda and P. E. Hart, *Pattern Classification and Science Analysis*. New York, NY: Wiley, 1973.

— A. Hanson and E. Riseman, "VISIONS: A computer system for

interpreting scenes," in *Computer Vision Systems*, A. Hanson and E. Riseman, Eds. New York, NY: Academic Press, 1978.
- B. P. McCune and R. Drazovich, "Radar with sight and knowledge," *Defense Electronics*, vol. 15, no. 8, Aug. 1983.
- H. Niemann and Y. T. Chien, Eds., "Knowledge based image analysis," *Pattern Recognition* (Special Issue) vol. 17, no. 1, 1984.
- L. Wesley and A. Hanson, "The use of an evidential-based model for representing knowledge and reasoning about images in the VISIONS system," in *Proc. of the Workshop on Computer Vision: Representation and Control*, Rindge, NH, Aug. 1982.

Further Reading for Maintenance and Troubleshooting

- A. L. Brown, "Qualitative knowledge, causal reasoning and the localization of failures," MIT AI Lab AI-TR-362, Ph.D. Thesis, Nov. 1976.
- R. Davis *et al.*, "Diagnosis based on description of structure and function," in *Proc. AAAI-82*, Aug. 1982.
- J. DeKleer and J. S. Brown, "Foundations of envisioning," in *Proc. AAAI-82*, Aug. 1982.
- M. R. Genesereth, "Diagnosis using hierarchical design models," in *Proc. AAAI-82*, Aug. 1982.
- J. J. King, "Artificial intelligence techniques for device troubleshooting," Computer Science Laboratory Technical Note Series CSL-82-9 (CRC-TR-82-004), Hewlett Packard, 1501 Page Mill Road, Palo Alto, CA 94304, Aug. 1982.
- W. R. Simpson and H. S. Balaban, "The ARINC research system testability and maintenance program (STAMP)," in *Proc. 1982 IEEE Autotestcon Conf.*, (Dayton, OH), Oct. 1982.

Jude E. Franklin (Senior Member, IEEE) received the B.S., M.S., and Ph.D. degrees in electrical engineering from Catholic University of America, Washington, DC.

He is the Senior Vice President/General Manager of the Technology Division and the Manager of Research and Development at PRC/GIS, McLean, VA. At PRC, he directs R&D programs in Expert Systems, Natural Language, Distributed Problem Solving, Computer Security, Feature Extraction, Software Engineering, and Information Systems Engineering. Prior to joining PRC, he was Manager of the Navy Center for Applied Research in Artificial Intelligence, located at the Naval Research Laboratory and a Vice President of Applied Engineering at MAR Inc. His experience includes the design, analysis, and management of AI programs, computer systems, sonar systems, communication systems, and acoustic signal processing systems.

Dr. Franklin is a Fellow of the Washington Academy of Sciences, member of American Association of Artificial Intelligence, Sigma Xi Honorary Research Society, and Acoustics Society of America. He is the author of over 40 technical papers, conference papers, and technical reports. He has been active in the organization of IEEE technical conferences including vice chairman for the 1987 and chairman of the 1989 Conference on AI Systems in Government, co-editor of an IEEE EXPERT special issue on Applications of AI for the Government. He serves on the AFCEA committee to investigate technology insertion into SDI and a special committee on Security & Integrity of SDI.

Cora Lackey Carmody (Member, IEEE) received the B.S. and M.A. degrees in mathematics from the Johns Hopkins University, Baltimore, MD, and the M.S. degree in computer science from Fairleigh Dickinson University, Rutherford, NJ.

She is Chief Scientist of PRC/GIS's Space Systems Technologies Division, McLean, VA, where she is currently supporting the Space Station Software Support Environment project for NASA's Johnson Space Center. She has been with PRC since 1978, primarily in military information systems, specifically command & control software.

Karl Keller received the B.S. degree in physics and engineering from Washington and Lee University, Lexington, VA, in 1983, and the M.E. degree in systems engineering from the University of Virginia, Charlottesville, VA, in 1985. He is currently a Ph.D. candidate in systems engineering at the University of Virginia.

He is a member of the technical staff in the Artificial Intelligence Technical Center of the Washington C^3I Division of the MITRE Corporation, McLean, VA. He is currently conducting applied research in machine learning for military intelligence analysis. His other research interests include knowledge representation, knowledge acquisition, and knowledge-based simulation.

Tod S. Levitt (Member, IEEE) received the B.S. degree from Case Western Reserve University, Cleveland, OH, and the M.A. and Ph.D. degrees in mathematics from the University of Minnesota.

He is currently a Principal Research Scientist at Advanced Decision Systems, Mountain View, CA. He is the principal investigator in the development of a knowledge-based image understanding surveillance system for the interpretation of synthetic aperture radar imagery under DARPA's Advanced Digital Radar Imagery Exploitation System (ADRIES) project. As part of this research, he has co-developed a probabilistic certainty calculus that performs model-based Bayesian inference to accrue evidence for the belief in system hypotheses. The system performs partial matching, accounts for and represents multiple (conflicting) interpretations of the military situation, and also models the influence of terrain and military tactics in the evidential accrual process. Other work in evidential reasoning includes the use of inductive learning techniques for eliciting rules and probabilistic estimates from domain experts, domain independent evidential accrual in model-based, open-ended systems, and machine learning for induction of rules and probabilities. He is also project supervisor for the Knowledge-Based Vision section of DARPA's Autonomous Land Vehicle (ALV) program. The objectives of this research are to provide the basic visual modeling, prediction, and recognition capabilities for the perceptual system of the ALV. This work involves research in machine vision for understanding natural terrain. In the course of this work, he has developed a mathematical framework for the representation of the visual memory of a mobile robot that allows navigation and guidance based on visual events without the need of precise metric information. This work has been implemented in the QUALNAV model for qualitative spatial reasoning. His current interests are in the fields of image understanding and computer vision, artificial intelligence, evidential reasoning, and digital signal and image processing.

Brandon L. Buteau (Associate, IEEE) was born in Boston, MA, on February 8, 1954. He received the B.S. degree cum laude in applied mathematics (computer science) from Harvard University, Cambridge, MA, in 1976.

Since then he has worked at Planning Research Corporation (PRC), McLean, VA, on the design and development of information systems and technology for a variety of national intelligence systems. His efforts have earned commendations from both the Defense Intelligence Agency and the Defense Communications Agency. In his current role as a systems applications scientist, he is leading a research project in the area of distributed problem solving for PRC's Government Information Systems group. This project involves the integration of several advanced technologies into a unified problem-solving architecture, including distributed expert systems, natural language understanding, active temporal data bases, and machine learning.

Reprinted from *Proceedings: IEEE National Aerospace and Electronics Conference [NAECON]*, 1988, pages 1200-1207.

AN EXPERT SYSTEMS APPROACH TO DECISION SUPPORT IN A TIME-DEPENDENT, DATA SAMPLING ENVIRONMENT (A BRIEF DISCUSSION)

Captain Peter G. Raeth
Air Force Systems Command
Armament Division
3246th Test Wing /TZWT
Electronic Warfare Test Division
Eglin AFB, FL 32542-5424

ABSTRACT

What framework can be devised to assist analysts in coping with the growing volume of data needed to support the analysis of today's complex systems? Such a framework should organize and use an analyst's knowledge in an automated environment and go beyond merely plotting raw data or charting statistics derived from raw data.

This paper discusses the author's approach to the use of raw data which greatly reduces the volume of data while increasing its value for analysis and decision-making. Consistent application of an analyst's knowledge in an automated environment readily transforms raw data into information and the information into recommendations. The approach discussed in this paper concentrates first on capturing the analyst's knowledge. The method permits the incremental, and fairly random, capture of knowledge at several levels of detail. The human view of the knowledge base is' transformed into the computer view which is then applied to the raw data. Two previous papers based on the partial implementation of these techniques are referenced.

INTRODUCTION

In order to make recommendations about a system or to determine how a system reacted during a given period of time, analysts must process a huge volume of raw data. The sheer volume of data can be staggering, especially in fields such as electronic warfare systems testing. The resulting difficulty in sifting useful information from the data leaves little time for coming to meaningful conclusions. The analyst needs a consistent method of applying knowledge about a system. This method should work well in an automated environment which converts raw data into meaningful information and recommendations. The intent is to quickly determine what a system did during a given period of time and make recommendations accordingly.

The flow of data from its raw form to suggested recommendations is illustrated in Figure 1. As the data move through this scenario, they become lower in volume but higher in value relative to the goal of the analysis.

Figure 1 illustrates three transformation phases: raw data collection, state analysis, and recommendation determination. The computer must pass the raw data through the state analysis and recommendation phases and deliver appropriate reports using the knowledge supplied by the analyst. Time periods which contain no active states and later, no offered recommendations, can be deleted from consideration. Each of the three phases will now be discussed.

RAW DATA COLLECTION

The collection and organization of raw data is a crucial step in any automated analysis. Data from each system in the test must be gathered, recorded, converted, and passed on to the state analysis phase (Figure 2).

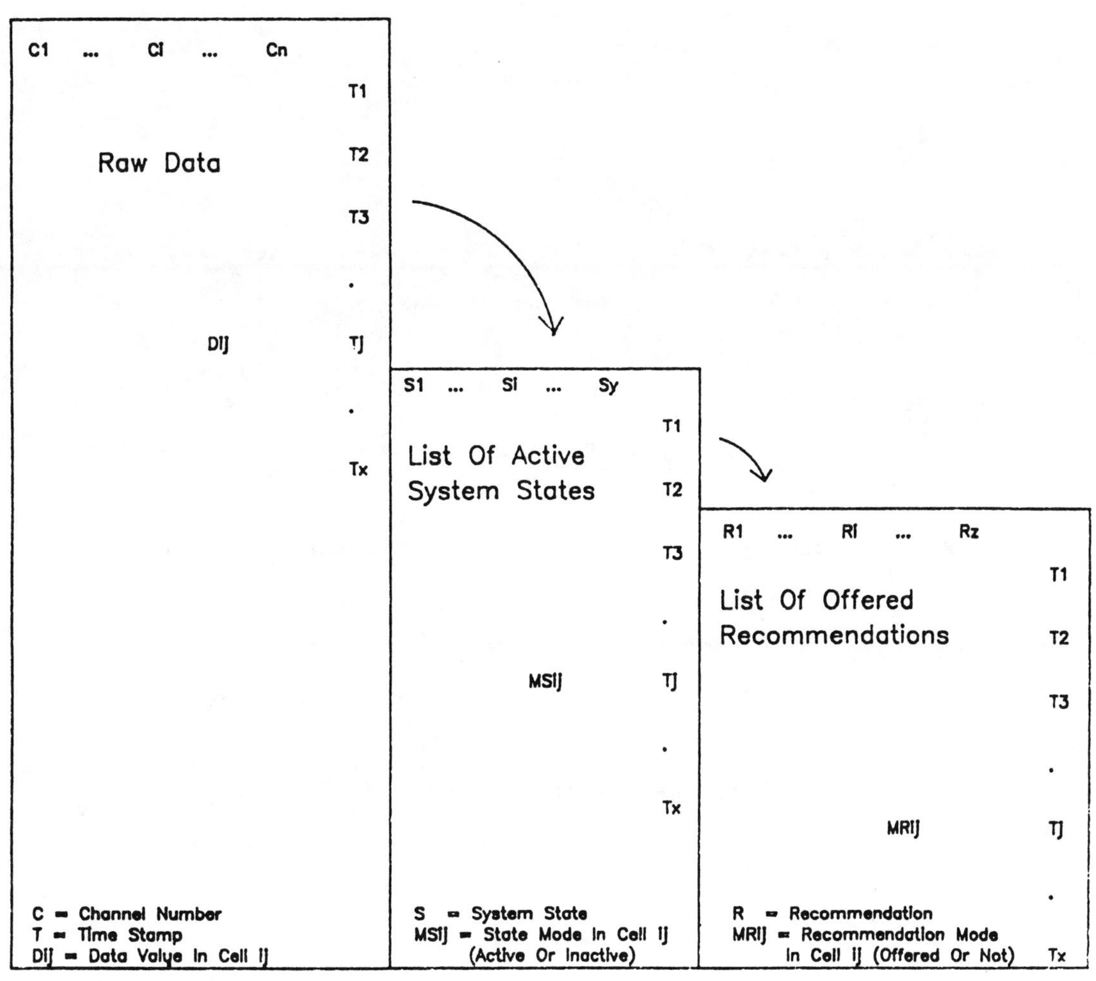

Figure 1

Transforming Raw Data Into Relevant Information

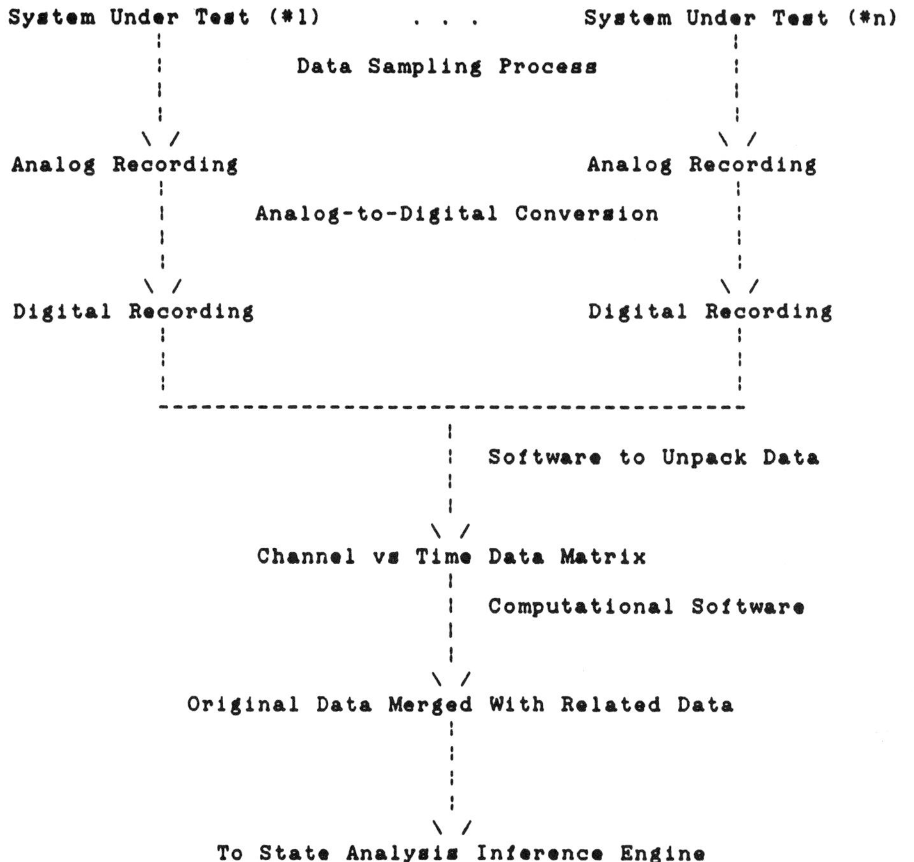

Figure 2

The Development of Raw Data for the State Analysis Inference Engine

There are two types of raw data, measured and derived. Measured data come directly from system sampling. Derived data are the result of computations performed on measured data. Examples of derived data are Fourier Transforms, averages, slopes, standard deviations, and other results which can be directly computed using measured data. Derived data are simply merged with measured data to create additional data channels. This results in a two-dimensional data matrix where the channels are the columns and the time stamps are the rows. The state analysis phase addresses this matrix, either directly or indirectly, to determine what the system did during any given period of time. The data to be collected during the raw data collection phase are determined by the requirements of the state analysis phase.

STATE ANALYSIS

Two things must happen to support the state analysis phase. First, the knowledge of the analyst must be captured. The supporting mechanism must permit incremental, and fairly random, knowledge capture. Second, this knowledge must be used in a consistent manner, along with the raw data, to determine what the system did during any given period of time.

Capturing the Analyst's Knowledge. The knowledge of the analyst must be gathered and organized into a detailed description of how system states are recognized in the raw data. Based on extensive interviews with experienced analysts, three basic categories make up this knowledge base:

States: internal or external conditions of the system which remain active for a given period of time. [See MacFarlane for a good discussion.]

Events: correlations of raw data which must be observed in time sequence before a state is determined to be active.

Specification Sets: triplets which identify a data channel, a data value, and a numerical comparison operator (<, >, =, etc). A list of these sets make up an event. The sets in an event must all be recognized at the same time before the event is determined to have occurred. Specification sets are used to identify events, and events are used to identify states. The comparison operators indicate integrity constraint ranges. Integrity constraints are restrictions on the real world for which a model is valid [Gal & Minker].

In addition to the three basic categories mentioned above, there are four categories of related knowledge which must also be collected:

Dwell Time: the minimum length of time an event must last.

Transition Time: the maximum length of time between the start of one event and the beginning of the next event.

Accept-Only-If Rules: rules which must be satisfied before a state, event, or set is determined to exist. These rules are tied to a particular state, event, or set and are invoked only if the given state, event, or set is indicated in the raw data. Acceptance is not accomplished unless all the rules are satisfied.

Reject-Only-If Rules: rules which must be satisfied before a state, event, or set is determined to not exist. These rules are tied to a particular state, event, or set and are invoked only if the given state, event, or set is not indicated in the raw data. Rejection is not accomplished unless all the rules are satisfied.

If no rules are given in the Accept-Only-If or Reject-Only-If categories, then the inference engine can proceed with the analysis without evaluating additional information. Accept-Only-If and Reject-Only-If rules allow the analyst to specify judgment factors which can, in specific instances, override the integrity constraints.

Accept-Only-If and Reject-Only-If rules take the form:

If Q1 then If Q2 then ... If Qn then accept/reject this state, event, or set.

Where Qx can be a state name or a specification set.

An example of a Reject-Only-If rule is:

If BAD-TIME then Reject. Means that the state, event, or set that the rule is tied to will be determined to not exist only if the time stamp is incorrect.

An example of an Accept-Only-If rule is:

If SUB-SYSTEM-B then Accept. Means that Sub-System-B must be active before the indicated state, event, or set can be meaningfully identified.

It is very important that an analyst-oriented mechanism be devised for the capture of these seven categories of knowledge. The mechanism chosen must use the language of the analyst and must be simple enough for the analyst to incrementally add knowledge in a fairly random fashion. One mechanism for doing this is the expanded spreadsheet shown in Figure 3.

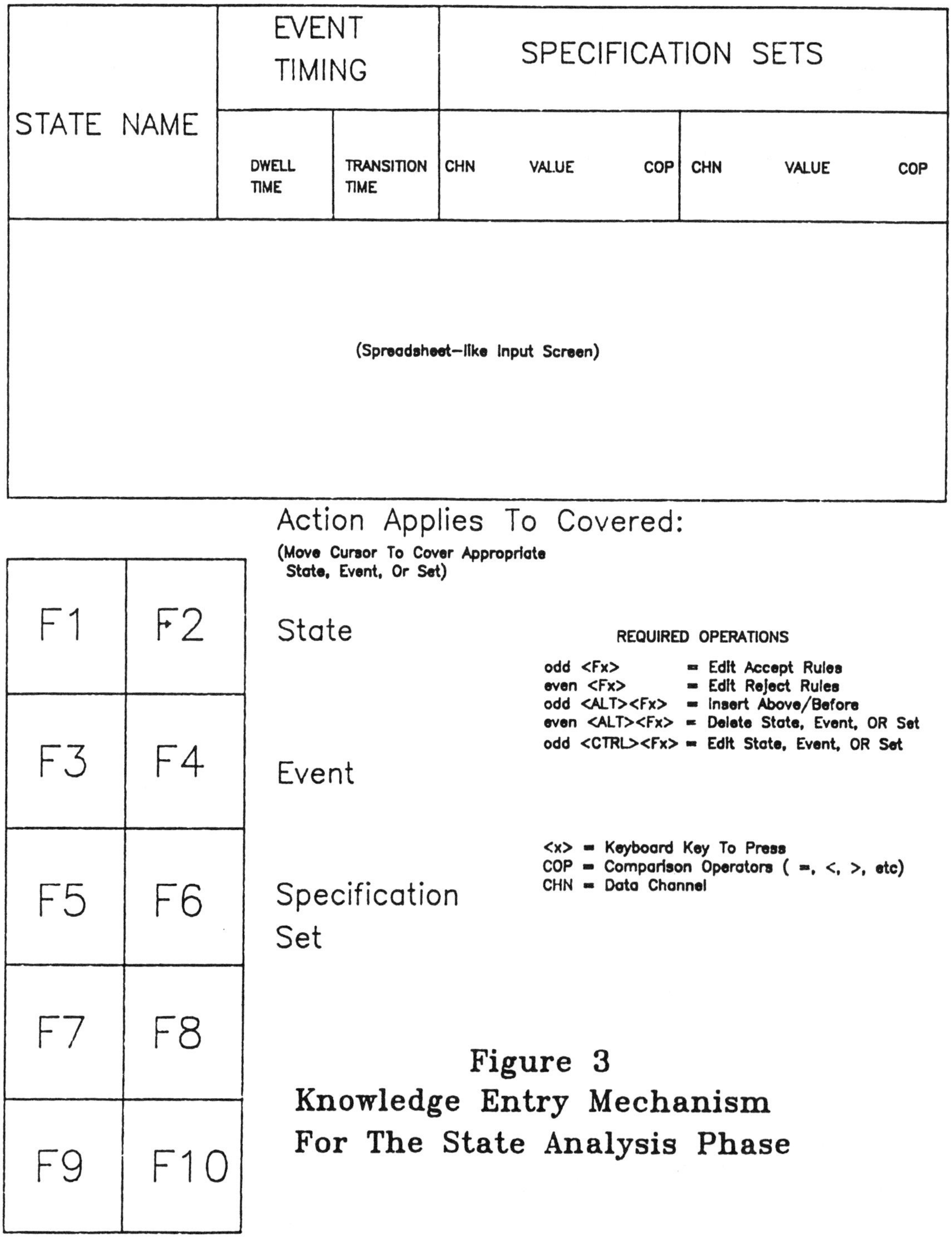

STATE NAME	EVENT TIMING		SPECIFICATION SETS					
	DWELL TIME	TRANSITION TIME	CHN	VALUE	COP	CHN	VALUE	COP

(Spreadsheet-like Input Screen)

Action Applies To Covered:
(Move Cursor To Cover Appropriate State, Event, Or Set)

F1	F2	State
F3	F4	Event
F5	F6	Specification Set
F7	F8	
F9	F10	

REQUIRED OPERATIONS

odd <Fx> = Edit Accept Rules
even <Fx> = Edit Reject Rules
odd <ALT><Fx> = Insert Above/Before
even <ALT><Fx> = Delete State, Event, OR Set
odd <CTRL><Fx> = Edit State, Event, OR Set

<x> = Keyboard Key To Press
COP = Comparison Operators (=, <, >, etc)
CHN = Data Channel

Figure 3
Knowledge Entry Mechanism For The State Analysis Phase

Using the extended spreadsheet shown in Figure 3, the analyst could easily capture state names, their related event, and specification set groupings, as well as their Accept-Only-If and Reject-Only-If rules. Functions which support the entry of this knowledge are: Edit-Accept-Rules, Edit-Reject-Rules, Insert, and Delete. These functions apply to states, events, and sets and can be easily activated via function keys. To indicate what to edit and where to insert or delete, the analyst simply moves the cursor to cover cells of the spreadsheet, then presses the appropriate function key.

The Computer View of the State Analysis Knowledge Base. The spreadsheet shown in Figure 3 is transformed by the computer into a multidimensional network database. An example of this database is shown in Figure 4. The states are listed vertically, the events for each state are listed horizontally, and the specification sets for each event are listed diagonally. The Accept-Only-If and Reject-Only-If rules are connected to the node for their respective state, event, or set. There is no logical limit to the number of states, events, sets, or rules. The only physical limitations are the amount of time and memory available to the analysis.

active search list. Once a specification set is found that does not compare as it should, the comparison process for that state stops and the comparisons for the next state start. If all the specification sets compare correctly and the dwell and transition time factors are within range, the current event is passed and the next event is made the current event for comparison purposes on the next sample. Once all events for a particular state are passed, that state is identified and placed on the active state list. The state report is also updated. An important point to note is that once a set, event, or state is found (or not found) in the raw data, it is not accepted (or rejected) unless all respective Accept-Only-If or Reject-Only-If rules are satisfied. In a previous paper, Raeth & Johnston discussed their partial implementation and use of this state analysis technique.

RECOMMENDATION DETERMINATION

Recommendations are based on the current and prior existence of states and other recommendations (taken either discretely or statistically accumulated). Each

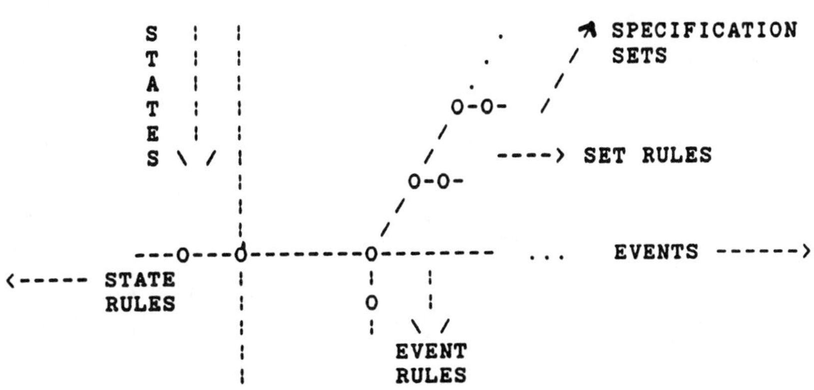

Figure 4

The Multi-Dimensional Network Database Representing the State Analysis Knowledge Base

The network shown in Figure 4 is traversed in a forward-chaining, breadth-first fashion. The inference engine uses the network to acquire the information it needs to tell when given sets, events, and states occur. The traversal of the network is such that each sample of data is compared to the specification sets of the current event of each state on the

recommendation is associated with a list of criteria which must be satisfied before that recommendation can be suggested. Each criteria has an asociated 'history' factor. This factor is a period of past time counting from the present time over which the given criteria may have been active. For instance, it may be important to know whether or not a switch was in the

"On" position during the past five seconds. The "history" factor can be specified as greater than zero if history is important or made equal to zero if only the present condition of the criteria is needed. A mechanism for capturing and organizing this type of knowledge can take the form of a relational database's input screen and the database management software's editing, deleting, and inserting functions. Figure 5 illustrates the input screen.

```
RECOMMENDATION1:  recommendation name
 CRITERIA1:  name of active state/recommendation    HISTORY1:  numeric factor
 CRITERIA2:  name of active state/recommendation    HISTORY2:  numeric factor
 etc

RECOMMENDATION2:  ______________________
 CRITERIA1:  ______________________                 HISTORY1:  __________
 CRITERIA2:  ______________________                 HISTORY2:  __________
 etc

RECOMMENDATIONx:
```

Figure 5

Input Screen for Capturing Recommendations and Their Associated Criteria

The database of recommendations and their associated criteria are transformed by the computer into a binary tree. Each recommendation's criteria are first sorted alphabetically. Then, the recommendations themselves are sorted according to the number of criteria associated with each recommendation (least number on top). The tree is constructed by placing the first recommendation on the tree with its associated criteria. The top criterion is the tree root and the suggested recommendation is the leaf. In a similar fashion, the other recommendations are placed on the tree with their associated criteria according to how they match recommendations and criteria already placed on the tree. This binary tree does not depend on any given order of recommendations or criteria. Each hangs on the tree regardless of whether it is a suggested recommendation or a criteria. An example recommendation knowledge base (after sorting) and its resulting binary tree is shown in Figure 6.

The tree shown in Figure 6 is traversed in a top-down, depth-first fashion. The left branch of each node is the branch taken if the recommendation or state making up the given criteria is currently active. The right side is taken if the recommendation or state is not currently active. This process continues until a suggested recommendation is achieved. Then, the recommendation report and the active recommendation list are updated. The process continues down the right branch of the node until the leaves of the tree are reached. Each time a recommendation or state is added to or removed from the active list, the binary recommendation tree must be traversed. In a previous paper, Raeth & Hardin discussed their partial implementation and use of this recommendation process.

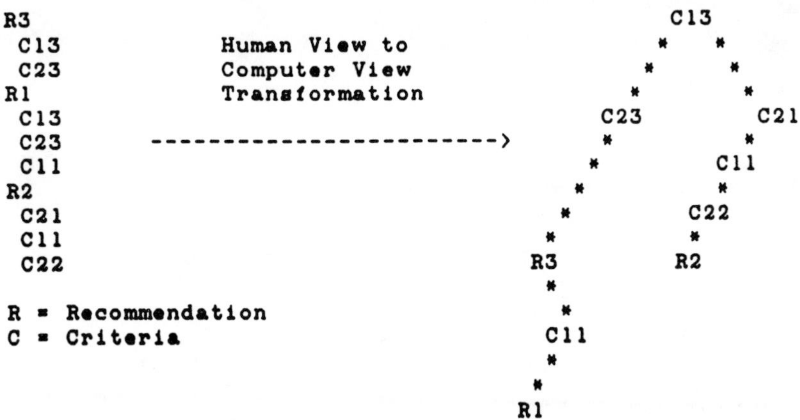

Figure 6

Sample Transformation of the Recommendation Knowledge Base Into a Binary Tree

SUMMARY

This paper has discussed a method of using raw data and an analyst's knowledge in an automated environment to turn data into information and information into suggested recommendations. [For a good tutorial on expert systems in general, see Gevarter.]

The contributions made by this paper are:

1) a way of getting meaning out of raw data samples.

2) a way of helping the analyst cope with the growing mounds of raw data while preserving the detail of the data.

3) a means of applying the analyst's knowledge consistently.

4) a means of identifying, structuring, and collecting [Waldron] the knowledge used to perform state analysis and to offer recommendations based on active states.

The current direction of this project is to document the software already produced for state analysis and to move that software into full production.

ACKNOWLEDGEMENTS

The author wishes to thank the following people:

Vicky Deiter, Doug Nation, and Marilyn Raeth for their review of this paper.

Ross Anderson for his encouragement and support.

Patricia Johnston and John Holland for their discussions on Goldenbird analysis.

James Graves (Georgia Tech Research Institute) for his insights into decision criteria.

BIBLIOGRAPHY

Gal, Annie and Jack Minker 'A Natural Language Database Interface that Provides Cooperative Answers' IEEE Conference on Artificial Intelligence Applications, 1985

Gevarter, William B. 'An Overview of Expert Systems' National Technical Information Service; Wash, DC; NTIS/DTIC # PB83-217562; May 82

MacFarlane, A.G.J. Engineering Systems Analysis Reading, MA; Addison-Wesley; 1964

Raeth, Peter G. and Patricia A. Johnston 'Using Expert Systems to Automate Goldenbird Stripchart Analysis' Association of Old Crows Joint Western-Mountain Region Technical Symposium, 1987

Raeth, Peter G. and James M. Hardin 'Using Expert Systems Technology to Develop a Statistical Strategist' 51st Meeting of the Mississippi Academy of Science, 1987

Waldron, Vincent R. 'Process Tracing as a Method for Initial Knowledge Acquisition' IEEE Conference on Artificial Intelligence Applications, 1985

Chapter 4: Knowledge Representation

The first task of any developer seeking to use expert systems techniques in new or existing software is to understand and develop a means to represent the knowledge appropriate to the problem domain. Usually, a means must be found to map the human view of the knowledge to something that the computer can use. The seven articles in this chapter discuss practical means of knowledge representation. Woods details a number of fundamental issues. Next, Adelson offers insight into several difficulties in knowledge representation that cut across many problem domains. Metaknowledge, what we know about what we know, and its use in expert systems is introduced by Aiello *et al.* A means of representing sequences of actions is given by Georgeff and Lansky. They are followed by Sandewall who talks about a way of representing objects and their features. The next paper, by Allen, offers ideas on working with time intervals. Finally, Francioni and Kandel discuss decision tables as applied to expert systems software.

Reprinted from *Proceedings of the IEEE*, Volume 74, Number 10, October 1986, pages 1322-1334.

Important Issues in Knowledge Representation

WILLIAM A. WOODS, MEMBER, IEEE

This paper discusses a number of important issues that drive knowledge representation research. It begins by considering the relationship between knowledge and the world and the use of knowledge by reasoning agents (both biological and mechanical) and concludes that a knowledge representation system must support activities of perception, learning, and planning to act. An argument is made that the mechanisms of traditional formal logic, while important to our understanding of mechanical reasoning, are not by themselves sufficient to solve all of the associated problems. In particular, notational aspects of a knowledge representation system are important—both for computational and conceptual reasons. Two such aspects are distinguished—expressive adequacy and notational efficacy.

The paper also discusses the structure of conceptual representations and argues that taxonomic classification structures can advance both expressive adequacy and notational efficacy. It predicts that such techniques will eventually be applicable throughout computer science and that their application can produce a new style of programming—more oriented toward specifying the desired behavior in conceptual terms. Such "taxonomic programming" can have advantages for flexibility, extensibility, and maintainability, as well as for documentation and user education.

INTRODUCTION

In computer science, a good solution often depends on a good representation. The first step in the development of most computer applications is the selection of a representation for the input, output, and intermediate results that the program will operate upon. For applications in artificial intelligence, this initial choice of representation is especially important. This is because the possible representational paradigms are diverse and the forcing criteria for the choice are usually not clear in the beginning. Yet the consequences of an inadequate choice can be devastating in the later stages of a project if it is discovered that critical information cannot be encoded within the chosen representational paradigm. When designing intelligent agents that can understand natural language, characterize perceptual data, or learn about their world, a suitable representational system for encoding their knowledge and the intermediate states of their reasoning is crucial. This is because the representational primitives, together with the system for their combination, effectively limit what such systems can perceive, know, or understand (see Fig. 1).

Fig. 1. *A priori* choices limit what a computer can perceive or know.

This paper discusses a number of issues that serve as goals for knowledge representation research. The objective is to discover general principles, frameworks, representational systems, and a good set of representational primitives for dealing with an open-ended range of knowledge. By "representational primitives" I mean to include not only primitive concepts but (especially) the primitive elements and operators out of which an open-ended range of learned concepts can be constructed. The focus of the paper will be on the problems that arise when designing representational systems to support any kind of **knowledge-based system**—that is, a computer system that uses knowledge to perform some task. I will first discuss some philosophical issues concerning the nature of knowledge and its relationship to the world and then consider a variety of issues that a comprehensive knowledge representation system must address.

The general case of a knowledge-based system can be thought of as a reasoning agent applying knowledge to

Manuscript received July 30, 1985; revised June 11, 1986. This paper was adapted from "What's Important about Knowledge Representation," by W. A. Woods, which was published in *IEEE Computer*, vol. 16, no. 10, pp. 22–27, October 1983. Copyright © 1983 IEEE. A slightly different version, entitled "Knowledge Representation: What's Important about It," will appear in a collection, *Knowledge Representation*, edited by G. McCalla and N. Cercone, to be published by Springer-Verlag.

The author is with Applied Expert Systems, Inc., Cambridge, MA 02142, and with Harvard University, Cambridge, MA 02138, USA.

achieve goals. Although the representational problems that need to be solved for many less ambitious knowledge-based systems may be simpler than those for a universal reasoning agent, the general principles are nevertheless the same.

Knowledge for Reasoning Agents

The role of reasoning for an intelligent agent is depicted in Fig. 2. Such an agent is perpetually engaged in an infinite loop of perceiving things, reasoning about them, and taking actions. The actions are determined by a set of internal

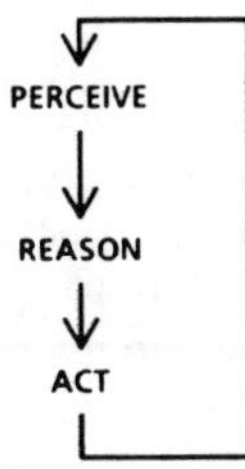

Fig. 2. The reasoning loop for an intelligent agent.

beliefs, goals, and objectives in interaction with what is perceived. A major task for such an agent is to acquire a model of the world in which it is embedded and to keep its model sufficiently consistent with the real world that it can achieve its goals. Knowledge of the world consists of two kinds of things—**facts** about what is or has been true (the known world state), and **rules** for predicting changes over time, consequences of actions, and unobserved things that can be deduced from other observations (the generalized physics/logic/psychophysics/sociology of the world).

An unavoidable characteristic of the internal model of the world is that it is, at best, an incomplete model of the real external world. In addition, it may be partly in error due to false assumptions and invalid reasoning. Differences between the internal model and the real world arise from a variety of sources, including (but not limited to):

- changes that have happened in the world since the agent recorded some fact about it (the world will not hold still);
- the inability of the agent to learn in a reasonable amount of time everything it could know in principle (there is too much to know);
- limitations of the knowledge representation system that preclude it from even conceptualizing certain things about the world (the world is richer than we can imagine).

(These are fundamental limitations that cannot be avoided simply by careful reasoning or avoiding unwarranted assumptions. There are, of course, other limitations that can arise from inadequate inferential mechanisms.)

Although it may not be apparent at first glance, this characterization is as accurate for most computer application systems as it is for a biological or artificial creature. It is a fair characterization of the function that a database system is intended to support, as well as a good model of the situation of an intelligent expert system. This perspective can be taken as a general characterization of the kinds of systems that require knowledge representation, and it can serve as a focus for the general problems that a knowledge representation system should solve. Specifically, it suggests that a knowledge representation system should support perception, reasoning, planning, and controlling actions with respect to some model of some world.

Modeling the External World

One theory of intelligence [5] conjectures that the essence of intelligence comes from a kind of internalized natural selection in which the intelligent agent has an internal model of the world against which it can try out hypothesized courses of action. This allows it to evaluate expected results without exposing it prematurely to the dangers of real actions in the real world. Whether this is the essence of intelligence or not, it is clearly a useful ability and an important role for knowledge representation—both in intelligent creatures and in computer artifacts.

Viewing the knowledge base of an intelligent agent as a **model** of the external world focuses attention on a number of problems that are not normally addressed in database systems or knowledge-based expert systems. For example, the questions of how the individuals and relationships in the model relate to the external world are not usually part of the competence of a database system. This role is filled at the time of input by a data entry specialist, after which the database behaves as if its contents **were** the world. If a person querying the database wonders what the terms in the database are supposed to mean (e.g. how is "annual coal production in Ohio" actually determined), there is nothing in the competence of the system to support this need. What is missing is a characterization of the relationship between the internal model and the world "out there."

Typical database systems support the storage and retrieval of "facts" about the world and certain aggregation and filtering operations on them, but do not perform operations of perception to map situations in the world into database contents. Nor do they perform planning operations leading to action in the world (unless you count report generators and graphics displays as actions in the world). The utility of a traditional database system rests in its support of an overall human activity in which the perception and action aspects are performed by the human users and maintainers of the database. However, the objectives of the overall combination of the database and its human "partners" are similar to those of our canonical intelligent agent. Thus an ideal database (or knowledge base) would be evaluated in terms of its support for the same kinds of reasoning, perception, and planning activities as performed by our intelligent agent (although, in some cases, parts of these activities are performed by human agents interacting with the system).

Perception and Reasoning by Machine

Increasingly, computer systems are being called upon to perform tasks of perception and reasoning as well as storage and retrieval of data. The problem of automatically perceiving larger patterns in data is a perceptual task that begins to be important for many expert systems applications. Recognition of visual scenes, natural language parsing and interpretation, plan recognition in discourse understanding, and continuous speech understanding are generic classes of perceptual problems which reasoning agents may be called upon to perform. Domain-specific instances include: medical diagnosis, mass spectrographic analysis, and seismic signal interpretation. The roles of a knowledge representation system in perception are important to all of these tasks.

Similarly, deductive reasoning applied to the information provided by perception is essential to the analysis of situations and the planning of actions in knowledge-based systems. Such reasoning is required to draw conclusions from facts, to connect known facts to enabling conditions for actions and expectations, to determine applicability of actions to situations, to evaluate alternative hypothetical actions, to determine when expectations and hypotheses are violated, to structure plans to achieve goals, and to detect inconsistencies among goals and to resolve them. Many of these activities involve the application of deductively valid rules of inference as in a formal predicate calculus. Other reasoning processes that can be applied to such information include: a) drawing "nonmonotonic" conclusions from limited amounts of information, b) determining degrees of likelihood or confidence of a supposition through probabilistic reasoning such as Bayesian analysis, c) conceptualizing and naming new entities through constructive actions such as planning and designing, and d) maximization or "satisficing" processes that seek the best choice among alternatives (e.g., the best fit to some specification or the best account of some data elements).[1] These reasoning processes do not consist of drawing deductively valid conclusions, and their operations differ in many respects from those of deductive inference.

All of the above processes make use of knowledge or information and draw conclusions. Some system of knowledge representation is implicit in all of these processes for recording and representing both the input information and the conclusions that are drawn. Whatever representational system is chosen it will play an important role in organizing the information and in supporting the internal substeps of the reasoning and/or perceptual processes.

It is important to realize that the reasoning involved here is dealing with a model of the world, not the real world. The model imposes an abstraction on the real world, segments the world into entities, and postulates relationships among them. Reasoning systems can draw conclusions about entities and propagate the consequences of assumptions, but do not deal directly with the world. Perceptual processes interact with the world to populate the internal model. Actions performed by the agent can affect both the external world and the internal model. These distinctions are fundamental for keeping track of what's going on in the knowledge base of a reasoning agent. Fig. 3 illustrates some elaborations of the basic reasoning loop to begin to take account of this distinction. Notice that at the point of action, there is an element of expectation introduced into the model to record the effect that the action is expected to have. Expectations have two roles: 1) the intended effect is modeled directly in the internal model of the world, and may be compared with the system's perceptions to determine if the action has succeeded, and 2) expectations condition perception by preparing the system to perceive what is expected (with consequent risks of percieving only what is expected).

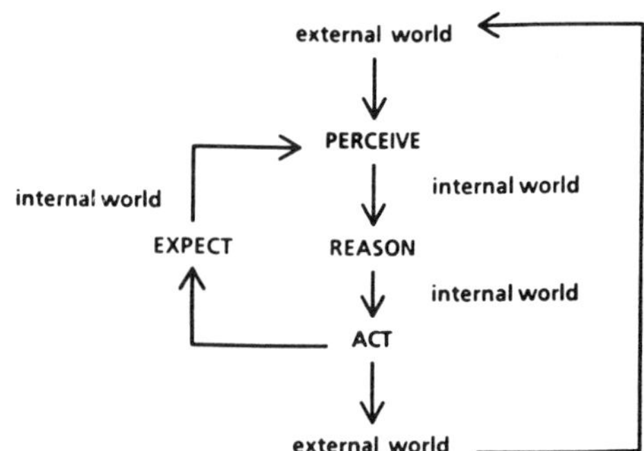

Fig. 3. Modeling the external world.

The Nature of the World and its Models

There are at least two ways to view the world and our perception of it. One way views the world as an ideal entity, perceived through sensors that are limited and prone to error, so that it is necessary to infer the hidden real world from the measurements of our sensors. A classic example of such a situation is that of a hidden Markov process, whose state transitions are only visible through their effects on a probabilistic output function that does not uniquely determine the underlying state. In this view, the nature of the "real world" can be inferred through reasoning or calculation (given enough data).

Another way to view the same situation is that the world is as it is, and is too complex and detailed to ever be captured or even fully conceived by any reasoning agent. In this view, the world as it presents itself to our senses is the ultimate reality, and the ideal models that we construct are just that—models—whose fidelity is traded off against the simplicity and efficiency of their use. From this latter perspective, it is not necessary to postulate objective reality for the idealized models, but only to consider them from the perspective of their utility in predicting the behavior of the world. All that is necessary is predictability within sufficient tolerances of precision and accuracy so that the reasoning agent can achieve its goals well enough for the species to survive. One model can be more "true" than another in that it predicts more than the other, and there can even be incomparable models that predict different things, each of which is equally "true" (faithful).

The second of the above perspectives seems to account best for the situation of a knowledge-based computer system (and probably our human situation as well)—the state of the world is far more detailed and its behavior more diverse than we could possibly replicate in a model with fewer states. In most mathematical models used by physicists and engineers, a deliberate approximation is used to eliminate inessential attributes of the problem and thereby focus on the essence. Hence the introduction of idealized concepts such as point masses and frictionless surfaces. From the perspective of our understanding of the world, we realize the nature of such approximations and can characterize the kinds of things that these idealized models leave out. When we look closely at our most complete perspective, however,

[1] "Nonmonotonic" conclusions are conclusions that are drawn in the face of explicit lack of knowledge and would not be made if certain information were known. Although the term is of relatively recent coinage, an early example is A. Collins' "lack of knowledge principle" [3], a principle for concluding that some property is not true of an object if it is not known to be true and enough is known about the object to be confident that the property in question would have become known if it were true. "Satisficing" is a term coined by H. Simon [9] for processes that are similar to optimization processes except that there is a criterion for what constitutes a sufficiently good solution to stop searching for something better—i.e., a process that seeks a "good enough" solution.

we find it likely that our best and most thorough understanding of the world is still, like the mathematical abstractions, only an approximate idealization, and there is no larger perspective available to us from which we can understand its limitations. Thus it appears that any reasoning we could do, and any understanding of the world we could have, must necessarily be an approximation.

An example of this limitation is weather simulation. The instabilities in the atmosphere that can be introduced by even a small temperature difference (below the levels of precision that we could expect to measure and use on the scale required to simulate the weather) can lead after a sufficient period of time to quite dramatic differences in the ensuing weather patterns. Hence it would be practically impossible to capture a fully accurate and precise initial state for a weather simulation, even if the model of the equations that govern the behavior of the weather were to be totally correct. Moreover, even if one could capture all the relevant information, it would be difficult to simulate the weather faster than it actually happens using any computational engine with fewer resources than the atmosphere itself. One can view the actual weather system as a highly parallel engine that determines its behavior more quickly and accurately than any existing computational artifact does. This line of reasoning leads to the possible conclusion that the most accurate, timely simulation of the weather may be the weather itself.

What is true of the weather is a part of what is true of the world. Hence, for purposes of an intelligent agent, models of limited fidelity appear to be a practical, if not logical, necessity if they are to be of any predictive use at all.

The Functions of a Knowledge Representation System

In the most general case, a knowledge representation system must support a number of different activities. Different techniques may be appropriate for representing different kinds of things and for supporting different kinds of activities, but there is a substantial overlap in the use of knowledge for different purposes. This forces us in some cases to either find a representational system that supports multiple uses of knowledge or else represent the same knowledge several times in different representational systems. All other things being equal, it would be desirable to have a representational system in which knowledge that has multiple uses will have a single representation that supports those uses. One reason for this is the cost of acquiring and maintaining independent representations. On the other hand, when issues of efficiency for different uses impose conflicting demands on a representation, and the benefits are sufficient to overcome the costs of acquiring, storing, and maintaining multiple representations, then one would prefer separate representations tailored to their use. In the next three sections, we will discuss three broad classes of use of knowledge that a knowledge representation system should support and some of the demands that they impose on a knowledge representation system.

The Knowledge Acquisition Problem

From the above perspective, it seems that the best any intelligent agent can hope for is to gradually evolve a more and more faithful and useful model of the world as its experience with the world and its experience with its own needs for prediction accumulate. This is true also for any computer knowledge base that is to perform some task or assist some human to do so. Thus a primary role of a knowledge representation system must be to support this evolutionary acquisition of more and more faithful models that can be effectively applied to achieving goals or tasks (at least up to the point where the additional complexity and cost of using a more faithful model outweighs the additional benefits to be gained).

Some of the issues that need to be addressed are as follows:

- how to structure a representational system that will be able to, in principle, make all of the important distinctions;
- how to remain noncommittal about details that cannot be resolved;
- how to capture generalizations so that facts that can be generalized do not have to be learned and stored individually;
- how to efficiently notice when new knowledge contradicts or modifies existing knowledge or hypotheses, and how to know/discover/decide what to do about it;
- how to represent values of time dependent attributes;
- how to acquire knowledge dynamically over the system's lifetime—especially to assimilate pieces of knowledge in the order in which they may accidentally be encountered rather than in a predetermined order of presentation.

The Perception Problem

In addition to the problems of acquiring a knowledge base, an intelligent agent must use the knowledge to advance its goals. This requires being able to perceive what is happening in the world in order to act appropriately. It is necessary for the agent to perceive that it is in a situation in which knowledge is applicable, and to find the knowledge that is relevant to the situation. It is also necessary to use knowledge to support the perception of what is the same and what has changed from a previously known state of the world. Among other things, knowledge will be used to perceive new individual entities that are present but formerly unknown and to identify new perceptions involving pre-existing concepts of known individuals.

Some of the issues that need to be addressed here are as follows:

- how to generate and search a space of possible hypotheses without combinatorial explosion;
- how to find the relationships between elements that have been identified and roles they could play in larger percepts;
- how to recognize when one perceptual hypothesis is a duplicate of another;
- how to find the best characterization of the situation;
- how to deal with errors in input or partially ill-formed perceptions.

Planning to Act

Like perception, planning to act can introduce new elements into the internal world that were not there before—namely, the planned actions and their expected results.

Planning is one of the kinds of action that the system can execute. It is one of a class of **internal actions** which differ from **overt actions** in that their effects happen in the internal, rather than the external, world. (We will ignore here the philosophical implications of the fact that the agent's internal world is a part of the real external world.) Internal actions require a knowledge representation system to be able to represent such things as plans, goals, hypotheses, and expectations.

Some of the issues that need to be addressed here are as follows:

- how to share large amounts of common knowledge between alternative hypotheses and different points in time;
- how to structure a plan to support monitoring for its successful execution;
- how to represent and trigger contingency plans and dynamically replan when something goes wrong;
- how to simulate and evaluate a plan;
- how to record the expectations and objectives that motivate a plan and recognize when a plan is no longer relevant;
- how to plan for multiple and possibly competing objectives.

The Role of a Conceptual Taxonomy for an Intelligent Agent

A fundamental problem for an intelligent computer agent that cuts across many of the above activities is analyzing a situation to determine what to do. For example, many expert systems are organized around a set of "production rules," a set of pattern-action rules characterizing the desired behavior of the system [4]. Such a system operates by determining at every step what rules are satisfied by the current state of the system, then acting upon that state by executing one of those rules. Conceptually, this operation entails testing each of the system's rules against the current state. However, as the number of rules increases, techniques are sought to avoid testing all of them.

One approach to this problem has been to assume that the pattern parts of all such rules are organized into a **structured taxonomy** covering all of the situations and objects about which the system knows anything. By a taxonomy, I mean a collection of concepts linked together by a relation of **generalization** so that the concepts more general than a given concept are accessible from it. By a structured taxonomy I mean that the concept descriptions have an internal structure so that, for example, the placement of concepts within the taxonomy can be computationally determined. An example of a structured taxonomy is the knowledge representation system KL-One [2]. A characteristic of such a taxonomy is that information can be stored at its most general level of applicability and indirectly accessed by more specific concepts said to "inherit" that information.

If such a taxonomic structure is available, the action parts of the system's rules can be attached to the concept nodes in the structure as pieces of "advice" that apply in the situations described by those concepts. The task of determining the rules applicable to a given situation then consists of classifying the situation within the taxonomy and inheriting the advice. Thus a principal role that a conceptual taxonomy can play is to serve as a conceptual "coat rack" upon which to hang various procedures or methods for the system to execute (see Fig. 4). A conceptual taxonomy can organize the pattern parts of a system's rules into an efficient structure that facilitates recognition as well as a number of other activities that a knowledge representation system must support.

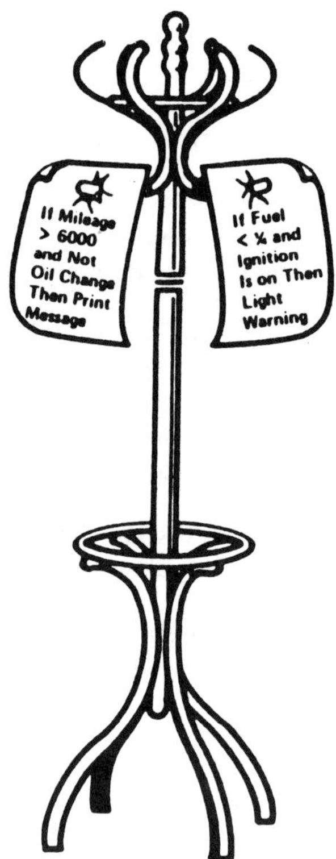

Fig. 4. A conceptual coat rack for organizing advice.

The Structure of Concepts

In building up internal descriptions of situations, one needs to use concepts of objects, substances, times, places, events, conditions, predicates, functions, individuals, etc. Each concept can be characterized as a configuration of attributes or parts, satisfying certain restrictions and standing in specified relationships to each other. A knowledge representation system that focuses on this type of characterization is the system KL-One.[2]

Space does not permit a complete exposition of KL-One in this paper. However, I want to use KL-One notations to illustrate the kinds of taxonomic organizations I am advocating. In this section, I will present a brief overview of the taxonomic structures in KL-One as a context for subsequent discussion. For a more complete exposition of KL-One, see Brachman and Schmolze [2].

[2]KL-One is the collaborative design of a number of researchers over an extensive period. Principal developers (besides myself) have been R. Brachman, R. Bobrow, J. Schmolze, and D. Israel. H. Levesque, B. Mark, T. Lipkis, and numerous other people have made contributions. Within this large group are different points of view regarding what KL-One is or is attempting to be. Over time those views have evolved substantially. What I say here represents primarily my own view, and may not be totally congruent with the views of my KL-One colleagues.

A **concept** node in KL-One has an associated set of **roles**—a generalization of the notions of attribute, part, constituent, feature, etc. In addition, it has a set of **structural conditions** expressing relationships among the roles. Concepts are linked to more general concepts by a relation called SUPERC. The more general concept in such a relationship is called the **superconcept** and is said to **subsume** the more specific **subconcept**. Some of a concept's roles and structural conditions are attached to it directly, while others are **inherited** indirectly from more general concepts.

The concepts and roles of KL-One are similar in structure to the general data-structure notions of record and field or to the "frame" (also called "schema" or "unit") and "slot" of AI terminology. However, there are several differences between a KL-One concept and these data structure notions. These differences include: the way that subsumption is defined and used, the presence of structural conditions attached to a concept, the explicit relationships between roles at different levels of generality, and the general intent of KL-One structures to model the semantics and conceptual structure of an abstract space of concepts (as opposed to being merely a data structure in a computer implementation).

This last point may require some elaboration. The goal of KL-One is not *per se* to produce a particular computer system, but rather to force the discovery and articulation of general principles of knowledge organization and structure. Expressive adequacy is an important driving force in KL-One research, emphasizing the semantics of the representation and its adequacy to make the kinds of subtle distinctions that people make when conceptualizing complex ideas. (The importance of the semantics of a semantic network is discussed elsewhere [10].) The KL-One effort has been more of an exercise in applied philosophical investigation of abstract conceptual structure than a design of computer data structures.

An Example of a Conceptual Taxonomy

The kind of taxonomic structure that I want to advocate is illustrated by the example in Fig. 5. Here, using KL-One notation, concepts are represented by ellipses and roles by circled squares. At the top of the figure is a high-level concept of Activity having roles for Time, Place, and Participants, which are inherited by all concepts below it. Below Activity, to the right, is the concept for a Purposive Activity, which differentiates (DIFFS) the general role for Participants into an Agent (the participant that has the purpose) and Other Participants. Purposive Activity introduces a new role called Goal to represent the purpose of the activity.

Below Purposive Activity is the fairly specific, but still generic, concept of Driving to Work. This concept modifies the Goal of Purposive Activity by adding Getting to Work as a value restriction (V/R), indicating that whatever fills the Goal must be an instance of Getting to Work. It also introduces a new role called Destination, with Place of Work as its value restriction. A structural condition (not shown) attached to the concept would specify how the Place of Work related to the Getting to Work goal—i.e., that it is the Destination of the Getting to Work goal. Driving to Work in Massachusetts is, in turn, a specialization of Driving to Work, with its Destination restricted (MODS) to a Place in Massachusetts. Driving to Work in Massachusetts is also a specialization of a Dangerous Activity with a Risk of Physical Harm.

This figure illustrates the kind of taxonomy one would expect to have in an intelligent computer agent, including both high-level abstractions and quite specific concepts. As

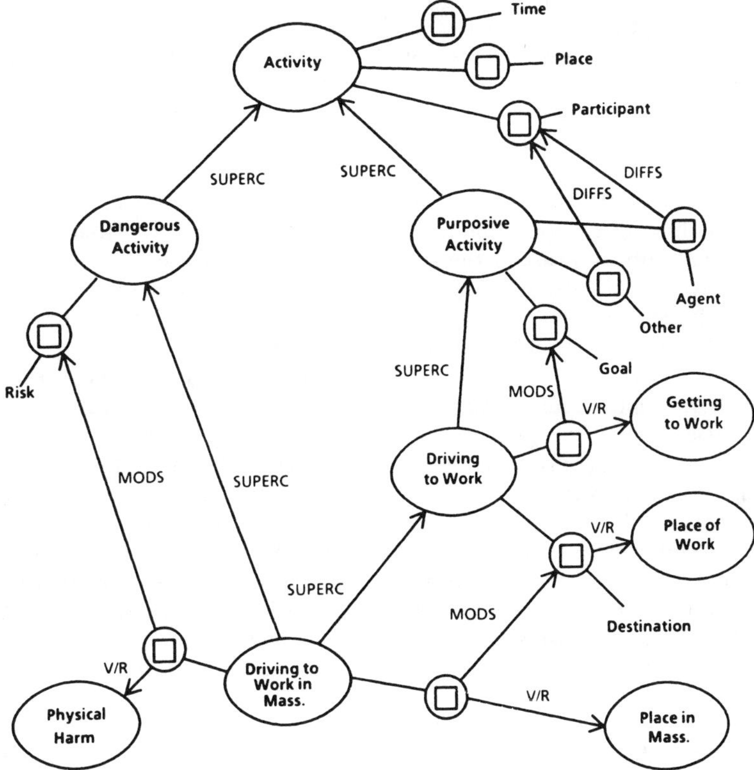

Fig. 5. An example of a conceptual taxonomy.

such a taxonomy is used and evolves, there is always room for inserting new levels of abstraction between existing ones. In fact, a well-defined classification procedure implemented in the KL-One system can automatically place a new description into a taxonomy, linking it by SUPERC connections to the concepts that most specifically subsume it and those that it in turn subsumes.

The Need for Taxonomic Organization

In most expert-system applications, a task description often satisfies several rules simultaneously, no one of which accounts for all of the task or supplants the relevance of the others. For example, adding an object to a display simultaneously changes the display and displays an object. Advice (the action parts of the rules) associated with both activities must be considered. Moreover, one description of a situation may subsume another more specific description, and their advice may supplement or contradict each other. Thus conventions are required to determine which advice takes precedence when conflicts arise.

For independent rules in a classical production-rule system, such conflicts may be discovered only when a conflicting situation occurs as input. In a taxonomic classification structure, however, the subsumption of the conditions of one rule by another can be discovered when the rule is assimilated into the taxonomy—at which time the person entering the rule can address the question of how the two rules should interact. The advice associated with the more specific rule then can explicitly include the information to override or supplement the more general rule.

Assimilating rules into a taxonomic knowledge structure not only facilitates the discovery of interactions at input time, but also promotes a compactness in the specification of the rules. By relying on the fact that concepts inherit information from more general concepts, one can usually create the concept for the pattern part of a new rule merely by adding a minor restriction to an existing concept.

In KL-One, when one wants to create a description of a situation that is more specific than a given one, it is only necessary to mention those attributes being modified or added; one does not have to copy all of the attributes of the general situation. Besides facilitating compact memory storage, this feature also facilitates updating and maintaining the consistency of the knowledge base by avoiding the creation of duplicate copies of information (which can then be independently modified and could accidentally be modified inconsistently).

The ability to assimilate new descriptions into an existing taxonomy at any level permits an evolutionary system design that can achieve the same standards of rigor as a top-down design without requiring concepts to be defined in a predetermined order. For most applications, even if one could get the initial design carefully laid out in a rigorous top-down mode, subsequent changes (e.g., required changes in accounting policies induced by new tax laws) will require the ability to modify the system in more flexible ways. A system's taxonomy of recognizable situations and situational elements should be viewed as an evolving knowledge structure that, like a person's world view, continues to be refined and developed throughout the lifetime of the system.

Taxonomic Programming

The use of a conceptual taxonomy to organize rules of behavior shows promise of becoming a productive discipline for organizing knowledge in a diverse range of activities—including general applications programming. For decades, standard compilers have routinely dealt with abstract arithmetic operators that have different implementation code depending on whether their arguments are fixed-point integers, floating-point numbers, complex numbers, etc. Each of these different *types* of operand have different representational conventions and require different treatment. There has been a steady evolution in the use of such "data types" in programming, including making programmer-defined types available to the applications programmer and the formalization of abstract data types that decouple the behavioral aspects of a data element from the details of its implementation. Both compile-time and run-time type checking are routinely used to detect and localize errors in programs. However, one can do more with the notion of a data type, beyond facilitating implementational flexibility and error checking, if we think of these types as nodes in a taxonomy of the kinds of objects that the system knows about.

Recently, so called "object-oriented" programming languages [7] have been developed to organize program behavior around a hierarchy of classes of objects. These classes play a role similar to the types in a typed programming language, but are used for quite another purpose. In object-oriented languages, behavior is invoked by sending a message to an object that results in a search up the class hierarchy for the most specific defined method that can handle the message. In essence, these languages facilitate the definition of abstract operations (analogous to the abstract arithmetic operators of addition and multiplication) that have different behavior for different classes of operand. Such languages have been highly effective in organizing complex programs for applications such as window-oriented graphics display, where operations such as PRINT, MOVE, SHAPE, REDISPLAY, etc., are to be applied to a variety of different kinds of objects—ranging over bitmaps (a rectangular array of bits), windows (rectangular regions of the display screen with further structure such as a header bar, scroll bars, redisplay functions, etc.), text windows (windows containing text), graphics windows (windows containing graphics such as line drawings), menus (windows containing active elements that can be selected and activated), etc.

The use of a taxonomy of different kinds of objects to organize the elements of such programs enables (and encourages) a programmer to share the specification of many aspects of the desired behavior over multiple classes of object. For example, most of the programs for printing, moving, and shaping windows can be reused for a variety of different subkinds of windows, while any aspect of a more specific kind of window that is not uniformly treated can be dealt with by a more specific method attached at a lower level in the taxonomy. This approach is to be contrasted with a traditional tendency to define a parallel and independent set of operations for each kind of operand to which they will be applied.

By identifying the *types* of a typed programming language and the object *classes* of an object-oriented programming language with the *concepts* of a taxonomic knowledge representation system, one can gain substantial improvements in the flexibility for expressing complex behaviors. The result could well produce a new style of programming, which Goodwin [6] termed "taxonomic programming." For example, if the types of the language are assumed to form a taxonomic lattice, complete with operations for conjunctions and disjunctions of types, then one of the classical difficulties with strongly typed programming languages—the difficulties with lists of mixed types—can be handed by specifying the type of the list as the most specific common subsumer of all of the types of its elements. Moreover, the concept of a list of elements of some type *x*, can itself be thought of as a structured type and might be considered a more specific type than say a *set* of elements of type *x*. This can be generalized to arbitrarily structured objects that are composed out of other objects, which are, in turn, composed out of still other objects, etc. An example of such a system is the AIPS system [13], a general-purpose graphics presentation system whose graphical presentation knowledge, coordinate system representations, display objects, etc., are all organized around a KL-One taxonomy of graphical display concepts.

Taxonomic programming has many advantages for creating systems that are flexible and extendable. When a programmer thinks of his task as specifying rules of behavior in terms of a taxonomy of kinds of operands and operators, he is encouraged to think at a level of generality beyond the immediate task at hand. This increases the likelihood that the procedure he defines will be automatically applicable to new kinds of object that may be introduced into the system at a later time. For example, why define a method that reshapes only menus when, with only a little more care, one can define a method that reshapes any window? One can identify, at the point of writing a method for an operator, what the most general kind of object is to which that operator could apply and then introduce that concept into the taxonomy at that point—even when only very specific uses of the operator in question are currently envisioned. The resulting system is much more likely to be easily extendable to a slightly different application. It also increases the likelihood that parts of one system can be reused in different systems or that two different systems might be combined into a system that can do more than a mere union of their separate abilities. The AIPS system, for example, contained extremely general concepts for display objects, coordinate systems, and coordinate system transformations. This enabled it to serve as a uniform graphical presentation interface for a wide variety of applications.

Combining programming methodologies with knowledge representation methodologies can have further advantages in the areas of programming documentation and automatic program analysis and synthesis. For example, the underlying knowledge structures in AIPS closely match the conceptual structures with which human programmers and users think of graphical entities and could easily serve as the basis for an on-line tutorial system or an automatic text generation system to produce documentation manuals.

Recognizing/Analyzing/Parsing Situations

One of the activities that a taxonomic structure can support is the process of situation recognition—recognizing that some of the elements currently perceived by an agent or in focus in some reasoning process constitute instances of a known situation. This process is the essence of many "diagnostic" expert systems applications such as medical diagnosis, spectrographic analysis, signal interpretation, and electronic trouble shooting. It also plays a central role in many aspects of more complex "synthesis" applications such as circuit design, machine configuration, financial planning, etc. It is the common abstract problem in tasks such as natural language understanding, continuous speech understanding, and the interpretation of visual scenes. Roughly, this process consists of discovering that some input data or some aspects of an intermediate state of reasoning can be interpreted as filling roles in some concept known to the system and therefore constitute an instance of that concept.

In general, merely characterizing an input situation as an instance of a single existing concept is not usually sufficient. Typically, a description of a situation will be a composite object, parts of which will be instances of other concepts assembled together in formally permitted ways, and the overall structure of these relationships must be recognized and characterized. Thus recognizing a situation in general is similar in many respects to parsing a sentence, although it can be considerably more complex. The similarity lies in the fact that the process creates a structure that integrates the perception of otherwise unconnected stimuli into a coherent, structured whole. On the other hand, whereas the grammatical relationships between parts of a sentence come from a fixed, rather small set of possibilities, the relationships among the "constituents" of a general situation may be arbitrary. These relationships may include events preceding one another in time; people, places, and physical objects in various spatial relationships with each other; people in physical or legal possession of objects; people in relationships of authority to other people; and people having certain goals or objectives.

One technique for efficiently recognizing situations is to use a "**factored**" knowledge structure [11] in which the common parts of different rules are merged so that their testing is done only once. Examples of factored knowledge structures include classical decision trees and typical ATN grammars. With such structures, one can effectively test a large set of rules without considering the rules individually. The taxonomic structures embodied in KL-One can provide a factored representation for parsing situations. Determining the most specific concepts that subsume the input situation can be done by using the chains of links from the elements of the situation to the roles of higher level concepts in which they can participate, using generalizations and extensions of the algorithms used to parse sentences.

The suitability of a representation for supporting algorithms of this sort is an important aspect of a knowledge representation system. A version of this technique using KL-One has been successfully applied in the PSI-Klone system, where an ATN parser is coupled with a KL-One taxonomy that organizes the semantic interpretation rules for a natural-language understanding system [1].

Two Aspects of Knowledge Representation

Two aspects of the problem of knowledge representation need to be considered. The first, **expressive adequacy**, has to do with the expressive power of the representation—that is, what it can say. Two components of expressive adequacy are the distinctions a representation can *make* (in order to capture subtleties) and the distinctions it can *leave unspecified* (in order to express partial knowledge). A second aspect, **notational efficacy**, concerns the actual shape and structure of the representation as well as the impact this structure has on the operations of a system. Notational efficacy, in turn, breaks down into such components as computational efficiency, conceptual clarity, conciseness of representation, and ease of modification.

It is important to distinguish expressive adequacy from notational efficacy, since the failure to clarify which issues are being addressed has exacerbated numerous arguments in this field. For example, an argument that first-order predicate calculus should be used because it has a well-understood semantics partially addresses expressive adequacy, but does not explicitly mention the issue of notational efficacy. The argument could be understood as advocating the use of the notations traditionally used by logicians, and in some cases this may even be what is meant. However, it is possible to invent many different notational systems, each having a first-order logic semantics, but having different attributes of notational efficacy.

To provide reasonable foundations for the practical use of knowledge in reasoning, perception, and learning, knowledge representation research should seek notational conventions that simultaneously address expressive adequacy and notational efficacy. A representational system is required that will be adequate for a comprehensive range of different kinds of inference and will provide computational advantages to inferences that must be performed often and rapidly. One class of inference that must be performed rapidly and efficiently is the characterization of one's current situation with respect to a taxonomically organized knowledge network.

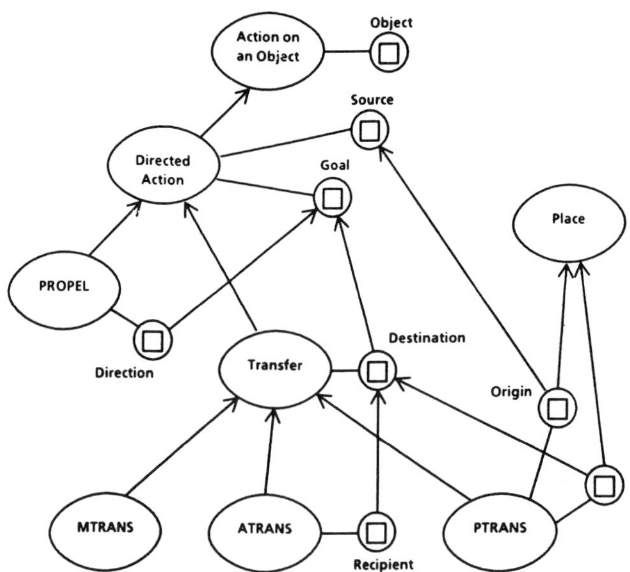

Fig. 6. Transfers and directed actions.

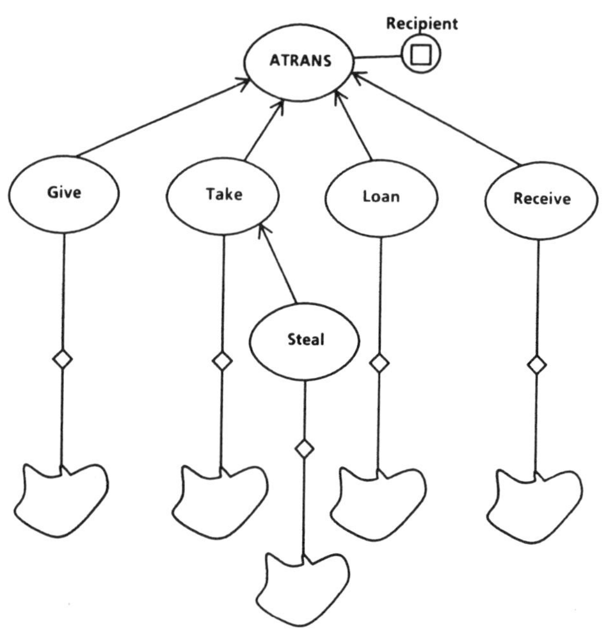

Fig. 7. Some kinds of ATRANS.

Expressive Adequacy

However efficient a representation may be for some purposes, it is all for naught if it cannot express necessary distinctions. In seeking a representation, one must avoid choosing a set of primitives that either washes out such distinctions as those among "walk," "run," "amble," "drive," and "fly," or overlooks the commonality between these specific concepts and the general concept "move." A structured inheritance network such as KL-One permits both benefits. As they become important, new distinctions can be introduced by refining or modifying existing concepts. Moreover, it is always possible to introduce more general concepts that abstract details from more specific ones. The explicit taxonomic structure allows one to move freely among different levels of generality, rather than being required to fix a single level of detail at which to characterize knowledge. Figs. 6–9 illustrate this kind of ability.

Fig. 6 illustrates how a number of Roger Schank's abstract transfer concepts (MTRANS, ATRANS, PTRANS, and PROPEL) [8] relate to each other and to some more general concepts. Fig. 7 illustrates how ATRANS relates to several more specific concepts (the balloons below the small diamonds signify structural conditions that characterize how the details of the more specific actions relate to the more general one). Fig. 8 illustrates how the specializations of one class of entity (Legal Entity in this case) can induce a corresponding space of specializations of an action that applies to such entities, presenting more specialized concepts for a variety of interesting classes of entity that can serve as the recipient of an ATRANS. Fig. 9 illustrates both kinds of subclassification of the abstract verb Like. This kind of taxonomic organization allows one to capture generalizations at the level of Transfer, Directed Action, or Action on an Object while still being able to store facts at the level of specific actions such as Loan or Give. The more specific concepts will automatically inherit information stored at the more general levels.

Common problems in axiomatizing a domain in the traditional predicate calculus are choosing the set of predi-

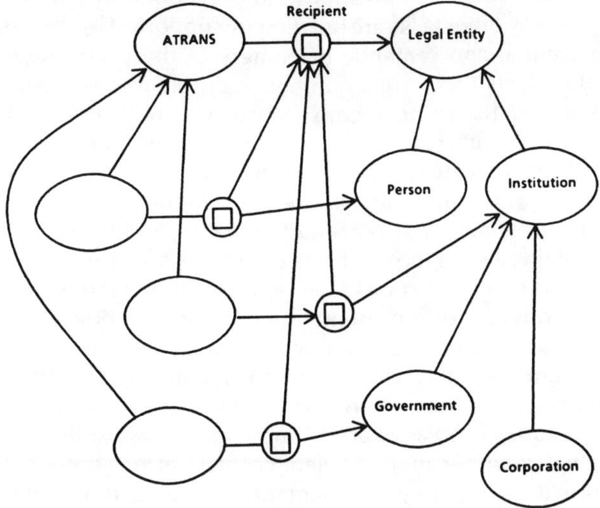

Fig. 8. More kinds of ATRANS.

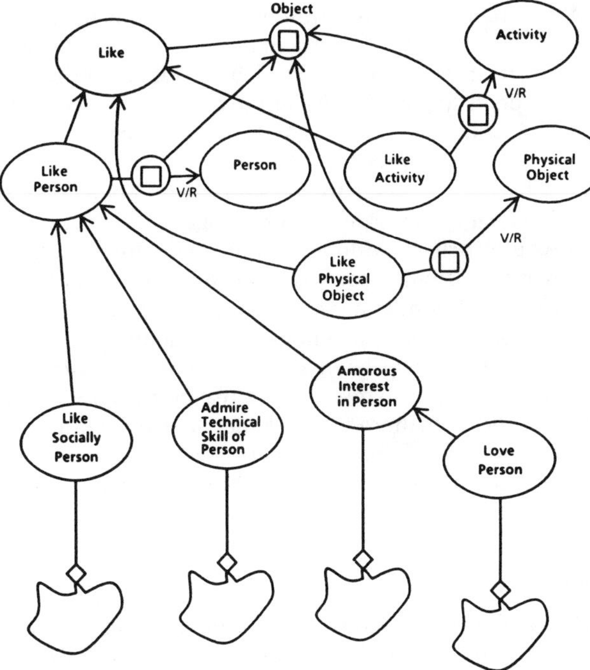

Fig. 9. Kinds of LIKE.

cates and deciding what arguments they will take. Inevitably, these decisions leave out distinctions that might be important for another purpose, such as time variables, situation variables, intermediate steps, and provisions for manner adverbial modification. Incorporating revisions of such decisions in a complex system could amount to redoing the axiomatization. One of the goals of KL-One, toward which some progress has been made, is to provide a terminological component for such axiomatizations (i.e., KL-One concepts provide the predicate inventory for the axiomatization) so that, for example, the time role of activities (see Fig. 5) can be virtually ignored in expressing an axiom in which time does not figure prominently, and yet remain present implicitly (or be added later) when a situation is encountered in which it is important.

Another need is an ability to refer to what we know, believe, suspect, or conjecture and to represent our varying states of knowledge over time. Often we need to record the sources of pieces of knowledge and evidence for that knowledge. The ability to refer to concepts as entities as well as to use them to refer to their instances, and the ability to represent propositions as entities (about which assertions can be made) as well as to believe them, are both critical to this ability.

Notational Efficacy

Expressive adequacy is a minimal requirement for a knowledge representation system. Eventually, one wants a framework in which the assimilation of arbitrary new information is not only possible but also is in some sense natural. For example, one would like small changes in knowledge to require small changes in the knowledge base, so that learning processes or even incremental debugging can be expected to eventually converge. Moreover, there must be operators for making fine adjustments as one gets close to the correct state. Thus an aspect of a knowledge representation system that is important for knowledge acquisition and learning is an analog of the mathematical property of compactness—that is, there should be points in the space of possible knowledge states that are arbitrarily close to the state of knowledge one wants to reach (in some suitable sense of "close"). It is necessary for the representation to be able to express not only the desired target knowledge, but also all of the intermediate states of knowledge that must be passed through along the way. If the knowledge representation system cannot represent some of the intermediate states or if the topology of the space of representations is such that you cannot get from one learning state to another without major detours or major reorganization, then learning will be difficult at best. This kind of support for knowledge acquisition is one element of notational efficacy.

Notational efficacy can be subdivided into issues of **computational efficiency** and **conceptual efficiency**. The aspects just discussed that support knowledge acquisition are primarily elements of conceptual efficiency. Elements of computational efficiency have to do with the utility of notations for supporting a variety of algorithms. Examples include some of the factoring transformations (which can reduce combinatoric search) and the use of links to access concepts related to other concepts (which can reduce searching and matching). Some aspects of a knowledge representation may support both computational and conceptual efficiency. For example, assimilating the pattern parts of rules into a taxonomic knowledge structure (to facilitate the discovery of interactions at input time) and the inheritance of common parts of those patterns (to provide a compactness in the rule specifications) are features that support both efficiency of operation and the conceptual efficiency of the knowledge acquisition and organization processes. The fact that sharing common parts of different rules can conserve memory storage and facilitate updating as well as minimizing duplicate testing of conditions is another such element.

In summary, a knowledge representation is called upon not only to express an adequate model of a domain of knowledge, but also to support a variety of computational and conceptual activities involved in acquiring and using that model.

The Relationship to Formal Logic

Many hold the opinion and have argued strongly that **formal logic**, by which they usually mean the **first-order predicate calculus** with its customary syntactic notations, provides all that one needs for knowledge representation. By this account, all that is necessary to represent knowledge is to axiomatize the appropriate information. Someone else, or perhaps the same person, will then produce an efficient general-purpose theorem prover or reasoning engine for using that axiomatization to perform tasks. Others, who perceive that something more than this is required for reasonable efficiency, nevertheless still view notations such as semantic networks as alternative encodings of things that can be expressed in predicate calculus, and they characterize the meaning of these notations in terms of equivalent predicate calculus terms and predicates.

While it is true that the first-order predicate calculus is able to axiomatize the behavior of any computational system, this does not mean that the properties of those systems follow from the predicate calculus. On the contrary, one can axiomatize any rule-governed behavior that one can rigorously specify, whether it's nature is deductively valid or not, logic-like or not. The difficult issues mostly stem from determining what the behavior should be and how to efficiently bring it about. This work remains whether the medium is a predicate calculus notation or some other representational formalism.

Thus the predicate calculus alone is not the solution. Moreover, the notations and style of activity associated with predicate calculus axiomatization can in some cases be misleading or distracting (or unnecessarily constraining) rather than helpful. For example, the semantic approach to belief modeling based on sets of possible worlds *a la* Hintikka and Kripke leads to models of belief with the undesirable consequences that the modeled agent must be assumed to believe all the logical consequences of his other beliefs (including all the truths of arithmetic—whether known to any mathematician or not). Likewise, although one can use the basic predicate calculus machinery to prove the existence of an individual satisfying certain properties, the predicate calculus machinery gives us no operator to name that individual and refer to it by name in subsequent reasoning. (Skolem functions and individual constants provide names for individuals conceived by the system designer in setting up the formal system, but there is no notion of process in the predicate calculus within which a name could be coined in the course of reasoning and then used later.) The classical Tarskian model of first-order logic provides a good way to characterize the necessary truths and the consequences of hypotheses in any situation (determined by the individuals in it and the predicates that are true of them), but it provides no machinery for characterizing the relationship between two situations. For example, it gives no leverage for characterizing what happens to a situation as a result of introducing a new individual.

The issues raised previously about sharing information across alternative hypotheses or points in time, the creation of new structures in response to perception or planning, and many other tasks are not directly supported by the basic inferential and semantic machinery of the predicate calculus. Thus there are a number of problems that need to be solved that are not inconsistent with a predicate calculus approach, but are nevertheless not solved merely by adopting the predicate calculus as a representation. These issues remain to be solved and addressed whether predicate calculus or some other representational system is used.

Finally, even if the predicate calculus is taken as the basis for part or all of a knowledge representation system, there remain issues of representation dealing with how best to structure the axioms and how to organize them to support efficient reasoning. Even within the family of predicate calculus approaches, there are different representational techniques such as Skolem-functions versus explicit existential quantification, Skolem-normal form, conjunctive normal form, clausal representations, etc.—all different notations sharing a basic semantic framework, but adopting different representations in order to support particular reasoning disciplines.

What I have said here is not meant to argue against the value of interpreting a representational system in terms of a predicate calculus perspective in order to understand those aspects that are equivalent to predicate calculus notions. I do, however, argue that the ordinary notations and semantics of the predicate calculus are insufficient by themselves—they need to be supplemented with additional machinery in order to solve many of the problems. This would include adding machinery for the modalities (necessity, possibility, contingent truth), machinery for the creation of referential names and attaching them to characterizations of their meanings, and machinery for non-monotonic reasoning, as well as coming to grips with some non-first-order problems such as reasoning about properties. In addition, there are other mechanisms required for dealing with perception and creative actions such as planning that are not customary in traditional formal logic. The next section will consider one such issue.

Concepts are More than Predicates

The search for a clean semantics for semantic networks has led some researchers to adopt an assumption that the nodes or arcs in a semantic network are simply alternative notations for the equivalent of predicate calculus predicates. In this section, I will present some counter arguments to that position. To illustrate the points, I will use the classical blocks-world arch example, illustrated in Fig. 10. In this figure, an arch is an assembly of three blocks, two of which are called uprights and the third is called a lintel. The balloon at the bottom signifies a structural condition about how the blocks have to be related to each other in order for them to constitute an arch—i.e., the lintel has to be supported by the uprights and there must be space between the uprights. The concept of such an arch specifies both that the arch is composed of three things which are individually blocks as well as the relationships that have to hold among them to be an arch.

We might consider identifying this concept of an arch with a predicate in the predicate calculus. However, there are a number of candidates, each of which might have equal

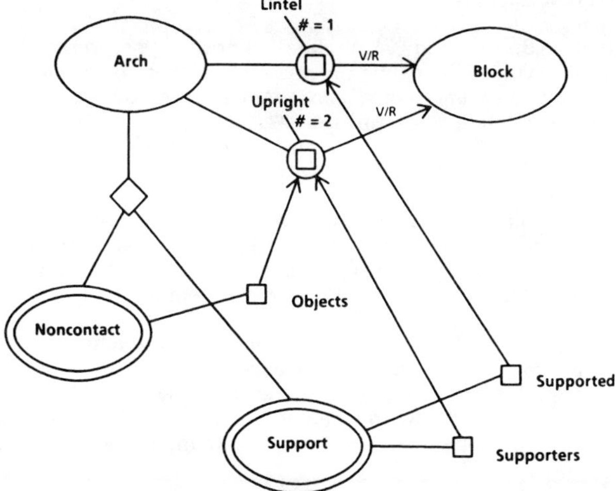

Fig. 10. A blocks-world arch.

claim to being the predicate represented by the arch concept. One of these is a predicate on three blocks that characterizes the relationship that the blocks must have in order to be an arch (i.e., the predicate of the structural condition). Let us call it ARCH(*X*, *Y*, *Z*). Another is a predicate of one argument that is true of an individual arch. Let us call it ARCH′(*X*). Still another is a predicate that is true of a situation *S* if there is an arch present in the situation. Suppose we call it ARCH″(*S*). Either of the first two, if not the third, appear to have a good claim to be the predicate that we would mean by the concept Arch (if that was all we meant by the concept). I would suggest, however, that the concept of an arch is something slightly different from any of these predicates, and yet has all three of these predicates (and more) associated with it. The concept is an abstraction that ties together a number of related predicates that characterize what an arch is. The concept also serves some other functions as well, such as serving as a place to keep records about one's experience with arches.

One of the elements of the arch concept that is missing from an identification of the concept with one of the simple arch predicates is the notion of what it would take to build one. The predicate ARCH(*X*, *Y*, *Z*), for example, is true of three blocks if they stand in a certain relationship. When viewed as an abstract set of ordered triples (as in a standard predicate calculus semantics) it does not provide a very useful object to serve as a plan for building an arch. If there do not happen to exist three blocks in an arch configuration, then the predicate ARCH(*X*, *Y*, *Z*) has the empty extension and there is not much structure there to use as a plan. Yet one of the things that we know about an arch is what its parts are and how they are arranged. The arch concept is at least a constellation of such predications (as uninstantiated, second-order, metalevel objects capable of being instantiated or as conceptual objects capable of being manipulated by reasoning and planning activities as entities in their own right—e.g., as subgoals to be achieved).

Another inadequacy of the concept-is-predicate view is the lack of anything that characterizes when two instances of a concept are to be considered the same or different. For example, two views of the same arch, one identifying a particular support with the *X* argument and the other identifying it with the *Y* argument should not count as two different arches. This is another piece of information that we know about arches and is part of the arch concept. I have suggested elsewhere [12] that a kind of object can be characterized by its recognition conditions and a "sameness" predicate that defines what counts as two instances versus one.

One can conceive of two different kinds of object that share the same recognition conditions but have different sameness predicates. For example, one can distinguish an "ordinary triangle" (which merely has three sides and three corners) from an "oriented triangle" (which in addition has a top and a bottom). Depending on one's perspective, a given instance of the former will be interpretable in different ways as instantiating the latter. When thinking in these terms, it becomes apparent that "same" is a relation not of two arguments but of three—we must consider whether two things *X* and *Y* are the same *Z*, where *Z* is a concept and *X* and *Y* are instances of it. We may have two distinct oriented triangles that are nevertheless the same triangle. Thus one of the things that may be associated with a concept is a sameness predicate, and this predicate stands in a qualitatively different relation to the concept from the predicates making up the conditions that any instance must satisfy.

Finally, there is nothing in either ARCH or ARCH′ (as abstract sets of ordered tuples) that characterizes the relationship between the arch as an individual and the individual blocks that make it up. There is nothing there to provide guidance to an inference engine that should conceptualize an individual arch when it perceives a constellation of three blocks in the appropriate relationship or to a planner than needs to know what subgoals to attempt in order to construct an arch.

The above discussion should convince the reader that there is something more to a concept than simply a predicate. A concept has an associated set of predicates as well as being a mental object that can support learning and planning. Note that I am not talking here about specific internal implementational structures. Rather, my notion of a concept should be interpreted as the abstraction that such implementational structures would stand for. The important issue is that a representational object representing a concept stands for more than a single predicate, and in the abstract world of concepts, there is structure that can be exploited for learning and planning.

Conclusions

This paper has discussed a number of issues of knowledge representation that go beyond merely being able to express a correct semantics. The discussion began with a consideration of the relationship between knowledge and the world and the use of knowledge by reasoning agents (both biological and mechanical). Specifically, a knowledge representation system is called upon to support activities of perception, learning, and planning to act. A number of issues are presented that a knowledge representation system must address to support these activities, and an argument is put forth that the mechanisms of traditional formal logic, such as the predicate calculus, while important to our understanding, are not by themselves sufficient to solve all of the problems. In particular, notational aspects of a knowledge representation system are important both for computational and conceptual reasons.

Two aspects of a knowledge representation system are important—expressive adequacy (the ability to make important distinctions and the ability to avoid undesired distinctions) and notational efficacy (the aspects of the notation that support both computational efficiency and conceptual activities such as knowledge acquisition and learning).

I have argued that taxonomic classification structures can advance both expressive adequacy and notational efficacy for intelligent systems. Such techniques, I believe, eventually will be applicable throughout computer science. Emphasizing the expressive adequacy and conceptual efficiency of a representation—rather than solely the computational efficiency of data structures—prepares the way for a general methodology of representation that will transcend different applications and different implementation techniques. Ultimately, this trend should lead to a way of specifying computational behavior in terms of high-level conceptual operators that match the conceptual structures of the human programmer, factoring out the issues of implementational efficiency for separate consideration or even automatic compilation.

The increased emphasis on abstract data types and "object-oriented" programming mark the beginning of this trend. The next logical step is to generalize the notion of abstract data types to the level of abstraction, inheritance, and expressive adequacy present in a sophisticated knowledge representation system. This step could produce a new style of programming—more oriented toward specifying the desired behavior in conceptual terms and characterizing the differences in behavior for different classes of operand. Such "taxonomic programming" can have enormous advantages in flexibility, extensibility, and maintainability, as well as for documentation, user education, error reduction, and software productivity. Moreover, such representations can make it possible to combine independently developed systems to produce integrated systems more powerful than the mere union of their parts.

References

[1] R. J. Bobrow and B. L. Webber, "Knowledge representation for syntactic/semantic processing," in *Proc. 1st Annu. Nat. Conf. on Artificial Intelligence*, American Association for Artificial Intelligence, pp. 316-323, 1980.

[2] R. J. Brachman and J. Schmolze, "An overview of the KL-One knowledge representation system," *Cogn. Sci.*, vol. 9, pp. 171–216, 1985.

[3] A. Collins, E. H. Warnock, N. Aiello, and M. L. MIller, "Reasoning from incomplete knowledge," in *Representation and Understanding: Studies in Cognitive Science*, D. G. Bobrow and A. Collins, Eds. New York, NY: Economic Press, 1975, pp. 383–415.

[4] R. Davis, B. Buchanan, and E. Shortliffe, "Production rules as a representation for a knowledge-based consultation program," *Artificial Intell.*, vol. 8, no. 1, pp. 15–45, 1977; also in *Readings in Knowledge Representation*, R. J. Brachman and H. J. Levesque, Eds. Los Altos, CA: Kaufman, 1985, pp. 371–387.

[5] D. C. Dennett, "Why the law of effect will not go away," in *Brainstorms*. Cambridge, MA: MIT Press, Bradford Books, 1978, pp. 71–89.

[6] J. W. Goodwin, "Taxonomic programming with KL-One," Tech. Rep. LiTH-MAT-R-79-5, Informatics Lab., Linköping Univ., Sweden, 1979.

[7] D. Robson, "Object-oriented software systems," *Byte*, vol. 6, no. 8, pp. 74–86, Aug. 1981.

[8] R. C. Shank and C. J. Rieger, III, "Inference and the computer understanding of natural language," *Artificial Intell.*, vol. 5, no. 4, pp. 373–412, Winter 1974; also in *Readings in Knowledge Representation*, R. J. Brachman and H. J. Levesque, Eds. Los Altos, CA: Kaufman, 1985, pp. 119–139.

[9] H. Simon, "Rational choice and the structure of the environment," *Psychol. Rev.*, vol. 63, p. 129, 1956.

[10] W. A. Woods, "What's in a link?: Foundations for semantic networks," in *Representation and Understanding: Studies in Cognitive Science*, D. G. Bobrow and A. Collins, Eds. New York, NY: Academic Press, 1975, pp. 35–82; also in *Readings in Knowledge Representation*, R. J. Brachman and H. J. Levesque, Eds. Los Altos, CA: Kaufman, 1985, pp. 217–241.

[11] —, "Cascaded ATN grammars," *Amer. J. Comput. Lingu.*, vol. 6, no. 1, pp. 1–15, Jan.-Mar. 1980.

[12] —, "Problems in procedural semantics," in *Meaning and Cognitive Structure*," Z. Pylyshyn and W. Demopoulos, Eds. Hillside, NJ: Ablex, forthcoming 1986.

[13] F. Zdybel, M. D. Yonke, and N. R. Greenfeld, "Application of symbolic processing to command and control," BBN Rep. 3849, Bolt, Beranek and Newman, Inc., Cambridge, MA, pp. 1-123, 1980.

Constructs and Phenomena Common to the Semantically-Rich Domains

Beth Adelson
Department of Psychology & Social Relations, Harvard University, Cambridge, MA 02138

This article looks at some phenomena that recur during expert problem solving across domains including computer science, physics, chess, and mathematics. Several of the mechanisms which have been hypothesized to underlie performance in one domain are then used to explain the results found in the others. Competing explanations, mechanisms which are not fully specified, and untested empirical issues are also discussed.

In this article I attempt to do four things. The first is to identify some phenomena that are common to several semantically-rich, problem-solving domains. The second is to cite some constructs which although proposed to account for a specific phenomenon in one domain, seem also to account both for other phenomena in that domain, and for similar phenomena in other domains. The third is to try to combine existing constructs so as to provide more complete explanations for existing data. And the fourth is to discuss issues which arise from competing classes of explanations and from incompletely specified mechanisms.

The goals motivating this attempt are two. The first is to support the position that studying semantically-rich domains can give us insights into general cognitive principles. In support of this position, the studies cited here deal with skills both developed in more widely disparate domains and studies using more highly varied paradigms and stimuli than is usually done in experimental psychology; nonetheless, similarities appear. A word on the related issue of the transfer of skill across domains is appropriate here. Previously, researchers have asked whether principles in problem solving are general by asking whether or not problem-solving skills transfer from one task to another. Despite the fact that the results of these transfer studies have frequently been negative,[1] arguments will be presented supporting the position that a lack of transfer does not necessarily mean a lack of generality. It may be that processes which were formed in conjunction with particular structures cannot be induced to function in isolation from those structures; however, that does not mean that similar processes do not develop around the structures in several distinct domains.

The second goal is to apply already existing explanatory mechanisms to new data, thereby giving the reader a sense of the capabilities and the limitations of the theoretical constructs that currently exist. This will allow us to see whether our currently existing explanations are headed towards power and parsimony or towards collections of special cases.

WHAT'S IN A CHUNK

Although the phenomenon of chunking has received a fair amount of attention, several interesting aspects of the concept have gone unattended. As background, let me first present the aspect of the phenomenon on which attention has

Reprinted with permission from *International Journal of Intelligent Systems,* Volume 1, 1986, pages 1-14.

been focussed. Generally, the results of chunking experiments in several domains indicate that experts use functional relationships which exist among the elements of a problem to structure the unitary elements of the problem into chunks. One example is found in the memory experiments of Chase and Simon, who have done extensive work on the cognitive skills of chess masters.[2–4] Chess Masters were shown mid-game boards for 5 s, after which the Masters were asked to select pieces from a full set and place them on an empty board attempting to reproduce what they had just seen. They found that those pieces that were placed within two seconds of the previous piece shared either an attack or defense relationship to the previous piece. They interpreted this to mean that functional relations among the items had been used to structure them into chunks.

Again using inter-response latencies in a recall task, Reitman[5] found that Master Go players also recall game boards as functional clusters. In her experiment, the chunks used by the masters again consisted of pieces that shared attack or defense relationships. A similar finding for the domain of electronics was reported by Egan and Schwartz,[6] who examined the recall clusters of electronics engineers for circuit diagrams: they found, once again, that an expert recalls clusters that are functionally based. This phenomenon, of sensing the structure of a problem before comprehending the details has long been remarked on in mathematics,[7] and Michener[8] has studied the issue empirically.

When the finding is formulated in this way, it focuses our attention on the internal structure of the chunks. But we can also ask what are the emergent properties of the chunks and, given such emergent properties, what is their utility? Three emergent properties will be discussed here: (1) the modular nature of chunks, (2) the abstract nature of chunks, and (3) the integration of chunks into larger knowledge structures.

Modularity

The results of several studies suggest that experts handle chunks as independent elements, to be combined in forming representations of problem statements. The work of Jeffries and Atwood illustrates this point.[9,10] They collected protocols from computer scientists as they were designing an indexing program which would create a list of all of the occurrences of a prescribed set of key terms in a text. Consider one representative protocol: The subject stated in her initial representation that the solution to the problem would consist of three main procedures. They were: read in the set of key terms, compare the key terms to the text, and store the resulting index. The way that the problem was formulated by this subject does suggest that the experts represented the problem statements as sets of modules. (Protocols were also collected from a slightly less expert group. These protocols were less well organized; for example, the level of detail of the elements of the solution at any one time was more uneven than in the expert group.) Another example of modular representation is provided by Atwood, Turner, Ramsey, and Hooper.[11] They showed specifications for to-be-designed programs to expert programmers who were then asked to summarize these specifications. Using a method of Kintsch,[12] the summaries were analyzed to reveal the structure of their contents. The results of the analyses suggested that the summaries were organized to reflect the main processes of the to-be-designed program. This type of organization seems to reflect a modular representation. Also in the domain of programming, Adelson[13] found that expert performance could be explained by the existence of modular representations. In mathematics, Lewis[14] documented the observation that mathematicians sometimes transform equations into more modular forms by replacing repeated subexpressions with temporary variables.

The utility of this type of representation is two-fold. First, since modules are distinct units the problem solver may consider them in isolation. This reduces the load on his or her working memory, since it is then unnecessary to hold all of the elements of the problem statement in mind at once. The effect here is that problem solvers can then be free to find the optimal solution for the particular element that they currently are considering. It seems that this potentially helpful strategy is in fact used. For example, in the protocol described above, from the Jeffries et al. study,[9] one of the procedures that the expert stated early on in her protocol was a means of comparing the key terms to the text. She later went back and compared alternative methods for doing this without making mention of the other aspects of the problem which she had initially identified as separate processes.

A second way in which modular representations are useful is in providing simple and coherent elements that allow a problem to be worked more easily. A simple but clear example comes from Lewis.[14] He found that three out of his five expert subjects (mathematicians at IBM Watson Research Center) rewrote

$$7(4x - 1) = 3(4x - 1) + 4$$

in a form such as

$$7(z) = 3(z) + 4.$$

We are lucky that his subjects explicity wrote out the substitution even though the problem is simple enough to have allowed them to perform the operations mentally since it allows us to see an interesting stage of the solution. As to why the calculation was written out, Lewis also found that his experts did not always rewrite an equation in the form which would allow the shortest solution method. He suggests that this may be the case because the expert can obtain a solution to an easy problem so quickly that even the method which is not the very fastest is still fast enough. For example, only one subject took advantage of the repeated subexpression "$x - 2$" that can be factored out in solving

$$6(x - 2) - 3(4 - 2x) = x - 12.$$

Lewis' suggestion seems particularly plausible in cases in which the time and effort involved in transforming the original equation would lead to an overall increase in solution time.

Abstract Information

The next emergent property is that expert problem solvers usually represent a problem statement in terms of chunks containing abstract information. This abstract information has a specific character, in that it is information which was not present in the original problem statement, but which was derived from the givens and is a reflection of the expert's understanding of both the particular problem and his domain as a whole.

A study done by Adelson[13] illustrates the phenomenon. In one set of conditions, expert programmers performed a detailed-level task on each of a set of programs and then they were questioned on some abstract property of the programs. Although the subjects had not been explicitly provided with any information about the abstract property, and in fact the task was purposely designed to draw their attention to the nonabstract properties of the programs, the expert subjects nonetheless were able to answer the abstract questions. In comparison, performance was significantly worse when expert subjects performed an abstract task and then answered a concrete question. The pattern of results was reversed for the novice subjects, for whom concrete questions were easier than abstract ones. That the experts but not the novices answered the abstract questions effectively, even after having performed a concrete task, supports the suggestions that experts' representations involve abstract information that is not explicitly present in the original problem statement. Evidence for the introduction of abstract information into expert representations also appears in the domains of physics,[15–18] computer programming,[19,20,11] baseball,[21,22] and mathematics.[14,23,24,8]

Turning to the utility of the abstracted information, the work of Simon and Simon[15] provides a clear example. Typically, they find that early on in the protocol of an expert physicist numerical values appear that both were necessary to solve the problem and were not given in the problem statement. For example, if velocity is needed to solve a problem, but only values for acceleration and time are presented, the expert derives the value for velocity, and ignores the original values for acceleration and time. The utility of recasting a problem representation so it contains just the concepts and numerical values needed to work towards the solution is obvious. The existence of abstract chunks highlights an important aspect of expertise: the expert has a deep understanding of the constructs in his field and this understanding is reflected in the way that he formulates a problem.

Organization of Chunks

As stated above, it is well known that experts group the simple elements of a problem into first-order chunks. However, the fact that then these first-order chunks are often further organized into interesting and complex knowledge structures has largely been overlooked. This is an aspect of cognitive skill that bears some attention. The data of Chase and Simon[2] provide a typical result suggestive of this phenomenon. They found that their chess Masters recalled more chunks than did their class A players. They also observed that the Masters would examine the partially recalled board systematically (in either a clockwise or counterclockwise pattern). They hypothesized that the Master may not simply be recalling unrelated chunks from memory, but rather is using some larger knowledge structure to guide his recall. (Chase and Simon do not exclude the alternative hypothesis that the Master is reconstructing rather than recalling the board; that issue, however, is outside of the scope of this article). Evidence that chunks are organized into larger knowledge structures appears again in the domain of chess in the results of Charness,[25] as well as for computer programming,[26,9] Go,[5] and physics.[18]

The notion of a retrieval structure[27] appears to help in accounting for these interesting data. A retrieval structure is a domain-specific, long-term memory structure. At the functional level of description it can be thought of as a data structure built up through practice and allowing direct access to the information that it holds. Direct access means that the information in a location can be obtained by going straight to the location, which is to be contrasted with serial access in which the only way to get to the nth item stored is by starting at the beginning and first accessing its $n - 1$ predecessors. The construct is presented in detail in Chase and Ericsson[27] who report a subject (SF) who developed a memory span of 80 digits. They suggest that the performance of their subject was due to the use of a hierarchical retrieval structure. Protocol analysis, computer simulation, and traditional experimental paradigms lend support to Chase and Ericsson's claim that a fixed, direct-access structure was being used. The subject's recall clusters were highly predictable even with varied stimuli, suggesting a fixed structure. His recall pauses suggested that the structure was hierarchical.

Before an experimental session would begin, SF and the experimenter would agree on a diagram of the retrieval structure that he would be using. After the digit sequence had been read, the experimenter would point to a location on the diagram and SF would produce the digits comprising that chunk. This suggests that the access was direct.

Chase and Ericsson suggest that a retrieval structure is generally useful for problem solving for two reasons: First, its invariance across situations allows series of subproblems to be performed without challenging the solver's capacity with keeping track of where he is in his solution (since part of knowing the retrieval structure is knowing how to traverse it). Second, the direct-access feature allows intermediate results to be quickly stored and retrieved, again without taxing working memory.

Although Chase and Ericsson did not mention it, the construct of a retrieval structure (including a nonhierarchical one) can clearly be used to explain existing data. For example, Charness (1976) found that chess Masters' memory for game boards did not decrease after a rehearsal suppression task. As Charness states, this suggests that long, rather than short term memory is involved. It is therefore possible that it was a retrieval structure that was being accessed by the chess Masters. The systematic recall observed in experts in Chess and Go[2,5] can also be explained by the existence of retrieval structures in those domains. The recall data of Spilich, Voss, and Chiesi[22] also suggests the use of a retrieval structure. In their article, expert baseball enthusiasts were able to recall detailed accounts of baseball episodes. The orderly structure of the information in the protocols seemed to reflect a well differentiated retrieval structure, rather than a haphazard dumping of the contents of short term memory. Jeffries, Turner, Polson, and Atwood[9] propose the concept of a "design schema" as an abstract framework which expert program designers use to generate the structure of a solution to a design problem; this can also be seen as a retrieval structure, in that both are abstract frames, with fixed structures, that can be used to guide the expert through his various steps in a problem.

THE MECHANISMS OF COMPOSITION AND PROCEDURALIZATION

Anderson, Greeno, Kline, and Neves[28] and Neves and Anderson[29] have constructed production systems which simulate problem-solving behavior in the domains of geometry and formal proof construction. Two of the mechanisms included in the production systems, called composition and proceduralization, can produce the smooth fast performance typical of the expert. Although this aspect of expert behavior would need to be recreated in any model hoping for sufficiency, these particular mechanisms are also interesting because the first provides an explanatory mechanism for the formation of chunks by experts, and the second yields the seemingly counterintuitive finding that experts are often unable to report the procedures that they successfully use (discussed at the end of this section).

A word on production systems may be useful here. They are sets of one or more rules, (called "productions"), each of which consists of an antecedent condition and a consequent action. If the antecedent condition of a production under consideration is met, then the consequent action of the production is carried out. Once a production has been carried out, the remaining productions in the set are no longer considered on that pass, i.e., only one production executes during a single pass. (There are several methods used by different researchers to break ties when more than one production can apply on a given pass.) Productions can become streamlined into "procedures" after they have been frequently executed during problem solving. Proceduralization turns general productions into specific productions; ones that execute specific actions under specific conditions. A procedure is an instruction which is directly executable, in an automatic, start to finish fashion. Proceduralization, therefore allows the observed smooth execution of the steps of a solution. The contrasting case exists for the novice, who can perform a task but only by "interpreting" the information he has in his knowledge base.*

*The terms "procedural" and "interpretive" and the analogy come from computer science. Procedures are instructions to the computer that can be written in a form which allows immediate execution without an intermediate translation stage, whereas interpretation involves first translating an instruction into executable form and then carrying it out. As a result of the extra step involved, it takes longer to carry out any one step in an interpretive than in a procedural fashion.

The composition mechanism acts to combine sets of simple procedures. The following example from Neves and Anderson[29] illustrates the process. Consider the composition process of productions P1 and P2:

P1: If you see a red light
Then assert danger.

P2: If there is danger
and another person is near you
Then warn that person.

Add together the If and the Then sides of the procedures to produce:

P3: If you see a red light
and there is danger
and another person is near you
Then assert danger
and warn that person.

However, the parts of the antecedent of P3 ("there is danger") that are dependent on action taken by P1 ("assert danger") should not be conditions for P3 or else P3 will never be satisfied. To remedy this situation delete clauses in the If side of the composite if they occurred in the Then side of the first production and the If side of the second. Following the two parts of this algorithm P4 would result from the composition of P1 and P2.

P4: If you see a red light
and another person is near you
Then assert danger
and warn that person.

Neves and Anderson[29] make the assumption that after two procedures have undergone composition, they will then be executed in one step whose execution time is less than the sum of the execution time of the original two (this is a common assumption in production systems). So, according to Neves and Anderson, composition accounts for the speed-up in expert performance because under their assumptions a production system with a few large productions will take less time to execute than a production system with many small productions. Larkin[30] also found evidence which supports the constructs of proceduralization and composition. She reported that the activities of (a) selecting and (b) applying principles takes only a small proportion of the problem solving time of expert physicists. Composition, which leads to the formation of larger pattern-matching conditions predicts the former result, whereas proceduralization predicts the latter.

Composition is a candidate for the mechanism accounting for expert chunking. The argument for this is as follows: One effect of composition is a production that has a greater number of antecedent conditions than did the original productions.[29] So, when we see evidence of chunking in a subject's protocol we may be seeing an effect of composition. That is, we may be seeing that the expert has recognized or matched the original problem statement against a single abstract concept rather than piece by piece, as a set of surface cues. And his response then, is the result of this matching. A direct test is needed for this hypothesis.

Proceduralization may account for an aspect of the abstraction phenomenon, that initially is counterintuitive. Both Brown and Burton[23] and Adelson[13] have found that in some situations experts have difficulty in providing information about the details of a task which they have just completed successfully. For example, Brown and Burton[23] found that grade school math teachers could of course solve simple artithmetic problems, but they were unable accurately to describe the procedures that they used to do so. Adelson[13] found that following comprehension tasks, expert computer programmers had more difficulty than did novice programmers in answering detailed questions about the programs that they had just seen. Proceduralization could account for the above two findings, since it is assumed that the contents of a procedure cannot be inspected.*

Lewis[14] suggested the possibility that the powerful problem-solving operators of the expert might not be the result of composition. His argument is that composition procedures new procedures which mimic not only the results of the original procedures, but also their step-by-step processes. He also pointed out that experts can develop shortcut operators, ones which do not mimic the processing involved in the original procedures. So although composition may produce levels of performance that are equal to those of the expert, it may not reproduce his behavior and therefore may not be the process through which experts develop their skills.

THE ROLE OF CATEGORIZATION IN PROBLEM SOLVING

The Phenomena: The Vertical and Horizontal Structure of Conceptual Categories

The work of Rosch and her colleagues[31,32] and of Murphy and Smith[33] suggests that concrete objects are categorized at several levels of abstraction. Furthermore, as these investigators have found, in each case one level of abstraction seems to be preferred by subjects; it has been named the "basic" level. The basic level of a taxonomy is defined as the level at which there is the greatest increase in the number of features listed when subjects are asked to list the features of members at varying hierarchical levels. Moreover, the basic level has certain properties which have implications for the use of labels for objects. For example, the basic level is the level at which items are first classified. (That is why it is considered the "preferred" level.) For example, a subject is most likely to call the roughly spherical, red object that he or she is holding an apple, although it is also understood to be both a fruit and a Macintosh.

Rosch and Mervis[31] have also found that, for any given level of abstraction, categories of concrete objects are highly structured in the sense that the members of the category vary in the degree to which they are judged to be good exemplars of the category. For example, when presented with an apple and a pomegranate, subjects will judge the apple to be more representative or prototypical of the category fruit.

The categorization of concrete objects is an accepted phenomenon; however, recent results suggest that abstract concepts may also show similar categorization effects.[34–36,18] Chi, Glaser, and Rees[18] suggested that concepts of physics may be subject to vertical classification. They found that experts can classify physics problems at various levels of abstraction, but that their first choice is a moderately abstract level. They also found that the first classification made by novices is less inclusive than that made by experts. Chi et al. suggested that these preferred categories are the "basic" categories for each group, a plausible suggestion since they conform to two of Rosch's criteria: items at the basic level are moderately abstract in that they have an intermediate number of features associated with them and they are the first classification to be used for an item. Chi et al. also suggest that experts have a more abstract basic level of conceptualization than novices since their first classification is more inclusive. This interpretation agrees with the finding that experts prefer to represent problems at a more abstract level

*The assumption is made because the construct of proceduralization is an analogy from computer science. Compilers in computers translate high level language instructions into procedures in lower level languages. And although the compiler understands the high level language (in the sense of being able to translate it) it does not understand the lower level language. Anderson and Neves[29] also assume that the only way of knowing what is in a procedural representation is through observing the performance that results from it.

than novices.[13] It also brings up the issue of how the basic level changes with expertise and context. This issue was only mentioned in the original Rosch experiments; however, as researchers explore the utility of the construct, the importance of these factors comes to light.

Adelson[36] directly addressed the question of whether abstract concepts in computer programming have a basic level of categorization. Using the methodology of Rosch et al.[32] asking subjects to list the features of the members of a well known concept of the domain (e.g., sorting algorithms which are well studied in computer science), Adelson found that the number of features listed by subjects increased greatly between the superordinate and the intermediate levels as compared to the increase between the intermediate and subordinate level. This result parallels that of Rosch et al.[32] Adelson[36] also found that the level identified as the basic one (using the feature listing task) was the level which subjects used first when classifying algorithms in a sorting task.

Turning to the horizontal structure of abstract categories, the results of Adelson[34,35] suggest that, within a given level of abstraction, familiar concepts in a domain are structured in a way similar to concrete objects. That is, the category members were found to vary in their prototypicality. Additionally, the frequency with which the category member's features appeared across the category predicted its prototypicality. This finding is again parallel to that of Rosch and Mervis.[31]

The Utility of Structured Categories

The existence of conceptual categories whose members differ in their degree of category membership may be useful in explaining the exceptional recall that is found in expert problem solvers. Although the data about this are very sketchy, the possibilities are interesting to consider.[6,5,2,25,37] Chase and Simon[2] and Charness[25] did find exceptional recall for chess boards, but also, at a finer level of detail, they found that such recall extended to novel meaningful as well as familiar chess patterns. Even if we accept chunking as a reliable phenomenon, we still need an explanatory mechanism for a process. At this point no data bear directly on this issue; however, Chase and Simon,[2] de Groot,[37] and Charness[25] all express the same intuition. Since novel patterns are well recalled, each researcher suggests the possibility that configurations of chess pieces are identified and labelled as members of a class of patterns, rather than only as instances of unique patterns. Adelson's finding[34,35] that in the domain of computer science, various sorting algorithms are perceived as members of a category that exhibit variations in their perceived prototypicality supports this possibility. That is, the finding suggests that subjects' representations in some sense consist of items related by category membership, rather than of a large number of unrelated concepts. Schneiderman's[19] finding that the recall distortions of computer programmers preserve the concepts, but not the specific form of previously seen material, also suggests that individual items are encoded as instances of categories.

Along with composition and proceduralization, Anderson, Greeno, Kline, and Neves[28] also describe a process called generalization. Although the mechanics of generalization are not described in detail by Anderson et al., the essence of the generalization process is that by building up general descriptions of problems, rules are formed to capture the similarities both between various problem statements and between the solutions for the problem statements. It is possible to use this same mechanism to explain the formation of the abstract categories that Chase and Simon[2] find their Masters using. That is, in both cases we have mechanisms whose function is to identify nonidentical stimuli as members of some equivalence class.

GETTING FROM THE MEANS TO THE ENDS: WORKING FORWARDS VERSUS WORKING BACKWARDS

One problem-solving strategy that is mentioned frequently in the literature is means-ends analysis.[38-40] When using means-ends analysis the problem solver first assesses the difference between the present state of the problem and the eventual goal state and then chooses an operator intended to decrease the difference. The process is repeated until the present and goal states are identical. The construct has been used successfully to model problem-solving behavior;[38-40] however, additional assumptions must be made in order to implement simulations of the strategy, since the difference between existing and goal states usually can be reduced in a number of ways.[39] The use of two problem-solving strategies, known as working forwards and working backwards, has been noted both in physics and geometry, and the two strategies have now been used in the implementation of means-ends analyses.[15,30,16,29,28] In working forwards to solve (for example) a physics problem, an equation is selected in which only one variable is currently unknown. This unknown variable is then solved for and the process is repeated until the variable being solved for is the answer to the problem.

In working backwards, an equation containing the desired unknown (i.e., the answer) is constructed; if this is the only unknown in the equation the problem is solved; if not, another equation is constructed containing the other unknown and the process is repeated until all unknowns can be solved for.

Several assumptions are usually made about working forwards, whereas several points that are needed to account for working backwards are usually left unexplained. The researchers cited above claim that working forwards is usually done by experts, whereas working backwards is usually done by novices. However, evidence of one strategy or the other having been chosen is usually gotten from protocol analysis and we can ask: are experts really working forwards or are they first working backwards and the reporting only the forward process? Since experts usually do not verbalize as much of their problem solving process as do novices, this is a real possibility.[15] Consider this example. Sam Spade keeps his scotch in a locked safe box in his bottom desk drawer. The key is in the top drawer. If Sam opens the top drawer before the bottom, one could very well describe him as working forwards. On the other hand, he may have opened the top drawer only after having imaged opening the bottom drawer with its locked box first, in which case he would really be working backwards. The same behavior occurs in both cases. Larkin (cited in Chi et al.[18]) finds that experts do give evidence of working backwards on novel or complex problems, and it is possible that they do so for shorter problems as well, but they manage to hold the backward chain in memory.

Even if the information in the protocol is an accurate record of the subject's solution path, the evidence is equivocal as to whether or not experts work forwards whereas novices work backwards. As stated above, evidence for the difference between the two groups is obtained by: Simon and Simon,[15] Larkin,[30] Larkin, McDermott, Simon, and Simon,[16] Neves and Anderson,[29] and Anderson, Greeno, Kline, and Neves[28] as well as by Atwood and Jeffries.[10] Nevertheless, some negative evidence can also be found. As mentioned above, Larkin finds that experts give evidence of working backwards when problems are novel or complex and Greeno, Magone, and Chaiklin[39] found that novices performed better on geometry problems which were structured so as to require working forwards rather than backwards. The data suggest that regardless of their level of expertise, subjects work forwards on problems which they do not find difficult. The data are less clear on problems that are ill-structured, unfamiliar, or in otherways complex. It is possible in these cases that both novices and experts work both backwards and forwards; backwards to find starting points and forwards once those starting points have become clear.

Whether the problem solver is working forwards or backwards, the selection of the steps in the solution still needs an explanation. Three possibilities suggest themselves. Although two were discussed previously in other contexts, they can also be put to work here. To the extent that we see unsuccessful attempts with various solution methods in a solver's protocol, we have evidence that a procedure of generating and testing may be all that is used in selecting steps. Novices do show this kind of searching around behavior. Experts, however, do not (at least when working on the kind of standard text book type of problems used in this research); Larkin (cited in Chi et al.[18]) finds that novices, but not experts, frequently generate clusters of equations with similar variables. What mechanism may be responsible for the efficient problem solving of the experts? Larkin[30] suggests that principles may become encoded as problem solving procedures and the conditions for the application of the principles may become encoded as antecedent conditions for their execution.* However, other mechanisms could also be directing problem solving. The solver could have a retrieval structure with whatever procedures are necessary for the solution embedded in it. This possibility would be supported to the extent that progress in the solution (i.e., pauses and bursts of activity) followed the hypothesized structure.

SUMMARY AND CONCLUSIONS

Most of the work that has been done in this domain has only been done fairly recently. As a result, we do not yet have theories of problem solving which are sufficiently specified. This article has therefore, attempted to examine constructs or principles which seem to play a frequent role in the phenomena in this domain as an intermediate step in the building of adequate theories.

Particular thanks to R. Duncan Luce. Also to Roger Brown, Stephen Kosslyn, E. E. Smith, Sheldon White, Miriam Schustack, and William Estes for their advice and guidance. Request for reprints should be set to Beth Adelson, Department of Psychology & Social Relations, Harvard Univeristy, Cambridge, MA 02138.

References

1. H.A. Simon and J.R. Hayes, "The understanding process: Problem isomorphs," *Cognitive Psychology,* **8**, 165–190 (1976).
2. W.C. Chase, and H.A. Simon, "Perception in chess," *Cognitive Psychology,* **4**, 55–81 (1973).
3. H.A. Simon and K. Gilmartin, "A simulation of memory for chess positions," *Cognitive Psychology,* **5**, 29–46 (1973).
4. H.A. Simon and Barenfeld, "An information processing analysis of perception in problem-solving," *Psychological Review,* **76**, 473ff (1969).
5. J.S. Reitman, "Skilled perception in go: Deducing memory structures from inter-response times," *Cognitive Psychology,* **8**, 336–356 (1976).
6. D.E. Egan and B.J. Schwartz, "Chunking in recall of symbolic drawings," *Memory and Cognition,* **7**(2), 149–158 (1979).

*Chi, Feltovich, and Glaser[17] examined the protocols of expert physicists and found that they would first associate the surface elements of a problem with physics principles and then they would mention the conditions under which these principles actually are relevant. For example, problems which contain inclined planes may sometimes be solved using Conservation of Energy and sometimes using Force Laws. Their experts tended to consider both laws and then state the conditions under which each would apply in order to choose the appropriate solution method.

Larkin[30] also suggests that what has been called "intuition" in experts may instead be their knowledge of when a particular principle applies, in contrast to novices who must search through their knowledge bases trying out various principles for their appropriateness.

7. R.D. Luce, Personal communication.
8. E.R. Michener, "Understanding understanding mathematics," *Cognitive Science,* **2**, 361–383 (1978).
9. R. Jeffries, A. Turner, P. Polson, and M. Atwood, "The processes involved in designing software," In *Cognitive Skills and Their Acquisition*, Anderson (Ed.), Lawrence Erlbaum Associates, Hillsdale, NJ, 1980.
10. M. Atwood and R. Jeffries, "Studies in plan construction," Technical Rep. SAI-80-028-DEN. Science Applications, Englewood, CO, 1980.
11. M. Atwood, A. Turner, R. Ramsey, and J. Hooper, "An exploratory study of the cognitive structures underlying the comprehension of software design problems," ARI Technical Rep. 392. SAI-79-100-DEN, 1979.
12. W. Kintsch, *The Representation of Meaning in Memory*, Hillsdale, NJ: Lawrence Erlbaum, Hillsdale, NJ, 1974, 279 pp.
13. B. Adelson, "When novices surpass experts: How the difficulty of a task may increase with expertise," submitted, 1983.
14. C. Lewis, "Skill in algebra," In *Cognitive Skills and Their Acquisition*, Anderson (Ed.), Lawrence Erlbaum Associates, Hillsdale, NJ, 1981.
15. D.P. Simon and H.A. Simon, "Individual differences in solving physics problems," In *Children's Thinking: What Develops?*, R.S. Siegler (Ed.), Lawrence Erlbaum Associates, Hillsdale, NJ, 1978.
16. J. Larkin, J. McDermott, H. Simon, and D. Simon, "Models of competence in solving physics problems," *Cognitive Science,* **4**, 317–345 (1980).
17. M. Chi, P. Feltovich, and R. Glaser, *Representation of Physics Knowledge by Novices and Experts*, in press.
18. M. Chi, R. Glaser, and E. Rees, "Expertise in problem solving," *In Advances in the Psychology of Human Intelligence*, Lawrence Erlbaum, Hillsdale, NJ, 1981, Vol. 1.
19. B. Schneiderman, "Measuring computer program quality and comprehension," Technical Rep. No. 16, University of Maryland, Department of Information Systems Management, College Park, MD, 1977.
20. N. Wirth, "Step-wise refinement," *Communication of the ACM,* **14** (1972).
21. Chiesi, Spilich, and Voss, *Journal of Verbal Learning and Verbal Behavior,* **18**, 257–273 (1979).
22. Spilich, Voss, and Chiesi, *Journal of Verbal Learning and Verbal Behavior,* **18**, 275–290 (1979).
23. J.S. Brown and R.R. Burton, "Diagnostic models for procedural bugs in basic mathematical skills," *Cognitive Science,* **2**, 155–192 (1978).
24. G. Polya, *Mathematical Discovery*, Wiley, New York, 1962.
25. Charness, "Memory for chess positions," *Journal of Experimental Psychology: Human Learning and Memory,* **2**, 641ff (1976).
26. B. Adelson, "Problem solving and the development of abstract categories in programming languages," *Memory and Cognition,* **9**(4), 422–433 (1981).
27. W.C. Chase and A. Ericsson, "Skilled memory," In *Cognitive Skills and Their Acquisition*, Anderson (Ed.), Lawrence Erlbaum Associates, Hillsdale, NJ, 1980.
28. J.R. Anderson, J.G. Greeno, P.J. Kline, and D.M. Neves, "Acquisition of problem-solving skill," In *Cognitive Skills and Their Acquisition*, Anderson (Ed.), Lawrence Erlbaum Associates, Hillsdale, NJ, 1980.
29. D.M. Neves and J.R. Anderson, "Knowledge compilation: Mechanisms for the automatization of cognitive skills," In *Cognitive Skills and Their Acquisition*, Anderson (Ed.), Lawrence Erlbaum Associates, Hillsdale, NJ, 1980.
30. J. Larkin, "Enriching formal knowledge," In *Cognitive Skills and Their Acquisition*, Anderson (Ed)., Lawrence Erlbaum Associates, Hillsdale, NJ, 1980.
31. E. Rosch and C. Mervis, "Family resemblances: Studies in the internal structure of categories," *Cognitive Psychology*, 1975, **7**, 573–605 (1975).
32. E. Rosch, C.B. Mervis, W.D. Gray, D.M. Johnson, and P. Boyes-Braem, "Basic objects in natural categories," *Cognitive Psychology,* **8**, 382–439 (1976).
33. G. Murphy and E.E. Smith, "Basic-level superiority in picture categorization," *Journal of Verbal Learning and Verbal behavior*, **21**(1), 1–20 (1982).
34. B. Adelson, "Cognitive spaces for computational categories," Paper Presented at the Fourth Annual Meeting of the Cognitive Science Society, Berkeley, CA, 1981.
35. B. Adelson, "Prototypicality in programming concepts," manuscript in preparation.
36. B. Adelson, "The basic level in programming concepts," manuscript in preparation.
37. A.D. de Groot, *Thought and Choice in Chess*, Mouton and Company, Paris, 1965.
38. H.A. Simon, "Information processing models of cognition," *Annual Review of Psychology,* **30**, 363–396 (1979).
39. J.G. Greeno, M.E. Magone, S. Chakin, "Theory of constructions and set in problem solving," *Memory and Cognition,* **7**(6), 445–461 (1979).
39. R. Bhaskar and H.A. Simon, "Problem solving in semantically-rich domains," *Cognitive Science,* **1**, 193–215 (1977).

40. A. Newell and H.A. Simon, *Human Information Processing*, Prentice-Hall, Englewood Cliffs, NJ. 1972.
40. B. Adelson, "Knowledge structures of computer programmers, *Proceedings of the Fourth Annual Meeting of the Cognitive Science Society*, 1981.
41. P. Jolicoeur, M. Gluck, S.M. Kosslyn, *From Pictures to words*, in press.
42. J.G. Greeno, "Cognitive objectives of instruction," In *Cognition and Instruction*, D. Klahr (Ed.), Lawrence Erlbaum, NJ, 1976.
43. J.G. Greeno, "Indefinite goals in well-structured problems," *Psychological Review*, **83**(6), 479–491 (1976).
44. K. McKeithen, J.S. Reitman, H. Rueter, and S.C. Hirtle, "Knowledge organization and skill differences in computer programmers," *Cognitive Psychology*, **13**(3), 307–325 (1981).
45. J.S. Reitman, and H. Reuter, "Organization revealed by recall orders and confirmed by pauses," *Cognitive Psychology*, **12**, 559–581 (1980).
46. G. Ryle, *The Concept of Mind*, Hutchinson, London, 1949.
47. R.M. Shiffrin and W. Schneider, "Controlled and automatic human information processing: II. Perceptual learning, automatic attending and a general theory." *Psychological Review*, **84**, 127–190 (1977).

Reprinted from *Proceedings of the IEEE,* Volume 74, Number 10, October 1986, pages 1304-1321.

Representation and Use of Metaknowledge

LUIGIA AIELLO, CARLO CECCHI, AND DARIO SARTINI

The need for expressing and using metalevel knowledge is emerging in the design of several kinds of AI systems.

The careful distinction between **object-level** *and* **metalevel** *notions and the formalization of the latter has first been carried out by logicians for foundational reasons; subsequently, the distinction has been exploited in Artificial Intelligence and Computation Theory, revealing itself to be of great relevance to Automated Deduction and Problem Solving.*

This paper concentrates on the use of metaknowledge in building knowledge-based systems. In order to introduce the issue, some motivating examples are presented. We then review various paradigms for **combining knowledge** *and* **metaknowledge**, *with the aim of abstracting general criteria that should underly the construction of viable AI systems, as far as metaknowledge is concerned. Furthermore, a general overview of the uses of metaknowledge in AI is provided and, among them, we concentrate on* **inference control**, *which can be conveniently exercised by formalizing control strategies at the metalevel and by letting the inference engine depend on metalevel descriptions. The technique is presented with the aid of some examples, chosen from practical AI applications, that are expressed in the formalism of* **Horn clause logic.**

The issue of **self-descriptive** *systems is then addressed. A system that embodies and can use an adequate description of itself allows for* **self-evaluation** *(e.g., the estimate of the resources needed to perform a given task) and for* **self-modification** *(e.g., the automatic improvement of deduction performance by profiting from experience gained in previous deductions).*

INTRODUCTION

The clear separation between **object language** and **metalanguage** has first been formally set by logicians in the study of the foundations of mathematical reasoning. Due to paradoxes caused by language/metalanguage confusion (as in "this sentence is false"), logicians have realized the importance of separating the language and the arguments used to describe a certain domain, from the language and the arguments used to characterize this activity. The clear distinction between the two levels (and the overcoming of paradoxes as the above mentioned one) has been rigorously treated by Tarski in the period 1923–1938. In the 1930s, Gödel developed a particular mathematical technique (called "gödelization") whereby the formulas of the theory of arithmetics can be **named** via a coding in the theory itself, and relevant metatheoretical notions such as well-formedness, substitution, and theoremhood can be represented in the theory as predicates defined on codes (gödel numbers). The construction was carried on in order to prove the incompleteness of "mechanical" deduction with respect to logical truth [26], which was achieved by exhibiting a formula that is true in the standard interpretation of arithmetics but that cannot be derived as a theorem with the deductive apparatus of first-order arithmetics.

Perhaps paradoxically, the technique involved in the proof of Gödel's incompleteness theorem, which establishes a limitation of mechanical deduction, has provided suggestive and useful hints in the search for enhancements of the deductive power of reasoning systems. Here, "deductive power" has to be interpreted in a pragmatic sense, whereby once the set of derivable theorems has been formally defined, the main concern is in **how easily** theorems can be derived. As stressed by Aiello and Weyhrauch in [1], in reading a book on algebra one realizes that most of the stated lemmas and theorems are actually metatheorems: if a system does not allow for the representation of metalevel knowledge they could not even be expressed; while the possibility of expressing and using them as subsidiary inference rules makes feasible the proof of complex facts. The capability of **representing** metaknowledge, deriving metatheorems, and **using** them in deduction is crucial in the extension of the deductive tools of a system.

One may argue at this point that computers, being able to perform thousands of inference steps per second, may not need to rely on the approach used by humans in order to carry on complex derivations, rather they can base inference on simple and uniform deduction primitives. However, complete deduction strategies for theorem proving based on the latter view, e.g., Robinson's **resolution** [38], which received great attention around the 1960s, turned out to be inefficient because of combinatorial explosion. The inefficiency does not depend upon the particular deduction strategy, rather it is intrinsic in the requirement that a whole logic should be treated by means of a uniform, indifferentiated strategy [21].

The immediate application of **meta** to theorem proving has been the use of metatheorems in order to shorten object-level proofs, for instance by asserting the validity of a fact by simply looking at its syntactic structure instead of repeatedly applying inference rules (to appreciate the difference between these two ways of proving, both in terms

Manuscript received April 11, 1985; revised May 17, 1986. This work was partially supported by the Italian Ministry of Public Education, by ENIDATA spa, and is a contribution to the COST13 project N.21 "Advanced Issues in Knowledge Representation."

The authors are with the Dipartimento di Informatica e Sistemistica, Università di Roma "La Sapienza," Rome, Italy.

of complexity and in perspicuity, the reader can consider the first example in Section I.)

The interest in metatheory—first manifested in the field of theorem-proving—soon extended to almost all the areas of Artificial Intelligence, among them Knowledge Representation, Problem Solving, Expert Systems, and Deductive Databases. In Knowledge Representation, the use of metatheoretical notions provides an alternative to the development of special-purpose languages (see, for instance, [6], and the third example in Section I). In this respect, metalevel knowledge and reasoning are advocated as a paradigm for knowledge representation, which can account for formalization of beliefs, default reasoning, inference in changing situations, etc., rather than simply as a means to facilitate the deduction process.

Both aspects of metalevel reasoning discussed above have been exploited in AI applications, by building on the ideas of the "pioneer" research work on *meta*, of which the best account is—in our opinion—Weyhrauch's paper on his FOL system [47].

An account of the uses of metaknowledge in the construction of AI systems, independently of the knowledge representation formalism they are based on (i.e., logic, production systems, semantic nets, etc.) has been provided in [5, Section 2], where the reader may also find references to the relevant literature. Here we add only a few more hints to those provided so far, by commenting on the following classification of such uses. It has been made on the basis of the objectives that have been pursued by using metaknowledge and metalevel reasoning in AI systems, which are

- to devise proof strategies in automated deduction systems;
- to control the inference in problem solving;
- to increase the expressive power of knowledge representation languages.

We have already said above that the basic idea underlying the use of metaknowledge in Automated Deduction is related to the possibility of defining at the metalevel new inference rules for the object-level (see examples 1 and 2 in subsection I-A). An inference rule (meta-axiom or metatheorem) is viewed as a high-level method for building formal proofs that resemble semiformal ones.

The area of inference control is probably the one in which the use of metalevel knowledge has more strongly been advocated in order to allow reasoning about control (a more detailed account of this problem will be given in Section II, in the framework of Logic Programming). Inference control may be viewed as a specific case of proof strategy definition: the basic idea is to allow the definition of both domain-independent and domain-dependent solution strategies by means of control knowledge (heuristics) with the goal of pruning the search tree and therefore increasing the efficiency of the search process.

In the area of Knowledge Representation we have seen a great proliferation of languages and formalisms specifically devised to take into account particular representation needs. The combination of object-level and metalevel representations has proved to be a very useful "abstraction mechanism" that allows a variety of representation problems to be coped with; these range from self-referential knowledge to awareness, knowledge about beliefs, nonmonotonicity, default reasoning, and reasoning about multiple views of objects. Within this area we may also include the problems that arise when different modules of knowledge have to know about each other, either in order to build interfaces between separate bodies of knowledge or in the construction of user interfaces.

The possibility of better exploiting in applied Artificial Intelligence the ideas outlined so far requires a **declarative** approach to **programming** or, even better, the possibility of **programming in logic**, where the construction of a system is viewed as the setting up of a formal theory. To this end, a crucial step has been Kowalski's **procedural interpretation** of logic, whereby first-order sentences of a particular form (i.e., definite Horn clauses) are regarded as procedures for achieving **goals** [32], [33], represented as atomic first-order formulas. The possibility of programming in logic has allowed the practical use of its computational paradigm in many AI applications, especially in the construction of Expert Systems, Problem Solvers, and Deductive Databases. Among the many advantages that Logic Programming shows, we point out that the distinction between **deduction** and **computation** is made very blurred (see also papers by Hayes [28], [29] in this respect), which allows for the experience gained in the use of metaknowledge within *interactive* theorem proving, to be transferred into the design of AI system which *automatically* accomplish computational tasks.

Organization of the Paper

In Section I we first provide, by means of simple examples, a flavor of what metalevel processing is. We then review some of the paradigms proposed in the literature for combining reasoning and metareasoning. Finally, we try to abstract general criteria that should underly such a combination in viable AI systems.

Section II focusses on the application of *meta* to the control of inference. The issue of control is introduced by means of two examples, chosen from practical AI applications. Horn clause logic and Logic Programming are used as a vehicle for presentation. The examples show the limitations of a uniform and undifferentiated deduction strategy, and suggest more effective control regimes. The control strategies presented, which to a great extent are domain-independent, can be formalized as metalevel knowledge. Hence we discuss in some detail the representation of provability in Horn clause logic in the logic itself; this involves naming relations, linking rules, and axiomatization of provability [8]. Then, having shown how control knowledge can be represented, we present two ways for using the metalevel descriptions, one based on a run-time change of context from theory to metatheory and *vice versa*, the other relying on partial instantiation and evaluation of meta-interpreters.

Section III is about self-modification and self-evaluation. The kind of self-modification we consider refers to the capability of a system to somehow profit from its experience by performing simple generalizations of what it has (or it has not) achieved within a task, remembering the generalizations, and eventually exploiting them—either in the same or in subsequent tasks—in order to prune the derivation of subgoals that are subsumed by the induced facts. By self-evaluation we mean the capability of a system to foresee its computational behavior in the accomplishment of a potential task. We discuss the importance of this feature and we relate it to the issue of inference, from the standpoint of the technical machinery needed to achieve it. Though we have no established results for self-evaluation, we have put forward the issue since we believe that it is an interesting application of metalevel reasoning that should receive particular attention.

I. Metaknowledge: Paradigms for Representation and Use

This section introduces the issue of metalevel knowledge representation by means of simple examples (I-A), then reviews some of the paradigms proposed in the literature for combining reasoning and metareasoning (I-B), and finally, profiting from the notions outlined in the review, it establishes a few principles that should underly the construction of viable AI systems which can effectively represent and use metaknowledge (I-C). Our position is for an explicit and uniform treatment of metaknowledge in AI systems, as opposed to *ad hoc* solutions, suited to particular applications, where metaknowledge is "hardwired" in the implementation code.

Many of the issues and notions raised in this section (e.g., naming functions, reflection principles, control strategies, partial evaluation, etc.) are then reconsidered and treated in more detail in Section II.

A definition of metaknowledge may be found in [30, ch. 7], where it is said that "meta-X" means "X about X": hence metaknowledge means knowledge about knowledge, metareasoning means reasoning about reasoning, metatheory means theory about theory. Given a theory, i.e., a language, a deductive apparatus, and a set of axioms, knowledge is what can be expressed in it: axioms are knowledge, theorems derivable from them are knowledge. Facts that can be asserted or proved **about** the theory are metaknowledge.

In order to better clarify this point we present some examples. They are intended to demonstrate what expressing knowledge at the metalevel means and how this enriches the capability of drawing conclusions at the theory level.

A. Examples

Consider the following axioms of propositional calculus:

A1. $A \leftrightarrow A$
i.e., any proposition is equivalent to itself;

A2. $(A \leftrightarrow B) \leftrightarrow (B \leftrightarrow A)$
i.e., equivalence is commutative;

A3. $(A \leftrightarrow (B \leftrightarrow C)) \leftrightarrow ((A \leftrightarrow B) \leftrightarrow C))$
i.e., equivalence is associative;

along with the inference rule

$$\text{R1}\ \frac{A[B],\quad B \leftrightarrow C}{A[C]}$$

which allows derivation from a proposition $A[B]$, i.e., A in which B occurs, of a proposition $A[C]$ where some or all of the occurrences of B have been replaced by C, provided that $B \leftrightarrow C$ holds. From this apparatus we can derive the proposition

$$S = (P \leftrightarrow Q) \leftrightarrow ((Q \leftrightarrow R) \leftrightarrow (R \leftrightarrow P))$$

by starting with $S \leftrightarrow S$, which is an axiom, then repeatedly applying R1 with the associativity and commutativity axioms (A2-3) until the following form is obtained:

$$((S \leftrightarrow (P \leftrightarrow P)) \leftrightarrow (Q \leftrightarrow Q)) \leftrightarrow (R \leftrightarrow R).$$

Finally, we can get S by successive eliminations of the tautologies (e.g., $R \leftrightarrow R$) to the right of the proposition, which is possible by R1 applied with axiom A1.

Writing the proof in all its detail takes about two pages! Indeed, the above proposition can be made to easily follow from the following metatheorem of propositional calculus:

MF1: *Each proposition S built out of sentential variables via the equivalence connective "↔" such that any sentential variable p occurs in S an even number of times is a theorem.*

Note that MF1 itself can be expressed in logic by introducing names for syntactic entities of the propositional calculus, as "↔," and by defining through axioms what is an occurrence of a variable in a proposition, what is an even number, etc. In this example, the propositional calculus is the object-theory, whereas its description in first-order logic along with MF1 is the metatheory. Note also that it is much easier to count the number of occurrences of variables in a formula than to construct an object level proof, hence if a deductive system, after having instantiated MF1 to S, is capable of procedurally interpreting the notions of occurrence, even number, etc., it can conduct the proof of S at the metalevel in a much more effective and elegant way than by object-level inference. Furthermore, the knowledge embodied in MF1 can be reused for the proof of other propositions similar to S.

As a second example, consider the following sentences of arithmetics, expressed in the language of predicate calculus:

F1. $\forall m\, n\, p.\ m = n \supset m - p = n - p$
F2. $\forall m\, n\, p.\ (m + n) - p = m + (n - p)$
F3. $\forall n.\ n - n = 0.$

The three facts above are derivable from the axioms of arithmetics and can be used in the resolution of linear equations of the form $n + a = b$. Consider now the sentence

∀ *fact x.* LINEAREQ(*wffof(fact), x)*
⊃ THEOREM(*mkequal(x, solve(wffof(fact), x)))*

extracted from a conversation in Weyhrauch's system FOL [47]. This is a piece of metaknowledge asserting that, if a fact at the theory level is a linear equation in x—i.e., it satisfies the metalevel predicate LINEAREQ—then a theorem is obtained by equating x to the solution of the equation (e.g., $n = b - a$), where the solution equation is obtained via the constructor *mkequal*. Note that x is a metalevel variable standing for object-level variables, e.g., n. This piece of metaknowledge allows us to assert that something is a theorem at the theory level by simply looking (within the metatheory) at the structure of a formula of the theory, rather than by a sequence of inferences within the theory, as one does in solving a linear equation via the facts F1-3 above. The theorem derived via the metafact above depends upon the same assumptions which the unsolved fact depended upon; indeed, the variable *fact* in the sentence above ranges over facts, which are FOL conclusions, consisting of a well-formed formula (accessible via *wffof*), a set of dependencies, and a justification for its derivation.

As a third example, consider the English sentence

"A person is innocent if he/she cannot be proved guilty"

which has been formulated in Horn clause logic (+ negation as failure) by Bowen and Kowalski [8] as follows:

$$Innocent(x) \leftarrow Person(x), \; not\ Demo(facts, \text{"}Guilty(x)\text{"}), \; Relevant(facts).$$

This is a mixed object/metalevel rule stating that the object-level fact that expresses the innocence of a person can be derived provided that the metalevel condition of provability does not hold between a set of relevant facts and the goal "*Guilty(x)*." The example shows that, by reasoning at the metatheory level and looking at the properties of the object level theory (is something provable or not?), one can extend the knowledge of the theory. Note, in fact, that a person can be guilty but not provably guilty, hence the innocence of a guilty but not provably guilty person could not be inferred by the object-level sentence

$$Innocent(x) \leftarrow Person(x), \; not\ Guilty(x).$$

B. Proposed Paradigms for Combining Reasoning and Metareasoning

Metaknowledge can be embedded in a system in various ways. The most primitive one consists in embedding chunks of metaknowledge as pieces of code in the system; in other words, metaknowledge is added by extending the implementation of the system with some routine that "does" what the piece of metaknowledge "says." For instance, in a LISP program that performs symbolic evaluation of arithmetic expressions, the metatheorem shown in the second example above about the solution of linear equations can be added to the system as a part of LISP code: a routine that applied to the data structure representing the equation $n + a = b$ builds the representation of the solution $n = b - a$. This way of embedding metaknowledge in a system may be considered as "hardwiring" metaknowledge, which suitably emphasizes the limitations of this approach.

A different view of the problem is the one underlying ML [27], a metalanguage designed for describing proofs and evaluations at the object level. Proof strategies can be expressed in ML for automatically carrying out proofs in the underlying object-level calculus, which is a typed λ-calculus with polymorphism. ML itself is based on that calculus, but, in addition, it has a mechanized notion of reduction (ML-evaluation). Unfortunately, the metalevel is not accessible for reasoning and proving, or at least only in a very restricted sense, hence meta-meta level and higher metalevels are not possible. However, many of the drawbacks of the "hardwiring" approach are overcome: the user has access to the metalevel for directly adding metaknowledge that expresses proof strategies, and the logical soundness of the extensions of the system is guaranteed by the strong type discipline, which allows (as theorems at the object level) only formulas that have been computed by ML with the type *theorem* to be added.

A third approach consists in allowing the user to access the object and the metalevel both for expressing and inferring new facts. This approach has been followed by various researchers: Weyhrauch [47], [4], Boyer and Moore [9], Bowen and Kowalski [8], Attardi and Simi [6]; we think it is the most promising for the construction of complex AI systems, in so far as object-level programming is the same as metalevel programming and the communication between the two levels allows for their combination.

We first examine the work of Boyer and Moore [9]. Their idea is to embed metafunctions—that have been proved correct—into their LISP theorem prover for properties of LISP programs, so as to use them as auxiliary deduction rules, in order to enrich the deductive power of the theorem prover. The expression of metafunctions is allowed in terms of an explicit denotation mapping: terms at the theory level are mapped into their canonical list representation. Metafunctions are then viewed as term transformations defined on their canonical representation.

The approach of amalgamating language and metalanguage has been explored by Bowen and Kowalski [8], in the context of Horn clause logic, and by Attardi and Simi [6] in OMEGA, an object-oriented formalism for knowledge representation. Both approaches rely on a language that is the same both for the object and the metalevel, though they are different in the two cases. Attardi and Simi extend OMEGA to deal with metatheoretical notions by including objects that describe OMEGA objects and the notion of derivability in OMEGA. Analogously, Bowen and Kowalski describe Horn clause syntax and provability in the logic itself: the provability relation is represented via a predicate *Demo* in the context of a set of clauses which are *de facto* a PROLOG interpreter written in PROLOG (the definition is reported in Section II-C2). The communication between the object level and the metalevel in both cases happens through linking rules, whose formulation is

$$\frac{Pr \mid-_M Demo(\text{"}T,\text{"}\ \text{"}A\text{"})}{T \mid-_L A} \qquad \frac{T \mid-_L A}{Pr \mid-_M Demo(\text{"}T,\text{"}\ \text{"}A\text{"})}$$

which specify how the results obtained at the metalevel are communicated to the object-level and *vice versa*. In the two rules, $\mid-_L$ means provability at the object level, $\mid-_M$ means provability at the metalevel; *Pr* is the set of axioms of the metatheory: a set of clauses providing the representation of Horn clause provability in Horn clause logic. The atomic formula *Demo("T," "A")* represents the relation that holds if and only if the set of clauses *T* whose metalevel name is "*T*" entails a formula *A* named by "*A*." The formula *A* is an existentially quantified conjunction of atomic formulas. A more detailed description of linking rules, naming functions and their use in problem solving is provided in Sections II-C and II-D.

The extension of a logic formalism with metalevel features has first been effectively demonstrated by Weyhrauch in his FOL system [47]. Weyhrauch's approach, differently from the above mentioned ones, is not for an amalgamation of knowledge and metaknowledge within a single context, rather it proposes a coexistence of them in different contexts.

An FOL context, sometimes called an *L/S* pair or *L/S* structure, consists of language *L*, a simulation structure *S*, attachments between the two, and a set of facts *F*. The language is first order, with sorts and conditional expressions both for formulas and for terms. The simulation structure *S* is a partial finite representation of some model; attachments are explicitations of the bindings between the symbols of *L* and their interpretation in *S*, which partially represents a model in so far as not all symbols of *L* have attachments and the counterparts for predicate and function symbols in *S* are not defined in general for all the individuals of the model. The set of facts *F* is finite and contains the axioms of the theory and the theorems that have been proved so far. As mentioned in Section I-A, facts are more complex objects than just formulas: they represent in addition dependencies on other facts and a justification for their assertion. A context in FOL can be seen as a snapshot in the construction of a theory/model pair, i.e., all the knowledge that has been gathered so far about a theory and a model for it, where such knowledge has been either provided to the system or derived within it.

An FOL context named META contains the general description of FOL contexts, i.e., a description of the aspects that are common to all FOL contexts, e.g., formulas, theorems, attachments, etc. META describes the proof theory and some of the model theory of an FOL context. The natural interpretation for META, i.e., the natural simulation structure, is the context for which META is the metatheory, plus FOL itself with its routines for dealing with formulas, facts, attachments, etc. The communication between a context and its metacontext is established via semantic attachment and reflection principles, a reflection principle being a relation between theoremhood in a theory and in its metatheory. So, similarly to the linking rules previously seen, provability at the theory level and at the metatheory level are tied by the following rule:

THEOREM("*W*")	in the metatheory
W	in the theory

where the double bar in the rule means applicability in both directions.

The strength of the FOL paradigm is in that it combines the power of programming and proving in first-order logic with the explicit representation and use of metalevel knowledge. The user can in fact access the metalevel, express knowledge in it, derive metatheorems, and use them as subsidiary deduction rules (whose correctness can be substantiated by a formal proof) in her/his subsequent proving activity. The weakness of FOL is primarily in the way the procedural interpretation of logic is dealt with. User-defined attachements are prone to inconsistency, since, for instance, one may attach a LISP program for a certain symbol that is incorrect with respect to the axioms that define that symbol in the theory. To overcome this problem, the automatic generation of LISP programs from first-order specifications has been explored and implemented (see [3]), however, due to the choice of ruling out nondeterminism, the technique can only deal with quite a constrained class of specifications.

C. Principles for Combining Reasoning and Metareasoning

In the previous subsection we have presented various ways of embedding and using metaknowledge, from the most naive (and almost useless) to the most effective and uniform paradigms. From the above review we can abstract a few essential notions we think should underly the realization of a system that smoothly and uniformly combines the use of knowledge and metaknowledge. They are the following:

1) A language and a metalanguage that have to be the same, so as to allow the expression of both knowledge and metaknowledge in the same form.

2) A deductive apparatus that has to be accessible at both levels and offers the possibility of proving (and not only of evaluating) at both levels, thus allowing the user to derive both theorems and metatheorems.

3) A naming relation between ground terms at the metalevel and linguistic expressions at the object level.

4) An axiomatization at the metatheory level of the relevant notions of the theory level, among them provability.

5) Reflection principles that connect theory and metatheory and allow the exportation of results from one level to the other.

Even though the use of metaknowledge has often been advocated with the aim of improving the efficiency of the proving process (and indeed Sections II and III of this paper demonstrate how this can be achieved), the doubt can arise that the setting up of a complex paradigm—as the one resulting from the above principles—can endanger efficiency, and that *ad hoc* solutions can be better in specific applications. We claim instead that, for the construction of viable AI systems, it is essential to provide them with uniform and general tools for the representation and use of metaknowledge (see also the positions emerging from the work in [17], [18], [25], [35]).

The system must also allow for the procedural use of declarative knowledge, which can be achieved either by compilation or by specialized proof procedures. An interesting technique in this context is that of **partial evaluation** [23], that "tailors" general rules to specific cases. Partial evaluation too can be obtained through a proving process to compute subsidiary rules and metarules. More details on partial evaluation and its application to metalevel processing can be found in Section II-E.

II. Metaknowledge and Inference Control

This section explores the application of metalevel reasoning to the **control of inference**. Examples and discussions are formulated in terms of Horn clause logic, the formalism underlying Logic Programming [32], [33]. However, the section is not a case for Logic Programming, rather, it uses its formal apparatus as a vehicle for presentation.

The problem of control raises from the **nondeterminism** present in a deduction process. In the framework of Logic Programming, where rules are expressed as clauses of the form A **if** B_1 **and** ... **and** B_n, AND-nondeterminism results from the freedom in the selection order of subgoals in solving the conjunction B_1 **and** ... **and** B_n and OR-nondeterminism raises from the presence of many clauses available for the solution of a goal A.

Standard implementations of Logic Programming (i.e., PROLOG systems [15], [10], [45], [37], [14]) deal with AND-nondeterminism by fixing a left-to-right selection order for the subgoals B_i in the conjunction B_1 **and** ... **and** B_n. OR-nondeterminism is dealt with by adopting a depth-first search strategy via backtracking, with a fixed try-order for the available clauses; namely, the order in which the clauses are written in the program. This strategy is often prone to inefficiency; in the sequel, we show in fact two examples that demonstrate the inadequacy of the standard resolution strategy and we elaborate on them in order to devise more flexible and efficient strategies. The first example comes from the field of **fault diagnosis** (it has been taken from [13]) and is particularly suited to discuss the control problem for OR-nondeterminism. The second example is in the context of **database query answering**, along the lines of [46], [42], and provides the basis for addressing the control problem for AND-nondeterminism.

A metalevel description of provability in Horn clause logic has been presented in [8]; as stressed in that paper, different control strategies can be explored by suitably defining the key metalevel predicates that describe the selection procedures. However, from the point of view of efficiency, it is not convenient to execute logic programs by running a metalevel interpreter on them, which by analogy corresponds to running LISP programs by calling the EVAL function, written in LISP, on their representation. Metalevel **descriptions**, in order to be effective, have to be **used** in an efficient way. In this section we show two methods by which this effective use can be achieved. The first one involves the interleaving of **upwards** and **downwards reflection** in deduction, along the lines of the work reported in [41], [36] in the field of functional evaluation. The second approach exploits **partial evaluation** in order to embed control—expressed as metaknowledge—in an object-level logic program, along the lines of [24], [40], [43], [44].

A. Control of OR-Nondeterminism

Before presenting the examples, we provide a concise background for Logic Programming.

A **logic program** is a set of **clauses** that are implications of the form $A \leftarrow B_1, \cdots, B_k$, $k \geq 0$, where $A, B_1, \cdots, B_k$ are atoms and the comma "," stands for logical conjunction.

An **atom** A is an atomic first-order expression $p(t_1, \cdots, t_n)$, where p is a predicate symbol of arity $n \geq 0$ and $t_1, \cdots, t_n$ are terms.

A **term** is either a constant, a variable, or a function name f—or arity $m > 0$—applied to m terms.

We reserve lower case starting identifiers for variables and uppercase starting identifiers for predicate, function, and constant symbols. For instance the term Cons(*A*, Cons(*B*, Nil)) contains no variables (such terms are called **ground**) and represents the list [*A*, *B*], which in infix notation is written as *A*.*B*.Nil.

The variables in a clause are meant to be universally quantified.

The **goal** is a conjunction of atoms $\leftarrow A_1, \cdots, A_k$, $k > 0$, to be derived from the clauses in the program. A goal is derived by a **backward chaining** use of the program clauses. The resolution is successfully completed when a **substitution** *s* has been found such that $A_1s, \cdots, A_ks$ (i.e., the goal list rewritten using the variable assignments in *s*) logically follows from the clauses.

1) Example: Fault Diagnosis: A simple system for finding faults in a car can be built out of a logical description of the components of a car, their parts, and their possible faults, along with a rule that relates a fault in a component to the faults in its subparts. The description is provided via a predicate *Part(x, y)*—meaning that *x* is a part of *y*—and a predicate *Fault(f, x)*, meaning that the fault *f* is present in component *x*. The clauses of the program are the following:

```
Part(Engine, Car) ←                 Fault(f, comp) ← Test(f, comp)
Part(Tank, Car) ←
Part(Carburetor, Engine) ←          Fault(In(f, part), comp) ←
Part(Filter, Carburetor) ←              Part(part, comp),
. . .                                   Fault(f, part)
```

The group of assertions to the left extensionally defines the *Part* relation between car components. The clauses to the right define the *Fault* predicate. The first states that *comp* has fault *f* if testing *comp* reveals fault *f*. Alternatively, the second clause states that *comp* has fault *In(f, part)* if *part* is part of *comp* and has fault *f*. For instance, if the assertion *Test(Empty, Tank)* ← is included in the program above, then the goal ← *Fault(f, Car)* succeeds with answer substitution {*f*: = *In(Empty,Tank)*}. If we include instead the predicate *Test(Dirty, Filter)*, then the goal succeeds with answer

{*f* := *In(In(In(Dirty, Filter), Carburetor), Engine)*}.

Of course, the small theory above has to be completed with facts coming from observation, i.e., assertions of the form *Test(f, comp)* ← must be embodied by querying the user or (this is more likely for fault diagnosis of electronic equipments) by actually performing the test of *comp* via dedicated input/output devices. Note however that no matter how *Test* is defined, the backward chaining procedure for resolution, applied to a goal ← *Fault(f, Car)*, has the effect of driving the series of tests that have to be performed in order to fix the car.

The order in which components are tested is crucial for the efficiency of the diagnosis process. Components that are *easy to access* and *likely to have a fault* must be tried first (the logic program above, which does not adopt this heuristics, behaves like a person who, after having realized that his/her car does not start, gets out of the car, takes all his/her tools, and starts dismounting the carburetor before having checked the tank!).

From the "easy-likely first" heuristics in fault diagnosis we can abstract a control strategy for OR-nondeterminism. In fact, we can relate the difficulty of accessing a certain component to its level of embedding in the hierarchy of components, defined by the *Part* relation (e.g., *Filter* is more embedded than *Tank* in the hierarchy of components). This in turn reflects into the complexity of deducing the goal ← *Fault(f,Car)* by using the assertion *Test(Dirty, Filter)* ←, which is greater than that arising by the use of the assertion *Test(Empty, Tank)* ←, in case this latter holds (the two deduction trees are shown below; the trees describe the proof of the goal from the assertions, though the search of the proofs in Logic Programming is performed backwards).

d1)
$$\cfrac{\cfrac{\mathit{Test(Empty, Tank)}}{\mathit{Fault(Empty, Tank)}} \quad \mathit{Part(Tank, Car)}}{\mathit{Fault(In(Empty, Tank), Car)}}$$

d2)
$$\cfrac{\cfrac{\cfrac{\cfrac{\mathit{Test(Dirty, Filter)}}{\mathit{Fault(Dirty, Filter)}} \quad \mathit{Part(Filter, Carbur)}}{\mathit{Fault(In(Dirty, Filter), Carbur)}} \quad \mathit{Part(Carbur, Engine)}}{\mathit{Fault(In(In(Dirty, Filter), Carbur), Engine)}} \quad \mathit{Part(Engine, Car)}}{\mathit{Fault(In(In(In(Dirty, Filter), Carbur), Engine), Car)}}$$

The complexity of solving a goal cannot be mechanically evaluated in general, as stressed in Section III in discussing self-evaluation. However, in particular cases (as in the example of this section) an estimate of the complexity can be calculated.

The likelihood of success in the resolution of a goal can be evaluated by distinguishing between **base** goals (e.g., ← *Test(f, Filter)*) and **derived** goals (e.g., ← *Fault(f, Carburetor)*). Base goals refer to knowledge that is either extensionally defined, or that is provided by the user on demand. Derived goals refer to relations intensionally defined via rules. The likelihood of success for a base goal must be provided as input to the control strategy, on a domain-dependent basis. The likelihood of success for derived goals can be inductively calculated by considering the form of the rules in the program.

In the sequel, we provide a formulation of the "easy-likely first" heuristics and we sketch the procedures for calculating the required parameters.

2) The Control Strategy: Consider again the previous example and suppose that the probability of a fault to be in *Tank* is P_1, while the one for *Engine* is P_2. If we assume, for the sake of simplicity, that *Tank* and *Engine* are the only parts of *Car* and that there is only one faulty component in *Car*, then the equality $P_2 + P_1 = 1$ holds. It is now necessary to quantify the expected complexity of proving a goal. Assume that the complexity of solving the goal ← *Fault((f, Car)* using the assertion *Part(Tank, Car)* ← is $C1_S$ in case of success, and $C1_f$ in case of failure. Similarly, $C2_S$ and $C2_f$ are the complexity coefficients when using *Part(Engine, Car)* ←. If the deduction strategy chooses to inspect *Tank* first, the expected complexity of proving ← *Fault((f, Car)* is given by

$$E_1 = P_1 * C1_S + P_2 * (C1_f + C2_S).$$

If the try-order is reversed, i.e., *Engine* is checked first, the expected complexity becomes

$$E2 = P_1 * (C2_f + C1_S) + P_2 * C2_S.$$

By simplifying the common terms we get

$$E_1 - E_2 = P_2 * C1_f - P_1 * C2_f$$

which leads to

$$E_1 < E_2 \text{ iff } P_1/P_2 > C1_f/C2_f.$$

Note that in our example, $P_1 > P_2$ (we assume that it is more likely to have an empty tank than a fault in the engine) and $C1_f < C2_f$, hence $E_1 \ll E_2$ and the control strategy strongly advises trying the rule *Part(Tank, Car)* ← first.

If it is assumed that the deduction exploits this control strategy, we can calculate the expected complexity in solving the top-level goal ← *Fault((f, Car)* as

$$E_0 = \min \{E_1, E_2\} + K$$

where K is a constant term. Note that E_0, E_1, and E_2 are expected complexities in case of success. The expected complexity of solving a goal in case of failure is expressed as the sum of the complexities in case of failure of all the alternatives. By letting

$$C0_S = E_0 \quad \text{and} \quad C0_f = C1_f + C2_f$$

we obtain values for the complexities of the top-level goal in terms of the complexities of alternative subgoals. This procedure can be recursively applied to $C1_S$, $C1_f$ and $C2_S$, $C2_f$, till base goals are reached, for which the expected complexities can be easily computed, since they reduce to the cost of a retrieval (for simplicity's sake, they can be assumed constant).

The inequality relation between the expected complexities E, previously derived for binary choice points only, can be extended in a straightforward way to deal with *n*ary choice points.

Once the parameters P_j, Cj_S, Cj_f have been determined for all the subgoals reachable from the top-level one, we can think of reordering the clauses in the program, so as to achieve an optimal try-order for them. Such a reordering is certainly possible for the program in the example of Section II-A1, however, in the general case, the process of deriving such parameters must take into account not only the static composition of the rule set, but also the dynamic behavior of the deduction till the values of the latter parameters are needed. Suppose in fact we wish to modify the program for fault diagnosis in order to better exploit the information acquired in the phase of testing. When a component is tested, there is a set of symptoms that lead to concluding that it is faulty. The symptoms generally give an indication of where to look next. This means that, for instance, the success likelihoods of subgoals cannot be determined in advance, rather they are available as the deduction proceeds; hence, the calculation leading to the optimal choice must be done dynamically.

The example of Section II-A1 for the discussion of the control strategy has been chosen so as to isolate the phenomenon of OR-nondeterminism, reducing AND-nondeterminism to a trivial instance. Note in fact that the search tree of the program in the example is essentially the tree of components defined by the *Part* relation. For programs that fall in this class, the control strategy can be formulated—in terms of the search tree—as follows:

- If node N can be alternatively expanded as N_1 and N_2 and if $\langle P_1, C2_f \rangle$ is the pair of probability/complexity parameters for N_1 and $\langle P_2, C2_f \rangle$ is the pair for N_2, then expand first node N_i for which the quotient P_i/Ci_f is maximum

(where for the sake of simplicity we have considered binary choice points only).

The above rule expresses metaknowledge about control. A metalevel description of provability is reported in Section II-C. The way metarules, such as the above one, can be incorporated in order to drive the deduction process is also discussed there.

B. Control of AND-Nondeterminism

Similarly to the approach followed in discussing OR-nondeterminism, we present a simple example where the phenomenon of AND-nondeterminism is isolated by reducing OR-nondeterminism to a trivial instance. Then we abstract from the example the criteria underlying the control strategy, which can be described at metalevel and incorporated into the definition of provability reported in Section II-C. The reported strategy for goal selection in a conjunction, developed by Smith in the framework of database querying, is described in detail in [42]. It is essentially based on the principle of choosing first the goal that causes the least branching of possibilities in the search tree (Warren [46]). Smith's strategy, as compared to Warren's, is more accurate but more complex.

1) Example: Querying Deductive Databases: Suppose we have a database for representing the employees of a firm. All information about one employee is stored as a ground unit clause of the form

Employee(name, level, partner-name) ←

where *name* is the name of the person, *level* is the salary level, and *partner-name* is the name of the employee's husband/wife to be used for tax accounting operations (in a realistic application there would be many more fields specifying other information, but we only concentrate on these three since our concern is on control issues). Suppose the database is queried to retrieve all employees *emp* such that their partners *partner* work in the same firm and their level *lev* is the highest one:

Query(*emp*) ← *Employee(emp, lev, partner),*
Employee(partner, __, emp),
Highest(lev)

(the symbol "__" is used for don't-care variables). According to the standard PROLOG proof procedure, answering the query amounts to the following generate-and-test computation: all employees *emp* are fetched in sequence with their level *lev* and partner-name *partner*, and for each of them, the database is looked up to see if *partner* is present in the firm (if that is not the case the employee is disregarded), then the level *lev* is checked to see if it is the highest one: in this case the employee is kept in the answer, otherwise it is disregarded. The generate-and-test computation takes place when all the answers to a query are required (or when only one answer is requested, but none exists). Indeed, even when a single answer is sufficient and it exists, the generate-and-test takes place until a solution is found, then it is suspended for outputting the answer.

Suppose now that in the firm there are 1000 employees, 10 of them only being employed at the highest salary level. The above query answering process causes a 1000-fold branching where, at each branch, a conjunction

← *Employee(partner, __, emp), Highest(lev)*

has to be evaluated, with all named variables fully instantiated. If, instead of following a left-to-right selection rule for the conjuncts in the query, the conjunction get processed as:

← *Highest(lev), Employee(emp, lev, partner),*
Employee(partner, __, emp)

then the answering process presents a ten-fold branching where at each branch the predicate *Employee(partner, __, emp)* has to be evaluated with *emp* already bound to an employee of highest level and *partner* to his/her partner. With the latter ordering, 990 activations of the goal ← *Employee(partner, __, emp)* are avoided.

2) The Control Strategy: The strategy for reordering a conjunctive query is formulated by Smith in [42] in terms of a metalevel function *AvgSize(a, b)* that expresses the average size of the set of answers to the atomic goal *b* instantiated by the answers to the goal *a*. In our example, if we number the atoms in the antecedent of the rule defining *Query* as A_1, A_2, A_3 according to the left-to-right order, and if we assume that on average, 1 person out of 100 has the highest salary level, and that 1 out of 50 persons has a partner in the same firm, we have the following figures for *AvgSize:*

- $AvgSize(\text{"true," "}A_1\text{"}) = 1000$
- $AvgSize(\text{"}A_1\text{," "}A_2\text{"}) = 1/50$
- $AvgSize(\text{"}A_1, A_2\text{," "}A_3\text{"}) = 1/100$
- $AvgSize(\text{"true," "}A_3\text{"}) = 1$
- $AvgSize(\text{"}A_3\text{," "}A_1\text{"}) = 10$
- $AvgSize(\text{"}A_3, A_1\text{," "}A_2\text{"}) = 1/50$

(where with "*E*" we denote the metalevel representation of an expression *E*, and the 0-place predicate *true* is defined by the clause *true* ←).

A normalized expression for the complexity of answering the query with the left-to-right ordering (where constant factors not influencing the comparison among different orderings have been disregarded) is the following:

$$
\begin{aligned}
&C_{1,2,3} = \\
&= AvgSize(\text{"true," "}A_1\text{"}) + && \text{(normalized cost of retrieving all employees)}\\
&\quad AvgSize(\text{"true," "}A_1\text{"}) * && \text{(for each of them)}\\
&\quad\quad (AvgSize(\text{"}A_1\text{," "}A_2\text{"}) + && \text{(normalized cost of retrieving the partner (if any))}\\
&\quad\quad AvgSize(\text{"}A_1\text{," "}A_2\text{"}) * && \text{(in case the employee has a partner in the firm)}\\
&\quad\quad AvgSize(\text{"}A_1, A_2\text{," "}A_3\text{"})) > && \text{(normalized cost of checking the salary level)}\\
&> 1000.
\end{aligned}
$$

Similarly, for the second ordering, we have

$$
\begin{aligned}
&C_{3,1,2} = \\
&= AvgSize(\text{"true," "}A_3\text{"}) + \\
&\quad AvgSize(\text{"true," "}A_3\text{"}) * \\
&\quad (AvgSize(\text{"}A_3\text{," "}A_1\text{"}) + \\
&\quad AvgSize(\text{"}A_3\text{," "}A_1\text{"}) * AvgSize(\text{"}A_3, A_1\text{," "}A_2\text{"}) = \\
&= 1 + 10 + 10 * 1/50 < 12.
\end{aligned}
$$

The above expressions for normalized costs rely on the assumption that the subgoals in the conjunction can be evaluated via a database retrieval, hence it makes sense to consider a unitary cost-per-answer, so as to express the cost of finding all solutions of A_1 (say) as *AvgSize("true," "A_1")*.

The expression C_π for the normalized cost of evaluating a conjunction in the order expressed by the permutation π can be used to drive a heuristic search of the optimal ordering (see [42] for details). The example in Section II-B1 suggests the approach of a static rearrangement of the subgoals in a query; however, the evaluation function C_π relies on estimates of average sizes, which cannot in general be provided on the basis of statical analysis. Consider in fact the clause *P(x, y) ← Q(x, z), R(z, y)* and assume that the goal *← Q(A, z)* admits 2 answers, while *← Q(B, z)* admits 1000. Suppose also that

AvgSize("Q(x, z)," "R(z, y)") = 2

and

AvgSize("true," "R(z, y)") = 100.

Now, for the query *← P(A, y)*, the best ordering for the antecedent of the rule is the written one, since it gives a normalized cost of 6, while the opposite ordering gives a cost greater than 100. With the query *← P(B, y)*, the situation is reversed: the written order gives a cost of 3000, while the opposite one gives less than 200 (note that the goal *← Q(x, z)*, when both arguments are fully instantiated, has an average size less than or equal to 1).

The conclusion, as for OR-nondeterminism, is that static reordering is not adequate in general. Rather there should be a control strategy for goal selection in conjunctions that, by dynamically determining the average sizes associated with the goals, chooses the next goal to be tried.

The method sketched above applies only to a very simple class of logic programs; namely, those where there is a single rule chaining and the subgoals of the top-level one are extensionally defined by ground unit clauses. In the more general case where subgoals too are defined via rules, the principle of constant cost-per-answer does not hold any longer. Indeed, the control strategy should access both average sizes of answers and estimates of the cost (in complexity) per-answer, which may not be the same for all subgoals.

As for the OR-nondeterminism example, we observe that reasoning and metareasoning must in general be interleaved. In the above example, for instance, with the query *← P(B, y)*, a metalevel computation occurs after the clause has been activated in order to choose the first conjunct in the antecedents to be solved; then the computation proceeds at the object level for solving the goal list *← R(z, y), Q(B, y)*.

C. Metalevel Representation of Provability in Logic Programming

In order for metaknowledge to be used in deduction, it must be **coded** in some formalism. In Section I we have argued that the metalevel formalism must be the same as the object-level one. This hypothesis forces a requirement on the **expressive power** of the object-level formalism: the language must allow the representation of its own syntax and the deductive apparatus must be powerful enough to represent significant notions of its own metatheory, among them **provability**. Horn clause logic possesses this expressive power, as demonstrated by Bowen and Kowalski in [8]. We briefly survey that work here since it will be frequently referred to in the remainder of the paper.

Provability in Horn clause logic can be represented in the logic itself via a three-place predicate *Demo(theory, goals, answer)* in the context of a set of clauses *Pr*. The representation condition that must hold for *Demo* is the following:

1) The Representation Condition:

- For each logic program *T*, goal-list *G*, substitution *s*, *G* succeeds in *T* with computed answer substitution *s* if and only if the goal *← Demo("T," "G," ans)* succeeds in *Pr* with computed answer substitution {*ans* := "*s*"},

where "*T*," "*G*," and "*s*" are names for *T*, *G*, and *s*, respectively.

The **naming function** "..." associates a ground term "*E*" with any syntactic entity *E* (a term, an atom, a clause, etc.). It involves the **coding** of symbols in *T* with positive integers, known as goedel numbers. For instance, the atom *P(x, A)* can be named by *Atom("P,"Var("x").Const("A").Nil)*, where "*P*," "*x*," and "*A*" are the gödel numbers for *P*, *x*, and *A*, respectively (say 3, 1, and 13).

2) Demo: The Representation of Provability: The following clauses define the *Demo* predicate; additional clauses are needed for the auxiliary predicates involved in the definition, which for the sake of conciseness have not been listed here. The top-level clauses for *Demo*, along with those for auxiliary predicates constitute the theory we have called *Pr*, i.e., the Horn clause formulation of Horn clause provability.

```
Demo(theory, goals, answer) ←
        Empty(goals), Empty(answer)
Demo(theory, goals, answer) ←
        Select(goals, goal, rest),
        Member(clause, theory),
        Rename(clause, goals, variant-clause),
        Parts(variant-clause, conclusion, conditions),
        Match(goal, conclusion, subst),
        Apply(conditions+rest, subst, newgoals),
        Demo(theory, newgoals, ans),
        Compose(subst, ans, answer)
```

(where we have chosen to name the set $x \cup y$ by the term "x" + "y").

The first clause for *Demo* deals with the trivial case of empty goal-list, which is achieved under any *theory* with the empty answer substitution.

The second clause deals with the case of nonempty goal list, in which

- an atom *goal* has to be *Selected* from *goals;*
- a *clause Member* of *theory* has to be chosen (and renamed) such that: its head *conclusion Matches* with *goal*, yielding the mgu *subst;*
- its tail *conditions*, merged with the *rest of goals* and rewritten according to *subst* by *Apply*, yields the goal-list *newgoals;*
- *newgoals* can be proved from *theory*, with answer *ans*.
- If such a clause can be found, the result answer is the composition *answer = subst answ*, represented in *Pr* by *Compose*.

We discuss now the way control knowledge is described at the metalevel via *Demo*. It is important to note that in the above definition of provability, the selection points for goals and clauses have been isolated and the choice has been left to the predicates *Select(goals, goal, rest)* and *Member(clause, theory)*, respectively.

3) Defining the Choice Predicates: Different strategies for goal and clause selection can be represented by suitably providing definitions for *Select* and by replacing *Member* with any other choice predicate. We can for instance define Warren's strategy (see the beginning of Section II-B) for goal selection as follows:

WarrenSelect(goals, theory, goal, rest) ←
    AvgSizes(goals, theory, sizes),
    Least(sizes, goals, goal, rest)

AvgSizes(Nil, __, Nil) ←
AvgSizes(goal.goals, theory, size.sizes) ←
    AvgSize(goal, theory, size),
    AvgSizes(goals, theory, sizes)

Least(s.NIL, goal.Nil, goal, Nil) ←
Least(s1.s2.sizes, g1.g2.goals, goal, g2.rest) ←
    s1 < s2, Least(s1.sizes, g1.goals, goal, rest)
Least(s1.s2.sizes, g1.g2.goals, goal, g1.rest) ←
    s1 ≥ s2, Least(s2.sizes, g2.goals, goal, rest).

Note that we have introduced a forth argument in the selection predicate, the set of program clauses *theory*, which is ultimatly passed to *AvgSize* in order to estimate the average size of the set of answers for a goal under the given theory. Smith's strategy for goal selection requires a slight adaptation of Section II-B, since it involves **reordering** rather than **selection**: the new version of the second clause for *Demo* is as follows:

Demo(theory, goals, answer) ←
  goals = goal.rest,
  Member(clause, theory),
  Rename(clause, goals, variant-clause),
  Parts(variant-clause, concl, conds),
  Match(goal, concl, subst),
  Apply(conds, subst. newconds),
  <u>*SmithReorder(newconds, theory, ordconds),*</u>
  Apply(rest, subst, newrest),
  Append(ordconds, newrest, newgoals),
  Demo(theory, newgoals, ans),
  Compose(subst, ans, answer)

where it is assumed that the top-level query is of the form ← *Query*($X_1, \cdots, X_n$) and the predicate symbol *Query* is defined by the clause whose head is the same as the above goal and tail is the conjunction that constitutes the query and $X_1, \cdots, X_n$ are the variables occurring in the conjunction. A reordering takes place each time a rule is applied, by invoking *SmithReorder* on the instantiated antecedent of the clause. This version of *Demo* explicitely refers to a list representation of a conjunction, in fact the operator + used in Section II-B is replaced here by *Append.*

Similar constructions can be made for clause selection, by defining procedures in *Pr* according to the criteria in Section II-A.

A remark on procedurality/declarativity is now in order. Note that no matter how the choice predicates are (reasonably) expanded, the **extension** of the *Demo* relation must remain unchanged. On the other hand, certain strategies perform better than others and the performances have to reflect in the metalevel description via *Demo*, i.e., different versions of the choice predicates must accordingly cause different **computational behaviors** in the procedural interpretation of *Demo*. This is to stress that, when control is involved, a metalevel description must refer to some underlying evaluator in order to make sense.

The definition provided in Section II-C2 for *Demo* does not make explicit the strategy for visiting the search tree of the proof: if it is run on a standard PROLOG intepreter, then it will inherit PROLOG backtracking, resulting in a depth-first strategy. If instead it relies on a breadth-first interpreter, it will inherit breadth-first search. In order for Section II-C2 to be of any significance to control issues, we have to specify a machine on which it has to run. For our purposes, we can assume that the underlying evaluator is standard PROLOG. We will have therefore a *Demo* based on backtracking, but with degrees of freedom in the choice of goals and clauses (while a PROLOG interpreter imposes a fixed left-to-right order). Note, however, that if we had wished a breadth-first search strategy, we could as well have programmed it in depth-first PROLOG, but we should have provided a different definition for *Demo* (see [12] for the representation of breadth-first computations in PROLOG). The breadth-first version of *Demo* written in PROLOG is however far less concise then Section II-B and, indeed, the advantages of Logic Programming seem not to be exploitable in writing such a program.

The next section shows two methods for allowing a logic programming interpreter to follow the control specified in *Pr.* The naive approach of simply running the programs via *Demo* is unacceptable from the point of veiw of efficiency, since it introduces an additional level of simulation into the evaluation process and ultimately, the efficiency of deduction is what control is aimed at.

D. Metalevel Control via Run-Time Reflection

As discussed in Section I, reflection principles are linking rules between object and metalevel, justified by the representation condition that must hold for the axiomatization of provability (e.g., see Section II-C1 for Horn clause logic).

We let |− stand for provability in Horn clause by writing $T \mid -^{S} G$ for *G* is provable in *T* with computed answer substitution *s*, for any set of clauses *T*, goal-list *G* (the variables in *G* are assumed to be existentially quantified), and substitution *s*. The linking rules are then formulated as follows:

$$\text{DWR} \quad \frac{T \mid -^{S}\, G}{Pr \mid -^{\{ans:=\text{“}s\text{”}\}}\ Demo(\text{“}T\text{,” “}G\text{,” } ans)}$$

Downwards Reflection

$$\text{UWR} \quad \frac{Pr \mid -^{\{ans:=\text{“}s\text{”}\}}\ Demo(\text{“}T\text{,” “}G\text{,” } ans)}{T \mid -^{S}\, G}$$

Upwards Reflection.

The reason for the terms **downwards** and **upwards** reflection is clarified by considering a backwards usage of the above inference rules. Rule DWR allows derivation of a *Demo* goal at the metalevel by a derivation at the object level, hence it allows a metalevel evaluation to proceed "down to the earth"; UWR, instead, requires a derivation at the metalevel in order to solve an object-level goal, hence it causes the object-level evaluation to proceed "up in the metalevel." Upwards reflection should occur when choices have to be exercised; for instance, when the underlying evaluator is about to execute the right-hand side of a rule, it can reflect upwards thus causing the activation of the metalevel procedure *SmithReflect* that reorders the conjuncts in the rule, then reflect downwards to continue the execution at the object level.

The implementation of evaluation with reflection has been given great attention in the framework of Functional Programming (e.g., [36], [41]). The issue—carried over to Logic Programming—remains in essence the same. Briefly, one starts with a procedure *HwDemo* (Hard-Wired Demo, as opposed to its description at metalevel that involves the predicate symbol *Demo*) that implements resolution. The procedure has two input parameters, the first for the data structure *R[T]* that represents a set of clauses *T* and the second for *R[G]*, the representation of a goal. A third parameter is of output type, used to communicate the answer in solving *G* from *T*: it is *R[s]* when there is a success with computed answer substitution *s* and the token **fail** if *G* finitely fails. In any other case, the execution of *HwDemo* does not terminate. For simplicity's sake we leave aside the problem of multiple answers to an interrogation which have to be provided on demand. The definition of *HwDemo* is recursive and is mirrored by a metalevel definition of *Demo* (say the one in Section II-C2). Now, each time the procedure is entered (either called from the system top-level or from itself), a choice point is met where there are the following options for executing *HwDemo(R[T], R[G], Ans)*:

i) Transform that call to *HwDemo(R[Pr], R[Demo("T," "G," ans)], Ans)* (upward reflect). If that call returns with *Ans* = *R[{ans:= "s"}]* then let *Ans* = *R[s]*.

ii) If *T* = *Pr* and *G* = *Demo("T," "Q," answ)*, then transform that call into *HwDemo(R[T], R[G], Ans')* (downward reflect). If *Ans'* = **fail** then let *Ans* = **fail**, else if *Ans'* = *R[s]* then let *Ans* = *R[{answ:= "s"}]*.

iii) Perform the operations of goal selection, clause selection, renaming, etc. (do not reflect).

Note that both choices (i) and (ii) involve a change of context, in so far as the theory from which the goal has to be derived changes. Therefore, efficient mechanisms for change of context must be available (essentially, a similar machinery to that for handling **packages** in conventional languages is required). Furthermore, choice (i) involves transformations *R[E]* → *R["E"]* for preparing the parameters of the new call, and *vice versa* *R["E"]* → *R[E]* for communicating the result substitution. Choice (ii) requires the opposite transformations. The distinction between *R[E]* and *R["E"]* can be made very blurred in a computer program, with a similar technique to quotation in LISP, whereby the representation of an expression *E* is marked in some way (by prefixing a QUOTE in LISP) in order to prevent substitutions to take place in *E*.

The crucial problem in the organization described so far is the choice among the options (i), (ii), and (iii) above. In particular, when the computation at metalevel is just a simulation of what would be performed by the lower level interpreter, the system proceeds in an unnecessarily inefficient way. Hence the proper timing of downward reflection is of primary concern. A first solution consists in leaving no freedom to the evaluator and having level jumps explicitly issued in the logic program. This is achieved by introducing two control constructs:

- $M\uparrow\{A_1, \cdots, A_n\}$ is treated as an atomic goal; when it is selected it causes the evaluation of $A_1, \cdots, A_n$ to take place at a higher level via *Demo("T," "$A_1, \cdots$)*, in the context of a set of clauses corresponding to the metatheory named by *M*.
- $\downarrow\{A_1, \cdots, A_n\}$ is treated as an atomic goal. When a goal of the form *Demo("T," "$\downarrow\{A_1, \cdots$)* is selected, it gets resolved as $\leftarrow A_1, \cdots A_n$, in the context of *T*.

Note that the possibility of explicitly naming the metatheory in the ↑-construct allows for chosing among different control strategies in any place of the program, thus relying on the most appropriate one for the goals at hand. Note also that a ↑-construct can be issued from a metalevel too, thus allowing an unbound tower of metalevels.

As an example, consider the representation of provability provided in Section II-C3 in terms of Smith's reordering of conjunctions and call it *Reorder*. Consider now the clause in Section II-B2 by which we demonstrated the need of dynamic reordering (for convenience the clause is repeated here): $P(x, y) \leftarrow Q(x, z), R(z, y)$. The desired behavior in resolution can be programmed, in terms of the above defined constructs as:

$$Query(x, y) \leftarrow Reorder\uparrow\{P(x, y)\}$$
$$P(x, y) \leftarrow \downarrow\{Q(x, z)\}, \downarrow\{R(z, y)\}.$$

Such a program achieves the optimal reordering of the two subgoals in the clause defining *P* according to the value to which *x* is instantiated in the query. We recall in fact from Section II-B2 that, with the assumptions on the sizes of $Q(x, z)$ and $R(z, y)$, the optimal reordering depends on the value bound to *x*.

E. Metaknowledge and Partial Evaluation

This section discusses the **embedding** of metalevel control knowledge into a logic program by means of the **partial evaluation** [23], [7] of a meta-interpreter which is given the logic program as input.

The partial evaluation of a program *P* consists in the specialization of *P* for a particular input ***I*** (which can be partially instantiated). Specialization can be applied both to meta- and object-level programs; the only difference consisting in the nature of the input ***I***.

In the sequel, for notational simplicity, in describing the functionality of higher order programs, we will not distinguish between the extension of a program *P* (i.e., the function computed by *P*) and its intension (i.e., the syntactic structure of *P*): the aspect of a program being discussed will be apparent from the context.

1) Partial Evaluation: We start with some definitions. Given a set *D* whose elements represent input/output data, and a program Prog in a programming language *L*, Prog represents the partial mapping

$$\mathbf{Prog}: D^* \rightarrow D$$

(where D^* denotes the set of all finite sequences of *D*).

Let $d_1, \cdots, d_n$ be values in *D*. A program $\text{Residual}_{\text{Prog}}$ is a residual program for Prog with respect to $\langle d_1, \cdots, d_m \rangle$ $(0 \le m \le n)$ iff

$\text{Prog}\langle d_1, \cdots, d_m, \cdots, d_n \rangle = \text{Residual}_{\text{Prog}}\langle d_{m+1}, \cdots, d_n \rangle$, for all $d_{m+1}, \cdots, d_n \in D$.

A program PE is a **partial evaluator** iff PE $\langle$Prog, $d_1, \cdots, d_m\rangle$ is a residual program for Prog with respect to $\langle d_1, \cdots, d_m\rangle$ for all L-programs Prog and values $d_1, \cdots, d_m \in D$. Note that PE is assumed to be a total function.

A program Int is an interpreter for L-programs iff Prog $\langle d_1, \cdots, d_n\rangle$ = Int $\langle$Prog, $d_1, \cdots, d_n\rangle$ for all L-programs Prog and sequences of data $d_1, \cdots, d_n$.

The compilation of a program Prog can be seen as the partial evaluation of an interpreter Int, which is given Prog as the only input PE (Int, Prog) = $\text{Comp}_{\text{Prog}}$. The resulting residual program being a function $\text{Comp}_{\text{Prog}}: D^* \to D$.

$\text{Comp}_{\text{Prog}}$ represents the compilation of Prog and, at the same time, it can be viewed as a specialized version of the interpreter which can deal with Prog only. This can be easily seen from the fact that Prog and $\text{Comp}_{\text{Prog}}$ have the same input–output behavior.

Running $\text{Comp}_{\text{Prog}}$ is more efficient than running the original program Prog, the partial evaluation process producing a less general version of Prog. In fact the symbolic manipulation (usually a source-to-source transformation) performed by a partial evaluator on a program induces some simplifications on it (e.g., propagation of input ground terms, reduction of instantiated terms, fold/unfold of procedures, elimination of conditionals that treat cases that do not hold for a particular program, etc.). We do not expand on simplification techniques, though important, since they are outside of the scope of this section.

In the particular case where partial evaluation is applied to meta-interpreters, it "wires" the control strategy defined by the meta-interpreter into logic programs provided as input I [22], [24]. These considerations provide a useful way for dealing with inference control in logic programming, since it is possible to specify different meta-interpreters, each of them embodying a different deduction strategy, hence, the compilation of a particular strategy into any object program can be easily achieved by means of the application of a general-purpose partial evaluator to logic (meta) interpreters.

Although the ultimate goal of the application of partial evaluation to inference control of logic programs is to obtain more efficient object-level programs that inherit the functionality of the meta-interpreter, while evaluated by a standard interpreter (e.g., what we called *HwDemo* in the previous section), it is, however, not always possible to eliminate calls to metapredicates in the program $\text{Comp}_{\text{Prog}}$. So the application of partial evaluation to inference control of logic programs can be interpreted as a way to implement upward reflection at compile time, while the run-time application of upward reflection has been described in Section II-D.

2) An Example: Coroutining: We present as an example (along the lines of [24]) a meta-interpreter that realizes a coroutining evaluation of logical queries. Cooperative evaluations of this kind are useful, for instance, in case of programs that specify a generate-and-test process where the generated objects present an incrementally built composed structure, which can be tested before it is completely generated.

The meta-interpreter will then be partially evaluated on a program for "samefringe," i.e., a program that checks the equality of the leaves of two binary trees.

The definition of the coroutining meta-interpreter is the following:

Coroutine(goal1, goal2, __) ←
Empty(goal1), Empty(goal2)
Coroutine(goal1, goal2, shared-vars) ←
Produce(goal1, rest-goal1, shared-vars, rest-vars),
Produce(goal2, rest-goal2, rest-vars, new-vars),
Coroutine(rest-goal1, rest-goal2, new-vars),

where *Produce* (*goal, rest-goal, vars, rest-vars*) is a meta-interpreter that evaluates *goal* till one of the variables *vars* gets instantiated: in which case it binds *rest-goal* and *rest-vars* to the remaining *goals* and *vars*, and stops the evaluation.

The definition of *Produce* is the following:

Produce(goal, rest-goal, vars, vars) ←
Empty(goal), Empty(rest-goal)
Produce((goal . goals), newgoals, vars, newvars) ←
SelectClause(goal, conditions),
NewInstance(vars, newvars),
Append(conditions, goals, newgoals).
Produce((goal . goals), newgoals, vars, newvars) ←
SelectClause(goal, conditions),
NoNewInstance(vars),
Append(conditions, goals, new),
Produce(new, newgoals, vars, newvars)

where *SelectClause* has the same behavior of the sequence of calls as *Select, Member, Rename, Parts, Match,* and *Apply* in the clause defining *Demo* in Section II-C.

Given the latter definition of *Produce*, the application of *Coroutine* alternates the evaluation of *goal1* and *goal2* till the empty goals are reached.

The definition of the leaves generator is the following:

Leaves (Leaf(x), (x . q) − q) ←
Leaves (Tree(left-tree, right-tree), leaves − q) ←
Leaves (left-tree, leaves − lr),
Leaves (right-tree, lr − q)

where a tree is either a leaf (e.g., *Leaf(x)*) or a left tree connected to a right tree (e.g., *Tree(left-tree, right-tree)*) and we represent lists as difference lists by means of the infix operation "−."

Now, if we want to know if the two trees have the same fringe we can either run a conjunctive goal like

← Leaves (Tree1, leaves − NIL), Leaves (Tree2, leaves − NIL)

or evaluate the meta-interpereter on the two latter subgoals as in

← Coroutine ((Leaves (Tree1, lvs − Nil)),
(Leaves (Tree2, lvs − Nil), (lvs)).

Running the former goal results in a complete generation of the leaves of Tree1 before checking the corresponding leaves of Tree2, while running the second suffers the overhead caused by the extra level of simulation of the computation.

Specializing the program of *Coroutine* for a goal in which *tree1* and *tree2* are left unspecified, leads to the following clauses for *Produce*:

```
Produce(Nil, Nil, vars, vars) ←
Produce((Leaves(Leaf(x), (x . q) − q). goals),
                    newgoals, (lvs), newlvs) ←
     Var (lvs),
     Produce(goals, newgoals, (lvs), newlvs)
Produce((Leaves(Leaf(x), (x . q) − q) . goals),
                    goals, (lvs), (q)) ←
     NonVar (lvs)
Produce((Leaves(Tree(lt, rt), leaves − q) . goals),
                    newgoals, (lvs), newlvs) ←
Produce((Leaves (lt, leaves − lr), Leaves (rt, lr − q) . goals),
                                   newgoals,
                                       (lvs),
                                   newlvs)
```

where the evaluation of *Produce* on the program *Leaves* is reduced almost only to a sequence of syntactic unifications (see also [2] for a coroutining solution to the samefringe problem, based on program transformations).

It is possible to think of several other useful applications for partial evaluation to Logic Programming. For instance, Shapiro presents in [40] some programs that inherit specific functionalities (e.g., termination control, deadlock handling, etc.) by means of the partial evaluation of different meta-interpreters.

Several effective applications have been presented in the literature, among them: a PROLOG compiler obtained by the partial evaluation of a LISP interpreter for PROLOG [31]; query optimization in a PROLOG interface for databases [44]; the efficient control of rule activation in expert systems [43].

III. Self-Modification and Self-Evaluation

In this section it is argued that metaknowledge should be used by a system also in order to a) profit from previous experience in the accomplishment of an actual task (this is a form of **self-modification**) and b) in order to foresee its computational behavior in the accomplishment of a potential task (**self-evaluation**).

After having discussed the two issues in Subsection III-A, we present in a fairly detailed way a self-modification scheme for Logic Programming, along with its overall metalevel description (this is the content of Subsection III-B).

A. Issues in Self-Modification and Self-Evaluation

Conventional knowledge-based systems do not improve their performance by repeating the same sequence of deduction steps on similar—or even identical—problems. Usually, it is the user who discovers general facts, either positive or negative, out of series of actual deductions, either successful or failed. A general fact can be used in subsequent deductions as a **lemma**, thus avoiding going through the sequence of steps which led to its generation. The knowledge embodied in the lemma is not necessary since it is entailed by the object-level knowledge, nonetheless, the process of checking that a fact is an instance of a lemma is generally much simpler than proving the fact itself. The **incorporation** of a lemma into a knowledge-based system is accomplished either by adding it to the object-level knowledge, by adding it as metaknowledge on control, or—not very actractively—by coding it explicitely in the implementation language of the inference engine.

The process of inducing general lemmas out of deduction instances and embodying them into the knowledge base, so as to be later used by the inference engine, can be carried on by the system **automatically**, with no intervention from the user. Subsection III-B demonstrates how self-modification can be achieved in terms of automatic **lemma generation** and **memoization.**

The capability of self-evaluation is as necessary as self-modification in order for a system to exhibit intelligent behavior. With such a feature, a user can know in advance an **estimate** of the computing time (computing resources, in general) consumed by a system in order to perform a given task. In situations where the response time is critical, the estimate could lead the user to decide not to rely on the system if its foreseen response time is unacceptable.

Self-evaluation can be achieved by letting the system reason on a metalevel **self-description.** The description involves the deductive tools of the inference engine, along with the computational complexities of the corresponding deductive processes. No matter how accurate is the self-description, the required estimates cannot be reliable in general, i.e., if the estimate has to be an **upper bound** of the actual computing complexity in solving a given problem, no total procedure can be devised, since such a procedure could be used to solve the halting problem for Turing machines, provided that the formal deduction system under consideration is Turing complete, i.e., it allows the expression of all general recursive functions. This theoretical limitation, however, does not exclude self-evaluation if the notion of estimate is interpreted in a less strict sense, or if the class of allowed rule sets is constrained.

Methods for self-evaluation have to rely on a solid theory of **complexity inference** for logical languages, which, at present, has only started to be taken into consideration in the literature (see, for instance, [39]). Some insights into the problem can be gained by considering the work in the related field of inference control, where the complexity of deduction has to be estimated in order to be optimized (see, for instance, [42], [46]). Note, in fact, that the methods sketched in Section II for driving the choices involved in (both OR and AND) nondeterminism are based in the evaluation of the complexity of deduction that results in following each of the open paths, and indeed provide procedures for estimating these complexity measures for rule-sets of a particular form. We are currently exploring possible generalizations of the class of rule-sets on which the above procedures would be applicable.

B. Self-Modification by Lemma Generation and Memoization

The method is very simple and intuitive. A **memory structure** $Memo_T$ is associated with any object level theory *T*. The memory structure is partitioned into two substructures, $Memo_T = PosMemo_T \cup NegMemo_T$, one for **positive** lemmas, the other for **negative** ones. When a goal *G* is proved (disproved), it can happen that a **more general** version $G^{\sim}$ can be induced from the steps performed in the proof (disproof) of *G*. The more general version $G^{\sim}$ is then embodied into $Memo_T$ according to its sign: it is incorporated into $PosMemo_T$ if *G* succeeded, into $NegMemo_T$ if *G* failed.

In the sequel, we say that a goal *G* has been **derived** from a set of clauses *T* meaning that either *G* succeeded in *T* or *G* finitely failed in *T*. An atom *A* is a more general version of B iff there exists a substitution s such that $As = B$. If A is a more general version of B it is also said to **subsume** B.

1) Embodying: A goal $G^{\sim}$ is recorded in $Memo_T$ by an operation that is not straight addition. Given an atom A and a set of atoms S, the operation of embodying A into S is defined as follows:

> If A is subsumed by some atom $A^{\sim}$ in S then A is not added to S; otherwise, A is added to S after the deletion of all atoms B in S subsumed by A.

Resolution with lemma generation and straight memoization is now described.

2) Resolution + Lemma Generation + Straight Memoization: Each time an atom A is going to be solved, first $Memo_T$ is inspected by nondeterministically looking-up both $PosMemo_T$ and $NegMemo_T$.

i) If an atom B is found in $PosMemo_T$ such that A and B **unify** with mgu σ then the invocation of A immediately stops with success.

ii) Similarly, if an atom $A^{\sim}$ is found in $NegMemo_T$ such that $A^{\sim}$ **subsumes** A, then the resolution of A immediately stops with failure.

iii) If none of the two cases above is given, the resolution goes on as described in Section II-C by clause selection, unification, etc. When and if resolution of A terminates, a more general version $A^{\sim}$ of A is obtained by **lemma generation** (described later) as a side-effect of resolution, and $A^{\sim}$ is embodied into $Memo_T$ according to its sign as described in Section III-B1.

In the procedure informally described above, it is intended that the inspection of $Memo_T$ (points (i) and (ii)) is performed each time an atomic goal has to be solved, not only at the top level. On the other hand, the recording of the lemma obtained as side-effect of the resolution (point (iii)) may be done either for all the intermediate subgoals generated by resolution, or just for the atoms in the top-level goal.

The consultation of a memory database storing both positive and negative facts before resolution has been introduced in [13] in the context of **query the user**. In that framework, the Memo structure records facts explicitely asserted by the user on demand.

The technique of **memoization** has been first advocated by Michie in [34] and exploited in functional programming. In this simpler computational paradigm, the technique amounts to recording during the evaluation the completed function calls, along with their arguments and the computed result. Memoization, applied for instance to recursive definitions like the one defining Fibonacci's numbers (i.e., $fib(0) = 0$, $fib(1) = 1$, $fib(n + 2) = fib(n + 1) + fib(n)$), reduces an exponential computation to a polynomial one.

A less general organization than that in Section III-B2 has been proposed in [19], [20]. There, the Memo structure records negative facts only and no attempt is made to generalize failures, i.e., the failed facts themselves are stored in $Memo_T$.

Note that if the Memo structure is not reset after a top-level invocation, the generated lemma can be used in subsequent invocations in order to speed up the resolution process. Obviously, it is not advantageous to embody **all** the lemmas generated during a top-level query, hence we will elaborate on the scheme proposed in Section III-B2 in order to achieve a **selective** memoization strategy. Selective memoization is discussed in Section III-B5.

In order for Section III-B2 to be complete, we need to discuss the process of lemma generation.

3) Generation of Negative Lemmas Upon Failure: A method for the induction of negative lemmas out of failure instances has been worked out in [12]. The method has been developed in the framework of intelligent backtracking and constitutes a complementary approach to those in [11] and [16]. Indeed, intelligent backtracking can be seen as self-modification restricted to one top-level invocation, i.e., with intelligent backtracking, the system modifies itself by learning negative lessons from failures occurred in the solution of subgoals generated from the top-level one, and exploits that knowledge to improve its performance in the solution of subsequently generated subgoals.

Since the negative lessons are represented as atomic (negative) assertions, rather than as minimal inconsistent deduction trees, as in [11], our method is better suited to the implementation of a general self-modification scheme.

Given the scope of the paper, the details of the generalization procedures are not provided, rather, the main ideas underlying them are shown by means of an example.

a) Example: Consider the following logic program and goal:

Parent(John, Tom) ← *Parent(Jim, Dave)* ←
Parent(Tom, Mary) ← *Parent(Jim, Mike)* ←

GrandParent(x, y) ←
  Parent(x, z), Parent(z, y)
Goal: ← *GrandParent(Jim, Mary).*

Note that the goal in the above example fails, since Jim has two sons (Dave and Mike), none of which has children. Indeed, from the failure of the goal, the following general assertion can be induced:

- *not GrandParent(Jim, x)* standing for "Jim has no grandchildren."

The synthesis of the above general assertion closely follows the resolution process: the goal ← *GrandParent(Jim, Mary)* unifies with the head of the rule defining GrandParent, with mgu $u = \{x := Jim, y := Mary\}$. The goal is reduced to

← *Parent(Jim, z), Parent(z, Mary).*

The first atom in the conjunction, *Parent(Jim, z)*, admits two solutions, i.e., $s_1 = \{z := Dave\}$ and $s_2 = \{z := Mike\}$. The second conjunct is instantiated by s_1 and s_2, thus obtaining ← *Parent(Dave, Mary)* and ← *Parent(Mike, Mary)*, which have to be tried in turn. Both subgoals finitely fail, since they do not unify with any of the available rules for *Parent*. In the analysis of the failure of unification between *Parent(Dave, Mary)* and the heads of the rules for *Parent*, it turns out that the "culprit" argument in the goal is *Dave*, since it mismatches with all the first arguments of the available heads. Hence, from the failure of the subgoal ← *Parent(Dave, Mary)*, the following general assertion is generated:

- *not Parent(Dave, w)* i.e., Dave has no children.

Similarly, from the failure of ← *Parent(Mike, Mary)*, the following is obtained:

- *not Parent(Mike, w')* i.e., Mike has no children.

Since s_1 and s_2 are the only answers to the goal ← *Parent(Jim, z)*, the following conjunction cannot be solved:

$\leftarrow$ *Parent(Jim, z), Parent(z, w")*, which implies that the following version of the head of the rule defining *GrandParent* cannot be solved:

GrandParent(Jim, w"), which, given that the tried rule is the sole available for *GrandParent*, implies that the following goal is bound to fail:

$\leftarrow$ *GrandParent(Jim, w")*.

Obviously, any instance of the negative assertion is bound to fail as well, hence the reason for the subsumption condition in the inspection of $NegMemo_T$ described in Section III-B2.

4) Generation of Positive Lemmas Upon Success: The procedure for generating positive lemmas in case of success is in principle similar to (and simpler than) the one for negative lemmas: base cases of success are treated by analyzing the behavior of unification, while general cases involving rule chaining are treated by recursively generating lemmas corresponding to the antecedents of the chained rule and composing them according to the shape of the rule. Again, the method is sketched with the aid of an example.

a) Example: Suppose we add the rules in Subsection III-B3a the following fact:

Parent(God, x) $\leftarrow$ i.e., God is everybody's parent.

The goal $\leftarrow$ *GrandParent(God, Jim)* succeeds, since the following conjunction does: $\leftarrow$ *Parent(God, God), Parent(God, Jim)*, which is an instance of the antecedent of the rule defining *GrandParent*, rewritten with the used mgus.

In solving both the two conjuncts *Parent(God, z)* and *Parent(z, Jim)*, the unit clause *Parent(God, x)* $\leftarrow$ has been used, hence the following conjunction would have been solved as well with answer substitution $s = \{z := God\}$:

$\leftarrow$ *Parent(God, z), Parent(z, y)*.

By considering the shape of the rule defining *GrandParent* it follows that the goal $\leftarrow$ *GrandParent(God, w)* succeeds with the empty substitution, hence the assertion *GrandParent(God, w)* (i.e., God is everybody's grandparent) is a logical consequence of the clauses in the example. Note that, again, the induced assertion is more general than the original goal. If an atom *B* unifies the induced assertion with mgu *u*, then the substitution *u* is a correct answer substitution for the goal $\leftarrow$ *B*. Hence the reason for the unifiability condition in the inspection of *PosMemo* in Section III-B2.

In general, both for the positive and for the negative case, it is not always possible to obtain a lemma strictly more general than the original goal (instantiated with the computed answer substitution, in the positive case). For the sake of uniformity, we let the resolution process return a lemma in any case, with the provision that the lemma $A^{\sim}$ for the goal $\leftarrow$ *A* can sometime be *A* itself in the negative case, or *As* in the positive case, where *s* is the computed answer substitution for $\leftarrow$ *A*.

The overall behavior of resolution with (both positive and negative) lemma generation can be described via a metalevel predicate

Solve&Generalize(theor, goal, answer, lemma, compl)

whose arguments have the following interpreation:

- *theor* is the set of clauses under which *goal* has to be proved;
- *goal* is an atom;
- *answer* is either a term *Yes(subst)*, where *subst* is a computed answer substitution for *goal* in *theor*, or the constant *No*, in which case *goal* finitely fails in *theor*.
- *lemma* is an atom more general than *goal* if *answer* = *No*, otherwise it is more general than *goal* instantiated with *subst*, if *answer* = *Yes(subst)*;
- *compl* is some measure of the complexity of deriving *goal* (involved in selective-memoization).

It is assumed that *goal* is an atom that cannot be derived upon inspection of the Memo structure, hence *Solve&Generalize* has to describe the usual process of clause selection, unification, etc.

We do not expand the definition of *Solve&Generalize*, rather we draw attention to the above input/output specification, since *Solve&Generalize* is involved in the overall metalevel description of the proposed self-modification scheme provided later in Section III-B6.

5) Selective Memoization: The self-modification scheme described in Section III-B2 is essentially meant to introduce redundancy in the overall knowledge base of the deduction system by storing **explicitely** at metalevel assertions that can be derived from the **intensional** information in the object-level knowledge base. Data retrieval, as compared to resolution, is simpler and faster; however, as the size of the Memo database increases, retrieval tends to loose convenience with respect to resolution, apart from the space consumption caused by straight memoization.

These problems can be overcome by adopting a selective memoization, whereby not all the generated lemmas are recorded, but only those whose derivation has been nontrivial. The nontriviality of a fact *A*—with respect to the rules from which it has to be derived—can be measured by the quotient $NT_A = C_A/S_A$, where C_A is the time complexity in deriving *A* and S_A is the structural complexity of *A*. The parameter NT_A can be interpreted as the *number of times A has to be visited in order to do the same computational effort as to derive it*, provided that S_A is measured as the complexity of visiting *A*. Obviously, the larger is NT_A, the more convenient is to retrieve *A* rather than to deduce it, while if $NT_A \leq 1$, it takes less or the same time to deduce *A* than to retrieve it, hence storing *A* is a waste of space and time.

The nontriviality quotient of an assertion *A* alone is not a sufficient information for driving a memoization strategy. The parameter that expresses the size increment (or decrement) Δ_A of the Memo database due to the embodying of *A* should also be taken into account.

If $NT_A \leq 1$, *A* must not be recorded; if $NT_A > 1$ and $\Delta_A \leq 0$, recording *A* is convenient; in the other case, i.e., $NT_A > 1$ and $\Delta_A > 0$, the choice is open and depends on the memoization strategy, which should weight NT_A against Δ_A and decide according to the space available in Memo. Any memoization strategy can be explored by suitably defining the metalevel predicate symbol *Embody*, by which the procedure of Section III-B1—improved by selective memoization—is described at metalevel. *Embody* should be defined as:

Embody(S(facts,size),A(atom,compl),NewS(newfacts, newsize))

with the following interpretation for its arguments:

- *facts* is the set of facts in which *atom* has to be embodied (*facts* is either PosMemo_T or NegMemo_T of some object-level theory *T*);
- *size* is the size of the set *facts*;
- *atom* is the atom to be embodied in *facts*;
- *compl* is the complexity of deriving *atom*;
- *newfacts* is the new set of facts after *atom* has been embodied in *facts* (possibly *facts* = *newfacts* if no incorporation has occurred);
- *newsize* is the size of *newfacts*.

As for the metalevel predicate symbol *Solve&Generalize*, we do not expand the definition of *Embody*, rather we keep in mind its input/output specification in the metalevel description of the self-modification scheme provided below.

6) A Metalevel Description of Self-Modification: The description is provided by means of a three-place predicate *Demo&Memo(theory, goals, answer)* where *theory* is the set of clauses under which *goals* has to be derived, *goals* is a list of atoms and *answer* is either a positive or a negative answer, as for *Solve&Generalize* introduced in Section III-B4. For the sake of clarity, we start with a simple preliminary version which restricts lemma generation and memoization to the subgoals in the top-level query, then we discuss the extensions to be performed in order to accommodate the general scheme as described in Sections III-B1–III-B5.

In order to provide a pure logical definition of *Demo&Memo*, two extra arguments would be required for the formalization of the transitions that take place in the Memo structure; namely, an argument **currmemo**, representing the current Memo structure, and **finmemo**, that is the final Memo structure after the derivation of the goal (if any). For simplicity's sake, we avoid this by allowing assertion and retraction of clauses in the metalevel knowledge base. Since side-effects are allowed, the order in which actions are performed is part of the specification, hence we will assume a particular deterministic resolution procedure; namely, the one adopted in standard PROLOG systems, i.e., depth-first search rule, left-to-right goal selection, and fixed try-order, imposed by the sequence of clauses written in the text of the program.

The Memo structure is represented by two ground unit clauses of the metatheory:

- *PosMemo("P", ps)* ←, where *"P"* is the metalevel coding of a set of atoms and *ps* is a positive integer representing the size of *"P,"*
- *NegMemo("N", ns)* ←, with similar interpretation for the arguments *"N"* and *ns*.

1) *Demo&Memo(theory, goals, answer)* ← *Empty(goals)*
2) *Demo&Memo(theory, goals, answer)* ←
   Select(goals, goal, rest),
   Demo&MemoAtom(theory, goal, ans),
   Demo&MemoRest(ans, theory, rest, answer)
3) *Demo&MemoAtom(theory,goal,No)*←
   NegSubsumed(goal)
4) *Demo&MemoAtom(theory, goal, Yes(subst))* ←
   not NegSubsumed(goal),
   PosRetrieved(goal, subst)
5) *Demo&MemoAtom(theory, goal, answer)* ←
   not negSubsumed(goal),
   not PosRetrieved(goal, subst),
   Solve&Generalize(theory, goal, answer, lemma, compl),
   MemoLemma(answer, lemma, goal, compl)
6) *Demo&MemoRest(No, theory, rest, No)* ←
7) *Demo&MemoRest(Yes(subst), theory, rest, answer)* ←
   Apply(rest, subst, newrest),
   Demo&Memo(theory, newrest, answ),
   ComposeAnswer(subst, answ, answer)
8) *NegSubsumed(goal)* ← *NegMemo(atoms, size),*
   Member(atoms, atom),
   Subsumes(atom, goal)
9) *PosRetrieved(goal, subst)* ← *PosMemo(atoms, size),*
   Member(atoms, atom),
   Unifies(atom, goal, subst)
10) *MemoLemma(No, lemma, compl)* ←
   Retract(NegMemo(atoms, size)),
   Embody(S(atoms,size),A(lemma,compl),
   NewS(toms1, size1)),
   Assert(NegMemo(atoms1, size1))
11) *MemoLemma(Yes(subst), lemma, compl)* ←
   Retract(PosMemo(atoms, size)),
   Embody(S(atoms,size),A(lemma,compl),
   NewS(atoms1, size1)),
   Assert(PosMemo(atoms1, size1))

A few remarks on the program are in order. Clauses (4) and (5) in the definition of *Demo&MemoAtom* contain negated atoms in their antecedents so as to have disjoint conditions among the three rules for the predicate. A better efficiency can be gained by using PROLOG control facilities so as to avoid multiple activations of the same predicate-instance.

The necessary renaming of variables of a *clause* in *theory* or of an *atom* in Memo is handled by *Rename(atom, variantatom)*. A pure logical definition of *Rename* requires an extra-argument *vars*, representing the set of variables that should not occur in the renamed *variantatom* (i.e., *vars* is the set of variables occurring in the goal-list from where the current goal has been selected). For simplicity's sake it is assumed that *Rename* always introduces "fresh" variables by recording an index to be suffixed to the variables in *atom*, to be incremented each time *Rename* is activated.

The predicate *ComposeAnswer(subst, answ, answer)* performs substitution composition, provided that *answ* is not reporting failure.

The predicate *Retract(atom)* deletes from the program database a clause whose head unifies with *atom*, hence, in clauses (10) and (11), it has the effect of both retrieving the representations of NegMemo and PosMemo, respectively, and of deleting them. Note by the way that the clause referred to by *Retract*—in this program—is always univocally determined.

The program, as it stands, performs the memoization of the atoms in the top-level query only. If an exhaustive memoization of all the intermediate lemmas generated during the resolution is wanted, it is sufficient to define *Solve&Generalize* as a predicate mutually recursive with *Demo&Memo*. In that case, however, *Demo&Memo* should have extra-arguments for lemma generation and for complexity measures, as discussed in Section III-B4 for *Solve&Generalize*. In the above definition of provability, the handling of OR-nondeterminism has not been explicit, rather it relies on backtracking in the underlying interpreter. If extra arguments are added to *Demo&Memo* as suggested above, then some additional machinery is needed for recording lemmas and complexities referring to failed branches of the computation, or alternatively, if side-effects are to be avoided, the definition of provability should be complicated so as to make explicit the handling of OR-nondeterminism.

Conclusions

In the paper we have extensively discussed the issues of metalevel knowledge representation and use. Different proposed paradigms for combining reasoning and metareasoning have been considered and some principles upon which the combination should be based have been abstracted.

Among the various applications of metalevel reasoning reviewed in the paper, we have focussed on inference control. The problem has been addressed with the aid of two paradigmatic examples, chosen from practical AI applications and expressed in Logic Programming. Two corresponding control strategies have been presented and their metalevel description has been discussed. We have then shown how control knowledge—declaratively expressed as metalevel knowledge on control—can be effectively used to drive the deduction process, either at run-time, via reflection, or at compile-time, via partial evaluation.

Finally, we have considered self-modification and self-evaluation: they are desirable features AI systems should possess and constitute interesting (somewhat unexplored) applications of metalevel processing. A self-modification scheme for resolution-based systems has been proposed and its metalevel description has been outlined.

References

[1] M. Aiello and R. W. Weyhrauch, "Checking proofs in the metamathematics of first order logic," in *Proc. of the 4th IJCAI*, pp. 1–8, 1975.

[2] L. Aiello, G. Attardi, and G. Prini, "Towards a more declarative programming style," in *Formal Description of Programming Concepts*. Amsterdam, The Netherlands: North-Holland, 1978, pp. 121–137.

[3] L. Aiello, "Automatic generation of semantic attachements in FOL," in *Proc. of the 1st AAAI Conf. on AI*, pp. 90–92, 1980.

[4] L. Aiello and R. W. Weyhrauch, "Using meta-theoretic reasoning to do algebra," in *LNCS 87*. New York: Springer Verlag, 1980, pp. 1–13.

[5] L. Aiello and G. Levi, "The uses of metaknowledge in AI systems," in *Proc. ECAI*, pp. 705–717, 1984.

[6] G. Attardi and M. Simi, "Metalanguage and reasoning across viewpoints," in *Proc. ECAI*, pp. 315–324, 1984.

[7] L. Beckman, "A partial evaluator and its use as a programming tool," *Art. Intell.*, vol. 7, no. 4, pp. 319–357, 1976.

[8] K. A. Bowen and R. A. Kowalski, "Amalgamating language and metalanguage in Logic Programming," in *Logic Programming*, K. Clark and S. Tarnlund, Eds. New York: Academic Press, 1982, pp. 153–173.

[9] R. S. Boyer and J. S. Moore, "Metafunctions: Proving them correct and using them efficiently as new proof procedures," SRI Tech. Rep. CSL-108, 1979.

[10] M. Bruynooghe, "An interpreter for predicate logic programs: Part I," Tech. Rep. CW 10, Applied Mathematics and Programming Division, Katholieke Universiteit, Leuven, Belgium, 1976.

[11] M. Bruynooghe and L. M. Pereira, "Deduction revision by intelligent backtracking," in *Implementations of PROLOG*, J. A. Campbell, Ed., Ellis Horwood, 1984, pp. 194–215.

[12] C. Cecchi, "Learning general lessons from failure," forthcoming.

[13] K. L. Clark and F. G. McCabe, "PROLOG: A language for implementing Expert Systems," in *Machine Intelligence 10*, J. E. Hayes, D. Michie, and Y-H Pao, Eds., Ellis Horwood, 1982, pp. 455–470.

[14] K. L. Clark and F. G. McCabe, *MICRO-PROLOG: Programming in Logic*. Englewood Cliffs, NJ: Prentice-Hall, 1984.

[15] A. Colmerauer, H. Kanoui, R. Pasero, and P. Roussel, "Une systeme de communication homme-machine en Français," Res. Rep., Artificial Intel. Group, Univ. of Aix-Marseille, France, 1973.

[16] P. T. Cox, "Finding backtrack points for intelligent backtracking," in *Implementations of PROLOG*, J. A. Campbell, Ed., Ellis Horwood, 1984, pp. 216–233.

[17] R. Davis, "Generalized procedure calling and content-directed invocation," in *SIGPLAN Notices*, vol. 12, no. 8, pp. 45–54, 1977.

[18] R. Davis and B. G. Buchanan, "Meta-level knowledge: Overview and applications," in *Proc. of 5th IJCAI*, pp. 920–927, 1977.

[19] M. Dincbas and J. Le Pape, "Metacontrol of logic programs in METALOG," in *Proc. Int. Conf. on Fifth Generation Computer Systems*, pp. 361–370, 1984.

[20] M. Dincbas, "The METALOG problem-solving system: An informal presentation," in *Proc. IFIP*, pp. 80–91, 1980.

[21] R. Ehrenfeucht and M. Rabin, "There is no perfect proof procedure," unpublished paper, 1972.

[22] P. Emanuelson and A. Haraldsson, "On compiling embedded languages in LISP," in *Proc. ACM LISP Conf.*, pp. 208–215, 1980.

[23] Y. Futamura, "Partial evaluation of computation process- an approach to compiler-compiler," in *Syst. Comput. Contr.*, vol. 2, no. 5, pp. 721–728, 1971.

[24] J. Gallagher, "Transforming logic programs by specialising interpreters," to appear in *Proc. of ECAI '86*.

[25] M. R. Genesereth, "An overview of meta-level architecture," in *Proc. of the AAAI Conf.*, pp. 119–124, 1983.

[26] K. Goedel, "Üeber formal unentscheidbare Saetze der Principia matematica und verwandter Systeme I," in *Monatschefte für Mathematik und Physic*, no. 38, pp. 173–198, 1931.

[27] M. Gordon, R. Milner, and C. Wadsworth, "Edinburgh LCF: A mechanized logic of computation," in *LNCS 78*. New York: Springer-Verlag, 1979.

[28] P. J. Hayes, "Computation and deduction," in *Proc. Symp. Math. Foundations of Computer Science*. Prague, Czechoslovakia: Czechoslovakian Acad. Sci., 1973, pp. 105–117.

[29] P. J. Hayes, "In defense of logic," in *Proc. 5th IJCAI*, pp. 559–565, 1977.

[30] F. Hayes-Roth *et alt.*, Eds. *Building Expert Systems*. Reading, MA: Addison-Wesley, 1983.

[31] K. Kahan, "A partial evaluator of Lisp written in Prolog," UPMAIL Tech. Rep. 17, Univ. of Uppsala, Sweden, 1985.

[32] R. A. Kowalski, "Predicate logic as programming language," in *Proc. IFIP*, pp. 569–574, 1974.

[33] —, *Logic for Problem Solving*. Amsterdam and New York: North Holland, 1979.

[34] D. Michie, "Memo functions and machine learning," *Nature*, vol. 218, pp. 19–22, 1968.

[35] A. Porto, "Two-level PROLOG," in *Proc. Int. Conf. on Fifth Generation Computer Systems*, pp. 356–360, 1984.

[36] J. des Rivières and B. C. Smith, "The implementation of procedurally reflective languages," in *Proc. ACM Symp. on LISP and Functional Programming*, pp. 331–347, 1984.

[37] G. M. Roberts, "An implementation of PROLOG," M.Sc. thesis, Univ. of Waterloo, Waterloo, Ont., Canada, 1977.

[38] J. A. Robinson, "A machine oriented logic based on the resolution principle," in *ACM J.*, vol. 12, pp. 23–41, 1965.

[39] E. Y. Shapiro, "Alternation and the computational complexity of logic programs," *J. Logic Programming*, no. 1, pp. 19–33, 1984.

[40] —, "Programming in Concurrent Prolog," Lecture Notes,

Advanced Course in AI, Vignieu, France, 1985.

[41] B. C. Smith, "Reflection and semantics in LISP," in *Proc. 11th ACM Conf. on Principles of Programming Languages*, pp. 23–35, 1984.

[42] D. E. Smith and M. R. Genesereth, "Ordering conjunctive queries," *Artificial Intell.*, no. 26, pp. 171–215, 1985.

[43] A. Takeuchi and K. Furukawa, "Partial evaluation of PROLOG programs and its application to meta-programming," Tech. Rep. ICOT Research Center, Tokyo, 1985.

[44] R. Venken, "A PROLOG meta-interpreter for partial evaluation and its application to source-to-source transformations and query-optimization," in *Proc. ECAI*, pp. 91–100, 1984.

[45] D. H. D. Warren, "Implementing PROLOG—Compiling predicate logic programs," Res. Rep. 39, 40, Dept. of Artificial Intelligence, Univ. of Edinburgh, Edinburgh, U.K., 1977.

[46] D. H. D. Warren, "Efficient processing of interactive relational database queries expressed in logic," in *Proc. VLDB Conf.*, pp. 272–281, 1981.

[47] R. W. Weyhrauch, "Prolegomena to a theory of mechanized formal reasoning," *AI J.*, no. 13, pp. 133–170, 1980.

Reprinted from *Proceedings of the IEEE,* Volume 74, Number 10, October 1986, pages 1383-1398.

Procedural Knowledge

MICHAEL P. GEORGEFF AND AMY L. LANSKY

Much of commonsense knowledge about the real world is in the form of ***procedures*** *or* ***sequences*** *of actions for achieving particular goals. In this paper, a formalism is presented for representing such knowledge using the notion of* ***process****. A declarative semantics for the representation is given, which allows a user to state* ***facts*** *about the effects of doing things in the problem domain of interest. An operational semantics is also provided, which shows* ***how*** *this knowledge can be used to achieve particular goals or to form intentions regarding their achievement. Given both semantics, our formalism additionally serves as an executable specification language suitable for constructing complex systems. A system based on this formalism is described, and examples involving control of an autonomous robot and fault diagnosis for NASA's space shuttle are provided.*

I. Introduction

There is an increasing demand for systems or artificial agents that can interact with a dynamic environment to achieve particular goals. Common applications of this type include robotic functions, construction and assembly tasks, navigation and exploration by autonomous vehicles, control and monitoring of systems, and servicing and maintenance of equipment. Agents capable of operating effectively in these kinds of domains must be able to reason about their tasks and determine how to act in given situations—i.e., they must be capable of *practical reasoning* [6]. To build such systems we need to be able to represent knowledge about the effects of actions and how these actions can be combined to achieve specific goals.

Within artificial intelligence (AI), there have been two approaches to this problem, with a somewhat poor connection between them. In the first category, there is work on theories of action, i.e., on what constitutes an action *per se* [1], [12], [17]. This research has focused mainly on problems in natural-language understanding concerned with the meaning of action sentences. Some attempts have also been made to indicate how these theories could be used for general reasoning about actions [2], [17]. Second, there is work on planning—that is, the problem of constructing a plan by searching for a sequence of actions that will yield a particular goal [2], [7], [19], [21], [23], [25], [27], [28].

Surprisingly, almost no work has been done in AI concerning the execution of *preformed plans* or *procedures*—yet this is the almost universal way in which humans go about their day-to-day tasks, and probably the only way other creatures do so. Actually searching the space of possible future courses of action, which is the basis of most AI planning systems, is relatively rare.

For example, consider the task of driving to work each day. For most of us, our plan of action has been worked out in advance. Once we establish a goal for ourselves to leave home and get to work, we follow some internal procedure or pattern of getting to the car, getting in, driving a certain route, searching the parking lot in a particular fashion once we get there, parking, and then walking to the office. Rarely do we ever derive this plan from first principles. In fact, we often seem to perform these actions without even thinking! This pattern applies to many of the tasks we perform everyday.

Of course, there are often situations in which our normal procedures or plans must be modified or reconsidered. Rather than derive completely new plans, we usually adjust to situations by operating in a piecemeal fashion; we keep an overall plan in mind and elaborate it as we proceed and acquire more knowledge of the world. For example, if we run into an obstacle on the road to work, a new route may need to be found for part of the journey. Although the top level plan of action may remain the same, different means of realizing pieces of the plan would be used, depending on the particular situation. For instance, one might be able to avoid the obstacle simply by driving around it. On the other hand, if the obstacle were large, one may have to use more complex avoidance procedures that involved turning onto side streets, continuing in the same general direction, and the like.

This strategy of operation might be called *partial hierarchical* planning. The idea is simply to intermix the formation of plans and their execution; i.e., form a partial overall plan, figure out near-term means, execute them, further

Manuscript received April 26, 1985; revised May 14, 1986. This research has been made possible in part by a gift from the System Development Foundation, by the Office of Naval Research under Contracts N00014-80-C-0296 and N00014-85-C-0251, and by the National Aeronautics and Space Administration (NASA) under Contract NAS2-11864. The views and conclusions contained in this paper are those of the authors and should not be interpreted as representative of the official policies, either expressed or implied, of the Office of Naval Research, NASA, or the United States government.

The authors are with the Artificial Intelligence Center, SRI International, Menlo Park, CA 94025, and the Center for the Study of Language and Information, Stanford University, Stanford, CA 94025, USA.

expand the near-term plan of action some more, execute, and so on. This approach has many advantages. First of all, systems generally lack sufficient knowledge to expand a plan of action to the lowest levels of detail—at least if the plan is expected to operate effectively in a real-world situation. The world around us is simply too dynamic to anticipate all circumstances. By finding and executing relevant procedures when they are truly needed, a system may stand a better chance of achieving its goals.

A combined planning/execution architecture can also be *reactive*. By reactive, we mean more than a capability of modifying current plans in order to accomplish given goals; a reactive system should also be able to completely change its focus and pursue new goals when the situation warrants it. This is essential for domains in which emergencies can occur and is an integral component of human practical reasoning. In a system that expands plans dynamically and incrementally, there are frequent opportunities to react to new situations and changing goals. Such a system is therefore able to rapidly modify its intentions (plans of action) on the basis of what it currently perceives as well as upon what it already believes, intends, and desires.

Of course, how we represent knowledge is just as important as how we use it. Representing knowledge of dynamic environments as procedures, rather than as sets of rules about individual atomic actions, has many advantages. Most obvious is the computational efficiency gained in not having to reconstruct particular plans of action over and over again from knowledge of individual actions.

Second, much expert knowledge is already procedural in nature: for example, consider the knowledge one might have about kicking a football, performing a certain dance movement, cooking a roast dinner, solving Rubik's cube, or diagnosing an engine malfunction. In such cases, it is highly disadvantageous to "deproceduralize" this knowledge into disjoint rules or descriptions of individual actions. To do so invariably involves encoding the control structure of the procedure in some way. Usually this is done by linking individual actions with "control conditions," whose sole purpose is to ensure that the rules or actions are executed in the correct order. This approach can be very tedious and confusing, destroys extensibility, and lacks any natural semantics.

By having direct access to the particular behavior that a procedure has or will follow, we can also make much more effective use of procedural knowledge. For example, suppose that an agent knows that in order to achieve G it will follow a course of action of the form: $X \rightarrow Y \rightarrow Z$. When performing X, the agent will know that it is setting up a situation for later performance of Z. When the agent finally does get around to Z, it will likewise know that X and Y have been performed, thereby allowing it to make various assumptions—for instance, that certain environmental conditions will be in place because X and Y tend to make them that way.

Having access to one's history of actions can also free an agent from stringent reliance on its sensors—the agent can derive much information about the state of the world simply from knowledge of its previous activities (see also [4], [20], [26]). For example, if cooks follow the steps of a recipe exactly, they can usually assume the food will turn out right—they do not have to perpetually taste it. They might not even know what it *should* taste like, especially at intermediate points in the recipe. Work by Lansky [14] has taken very seriously the notion of encoding domain requirements primarily in terms of actions and their interrelationships and deriving knowledge from past activity. Such an approach appears particularly advantageous for describing the properties of multiagent domains.

Another advantage of representing knowledge as procedures is that we are able to reason about those procedures as *whole entities*. For example, given two procedures for fixing a broken pipe, we can evaluate those procedures in their entirety and decide which has the best cost/benefit properties in a particular situation. This type of "metalevel" or reflective reasoning about our own internal procedures enables us to perform effectively and would be intractible if the steps of the procedure had been broken down into seemingly unrelated rules.

The primary aim of this paper is to provide a basis for representing and reasoning about procedural forms of knowledge in a way that allows an agent to deal effectively with a dynamically changing world. It is important that the agent be able to use these procedures to form intentions to achieve given goals, to react to particular events, to modify intentions in the light of new beliefs or goals, and to reason about these things in a timely way.

To do this, we first introduce the notions of action and process. We then provide a means for describing these and define a declarative and operational semantics for our formalism. Together, these provide a way of both stating procedural facts about the problem domain and a method of practical reasoning about how to use this knowledge to achieve given goals.

More generally, the formalism may also be viewed as the basis for an executable specification language. Just as for Prolog [5], the declarative semantics provides a means for stating *facts* about the problem domain, and the operational semantics yields a means of *using* these facts to achieve given goals.

A system based on the proposed representation has been implemented and is currently being used for the control of an intelligent robot and for fault isolation and diagnosis on NASA's space shuttle. An early version of an implemented system is described in Georgeff and Bonollo [8] and the latest work in Georgeff and Lansky [11].

II. Processes and Actions

We assume that, at any given instant, the world is in a particular *world state*. This state embodies not only the external environment, but also the world inside an agent—its internal cognitive world. As time progresses, the world state changes through the occurrence of *actions* (or *events*).[1]

Most early work in AI represented actions as mappings from world states to world states [7], [16], [19]. However, these models can describe only a limited class of actions and are too weak to be used in dealing with multiagent or dynamic worlds. Some attempts have recently been made to provide a better underlying theory for actions. McDermott [17] considers an action or event to be a set of sequences of states, and describes a temporal logic for reasoning about such actions and events. Allen [1] also con-

[1]Although some authors make a distinction between actions and events, in this paper we treat the two terms synonymously.

siders an action to be a set of sequences of states and specifies an action by describing the relationships among the intervals over which the action's conditions and effects are assumed to hold. However, although it is possible to state arbitrary properties of actions and events, it is not obvious how these logics can be used effectively for practical reasoning.[2]

Our notion of action is essentially the same as that of McDermott and Allen; namely, we consider actions to be sets of sequences of world states. However, to reason about how to achieve given goals or test certain properties of the world, the concept of action alone is not sufficient. In particular, we need to know *how* the actions are actually generated.

To do this, we introduce a notion of *process*. Informally, a process can be viewed as an abstract mechanism that can be executed to generate a sequence of world states, called a *behavior* of the process. The set of all behaviors of a process constitutes the *action* (or *action type*) generated by the process.[3]

Having a notion of process allows us to make a distinction that is critical for practical reasoning—we can distinguish between behaviors that are *successful* executions of the process (i.e., successful instances of the action) and those that are unsuccessful (those that have *failed*). The ability to represent both successful *and* failed behaviors is very important in commonsense reasoning and is critical in multiagent and dynamic environments (e.g., see [13]).

The need for representing both failed and successful behaviors is most clearly seen in problems that involve multiple agents. In such worlds, the potential side effects of an agent's activities can have a dramatic influence upon other agents. In a system that uses a combined planning/execution framework as described earlier, and in which knowledge of the world is incomplete or uncertain, it is usually not possible to predict whether a process will succeed or fail. (Of course, even if we were fully planning everything out in advance, we still could not realistically anticipate all of the consequences of executing a process.) Given this, it is clear that both failed and successful process executions will occur, and thus both must be available for reasoning about potential process interactions.

Using a notion of process is particularly important for reasoning about failed attempts to accomplish given goals. For example, suppose that we have two ways to get to the airport to catch a plane; one involving making a bus connection and the other, although less convenient, by car along a busy route frequented by numerous taxis. Clearly, the successful behaviors of both methods (processes) will result in catching the plane. However, failure to make the bus connection could leave us in a state from which we could not recover (because next available means of transport to the airport will arrive too late), whereas failure of the other method, given the availability of taxis, would not be so catastrophic. We might, therefore, decide to use the second, less convenient method, if we compare their possible modes of *failure*. And this can only be done with a notion of process—an intrinsic part of deducing a failed behavior is knowing exactly how that behavior was generated in the first place.

The need for representing failed behaviors also arises in natural-language understanding. For example, it is important to have a denotation for action sentences (such as "she was painting a picture") that allows for action failure, even in midperformance ("she was painting a picture when her paints ran out"). The action referred to in the second sentence must be one of the failed behaviors of the picture-painting process, and there is no way to derive this solely from a set of successful picture-painting behaviors.[4]

Finally, the notion of process failure also allows us to represent tests on world states in a particularly simple way without the introduction of knowledge or belief structures (cf. [18]). To do this, we let certain successful process behaviors stand as tests for a given condition. This can only be done if we are guaranteed that the process will only succeed when the condition is indeed true. (If the process fails, we can, of course, assume nothing about the condition. Although we might usually be happy to equate process failure with the negation of the condition being tested, we may not always wish to do so. In such cases, we might need one process to test for a given condition and another process to test for its negation.)

III. Process Descriptions

Abstractly, a process can be modeled by two sets of behaviors, one set representing the successful behaviors of the process and the other the failed behaviors. However, to reason about these behaviors we need some means for describing them; moreover, whatever descriptions we use need to be amenable to efficient reasoning techniques.

In this section we present a means of describing processes as sequences of particular *subgoals* or behaviors to be achieved. Each process is represented by a labeled transition network with distinguished start and finish nodes and arcs labeled with subgoal descriptions (see Fig. 1). Any realizable behavior that achieves each of the subgoals labeling some path through the network from the start node to a final node constitutes a successful behavior of the process; a failed behavior is one which terminates with failure to achieve a subgoal on the path.

Operationally, we may view a process description as being *executed* in the following manner. At any moment during execution, control is at a given node n. An outgoing arc a may be transitted by *successfully* executing a process that achieves the subgoal (behavior) labeling a. If no outgoing arc from n can be transitted, process execution *fails*. Execution begins with control at the start node and *succeeds* if control reaches the finish node.

We now give a more formal definition of our process de-

[2]Allen [2] does, however, propose a method of forming plans that is based on a restricted form of his interval logic.

[3]In this paper we restrict our attention to sequential, nonconcurrent processes. Our work on implementing a system based on this theory, however, has incorporated the notion of concurrently active, communicating processes.

[4]The reason is that different processes or procedures can generate the same set of successful behaviors yet have different failure modes. For example, consider two procedures for jumping across a stream. In the first procedure, we check that the stream is sufficiently narrow to jump and, if so, jump it; otherwise, we do not even attempt to cross. In the second procedure we perform the same test but attempt the jump irrespective of the outcome of the test. Assuming the test accurately determines whether or not the stream can be jumped, both procedures yield the same set of successful behaviors, but the failed behaviors are quite different!

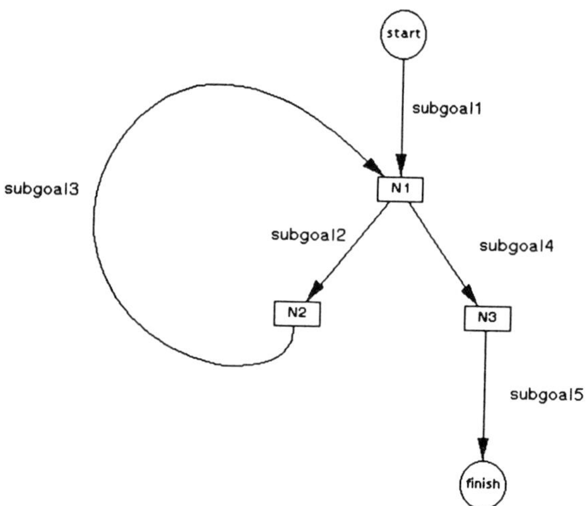

Fig. 1. A process description.

scription language. First, we assume a fixed set S, possibly infinite, of *state descriptions* and a fixed set A, also possibly infinite, of *action descriptions*. A *process description* can then be represented as a tuple $P = \langle N, E, \delta, n_I, N_F, \alpha, \rangle$, where

- N is a set of *nodes*
- E is a set of *arcs*
- $\delta: N \times E \rightarrow N$ is the *process control function*
- $n_I \in N$ is the *start node*
- $N_F \subset N$ is a set of *final nodes*
- $\alpha: E \rightarrow A$ associates an action description with each arc; these action descriptions are called *goal descriptions*.

Rather than represent process descriptions in this formal mathematical way, we use a graphical form as typified in Fig. 1.

Both the state and action description languages are based on predicate calculus. Each state description is a first-order predicate-calculus formula and can be viewed as denoting a set of states; namely, those in which it is true. For example, a formula of the form ((**on a b**) ∧ (**on b c**)) could be used to denote the world states in which block **a** is on top of block **b**, which in turn is on top of block **c**.

An action description consists of an action predicate applied to an *n*-tuple of terms. Each action description denotes an *action type* or set of *behaviors*. For example, an expression like (**walk a, b**) could be taken to denote the set of behaviors in which an agent walks from point **a** to point **b**.

It is also desirable to allow a class of action descriptions that relate directly to world states. We thus extend the action description language to include actions that achieve a given state condition p (represented by an action description of the form (! p)), actions to test for p (represented as (? p)), and actions that preserve p (represented as (# p)). We call these *temporal action descriptions*. In each of these, p is a state description—i.e., a description of the type of state to be achieved, tested for, or preserved. For example, an action that achieves a state in which block **a** is on block **b** might be described by a temporal action description of the form (! (**on a b**)).

Action descriptions may also be combined into *action expressions*. These are composed in the usual way using conjunctive and disjunctive operators. Thus an action expression of the form (! p) ∧ (? q) denotes an action that both tests for q and achieves p.

Having a means for describing processes, we now need a way to state properties about them. In this paper, we are interested primarily in describing the effects of successful behaviors of a process; that is, we want to be able to express the fact that, under certain conditions, successful execution of the process will result in a certain behavior being achieved. We will call such facts *process assertions*.

A process assertion consists of a *process description*, P, describing a process; a *precondition*, c, denoting a set of world states in which the process is applicable; and an *effect*, g, characterizing the set of successful behaviors the process can actually generate when commenced in a state satisfying c. We will write such an assertion as $c \langle P \rangle g$.

The intent or meaning of this assertion is that any successful behavior of process P whose first state satisfies precondition c will also satisfy the effect g. From an operational viewpoint, if c holds at the commencement of execution of process P, g will be realized by a successful execution of the process.

Process assertions may also use variables. Such variables may appear in the precondition c, in the process description P, as well as in the effect g. We make a distinction between *local* variables (prefixed by %) and *global* variables (prefixed by $). All global variables must have a fixed interpretation over the entire assertion and are taken to be universally quantified. In contrast, local variables must have a fixed intepretation in the interval of states during which a given arc is transitted, but can otherwise vary. They cannot appear in the preconditions or the effect of a process assertion, and are existentially quantified over the scope of the arc on which they appear. (Local variables are often needed in loops where it is necessary to identify different elements from one iteration to the next.)[5]

A typical process assertion is shown in Fig. 2.

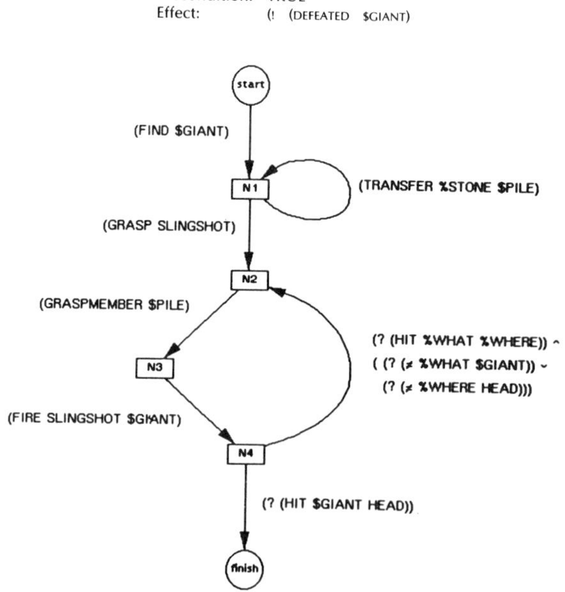

Fig. 2. David and Goliath.

[5]In fact, because we want to allow local variables to denote different objects on different *transitions* of the *same* arc, we strictly have to interpret variables with respect to the underlying tree structure of a process obtained by "unwinding" all loops appearing in the process description.

IV. Declarative Semantics

The declarative semantics of process assertions is intended to describe what is *true* about the underlying system of processes and the world in which they operate. Such a semantics says nothing about *how* such knowledge can be used to achieve particular goals—rather, it simply allows one to state *facts* about certain behaviors. In the preceding section we provided a preliminary, intuitive meaning for process assertions. In this section we present a more formal declarative semantics. First, however, we begin with a closer examination of the process assertion depicted in Fig. 2.

The "David and Goliath" procedure can be viewed as a plan for defeating a giant with a slingshot. The procedure involves gathering stones, placing them in a pile, getting a slingshot, and then repeatedly taking up a stone and shooting it until the giant is hit on the head. In this particular domain, hitting a giant on the head with a stone hurled by a slingshot always results in the giant's defeat. The procedure is nondeterministic and allows agents to gather as many stones as they wish, limited only by their ability to continue gathering them. The procedure is not guaranteed to be successful—it may fail if any one of the actions labeling the arcs of the network cannot be accomplished (and no other alternative path can be taken).

It is important to note how the process assertion captures *implicit* knowledge of the problem domain. This knowledge is of two kinds: one concerning the validity of the procedure, the other heuristic. For example, hitting giants on the head with an object propelled from a slingshot will not always defeat them (e.g., if it is a cotton ball), but will if it is a stone. Thus the validity of the effects of the procedure depends critically on the structure of the procedure itself, which ensures the only stones are placed in the pile. (Strictly, the procedure should also ensure that the pile is initially empty or contains nothing but stones.)

The procedure also captures heuristic knowledge in that earlier actions may make subsequent actions more likely to succeed. For example, the slingshot may require a certain size and weight of stone; however, instead of this being represented as an explicit test that precedes the shooting action, it is represented implicitly by the context established by the procedure. In this case, the assumption is that any stone that can possibly be gathered will most likely possess the appropriate characteristics. Note that this does not affect the validity of the procedure; if a stone does not have the necessary properties, the action of shooting the slingshot will fail.

At first glance, it seems that the semantics of a process description could be determined solely on the basis of successful behaviors which satisfy each of its subgoals. On closer examination, however, it becomes clear that this will not quite do. For example, if a node has multiple outgoing arcs (such as nodes *N*1 and *N*4 in Fig. 2) we need to allow several of these arcs to be tried until one is found successful. This is exactly the sort of behavior required of any useful conditional plan or program; if a test on one branch of a conditional fails (returns false), it is necessary to try other branches of the conditional. Similarly, in many real-world situations, it is often desirable to allow multiple attempts to achieve a goal before relinquishing that goal (for example, if a stone is accidently dropped when trying to pick it from the pile). The problem with failed attempts, however, is that they may change the state of the world. Thus to obtain a proper semantics, paths through a process network must allow behaviors that explicitly include failed attempts at realizing tests and actions as well as successful ones.

Moreover, it is important to realize that the goal descriptions labeling process arcs actually refer to other *processes*; namely, those whose successful behaviors realize the described goals. If this were not the case, we could not talk sensibly about failed attempts to achieve goals—failures can only be understood relative to the processes or methods used to generate actions, and not with respect to actions alone. Second, from a practical point of view, it would be very difficult to say anything useful about process assertions that were *not* grounded in performable actions—the resulting processes would be too unconstrained.

We now give a more formal definition of the semantics of process assertions. We first need to specify the interpretation for the symbols appearing in our description language. We will assume a fixed domain *D* of objects and a possibly infinite set of states. Given a particular state, a *state interpretation* associates with each constant symbol and variable an object from *D*, with each predicate symbol a relation over *D*, and with each function symbol a function on *D*. The meaning of a given state description is then defined under the usual semantics for first-order predicate calculus.

Similarly, we can define a *behavior interpretation* that associates a set of behaviors with each action predicate. The meaning of an action predicate is then taken to be the corresponding set of behaviors in the interpretation of that predicate. We also need to specify the meaning of temporal action descriptions. If *p* is a state description, then

- (! *p*) denotes those behaviors whose last state satisfies *p*.
- (? *p*) denotes those behaviors whose first state satisfies *p*.
- (# *p*) denotes those behaviors all of whose states satisfy *p*.

Conjunction and disjunction of action descriptions denote behavior-set intersection and union, respectively.

Finally, we are in the position to give a meaning to process descriptions. Each process description will be taken to denote a set of *successful* behaviors *and* a set of *failed* behaviors. To build a description of these behaviors, we first introduce the notion of process applicability. A process *P* is said to be *applicable* to a goal (i.e., an action type) *B* for a set of states *S*, if every behavior in the success set of *P* that begins in a state $s \in S$ is also in *B*.

Now let *n* be a node in a process description *P*. An *allowed behavior* starting at node *n* is a sequence composed of behaviors of processes applicable to the goals labeling the arcs emanating from *n*. Each allowed behavior represents a series of attempts by applicable processes to transit an arc leaving *n*, until one succeeds or they all fail. Thus the set of allowed behaviors starting at a node *n* can be partitioned into two sets: those representing successful transits to a succeeding node, and those that represent failures to leave the node. The first set is denoted by *succ*(*n*, *a*). Each of its behaviors must be a sequence of unsuccessful attempts by processes applicable to goals on arcs emanating from *n*, followed by a behavior of an applicable process that succeeds for some arc *a*. The second set, *fail*(*n*), consists of the null behavior along with those behaviors composed only of failed attempts of applicable processes.

Given these two types of allowed behaviors, we can then recursively define the *success* and *failure sets* for a node n, denoted $S(n)$ and $F(n)$, respectively, as follows:[6]

1) If n is a final node, then $S(n)$ and $F(n)$ are both empty.
2) If n has arcs a_i to nodes m_i, $1 \le i \le k$, then
   $S(n) = \cup_i succ(n, a_i).S(m_i)$
   and
   $F(n) = \cup\{ fail(n), \cup_i succ(n, a_i).F(m_i)\}$.

The success and failure sets of a process description P are then taken to be the success and failure sets, respectively, of the initial node of P.

As an example, consider the process networks shown in Fig. 3 where the arcs are labeled with applicable processes.

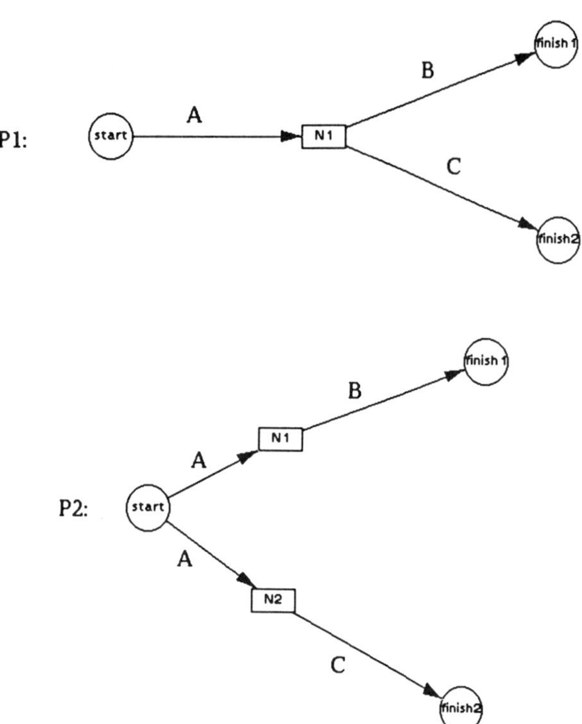

Fig. 3. Sample process networks.

For a process A, let A denote the set of its successful behaviors, and A_F the set of its failed behaviors. Then the success and failure sets for each of the process networks in Fig. 3 may be described as follows:[7]

$P1$:	$(A_F)^*.A.(B_F\|C_F)^*.B$	$P1_F$:	$(A_F)^+$
	$(A_F)^*.A.(B_F\|C_F)^*.C$		$(A_F)^*.A.(B_F\|C_F)^+$
$P2$:	$(A_F)^*.A.(B_F)^*.B$	$P2_F$:	$(A_F)^+$
	$(A_F)^*.A.(C_F)^*.C$		$(A_F)^*.A.(B_F)^+$
			$(A_F)^*.A.(C_F)^+$.

[6]If $w_1 = s_1, \cdots, s_k$ and $w_2 = s_k, \cdots, s_n$, then $w_1 \cdot w_2 = s_1, \cdots, s_{k-1}, s_k, s_{k+1}, \cdots, s_n$. This operation is extended to sets of sequences in the usual way. Note that this formulation allows a single state to satisfy a sequence of goals.

[7]The notation used is that for standard regular expressions. The symbols * and + denote zero or more and one or more repetitions, respectively. The symbol | is used to denote a choice between alternatives, e.g., $(A|B)$ denotes a behavior of form A or B.

Notice that, while the semantics given above allows for multiple attempts to achieve the goals exiting any given mode, it does not allow for backtracking to previous nodes in the net.

Now that we have given an interpretation for process descriptions, we are finally in a position to specify the meaning of a set of process assertions. Consider a process assertion $c\langle P\rangle g$. This assertion is actually a requirement of the following form: for each behavior b in the success set denoted by P, if the first state of b satisfies c, then b satisfies g. This requirement must be met by all process assertions.

V. Operational Semantics

Process assertions provide a way of describing the effects of actions in some dynamic problem domain. But how can a system or "agent" *use* this knowledge to achieve its goals? That is, we currently have a knowledge representation that allows us to state certain properties about actions and what behaviors constitute what actions. We have not explained, however, how an agent's *wanting* something can provide a rationale for or *cause* an agent to *act* in a certain way.

One way to view the causal connection between reasoning and action is as an *interpreter* that takes goals as well as knowledge about the state of the world as input and, as a result, forms intentions to perform certain actions and then acts accordingly. An abstract representation of such an interpreter may be considered to be the *operational semantics* of the knowledge representation language. In this section we provide a description of such an interpreter.

To ground our interpreter in some executable framework, we must make certain assumptions. First of all, if a system is to be able to achieve its goals, it must be able to bring about certain actions, and thus be able to affect the course of behavior. Thus we assume a system containing certain *primitive processes* capable of activating various external effectors. The system must also be able to sense the world to the extent of determining the success or failure of primitive processes—indeed, this is the only way it can sense the state of the world.

Given these capabilities, the system tries to achieve its goals by applying the following interpreter to applicable processes.[8] The interpreter works by exploring paths from a given node **n** in a process description **P** in a depth-first manner. To transit an arc, the interpreter unifies the corresponding arc assertion with the effects of the set of all process descriptions, and executes a set of the unifying processes, one at a time, until one terminates satisfactorily. If there are no matching processes, or none of the matching processes on any of the outgoing arcs are successful, the execution of **P** fails. Note that the precondition of each process must be satisfied when it is applied, in order for it to be truly applicable.

```
function successful (P n)
  if (is-end-node n) then
    return true
  else
```

[8]This interpreter is very similar to the parsers and generators used for Augmented Transition Networks [29]. It differs in the amount of backtracking allowed and in the use of unification to match arc labels and networks.

```
    arc-set := (outgoing-arcs n)
    pr-a-set := (processes-that-unify arc-set)
    do until (empty pr-a-set)
      pr-a := (select pr-a-set)
      pr := (process pr-a)
      a := (arc pr-a)
      if (satisfied (precondition pr)) then
        if (successful pr (start-node pr)) then
          return (successful P (terminating-node a))
      pr-a-set := (processes-that-unify arc-set)
    end-do
    return false
  end-function
```

The function **processes-that-unify** takes a set of arcs and returns the set of processes that unify with some arc in the set, along with the specific arc with which each unifies. The functions **process** and **arc** select out the process instance and corresponding arc from each element of this set. The function **select** selects an element from a set. The order in which selections are made is called the *selection rule*. We call the rule governing the number of times a process may be tried the *application rule* (for this particular interpreter, the application rule is embodied in the function **processes-that-unify**).[9] The function **return** returns from the enclosing **function**, not just the enclosing **do**. The system starts by executing a process description with a single arc labeled with the initial goal.

Of course, it is important that the operational and declarative semantics be consistent with each other. The declarative semantics defines a set of behaviors for each process. The operational semantics also defines a set of behaviors for each process, but this set depends on the selection and application rules used in the above algorithm. Let P_D be the set of successful behaviors for a process P as given by the declarative semantics, and let $P_{O,R,A}$ be the set of successful behaviors for P as given by the operational semantics for selection rule R and application rule A. It is not difficult to show that $P_{O,R,A} \subset P_D$. This means that any behavior generated by the interpreter given above will satisfy the declarative semantics. The proof involves showing that both the success set and failure set of a process under the operational semantics are each a subset of the success set and failure set, respectively, of the process under the declarative semantics. This can be done using double induction, first, on the number of processes that are applied at a node, and second, on the length of a particular path through the process (where length is measured in number of nodes in the path). The proof is straightforward once it is recognized that any path resulting from use of a selection rule R and an application rule A will automatically be one of the paths covered by the declarative semantics, and that any sequence of process attempts (as well as primitive actions) will be considered successful (or a failure) both declaratively and operationally.

[9] In a practical implementation of the operational semantics, it is usually best to use an application rule that tries each matching process exactly once. This allows the realization of all the control constructs of standard programming languages while meeting reasonable bounds on time resources. However, variations in which each process is tried multiple times could be incorporated without conflicting with the declarative semantics of process characterizations.

Note that we have made no assumptions about whether a process will succeed or fail—this is determined solely by the environment. As discussed earlier, in the real world, the success or failure of processes simply cannot always be predicted. Thus the above interpreter must be embedded in an environment in order to be truly useful. Without this, the operational semantics given above would be of little interest: it would produce just one possible success set for a given process or goal without any expectation that this behavior could be realized. However, because the interpreter is actually operating using a mixed planning/execution strategy, the environment itself determines process success or failure. This is quite different from standard AI planning systems, where success of primitive actions is assumed. It is also quite different from the operational semantics of pure Prolog, though would be similar to a semantics for Prolog with input and output streams.

Of course, if we did have additional knowledge about the state of the environment and the success or failure of the primitive actions, we could use the above interpreter in a pure planning mode. However, the inclusion of $P_{O,R,A}$ in P_D would be, in general, strict. That is, the interpreter may not achieve some given goal even when, according to the declarative semantics, there exists a way to achieve it. This is partly because the interpreter fixes the selection rule and application rule. Even by allowing all possible selection and application rules, however, we would still not attain completeness. The problem is that the interpreter does not have the machinery to deduce facts about world state that can be inferred using the declarative semantics. If an interpreter were capable of deducing all possible behaviors of a process from its description, and if it could also arbitrarily combine processes to generate any achievable behavior, that interpreter would also be able to generate any behavior in P_D. It is clear that such an interpreter would be extremely difficult (if not impossible) to construct. However, in the next section we provide a limited set of proof rules for deducing facts about process behaviors as well as for combining processes to achieve particular effects.

VI. Action Decomposition Rules

As described above, the operational semantics we have provided is actually not as strong as it could be. For example, if an arc is labeled with a goal of the form $(!\,(p \vee q))$, we can determine from the declarative semantics that a process with effect p (or effect q) will be applicable (assuming its preconditions are satisfied). The interpreter given above, however, cannot make this determination.

One way of strengthening our interpreter is to provide it with additional proof rules about the behavior of processes. For example, we might use standard rules of logic along with proof rules such as the following:

$$c1\langle P\rangle g1 \wedge c2\langle P\rangle g2 \equiv (c1 \wedge c2)\langle P\rangle(g1 \wedge g2)$$

$$c1\langle P\rangle g1 \vee c2\langle P\rangle g2 \supset (c1 \wedge c2)\langle P\rangle(g1 \vee g2).$$

We can also devise additional rules for *combining* processes. As before, we use the notation $c\langle P\rangle g$ to mean that every successful behavior of P whose first state satisfies c also satisfies the temporal assertion g. However, we extend the notation to failed behaviors as well, using assertions of the form $c\langle P\rangle_F\, g$ to describe the effects of failed behaviors.

The symbols ";" and "|" represent sequential composition and [nondeterministic] branching, respectively.

Conjunctive Testing

$$\frac{\begin{array}{c} c\langle P_1\rangle(?\ p \wedge !\ c' \wedge (\#\ q \vee \#\,(\neg q))) \\ c'\langle P_2\rangle(?\ q) \end{array}}{c\langle P_1\,;\,P_2\rangle(?\ (p \wedge q))}$$

Conjunctive Achievement

$$\frac{\begin{array}{c} c\langle P_1\rangle(!\ p \wedge !\ c') \\ c'\langle P_2\rangle(!\ q \wedge \#\ p) \end{array}}{c\langle P_1\,;\,P_2\rangle(!\ (p \wedge q))}$$

Disjunctive Testing

$$\frac{\begin{array}{c} c\langle P_1\rangle(?\ p) \\ c\langle P_2\rangle(?\ q) \\ c\langle P_1\rangle_F((\#\ q \vee \#\,(\neg q)) \wedge !\ c) \\ c\langle P_2\rangle_F((\#\ p \vee \#\,(\neg p)) \wedge !\ c) \end{array}}{c\langle P_1|P_2\rangle(?\ (p \vee q))}$$

Disjunctive Achievement

$$\frac{\begin{array}{c} c\langle P_1\rangle(!\ p) \\ c\langle P_2\rangle(!\ q) \\ c\langle P_1\rangle_F(\#\ c) \\ c\langle P_2\rangle_F(\#\ c) \end{array}}{c\langle P_1|P_2\rangle(!\ (p \vee q))}$$

The above rule for disjunctive achievement, for example, can be read as follows: if process P_1 achieves p when begun in state c, and if process P_2 achieves q when begun in state c, and if, in addition, failures of processes P_1 and P_2 leave c intact, then processes P_1 and P_2 can be combined into a disjunctive process that generates a behavior of the form $(!\ (p \vee q))$ when begun in state c.

To obtain a new operational semantics that incorporates these proof rules, the interpreter provided in the previous section would have to be modified to allow application of the proof rules when necessary.

VII. System Description

We now describe a procedural reasoning system (PRS) based on the theory described in the previous sections. It goes beyond the theory in several ways, including the addition of a database of beliefs and an enhancement of the way procedures can be invoked. These additions enable the system to exhibit not only goal-directed behavior, but behavior that is *reactive* to particular situations.

The PRS system also makes extensive use of world states and actions that refer to an agent's internal cognitive components. These "metalevel" states and actions are manipulated by the system in the same way it handles states and actions that deal solely with the outside world. They enable the system to form goals or react to situations that deal with the system's internal workings–for example, to figure out how to choose between multiple applicable procedures for a particular task, how to establish new desires, beliefs, or intentions based on particular situations, and so on. All of this metalevel reasoning, however, is done within the same formal context in which reasoning about the external world is done—i.e., in the context of the formalism presented in this paper.

The overall structure of a procedural reasoning system is shown in Fig. 4. The system consists of a *database* containing currently known *facts* (or beliefs) about the world, a set of current *goals* (tasks) to be accomplished, a set of *process assertions* (plans) that described procedures for achieving given goals or reacting to particular situations, and an *interpreter* (inference mechanism) for manipulating these components. At any moment in time, the system will also have a *process stack* that contains all currently active processes. This stack can be viewed as the system's current *intentions*. We now look at these components in more detail.

The database is intended to describe the state of the world at the current instant, and thus contains only state descriptions. Its primary function is to keep track of facts about the world state that can be inferred as consequences of process executions. We also use the database to provide the system with knowledge of the initial world state. A STRIPS-like rule is used for determining the full effects of processes; that is, facts are assumed to remain unchanged throughout a process unless we can infer otherwise. Updates to the database require the use of consistency maintenance procedures.

As in the preceding sections, goals are represented by action descriptions, and can be viewed as specifying a desired behavior of the system. This view of goals as behaviors is unlike that used by most planning systems. In such systems, goals can only be represented as descriptions of state

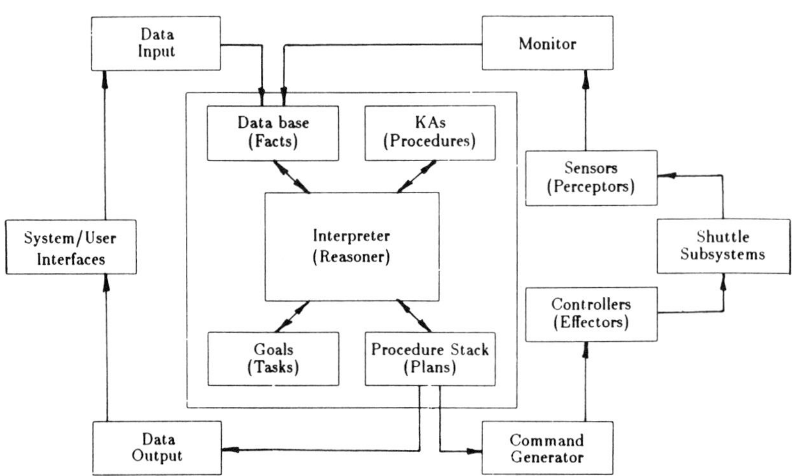

Fig. 4. System structure.

conditions to be achieved. The scheme adopted here allows us to express a much wider class of goals, including goals of maintenance (e.g., "achieve *p* while maintaining *q* true") and goals with resource constraints (e.g., "achieve *p* without using more than one tool").

As indicated earlier, we have also enhanced the way procedures supplied to PRS may be invoked. Rather than just being applicable to given goals, some procedures become applicable when certain facts become known to the system—i.e., they are *fact invoked*. Fact-invoked procedures are particularly useful for creating a *reactive* system—i.e., one that can change focus in reaction to particular situations. Such procedures are usually associated with some *implicit* goal. For example, a fact-invoked procedure for putting out a fire might become applicable whenever the system notices that there is a fire. Although this procedure does not respond to any explicit goal *per se*, it actually achieves an underlying implicit goal of all organisms—to stay alive. Within the context of the formalism presented in the preceding sections, a process assertion for a fact-invoked procedure has a precondition that describes the condition under which it is applicable and an effect part that matches all implicit goals of the system. This guarantees that each fact-invoked procedure becomes applicable whenever its precondition becomes true.

The PRS interpreter runs the entire system. From a conceptual viewpoint, it operates in a relatively simple way. At any particular point in time, certain goals are active in the system, and certain facts or beliefs are held in the system database. Given these goals and facts, a subset of the procedures in the system (both fact-invoked and goal-invoked) will be applicable. One of these procedures will then be chosen for execution. In the course of transitting the body of the chosen procedure, new goals will be formulated and new facts will be derived. When new goals are added to the goal stack, the interpreter checks to see if any new procedures are relevant, selects one, and executes it. Likewise, whenever a new fact is added to the database, the interpreter will perform appropriate consistency-maintenance operations on the database and possibly activate newly applicable procedures.

Because the system is repeatedly assessing its current set of goals, beliefs, and the applicability of procedures, the system exhibits a very reactive form of behavior, rather than being merely goal-driven. For example, when a new fact enters the system database, execution of the current process network might be suspended, with a new relevant process network taking over. One of the ways the system resolves which procedures to execute at any given time is by using other *metalevel* process networks. These metalevel procedures are manipulated and invoked by the system in the same way as any other procedure. However, they respond to facts and goals pertaining to the system itself, rather than just those of the application domain. For example, one typical metalevel procedure might respond to a goal of the form: (CHOOSE-BEST-PROCESS **$goal $list-of-procedures**). Besides the task of process selection, metalevel procedures can also be used to combine process networks in order to achieve composite goals. For example, metalevel procedures are used in the current system to apply some of the proof rules described in Section VI.

As in the interpreter of Section V, unification is used to determine whether or not a given process matches a given goal or database fact. Just as in Prolog, and unlike standard programming languages, this means that it is not necessary to decide before execution which process variables are to count as input variables and which are to count as output variables. This is important for providing flexibility and ease of verification. It also means that variables will not be unnecessarily bound, which can often be advantageous in allowing difficult decisions to be avoided or deferred. For example, if we had a goal of the form (P $x) (where $x is not bound), and a procedure for achieving (P $x) for all objects $x, we could use this procedure for achieving the goal without having to select a particular object to apply it to.

PRS also differs from conventional programming languages in its more flexible means of representing and using procedural information. For example, procedures are not "called" as in standard programming languages. Instead, they are invoked whenever they can contribute to accomplishing some goal or reacting to some situation. Just as procedures cannot be called, neither can they call any other procedure; they can only specify what goals are to be achieved and in what order. This makes PRS procedures much more amenable to modular verification techniques. Another difference that sets PRS apart from standard programming languages is that it is not deterministic. For example, several procedures may be relevant to a goal at any one time, and the order in which they are chosen for execution may, in general, be nondeterministic.

We have implemented an experimental system based on the ideas presented above. The implemented system is written in LISP and runs on a Symbolics 3600 machine. User interaction occurs via a graphical package that allows direct entry and manipulation of process networks.

VIII. Sample Problem Domains

A. Space Shuttle

One area in which advanced automation can be of particular practical benefit is fault isolation and diagnosis in complex systems. PRS is particularly suited to this kind of application not only because a diagnostic domain is dynamic, requiring quick response to faults, but also because much of diagnostic knowledge is specifically procedural in nature. In this section we describe one such application—diagnosis of the Reaction Control System (RCS) of NASA's space shuttle. The structure of the RCS module is depicted in Fig. 5. Sample malfunction procedures from the NASA diagnostic manuals are shown in Fig. 6.

The manner in which we have represented procedures for our RCS application reflects what we have said in the preceding sections—i.e., that actions and tests must be represented by whatever condition they achieve or test for, rather than by some arbitrary name. For example, there exist several malfunction procedures for lowering the pressure in tanks that have high pressure readings, and likewise, raising the pressure in tanks with low pressure. Currently, the NASA diagnostic procedures for these tasks make explicit calls to particular procedures that raise and lower tank pressure. Whenever the desired methods for altering tank pressure change, these procedure calls also have to be explicitly changed.

Our methodology for invoking procedures obviates the need for such changes; goals to lower or raise tank pressure may be posted as such—any applicable procedure may then respond. In fact, for this particular example, the goal un-

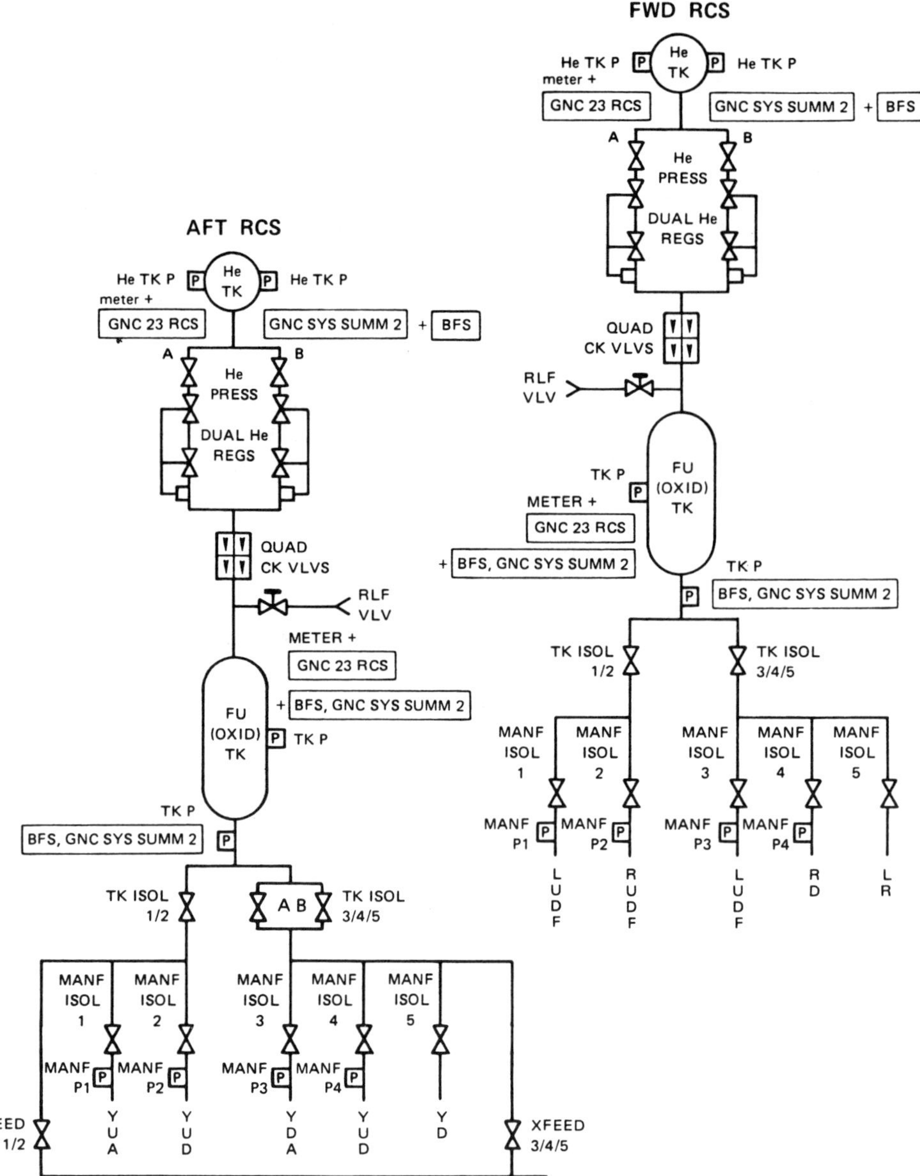

Fig. 5. The reaction control system.

derlying the pressure alterations may simply be to "normalize" the pressure of the tank. Invoking procedures on the basis of desired effects results in a much more modular and useful way of constructing large systems. Given a set of procedures associated with the actual goal that they achieve, the procedures may be reused in other circumstances in which they might be useful, or easily replaced by other procedures that achieve their particular goal in a better way.

1) RCS Database: Our first task in encoding the RCS application was to capture the structure of the physical RCS system (depicted in Fig. 5) as a set of database facts. Once inserted into the system database, these facts are used during fault diagnosis to identify particular components of the system and their properties. For example, a sample set of structural facts are given below.

```
(TYPE RCS F RCS.1)
(TYPE HE-PRESSURIZATION OX HEP.1.1)
(TYPE HE-PRESSURIZATION FUEL HEP.1.2)
(PART-OF HEP.1.1 RCS.1)
(PART-OF HEP.1.2 RCS.1)
(TYPE HE-TANK HET.1.1.1)
(PART-OF HET.1.1.1 HEP.1.1)
(TYPE HE-TANK HET.1.2.1)
(PART-OF HET.1.1.1 HEP.1.2)
```

Two types of structural facts have been used for this application—TYPE facts, which declare specific components or subsystems and assign to them unique identifiers, and PART-OF facts, which state which components are part of which subsystems. For example, (TYPE RCS F RCS.1) declares the entire front RCS and assigns it the identifier RCS.1. Each RCS contains two helium pressurization subsystems, one for the oxidant part of the system, the other for the fuel subsystem. For the system RCS.1 these are labeled as HEP.1.1 and HEP.1.2, respectively. Finally, each helium pressurization system contains its own helium tank.

Once we encode the structure of the RCS in this fashion, the diagnostic procedures can make use of this information to perform what might be considered simple commonsense tasks for an astronaut. For example, if a malfunction procedure has the test "Is the oxidant helium tank pressure greater than the fuel helium tank pressure for the front

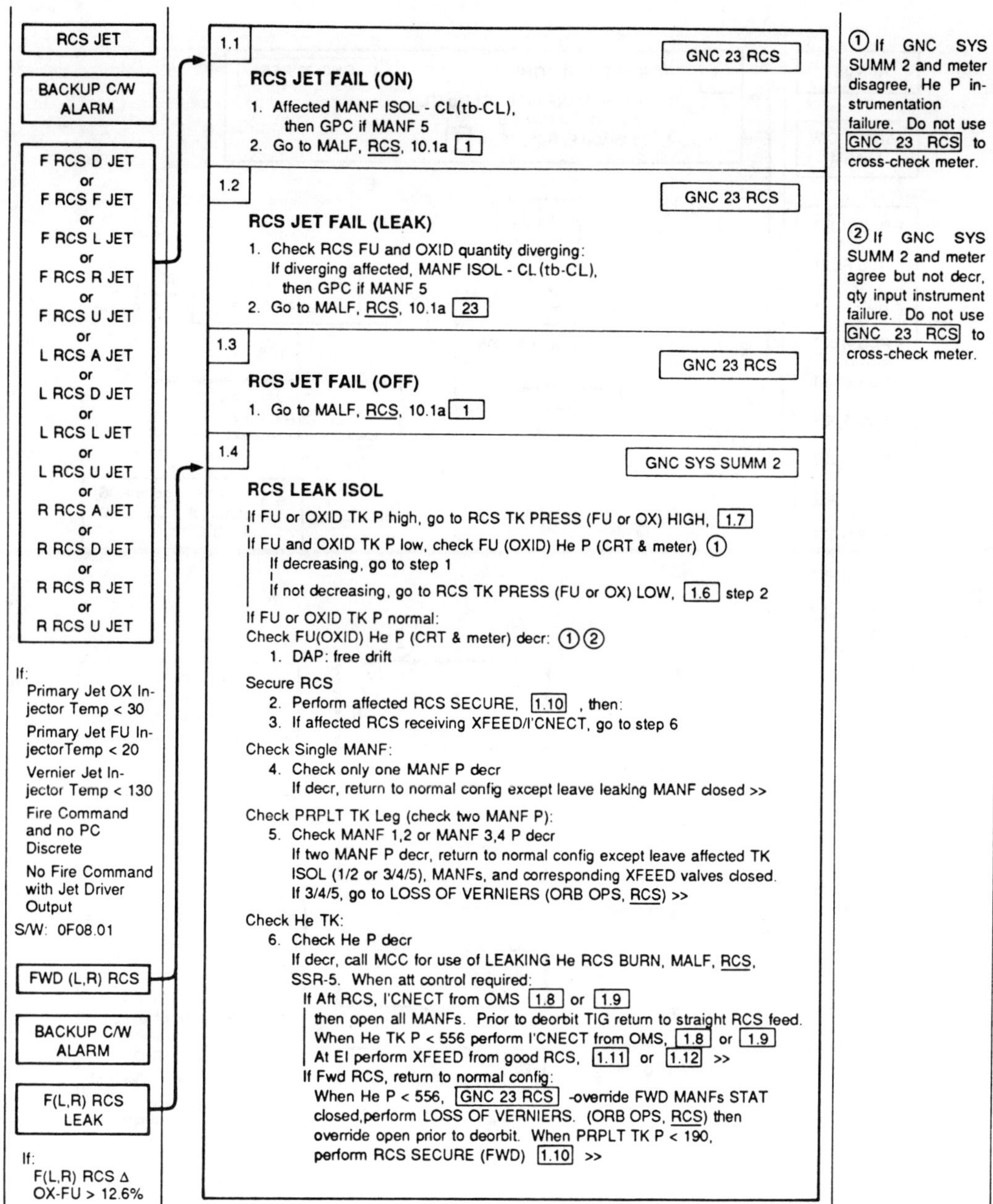

Fig. 6. Some RCS malfunction procedures.

RCS?" the test can be represented in a way that is impervious to system reconfiguration, is not hard-wired to particular identifiers, and can be used for any RCS. This is done using unification—matching database facts against queries composed as logical combinations of atomic formulas. In this case, the query would have the following form:

```
(? ((TYPE RCS F $rcs-id) ∧
    (TYPE HE-PRESSURIZATION OX $hep-ox) ∧
    (PART-OF $hep-ox $rcs-id) ∧
    (TYPE HE-PRESSURIZATION FUEL $hep-fuel) ∧
    (PART-OF $hep-fuel $rcs-id) ∧
    (TYPE HE-TANK $he-ox-tank) ∧
    (PART-OF $he-ox-tank $hep-ox) ∧
    (TYPE HE-TANK $he-fuel-tank) ∧
    (PART-OF $he-fuel-tank $hep-fuel) ∧
    (PRESSURE $he-ox-tank $ox-press) ∧
    (PRESSURE $he-fuel-tank $fuel-press) ∧
    (> $ox-press $fuel-press))).
```

2) RCS Process: We will now concentrate on the procedure called RCS JET FAIL (ON), which can be seen as Step 1.1 of Procedure 10.1, as well as 10.1a (a portion of the entire malfunction procedure is shown in Fig. 7). Notice how diagnostic conclusions (such as "JET DRIVER FAILED-ON ELECTRICALLY") are displayed in highlighted boxes.

PRS uses several processes to implement this diagnostic procedure. The main top-level process for dealing with the "JET FAIL (ON)" failure is called JET-FAIL-ON and is shown in Fig. 8. This process is fact invoked—that is, it responds when the system notices that certain lights, alarms, and computer monitor readings appear. Its precondition has the form:

```
(LIGHT RCS-JET) ∧ (ALARM BACKUP-CW) ∧
(FAULT $rcs-id RCS $jet-id JET) ∧
(JETFAIL-INDICATOR ON $manf-id).
```

In order to get JET-FAIL-ON running, these four facts (with instantiations of the three variables $rcs-id, $jet-id, $manf-

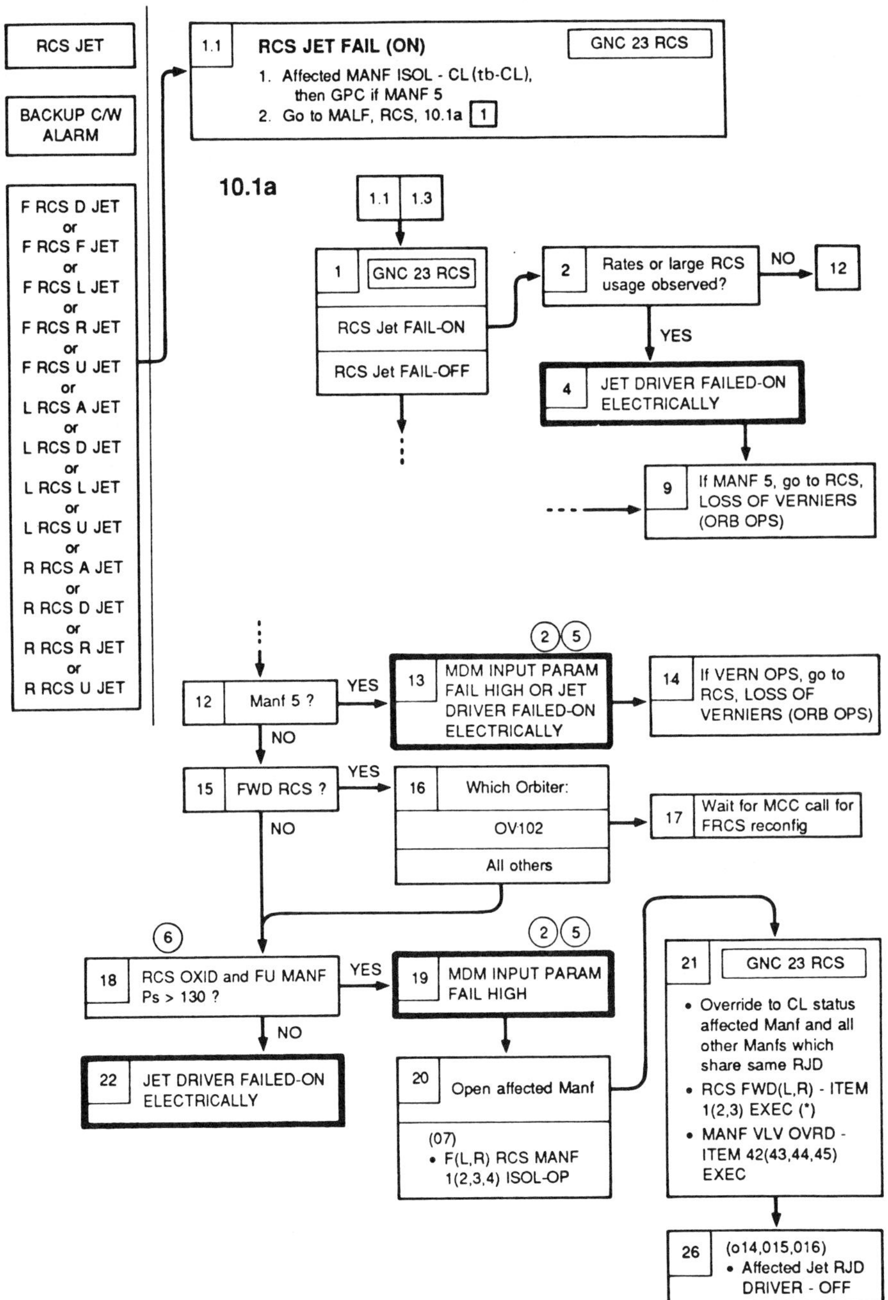

Fig. 7. RCS JET-FAIL-(ON) malfunction procedure.

id) must be added to the system database. For example, we might add the following facts:

(LIGHT RCS-JET)
(ALARM BACKUP-CW)
(FAULT RCS.1 RCS THR.1.1 JET)
(JETFAIL-INDICATOR ON MIV.1.1.1).

This tells the system that *there is* an actual malfunction in a specific RCS subsystem, jet, and manifold. The system will then react and apply the JET-FAIL-ON process.

Starting at its START node, JET-FAIL-ON will proceed and try to traverse its first edge, labeled with the goal expression (! (CLOSED-MANIFOLD $manf-id)). In other words, the system must find some way to close the given manifold. This corresponds to the first step of the malfunction procedure in Fig. 7, which reads: "Affected MANF ISOL-CL (tb-CL), then GPC if MANF 5." Notice how we have abstracted the overall *goal* or *intent* of this step (to close the manifold) from a particular instruction in the malfunction book which only states *how* to achieve the goal.

In this case, there are actually two different ways of achieving a closed manifold: for all manifolds, a talk-back switch is set to the closed position, but for vernier manifolds (type 5 manifolds), a setting must also be made on the computer console. These two ways of achieving a behavior of form (! (CLOSED-MANIFOLD $manf-id)) are reflected in the two procedures shown in Fig. 9, CLOSED-MANIFOLD and CLOSED-MANIFOLD-VERNIER. Each responds to a goal of the form (! (CLOSED-MANIFOLD $manf-id)). However, their preconditions constrain their applicability further—CLOSED-MANIFOLD will only be truly applicable if the manifold in question is

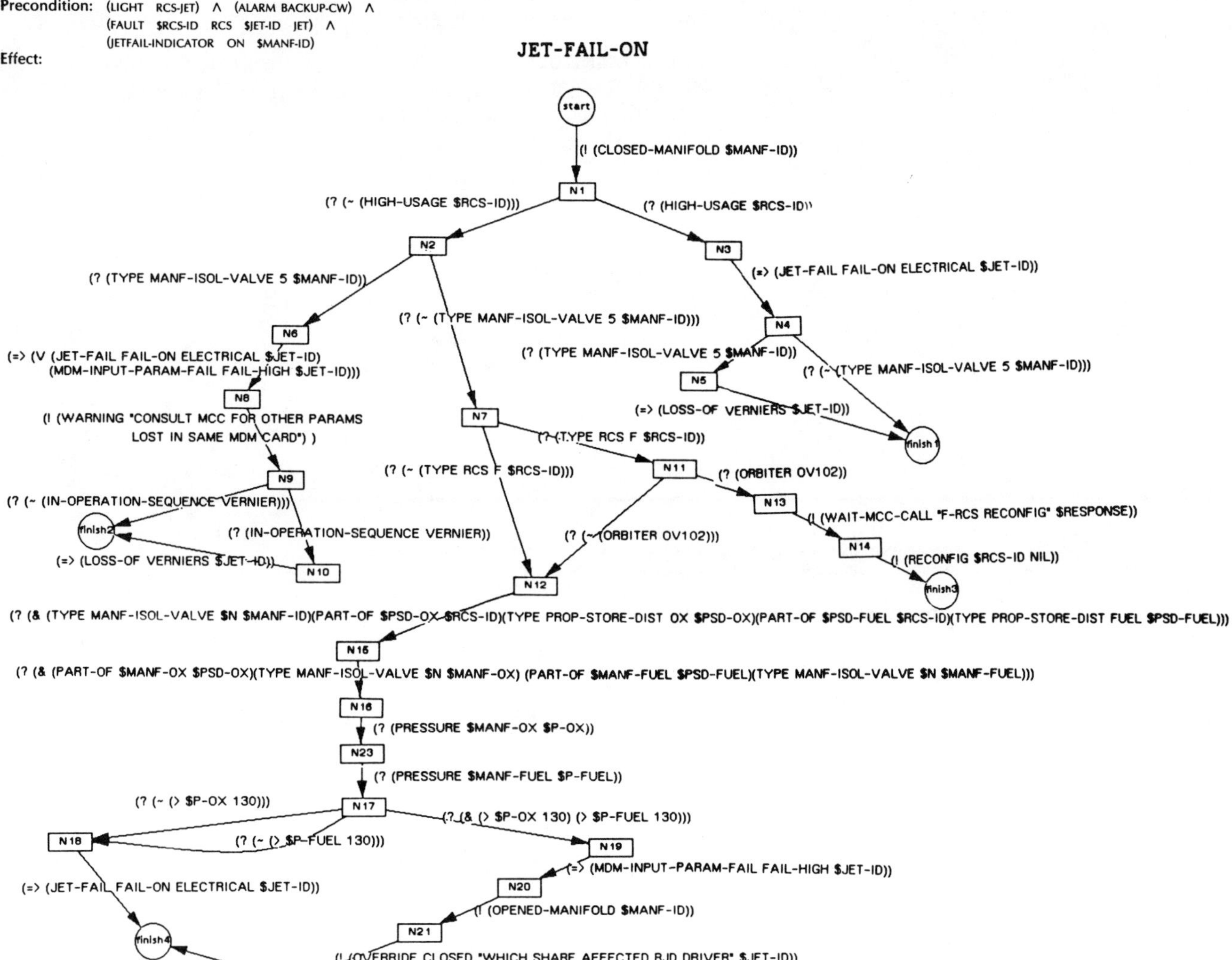

Fig. 8. JET-FAIL-ON process.

not of type 5, and CLOSED-MANIFOLD-VERNIER will only be applicable if the manifold *is* of type 5. In situations in which more than one process is truly applicable to a given goal, metalevel processes are used to resolve which is most useful.

Of course, given the semantics of process assertions, there is yet another way to achieve (! (CLOSED-MANIFOLD $manf-id)). In particular, a goal of the form (! P) will automatically be achieved if the system already believes that P is true. For this case, if the system already has in its database a fact of the form (CLOSED-MANIFOLD MIV.1.1.1), a goal of the form (! (CLOSED-MANIFOLD MIV.1.1.1)) will automatically succeed—no executions of the processes CLOSED-MANIFOLD and CLOSED-MANIFOLD-VERNIER need be undertaken.

It is precisely the lack of this kind of goal semantics and reasoning ability that caused a recent space shuttle flight to abort. Although the shuttle system knew that a particular manifold was closed, it found itself unable to proceed when an instruction of the form "close the manifold" was given to it. This is because all of the manifold-closing procedures available *presumed* an open manifold—they could not close a manifold that was already closed! If the procedures had been written in terms of the goals to be accomplished, rather than as fixed hard-wired procedure calls, the shuttle system could have realized that its goal to close the manifold had already been achieved.

We continue now with one more step in the execution of the JET-FAIL-ON process. If the goal to close the manifold actually succeeds, the system will then move on to the next node and choose a new outgoing arc to traverse. One possible choice might be the arc labeled (? (¬ (HIGH-USAGE $RCS-id)))—i.e., our goal is to determine whether there is *not* high usage in the affected RCS. To handle a goal of this form the system will first check to see if there are any database facts or processes that match this goal precisely. Because we can have negated facts in the system database, it is possible that a fact of form (¬ (HIGH-USAGE RCS.1)) *is* present in the database (for the sake of argument we have assumed that $rcs-id is bound to RCS.1). Similarly, there may be a process with an invocation part that indicates it is useful for precisely a goal of the form (? (¬ (HIGH-USAGE $rcs-id))). If a matching database fact or a successful matching process is found, then the system will attempt to satisfy the goal in these ways. However, if no such fact or matching process is present, the system will try to achieve the goal using any other means at its disposal.

For goals composed of certain negated predicates, a metalevel process is available which tries to achieve the goal

Precondition: (TYPE MANF-ISOL-VALVE 5 $MANF-ID)
Effect: (! (CLOSED-MANIFOLD $MANF-ID))

CLOSED-MANIFOLD

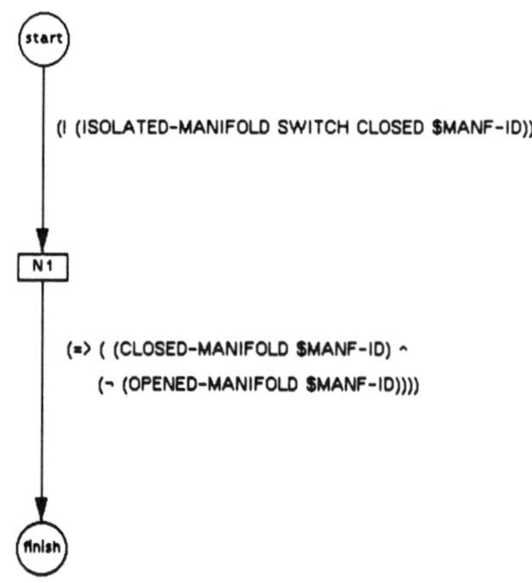

Precondition: (TYPE MANF-ISOL-VALVE $N $MANF-ID) ∧ (≠ $N 5)
Effect: (! (CLOSED-MANIFOLD $MANF-ID))

CLOSED-MANIFOLD-VERNIER

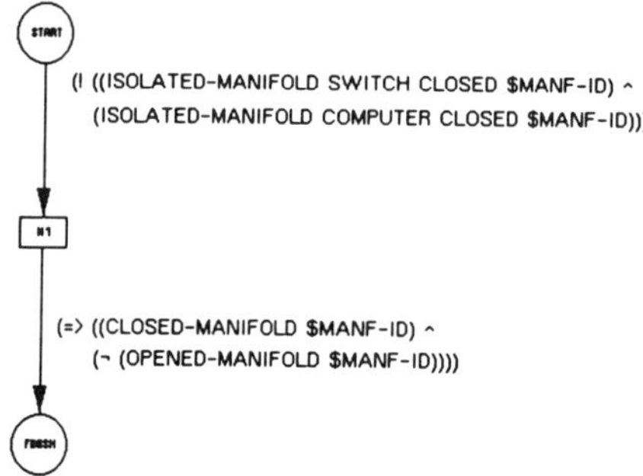

(! ((ISOLATED-MANIFOLD SWITCH CLOSED $MANF-ID) ^
(ISOLATED-MANIFOLD COMPUTER CLOSED $MANF-ID)))
(=> ((CLOSED-MANIFOLD $MANF-ID) ^
(¬ (OPENED-MANIFOLD $MANF-ID))))

Fig. 9. Process for closing a manifold.

using the "negation as failure" rule [15]. In other words, for a goal of form (! (¬ P)) or (? (¬ P)), the metalevel process will try to achieve (! P) (or (? P)), and if it fails to do so, will assume that the original negated goal has succeeded. In our current system, this is precisely how the goal (? (¬ (HIGH-USAGE $rcs-id)) is handled. Other metalevel processes also exist for achieving a conjunct of goals or a disjunct of goals.

B. Autonomous Robot

A second interesting application of PRS is in the control of autonomous robots. Our experimentation in this domain is being done using SRI International's new robot, Flakey. To effectively tackle this problem, we had to use multiple, concurrently active PRS modules, each consisting of its own database, goal stack, and processes that monitor and control different aspects of the robot's activity.

As our objective, we envisaged the robot in a space station acting as an astronaut's assistant. When asked to get a wrench, for example, the robot works out where the wrench is kept, plans a route to get it, and goes there. If the wrench is not there the robot reasons further about how to obtain information on its whereabouts, and finally returns to the astronaut with the wrench or explains why it could not be retrieved. In another scenario, the robot may be in the process of retrieving the wrench when it notices a malfunction light for one of the jets in an RCS module of the space station. It reasons that this is of higher priority than retrieving a wrench and sets about diagnosing the fault and correcting it. After having done this, it continues with its original task, finally telling the astronaut what has happened.

To accomplish these tasks the robot must not only be able to create and execute plans, but must be willing to interrupt or abandon a plan when circumstances demand it. Since other agents can move obstacles and issue demands even as the robot is planning, and since its view of the world can change as fast as the robot itself is moving, performance of the task requires a robot which is perceptive and highly reactive as well as goal directed.

The way we have structured the processes for this domain has actually conformed somewhat to Brooks' notion [3] of a *vertical* decomposition of robot functions (in contrast to the traditional *horizontal* decomposition into functional modules). The top-level robot module is used to perform higher level cognitive functions: overall route planning and high-level guidance. The lower the level of a module, the more primitive its function. Below the highest level are modules which put together sonar sensory data and figure out where "walls" and "doors" are. Even lower level modules reactively monitor the more rudimentary aspects of the navigation process—reacting to obstacles, maintaining a parallel bearing to the wall, getting back on course when veering takes place, etc.

Our present version of the robot application system is more fully described elsewhere [22]. Currently, the robot's model of the external world is particularly simple: apart from topological knowledge about hallways and rooms, the beliefs of the robot consist solely of its most recent sonar readings, various velocities and accelerations, and some indicators regarding the status of simulated external systems (such as the RCS module). Of course, any realistic application of the system would require that the robot be capable of building and storing much more complex models of the world around it.

IX. Conclusions

We have presented a model for action and a means for representing knowledge about procedures. The importance of reasoning about *processes* rather than simple histories or state sequences was stressed. In particular, we have indicated the role that process *failure* plays in practical reasoning.

A declarative semantics for the representation was provided that allows a user to specify facts about processes and their behaviors. This semantics is important for providing a model-theoretic basis to the knowledge representation. We have also given an operational semantics that shows how these facts can be used by an agent to achieve (or form intentions to achieve) its goals. A critical feature of the interpreter, and one that distinguishes it in kind from most existing AI planners, is that it is situated in an environment with which it interacts during the reasoning process. We consider the partial hierarchical planning that results to be an essential component of effective practical reasoning.

The knowledge representation we have described can also be used for symbolic planning in the traditional sense, although we would need to provide additional axioms stating under what conditions primitive processes would be successful. Indeed, the operators of many standard planning systems (such as NOAH [21], DEVISER [27], and SIPE [28]) can be viewed as restricted forms of process assertions.

Our formalism can also be viewed as an *executable specification language*—that is, as a programming language that allows a user to directly describe the behaviors desired of the system being constructed. The fact that the language has a declarative semantics allows *facts* about the behavior of the system to be stated and verified independently. The operational semantics provides a means for directly *executing* these specifications to obtain the desired behavior. In this sense, the language has much in common with Prolog, except that is applies to dynamic domains instead of static domains.

We have described a practical implementation of a system based on this model, and have shown how it can be applied for fault diagnosis and in the control of autonomous robots in highly dynamic situations. Although we have used parallel instances of PRSs within our implementation, we have yet to extend our formal model to deal with it. Some work in this direction is described by Georgeff [9], and work on synchronizing the activities of multiple agents has been done by Lansky [14] and Stuart [24].

Acknowledgment

There have been a number of people involved in the design, implementation, and testing of PRS, including P. Bessiere, M. Schoppers, J. Singer, and M. Tyson. We are most grateful for their contributions.

References

[1] J. F. Allen, "A general model of action and time," Computer Science Rep. TR 97, Univ. of Rochester, Rochester, NY, 1981.

[2] —, "Maintaining knowledge about temporal intervals," *Commun. ACM*, vol. 26, pp. 832–843, 1983.

[3] R. A. Brooks, "A robust layered control system for a mobile robot," A.I. Memo864, MIT Artificial Intell. Lab., Cambridge, MA, 1985.

[4] K. M. Chandy and J. Misra, "How processes learn," in *Proc. 4th ACM Symp. on Principles of Distributed Computing*, 1985.

[5] W. F. Clocksin and C. S. Mellish, *Programming in Prolog*. Berlin, FRG: Springer-Verlag, 1984.

[6] D. Davidson, *Actions and Events*. Oxford, England: Clarendon, 1980.

[7] R. E. Fikes and N. J. Nilsson, "STRIPS: A new approach to the application of theorem proving to problem solving," *Artificial Intell.*, vol. 2, pp. 189–208, 1971.

[8] M. P. Georgeff and U. Bonollo, "Procedural expert systems," in *Proc. 8th Int. Joint Conf. on Artificial Intelligence* (Karlsruhe, West Germany, 1983).

[9] M. P. Georgeff, "A theory of action for multiagent planning," in *Proc. 4th Nat. Conf. on Artificial Intelligence* (Austin, TX, 1984).

[10] M. P. Georgeff, A. L. Lansky, and P. Bessiere, "A procedural logic," in *Proc. 9th Int. Joint Conf. on Artificial Intelligence* (Los Angeles, CA, 1985).

[11] M. P. Georgeff and A. L. Lansky, "A system for reasoning in dynamic domains: Fault diagnosis on the space shuttle," Tech. Note 375, Artificial Intelligence Center, SRI Int., Menlo Park, CA, 1985.

[12] G. G. Hendrix, "Modeling simultaneous actions and continuous processes," *Artificial Intell.*, vol. 4, pp. 145–180, 1973.

[13] C. A. R. Hoare, S. D. Brookes, and A. W. Roscoe, "A theory of communicating sequential processes," Tech. Monograph PRG-16, Computing Lab., Oxford Univ., Oxford, England, May 1981.

[14] A. L. Lansky, "Behavioral specification and planning for multiagent domains," Tech. Note 360, Artificial Intelligence Center, SRI Int., Menlo Park, CA, 1985.

[15] J. W. Lloyd, *Foundations of Logic Programming* (Symbolic Computation Series). Berlin, FRG: Springer-Verlag, 1984.

[16] J. McCarthy, "Programs with common sense," in *Semantic Information Processing*, M. Minsky, Ed. Cambridge, MA: MIT Press, 1968.

[17] D. McDermott, "A temporal logic for reasoning about plans and processes," Computer Science Res. Rep. 196, Yale Univ., New Haven, CT, 1981.

[18] R. C. Moore, "Reasoning about knowledge and action," Tech. Note 191, Artificial Intelligence Center, SRI Int., Menlo Park, CA, 1980.

[19] S. J. Rosenschein, "Plan synthesis: A logical perspective," in *Proc. 7th Int. Joint Conf. on Artificial Intelligence*, pp. 331–337, 1981.

[20] —, "Formal theories of knowledge in AI and robotics," Tech. Note 362, Artificial Intelligence Center, SRI Int., Menlo Park, CA, 1985.

[21] E. D. Sacerdoti, *A Structure for Plans and Behavior*. New York, NY: Elsevier-North Holland, 1977.

[22] M. P. Georgeff, A. L. Lansky, and M. Schoppers, "Reasoning and planning in dynamic domains: An experiment with an autonomous robot," Tech. Note 380, Artificial Intelligence Center, SRI Int., Menlo Park, CA, 1986.

[23] M. Stefik, "Planning with constraints," *Artificial Intell.*, vol. 16, pp. 111–140, 1981.

[24] C. Stuart, "An implementation of a multi-agent plan synchronizer using a temporal logic theorem prover," in *Proc. 9th Int. Joint Conf. on Artificial Intelligence* (Los Angeles, CA, 1985).

[25] A. Tate, "Goal structure—Capturing the intent of plans," in *Proc. 6th European Conf. on Artificial Intelligence*, pp. 273–276, 1984.

[26] V. Nguyen and K. J. Perry, "Do we really know what knowledge is," Tech. Note, IBM T.J. Watson Res. Cen., Yorktown Heights, NY, 1986.

[27] S. Vere, "Planning in time: Windows and durations for activities and goals," Tech. Rep., Jet Propulsion Lab., Pasadena, CA, 1981.

[28] D. E. Wilkins, "Domain independent planning: Representation and plan generation," *Artificial Intell.*, vol. 22, pp. 269–301, 1984.

[29] A. T. Woods, "Transition network grammar for natural language analysis," *Commun. ACM*, vol. 13, pp. 591–606, 1970.

Reprinted from *Proceedings of the IEEE*, Volume 74, Number 10, October 1986, pages 1345-1353.

Nonmonotonic Inference Rules for Multiple Inheritance with Exceptions

ERIK SANDEWALL

The semantics of inheritance "hierarchies" with multiple inheritance and exceptions is discussed, and a partial semantics in terms of a number of structure types is defined. Previously proposed inference systems for inheritance with exceptions are discussed. A new and improved inference system is proposed, using a fixed number of nonmonotonic inference rules. The hierarchy is viewed as a set of atomic propositions using the two relations isa (subsumption) and nisa (nonsubsumption). General results concerning systems of nonmonotonic inference rules can immediately be applied to the proposed inference system.

I. Multiple Inheritance with Exceptions: Surprisingly Difficult

Inheritance "hierarchies" are a classical mechanism in artificial intelligence. They allow information to be stated about general concepts, and to be "inherited" by their specializations and instantiations, through one or more layers. Unlike the block structures, and other inheritance mechanisms in other branches of computer science outside AI, artificial intelligence research has often emphasized the need for multiple inheritance, where a more specific concept may inherit information from several more general concepts. For example, the concept "boy" should be able to inherit information both from "male person" and from "child," neither of which is a specialization of the other.

A second requirement on practical inheritance hierarchies is that they should allow for exceptions. For example, the concept "ostrich" is a specialization of "bird," and should inherit the properties of birds, except certain properties such as being able to fly.

Exceptions are fairly easy to deal with in single-inheritance systems, and can be obtained through conventional block structures. Multiple inheritance without exceptions is also easy to deal with theoretically. The combined structure, multiple inheritance with exceptions, offers, however, a number of unpleasant and challenging surprises. The purpose of the present section is to discuss them. In the sequel, we shall use the shorter term "inheritance systems" for our topic, since the precise term "systems with multiple inheritance and exceptions" is too elaborate.

Manuscript received May 6, 1985; revised May 6, 1986. This research was supported by the Swedish Board of Technical Development.

The author is with the Department of Computer and Information Science, Linköping University, 58183 Linköping, Sweden.

We shall represent exceptions through separate exception links in the structure, as previously used by Fahlman [2], Etherington and Reiter [1], and Touretzky [7]. One then has a structure of nodes, and two kinds of links between nodes, which will here be written *isa*(*x*, *y*) and *nisa*(*x*, *y*). The intended meaning is "any *x* is usually a *y*" and "any *x* is usually not a *y*," respectively. The inclusion of "usually" means that those statements may be qualified by other links in the hierarchy. For example, some of the peculiar properties of ostriches and penguins, relative to other birds, may be expressed as follows:

isa(Ostrich, Bird)
isa(Penguin, Bird)
isa(Bird, AbleToFly)
nisa(Ostrich, AbleToFly)
nisa(Bird, UprightWalker)
isa(Penguin, UprightWalker)

Fig. 1 illustrates this structure, and it also illustrates the graphical notation that we shall use in the sequel (and which

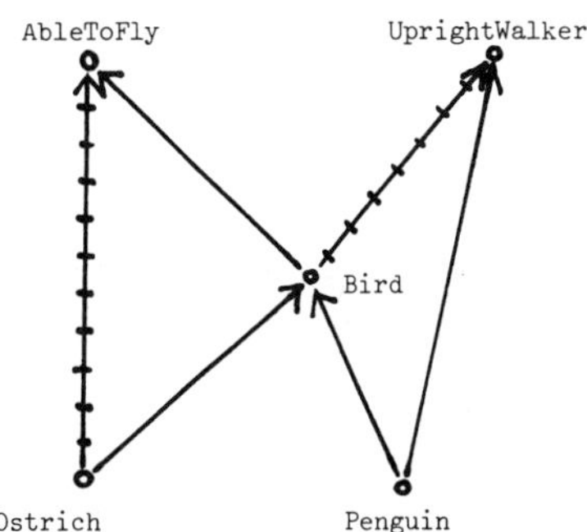

Fig. 1.

is standard in the literature); namely, to represent *isa* by an ordinary arrow and *nisa* by an arrow with crossbars.

The primary use of such a structure is to "answer questions," or more specifically, for answering questions about explicit and implicit (derivable) links between nodes. If *I* is an inheritance structure, then an extension of *I* is a larger structure which contains the same nodes and links as *I*, but where some additional links may have been added; namely, all those links that can be derived from the links in *I*. (No

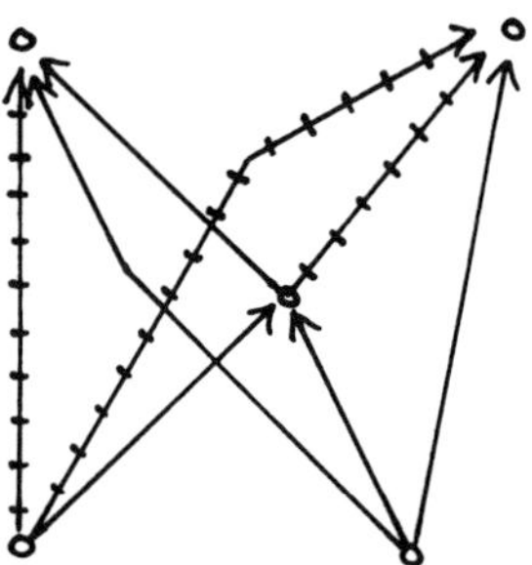

Fig. 2.

extra nodes may be added.) Fig. 2 shows the extension of the structure in Fig. 1.

We shall now consider three inheritance structures which are particularly interesting, and which we shall refer to as Type 1, Type 2, and Type 3 structures. Each of them is a configuration of a few nodes and arcs, so a practical inheritance structure may contain one or more substructures of each type.

The following often-used example, also shown in Fig. 3(a), is a Type 1 structure:

isa(Clyde, RoyalElephant)
isa(RoyalElephant, Elephant)
isa(Elephant, Gray)
nisa(RoyalElephant, Gray)

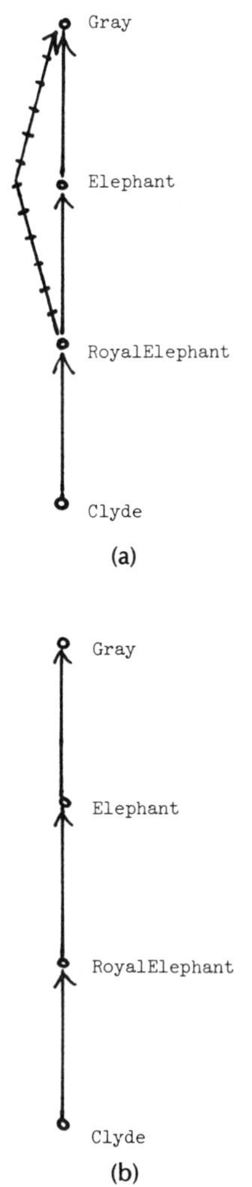

Fig. 3. (a) Type 1.

In this example, we wish to be able to infer

nisa(Clyde, Gray)

because he is a *RoyalElephant*, which are known to be non-*Gray*. Thus the *nisa* link from *RoyalElephant* to *Gray* inhibits the transitivity of *isa* links, which would otherwise imply

isa(RoyalElephant, Gray)

causing a contradiction (or implying that there are no *RoyalElephants*). Furthermore, we wish to be able to infer, in this example,

isa(Clyde, Elephant)

but still the transitive conclusion from that statement plus

isa(Elephant, Gray)

should be blocked.

The particular problem with Type 1 structures is that they introduce nonmonotonicity. Consider the structure in Fig. 3(b), which is a part of Fig. 3(a). The extension of Fig. 3(b) naturally contains the link

isa(Clyde, Gray)

Thus as we go from the structure in Fig. 3(b), to the larger structure in Fig. 3(a), some of the possible conclusions are lost. We say that a logical system is monotonic if it satisfies

$$A \subseteq B \rightarrow Th(A) \subseteq Th(B)$$

where A and B are sets of propositions, and $Th(A)$ is the set of conclusions that can be obtained from A, i.e., the extension. If in our case we let A and B be sets of links, then Type 1 structures fail to satisfy the property of monotonicity.

The following is the standard example of a Type 2 structure, also illustrated in Fig. 4:

isa(Nixon, Republican)
isa(Nixon, Quaker)
isa(Quaker, Pacifist)
nisa(Republican, Pacifist)

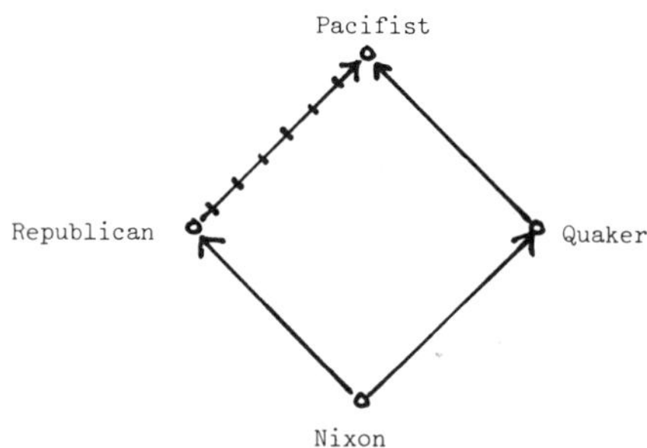

Fig. 4. Type 2.

The puzzling question is of course: is Nixon a Pacifist or is he not? The stance that one usually takes is that this structure has two distinct extensions, as shown in Fig. 5(a) and (b). In one extension, Nixon is a Pacifist, and in the other extension Nixon is not a Pacifist.

If such a structure arises in a practical situation, one would, of course, expect the software to look for additional information which allows it to choose between those two alternatives. But as long as we only study the inheritance

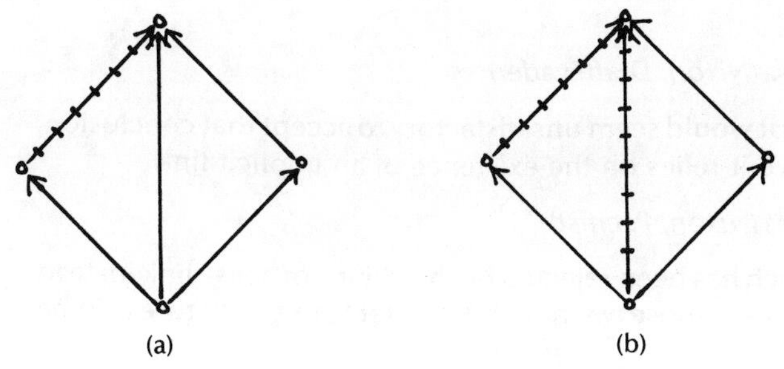

Fig. 5.

structures themselves, the additional information is not available, and the only possible approaches are to allow both alternatives, or to ban such structures altogether.

Although there are two possible extensions, it is not permitted to mix them arbitrarily. For example, if the original structure also contains the link

isa(RichardNixon, Nixon)

which makes sense if the node *"Nixon"* refers to any member of the Nixon family, then the first extension would contain the link

isa(RichardNixon, Pacifist)

and the second extension would instead contain the link

nisa(RichardNixon, Pacifist).

Fig. 6(a) and (b) shows the revised extensions. Notice how the relation between the nodes *RichardNixon* and *Pacifist*

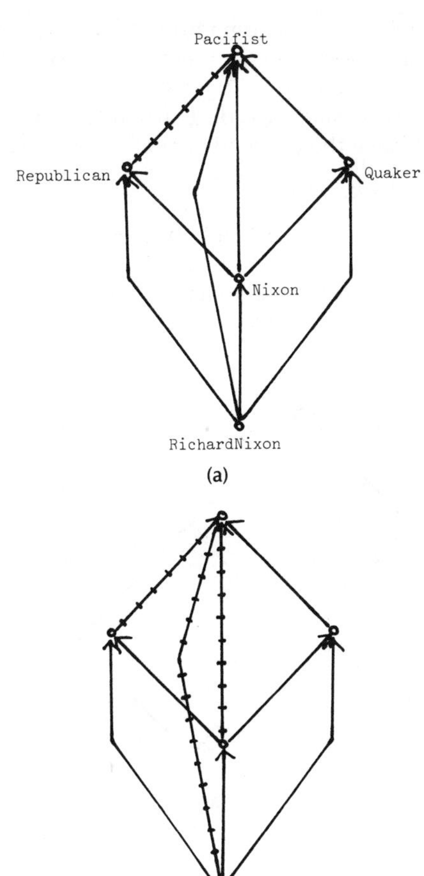

Fig. 6.

are obtained from a Type 1 structure in both cases. An incorrectly designed inference system would produce four extensions, including one where *Nixon* is a *Pacifist* and *RichardNixon* is not, and also the reversed one. Touretzky uses the term decoupling for that anomaly.

It should not come as a surprise that multiple extensions arise in the same context as nonmonotonicity. Sandewall [4] describes the nonmonotonic operator Unless, and the observation that it leads to multiple extensions.

The following is a Type 3 structure, also illustrated in Fig. 7:

isa(Whale, Mammal)
isa(Mammal, LandAnimal)
nisa(Whale, LandAnimal)
isa(LandAnimal, x).

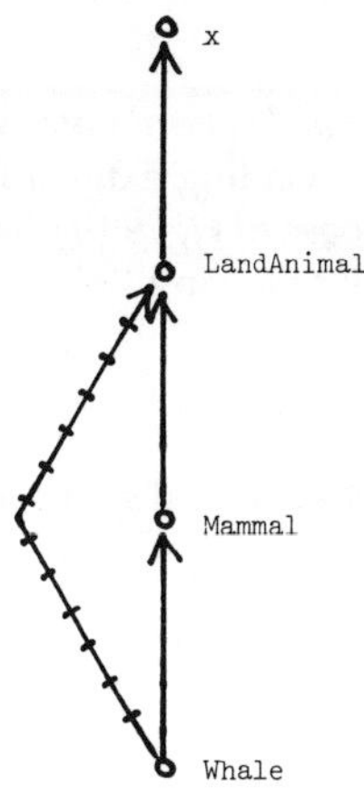

Fig. 7. Type 3.

The issue is whether the link

isa(Whale, x)

should be allowed in the extension(s) of this structure. It seems natural that the answer must be "no," since the only reason why a *Whale* would be an *x* is by the three-step *isa* chain, and the *nisa* link specifies an exception to its lower two links. However, if the link

isa(Mammal, x)

is added to the original structure as an independent fact, then there does not seem to be any reason for not drawing the conclusion that *Whales* are *x*'s. This means that different facts have a different derivational "power." The distinction is not simply between axioms and derived facts, however. Suppose that the two links

isa(Mammal, y)
isa(y, x)

are both added, then again one should be allowed to conclude that *Whales* are *x*'s. The difference in derivational "power" is, therefore, made on the basis of derivational dependency, or which links are derived from, and therefore dependent on which other links.

The following example also illustrates this point. Suppose we augment the *"Nixon"* example of Type 2 structures, with the additional link

isa(Pacifist, DraftEvader)

as in Fig. 8. Like before, we have two choices for the relation

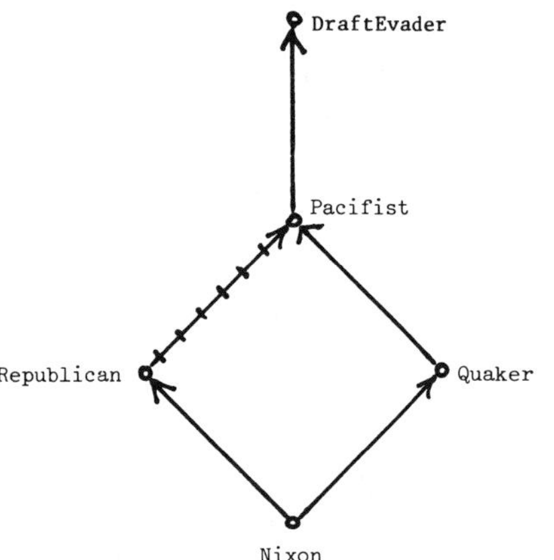

Fig. 8. Type 3B.

between the *Nixon* and *Pacifist* concepts, as shown in Figs. 9(a) and 10(a). The corresponding extensions are shown in Figs. 9(b) and 10(b). The case of Fig. 9(b) offers no surprises, but in Fig. 10(b) both the following links are derived ones:

nisa(Nixon, Pacifist)
isa(Quaker, DraftEvader)

and if those links would have equal status, then one should be allowed to derive

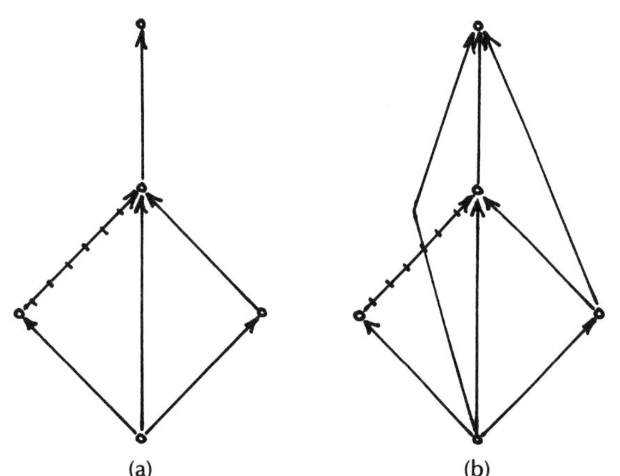

Fig. 9.

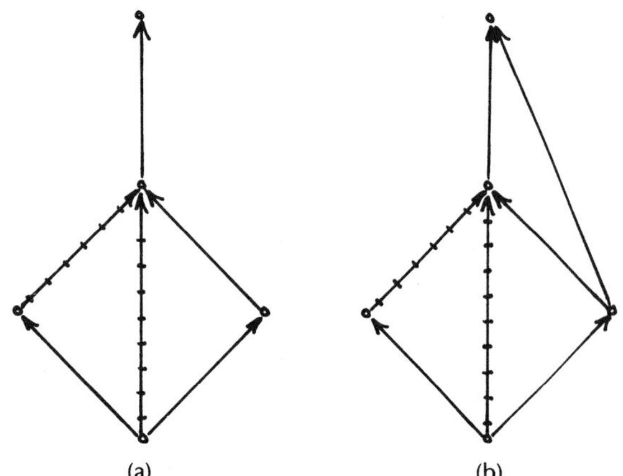

Fig. 10. (b) Type 3B.

isa(Nixon, DraftEvader)

but it would seem unsatisfactory to accept that conclusion, since it relies on the existence of an implicit link

isa(Nixon, Pacifist)

which has been rejected by the choice of a *nisa* link instead between those two nodes. We refer to this structure as Type 3B.

In the context of derivational history, we must also consider the structure shown in Fig. 11, which we shall refer

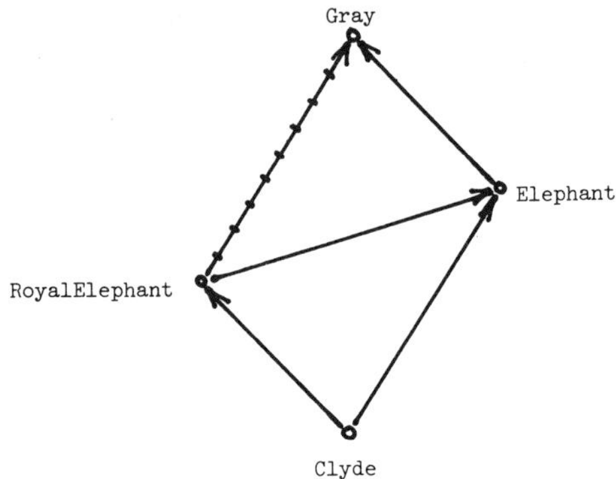

Fig. 11. Type 1B.

to as Type 1B. It is like Type 1, except that the link

isa(Clyde, Elephant)

in the example is now an axiom rather than a derived link. We require that also in this case there shall be only one extension, where *Clyde* is non-*Gray*, just like in Type 1.

Type 1B is also similar to Type 2 structures, except that in Type 1B we have the extra "diagonal" link, namely

isa(RoyalElephant, Elephant)

in the example. It is this extra link that restricts one of the two extensions that Type 2 structures have.

Finally, we give in Fig. 12 a slightly more complex example, which does not introduce any new situation type,

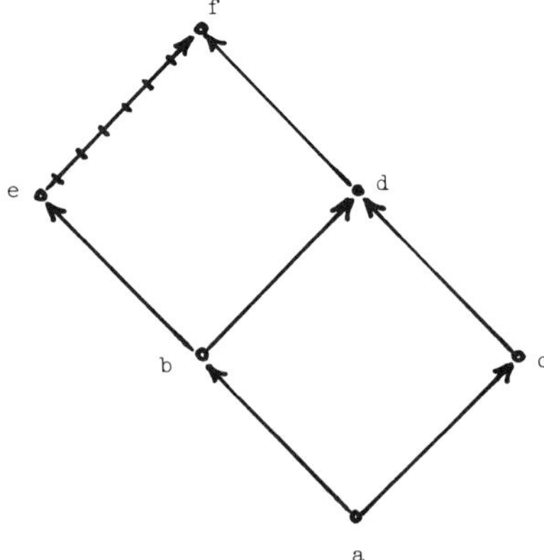

Fig. 12.

but which illustrates the applications of the given three types. (The representation by formulas is omitted.) Fig. 13(a) and (b) shows structures which must certainly be admitted

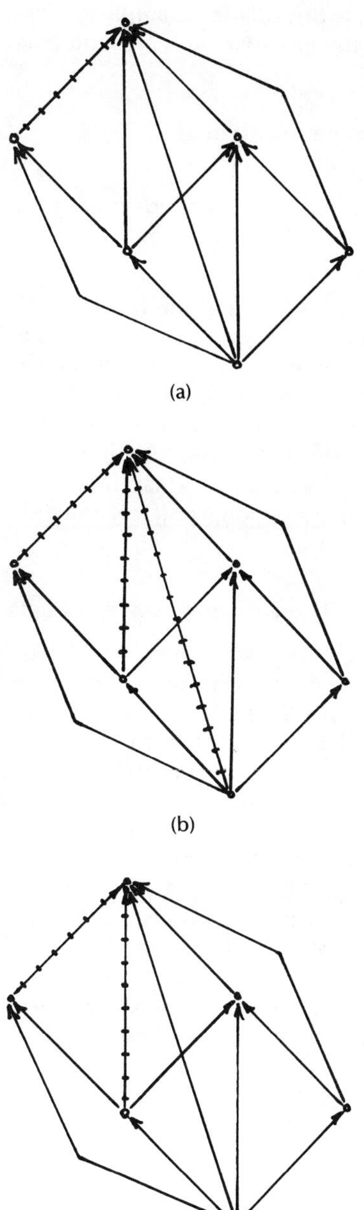
(a)

(b)

(c)

Fig. 13.

as extensions. But what about the structure in Fig. 13(c)? The link *isa(a, f)* may not be derived from *isa(a, d)* and *isa(d, f)*, since we have a Type 1 substructure with *a, b, d, f*. It could, however, be obtained from *isa(a, c)* and *isa(c, f)*, since the nodes *a, b, c, f* form a Type 2 substructure.

At this point it is clear that the structures of multiple inheritance with exceptions have a very nonstandard logic. Their analysis requires the use of concepts such as "conclusions" and "derivation," but the three types of special structures all run counter to the traditional rules and intuitions of logic. Type 1 structures show that we have nonmonotonicity; Type 2 structures show that we have to deal with multiple extensions; and Type 3 structures show how the logical dependency between propositions affect the possible conclusions.

Before we proceed to the formal treatment of these structures, we must, therefore, discuss in some more depth how they relate to ordinary logic.

II. Philosophical Aspects of Inheritance Systems

The concept of having several extensions in a logical system is an unusual one, and may be considered hard to accept. That depends, however, on what we think of as the goal for logic. If the goal is to formulate what are the admissible conclusions, one at a time, from a given set of axioms, then one will expect to have a unique extension; namely, the set of all (individually) admissible conclusions. If, on the other hand, one takes as the goal to specify what are admissible belief structures, when a certain set of "base beliefs" is given, then one should not be so disturbed by the existence of several distinct, but admissible belief structures or extensions. Thus the quest for monotonicity is related to the yearning for the single truth.

We proceed now to the issue of semantics. In order to develop a logical theory, one must first identify the structure of the phenomena that one wishes to describe, and then define the language and inference machinery of the logic, i.e., the syntax of formulas, the choice of axioms and inference rules, and so forth. The semantics of the logical theory is then the formal expression of the "structure of the phenomena," together with an account of how it relates to the selected language.

In the case of inheritance systems, it is not so easy to define semantics. One might like to base semantics on the notion of individual objects, which are "members of" or "included in" the concepts represented by nodes in the inheritance structure. Then *isa(x, y)* would mean something along the lines of "all *x* are *y*" or, since exceptions are allowed, "most *x* are *y*." Exactly what does "most" mean? Suppose 90 percent is sufficient. But then if we have

$$isa(x_0, x_1),\ isa(x_1, x_2), \ldots,\ isa(x_9, x_{10})$$

we might in the worst case obtain that only about 35 percent of the x_0 are x_{10}!

Another annoying problem is that when we analyze fine points of the proposed system (e.g., whether to admit the extension in Fig. 13(c)), proposed solutions cannot be refuted by concrete counter-examples, exactly because it is the purpose of the system to accomodate occasional counter-examples. Consider the example of a Type 3 structure that was given above, where the issue was whether

isa(Whale, x)

should be an admitted conclusion. If we select

x = AnimalWithLegs

then all the links in the structure are in accordance with our common sense knowledge, and *isa(Whale, x)* should not follow. If, on the other hand, we select

x = AnimalThatObtainsOxygenFromAtmosphere

then again all the links in the structure agree with common sense knowledge, and it would be correct for *isa(Whale, x)* to follow. (We cannot choose *x = AnimalWithLungs* since insects do not have lungs.) Regardless whether *isa(Whale, x)* is in the extension or not, we can accomodate both choices of *x* in an application, simply by adding an extra *isa* or an extra *nisa* link, for the case where the deduction machinery does not yield the desired result.

The task of defining an inheritance system might, therefore, be thought of as a design task, rather than as the task of identifying the correct forms of reasoning. The system should be designed so that reasonable consistency is obtained, and so that one minimizes the number of extra *isa* or *nisa* links that have to be introduced in order to adapt the results of the inference machinery to reality.

The published attempts to define the formal properties of inheritance systems have, therefore, used a nonstandard inference process for defining extensions. One then defines an inference operator that will add one more link (or other, similar construct) to an inheritance structure. The admissible extensions of an inheritance structure *I* are defined to be those supersets of *I* which are closed with respect to the inference operator (i.e., everything that the operator would add to the closure, is already a member there), and which are well founded (i.e., all members of the extension must either be members of *I*, or have been obtained through the inference process).

By the standards of traditional logic, it is debatable whether such an inference process qualifies as a satisfactory semantics. One would like to have a semantics which defines the underlying meaning, and which the inference process could be measured against. But as we have already seen, such an underlying meaning is hard to identify and make precise. Inheritance systems were introduced because of their utility, not because of elegant formal properties.

Another, although related, problem is that the inference processes for inheritance systems that have been proposed so far, are not particularly transparent. The reader is not likely to accept them as they stand, and in fact the only reasonable way to approach them is through specific cases, like our Type 1, 2, and 3 structures. I, therefore, propose that we consider such collections of structure types as the definition of the semantics, for the time being. They do satisfy two basic requirements for a semantics: they can easily be grasped intuitively, and one can effectively determine whether a proposed inference process will deal with them correctly. The remaining problem is that they are only partial semantics, so it would be possible to have two inference processes, both of which agree with the semantics, i.e., they treat the specified structure types correctly, but still they can be proven to be nonequivalent. If and when that happens, one will have to identify additional structure types for which those processes differ, and add them to the semantics, thereby making it more precise.

If we think of logic and/or information science as an empirical endeavor, where information structures which actually occur in the real world are identified and formally characterized, then such an iterative strategy should be quite acceptable.

III. Current Theories for Inheritance Systems

A. Etherington and Reiter: Default Logic

Etherington and Reiter [1] express each link in the inheritance structure as an inference rule in default logic [3]. They give criteria and proofs for the existence of extensions, and prove the correctness of certain inference algorithms. They claim that the resulting system defines a semantics for inheritance systems.

Their approach, however, suffers from the problem that the transformation from links to inference rules is not a local operation. In the "Clyde" example for Type 1 structures, for instance, the link that we represented as

isa(Elephant, Gray)

would have to be rewritten as

if Elephant(x)
and unless it leads to a contradiction to assume
$Gray(x) \wedge \neg RoyalElephant(x)$
then Gray(x)

As Touretzky [6] pointed out, the need to write exceptional cases explicitly into the inference rules that may be affected by them, is very impractical. Also, of course, the very point with nonmonotonic reasoning and exception links is that we should not have to perform that chore.

In order to deal with Type 3 structures, where derivational dependency comes into play, one would presumably have to introduce additional artifacts into the inference machinery.

B. Touretzky's Theory for Inheritance Systems

Touretzky [6], [7], uses an approach which is more remote from ordinary logic: he considers an inheritance system as a set of paths, where each path is an annotated sequence of nodes. A link is a path of length 2, and in the originally given structure all paths are links. The inference operator combines two paths of length *k* into one new path of length $k + 1$. Therefore, unlike ordinary logic, each derived path contains its entire derivation history. This allows a correct treatment of Type 3 structures. Also, although the inference operator primarily operates on two existing paths, it may be inhibited by other paths in the structure. This allows a correct treatment of Type 1 and Type 2 structures as well.

The following summarizes the concepts and notation that are used in [7, ch. 2], together with some additional notation that we need.

We consider sequences $\langle u_1, \ldots, u_n \rangle$, called paths, where each u_i has either of the following forms

$+x_i$
$-x_i$

where again each x_i is a node in the network, and where the sign must be + in each u_i, except u_n. (Minus signs in other positions would be used, e.g., for "non-*x* objects are *y*'s," expressed as

$$\langle -x, +y \rangle$$

but Touretzky [7] does not analyze such constructs.)

Touretzky [7] also allows u_i of the form

$$\#x_i$$

but we ignore that case here since it does not add to the difficulty of the analysis. A path characterizes the conjunction of the following propositions:

$isa(x_1, x_2)$
$isa(x_2, x_3)$
...
$isa(x_{n-1}, x_n)$, if $u_n = +x_n$
$nisa(x_{n-1}, x_n)$, if $u_n = -x_n$.

We write $u \pm v$ when *u* and *v* contain the same node but with opposite signs. (Thus $\pm$ is a relation symbol here.) If

Φ is a set of paths, $C(\Phi)$ is defined by

$$C(\Phi) = \{\langle u_1, u_n\rangle \mid \langle u_1, u_2, \ldots, u_n\rangle \in \Phi\}.$$

A path $\sigma = \langle u_1, u_2, \ldots, u_{n-1}, u_n\rangle$, where $n > 2$, is said to be inheritable in a set Φ of paths iff the following holds:

$\langle u_1, \ldots, u_{n-1}\rangle \in \Phi$
$\langle u_2, \ldots, u_n\rangle \in \Phi$
Φ does not contradict σ
Φ does not preclude σ

where the last two conditions are defined as follows:

Φ contradicts σ iff some $\langle u_1, v\rangle \in C(\Phi)$ where $v \pm u_i$ for some i.
Φ precludes σ iff some $\langle v, w\rangle \in \Phi$ where $w \pm u_k$ for some k, where
$1 < k \leq n$, and either
$v = u_i$ for some $i \leq k$, or:
$\langle u_1, \ldots, u_i, v_1, \ldots, v_m, u_{i+1}\rangle \in \Phi$
and $v = v_j$ for some j
and $1 \leq i < k$.

Actually, Touretzky [7] first introduces this definition with the stronger restriction $k = n$, and then relaxes it to the above formulation. He hypothesises that the two variants result in the same extensions.

The set Φ of paths is a grounded expansion of a set S iff:

$S \subseteq \Phi$
Φ is closed under inheritance
every path in $\Phi - S$ is inheritable in Φ.

This is what we have called an extension above, and we shall use the two terms synonymously.

Since an inherited path is longer than the two paths that were used to form it, for every grounded expansion Φ of S there must be some sequence of sets

$S_0, S_1, \ldots, S_m$

where $S = S_0$, $S_m = \Phi$, and each S_i is obtained from its predecessor by adding one or more paths that is inheritable in S_{i-1}.

It is easy to verify that this definition treats our structure types correctly. In Type 1 structures there are two potential and contradictory derived links (*Clyde* is *Gray* or non-*Gray*, in the example), and one of them has to be inhibited. This is done through the preclusion condition. The two "either" cases in the definition of preclusion deal with Type 1 and Type 1B, respectively.

In Type 2 structures, the two extensions are again obtained through the preclusion condition, which guarantees that whichever of the competing links is derived first, that will inhibit the other.

Type 3 structures, finally, are handled correctly by virtue of the contradiction requirement. For Type 3B, the last clause in the definition of a grounded expansion plays a key role for avoiding the incorrect conclusion

isa (Nixon, DraftEvader)

in the example in Fig. 10(b). That proposition is, in fact, derivable in some subsets of the grounded expansion, but it is not rederivable in the full expansion.

As we have already discussed, it is difficult to understand the full consequences of Touretzky's definitions. For example, although Type 1B is handled correctly, the slightly

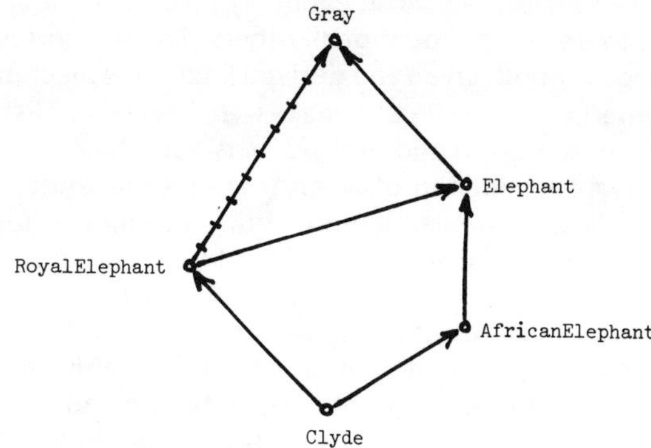

Fig. 14. Type 1C.

modified structure shown in Fig. 14 will result in two extensions, i.e., one where *Clyde* is *Gray* and one where he is non-*Gray*. The reason is that the definition of preclusion only allows the sequence of nodes $v_1, \ldots, v_m$ to be inserted between two successive nodes u_i and u_{i+1}, so the extra node "*AfricanElephant*" in Fig. 14 results in a loss of the preclusion condition. We refer to this structure as Type 1C.

This apparent anomaly supports our suggestion that the collection of structure types should be seen as the currently available, partial semantics for inheritance systems. New, proposed theories for inheritance should first be calibrated against that semantics, and only then should one maybe investigate whether they are equivalent to Touretzky's or other existing theories.

IV. An Alternative Inference Machinery, Using Nonmonotonic Logic, for Multiple Inheritance with Exception

Touretzky chose to use paths, or sequences of nodes, as the "propositions" in his inference machinery. This allows him to capture the special properties of the structure types, but it disassociates him from ordinary logic, which makes it particularly difficult to extend the system, or combine it with other logic. Etherington and Reiter, on the other hand, stayed within a nonmonotonic logic with considerable generality, but at the expense of sacrificing the spirit of exception statements.

In this section we shall show that it is possible to obtain the best of both worlds. We use propositions of the following forms:

isax (x, y, s)
isa (x, y, s)
precl (x, y, z, s)
cntr (x, y, z, s)

where *x*, *y*, and *z* are nodes, and *s* is either of the symbols + or −. The ternary *isa* relation is defined by

isa (x, y, +) = isa (x, y)
isa (x, y, −) = nisa (x, y).

If we wish to stay within a single-sorted logic we can, of course, view ternary *isa* as an abbreviation. Type 3 structures, and the dependency conditions they imply, are handled by the separate relation *isax* which is a stronger form of *isa*. Thus *isax (x, y, s)* means that *isa (x, y, s)* independently of other links. In practice one would use *isax* for stating the explicitly given facts, or axioms (hence the relation's name), or at least for stating those explicitly given facts that are known to be independent of all others. The relation *isa* is then used for both given and derived links. This machinery is counterintuitive, since it makes an irrelevant distinction—between given and derived facts—and fails to make the relevant distinction of when there are dependencies between facts. However, it does work, since independent, derived facts are treated in a different way, and as we do not view the inference machinery as a way of clarifying the semantics we do not have any qualms about it.

The relations *precl* and *cntr* are used for technical purposes, which will be apparent in the inference rules. They are vaguely related to Touretzky's preclusion and contradiction conditions, respectively.

If *s* is + or −, then −*s* is the opposite sign as *s*.

Inference rules are written in the general form "If Φ_1 is known, and Φ_2 is not known, then infer Φ_3." As we showed in [5] one may determine extensions for such rules as follows: Construct a sequence of successively increasing sets of propositions,

$$\Gamma_0, \Gamma_1, \Gamma_2, \ldots, \Gamma_n, \ldots$$

where each Γ_{i+1} is constructed from Γ_i by selecting an instantiation of a rule where Φ_1 is a subset of Γ_i, and Φ_2 is disjoint from Γ_i, and then choosing Γ_{i+1} as $\Gamma_i \cup \Phi_3$. The process is continued to its (possibly infinite) limit, where we obtain an extension. Depending on the order in which the rules are applied, we may obtain several different extensions. Among them, all consistent extensions are accepted, but not those extensions containing both the propositions *p* and $\neg p$ for some *p*.

Intuitively speaking, this makes it possible for an inference rule to generate "mines" which will later on create a contradiction in the proof sequence, and thereby render it invalid. The relations *precl* and *cntr* are used for exactly this purpose.

The following is our set of inference rules:

1) if *isax(x, y, s)* $\in \Gamma$
   then add to Γ:
      isa(x, y, s)
2) if *isa(x, y, s)* $\in \Gamma$
   then add to Γ:
      $\neg$*isa(x, y, −s)*
3) if the following are members of Γ:
      isa(x, y, +)
      isa(y, z, s)
   and the following are not members of Γ:
      isa(x, z, −s)
      cntr(x, y, z, s)
   then add to Γ:
      isa(x, z, s)
      precl(x, y, z, s)
      $\neg$*cntr(x, y, z, s)*
4) if the following are members of Γ:
      precl(x, y, z, s)
      isa(x, v, +)
      isa(v, y, +)
   then add to Γ:
      precl(x, v, z, s)
5) if the following are members of Γ:
      precl(x, y, z, s)
      isa(y, z, −s)
   then add to Γ:
      $\neg$*isa(y, z, −s)*
6) if the following are members of Γ:
      isa(x, y, +)
      isa(y, z, +)
      isa(x, z, −)
      isa(z, v, s)
   and the following is not a member of Γ:
      isax(y, v, s)
   then add to Γ:
      cntr(x, y, v, s).

In order to understand how these rules work, let us first ignore Rule 6 and the *cntr* literals in Rule 3. The thus simplified rules are sufficient for dealing with Type 1 and Type 2 structures, and their variants.

Rule 3 is set up so that it will not produce combinations of the form

isa(x, y, +) and *isa(x, y, −)*

which would immediately lead to a contradiction through Rule 2. Through Rule 3 we, therefore, obtain a correct treatment of Type 2 structures. Also, when Rule 3 is applied, the auxiliary proposition

precl(x, y, z, s)

is obtained. Rule 4 then "slides" the second argument of the *precl* proposition down through the structure, which is used for spotting Type 1, 1B, and 1C structures. Rule 5 is used to invalidate undesirable extensions. (This is a counterpart of backtracking, in the logical system.)

Consider, for example, the contents of Fig. 3(a), which would be represented in our notation as follows:

isa(Clyde, RoyalElephant, +)
isa(RoyalElephant, Elephant, +)
isa(Elephant, Gray, +)
isa(RoyalElephant, Gray, −).

There are two main alternatives for the order in which the inference rules are applied. If we start by inferring that *Clyde* is non-*Gray*, we obtain using Rule 3:

isa(Clyde, Gray, −)
precl(Clyde, RoyalElephant, Gray, −).

After that we also conclude that Clyde is an elephant, and obtain:

isa(Clyde, Elephant, +)
precl(Clyde, RoyalElephant, Elephant, +)

An attempt to deduce that *Clyde* is *Gray* at this point is prevented by the inhibiting conditions in Rule 3, since we have already concluded that he is non-*Gray*.

In the other main alternative, we do not first conclude that *Clyde* is non-*Gray*. Instead we start with

isa(Clyde, Elephant, +)
precl(Clyde, RoyalElephant, Elephant, +)

using Rule 3, and then

isa(Clyde, Gray, +)
precl(Clyde, Elephant, Gray, +).

At this point we cannot conclude that *Clyde* is non-*Gray*, because of the symmetrical character of Rule 3. However,

Rule 4 allows us to conclude

precl(Clyde, RoyalElephant, Gray, +)

and then Rule 5 identifies the problem and reacts by intentionally creating a contradiction, namely

$\neg$ *isa(RoyalElephant, Gray, −)*

which contradicts what was originally stated as

isa(RoyalElephant, Gray, −).

In summary, two extensions are created, but one of them develops an inconsistency, and only the other one is admissible. For simplicity, in this example we have started from the *isa* propositions, and ignored the original step from *isax* to *isa*.

This machinery for Type 1 structures also prevents decoupling.

For Type 3 structures, we need the full set of rules. Rule 6 identifies the Type 3 structure. For example, in Fig. 7, we would have x = *Whale*, y = *Mammal*, z = *LandAnimal*, and $v = x$ in Rule 6. Also Rule 6 can be applied if *isa(Mammal, x)* is only a derived link, but not if it is a given link, because in the latter case there will be an *isax* proposition. The conclusion from Rule 6,

cntr(x, y, v, s)

indicates that it is not allowed to combine

isa(x, y, +)
isa(y, v, s)

using Rule 3. Rule 3, therefore, checks for this condition in a nonmonotonic assumption, so if Rule 6 has been used then Rule 3 will not be applied. On the other hand, if Rule 3 has been applied first, then Rule 6 will be applied anyway, and a contradiction will occur which will invalidate the present derivation.

If the "bypass link" *isa(Mammal, x)* is derived but independent, because there is some separate node *r* for which *isa(Mammal, r)* and *isa(r, x)*, then it will not be possible to chain *isa(Whale, Mammal)* with *isa(Mammal, x)*, to obtain the conclusion *isa(Whale, x)* that we should be allowed to have in this case. However, the desired conclusion can be obtained in another order, namely

isa(Whale, Mammal)
isa(Whale, r)
isa(Whale, x)

without being inhibited by Rule 6.

In summary, the proposed inference machinery handles the three structure types and their variants correctly. It is clearly not equivalent to Touretzky's machinery since they handle Type 1C differently. It would probably not lead to any complications to change his system in that respect, but we have not tried to prove the equivalence between a revision of his system and our system.

It is certainly too early to claim that the present system, or any other proposed inference machinery, deals correctly with all situation types that one can think of. (In particular, the literature rarely deals with the case of circular *isa* structures, and they are also not represented in the structure types in this paper.) In fact, it is not even clear that there will always be a good common-sense answer as to what is the correct way of dealing with a situation type. The present contribution is only claimed to be a step in the iterative discovery process that we must pursue at present. Meanwhile, it has also pointed towards a way of accomodating the peculiar characteristics of inheritance systems semantics in nonmonotonic logic, for which some results at least are known. Thus it is not necessary to use the inheritance paths, containing the trace of the derivation, as Touretzky does.

In a previous paper about nonmonotonic logic [5], we proved results for systems of nonmonotonic inference rules which are directly applicable here. The following results follow directly, with the requirement on extensions that they must be consistent as we have assumed in this section:

Proposition: Let Φ and Φ' be two extensions of an inheritance structure S, for which $\Phi \subseteq \Phi'$. Then $\Phi = \Phi'$.

Proposition. Every union of distinct extensions of an inheritance structure, is inconsistent.

Touretzky proved these results for his system, but for the latter proposition his proof only applies if the *isa* relation in S is acyclic. By relying on nonmonotonic logic we obtain the same results as special cases of generally applicable theorems, and we obtain a more general result since we do not need the limitation of an acyclic *isa* relation.

References

[1] D. W. Etherington and R. Reiter, "On inheritance hierarchies with exceptions," in *Proc. (U.S.) Nat. Conf. on Artificial Intelligence*, pp. 104–108, 1983.

[2] S. E. Fahlman, *NETL: A System for Representing and Using Real-World Knowledge.* Cambridge, MA: MIT Press, 1979.

[3] R. Reiter, "A logic for default reasoning," *Artificial Intell.*, vol. 13, no. 1, pp. 81–132, 1980.

[4] E. Sandewall, "An approach to the frame problem, and its implementation," in B. Meltzer and D. Michie, Eds., *Machine Intelligence 7.* New York, NY: Wiley, 1972, pp. 195–204.

[5] ——, "A functional approach to non-monotonic logic," in *Proc. 1985 Int. Joint. Conf. on Artificial Intelligence*, pp. 00–00, 1985, and *Computat. Intell.*, vol. 1, no. 2, pp. 00–00, 1985.

[6] D. S. Touretzky, "Implicit ordering of defaults in inheritance systems," in *Proc. (U.S.) Nat. Conf. on Artificial Intelligence*, pp. 322–325, 1984.

[7] ——, "The mathematics of inheritance systems," Rep. CMU-CS-84-136, Dep. Comput. Sci., Carnegie-Mellon Univ., Pittsburgh, PA, 1984; also London, UK: Pitman, 1986.

Artificial Intelligence and Language Processing

David Waltz Editor

Maintaining Knowledge about Temporal Intervals

JAMES F. ALLEN *The University of Rochester*

ABSTRACT: ***An interval-based temporal logic is introduced, together with a computationally effective reasoning algorithm based on constraint propagation. This system is notable in offering a delicate balance between expressive power and the efficiency of its deductive engine. A notion of reference intervals is introduced which captures the temporal hierarchy implicit in many domains, and which can be used to precisely control the amount of deduction performed automatically by the system. Examples are provided for a database containing historical data, a database used for modeling processes and process interaction, and a database for an interactive system where the present moment is continually being updated.***

James F. Allen's main interests are in artificial intelligence in particular natural language processing and the representation of knowledge.

Author's Present Address: James F. Allen, Computer Science Department, University of Rochester, Rochester, NY 14627.

The research described in this paper was supported in part by the National Science Foundation under Grants IST-80-12418 and IST-82-10564, and in part by the Office of Naval Research under Grant N00014-80-C-0197.

1. INTRODUCTION

The problem of representing temporal knowledge and temporal reasoning arises in a wide range of disciplines, including computer science, philosophy, psychology, and linguistics. In computer science, it is a core problem of information systems, program verification, artificial intelligence, and other areas involving process modeling. (For a recent survey of work in temporal representation, see the special sections in the April 1982 issues of the ACM SIGART and SIGMOD Newsletters.)

Information systems, for example, must deal with the problem of outdated data. One approach to this is simply to delete outdated data; however, this eliminates the possibility of accessing any information except that which involves facts that are presently true. In order to consider queries such as, "Which employees worked for us last year and made over $15,000," we need to represent temporal information. In some applications, such as keeping medical records, the time course of events becomes a critical part of the data.

In artificial intelligence, models of problem solving require sophisticated world models that can capture change. In planning the activities of a robot, for instance, one must model the effects of the robot's actions on the world to ensure that a plan will be effective. In natural language processing, researchers are concerned with extracting and capturing temporal and tense information in sentences. This knowledge is necessary to be able to answer queries about the sentences later. Further progress in these areas requires more powerful representations of temporal knowledge than have previously been available.

This paper addresses the problem from the perspective of artificial intelligence. It describes a temporal representation that takes the notion of a *temporal interval* as primitive. It then describes a method of representing the relationships between temporal intervals in a hierarchical manner using constraint propagation techniques. By using *reference intervals,*

"Maintaining Knowledge about Temporal Intervals" by J.F. Allen from *Communications of the ACM,* November 1983, Volume 26, Number 11, pages 832-843. Copyright © 1983, Association for Computing Machinery, Inc., reprinted by permission.

the amount of computation involved when adding a fact can be controlled in a predictable manner. This representation is designed explicitly to deal with the problem that much of our temporal knowledge is relative, and hence cannot be described by a date (or even a "fuzzy" date).

We start with a survey of current techniques for modeling time, and point out various problems that need to be addressed. After a discussion of the relative merits of interval-based systems versus point-based systems in Section 3, a simple interval-based deduction technique based on constraint propagation is introduced in Section 4. This scheme is then augmented in Section 5 with reference intervals, and examples in three different domains are presented. In the final sections of the paper, extensions to the basic system are proposed in some detail. These would extend the representation to include reasoning about the duration of intervals, reasoning about dates when they are available, and reasoning about the future given knowledge of what is true at the present.

The system as described in Section 5 has been implemented and is being used in a variety of research projects which are briefly described in Section 6. Of the extensions, the duration reasoner is fully implemented and incorporated into the system, whereas the date reasoner has been designed but not implemented.

2. BACKGROUND

Before we consider some previous approaches to temporal representation, let us summarize some important characteristics that are relevant to our work:

- The representation should allow significant imprecision. Much temporal knowledge is strictly relative (e.g., A is before B) and has little relation to absolute dates.
- The representation should allow uncertainty of information. Often, the exact relationship between two times is not known, but some contraints on how they could be related are known.
- The representation should allow one to vary the grain of reasoning. For example, when modeling knowledge of history, one may only need to consider time in terms of days, or even years. When modeling knowledge of computer design, one may need to consider times on the order of nanoseconds or less.
- The model should support *persistence*. It should facilitate default reasoning of the type, "If I parked my car in lot A this morning, it should still be there now," even though proof is not possible (the car may have been towed or stolen).

This does not exhaust all the issues, and others will come up as they become relevant. It provides us with a starting criteria, however, for examining previous approaches. Previous work can be divided roughly into four categories: state space approaches, date line systems, before/after chaining, and formal models.

State space approaches (e.g., [7, 17]) provide a crude sense of time that is useful in simple problem-solving tasks. A state is a description of the world (i.e., a database of facts) at an instantaneous point in time. Actions are modeled in such systems as functions mapping between states. For example, if an action occurs that causes P to become true and causes fact Q to be no longer true, its effect is simulated by simply adding fact P to the current state and deleting fact Q. If the previous states are retained, we have a representation of time as a series of databases describing the world in successive states. In general, however, it is too expensive to maintain all the previous states, so most systems only maintain the present state. While this technique is useful in some applications, it does not address many of the issues that concern us. Note that such systems do provide a notion of persistence, however. Once a fact is asserted, it remains true until it is explicitly deleted.

In datebase systems (e.g., [4, 5, 12, 13]), each fact is indexed by a *date*. A date is a representation of a time such that the temporal ordering between two dates can be computed by fairly simple operations. For example, we could use the integers as dates, and then temporal ordering could be computed using a simple numeric comparison. Of course, more complicated schemes based on calendar dates and times are typically more useful. Because of the nice computational properties, this is the approach of choice if one can assign dates for every event. Unfortunately, in the applications we are considering, this is not a valid assumption. Many events simply cannot be assigned a precise date. There are methods of generalizing this scheme to include ranges of dates in which the event must occur, but even this scheme cannot capture some relative temporal information. For instance, the fact that two events, A and B, did not happen at the same time cannot be represented using fuzzy dates for A and B. Either we must decide that A was before B, or B was before A, or we must assign date ranges that allow A and B to overlap. This problem becomes even more severe if we are dealing with time intervals rather than time points. We then need fuzzy date ranges for both ends of the interval plus a range for the minimum and maximum duration of the interval.

The next scheme is to represent temporal information using before/after chains. This approach allows us to capture relative temporal information quite directly. This technique has been used successfully in many systems (e.g., [4, 13]). As the amount of temporal information grows, however, it suffers from either difficult search problems (searching long chains) or space problems (if all possible relationships are precomputed). This problem can be alleviated somewhat by using a notion of reference intervals [13], which will be discussed in detail later. Note that a fact such as "events A and B are disjoint" cannot be captured in such systems unless disjunctions can be represented. The approach discussed in this paper can be viewed as an extension of this type of approach that overcomes many of its difficulties.

Finally, there is a wide range of work in formal models of time. The work in philosophy is excellently summarized in a textbook by Rescher and Urquhart [16]. Notable formal models in artificial intelligence include the situation calculus [14], which motivates much of the state space based work in problem solving, and the more recent work by McDermott [15]. In the situation calculus, knowledge is represented as a series of situations, each being a description of the world at an instantaneous point of time. Actions and events are functions from one situation to another. This theory is viable only in domains where only one event can occur at a time. Also, there is no concept of an event taking time; the transformation between the situations cannot be reasoned about or decomposed. The situation calculus has the reverse notion of persistence: a fact that is true at one instance needs to be explicitly reproven to be true at succeeding instants.

Most of the work in philosophy, and both the situation calculus and the work by McDermott, are essentially point-based theories. Time intervals can be constructed out of points, but points are the foundation of the reasoning system. This approach will be challenged in the upcoming section.

One other formal approach, currently under development, that is compatible with an interval-based temporal representation is found in the Naive Physics work of Hayes [10, 11]. He proposes the notion of a *history*, which is a contiguous block of space-time upon which reasoning can be organized. By viewing each temporal interval as one dimension of a history, this work can be seen as describing a reasoning mechanism for the temporal component of Naive Physics.

3. TIME POINTS VS. TIME INTERVALS

In English, we can refer to times as points or as intervals. Thus we can say the sentences:

We found the letter at twelve noon.
We found the letter yesterday.

In the first, "at twelve noon" appears to refer to a precise point in time at which the finding event occurred (or was occurring). In the second, "yesterday" refers to an interval in which the finding event occurred.

Of course, these two examples both refer to a date system where we are capable of some temporal precision. In general, though, the references to temporal relations in English are both implicit and vague. In particular, the majority of temporal references are implicitly introduced by tense and by the description of how events are related to other events. Thus we have

We found the letter while John was away.
We found the letter after we made the decision.

These sentences introduce temporal relations between the times (intervals) at which the events occurred. In the first sentence, the temporal connective "while" indicates that the time when the find event occurred is during the time when John was away. The tense indicates that John being away occurred in the past (i.e., before now).

Although some events appear to be instantaneous (e.g., one might argue that the event "finding the letter" is instantaneous), it also appears that such events could be decomposed if we examine them more closely. For example, the "finding the letter" might be composed of "looking at spot X where the letter was" and "realizing that it was the letter you were looking at." Similarly, we might further decompose the "realizing that it was the letter" into a series of inferences that the agent made. There seems to be a strong intuition that, given an event, we can always "turn up the magnification" and look at its structure. This has certainly been the experience so far in physics. Since the only times we consider will be times of events, it appears that we can always decompose times into subparts. Thus the formal notion of a time point, which would not be decomposable, is not useful. An informal notion of time points as very small intervals, however, can be useful and will be discussed later.

There are examples which provide counterintuitive results if we allow zero-width time points. For instance, consider the situation where a light is turned on. To describe the world changing we need to have an interval of time during which the light was off, followed by an interval during which it was on. The question arises as to whether these intervals are open or closed. If they are open, then there exists a time (point) between the two where the light is neither on nor off. Such a situation would provide serious semantic difficulties in a temporal logic. On the other hand, if intervals are closed, then there is a time point at which the light is both on and off. This presents even more semantic difficulties than the former case. One solution to this would be to adopt a convention that intervals are closed in their lower end and open on their upper end. The intervals could then meet as required, but each interval would have only one endpoint. The artificiality of this solution merely emphasizes that a model of time based on points on the real line does not correspond to our intuitive notion of time. As a consequence, we shall develop a representation that takes temporal intervals as primitive.

If we allowed time points, intervals could be represented by modeling their endpoints (e.g., [4]) as follows: Assuming a model consisting of a fully ordered set of points of time, an interval is an ordered pair of points with the first point less than the second. We then can define the relations in Figure 1 between intervals, assuming for any interval t, the lesser endpoint is denoted by t− and the greater by t+.

We could implement intervals with this approach, even given the above argument about time points, as long as we assume for an interval t that $t- < t+$, and each assertion made is in a form corresponding to one of the interval relations. There are reasons why this is still inconvenient, however. In particular, the representation is too uniform and does not facilitate structuring the knowledge in a way which is convenient for typical temporal reasoning tasks. To see this, consider the importance of the *during* relation. Temporal knowledge is often of the form

event E′ occurred during event E.

A key fact used in testing whether some condition P holds during an interval t is that if t is during an interval T, and P holds during T, then P holds during t. Thus *during* relationships can be used to define a hierarchy of intervals in which propositions can be "inherited."

Furthermore, such a *during* hierarchy allows reasoning processes to be localized so that irrelevant facts are never considered. For instance, if one is concerned with what is true "today," one need consider only those intervals that are *during* "today," or above "today" in the *during* hierarchy. If a fact is indexed by an interval wholly contained by an interval representing "yesterday," then it cannot affect what is true now. It is not clear how to take advantage of these properties using the point-based representation above.

4. MAINTAINING TEMPORAL RELATIONS

4.1. The Basic Algorithm

The inference technique described in this section is an attempt to characterize the inferences about time that appear to be made automatically or effortlessly during a dialogue, story comprehension, or simple problem-solving. Thus it should provide us with enough temporal reasoning to participate in these tasks. It does not, however, need to be able to account for arbitrarily complex chains of reasoning that could be done, say, when solving a puzzle involving time.

We saw above five relations that can hold between intervals. Further subdividing the *during* relation, however, pro-

Interval Relation	Equivalent Relations on Endpoints
t < s	t+ < s-
t = s	(t- = s-) & (t+ = s+)
t overlaps s	(t- < s-) & (t+ > s-) & (t+ < s+)
t meets s	t+ = s-
t during s	((t- > s-) & (t+ =< s+)) or ((t- >= s-) & (t+ < s+))

FIGURE 1. Interval Relation Defined by Endpoints.

vides a better computational model.[1] Considering the inverses of these relations, there are a total of thirteen ways in which an ordered pair of intervals can be related. These are shown in Figure 2.

Sometimes it is convenient to collapse the three *during* relations (d, s, f) into one relationship called *dur*, and the three *containment* relations (di, si, fi) into one relationship called *con*. After a quick inspection, it is easy to see that these thirteen relationships can be used to express any relationship that can hold between two intervals.

The relationships between intervals are maintained in a network where the nodes represent individual intervals. Each arc is labeled to indicate the possible relationship between the two intervals represented by its nodes. In cases where there is uncertainty about the relationship, all possible cases are entered on the arc. Note that since the thirteen possible relationships are mutually exclusive, there is no ambiguity in this notation. Figure 3 contains some examples of the notation. Throughout, let N_i be the node representing interval i. Notice that the third set of conditions describes disjoint intervals.

Throughout this paper, both the above notations will be used for the sake of readability. In general, if the arc asserts more than one possible relationship, the network form will be used, and in the case where only one relationship is possible, the relation form will be used.

For the present, we shall assume that the network always maintains complete information about how its intervals could be related. When a new interval relation is entered, all consequences are computed. This is done by computing the transitive closure of the temporal relations as follows: the new fact adds a constraint about how its two intervals could be related, which may in turn introduce new constraints between other intervals through the transitivity rules governing the temporal relationships. For instance, if the fact that i is *during* j is added, and j is *before* k, then it is inferred that *i* must be *before* k. This new fact is then added to the network in an identical fashion, possibly introducing further constraints on the relationship between other intervals. The transitivity relations are summarized in Figure 4.

The precise algorithm is as follows: assume for any temporal relation names r1 and r2 that T(r1, r2) is the entry in the transitivity table in Figure 4. Let R1 and R2 be arc labels, assume the usual set operations (∩ for intersection, ∪ for union, ⊂ for proper subset), and let ε be the empty set. Then *constraints (R1, R2)* is the transitivity function for lists of relation names (i.e., arc labels), and is defined by:

```
Constraints (R1, R2)
    C ← ε;
    For each r1 in R1
        For each r2 in R2
            C ← C ∪ T(r1, r2);
    Return C;
```

Assume we have a queue data structure named *ToDo* with the appropriate queue operations defined. For any two intervals i, j, let N(i, j) be the relations on the arc between i and j in the network, and let R(i, j) be the new relation between *i* and j to be added to the network. Then we have the following algorithm for updating the temporal network:

```
To Add R(i, j)
    Add ⟨i, j⟩ to queue ToDo;
    While ToDo is not empty do
begin
    Get next ⟨i, j⟩ from queue ToDo;
    N(i, j) ← R(i, j);
    For each node k such that Comparable(k, j) do
    begin
        R(k, j) ← N(k, j) ∩ Constraints(N(k, i), R(i, j))
        If R(k, j) ⊂ N(k, j)
            then add ⟨k, j⟩ to ToDo;
    end
    For each node k such that Comparable(i, k) do
    begin
        R(i, k) ← N(i, k) ∩ Constraints(R(i, j), N(j, k))
        If R(i, k) ⊂ N(i, k)
            then add ⟨i, k⟩ to ToDo;
    end
end;
```

[1] This fact was pointed out to me by Marc Vilain and was first utilized in his system [18].

Relation	Symbol	Symbol for Inverse	Pictoral Example
X *before* Y	<	>	XXX YYY
X *equal* Y	=	=	XXX YYY
X *meets* Y	m	mi	XXXYYY
X *overlaps* Y	o	oi	XXX YYY
X *during* Y	d	di	XXX YYYYYY
X starts Y	s	si	XXX YYYYY
X finishes Y	f	fi	XXX YYYYY

FIGURE 2. The Thirteen Possible Relationships.

Relation	Network Representation
1. i *during* j	N_i --(d)→ N_j
2. i *during* j or i *before* j or j *during* i	N_i --(< d di)→ N_j
3. (i < j) or (i > j) or i *meets* j or j *meets* i	N_i --(< > m mi)→ N_j

FIGURE 3. Representing Knowledge of Temporal Relations in a Network.

We have used the predicate *Comparable(i, j)* above, which will be defined in Section 5. For the present, we can assume it always returns true for any pair of nodes.

4.2. An Example

Consider a simple example of this algorithm in operation. Assume we are given the facts:

S *overlaps or meets* L

S is *before, meets, is met by,* or *after* R.

B r2 C / A r1 B	<	>	d	di	o	oi	m	mi	s	si	f	fi
"before" <	<	no info	< o m d s	<	<	< o m d s	<	< o m d s	<	<	< o m d s	<
"after" >	no info	>	> oi mi d f	>	> oi mi d f	>	> oi mi d f	>	> oi mi d f	>	>	>
"during" d	<	>	d	no info	< o m d s	> oi mi d f	<	>	d	> oi mi d f	d	< o m d s
"contains" di	< o m di fi	> oi di mi si	o oi dur con =	di	o di fi	oi di si	o di fi	oi di si	di fi o	di	di si oi	di
"overlaps" o	<	> oi di mi si	o d s	< o m di fi	< o m	o oi dur con =	<	oi di si	o	di fi o	d s o	< o m
"over-lapped-by" oi	< o m di fi	>	oi d f	> oi mi di si	o oi dur con =	> oi mi	o di fi	>	oi d f	oi > mi	oi	oi di si
"meets" m	<	> oi mi di si	o d s	<	<	o d s	<	f fi =	m	m	d s o	<
"met-by" mi	< o m di fi	>	oi d f	>	oi d f	>	s si =	>	d f oi	>	mi	mi
"starts" s	<	>	d	< o m di fi	< o m	oi d f	<	mi	s	s si =	d	< m o
"started by" si	< o m di fi	>	oi d f	di	o di fi	oi	o di fi	mi	s si =	si	oi	di
"finishes" f	<	>	d	> oi mi di si	o d s	> oi mi	m	>	d	> oi mi	f	f fi =
"finished-by" fi	<	> oi mi di si	o d s	di	o	oi di si	m	si oi di	o	di	f fi =	fi

FIGURE 4. The Transitivity Table for the Twelve Temporal Relations (omitting "=").

These facts might be derived from a story such as the following:

John was not in the room when I touched the switch to turn on the light.

where we let S be the time of touching the switch, L be the time the light was on, and R be the time that John was in the room. The network storing this information is

R ←-(< m mi >)- - S - -(o m)→ L.

When the second fact is added, the algorithm computes a constraint between L and R (via S) by calling the function *Constraints* with its two arguments, *R1* and *R2*, set to {oi mi} and {(m mi)}, respectively. Note that we obtained the inverse of the arc from S to L simply by taking the inverse of each label. *Constraints* uses the transitivity table for each pair of labels and returns the union of all the answers. Since

T(oi, <) = (< o m di fi)

T(oi, m) = (o di fi)

T(oi, mi) = (>)

T(oi, >) = (>)

T(mi, <) = (< o m di fi)

T(mi, m) = (s si =)

T(mi, mi) = (>)

T(mi, >) = (>)

we compute (< > o m di s si fi =) as the constraint between L and R and thus obtain the network

```
R ←(< m mi >)- - S - -(o m)→  L
↑                             |
- - -(< > o m di s si fi =)- - - - -
```

Let us consider what happens now when we add the fact

L *overlaps*, *starts*, or is *during* R

This fact might arise from a continuation of the above story such as

But John was in the room later while the light went out

Taking the intersection of this constraint with the previously known constraint between L and R to eliminate any impossible relationships gives

L - -(o s)→ R

To add this constraint, we need to propagate its effects through the network. A new constraint between S and R can be calculated using the path:

S - -(o m)→ L - -(o s)→ R

From the transitivity tables, we find:

T(o, o) = (< o m)

T(o, s) = (o)

T(m, o) = (<)

T(m, s) = (m)

Thus the inferred constraint between S and R is

S - -(< o m)→ R.

Intersecting this with our previous constraint between S and R yields

S - -(< m)→ R.

With respect to the example story, this is equivalent to inferring that John entered the room (i.e., R started) either after I touched the switch or at the same time that I finished touching the switch. Thus the new network is:

```
R ←(< m)- - S - -(o m)→ L
↑                       |
- - - - - - - - - - (o s) - - - - - - - -
```

Of couse, if there were other nodes in the network, there would be other constraints derived from this new information. Thus, if we added a new interval D, say with the constraint D - -(d)→ S, we would infer the following new relationships as well:

D - -(<)→ R

D - -(< o m d s)→ L.

4.3. Analysis

A nice property of this algorithm is that it only continues to operate as long as it is producing new further constrained relationships between intervals. Since there are at most thirteen possible relationships that could hold between two intervals, there are at most thirteen steps that could modify this relationship. Thus for a fixed number of nodes N, the upper limit on the number of modifications that can be made, irrespective of how many constraints are added to the network, is 13 × the number of binary relations between N nodes, which is:

$$13 \times \frac{(N-1)(N-2)}{2}$$

Thus, in practice, if we add approximately the same number of constraints as we have nodes, the average amount of work for each addition is essentially linear (i.e., N additions take $O(N^2)$ time; one addition on average takes $O(N)$ time).

The major problem with this algorithm is the space requirement. It requires $O(N^2)$ space for N temporal intervals. Methods for controlling the propagation, saving time and space, will be discussed in the next section.

It should be noted that this algorithm, while it does not generate inconsistencies, does not detect all inconsistencies in its input. In fact, it only guarantees consistency between three node subnetworks. There are networks that can be added which appear consistent by viewing any three nodes, but for which there is no consistent overall labeling. The network shown in Figure 5 is consistent if we consider any three nodes; however, there is no overall labeling of the network.[2] To see this, if we assign the relationship between A and C, which could be f or fi according to this network, to either f alone, or fi alone, we would arrive at an inconsistency. In other words, there is no consistent labeling with A - -(f)→ C, or with A - -(fi)→ C, even though the algorithm accepts A - -(f fi)→ C.

To ensure total consistency, one would have to consider constraints between three arcs, between four arcs, etc. While this can be done using techniques outlined in Freuder [9], the computational complexity of the algorithm is exponential. In practice, we have not encountered problems from this deficiency in our applications of the model. We can verify the consistency of any subnetwork, if desired, by a simple backtracking search through the alternative arc labelings until we

[2] This network is due to Henry Kautz, personal communication.

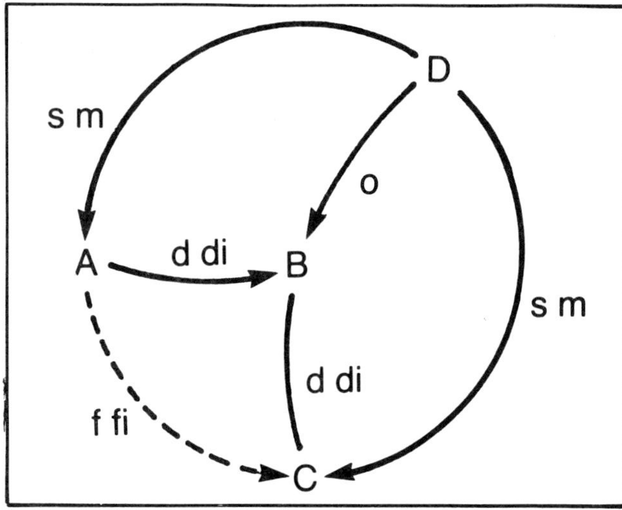

FIGURE 5. An Inconsistent Labeling.

arrive at a labeling for the whole subnetwork in which every arc has only one label.

5. CONTROLLING PROPAGATION: REFERENCE INTERVALS

In order to reduce the space requirements of the representation without greatly affecting the inferential power of the mechanism, we introduce *reference intervals.* Formally, a reference interval is simply another interval in the system, but it is endowed with a special property that affects the computation. Reference intervals are used to group together clusters of intervals for which the temporal constraints between each pair of intervals in the cluster is fully computed. Such a cluster is related to the rest of the intervals in the system only indirectly via the reference interval.

5.1. Using Reference Intervals

Every interval may designate one or more reference intervals (i.e., node clusters to which it belongs). These will be listed in parentheses after the interval name. Thus the node names

I1(R1)

I2(R1, R2)

describe an interval named I1 that has a reference interval R1, and an interval named I2 that has two reference intervals R1 and R2. Since I2 has two reference intervals, it will be fully connected to two clusters. An illustration of the connectedness of such a network is formed in Figure 6.

The algorithm to add relations using reference intervals is identical to the previous addition algorithm except that the comparability condition is no longer universally true. For any node N, let *Refs*(N) return the set of reference intervals for N. For any two nodes K and J, *Comparable(K, J)* is true if

1) *Refs*(K) ∩ *Refs*(J) is not null, that is, they share a reference interval; or
2) K ε *Refs*(J); or
3) J ε *Refs*(K).

Since reference intervals are simply intervals themselves, they may in turn have their own reference intervals, possibly defining a hierarchy of clusters. In most of the useful applications that we have seen, these hierarchies are typically tree-like, as depicted in Figure 7.

If two intervals are not explicitly related in the network, a relationship can be retrieved by finding a path between them through the reference intervals by searching up the reference hierarchy until a path (or all paths) between the two nodes are found. Then, by simply applying the transitivity relationships along the path, a relationship between the two nodes can be inferred. If one is careful about structuring the reference hierarchy, this can be done with little loss of information from the original complete propagation scheme.

To find a relationship between two nodes I and J, where N(i, j) represents the network relation between nodes i and j as in Section 4.1, we use the algorithm:

If N(I, J) exists
then return N(I, J)
else do
  Paths := Find-Paths(I, J)
  For each *path* **in** *Paths* do
    R := *R* ∩ *Constrain-along-path(path)*
  return *R*;
end;

The function *Find-Paths* does a straightforward graph search for a path between the two nodes with the restriction that each step of the path must be between a node and one of its reference intervals except for the one case where a direct connection is found. Thus, a path is of the general form

$$n_1, n_2, \ldots, n_k, n_{k+1}, \ldots, n_m$$

where all of the following hold:

- - for all i from 1 to k − 1, n_{i+1} is a reference interval for n_i;
- - n_k and n_{k+1} are connected explicitly;
- - for all i from k + 1 to m − 1, n_i is a reference interval for n_{i+1};

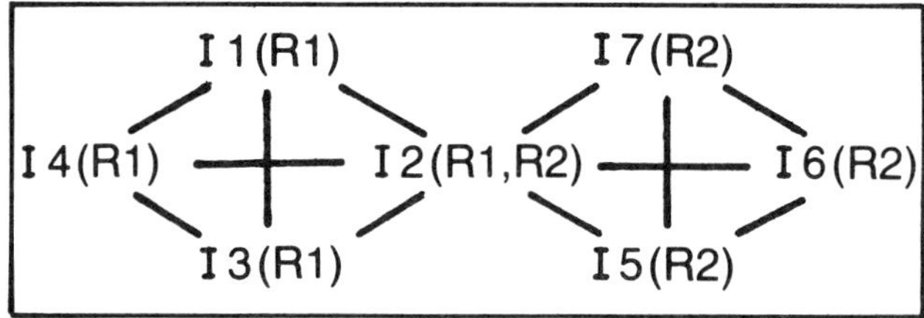

FIGURE 6. The Connectness of a Network with Two Reference Intervals.

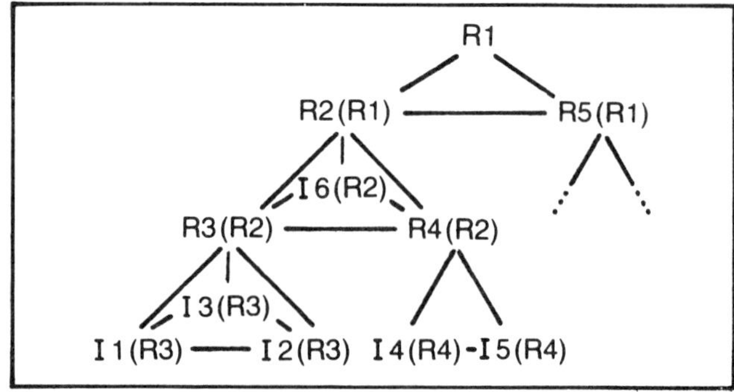

FIGURE 7. A Tree-Like Hierarchy Based on Reference Intervals.

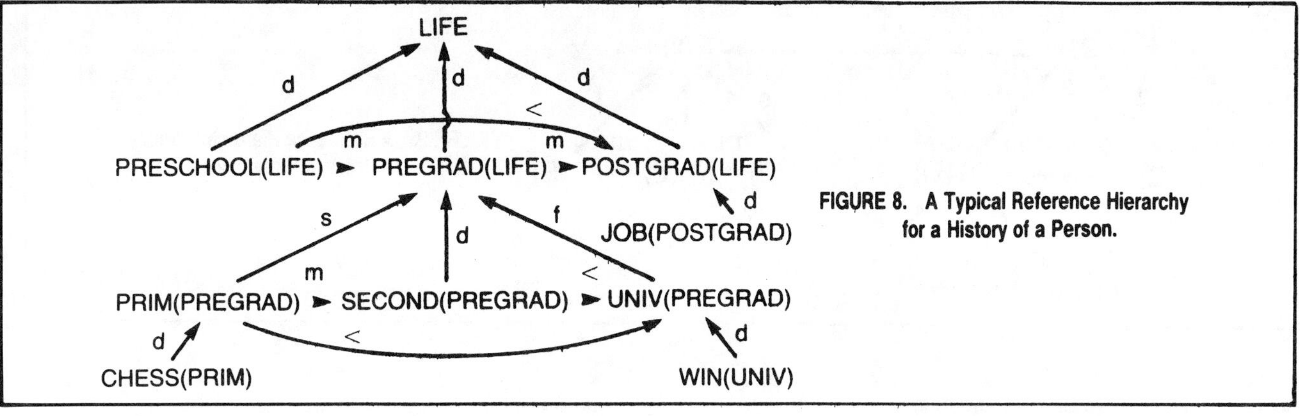

FIGURE 8. A Typical Reference Hierarchy for a History of a Person.

The function *Constrain-along-path* simply takes a path and computes the transitivity constraints along it. Thus if a path consisted of the nodes $n_1, n_2, n_3, \ldots, n_m$, we compute the relation between n_1 and n_m as follows:

$$R := N(n_1, n_2)$$

$$R := \text{Constraints}(R, N(n_2, n_3))$$

$$R := \text{Constraints}(R, N(n_3, n_4))$$

$$\ldots$$

$$R := \text{Constraints}(R, N(n_{m-1}, n_m))$$

where *Constraints* was defined in Section 4.1.

5.2. Examples

There are no restrictions imposed by the system on the use of reference intervals. Their organization is left up to the system designer. Certain principles of organization, however, are particularly useful in designing systems that remain efficient in retrieval, and yet capture the required knowledge. The most obvious of these is a consequence of the path search algorithm in the previous section: the more tree-like the reference hierarchy, the more efficient the retrieval process. The others considered in this section exploit characteristics of the temporal knowledge being stored.

With domains that capture historical information, it is best to choose the reference intervals to correspond to key events that naturally divide the facts in the domain. Thus, if modeling facts about the history of a particular person, key events might be their birth, their first going to school, their graduation from university, etc. Kahn and Gorry [13] introduced such a notion of reference events in their system. Other times in their system were explicitly related to these reference events (i.e., points). In our system, the intervals between such key events would become the reference intervals. Other time intervals would be stored in the cluster(s) identified by the reference intervals that contain them. Thus, we could have a series of reference intervals for the time from birth to starting school (PRESCHOOL), during school (PREGRAD), and after graduation (POSTGRAD). In addition, certain reference intervals could be further decomposed. For example, PREGRAD could be divided into primary and secondary school (PRIM and SECOND) and the time at university (UNIV). The times of the rest of the events would be stored with respect to this reference hierarchy. Figure 8 depicts this set of facts including its reference hierarchy, plus intervals such as the time spent learning chess (CHESS), the time the person won the state lottery (WIN), and the time of the first job (JOB). If an event extended over two reference intervals, then it would be stored with respect to both. For example, if learning to play chess occurred during primary and secondary school, the interval CHESS would have two reference intervals, namely, PRIM and SECOND.

We can now trace the retrieval algorithm for this set of facts. Let us find the relationship between CHESS and WIN. There is no explicit relationship between the intervals, so we must search up the reference hierarchy. Only one path is found, namely:

CHESS(PRIM) - -(d)→ PRIM(PREGRAD) - -(<)→
UNIV(PREGRAD) - -(di)→ WIN(UNIV)

Applying the transitivity relations along the first path, we infer first that

CHESS *before* UNIV

and then

CHESS *before* WIN.

The fact that CHESS is *before* JOB can be inferred similarly from the path

CHESS - -(d)→ PRIM - -(s)→ PREGRAD - -(m)→
POSTGRAD - -(di)→ JOB.

Consider another domain, namely, that of representing information about processes or actions. Such knowledge is required for problem-solving systems that are used to guide the activity of a robot. Each process can be described as a partial sequence of subprocesses. Such a decomposition is not described in absolute temporal terms (i.e., using dates), but by the subprocess's relation to its containing process. Thus a natural reference hierarchy can be constructed mirroring the process hierarchy. For example, consider a process P consisting of a sequence of steps P1, P2, and P3 and another process Q consisting of subprocesses Q1 and Q2 occurring in any order, but not at the same time. Furthermore, let Q2 be decomposed into two subprocesses Q21 and Q22, each occurring simultaneously. To simulate a world in which process P begins before Q begins, we can construct the reference hierarchy in Figure 9. With this organization we can infer relationships between subprocesses of Q and subprocesses of P in the same manner as above. As long as the decomposition of

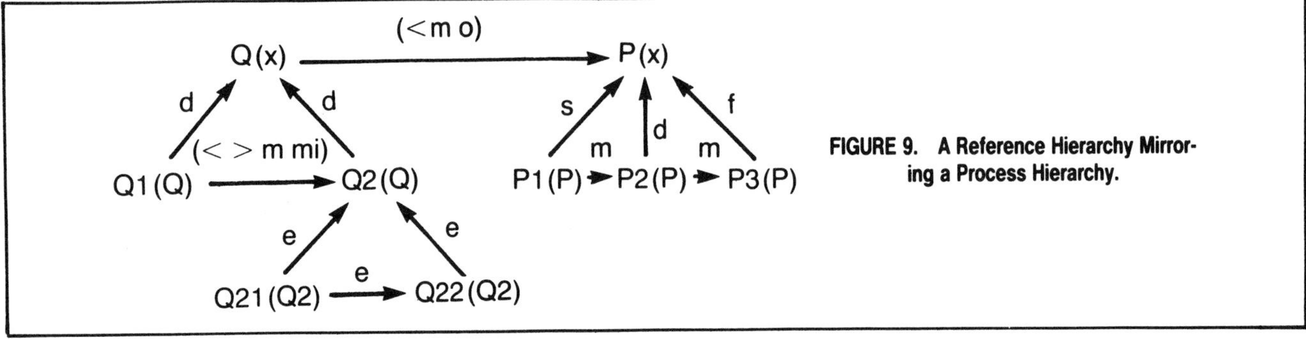

FIGURE 9. A Reference Hierarchy Mirroring a Process Hierarchy.

processes or actions can be done independently (such as in the NOAH system [17]), this organization will capture all the relevant temporal knowledge.

More interesting cases arise when there may be interactions among subprocesses. For instance, we might want to add that Q1 must occur before Q21. Note that, in adding Q1 *before* Q21, we can infer a new relationship between Q1 and Q2 from the path

Q1(Q) - -(<)→ Q21(Q2) - -(e)→ Q2(Q)

because Q1 and Q2 share the reference interval Q. It does not matter that Q21 does not share a reference interval with Q1. In more complicated cases, we will find relationships between subprocesses such that an important relationship between the processes containing the subprocesses will not be inferred because they do not share a reference interval. For instance, if we learn that Q2 overlaps P1, adding this will not cause the relationship between Q and P to be constrained to simply the *overlaps* relation even though that would be a consequence in the system without reference intervals. There is no path consisting of two arcs from Q to P that is affected by adding Q2 *overlaps* P1.

To allow this inference, we need to reorganize the reference hierarchy. For example, we could, when adding a relation between two noncompatible nodes, expand one of the node's reference intervals with the other node's reference intervals. In this scheme, to add Q2 *overlaps* P1, we would first add P to Q2's reference interval list. Then adding the relation will allow the appropriate changes. In particular, among others, we would infer that

Q2(Q, P) - -(o)→ P(X)

from the path

Q2(Q, P) - -(o)→ P1(P) - -(s)→ P(X),

and then infer

Q(X) - -(o)→ P(X)

from the path

Q(X) - -(di)→ Q2(Q, P) - -(o)→ P(X)

and the previous constraints between Q and P. The final state of the processes after these two additions is summarized in Figure 10.

Manipulating the reference hierarchies as in this example can be effective if used sparingly. With overuse, such tricks tend to "flatten out" the reference hierarchy as more intervals become explicitly related. In domains where such interactions are rare compared with the pure decompositional interactions, it can be very effective.

5.3. Representing the Present Moment

The technique of reference interval hierarchies provides a simple solution to the problem of representing the present moment. In many applications, such as natural language processing and process modeling, the present is constantly moving into the future. Thus a representation of *NOW* must allow for frequent updating without involving large-scale reorganization of the database each time.

Suppose we have a database in which all assertions are indexed by the temporal interval over which they hold. As time passes, we are interested in monitoring what is true at the present time, as well as in the past and future. The method suggested here is to represent *NOW* as a variable that, at any specific time, is bound to an interval in the database. To update *NOW*, we simply reassign the variable to a new interval that is *after* the previous interval representing the present moment. The key observation is that while the present is continually changing, most of the world description is remaining the same. We can exploit this fact by using refer-

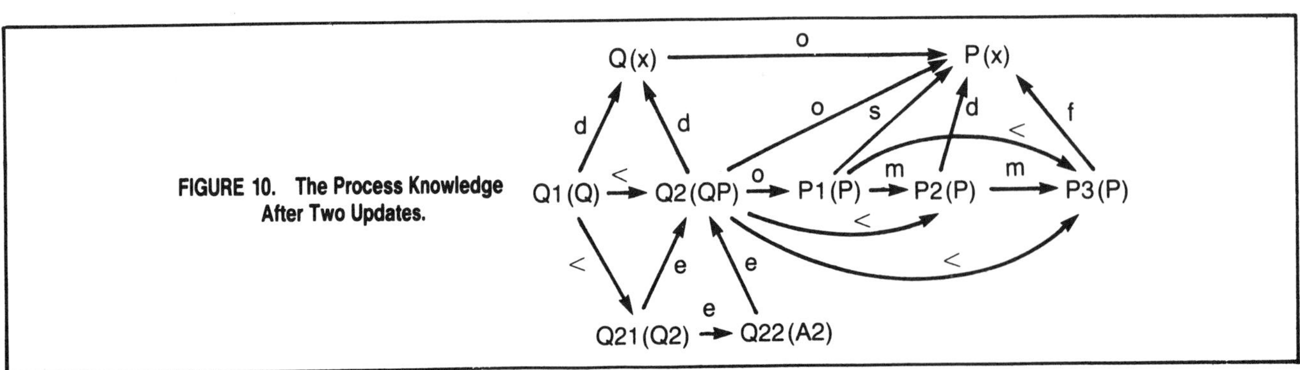

FIGURE 10. The Process Knowledge After Two Updates.

ence intervals to control the inferences resulting from updating *NOW*.

For example, let *NOW* be interval N1, which is *during* its reference interval R1. An example state of the database would be

N1(R1) *during* R1

R1 *before* I1, R1 *after* I2, R1 *during* I3

From this we can infer easily that the present (i.e., N1) is during I3, before I1, and after I2. If *NOW* then is updated (slightly), N2 can be defined as the new *NOW* using the same reference interval by adding the facts

N2(R1) *during* R1, N2(R1) *after* N1(R1)

Thus, *NOW* has been updated but most of the relations in the database have been unaffected, for the effects of N2 will only propagate to intervals referenced by R1. The reference interval R1 has "protected" the rest of the database from a minor change in the present moment.

Of course, eventually *NOW* will cease to be during R1 and a new reference interval will be needed. This will involve a more major update to the database, but the amount of work can be reduced if R1 itself has a reference interval that "protects" much of the database from it.

Thus we need a hierarchy of reference intervals, each containing the present moment. This hierarchy could be designed to mirror the set of English terms that can be used to refer to the present. For example, in English we can refer to the exact moment of an utterance (e.g., at a race, the starter may say "Go now!"), as well as to larger intervals such as "this morning," "today," and "this year." We can also refer to more event-oriented intervals such as "during this lecture" and "while at this bar." These are the types of intervals that should be maintained in the hierarchy representing the present. Furthermore, these intervals typically have well defined starting and termination points. Thus it is reasonable to assume that the temporal database will receive explicit notification when one of them ceases to contain the present. This allows the following important assumption:

> When updating the *NOW* interval, unless otherwise stated, its relationship to its reference interval(s) remains constant.

When one of the reference intervals in the hierarchy ceases to contain the present moment, a new reference interval is selected. (This new interval should usually be provided by the user.) This update is done in the identical fashion as described above with *NOW*. In particular, the relationship with the higher-level reference interval remains constant. A new *NOW* interval, below the new reference interval in the hierarchy, must be introduced. For example, the beginning of a new day would make much of the old hierarchy part of the past (i.e., "yesterday").

While many intervals will be generated by this succession of intervals for *NOW*, many of them can be garbage collected when the reference intervals are updated. In particular, any interval that is not used to index any events or facts may be removed from the database. In a system modeling a natural language dialogue, a large number of these intervals would be used only to index the time of an utterance: These generally can be deleted without harm.

6. DISCUSSION

The temporal representation described is notable in that it is both expressive and computationally feasible. In particular, it does not insist that all events occur in a known fixed order, as in the state space approach, and its allows disjunctive knowledge, such as that event A occurred either before or after event B, not expressible in date-based systems or simple systems based on before/after chaining. It is not as expressive as a full temporal logic (such as that of McDermott [15]), but these systems do not currently have viable implementations.

This balance between expressive power and computational efficiency is achieved by the restricted form of disjunctions allowed in the system. One can only assert disjunctive information about the relationship of two intervals. In other words, we can assert that A is *before* or *meets* B, but not that (A *meets* B) or (C *before* D). This limited form of disjunction is ideal for the constraint propagation algorithm.

The system has been implemented and is being used in a variety of applications. Both FRANZ LISP [8] and INTERLISP versions are running on a VAX 11/780 under UNIX. The system presently also includes the duration reasoner described below. It is currently being used in research in representing actions, events, and interactions that arise in natural language dialogues [1]. We are also using the representation as a world model for research in automatic problem-solving systems [3]. Such systems have long been constrained by their inadequate temporal models.

Vilain [18] has implemented a version of this system which, at the cost of greater space requirements, can perform consistency maintenance. In other words, in his system, when an inconsistency is found, the set of facts that caused the inconsistency can be identified. This system also explicitly allows time points in the representation and has a larger transitivity table, including all interval/point and point/point interactions. This violates the semantics of the interval representation, and so has not been adopted in our present system.

Let us consider why we would like time points, however. They seem to be referred to in English. We can, for instance, talk about the beginning and ending of events. There is no reason to assume, however, that these "endpoints" are truly zero-width points rather than intervals small enough so that they appear to be instantaneous. What this suggests is that there might be a minimum duration ε, such that all intervals of duration less than ε would be viewed as points. This would simplify our reasoning about such times for we would not have to worry about the possibility of two such intervals overlapping. It would be assumed either that these small intervals are equal or that one is before the other.

But this minimum size cannot be fixed in advance. A historian, for instance, may be happy to consider days as points, whereas the computer engineer, when reasoning about a logic circuit, would consider a day to be an eternity. Thus the interval size, where it is appropriate to simplify reasoning by assuming point-like times, varies with the reasoning task.

7. FUTURE RESEARCH AND EXTENSIONS

There are many areas in which this system is being extended. In particular, an interface to a *duration* reasoner has been incorporated into the system, and a system for reasoning about dates will be implemented in the near future. Finally, we are investigating reasoning systems that depend on the notion of persistence.

7.1. Duration Reasoning

We have designed a duration reasoning system based on the same principles as the interval relation reasoner described above. In particular, it is designed to allow relative information (e.g., interval A took longer than interval B) as well as representing uncertainty. The reasoner is again based on constraint propagation and a notion of reference durations can be defined.

Very briefly, the duration relationship between two intervals is expressed by outlining a range that includes the multiplicative factor which the duration of the first would be multiplied by to get the duration of the second. For example, the fact that the duration of A is less than the duration of B, expressed as *dur*(A) < *dur*(B), is represented by the relation A - -(0(1))→ B. In other words, *dur*(A) >= 0**dur*(B) and *dur*(A) < 1**dur*(B). The parentheses about the factor 1 indicate an open endpoint; thus the durations of A and B could not be equal. Both the upper and lower duration limits may be open or closed.

Duration information is encoded in a network orthogonal to the relationship network. Propagation across two duration restrictions is accomplished simply by multiplying the respective upper and lower duration limits. For example, if we have the facts

$$dur(A) <= dur(B)$$

$$dur(C) <= dur(B)$$

$$dur(B) < 2*dur(C)$$

which in network form would be

A - -(01)→ B - -(1(2))→ C

we derive the relation

A - -(0(2))→ C.

The duration reasoner and the interval reasoner are not independent of each other, however. They constrain each other by rules such as the following:

If I - -(dsf)→ J then dur(I) < dur(J).

Using this rule, constraints introduced in one network may introduce constraints in the other. In many examples, the networks may exchange information back and forth multiple times before the propagation terminates.

Reference durations correspond to the notion of scales, or common units. Constraints do not propagate through a reference duration. Thus, if the duration HOUR is a reference duration, and we add that dur(A) is between 1 and 2 hours, and dur(B) is less than one half an hour, no relation between dur(A) and dur(B) will be inferred. It will be derived at retrieval time via the reference duration HOUR. Further details on the duration reasoner can be found in [2].

7.2. Date Lines

Having considered the maintenance of relative temporal information in detail, we now consider how to exploit date information when available. Let a *date line* be any representation consisting of a fully ordered set of values taken to correspond to times. A date line corresponding to a simple calendar could be defined as follows:

values: ordered triples of integers, representing year, month (1–12), and day (1–31) (for example, (50 3 25) represents March 25, 1950)

comparison operation: orders triples in the obvious manner (for example, (50 3 25) < (75 1 1))

With date lines, the comparison operation between two times on the same date line is relatively inexpensive compared to searching the network of temporal relations.

Date line information could be incorporated into the present system by allowing any interval in the network to have date line information associated with it which identifies the dating system and dates associated with its start and end. The name of the date line is necessary to identify the operations for comparing values. A new interval, added with date line information specified, may affect the relationship to its reference interval and to the other intervals in its "reference class." For example, if two intervals are dated by the same date line, and have date values specified, those values can be used to calculate the exact relation between the intervals. If this relation is more specific than the information stored in the relational network, the network is updated and its effects propagated as usual.

When retrieving a relationship between two intervals dated by the same date line, the date information should be considered first before applying the usual network retrieval mechanism. Sometimes, however, the date line information will not be specific enough to pinpoint a specific relationship, and a network search will still be necessary. It may occur that one of the intervals being considered is dated but the other is not. In this case, the date information may be used only if a relationship can be found between the nondated interval and another interval dated by the same date line. In general, this may be too expensive to consider. A specific case that could be very useful, however, occurs when a reference interval involved in the search is dated by the appropriate date line. We can then compare the two dated intervals to obtain a relationship, which can be propagated back to the nondated interval.

A useful date line for dialogue systems is the time-of-day line. A reasonable implementation of this might have the basic duration of one minute, and have values consisting of an hour-minute pair. If the system were given access to a clock, this date line could be used extensively in the *NOW* hierarchy. Of course, the relative time database is still required to store the facts that are acquired during the dialogue as facts typically hold for much longer than the time that they are being talked about.

If the system does not have such easy access to an internal clock, it may still get time-of-day information occasionally during a dialogue. In this case, some of the *NOW* intervals will map onto the time-of-day line, while others will only be related to it by some relation (e.g., after 10 o'clock). In such a scheme, a new reference interval for the *NOW* interval would be created each time a precise time-of-day value was identified. For example, if the system learns that it is presently 10 o'clock, it can create an "after 10 o'clock" reference interval in which the *NOW* intervals will be contained until the next specific time is acquired. Whether such a technique is feasible requires further search.

7.3. Persistence of Intervals

The last requirement described in the introduction was that the representation should facilitate plausible inferences of the form "if fact P is true now, it will remain true until noticed otherwise." Most of the issues concerning this fall outside the range of this paper, as this system only knows about time intervals. However, a simple trick using this representation makes inferences of the above form easy to implement.

Typically, when a new fact is learned, its exact extent in time is not known. For instance, when I parked by car in the parking lot this morning I knew its location. Sitting at my desk now, I assume it is still there, though I have no proof of that fact. In general, I assume it will remain where it is until I pick it up. Thus, although I do not know the extent of the interval in which my car is parked, I want to be able to assume that this fact holds later in the day. The temporal representation is already based on the observation that most time intervals do not have precisely defined limits. If we allow the user to specify that some intervals should be assumed to extend as far as possible given the constraints, then we can use such intervals to index facts that are assumed to persist until discovered otherwise.

Thus, if we let a fact P be indexed by a *persistent interval* Tp, then testing P later during an interval t will succeed (by assumption) if it is possible that t is *during* Tp. Checking whether relationships between intervals are possible is easy, since the representation explicitly maintains this information.

For example, let Tp represent the interval in which my car is in the parking lot. I know that Tp is *met by* Tarrive, where Tarrive is the time that I arrived at school today. Then, if *NOW* is represented as interval Tnow, where Tnow *after* Tarrive, we can test if my car is on the parking lot. Since it is there during Tp, we are interested in whether it is possible that Tnow is *during* Tp. The known constraints allow us to infer the following:

Tp *met by* Tarrive, Tarrive *before* Tnow

=> Tp --(< odim)→ Tnow

Thus it is possible that Tnow is *during* Tp, since it is possible that Tp contains ("di") Tnow. So the test succeeds.

Of course, if it is later learned that the car was found to be missing during time interval Tmiss, then Tp is constrained to be *before* Tmiss (even though it is still persistent). If Tnow is then after or during Tmiss, then it is not possible any longer that Tnow is during Tp.

Managing a system such as this is a difficult problem that requires some form of truth maintenance (e.g., see [6]). These issues, however, are independent of the temporal representation. All that is shown here is that the necessary temporal calculations are easily done within this framework.

An interesting technique suggested by the above may simplify much of the computation required for truth maintenance for this type of assumption. In particular, let us assume that P holds during interval Tp, where Tp is a persistent interval. Furthermore, assume that for any time, P implies Q. Then if we test P at time t, and find it is true by assumption, so we can infer Q during time t. However, if we index Q by Tp rather than by t, then we still can use the fact that Q is true during t (by assumption), but if we ever discover further constraints on Tp that then eliminate the possibility that t is *during* Tp, then both P *and* Q will cease to be true (by assumption) during t. Thus, by indexing all the consequences of P by the same interval, Tp, we can revise our beliefs about P and all its consequences simply by adding constraints about Tp. While this idea obviously requires further investigation, it appears that it may allow a large class of assumption-based belief revision to be performed easily.

8. SUMMARY

We have described a system for reasoning about temporal intervals that is both expressive and computationally effective. The representation captures the temporal hierarchy implicit in many domains by using a hierarchy of reference intervals, which precisely control the amount of deduction performed automatically by the system. This approach is partially useful in domains where temporal information is imprecise and relative, and techniques such as dating are not possible.

Acknowledgments. Many people have provided significant help during the design and development of this work. In particular, I would like to thank Marc Vilain and Henry Kautz for work on developing and extending the system. I have also received many valuable criticisms from Alan Frisch, Pat Hayes, Hans Koomen, Drew McDermott, Candy Sidner, and the referees. Finally, I would like to thank Peggy Meeker for preparing the manuscript, and Irene Allen and Henry Kautz for the valuable editorial criticism on the final draft.

REFERENCES

1. Allen, J. F., Frisch, A. M., and Litman, D. J. ARGOT: The Rochester dialogue system, Proc. Nat. Conf. on Artificial Intelligence, AAAI 82, Pittsburgh, Pa., Aug. 1982.
2. Allen, J. F., and Kautz, H. A. "A model of naive temporal reasoning," to appear in J. R. Hobbs and R. Moore (Ed)., *Contributions in Artificial Intelligence, Vol. 1*, Ablex Pub. Co., Norwood, N.J., 1983.
3. Allen, J. F., and Koomen, J. A. Planning using a temporal world model. Submitted to 8th Int. Joint Conf. Artificial Intelligence, Aug. 1983.
4. Bruce, B. C. A model for temporal references and its application in a question answering program. *Artificial Intelligence 3* (1972), 1–25.
5. Bubenko, J. A., Jr. Information modeling in the context of system development. Proc. IFIP Congress 80, Oct. 1980, North-Holland, Amsterdam.
6. Doyle, J. A truth maintenance system. *Artificial Intelligence 12*, 3, (Nov. 1979), 231–272.
7. Fikes, R. E., and Nilsson, N. J. STRIPS: A new approach to the application of theorem proving to problem solving. *Artificial Intelligence 2*, (1971), 189–205.
8. Foderaro, J. K. *The FRANZ LISP Manual*. Dept. of Computer Science, U. of California, Berkeley, 1980.
9. Freuder, E. C. A sufficient condition for backtrack-free search. *J. ACM 29*, 1 (Jan. 1982), 24–32.
10. Hayes, P. J. The Naive Physics manifesto. In *Expert Systems*, D. Michie (Ed.), Edinburgh U. Press, 1979.
11. Hayes, P. J. Naive Physics I: Ontology for liquids. Working Paper 63, Institut pour les Etudes Semantiques et Cognitives, Geneva, 1978.
12. Hendrix, G. G. Modeling simultaneous actions and continuous processes. *Artificial Intelligence 4*, 3 (1973), 145–180.
13. Kahn, K. M., and Gorry, A. G. Mechanizing temporal knowledge. *Artificial Intelligence 9*, 2 (1977), 87–108.
14. McCarthy, J., and Hayes, P. Some philosophical problems from the standpoint of artificial intelligence. *Machine Intelligence 4*, Edinburgh U. Press, 1969.
15. McDermott, D. A temporal logic for reasoning about processes and plans. *Cognitive Science 6*, (1982), 101–155.
16. Rescher, N., and Urquhart, A. *Temporal Logic*. Springer-Verlag, New York, 1971.
17. Sacerdoti, E. D. *A Structure for Plans and Behavior*. Elsevier North-Holland, New York, 1977.
18. Vilain, M. A system for reasoning about time. Proc. AAAI 82, Pittsburgh, Pa., Aug. 1982.

CR Categories and Subject Descriptors: I.2.4 [**Knowledge Representation Formalisms and Methods**]: Representations—*time, temporal representation*; I.2.3 [**Deduction and Theorem Proving**]: Deduction—*constraint propagation, temporal reasoning*; H.3.3 [**Information Search and Retrieval**]: Clustering, Retrieval Methods

General Terms: Algorithms

Additional Key Words and Phrases: temporal interval, interval reasoning, interval representation.

Received 12/81; revised 3/83; accepted 5/83

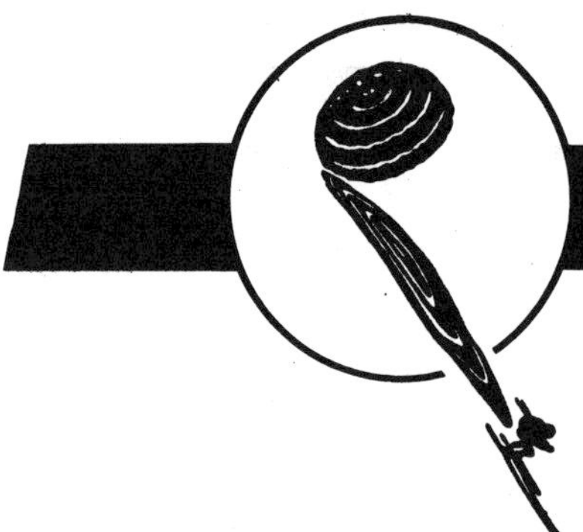

A Software Engineering Tool for Expert System Design

Joan M. Francioni, Michigan Technological University

and

Abraham Kandel, Florida State University

Tools and Techniques

Editor's note: We are pleased to begin with this issue our Tools and Techniques track, devoted to major technological tools and programming techniques currently under research and development. We hope readers will find these tools and techniques widely applicable to their own pursuits, and that they will gain insights into their own alternative approaches.

In this issue, Francioni and Kandel present a method for creating decision tables based on fuzzy logic, for use in designing expert systems and verifying and managing knowledge bases. Bard's "Rule-Based Inferencing Functions for APL Applications" describes how a library of inferencing routines can easily be added to applications written in procedural languages like APL.

Future issues will contain additional Tools and Techniques articles. Please send us your comments.

—Doug DeGroot
Associate Editor

Expert systems are computer programs, or sets of programs, using domain knowledge and reasoning techniques to solve problems normally requiring expertise—either from a human expert, a number of experts, written information, or some combination thereof. Although no standard format exists for expert systems, they minimally comprise a knowledge base, an inference procedure, and a working memory.[1] Knowledge bases contain domain facts and heuristics representing human expert domain knowledge. Inference procedures are control structures using knowledge bases for solving problems posed to expert systems. As expert systems work on problems, working memory serves as a global database keeping track of problem status. We will focus on knowledge bases.

Expert systems can represent knowledge in many different ways; a common example is as production type rules, generally written in "if-then" format.[2] Such production type knowledge bases can be processed easily. However, designing knowledge bases presents a difficult problem.[3] Although working expert systems exist today, no general methodology exists for designing production type knowledge bases for new expert systems—due primarily to knowledge base characteristics. First, they typically contain a large domain of both precise and imprecise data qualified and quantified in either a deterministic way (such as, "If there are five colors, then . . .") or in a nondeterministic way (such as, "If there are a few colors, then . . ."). Having both types of information present makes it hard to represent knowledge in a way that

Table 1. The parts of a decision table.

condition stub	condition entry
action stub	action entry

Table 2. A decision table to sort N records.

	1	2	3	4	5	6
Number of records ≤ 10	T	T	F	F	F	F
Records have ≤ 3 fields	T	-	-	-	-	-
Records have > 3 fields	-	T	-	-	-	-
Records have > 100 fields	-	-	T	F	F	F
Alphabetizing	-	-	-	T	F	F
Recursion available	-	-	-	-	F	T
Call insertion sort	X					
Call selection sort		X				
Call heapsort					X	
Call quicksort						X
Call bucket sort				X		
Sort pointer array to records			X			
Go again			X			
Stop	X	X		X	X	X

Table 3. A simple decision table.

	1	2	3
C1	T	-	F
C2	-	F	T
C3	F	F	T
A1	X		X
A2		X	

ensures consistency throughout the system. Second, due to the size of a typical knowledge base, keeping track of information is difficult as design progresses. As a result, redundant and contradictory information ends up being stored in the knowledge base.

These characteristics require that design tools, as well as a straightforward methodology, be developed for designing production type knowledge bases. In particular, the tool should have the following properties:

(1) The tool should be well defined and application independent. We want a general tool that can be used for any production type knowledge base.

(2) The tool should facilitate meaningful documentation as design progresses. It would be foolish and time consuming to go back at the end and try to document a design.

(3) The tool should ensure design consistency as it progresses. Specifically, we want to detect contradictions and redundancies in the data. Such detection must be possible during the design's entire phase, including all modifications, enhancements, and changes.[4]

(4) The tool should be able to model the approximate reasoning involved in human knowledge.

This article proposes a design tool for constructing production type knowledge bases that provides a straightforward methodology and adheres to the above properties. We first describe an existing software engineering tool—the decision table (DT). The DTs-for-programming concept appeared in the early 1960s.[5-7] However, DT use in software design became widespread much later.[8,9] Although we can use these tables as software design tools, they do not work well when designing knowledge bases as presently defined. Specifically, they cannot handle imprecise conditions; hence, they cannot model approximate reasoning.

We will present a modified DT that can be used for designing a production type knowledge base. Some previous work along these lines modified DTs so that all conditions were defined as fuzzy variables.[10] Rules then became combinations of these variables, represented as fuzzy switching functions. Although this technique can handle imprecise conditions, the method of processing the table is somewhat restrictive. This article develops a more general imprecise DT and presents a definite method for constructing it.

Decision tables in software design

Simply put, a DT is a special form of table that determines decision rules based on a clearly identified set of conditions and resulting actions.[11] DTs contain four major parts—the condition stub and entry, and the action stub and entry. Table 1 shows the relative positions of these sections. The condition stub contains a row for each condition to be evaluated. Similarly, the action stub has a row for each action. The condition and action entry sections divide into columns called rules. Each column specifies values for certain conditions and resulting actions to be taken when those conditions meet the specified values. We can interpret a rule as "If (C_1 and C_2 and . . .) then (A_1 and A_2 and . . .)" where C_i is the condition, either true or false as specified by the condition entry, and A_i is the action to be taken as specified by the action entry.

Table 2 shows a simple DT to determine which sort to use when sorting a record array. In this DT, according to rule 4—if the number of records exceeds 10 and the size of the records is less than 100, and if you want to alphabetize, then call the Bucket Sort procedure to sort the records and exit the table. In this case, we "don't care" whether recursion is available or not since the bucket sort is not recursive.

A DT's condition entries and action entries are deterministic. Actions are taken if—and only if—a mark is present in the action entry and the rule's conditions are satisfied. If actions are numbered, then they should be processed in that order; otherwise, they are processed in their order of occurrence. Condition entries must be explicitly stated. A condition entry's possible values may be Boolean (as in the previous example) or they may be multivalued, but they cannot be infinite in number. For example, Table 2's first condition could not be "Number of Records" since possible condition entries are unlimited. This restriction exists because the DT must be able to be directly programmable.

DTs can be set up for each level of a top-down design. The top-level table would be general and would call other tables for more specific processing. Besides a DT's traditional condition and action entries, specific exit actions should also exist for each rule of a table. Hurley defines four permanent exit actions and one temporary exit action as follows.[8]

Permanent: 1) Return
2) Goto (table name)
3) Go Again
4) Stop

Temporary: 1) Call (table name)

Unlike other permanent exit actions, the Go Again exit does not really cause processing to leave a table. Rather, it facilitates the loop concept. For example, Table 2's rule 3 has a Go Again action entry. This means that an array of pointers to the records should be set up and sorted, instead of the original data array, if record number exceeds 10 and record size exceeds 100. To determine which sort to use on this new array, the DT is processed again—hence, the Go Again action. This article will consider the Go Again to be a temporary exit action like the Call (table name) exit action. Also, since these exits are only temporary, they will both be treated as nonexit actions in the table.

In keeping with general principles of structured programming, we will put the following restrictions on DT use:

(1) Any table entered by a Call from another table must have a Return as its only exit.
(2) Goto exits should be avoided at all costs.
(3) The top-level DT must have a Stop exit.

Table 2's DT is considered top level; hence, it has the Stop exit. DTs for the particular sorts called from this DT would have Return exits instead.

A DT's rule selection deals with the number of rules that can be selected in a single DT activation (that is, when the DT is called and processed). The following four conventional disciplines for rule selection exist:[12]

(1) Zero or more rules;
(2) At least one rule;
(3) Exactly one rule; and
(4) At most one rule.

In the first and second cases, in which more than one rule can be selected, rules themselves must be ordered to order actions to be taken. For example, consider Table 3's simple DT. Rules 1 and 2 will be selected if the three condition values are C1 = T, C2 = F, and C3 = F. In this case, it should be specified whether to do action A1 or action A2 first.

If it is possible that no condition set is satisfied in rule selection's second and third cases, then no rule would be selected. In such cases, the table is incomplete. Under either of these rule selection disciplines, Table 3's DT is considered incomplete since—when C1 = F, C2 = T, and C3 = F—no rule is satisfied.

In the last two cases of rule selection, contradictions and redundancies may arise in the DT. We formally define these as follows with respect to DTs.

(1) Contradiction exists when two or more rules in a table, where at most one rule can be selected, have equivalent condition entries but different action entries.
(2) Redundancy exists when two or more rules in a table, where at most one rule can be selected, have equivalent action entries.

Again, using Table 3's DT, rules 1 and 2 are contradictory since their conditions are equivalent (a "don't care" is equivalent to either a True or a False) but they have different actions. When contradiction occurs, rules should be reevaluated to determine which is correct. When redundancy occurs, rules should be combined when condition entries are different, and one rule should be omitted when condition entries are the same. In the case of Table 3's rules 1 and 3, the two rules should be combined into one where C1 = -, C2 = T, and C3 = -.

Neither contradictions nor redundancies should exist in a knowledge base—an important point. If contradictions exist in a knowledge base, then erroneous information is actually being represented. The contradiction would then have to be resolved by expert system users. This is undesirable. Redundant information in the knowledge base takes up valuable space; in addition, it necessitates extra processing by the inference procedure. Due to the normally large size of knowledge bases, this is undesirable as well. Knowledge base design using a DT checks for these conditions automatically, thereby ensuring that knowledge bases are contradiction and redundancy free.

Nondeterministic decision tables

DTs work well for all but one of the desirable knowledge base design tool properties discussed above. The major problem with using DTs, however, results from their completely deterministic nature. In each rule, each condition value must be precisely one of a finite possible-value set. But according to property (4)—presented in the introduction, and dealing with consistency—design tools must be able to model approximate reasoning. To force DTs to handle imprecise conditions results in an infeasible number of conditions. The basic problem here is that DTs must be directly programmed; hence, they are based on how computers work rather than how people think. For example, in Table 2's Sorting DT, arbitrary cutoffs were set up for the size and number of records to be sorted.

As another example, consider the Sorting DT's rules 1 and 2. In rule 1, the condition entry for "Records have > 3 Fields" is a "don't care" since, if the earlier condition is true, this one will automatically be false. A similar situation exists in rule 2 for the condition "Records have > 100 Fields." As humans, we understand this. However, these rules will be considered contradictory if checked by an automatic program verifier because condition entries are equivalent and action entries are different.

Using a nondeterministic form of the table is one way to handle imprecision in a DT. We define a *nondeterministic DT* (NDT) as a DT whose condition entries may be nondeterministic. Table 4 shows Table 2's modified DT to exemplify an NDT. We have generalized Table 2's first condition in this NDT and have combined the second, third, and fourth conditions into one. The table is now much more representative of how people think rather than based on how computers work. The next section details how to construct and use such an NDT.

Constructing an NDT for a knowledge base. Because of nondeterministic condition entries, we can generate NDTs quite simply. First, we need to identify variable conditions and possible actions. Next, we must set up relations between the two. Since we need not force conditions to be deterministic, NDTs can be generated in parallel during the design's initial thinking phase. Then, we must resolve contradictions and redundancies and determine any unnecessary solution parts. Human thought processes become unreliable in this second design phase.

To generate knowledge bases with no contradictions or redundancies, we need a formal method for verifying and managing the NDT. A *fuzzy DT,* based on the fuzzy-logic theory,[13] facilitates such a formal method. We discuss verification and management methods below.

Definition: A fuzzy DT (FDT) is an NDT where

(1) Each variable in the condition stub is a fuzzy variable, x_i within [0,1];

(2) Condition entry values are some fuzzy switching function of the x_i's; and

(3) Action entry values are intervals for accepted grades of membership.

Fuzzy switching functions of the x_i's for the condition entry are

x_i;
$1-x_i$: represented as $\overline{x_i}$;
$\max(x_i, x_j)$: represented as $x_i + x_j$;
$\min(x_i, x_j)$: represented as x_ix_j; and
any combination of the above.

Table 5 shows an FDT version of the previous Sorting NDT (← and → symbols are for scaling and are explained in the next section).

Table 4. A nondeterministic decision table to sort N records.

	1	2	3	4	5	6
Number of records	low	low	-	high	high	high
Size of records	small	med/large	huge	-	-	-
Alphabetizing	-	-	-	yes	no	no
Recursion available	-	-	-	-	no	yes
Call insertion sort	X					
Call selection sort		X				
Call heapsort					X	
Call quicksort						X
Call bucket sort				X		
Sort pointer array to records			X			
Go again			X			
Stop	X	X		X	X	X

Table 5. A fuzzy decision table to sort N records.

	1	2	3	4	5	6
Number of records (x_1)		$(x_1 \rightarrow .2)$			$x_1\ x_3\overline{x_4}$	
Size of records (x_2)		+				
Alphabetizing (x_3)	$x_1 + x_2$	$(\overline{x_2} \leftarrow .2)$	x_2	x_1x_3		$x_1\ \overline{x_3}x_4$
Recursion available (x_4)						
Call insertion sort	[0,.3]					
Call selection sort		[0,.5]				
Call heapsort					[.6,1]	
Call quicksort						[.6,1]
Call bucket sort				[.6,1]		
Sort pointer array to records			[.9,1]			
Go again			[.9,1]			
Stop	[0,.3]	[0,.5]		[.6,1]	[.6,1]	[.6,1]

Table 5 represents even the deterministic conditions as fuzzy variables. This causes no problems since their possible values will just be 0 or 1, which are within the necessary range. The following section presents a formal method for generating FDT rules from NDT rules. But first, let's look at one rule as an example. We see from rule 1 of the Sorting NDT (Table 4) that, when the record number is low and record size is small, we want to call the insertion sort. For the Sorting FDT (Table 5), x_1 represents record number and x_2 represents record size. By rule 1, if the max(x_1, x_2) is within [0,.3], we call the insertion sort. Since this will be true only if both x_1 and x_2 are low, the FDT's rule 1 matches the NDT's rule 1.

Since x_i condition variable values will be within [0,1], action entry acceptance intervals must also be within this range, meaning that specific value ranges for the variables must be mapped to the interval [0,1] so that an appropriate acceptance interval can be set. Specific action intervals would be determined by the expert or experts as the case may be. The references present a technique for determining such intervals.[14]

Specifically, FDTs are executed as follows.

(1) Rules are processed in order from left to right. On Go Again actions, table processing should be started over from the first rule.

(2) If none of a table's conditions are satisfied, then Return action becomes automatic.

(3) Some rules in the table can be specified as conditional rules, meaning that they should be considered only if directed by a table action. These rules should be set off from the others by a dashed vertical line (see Example 1 below).

(4) Unless processing a conditional rule, rule selection is at most one rule.

Table 6. Adjusting CPU utilization in a fuzzy decision table.

	1	2	3
Average number of processes (x_1)			
Average page fault frequency (x_2)	x_1x_2	$x_1 + x_2$	x_3
$x_1 > x_2$ (x_3)			
Call decrease-load	$\geq .8$		1
Call increase-frames	$\geq .8$		0
Process rule 3		$\geq .6$	
Return	$\geq .8$	$< .6$	X

Because of number (4) above, both contradictions and redundancies are possible in an FDT. Since this must not be allowed for a knowledge base, these must be detected and removed during the design stage.

Example 1. Table 6 gives a different FDT example, representing an intermediate-level table for choosing one of two ways to adjust a virtual memory system's CPU utilization. Rule 1 states that, if both the average number of system processes and average page fault frequency are high, then the system load should be decreased and the number of page frames allocated to each process should be increased. Because of rule 1's Return action, rule 2 is processed only if the conditions of rule 1 are not satisfied. In this case, either one or both of the averages are $< .8$. If at least one of them is high enough (that is, $\geq .6$), then the choice depends on whichever is higher according to conditional rule 3. If neither is high, no action is taken at this time.

Constructing FDT rules from NDT rules. For a given condition variable or expression, the acceptance range can be any interval within [0,1]. However, different ranges are sometimes more appropriate for the table. A simple technique for changing the range is to *scale the acceptance range*; that is, a scaling factor is added to or subtracted from both the condition variable value and acceptance range end points. We use an arrow (← or →) to represent the scaling direction.

Example 2—scaling an acceptance range. Assume that the acceptable range for condition x is [0, .2]. This means $x = .1$ and $x = .2$ are acceptable, but $x = .3$ is not. If we scale the acceptance range of x up by .2, then the scaled acceptance range is [.2, .4] and $x = .1$ implies ($x \rightarrow .2 = .3$), which is acceptable; $x = .2$ implies ($x \rightarrow .2 = .4$), which is acceptable; but $x = .3$ implies ($x \rightarrow .2 = .5$), which is not acceptable.

It is useful to note that, if one of the acceptance range end points is 0 and the range is being scaled up, the 0 end point does not have to be changed. The same is true for 1 when scaling down (true because a condition variable's initial value cannot be less than 0 nor greater than 1). In the above example, therefore, the scaled acceptance range could also have been [0, .4]. The only restrictions on scaling acceptance ranges are (1) if the acceptance range already has one end point = 0, it cannot be scaled down and, (2) if one end point = 1, it cannot be scaled up. We will discuss applications for scaling shortly.

For a given NDT rule, the action marked is to be taken only if each condition in the condition entry holds. Conditions can be represented by Boolean switching functions with deterministic DTs, which corresponds to "anding" different condition entries. With an FDT's fuzzy switching functions, however, how we combine two condition entries to "and" them depends on their acceptable ranges. For example, if we want to take an action only when x_1 is within [0,.3] and x_2 is within [0,.3], then (to combine these conditions) we would take the action only when the max(x_1, x_2) is within [0,.3] (which is what was done for Table 5's rule 1). But if we want to take an action only when x_1 is within [.7,1] and x_2 is within [.7,1]—and we again combine conditions so that we take the action when the max(x_1, x_2) is within [.7,1]—then it is possible that x_1 could be .6 and x_2 could be .8, and we would still take the action. Instead, the conditions should be combined so that we take the action only when the min(x_1, x_2) is within [.7,1].

Combining conditions of the same range is straightforward. However, combining conditions of different ranges poses some problems—problems avoided by scaling acceptance ranges for variables or expressions to be one of the following:

(1) Low range—one end point = 0, one end point $\leq .5$;

(2) High range—one end point $\geq .5$, one end point $= 1$; or

(3) Middle range—end points are complementary (that is, they add up to 1).

Algorithm 1—constructing FDT rules. Using this scaling convention, the following algorithm details the steps involved in constructing FDT rules from NDT rules:

Begin

(1) Determine acceptance range for each condition variable in rule.

(2) Scale the variables to a low, middle, or high range. For example,

If $x_1 \varepsilon [.2,.5]$, let $x_1' = x_1 \leftarrow .2 \varepsilon [0,.3]$.
If $x_2 \varepsilon [0,.2]$, let $x_2' = x_2 \rightarrow .3 \varepsilon [0,.5]$.

(3) Combine scaled variables according to range as follows:

(a) If both ranges are low, set the max of the variables to be in the low range. For example,

$x_1 \varepsilon [0.,3]$ and $x_2 \varepsilon [0.,3]$ implies $x_1 + x_2 \varepsilon [0,.3]$.

(b) If both ranges are high, set the min of the variables high to be in the high range. For example,

$x_1 \varepsilon [.6,1]$ and $x_2 \varepsilon [.6,1]$ implies $x_1x_2 \varepsilon [.6,1]$.

(c) If x_i is in a low range and x_j is in a high range, set either $(x_i + \overline{x_j})$ to be in the low range, or $\overline{x_i}x_j$ to be in the high range (by DeMorgan's law, these two are equivalent choices). For example,

$x_1 \varepsilon [0.,4]$ and $x_2 \varepsilon [.6,1]$ implies $x_1 + \overline{x_2} \varepsilon [0,.4]$ or $\overline{x_1}x_2 \varepsilon [.6,1]$.

(d) If both ranges are middle, set$(x_i\overline{x_j})(\overline{x_i}x_j)$ to be in the lower part of the middle range. For example,

$x_1 \varepsilon [.3,.7]$ and $x_2 \varepsilon [.3,.7]$ implies $(x_1\overline{x_2})(\overline{x_1}x_2) \varepsilon [.3,.5]$. (Notice that the new range is neither a low, high, nor middle range. This makes no difference as long as it is not necessary to combine this rule with another. If it does become necessary, the range can then be scaled accordingly.)

(4) If the scaled variables do not fit any of 3a-d, then set up the rule as a conditional rule.

End.

Example 3. As an example, here follows the construction of the Sorting FDT's rule 2 (see Table 5).

Given, $x_1 \varepsilon [0,.3]$ and $x_2 \varepsilon [.3,.8]$
Scaling both variables up by .2, $x_1 \rightarrow .2 \varepsilon [0,.5]$ and $x_2 \rightarrow .2 \varepsilon [.5,1]$.
Then, since x_1 is low and x_2 is high,
$(x_1 \rightarrow .2) + \overline{(x_2 \rightarrow .2)} \varepsilon [0,.5]$.
Based on the laws of fuzzy logic, this is equivalent to
$(x_1 \rightarrow .2) + (\overline{x_2} \leftarrow .2) \varepsilon [0,.5]$.

Verifying and managing FDTs. As stated earlier, we want a formal method for verifying that FDTs have no contradictions or redundancies. We must also ensure that no rule exists where the rule's condition entries can never satisfy action entry intervals. Since we have set up condition entries as fuzzy switching functions, we can use fuzzy logic's structure to provide the formalism we require.

A contradiction exists in an FDT when two or more rules have equivalent condition entries and different action entries. By comparing the functions of two rules using fuzzy switching logic, we can determine their equivalency. For example, consider two rules of an FDT whose condition functions are x_1 and $x_1 + (x_1 + x_2)$. Although these appear to be different functions, they are actually the same by the absorption axiom in fuzzy logic. If these two rules had different action entries, they would represent a contradiction. The easiest way to compare two fuzzy switching functions is to first minimize the two functions and then compare them directly. Various algorithms exist in the literature that can be used for minimizing fuzzy switching functions.[15-17] Therefore, we can automatically determine if two rules have equivalent condition functions.

A redundancy occurs in an FDT when the action entries of two rules are exactly the same intervals for the same actions. This can be checked by comparing the intervals directly. If two rules are redundant and condition functions are different, we will want to combine condition entries. In a deterministic DT, we would "or" condition entries of the two rules to form the new rule. For FDTs, it will depend on the ranges of variables as to how we combine rules. Resembling the steps for constructing FDT rules when condition entries were being "and" combined, the following are steps for "or" combining two FDT rules with the same action entries (we assume that variables have been previously scaled to an acceptable range):

(1) If both ranges are low, set the min of variables to be in the low range.

(2) If both ranges are high, set the max of variables to be in the high range.

(3) If x_i is in the low range and x_j is in the high range, set either $(x_i\overline{x_j})$ to be in the low range or $(\overline{x_i} + x_j)$ to be in the high range.

(4) If both ranges are middle, set $x_i\overline{x_j} + \overline{x_i}x_j$ to be in the lower part of the middle range.

When FDT rules combine, certain functions can arise that represent an impossible set of circumstances. For instance, the complementary laws $x_i + \overline{x_i} = 1$ and $x_i\overline{x_i} = 0$ do not hold in fuzzy logic. However, the relations $x_i + \overline{x_i} \geq .5$ and $x_i\overline{x_i} \leq .5$ always hold true. Therefore, if an FDT has the function $x_i + \overline{x_i}$ and the action interval is within [0,.5), then the conditions can never be satisfied and the rule is superfluous. The same is true for a condition function of $x_i\overline{x_i}$ and an action interval within (.5,1]. This can also be checked automatically.

Algorithm 2—constructing an FDT. We can now present an algorithm for constructing an FDT that contains no contradictory, redundant, or superfluous rules. This algorithm can also be used to modify an existing table by initializing *r-set* to be the set of all rules that will not be modified.

Let *r-set* be a set of FDT rules.

Begin
    Construct FDT rules according to Algorithm 1.
    Initialize *r-set* to be empty.
    While (there are rules not yet in *r-set*) Do
    Begin
        1. Consider a rule not in the *r-set*, rule′.
        2. If rule′ is not superfluous, Then
            Begin
                3. Compare rule′ with all the rules in *r-set* for contradictions or redundancies.
                4. If a contradiction exists between rule′ and a rule in *r-set*, Then
                    (a) delete one of the rules completely, or
                    (b) delete both rules and form a new rule.
                    (Do not add the new rule to *r-set* automatically.)
                5. Else
                    If a redundancy exists between rule′ and a rule in *r-set*, Then
                        combine the condition entry functions of the two rules (as per Algorithm 1);
                        delete the previous two rules.
                        (Do not add the new rule to *r-set* automatically.)
                6. Else {rule′ forms no contradiction or redundancy with any rule currently in *r-set*}
                        add rule′ to *r-set*.
            End (If);
    End (While Do);
End.

At first, it may seem that step 3 in the above algorithm could take considerable time. However, remember that these FDTs are design tools and should be constructed in a top-down manner. When FDTs are too large, another design level (and, therefore, another FDT) is usually warranted.

We have shown that FDTs are software engineering tools that designers can use to create production type knowledge bases. As DTs, they adhere to structured design techniques, are well defined and application independent, and provide meaningful documentation of designs throughout the entire design process. As NDTs, they can model approximate reasoning. And they can detect contradictory, redundant, and superfluous information in the knowledge base automatically because of fuzzy logic's formalism.

To set up knowledge bases, design engineers work with experts to generate NDTs representing knowledge. Using Algorithms 1 and 2, they then convert these NDTs to FDTs that are automatically verified free of contradictions, redundancies, and superfluous rules. If modifications are warranted, either during the design or after verification, engineers and experts consult before changing NDTs. Conversion to FDTs and verification procedures are then repeated.

FDTs work well for designing production type knowledge bases. How they can be used for relational and semantic-network-type knowledge bases remains to be seen. It should be noted that, since FDTs are just tools, we cannot expect them to guarantee correct designs. Rather, they can guarantee design consistency as we form and refine designs.

Acknowledgments

This work was assisted by research equipment supported by NSF grant number DCR-8404909. We gratefully acknowledge *IEEE Expert* referees, and Managing Editor Henry Ayling, whose combined suggestions have improved this article's presentation.

References

1. W.B. Gevarter, "An Overview of Expert Systems," U.S. Dept. of Commerce Report No. NBSIF 82-2505, Washington DC, Oct. 1982.
2. F. Hayes-Roth, D.A. Waterman, D.B. Lenat, *Building Expert Systems,* Addison-Wesley, Reading, Mass., 1983.
3. E.A. Feigenbaum, "Knowledge Engineering for the 1980s," tech. report, Computer Science Dept., Stanford Univ., Stanford, Calif. 94305, 1982.
4. R.S. Pressman, *Software Engineering: A Practitioner's Approach,* McGraw-Hill, New York, N.Y., 1982.
5. H.N. Cantrell, J. King, and F.E.H. King, "Logic Structure Tables," *Comm. ACM,* June 1961, pp. 272-275.
6. "Decision Tables: A System Analysis and Documentation Technique," IBM General Information Manual, IBM, White Plains, N.Y., 1962.
7. T.F. Kavanagh, "TABSOL—The Language of Decision Making," *Computers and Automation,* Sept. 1961, pp. 18-22.
8. R.B. Hurley, *Decision Tables in Software Engineering,* Van Nostrand Reinhold, New York, N.Y., 1983.
9. D.F. Langenwalter, "Decision Tables: An Effective Programming Tool," *Proc. First ACM SIGMINI Symp.,* ACM, New York, N.Y., 1978, pp. 77-85.
10. J.M. Francioni, A. Kandel, and C.M. Eastman, "Imprecise Decision Tables and Their Optimization," in *Approximate Reasoning in Decision Analysis,* M.M. Gupta and E. Sanchez, eds., North-Holland, New York, N.Y., 1982.
11. S.L. Pollack, H.T. Hicks, and W.J. Harrison, *Decision Tables: Theory and Practice,* Wiley Interscience, New York, N.Y., 1971.
12. J.R. Metzner and B.H. Barnes, *Decision Table Language and Systems,* Academic Press, New York, N.Y., 1977.
13. L.A. Zadeh, "Fuzzy Sets," *Information and Control,* Aug. 1965, pp. 338-353.
14. L.O. Hall, S. Szabo, and A. Kandel, "The Construction of Membership Functions of Fuzzy Sets for Use in Expert Systems," *Proc. First IFSA Congress,* Palma de Mallorca, Spain, June 1985.
15. A. Kandel, "On the Minimization of Incompletely Specified Fuzzy Functions," *Information Control,* Oct. 1974, pp. 141-153.
16. E.T. Lee, "Comments on Two Theorems by Kandel," *Information Control,* 1977, pp. 106-108.
17. M. Mukaidono, "An Improved Method for Minimizing Fuzzy Switching Functions," *Proc. Int'l Symp. Multivalued Logic,* IEEE Service Center, 445 Hoes La., Piscataway, N.J. 08854-4150, 1984, pp. 196-201.

Joan M. Francioni received her BS in mathematics from the University of New Orleans in 1977, and her MS and PhD in computer science from Florida State University in 1979 and 1981, respectively. From 1983 to 1985, she was a visiting assistant professor at Florida State University, during which time she researched fuzzy sets. Since then, she has been an assistant professor at Michigan Technological University, and has shifted her research interests to operating systems for parallel processing machines—focusing on load balancing techniques in hypercube multiprocessors, virtual memory for distributed memory systems, and operating system support for parallel program debuggers. A member of the IEEE, ACM, and Sigma Xi, she can be reached at the Dept. of Computer Science, Michigan Technological University, Houghton, MI 49931.

Abraham Kandel is a professor, and chair of the Computer Science Department, at Florida State University. In addition, he is director of FSU's Institute for Expert Systems and Robotics. He received his PhD in EECS from the University of New Mexico, his MS in EE from the University of California, and his BS in EE from the Technion-Israel Institute of Technology. He is a senior member of the IEEE and a member of NAFIPS, the Pattern Recognition Society, and the ACM. An advisory editor to the international journals *Fuzzy Sets and Systems, Information Sciences,* and *Expert Systems,* he has written numerous books and articles on fuzzy set theory and fuzzy techniques—including *Fuzzy Techniques in Pattern Recognition* (1982), *Fuzzy Relational Databases—A Key to Expert Systems* (1984), *Designing Fuzzy Expert Systems* (1986), and *Fuzzy Mathematical Techniques with Applications* (1986). His address is the Dept. of Computer Science, Florida State University, Tallahassee, FL 32306.

Chapter 5: Knowledge Acquisition

Once the developer has determined the knowledge appropriate to the problem domain and developed a means of representing it in the computer, the next step is to have a scheme for acquiring the knowledge. Knowledge acquisition is widely recognized as a serious bottleneck in expert systems development. To develop an expert system requires a human domain expert; getting the knowledge out of a human and into a computer is no trivial task. The five papers in this section are a big help in accomplishing this task. Cooke and McDonald lead with an excellent overview discussing a means of overcoming the weaknesses of classical knowledge acquisition methods. Finin *et al.* discuss the use of natural language in expert systems. Finin goes on, in the next article, to develop a way of building and maintaining large knowledge bases. Sharman and Kendall provide a case study showing how they applied acquisition principles. Lastly, Kitto and Boose discuss a way of choosing acquisition methods depending on the situation.

EH0303-8/90/0000/0269$01.00 © 1990 IEEE

Reprinted from *Proceedings of the IEEE*, Volume 74, Number 10, October 1986, pages 1422-1430.

A Formal Methodology for Acquiring and Representing Expert Knowledge

NANCY M. COOKE AND JAMES E. McDONALD

The process of eliciting knowledge from human experts and representing that knowledge in an expert or knowledge-based system suffers from numerous problems. Not only is this process time-consuming and tedious, but the weak knowledge acquisition methods typically used (i.e., interviews and protocol analysis) are inadequate for eliciting tacit knowledge and may, in fact, lead to inaccuracies in the knowledge base. In addition, the intended knowledge representation scheme guides the acquisition of knowledge resulting in a representation-driven knowledge base as opposed to one that is knowledge-driven. In this paper, a formal methodology is proposed that employs techniques from the field of cognitive psychology to uncover expert knowledge as well as an appropriate representation of that knowledge. The advantages of such a methodology are discussed, as well as results from studies concerning the elicitation of concepts from experts and the assignment of labels to links in empirically derived semantic networks.

INTRODUCTION

Expert systems are computer programs that perform a variety of complex problem-solving and decision-making tasks in well-specified areas of expertise. The successful performance of these systems relies heavily on human expert knowledge, as opposed to elaborate search strategies or general problem-solving methods. Some of the tasks performed by expert systems include diagnosis, prediction, design, interpretation, and instruction. For instance, MYCIN [1] is an expert system that was designed to perform diagnosis of infectious blood diseases and has been successful in producing judgements comparable to or better than, experienced diagnosticians. Another system, PROSPECTOR [2], has been successful in advising on the location of deposits of ore.

Reliance on a knowledge base is one feature that distinguishes expert systems from other programs which exhibit intelligent behavior. The expertise embodied in the knowledge base of an expert system consists of numerous facts, rules, procedures, and heuristics (i.e., rules of thumb) relevant to a particular domain. Expert system technology developed as a result of a new focus on specific knowledge is the driving force behind intelligent programs. Initially, the focus had been on general problem-solving methods and efficient knowledge representation and search strategies.

Knowledge engineering refers to the procedures involved in the development of expert systems. A major part of the knowledge engineer's task involves the acquisition of knowledge from experts in the form of facts and rules and the transformation of this information into a form that can be used by a computer program. The assumption that expert knowledge is mandatory in successful expert systems makes the acquisition and representation of knowledge critical to the design of these systems. Unfortunately, the transfer of knowledge from the human to the artificial expert is also the major bottleneck in expert system design.

THE KNOWLEDGE TRANSFER BOTTLENECK

Knowledge Acquisition Problems

There are several problems inherent in the current knowledge engineering methodology that contribute to the knowledge transfer bottleneck. The process of extracting expert knowledge is not well defined, but generally involves interaction between a knowledge engineer and a domain expert. The knowledge elicitation process is often viewed as an art (sometimes black magic) because there are no formal procedures or techniques available. However, there are some less formal methods that are commonly employed by knowledge engineers for this purpose. The two most popular of these are interviews and protocol analysis. Whereas both of these methods share many common elements, protocol analysis can be considered a structured interview and is, in this sense, more formal than an interview. Generally, the interview is conducted by a knowledge engineer who poses questions or problems to the expert. The expert in turn, is expected to provide answers or solutions that hopefully reveal some of the facts, rules, and heuristics relevant to the domain in question. In a similar fashion, protocol analysis involves the observation of a domain expert by a knowledge engineer as the expert solves a problem within his domain. Often the domain expert is asked to think aloud, to try to express his mental processes

Manuscript received June 16, 1985; revised March 21, 1986. This research was partially funded by Sperry Corporation, Aerospace & Marine Group, Albuquerque, NM.

The authors are with the Computing Research Laboratory and the Department of Psychology, New Mexico State University, Las Cruces, NM 88003, USA.

verbally. The protocol, consisting of the knowledge engineer's observations and the expert's thoughts, is later analyzed for particular features. It should be noted that not only are both of these techniques ill-defined, but they also involve introspection on the part of the expert, subjective interpretation by the knowledge engineer, and considerable communication between the participants. These characteristics of current knowledge acquisition techniques can lead to numerous problems that contribute to the knowledge transfer bottleneck.

One of the greatest problems with traditional interview and protocol techniques is that they involve introspection and verbal expression of knowledge, which is a difficult task for humans, especially experts. In fact, Johnson [3] has noted that, paradoxically, an increase in expertise seems to result in a decline in the ability to express knowledge. One of the major reasons for the rise of behaviorism in psychology was the failure of introspectionists to reliably report mental states. There is considerable psychological evidence suggesting that humans are not always able to accurately report on mental processes [4]. Thus the domain expert may not be able to convey the heuristics that he actually uses to solve problems. In fact, when asked to provide explanations for their behavior, experts often produce reasons, rules, or strategies that do not correspond to their actual behavior [3]. In summary, both the inaccuracies and inadequacies of verbal reports have important ramifications for knowledge engineering.

Some characteristics of human expertise add to the difficulty of verbally expressing knowledge. Much expert knowledge takes the form of heuristics, rules, and strategies that are procedural in nature. It seems especially difficult to convey procedural knowledge. To illustrate this difficulty, consider attempting to enumerate the steps required to tie a shoe. If you are like us, you will find it a difficult task, partly because like many skills, it is acquired by doing or watching and not by verbal instruction, just as the apprentice learns from the expert. Thus some knowledge is difficult to express verbally because it was not taught using language. Minsky [5] has stressed that "... self-awareness is a complex, but carefully constructed illusion" and that "... only in the exception, not the rule, can one really speak what he knows."

A related problem is that the knowledge that is involved in human expertise is often characterized as consisting of mental processes that are automatic or compiled. According to Shiffrin and Schneider [6], automatic processes do not require attention, are obligatory in nature, are unlimited in capacity, and are usually unconscious. Many of the processes or strategies that experts employ to solve problems are automatic and, therefore, not likely to be available for introspection. Knowledge compilation is the process of combining related productions into chunks of productions in order to speed up performance and decrease step-by-step processing, resulting in smoother performance [7]. Much expert knowledge appears to be compiled and, consequently, it is often difficult for the expert to reconstruct the original steps. Again, it is paradoxical that many of the traits of expertise, such as automatic processes and compiled knowledge, lead to difficulties in the explicit expression of knowledge.

Frequently, it appears that much human expertise is nothing more than intuition which is, by definition, hard to code into condition-action rules. Psychological studies of expertise have shown that what appears as intuition may actually be skilled pattern recognition ability. That is, the expert is able to quickly make sense of the environment by partitioning it into meaningful segments. For example, results of research on chess skill [8], [9] indicated that chess masters were able to quickly recognize and reconstruct meaningful chess board configurations, but not random configurations. These results were attributed to the experts' ability to perceptually chunk related pieces on a meaningful board. As with intuition, however, pattern recognition abilities are difficult to verbalize.

Even if experts were able to accurately introspect about their knowledge, they would still face the problem of communicating the knowledge to the knowledge engineer. Typically, the knowledge engineer is unfamiliar with the expert's domain and the domain expert is unfamiliar with programming and building expert systems. Ideally, the domain expert is also an experienced knowledge engineer, although that is rare. More often the domain expert tries to communicate his knowledge in a form understandable to the knowledge engineer and the knowledge engineer tries to convey to the domain expert the type, level, and form of knowledge that is needed to develop the expert system. Communication problems can make the knowledge acquisition process very tedious and inefficient. Often, this process results in either the knowledge engineer becoming an expert in the domain of interest or the domain expert becoming a knowledge engineer.

Another problem related to communication has to do with subjectivity on the part of the knowledge engineer. The expert's behavior is observed, interpreted, and transformed into a formalized version by the knowledge engineer. Likewise, whatever the expert says in an interview or protocol is interpreted and transformed by the knowledge engineer. These subjective procedures can be made more objective by using several knowledge engineers to obtain a consensus, but the problem still exists. Furthermore, when a knowledge engineer misinterprets an expert's knowledge, it might lead to the inclusion of faulty or conflicting rules in the system. Such errors could lengthen the expert system development process or might result in a faulty expert system. In summary, communication barriers inherent in traditional knowledge acquisition techniques contribute to their inefficiency.

Knowledge Representation Problems

Not only are the techniques for eliciting expert knowledge weak, but techniques for constructing a representation of the expert's knowledge are, for the most part, nonexistent. A knowledge engineer may derive a list of 100 or so "if-then" rules from an interview with an expert, without any indication of the relative importance of the rules, the number of rules that remain, or how the rules are organized or represented in the expert's memory. Typically, the knowledge (e.g., the set of rules) is represented in the expert system according to the selected system architecture. This tendency is probably magnified by the proliferation of expert system shells that embody a standard system architecture. The knowledge engineer must make the expert's knowledge conform to the scheme inherent in the system. Consequently, the acquisition of knowledge has generally been guided by the intended use of that knowledge in the expert system. If a system is to consist of production rules (condition–action pairs), then the knowledge engineer will try to extract knowledge from the expert in an "if–then" format. Other system architectures (e.g., frames, semantic networks) encourage different knowledge acquisition strategies.

There has been little concern expressed for the degree of fit between knowledge representation in the system and that possessed by the human expert. There is psychological evidence that indicates that experts differ from novices not only in terms of quantity of knowledge, but also in terms of how that knowledge is organized [10]. It is reasonable to assume that an expert system might benefit from the same type of knowledge representation that is associated with human expertise.

In addition, an important feature of expert systems is the ability to explain the line of reasoning involved in reaching a decision to an expert. This information is necessary for the training of potential experts and the debugging of faulty rules in the system. In many cases, the explanation consists of the series of rules that were used to reach a conclusion. Often these rules do not resemble the rules that the expert actually uses and, consequently, they are difficult for the human expert to understand [11]. The expert's need to understand the expert system is another argument for matching the content and organization of the expert system's knowledge base to that of a human expert.

It is unlikely that a single type of knowledge representation can efficiently capture expertise. Expert knowledge consists of concepts, relations, features, chunks, plans, heuristics, theories, mental models, etc. Whereas production rules are well-suited for the representation of some types of knowledge (e.g., heuristics), they are less suited for representing others. Hayes-Roth, Waterman, and Lenat [12] refer to the discrepancy between what an expert says and how it is represented in the program as the "representation mismatch." In order to avoid such a mismatch, we believe that the knowledge representation of the system should be driven by a formal knowledge acquisition process which would reveal the contents and organization of expert knowledge.

In summary, we contend that the selection of a knowledge representation scheme prior to the acquisition of the knowledge to be represented is putting the cart before the horse. The pre-selection of a knowledge representation scheme is necessarily independent of the knowledge and of the specific domain and, consequently, may misguide or wrongly constrain the knowledge acquisition process.

Knowledge Engineering Aids

A variety of knowledge engineering languages and tools have been developed to aid in expert system development. Some of these tools are stripped-down expert systems that contain inference engines and empty knowledge bases. EMYCIN is an example of such a system, based on MYCIN. These skeletal systems are, in theory, domain-independent (in the sense that the knowledge base is empty), but they are still designed for a certain class of problem solving activity, such as diagnosis. Other knowledge engineering languages (e.g., KEE, OPS5, ART) are more general and provide the user with a prespecified format for representing knowledge. These knowledge engineering languages, or environments, are basically high-level programming languages that aid in the formalization of knowledge once it is acquired. However, they do not address the problems involved in eliciting knowledge from experts or representing that knowledge in appropriate ways.

Some knowledge engineering tools aid in refining the knowledge base once it has been established. For example, TEIRESIAS interacts with the domain expert to refine the knowledge base by locating and debugging errors (i.e., incomplete or inconsistent rules) and by adding new rules as needed [13]. Systems like TEIRESIAS aid in automating the later phases of knowledge engineering involved with knowledge base testing and refinement, but again, they do not facilitate the initial knowledge acquisition process or provide freedom to flexibly represent knowledge.

An alternative approach has been to have the system acquire knowledge automatically, often referred to as machine learning. This approach assumes that by automating the learning process, the program can acquire expertise through learning in the same way human experts do. One example of this approach is seen in the Meta-DENDRAL system that induces rules about chemical mass spectrograms from examples. Michie [11] has discussed work on induction algorithms that allow learning by example, resulting in the generation of new rules for knowledge bases. He mentioned that such algorithms have often resulted in rules that were accurate and efficient, yet not understandable by human experts and proposed a structured induction technique in an attempt to overcome this limitation. Other work in machine learning has focused on learning by being told, as well as learning by discovery [14]. The "machine learning" approach to knowledge engineering problems is promising in that it removes the knowledge engineer from the loop and possibly, the domain expert as well. However, caution needs to be exercised in producing an adequate match between human expert knowledge and expert system knowledge.

In summary, these developments have not solved the basic knowledge engineering problems outlined in the beginning. There are tools that help to represent and formalize knowledge and there are tools for knowledge base refinement, but there are no satisfactory tools to aid in the elicitation of expert knowledge and no tools to aid in the selection of an appropriate knowledge representation. The machine learning approach is a step toward the initial generation of rules, but these rules often differ from those actually used by the experts, resulting in decrements in system comprehensibility. In the remainder of this paper we discuss a formal methodology for knowledge acquisition and representation which is based on techniques from cognitive psychology and which addresses the problems mentioned above.

A Formal Methodology for Knowledge Acquisition and Representation

In light of the limitations of current knowledge engineering methods, it seems that a new approach to the problem is needed. We propose a set of techniques which together comprise a formal knowledge acquisition and representation methodology. These techniques have been developed, applied, and studied in the field of cognitive psychology and have promising applications in knowledge engineering.

Related Research in Cognitive Psychology

The issues of knowledge representation and expertise have long interested researchers in cognitive psychology and artificial intelligence. The basic tenet of the information processing paradigm that underlies most work in cognitive psychology is that the human mind operates as a general-purpose symbol manipulator [15]. This emphasis on symbols has led to much of the research on knowledge representation. For instance, Collins and Loftus [16] developed a theory of semantic memory based on network representations of stored concepts. Other theories that attempt to explain empirical findings in language, perception, memory, and problem solving have been based on knowledge representations of various forms such as frames, scripts, features, propositions, and production systems.

Recently, psychological scaling algorithms have been developed that empirically generate specific types of knowledge representations (e.g., spatial, hierarchical, network) from data collected from human subjects. For example, given a set of pairwise estimates of proximity for a set of concepts, the Pathfinder algorithm produces a network representation in which concepts are represented as nodes and relations between pairs of concepts are represented as links between nodes [17], [18]. There are several means by which the proximity estimates for these algorithms can be derived, all of which are less direct than interviews and verbal reports. For instance, proximity estimates for sets of concepts can be derived from pairwise relatedness ratings, the location of items in a recalled list, frequency of co-occurrence in a sorting task, and confusion probabilities.

Along with the focus on knowledge representation, there has been a great deal of interest in the study of human expertise. Research has addressed differences between experts and novices in terms of pattern recognition abilities, memory organization, problem solving, decision making, and learning. Much of this research can be applied to work in expert systems [19].

Of particular relevance to the knowledge engineering issues discussed above is research in cognitive psychology that combines the representation of knowledge using psychological scaling algorithms with expertise. The goal of such studies has been to use various psychological scaling algorithms to generate representations of a well-specified domain of knowledge for subjects who vary in their level of expertise within that domain. The derived cognitive structures of novices are then compared with those of experts in an attempt to determine the cognitive features of expertise. Two examples of such studies will be discussed.

Schvaneveldt, Durso, Goldsmith, Breen, Cooke, Tucker, and DeMaio [20] derived cognitive structures of expert and novice fighter pilots using two different scaling techniques: multidimensional scaling (MDS-Alscal) and link-weighted networks (Pathfinder). Both of these techniques operate on estimates of psychological proximity. In this experiment, expert and novice fighter pilots rated all possible pairs of 30 flight-related concepts by assigning values from zero to nine to each pair, where zero indicated that the pair of concepts was highly unrelated and nine indicated that the pair was highly related. These ratings were then submitted to the two scaling algorithms.

MDS [21]–[24] positions each concept in a K-dimensional space where the distance between points reflects the psychological proximity of the corresponding concepts. As previously mentioned, Pathfinder produces a network with concepts represented as nodes and relations between concepts represented as links connecting some of the nodes [17], [18]. Links may be either directed (allowing traversal in only one direction) or undirected (allowing traversal in either direction). Thus relations between concepts may be either symmetrical or asymmetrical. With symmetrical proximity estimates, only undirected links can be included in the network representation. Both MDS and Pathfinder reduce a large amount of data, in the form of pairwise estimates, to a smaller set of parameters, but these two techniques tend to highlight different aspects of the underlying structure. Pathfinder focuses on the local relations among concepts, whereas MDS provides a more global understanding of the dimensionalized concept space.

Resulting cognitive structures were similar for fighter pilots with the same degree of experience, but differed across groups. Schvaneveldt *et al.* also found that a pilot could be classified as an expert or novice fighter pilot on the basis of his cognitive structure. Finally, an examination of the cognitive structures generated by the Pathfinder and MDS scaling algorithms highlighted concepts and relations that were common to both expert and novice representations, as well as those that appeared in only one of the representations. Schvaneveldt *et al.* discussed applications of these techniques to selection, training, and knowledge engineering. A direct extension of this work has been the development of, ACES, an air-to-air combat expert system [25].

Similar research has been done in the domain of computer programming expertise [26]–[28]. By exploring the cognitive organization of programming concepts using scaling techniques, similar to those described above, it has been determined that one aspect of programming expertise involves the organization of programming knowledge according to program meaning, or semantics, instead of syntax or surface structure. Cooke [27] derived Pathfinder network representations of abstract programming concepts based on relatedness ratings from programmers of varying levels of expertise. The resulting expert network is shown in Fig. 1. Cooke found that programmers could be correctly classified into either naive, novice, intermediate, or expert groups based on their cognitive representation. In addition, Cooke investigated the evolution of cognitive structures as programmers progressed from novices to experts. As a person learns to program, certain relations are formed, such as the relation between the concepts "algorithm" and "program." Likewise, certain misconceptions about programming are removed from the conceptual organization. This identification of expert relations and novice misconceptions can be used to guide education and training programs.

These two examples make use of only some of the psychological techniques that are available to elicit knowledge. Multidimensional- and network-scaling techniques have already been mentioned. Cluster analysis is a related technique that represents knowledge in the form of clus-

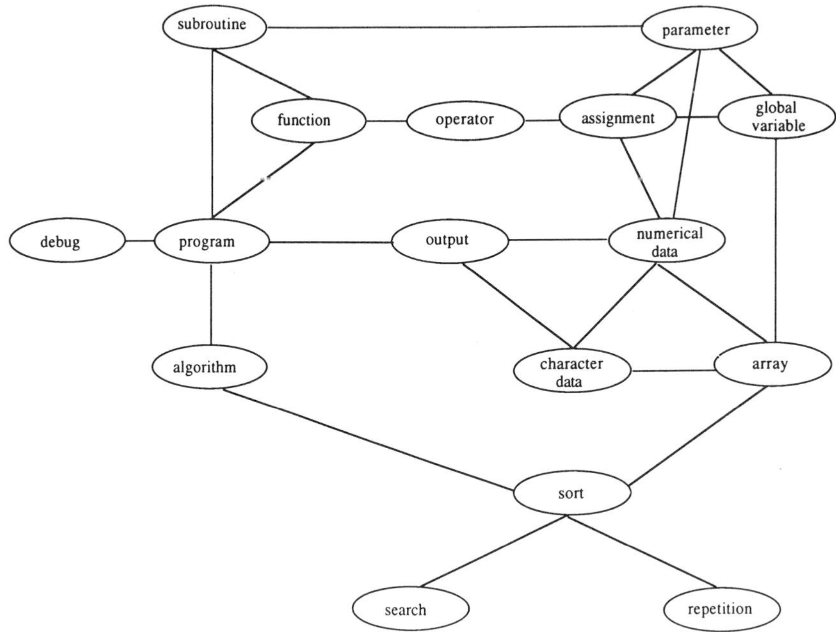

Fig. 1. Pathfinder network representation of expert programmer conceptual structure.

ters or categories. These three techniques: MDS, network, and cluster analysis fall under the general heading of psychological scaling techniques. In general, they are algorithms that transform complex data into a more understandable format that is assumed to have psychological validity. The resulting representation is dependent on the technique chosen to scale the data. For instance, the Pathfinder algorithm generates a network representation of knowledge, whereas MDS generates a spatial representation. However, it is possible to compare various representations to other behavioral measures such as recall performance or categorization in order to select the "best" representation [29]. One representation is better than another to the extent that it more closely matches the cognitive structure assumed to underly performance. Other psychological techniques that have been used to elicit knowledge of various forms include task analysis [30], repertory grid technique [31], and the critical incident technique [32]. Consequently, the field of cognitive psychology provides a number of alternatives to the weak knowledge acquisition and representation methods currently used by knowledge engineers.

The Application of Cognitive Techniques to Knowledge Engineering

To the extent that various cognitive techniques have been successful in representing expertise, they have potential application as knowledge acquisition and representation tools. Others [33], [34] have supported such an approach to knowledge engineering. Gammack and Young [34] argue that expert knowledge cannot always be represented in the form of "if-then" rules and that the interview is not suitable for eliciting some types of knowledge. In fact, it is likely that the belief that most expert knowledge can be represented in the form of rules stems from the fact that the interview methodology most readily elicits this type of knowledge. They concluded that knowledge types should be matched with the appropriate knowledge elicitation technique.

We are in the process of developing a knowledge acquisition "tool kit," consisting of cognitive scaling techniques, that could be used to aid the knowledge engineer in extracting relevant expert knowledge and representing that knowledge. This tool kit would become part of a formal methodology for expert system development. Some of the tools would be appropriate in certain applications, but not others. That is, multidimensional scaling might be better suited to extracting the important features underlying expert classification, whereas network scaling might be used to elicit conceptual relations. In the next section, some work is discussed that maps techniques for eliciting a set of domain-related concepts onto the type of concepts that are elicited by each.

The ability to automate the knowledge acquisition process is an important consequence of developing a formal methodology. An automated knowledge engineering system could be developed to take the place of the knowledge engineer in the preliminary stages of knowledge acquisition. In our conception, the domain expert would interact with the automated system which would direct the expert through a series of tasks (e.g., listing concepts, sorting concepts, rating concepts for relatedness, labeling links on a generated network, etc.) and apply psychological techniques to the expert's behavior to generate a "cognitive profile" of the expert. This profile might take the form of one or more network representations of the expert's knowledge with nodes representing specific concepts and links labeled with various relations. The knowledge engineer could use this information as a starting point for further knowledge acquisition or the cognitive profile might automatically be submitted to a knowledge refinement system along the lines of TEIRESIAS. Thus the automated system would facilitate the initial knowledge acquisition processes and possibly suggest a suitable knowledge representation scheme.

This methodology overcomes some of the previous problems inherent in the traditional weaker methods of knowledge engineering. First, unlike other knowledge engineering tools, this formal methodology can aid in the initial stages of knowledge engineering in which expert knowledge is elicited. In addition, the techniques we propose elicit knowledge from the expert indirectly by asking relatively simple questions (e.g., "how similar are these concepts?" versus "tell me what you know"). Even though experts may not be fully aware of their knowledge, the knowledge is often revealed through judgements, much as it is applied in their domains of expertise. Furthermore, because the proposed techniques suggest questions that need to be asked of the expert and because the techniques do not require answers from the expert in the form of "if-then"rules, the amount of interaction between the knowledge engineer and the expert is reduced and the type of knowledge that is elicited from the expert is not constrained by a particular knowledge representation. The data obtained are rich in information and various scaling techniques can each be applied to the same set of data to generate different knowledge representations. Finally, the comparison of a variety of psychological scaling techniques may provide clues as to the most compatible form, of knowledge representation for the expert system. It should be noted that different types of representations (e.g., network, spatial, production system) are not necessarily incompatible with each other, but might differ in the extent to which they match the expert's internal representation of knowledge.

Research Efforts Toward a Formal Methodology

Two issues that are critical to the application of scaling techniques to knowledge elicitation involve the initial elicitation of a set of concepts and the interpretation of the resulting representation. Many of the psychological scaling techniques discussed in this paper require a set of concepts, or items, that are to be scaled. In most cases, the concept set is intuitively selected, although in a few cases it is predefined (e.g., the set of all UNIX commands) [35]. The composition of the concept set is critical in applications of these techniques to knowledge engineering because missing or irrelevant concepts could result in gaps in the knowledge base. Another issue centers around the interpretation of scaling representations. In particular, the Pathfinder algorithm generates a network of nodes and unlabeled links. Whereas the links have weights, or costs, associated with the strength of each relationship, they are not differentiated by meaning. That is, there is no built-in way to distinguish the meaning of the relation between "bird" and "worm" from that between "bird" and "nest" if both pairs are connected with equal strength. In the application of these techniques to knowledge engineering, it becomes important to determine not only the presence of a relation, but the specific nature of the relation as well. In the following sections, research that addresses the concept elicitation and link interpretation issues is discussed.

Concept Elicitation Study

We have recently conducted an experiment [36] to address the question of how to elicit from experts, the domain-related concepts that are needed to perform many of the scaling techniques. The domain of "driving a car" was selected becaused it was broad enough to require a variety of types of knowledge and because "experts" in this domain were readily available in the subject pool of introductory psychology students.

Method: Thirty-two introductory psychology students from New Mexico State University volunteered in partial fulfillment of a research familiarization requirement. Thirty of the subjects participated in the experiment and were required to have a current valid driver's license. The remaining two subjects served as judges in the data analysis phase of the experiment. Four methods for eliciting concepts were considered and constituted the four conditions in a between subjects experimental design. These conditions were 1) concept listing, 2) interview, 3) step listing, and 4) chapter listing. Subjects were randomly assigned to these conditions and performed the task corresponding to their condition for a period of 20 min.

In the concept listing condition each of five subjects was asked to list on a sheet of paper all of the concepts that were relevant to driving a car. In the interview condition 15 subjects were divided into five teams of three. Within each team subjects were assigned to either an instructor role, a student role, or to the task of extracting concepts from the dialogue. The subject in the instructor role was told to act the part of a driver education instructor and to teach a student who has never driven before how to drive. The subject in the student role was told to act the part of a driver education student who had never driven before. This condition was designed to simulate a typical knowledge engineering interview. The dialogue was tape recorded and afterwards, the third subject listened to the tape and wrote down any concepts that were mentioned by either the student or the teacher concerning driving a car. In the step listing condition each of five subjects was told to list on a sheet of paper all of the steps that were relevant to driving a car, but not to worry about the order of the steps. Finally, in the chapter listing condition, each of five subjects was told to imagine that he was a driver education expert that had been asked to write a book on how to drive. The subject was then told to list on a sheet of paper the proposed chapter titles and subtitles for such a book. In general, the four conditions differed in the amount of structure or guidance inherent in the task with the first condition being least structured and the fourth being most structured.

Results and Discussion: Concepts generated in each of the conditions were transferred to 3 × 5 inch index cards (one concept per card). There was a total of 702 concepts (separate index cards) generated in this experiment. Of these, 17 percent were generated in the concept listing condition, 35 percent in the interview condition, 18 percent in the step listing condition, and 30 percent in the chapter listing condition. These values served as productivity measures in that they were indicative of the number of concepts that were generated, but did not take redundancy into account.

In order to determine the type of concepts elicited by each task, two additional subjects were recruited to assign each of the 702 concepts to one of seven categories. Each subject independently classified the concepts. They were presented with the total set of cards in a random arrangement, along with a list of categories. The categories were selected on the basis of the types of concepts that appeared to be present in the set of cards, but an attempt was also made to represent a variety of knowledge types in the categories (e.g., declarative and procedural). The seventh category, labeled "other" was a catch-all category and was to serve as an indicator as to whether or not the other six categories were sufficient. The category labels for all except the "other" category were also accompanied by one or more examples of typical concepts that might be assigned to each. The seven categories and accompanying examples are presented in Table 1.

Table 1. Concept Categories and Examples

Category	Examples
1. Explanation	*Watch for other drivers at intersections because they may not stop when they're supposed to.*
2. Rule: General	*Know your car. Wear seat belts.*
3. Rule: Condition	*When it rains, use your wipers. If the oil is low, then add more.*
4. Concept	*Skidding, brakes*
5. Procedure	*Put your foot on the brake, turn the key, and shift gears.*
6. Fact	*Drunk driving is dangerous.*
7. Other	—

After the sorting was completed, the number of cards in each category from each of the four conditions was recorded for each subject. Because there was considerable sorting agreement between the two subjects ($r(26) = 0.864$, $p < 0.001$) their data were combined. The percentage of concepts sorted in each of the seven categories is presented in the first column of Table 2. Most cards were sorted into the "concept" category, while the "other" category was least used, suggesting that the six categories were sufficient. Examples of concepts that were sorted into the "other" category include: "no suspension," "automatic driving," and "written." In the remainder of Table 2, the sorting distribution is presented for each condition. These percentages were based on the total number of concepts generated in each condition.

In general, these results indicated that the interview and chapter listing conditions elicited the largest number of concepts. However, this measure did not take into account redundancy. It is possible that the interview and chapter listing conditions elicited the same number of unique concepts as the concept and step listing conditions. An extension of this study which is currently underway involves the evaluation of the four concept elicitation tasks in terms of redundancy, completeness, and relevancy. That is, which task generates the most complete set of unique, domain-related concepts, but does not generate many concepts that are irrelevant to the domain?

Classification results revealed that each of the techniques differed in terms of the major types of concepts that they elicited. For instance, the interview and chapter listing techniques generated mostly concepts, whereas the concept listing and step listing technique generated mostly general rules. This type of information should be useful in matching a particular knowledge elicitation technique to the type of knowledge that is desired. For instance, on the basis of this study, if procedural knowledge in a particular domain was desired then the step listing task would be most appropriate for eliciting this type of knowledge. On the other hand, it might be useful to use a combination of tasks in cases in which the form that the knowledge is to take is not specified.

Interestingly, the interview technique did not yield as many "if-then" rules as was anticipated. In fact, none of the four techniques generated a large number of conditional rules. Perhaps this was due to the particular domain used in this study. Replications of this experiment in a variety of domains would serve to clarify this issue. At any rate, it is interesting that at least in this domain, little knowledge seems to be naturally expressed by experts in the form of "if-then" rules.

Link Labeling Study

In the previously mentioned study by Cooke [27], Pathfinder network representations were generated for a set of abstract programming concepts. The networks were generated from a set of relatedness ratings for all possible pairs of the concepts. The Pathfinder network presented in Fig. 1 was based on the average ratings from ten expert programmers. (Expertise was defined for the purpose of this study as three or more years of programming experience.) This network has 16 concepts (or nodes) and 24 links. In general, the network represents the experts' conceptual organization of the set of programming concepts. Linked concepts are assumed to be closely associated in memory. Such a representation might be applied to knowledge engineering as an organizational scheme for rules in a production system. An organization of rules based on such a

Table 2. Percent Cards Sorted in Each Category.

	Overall	Breakdown by Condition			
Category		Concept Listing	Interview	Step Listing	Chapter Listing
Explanation	6.0%	5.5%	7.2%	7.0%	4.9%
Rule: General	22.0	43.6	18.9	36.8	4.9
Rule: Condition	9.0	9.8	8.4	11.6	6.6
Concept	42.0	18.4	38.9	8.5	77.7
Procedure	9.0	5.1	9.5	25.6	1.2
Fact	9.0	15.0	12.5	8.9	1.2
Other	3.0	2.6	4.6	1.6	3.5

Table 3. Link Labels for Expert Programming Network in Fig. 1

Linked Concept Pair	Labeled Relation	Number of Subjects
subroutine–program	is part of	4
character data–output	is a type of	4
parameter–subroutine	is used with	3
program–output	produces	4
sort–search	involves	3
numerical data–parameter	can be a	4
function–operator	is a	3
function–program	is part of	4
debug–program	is done to	3
output–numerical data	consists of	3
array–character data	can consist of	3
function–subroutine	is a	4
assignment–parameter	is done to	4
array–global variable	can be	5
array–numerical data	can consist of	4
repetition–sort	is part of	4
program–algorithm	is the implementation of	3
sort–algorithm	is a	4
sort–array	is done to	5
assignment–operator	is a	4

Note: The four links that were not successfully identified were, *parameter–global variable, global variable–assignment, assignment–numerical data,* and *character data–numerical data.*

representation might then reduce the search time through the set of production rules. An additional application, as advocated in this paper, involves the application of the network representation to problems of knowledge acquisition and representation. It has been suggested that such a representation could serve as a cognitive profile of an expert or group of experts. However, a limitation of this approach concerns the fact that links in Pathfinder networks are unlabeled. It would be of interest to know not only *that* the concepts, "subroutine" and "parameter" are related, but also, *how* they are related. Therefore, in the following study, expert programmers were asked to assign meaning to each of the links of the network in Fig. 1.

Method: Subjects consisted of five programmers from New Mexico State University who had a variety of experience with different programming languages. Each subject was shown a hard copy of the expert programming network (Fig. 1) and asked to describe the relation between pairs of linked concepts in the network. The 24 pairs of concepts were listed on a separate sheet in a random order for each subject. The position of items in each pair was counterbalanced across subjects. Subjects were told to write a short sentence or phrase next to each pair of concepts in the list. The sentence or phrase was to take the following format: *concept-relationship-concept.* The instructions stated that the concepts within the pairs did not have to be ordered in the same way that they occurred in the list. They also stated that specific relations should be used as opposed to vague relations such as, "is similar to" or "is related to." Subjects were given as much time as they needed to complete the task.

Results and Discussion: For each pair of linked concepts, the relation and direction of the relation identified by each subject were noted. The identification of a link label was assumed to be successful in those cases in which three or more of the five subjects agreed on a label. In some cases, relations were equated that differed in the direction of the relation, but not the meaning. For instance, "function" *is part of a* "program" was judged to be equivalent in meaning to "program" *contains a* "function." According to these criteria, 20 of the 24 links were successfully identified. The resulting link labels are shown in Table 3, along with the number of subjects that agreed upon each label. The order of the concepts in each pair is indicative of the direction of the relation.

Of the 20 links that were identified, 11 were unique. The addition of link labels to the Pathfinder network transforms a purely associational network into a semantic network. The information that is conveyed in the links adds necessary detail to the knowledge representation. Also, the link labels can be differentiated according to the inferences that correspond to a particular relation. For instance, the presence of an "is a" relation might lead one to infer that properties of the superordinate concept could be inherited by the subordinate concept.

It should be noted that even when subjects felt that a relation existed between two concepts, it was often difficult to attach a label to the relation. An extension of this research involves the development of a technique for labeling links that might facilitate this process. Using this technique, subjects are not required to directly attach verbal labels to the links, but rather to judge the relatedness of pairs of links. We hope that these data can be used to classify links into groups with similar meanings. The resulting groups of linked pairs should be easier to label than individual links.

Conclusions

The concept elicitation and link labeling work discussed above have been aimed at adding to, or refining, some existing psychometric techniques in order to apply them to knowledge engineering issues. The elicitation of a set of domain-related concepts is an important first step in many of these techniques. The concept elicitation study also illustrated the variety of types of knowledge that can be associated with a single domain. Whereas it might be possible to code all of the knowledge about driving a car in the form of condition-action rules, some knowledge might be more efficiently or naturally represented in other forms. In the link labeling study the meaning of relations was incorporated into the Pathfinder network representation of programming concepts. The interpretation of a particular relation seems critical to the application of these techniques to knowledge engineering.

Future work includes assessing the psychological validity of various scaling techniques and determining which techniques are best suited for different types of knowledge. Our long-term goal is to develop a formal methodology in which different types of knowledge are acquired and represented using appropriate techniques. As techniques are determined to be successful, they will be incorporated into the automated knowledge engineering system.

We believe that the proposed methodology is a promising alternative to traditional knowledge engineering techniques. It overcomes many of the problems inherent in protocol analysis and interviews. In addition, it is an approach that advocates a knowledge-driven architecture, instead of representation-driven knowledge acquisition.

References

[1] E. H. Shortliffe, *Computer-Based Medical Consultations: MYCIN.* New York: American Elsevier, 1976.

[2] R. O. Duda, J. G. Gaschnig, and P. E. Hart, "Model design in the PROSPECTOR consultation system for mineral exploration," in *Expert Systems in the Micro-Electronic Age*, D. Michie, Ed. Edinburgh, UK: Edinburgh Univ. Press, 1979.

[3] P. E. Johnson, "What kind of expert should a system be?" *J. Med. Phil.*, vol. 8, pp. 77–97, 1983.

[4] R. E. Nisbett and T. D. Wilson, "Telling more than we can know: Verbal reports on mental processes," *Psychol. Rev.*, vol. 8, pp. 231–259, 1977.

[5] M. Minsky, "K-Lines: A theory of memory," in *Perspectives in Cognitive Science*, D. A. Norman, Ed. Norwood, NJ: Ablex, 1981.

[6] R. M. Shiffrin and W. Schneider, "Controlled and automatic human information processing: II. Perceptual learning, automatic attending, and a general theory," *Psychol. Rev.*, vol. 84, pp. 127–190, 1977.

[7] J. R. Anderson, "Acquisition of cognitive skill," *Psychol. Rev.*, vol. 89, pp. 369–406, 1982.

[8] W. G. Chase and H. A. Simon, "Perception in chess," *Cogn. Psychol.*, vol. 5, pp. 55–81, 1973.

[9] A. D. De Groot, *Thought and Choice in Chess.* The Hague, The Netherlands: Mouton, 1965.

[10] M. T. H. Chi, P. J. Feltovich, and R. Glaser, "Categorization and representation of physics problems by experts and novices," *Cogn. Sci*, vol. 5, pp. 121–152, 1981.

[11] D. Michie, "The state of the art in machine learning," in *Introductory Readings in Expert Systems*, D. Michie, Ed. London, UK: Gordon and Breach, 1982.

[12] F. Hayes-Roth, D. A. Waterman, and D. B. Lenat, Eds., *Building Expert Systems.* Reading, MA: Addison-Wesley, 1983.

[13] A. Barr and E. A. Feigenbaum, Eds., *The Handbook of Artificial Intelligence*, vol. II. Los Altos, CA: Kaufmann, 1982.

[14] R. S. Michalski, J. G. Carbonell, and T. M. Mitchell, *Machine Learning: An Artificial Intelligence Approach.* Palo Alto, CA: Tioga Publ. Co., 1983.

[15] R. Lachman, J. L. Lachman, and E. C. Butterfield, *Cognitive Psychology and Information Processing: An Introduction.* Hillsdale, NJ: Erlbaum, 1979.

[16] A. M. Collins and E. F. Loftus, "A spreading activation theory of semantic processing," *Psychol. Rev.*, vol. 82, pp. 407–428, 1975.

[17] R. W. Schvaneveldt and F. T. Durso, "Generalized semantic networks," presented at the meetings of the Psychonomic Society, 1981.

[18] R. W. Schvaneveldt, F. T. Durso, and D. W. Dearholt, "Pathfinder: Scaling with network structures," Memo. in Comput. and Cogn. Sci., MCCS-85-9, Comput. Res. Lab., New Mexico State Univ., 1985.

[19] N. M. Cooke, "Modeling human expertise in expert systems," Memo. in Comput. and Cogn. Sci., MCCS-85-12, Comput. Res. Lab., New Mexico State Univ., 1985.

[20] R. W. Schvaneveldt, F. T. Durso, T. E. Goldsmith, T. B. Breen, N. M. Cooke, R. G. Tucker, and J. C. DeMaio, "Measuring the structure of expertise," *Int. J. Man Machine Studies*, vol. 23, pp. 699–728, 1985.

[21] J. B. Kruskal, "Multidimensional scaling and other methods for discovering structure," in *Statistical Methods for Digital Computers*, Enslein, Ralston, and Wilf, Eds. New York: Wiley, 1977.

[22] J. B. Kruskal and M. Wish, "Multidimensional scaling," *Sage University Paper Series on Quantitative Applications in the Social Sciences*, no. 07-011. London, UK: Sage Publ., 1978.

[23] R. N. Shepard, "Analysis of proximities: Multidimensional scaling with an unknown distance function. I," *Psychometrika*, vol. 27, pp. 125–140, 1962.

[24] ——, "Analysis of proximities: Multidimensional scaling with an unknown distance function. II," *Psychometrika*, vol. 27, pp. 219–246, 1962.

[25] T. E. Goldsmith and R. W. Schvaneveldt, "ACES: Air Combat Expert Simulation," Memo. in Comput. and Cogn. Sci., MCCS-85-34, Comput. Res. Lab., New Mexico State Univ., 1985.

[26] B. Adelson, "Problem solving and the development of abstract categories in programming languages," *Memory and Cogn.*, vol. 9, pp. 422–433, 1981.

[27] N. M. Cooke, "Computer programming expertise: Variations in cognitive structures," in F. T. Durso, Chair, *Human Expertise: Papers in Honor of William Chase*, meetings of the Southwest Psychological Association, Austin, TX, 1985.

[28] K. B. McKeithen, J. S. Reitman, H. H. Rueter, and S. C. Hirtle, "Knowledge organization and skill differences in computer programmers," *Cogn. Psychol.*, vol. 13, pp. 307–325, 1981.

[29] N. M. Cooke, F. T. Durso, and R. W. Schvaneveldt, "Recall and measures of memory organization," *J. Exper. Psychol.: Learning, Memory, Cogn.*, in press.

[30] R. D. Huchingson, *New Horizons for Human Factors in Design.* New York: McGraw-Hill, 1981.

[31] F. Fransella and D. Bannister, *A Manual for Repertory Grid Technique.* London, UK: Academic Press, 1977.

[32] F. J. Landy and D. A. Trumbo, *Psychology of Work Behavior.* Homewood, IL: Dorsey Press, 1976.

[33] K. A. Butler and J. E. Corter, "The use of psychometric tools for knowledge acquisition: A case study," in *Artificial Intelligence and Statistics*, W. Gale, Ed. Reading, MA: Addison-Wesley, 1986.

[34] J. G. Gammack and R. M. Young, "Psychological techniques for eliciting expert knowledge," in *Research and Development in Expert Systems*, M. A. Bramer, Ed. London, UK: Cambridge Univ. Press, 1985, pp. 105–112.

[35] J. E. McDonald and R. W. Schvaneveldt, "The application of user knowledge to interface design," in *Cognitive Science and its Application for Human-Computer Interaction*, R. Guindon, Ed. Hillside, NJ: Erlbaum (in press).

[36] N. M. Cooke, "The elicitation of concepts and association in empirically derived semantic networks," unpublished manuscript, New Mexico State Univ., Las Cruces, NM, 1986.

Reprinted from *Proceedings of the IEEE*, Volume 74, Number 7, July 1986, pages 921-938.

Natural Language Interactions with Artificial Experts

TIMOTHY W. FININ, MEMBER, IEEE, ARAVIND K. JOSHI, FELLOW, IEEE, AND BONNIE LYNN WEBBER, MEMBER, IEEE

Invited Paper

The aim of this paper is to justify why Natural Language (NL) interaction, of a very rich functionality, is critical to the effective use of Expert Systems and to describe what is needed and what has been done to support such interaction. Interactive functions discussed here include defining terms, paraphrasing, correcting misconceptions, avoiding misconceptions, and modifying questions.

I. INTRODUCTION

Natural Language (NL) interfaces to database systems are already proving their worth. They allow users to get at the database facts they want, without the need to become system wizards. In this paper we are primarily concerned with NL interfaces for systems that do more than identify and retrieve facts. Such systems, often called knowledge based systems, expert systems, or advisory systems, are expected to provide analyses and/or advice to users faced with real problems. Our main goals in this paper are to justify why NL interaction, of a very rich functionality, is critical to the effective use of these systems and to demonstrate what is needed to support it. Even if one wanted to employ a formal language or menu and pointer-based system for this role, it would have to have many of the features of NL that allow the kinds of interactive functions that a system must support.

Returning for a minute to database systems, to naive or infrequent users of these systems, NL can mean getting their information simply by asking for it. In many cases, NL can even mean a shorter, simpler query than its formal counterpart [51]. "Smarts" built into NL front ends can increase their tolerance for typing errors, spelling errors, and divergences from polished grammar, or eliminate such errors entirely [67]. Other "smarts" can increase the scope of acceptable NL input beyond "syntactically sugared" formal queries by allowing the user to talk about things previously mentioned in either question or answer without previously having named them (via the use of anaphoric and deictic pronouns and noun phrases such as "them," "this," "those students," "the seniors," etc.). On the basis of such improvements in performance characteristics, NL interfaces for databases are becoming more attractive, even where the interactive behaviors required are not very sophisticated.

However, our claim is that for effective use of Expert Systems, NL is not just attractive but critical. This is because advice and diagnosis cannot be trusted blindly. From what we know of human–human advisory (problem-solving) interactions, a responsible person will not accept advice if s/he cannot request and receive clarification, e.g.,

System: In your agreement of sale, make sure that you have an engineering inspection clause.
User: Engineering inspection clause?
System: Right. You want that house inspected by an engineer just to make sure that there's nothing wrong with it that you might have missed. The clause stipulates that an engineer be permitted at your expense to inspect the house and that if there is anything in that house that requires a repair (and you can set the amount, $250 or $500), the owner will be responsible for it, or else you will be permitted to get out of the deal.

if s/he cannot verify that s/he has understood the advice or diagnosis correctly, e.g.,

Expert: OK, under those circumstances, I would not object to see you go back into that for another six months.
User: So you roll it over, in other words?
Expert: Right.

if s/he cannot get justification as to why that advice or diagnosis should be accepted, e.g.,

Expert: So you would still file as a single person.
User: Even though I'm married?

Manuscript received January 27, 1986; revised March 7, 1986. This work was supported by the U.S. Army under Grants DAAG-29-84-K-0061, DAAB-07-84-K-FO77, DAAG-29-84-9-0027, DAAG-29-85-K-0061; by DARPA under Grant N00014-85-K-0018, and by the National Science Foundation under Grants MCS-82-19116-CER, MCS-82-07294, MCS-83-05221, and DCR-84-10413.

The authors are with the Department of Computer and Information Science, University of Pennsylvania, Philadelphia, PA 19104, USA.

Expert: Right. Because you weren't married during the filing year, 1984.

or if s/he cannot add or delete parts of the problem specification (i.e., symptoms of the problem or constraints on its solution), e.g.,

User: What if I decided to postpone my wedding until January 1?

The complexity of this whole enterprise requires a rich communicative system such as NL (and currently nothing else) provides.

Moreover, there is another problem with achieving successful communication that makes NL not just attractive but critical. Suchman [63] describes the situation as follows:

> Successful communication under ordinary circumstances is not a product of the absence of trouble but of its repair. That is, the extent to which we are able to communicate effectively is not a function of our planning for all of the contingencies that might arise. It is the function of the locally managed, collaborative organization of the interaction itself—an organization designed to maximize efficiency by providing for the ongoing identification and repair of troubles that, when we are constructing our actions together, in real time, inevitably arise.

The kinds of interactions we will be concerned with are well-characterized by this observation. One reason for this is that we do not believe that we can count on users having well-formed problems, or an accurate understanding of the system's capabilities and terminology, or even a good understanding of the domain. Under such conditions, a formulaic, trouble-free interaction through a menu/pointer system or a formal language is inconceivable, assuming as it does that users can state their problems right off, receive a response, and leave, confident that applying this response to their problem will result in success. In what follows, we discuss specific NL capabilities that can help user and system to come to terms with each other. If such capabilities are not supported, any "user friendliness" intended in the system design (through NL input and output, for example) will be irrelevant. The appealing "syntactic sugar" will turn out to be "syntactic arsenic" in disguise.[1] Our position is that support for these NL capabilities cannot come solely from the NL interface itself: that in many cases, the underlying reasoning system and knowledge base will have to be adapted or designed *ab ovo* to support them. On any grounds, the latter is clearly preferable.

In this paper, we will not discuss basic issues involved in developing NL interfaces for database fact-retrieval systems (e.g., the syntactic and semantic analysis of user requests, resolving anaphora and ellipses, mapping between user concepts and an underlying database query language, etc.). One reason is that by now there exist some fine surveys of such work, which are quite up-to-date and easily accessible. For example, Perrault and Grosz [52] survey NL interface systems from the point of view of system architecture and development tools. Bates and Bobrow [4] describe the state of the art of NL interfaces for databases, circa 1984.

The other reason for ignoring these more basic issues is that we want to focus on other aspects of cooperative interactions where NL interfaces will really pay off. Some of the extensions to NL interfaces that we will be discussing in the following sections have, in fact, been developed in the context of database systems. However, the reason we have included them is the payoff they will have for interactions with knowledge based systems, expert systems, etc. (It is worth noting that as database systems are augmented with more and more general knowledge about their domains, as well as specific facts, the line between database system and knowledge based system becomes faint indeed.) For the reader's benefit though, we list here some of the major NL interface systems for databases: LUNAR [72], REQUEST and TQA [15], [53], REL [67], PLANES [70], NL Menu System [67], EXPLORER [37], RUS [5], [4], and a (nameless) system built by Ginsparg [22]. Some of the commercial systems are INTELLECT (Artificial Intelligence Corporation), THEMES (Frey Associates), RAMIS (Mathematica Products Group), CLOUT (Microrim), PLUME (Carnegie Associates), EXPLORER (Cognitive Systems), among others as reported in TARGET (*Artificial Intelligence Newsletter*, vol. 1, no. 3, May 1985).

There is one significant type of cooperative behavior needed for interacting with expert systems and knowledge based systems that we do not discuss here: that is, explanation, in the sense of justification: e.g.,

Tax Advisor: So you would still file as a single person.
Taxpayer: Even though I'm married?
Tax Advisor: Right, because you weren't married during the filing year, 1984.

Systems must be able to justify their actions—whether they act to ask a question or to offer some conclusion or advice (as above)—or users may not be sufficiently convinced to respond appropriately (i.e., to answer the system's question or to take its advice).[2] Here we just point the reader to some work on enabling systems to produce explanations which are accurate (i.e., they correctly specify the reason for the system's behavior) [64], [50], [16], clear [26], [58], and appropriate to their users' interests and/or level of expertise [44] [69].

The next section (Section II) describes two preliminary efforts aimed at giving systems the ability to automatically construct useful definitions of the terms they use. It also makes recommendations for further work in this area. Section III describes work that has been done on giving systems the ability to paraphrase their users' queries, allowing their users to verify how their queries have been understood. It also makes recommendations for further work. Section IV deals with recognizing and responding to disparities between the user's view of the world and that of the system. We describe some work that has been done in the context of database interfaces and some more applicable to interfacing with expert and knowledge based systems. Section V discusses research on the complementary issue of avoiding misleading remarks which may lead to new misconceptions on the user's part. And finally, Section VI discusses how troubles can be avoided or lessened in

[1]Note that this research also helps in isolating aspects of cooperative interaction which may be incorporated in a more formal system, to provide some of the flexibility available in NL, if that is the path one wishes to follow.

[2]Notice that this sense of "explanation" is different from its sense of "clarification," as in "Explain what you mean by X." We discuss clarifying terminology in Section II.

determining the user's problem, by having systems adapt their questions and interpretations to the user. We conclude in Section VII with a summary prognosis and recommendations.

II. Clarifying Terminology

To avoid troubles or as one way to repair them, user and system must establish a single terminology for interacting with each other. Without a single mutually understood terminology, the user will not be able to understand what the system is requesting or recommending; e.g.,

System: Is renal function stable?
User: What do you mean by 'stable'?

System: In your agreement of sale, make sure that you have an engineering inspection clause.
User: Engineering inspection clause?
System: Right. You want that house inspected by an engineer just to make sure that there's nothing wrong with it that you might have missed. The clause stipulates that an engineer be permitted at your expense to inspect the house and that if there is anything in that house that requires a repair (and you can set the amount, $250 or $500), the owner will be responsible for it, or else you will be permitted to get out of the deal.

nor will the expert system be able to understand what the user is requesting or reporting, comparable to the following exchange:

Medical Student: These are just stasis changes.
Expert Diagnostician: What do you mean by that?
Medical Student: Just from chronic venous insufficiency.

If the user and expert system are to establish a single terminology for the interaction, there are three options—either the system understands and accepts the user's terminology, the user understands and accepts the system's, or together they accept a third. Here we only want to consider the first two options.

Now it has been claimed [10] that in the case of a physician (i.e., a "natural" expert) interacting with a patient to take a history, it is an error for the physician to use his or her term for a symptom rather than the patient's—that patients seem to have difficulty characterizing their experiences in the physician's terms.[3] The physician must accommodate to the patient's terminology and, where necessary, seek clarification from the patient. For example, in the following exerpts from a history-taking session, the physician first attempts to verify that he and the patient mean the same thing by 'hemorrhoids' and later asks the patient directly what he means by 'diarrhea' [10].

Physician: How can I help you?
Patient: OK. I've had bleeding right before my bowel movement. Some diarrhea and then bleeding.
Phys: Mm-hm. And when is the first time that ever happened?
Pat: I really don't remember because I didn't make any note of it. It was just like a little diarrhea and no blood—I think I had hemorrhoids.
Phys: I see. And how did you know it that time that you had hemorrhoids?
Pat: I thought they were because I felt something there.
Phys: Did you feel something? A lump or something like that?
Pat: Yeah, like two little lumps.
Phys: Right. Now. Did the diarrhea come first?
Pat: Yeah.
Phys: And when was that? Roughly speaking.
Pat: About a month ago, maybe—but then it was on and off I—and I did—
Phys: You mean for one day you would have dia —a loose—*What do you mean by diarrhea?*
Pat: Not diarrhea. I don't even know. Like, I can't describe it—Like I would feel very gassy. And, like, I guess expel—I don't know if it was mucus or whatever.

[3]On the other hand, if the patient seems to lack a way to describe his/her experience (say, of pain), Cassell would have the physician suggest terms which may evoke a useful response from the patient.

This would seem to argue for the expert system adapting to the user's terminology—at least in attempting to acquire information from the user. Moreover, in later explaining any of its conclusions, it has been argued that a system must make its explanation understandable to the user—one element of which is that the explanation not contain terms that the user will not understand.

On the other hand, others have noted [12], [48], [39] that people are notoriously poor at defining terms in a way that a system can understand and use. This is well illustrated in the case of NL interfaces confronted with new terms that appear in database queries, that are not in their lexicons or grammars. There are currently four viable methods for dealing with such 'unknown' terms: the first, used in [27], is to let the user equate the new term or phrase with one the system already knows (which assumes solid knowledge of the system on the user's part); e.g.,

User: Who commands the *Kennedy*?
System: I don't know the word "command."
User: Let 'who commands the *Kennedy*' be the same as 'who is the captain of the *Kennedy*'
System: OK
User: Who commands the *Kennedy*?

The second method is to guide the user step by step through a simple defining process [48], [39] to acquire both syntactic and semantic information about the new term. If the word cannot be defined by this process, the user's only recourse is asking the system programmer to do it or forgetting about the term. It would be foolish to ask users

System: What do you mean by 'concerning'?
System: What do you mean by 'often'?
System: What do you mean by 'purple'?

and expect them to come up with a response that makes sense and is usable by the system.

The third method is to assume that the unknown term is a database value (a reasonable practice in large, rapidly changing systems where one does not want to put all database values with their syntactic and semantic features into the lexicon) and either wait until after query evaluation or verify understanding through query paraphrase (see next section) to find out whether this is a reasonable assumption. This is the method used in INTELLECT [25].

The fourth option is to punt entirely and ask the user to rephrase the query without the offending terms.

The upshot of all this is that it is not clear to us who should adapt to whom. If the expert system is to adapt to users' terminology (both in acquiring information from them and in explaining its conclusions to them), it will need to develop better ways of figuring out what the user means by particular terms. If a user is to adapt to the system's terminology, the system becomes responsible for explaining its terms to the user in ways that s/he will understand and find relevant. In most cases, this will be impossible to do without the use of freely generated language. So far, two attempts have been made at enabling systems to take this responsibility through the incorporation of domain-independent NL definitional capabilities. Both of these are described below.

A. *TEXT*

TEXT [45] was a first attempt at providing a domain-independent NL definitional capability which can explain to a user the system's understanding of the terms it uses. It does this by piggybacking its definitional machinery on an augmented version of the system's domain model. If the user requests the definition of a term or an explanation of the difference between two terms, TEXT

- circumscribes an area of relevant information that it might include in its response. (Not all this information need be included in the response, but no information other than this will be included.);
- chooses an appropriate, fairly flexible format for presenting the information (called a "schema");
- instantiates the schema with a subset of the relevant information, guided both by the options allowed in the schema and a desire to allow the 'focus' of the text to change in a natural way;
- produces NL text from the instantiated schema.

The kind of information represented in TEXT's augmented data model in order to produce useful definitions of and comparisons between terms includes

- simple taxonomic information (e.g., a SHIP is a WATER_VEHICLE);
- for each split in the taxonomy, the criterion for the split (called the "distinguishing descriptive attribute" or DDA). Each subtype in a partition has the same DDA name, but a different value for the attribute (e.g., WATER_VEHICLE has two subtypes SHIP and SUBMARINE. SHIP has its DDA, TRAVEL_MODE, equal to SURFACE, while SUBMARINE has TRAVEL_MODE equal to UNDERWATER);
- for each subtype in a partition, the database attributes that support its DDA (e.g., SHIP has DRAFT and DISPLACEMENT). These are called "supporting database attributes";
- for each subtype, a list of its other database properties (e.g., SHIP has additional properties PROPULSION, MAST_HEIGHT, MAXIMUM_SPEED, etc.).

TEXT's presentational schema were derived from empirical observation of short (one paragraph) definitions and term comparisons. (Current schema include IDENTIFICATION, CONSTITUENCY, COMPARE_&_CONTRAST, ATTRIBUTIVE.) Each schema corresponds to a regular grammar whose terminals are *rhetorical predicates* which themselves specify the kind of information from the system's data model that can instantiate them. For example, the IDENTIFICATION schema has the following structure:

Identification (Class & Attribute/Function)
{Analogy/Constituency/Attributive/Amplification}*
Particular_Illustration/Evidence +
{Amplification/Analogy/Attributive}
{Particular_Illustration/Evidence}

where "{ }" indicates optional constituents, "/" indicates alternatives, "+" indicates that the item may appear one or more times, while "*" indicates that the item is optional, but if present, may appear one or more times. With respect to the information in the system's data model, instantiating the rhetorical predicate "Attributive" with respect to a concept can be done via its DDA or its other database attributes. "Particular_Illustration" can be instantiated by any of the terminal elements below a concept in the data model. The following definition of AIRCRAFT_CARRIER produced by TEXT instantiates this IDENTIFICATION schema as the sequence of rhetorical predicates: Identification, Analogy, Particular_Illustration, Amplification, Evidence.

User: What is an AIRCRAFT_CARRIER?
TEXT: An aircraft carrier is a surface ship with a DISPLACEMENT between 78000 and 80800 and a LENGTH between 1039 and 1063. Aircraft carriers have a greater LENGTH than all other ships and a greater DISPLACEMENT than most other ships. Mine warfare ships, for example, have a DISPLACEMENT of 320 and a LENGTH of 144. All aircraft carriers in the ONR database have REMARKS of 0, FUEL_TYPE of BNKR, FLAG of BLBL, BEAM of 252,... A ship is classified as an aircraft carrier if the characters 1 through 2 of its HULL_NO are CV.

The original version of TEXT [45] used a unification grammar [34] to translate an instatiated schema into NL text. The current version uses McDonald's MUMBLE [42] as its tactical component [57] and runs 60 times faster. TEXT is now being provided with a richer representation language for its domain model, which should allow it to make better judgements in its selection of relevant material and in how it instantiates its response.

B. *CLEAR*

While TEXT attempts to provide a portable, domain-independent NL definitional capability for users of database systems, CLEAR [57] attempts to do the same for users of expert systems. It uses the information intrinsic to an expert system—i.e., its rule base—to explain to users how terms are used in the system. (There is really no intrinsic information that corresponds to what a term *means*.)

These term definitions should be of interest to both the system's intended users and its developers. That is, the larger the rule base, the greater the chance that the same term may be used in more than one way (especially with rules being developed by more than one person or rules changing over a period of time). This is something that system developers must avoid. CLEAR's NL definitions can help them avoid this.

For a user, it is important to understand the system's terminology when s/he must answer the system's questions. Asking "why" the system asked a question (an option already allowed in many expert systems, following the work of Davis [16]) is unlikely to help him/her understand its terms: e.g., "filing status," "single," "married filing jointly," etc, in the following question:

System: What is your filing status?
1—Single
2—Married filing separately
3—Married filing jointly
4—Unmarried head of household
User: WHY?
[i.e., WHY is it important to determine your filing status?]
System: ...in order to compute total number of exemptions.

IF the filing status of the client is single
AND the number of special exemptions claimed by the client is N
THEN the total number of exemptions on the tax form is $N + 1$.

Of course, one could manually store definitions of all the terms used in the system, but again, this would require an enormous amount of manual labor, with the additional difficulty of keeping the definitions up to date as the system develops.

In deciding whether to use a rule in explaining a given concept, CLEAR ranks it according to how it uses the concept. For example, a rule in which a concept appears in the premise as a property or its value (provided it does not also appear in the conclusion) is ranked HIGH with respect to that concept, while a rule in which that concept appears as a property whose value is inferred is ranked MEDIUM, and one in which the concept appears as a concluded value is ranked LOW. This means that a rule such as R1 below will be rated higher than R2, with respect to explaining the term "filing status," since in R1, it appears as a premise property and in R2, as an inferred property.

R1: IF the filing status of the client is single,
AND the number of special exemptions claimed by the client is N,
THEN the total number of exemptions on the tax form is $N + 1$.

R2: IF the marital status of the client is unmarried or separated,
THEN the filing status on the tax form is single.

The actual translation of rules from their internal format into English is done through the use of pre-stored templates in the manner of MYCIN's translation methods [60]. However, if there are many rules to present, CLEAR provides two ways of condensing their presentation. First, if the rules have been organized into a generalization hierarchy, then each subset of the rules to be presented which falls under a single node of the hierarchy can be presented in terms of the single general rule; e.g.,

IF the filing status of the client is X
THEN the total number of exemptions of the client is determined from X.

The second method involves automatic summarization of subsets of the rules to be presented which share common concepts in both premise and conclusion—for example, if in explaining "filing status," several rules have "filing status" as a property in the premise and "total number of exemptions" as an inferred property in the conclusion, this would be summarized as

Filing status can be used to determine the total number of exemptions.

(At the time of this writing, this summarization facility had not yet been implemented.)

A system such as CLEAR can only be as effective as its ability to recognize how terms are used in rules, without requiring additional knowledge of the domain.[4] In a MYCIN-like system, this can be done to some extent, but for really effective responses, more explicit rule structures are needed. For example, if the set of premise clauses in a rule *define* what it means for the concluded property to have a particular value, then that rule should actually be rated HIGH with respect to explaining that concluded value, as in

User: EXPLAIN head_of_household
System:
IF the client is unmarried on the last day of the year,
AND the client paid < 50% of the cost of keeping up the home of a parent or child who is a dependent of the client,
AND the client selects filing status head of household,
THEN the filing status of the client is head of household.

On the other hand, if the set of premise clauses set out symptomatic ways of *recognizing* whether the concluded property has a particular value, then that rule should be ranked LOW with respect to explaining the concluded value, as CLEAR currently does; e.g.,

User: EXPLAIN enterobacteriaceae
System:
IF the identity of the organism is not known with certainty,
AND the stain of the organism is gramneg,
AND the morphology of the organism is rod,
AND the aerobicity of the organism is aerobic,
THEN there is strongly suggestive evidence that the class of the organism is enterobacteriaceae.

[4]CLEAR is designed to be used with any expert system built using the HP-RL system [36] without altering the structure or content of the expert system's knowledge base.

Because CLEAR cannot distinguish between these two very different types of rules, it will make mistakes. But notice that the consequence of such mistakes is just a suboptimal definition, which includes material that the user may find irrelevant and omits some material that s/he may find useful. Thus a desire to be able to better clarify terms for users of expert systems argues for both an NL generation capability and more explicit structure to the system's knowledge base. Additional arguments for more explicit rule structures appear in [11] and have led to the development of MYCIN's successor NEOMYCIN.

C. Further Work

If systems are to be truly adept at clarifying the meaning of terms for their users, they will have to have the ability to

- adapt definitions to the current discourse context
- adapt definitions to users' previous knowledge and/or to their goals
- make use of pictures as well as text

With respect to adapting definitions to the current discourse context, McKeown [45] discusses the value that can be gained from discourse context defined as either 1) the user's previous requests for definitions or 2) those previous requests and their answers. (Neither sense of discourse context is stored or used in the original version of TEXT.) By just storing the user's previous requests for definitions, the following adaption can be made: if the user has previously asked about term A and now asks the difference between A and B, the system can omit information about A in its response (which it can presume to have given earlier) and just describe B with some concluding comparison between A and B. If the answer to the user's request for clarification of A had been stored as part of the discourse context as well, B can be presented in a way that mirrors the structure and content of the system's presentation of A. Also, if the user has asked about A and now asks about a similar concept B (where similarity can be defined with respect to the generalization hierarchy), parallels can be drawn between them and their similarities and differences discussed. If the answers to the user's requests have been stored as well, contrasts can be limited to those properties of A discussed previously. Otherwise any contrasts may be generated.

With respect to adapting definitions to users' previous knowledge and/or to their goals, an example of a goal-sensitive definition was given earlier:

System: In your agreement of sale, make sure that you have an engineering inspection clause.
User: Engineering inspection clause?
System: Right. You want that house inspected by an engineer just to make sure that there's nothing wrong with it that you might have missed. The clause stipulates that an engineer be permitted at your expense to inspect the house and that if there is anything in that house that requires a repair (and you can set the amount, $250 or $500), the owner will be responsible for it, or else you will be permitted to get out of the deal.

In this interaction, the user is asking the expert for advice on purchasing a house. In defining 'engineering inspection clause' here, the expert first shows how the concept is significant with respect to purchasing a house and then gives its formal definition. Depending on the current situation and the experience of the user, the significance of the concept to the current situation may be more or less evident from its formal definition alone, but to ensure that its significance is recognized, it should be included explicitly in the definition or the definition should be tailored explicitly to the goal. For example, compare the following three definitions:

User: DEFINE stainless_steel.

S1: Stainless steel is an alloy consisting of iron mixed with carbon, chromium, and often, nickel.

S2: Stainless steel is an alloy consisting of iron mixed with carbon, chromium, and often, nickel. It resists rust, stains, and corrosion caused by water and acid. On the other hand, a stainless steel blade does not take and keep a very sharp edge.

S3: Stainless steel is an alloy consisting of iron mixed with carbon, chromium, and often, nickel. It does not conduct heat efficiently, and the resulting hot and cold spots on the bottom of a stainless steel pan will not allow food to cook properly. Copper is a fast heat conductor and stainless steel pots are sometimes given copper bottoms to eliminate hot and cold spots.

Here, the user comes across the term "stainless steel," which s/he realizes s/he does not understand. If the user's goal in this situation is to buy kitchen knives, S1 and S3 would not be helpful definitions. If his/her goal is to buy pots, S1 and S2 would be equally as useless. In any goal-oriented interchange, a simple definition like S1 is rarely sufficient, and the significance of the term with respect to the goal must be made evident.

The importance of identifying and taking account of the speaker's plans and goals is becoming widely recognized both for understanding the speaker's utterances [1], [9], [38] and for generating cooperative responses [1], [31], [54], [61]. It should be applied to attempts to clarify meaning automatically as well.

Finally, systems should be able to make use of pictures (either pre-stored or automatically generated) in explaining terms or phrases to users. This is essential for 1) terms which name visual patterns, such as "spindle cell component" or "pseudopalisading" in neuropathology; 2) terms for objects and their parts which should be defined both visually (conveying structural information) and textually (conveying functional information); e.g., terms like "twing lines" and "boom vang," which refer to optional but useful parts of a racing dinghy;[5] or 3) terms or phrases which refer to actions; e.g., a term like "luff".[6]

[5] The 'boom vang' is a lever or block and tackle system used to hold the boom down and take the twist out of the mainsail. It should be anchored on the heel of the mast, as low as possible [14]. Notice that without a visual illustration, one would probably still not recognize a boom vang if it hit him/her on the head (and it usually does.)

[6] Luffing is when a leeward boat starts to turn into the wind, pushing up any windward boats beside it. One cannot luff beyond head to wind.

There has been some work in this area already. The APEX system [19] is an experimental system which creates pictorial explanations of action specifications automatically, as they are to be carried out in a particular context.[7] In addition, work in the Pathology Department at the University of Pennsylvania on a diagnostic expert system in neuropathology [62] projects the use of images (stored on an optical disk) to give users visual experience in concepts such as those mentioned above. If in working with the pathologist to diagnose a specimen, the system asks whether the specimen shows evidence of phenomenon *X*, the pathologist has the opportunity of seeing examples of *X* before answering. This is expected to enhance the reliability of the pathologist's response.

To summarize, we feel that it is critical in interactions with expert systems that user and system understand each other's terminology. Since this can rarely be established outside the context of the interaction—it is too much to ask of users that they learn the system's terminology beforehand—the interaction itself must provide for this functionality.

III. Paraphrase

The second place we see NL playing a critical role in users interacting with expert systems is in the use and understanding of paraphrase. Paraphrase—restating a text so as to convey its meaning in another way—is used in human–human interactions for three reasons, all attempting to ensure reliable communication.

- Where the listener notices an ambiguity or vagueness in the speaker's utterance, the listener can produce multiple paraphrases of the utterance, each corresponding to and highlighting a different alternative. (The set, of course, can be presented in an abbreviated way, to focus on the different options even more.) Given these presented alternatives, the speaker can easily indicate what was intended.
- Even without noticing an ambiguity or vagueness, the listener may attempt to paraphrase the speaker's utterance to show how it has been understood, looking for confirmation or correction from the speaker.
- The listener may paraphrase the speaker's utterance in order to confirm his or her belief that there is common understanding between them.

In each of these cases, a paraphrase must present a particular interpretation without itself introducing additional ambiguity.

The important thing about a paraphrase, whether to verify understanding or to point out the perceived ambiguity of the original utterance, is that it must be unambiguous, different in some obvious way from the original, and its meaning easily recognized. An early effort at paraphrasing wh-questions that conforms to these goals can be found in [43]. McKeown based her paraphrases on the linguistic notion of *given* versus *new* information. *Given* information is information either in the previous discourse or part of the shared world knowledge of the participants. In her database question/answering application (as part of the CO-OP system [33]), McKeown takes this to be information which the speaker assumes to be in the database. *New* information is of two types, that specifying what the speaker directly wants and that specifying the angle from which that want is to be satisfied. McKeown presents a simple algorithm for dividing the parse tree (syntactic analysis) of a question into three parts: 1) that conveying the new information directly wanted (the simple question, stripped of modifiers, containing the wh-phrase); 2) that conveying the interest angle of the new information (any syntactic modifiers of the wh-phrase); and 3) the assumed information (all the other modifiers). These parts are then rearranged and presented as an NL paraphrase to the user—assumptions, if any, first; direct question, second; and interest angle, if any, third. For example,

Question: Which division of the computing facility works on projects in oceanography?
Paraphrase: Assuming there are projects in oceanography, which division works on those projects? Look for a division of the computing facility.

If there is more than one modifier in a particular category, that information is presented in ways that avoid ambiguity. Additional assumptions are presented in parentheses after the first one; e.g.,

Question: Which division works on projects in oceanography sponsored by NASA?
Paraphrase: Assuming there are projects in oceanography (*those projects sponsored by NASA*), which division works on those projects?

Additional interest angles are presented in separate sentences at the end; e.g.,

Question: Which division of the computing facility in superdivision 5000 works on projects in oceanography?
Paraphrase: Assuming there are projects in oceanography, which division works on those projects? Look for a division of the computer facility. *That facility must be in superdivision 5000.*

Notice that if the system had parsed either question such that the additional modifiers were attached at other syntactically possible places, this would be evident from the paraphrase:

Paraphrase: Assuming there are projects in oceanography (*that oceanography sponsored by NASA*), which division works on those projects?[8]

Paraphrase: Assuming there are projects in oceanography, which division works on those projects? Look for a division of the computer facility. *That division must be in superdivision 5000.*

Thus although this system only produces paraphrases of one type of utterance—wh-questions, those paraphrases

[7] The STEAMER project makes excellent use of graphics to show cause–effect relationships in the control of a ship's steam plant, but not to explain terms [29].

[8] If this sounds like an impossible analysis, substitute the word "rockets" for oceanography.

are guaranteed both to differ from the original question, if the original contained any modifiers in the form of prepositional phrases or relative clauses (due to its separation of given and new information), and to be unambiguous (due to the way it presents multiple modifiers). McKeown's system produces paraphrases of the syntactic analysis of the user's question. However, McKeown notes that the paraphraser can also be made to generate an NL version of the system's interpretation and that this feature is used by CO-OP to produce a corrective response in case the user's question shows a misconception (cf. Section IV).

The point here is that there are places further down the line where the system's understanding of the user's query can become something other than the user intended. This can happen when relationships between items mentioned in the query are introduced explicitly. For example, in paraphrasing the following user query, Codd sees his proposed system RENDEZVOUS [12] as making explicit its interpretation of a dependency (grouping) relation among the items requested, as well as its interpretation of the relation implicit in the noun–noun compound "Houston parts":

User: Give me a list of all the part numbers, quantities, and suppliers of Houston parts.

...intervening clarifying interaction...

System: This is what I understand your query to be: "Find all combinations of part number, quantity on order, supplier name, and supplier city such that the supplier supplied the part to a project located in Houston."

Notice that a dependency should not always be assumed—for example, in the similarly structured request like "Give me a list of the seniors and graduate students of the Computer Science Department." If the system has not previously established the validity of these interpretations through a clarification process, they may not turn out to be the ones intended by the user.

Also during subsequent interpretation, a term may be understood in ways the user does not expect or intend—for example, "rock" in the LUNAR system [71] was understood in a strict geological sense as meaning "basalt" rather than in its everyday sense.

Or terms may be particularized in ways the user does not intend or expect; e.g., the query

User: What do you have written by Jones?

may be paraphrased as

Paraphrase: What reports do you have written by Jones?

whereas the user meant any reports, books, articles, etc.

Or discourse context-sensitive items such as definite pronouns, definite noun phrases, and ellipses may be resolved in ways that the user has not intended.

User: Do you have the report that Finin published in 1984?
Paraphrase: *Test whether there are any reports whose author is Finin and whose publication date is 1984.*
System: No.
User: What reports do you have by Joshi?
Paraphrase: *List reports whose author is Joshi and whose publication date is 1984.*

Here the user's second request has been interpreted[9] as if the publication date had been ellipsed, and the system has supplied it explicitly as 1984 (i.e., the publication date from the previous request). This may or may not be what the user intended by his/her query, but the paraphrase makes the system's interpretion clear.

All of these argue for at least one additional point of contact where users can verify that the system has understood their queries as they intended them. At least two researchers have reported on systems which generate a paraphrase of the system's understanding of the user's utterance rather than a paraphrase of its syntactic analysis: an NL front-end to databases developed at Cambridge by Sparck-Jones and Boguraev [7], and an extension to IBM's TQA system developed by Mueckstein [49].

In Sparck-Jones and Boguraev's system, NL paraphrases are generated at two points in the understanding process: first, after the system has produced an interpretation which reflects lexical and structural aspects of the user's query, and second, at a later stage in the processing, after the specific characteristics of the data to be sought in the database in answer to the user's query have been identified. Both of these paraphrases are produced by the same NL generator.

The TQA system is an NL front-end to databases, which translates NL queries into SQL expressions, which are then evaluated against the database. To allow the user to check the accuracy of the SQL expression, Mueckstein's system QTRANS produces a paraphrase of the SQL expression before it is evaluated, and does so in an interesting way. It parses the SQL expression using SQL's unambiguous context-free grammar.[10] It then maps this SQL parse tree into an English surface structure parse tree using application-independent translation rules in the form of a Knuth Attribute Grammar. During this process, nonterminal nodes of the SQL parse tree are *renamed* (receiving NL grammar labels like Noun_Phrase, Relative_Clause, etc.) *reordered* into pre- and post-modifiers within a Noun_Phrase, *deleted* (suppressing SQL artifacts like join conditions), or *inserted* (providing English function words). In its final step, QTRANS translates the terminal nodes into English and prints out the resulting string.

For example, consider the database query:

What parcels in the R5 zone on Stevens Street have greater than 5000 square feet?

(The database for one TQA application is zoning information for the city of White Plains, NY.)
TQA produces the following SQL expression as its interpretation of the query:

```
SELECT UNIQUE A.JACCN, B.PARAREA
  FROM ZONEF A, PARCFL B
```

[9]Following a technique used in PLANES [70].

[10]Mueckstein has made minor, domain-independent changes to the published SQL grammar, to make it more compatible with the English structures to be generated.

```
WHERE A.JACCN = B.JACCN
  AND B.STN ='Stevens St'
  AND B.PARAREA > 5000
  AND A.ZONE ='R5'
```

This, QTRANS paraphrases as

> Find the account numbers and parcel areas for lots that have the street name Stevens St, a parcel area of greater than 5000 square feet, and zoning code R5.

The important things to notice about this paraphrase are the following:

- It shows the user that the system will be responding with not just a direct answer to the query (i.e., noting just the appropriate parcels) but with additional information as to the area of each.
- It shows the user that the system only knows 'parcels' in terms of their account numbers.
- The phrase 'for lots' is derived from the FROM phrase and represents the 'real world' entity to which the tables ZONEF and PARCFL apply. This must be recorded by hand explicitly for each table and column name in the database.
- The first conjunct in the WHERE clause is suppressed because it is recognized as specifying only the keys compatible for joining.

Mueckstein notes that the difficulties in this approach to paraphrase stem from the need to rearrange the SQL query structure to bring it closer to the user's conceptualization of the domain without being too repetitive or introducing ambiguities. This requires the attribute grammar to be able to compare units across nonadjacent phrases in the SQL query and act accordingly.

Related to this use of paraphrase of a formal language query is the use of NL paraphrase in an automatic programming system developed at ISI [65]. Here the programmer's high-level specifications (which the system will turn into an executable program) are first paraphrased in English so that the programmer can check that s/he has said what s/he intended. (That is, constructs in this high-level language imply particular mapping/dependency relations which the programmar may not be aware of or may have forgotten.) So paraphrase is being used to catch errors before they get in the code—very similar to its use in query systems, to catch bad interpretations before they undergo expensive evaluation against a database.

Before concluding this section, we want to comment on the opposite side of this verification process—allowing the user to verify his/her own understanding of the system's requests or advice. This aspect of user–system interactions has long been overlooked, but is critical for the responsible use of expert systems. Paraphrase, the method employed by systems in resolving ambiguity and verifying their formal interpretation of users' queries, is only one of several methods that are used interactively between people to verify understanding. On the advice giver's side, s/he can explicitly test the recipient's understanding by asking questions or asking the recipient to repeat what s/he (the advice giver) has said. On the recipient's side, s/he may ask the advice giver to verify his/her understanding by *inter alia* paraphrasing the question or advice,

> Expert: OK, under those circumstances, I would not object to see you go back into that for another six months.
> User: So you roll it over, in other words?
> Expert: Right.

by offering back a conclusion s/he draws from the material to be understood

> User: Where is Schedule B?
> Expert: Do you have Schedule A?
> User: Yeah.
> Expert: OK. If you turn it over, there is Schedule B.
> User: Oh. It's on the back of Schedule A?
> Expert: Right.

or by offering back how s/he sees the (general) advice as instantiated to his/own situation

> Expert: You shouldn't mix drugs and alcohol.
> User: You mean I shouldn't take my vitamins with a Bloody Mary?

In user interactions with expert systems, who takes responsibility for initiating this verification will depend on the character of the user population, the goal of the system, and what is being verified. We believe that systems must always provide for users' attempts to verify their understanding of the system's queries. With respect to the system's conclusions and advice, some systems will have to build in tests of users' understanding; others will have to build in additional NL understanding machinery to allow users to initiate verification themselves.

IV. Correcting Misconceptions

Another aspect of the user and system coming to terms with each other for successful interactive problem solving involves their individual world views. Many of the troubles in interaction come from the fact that participants hold different views about the world and either do not realize it or fail to do anything about it. As a result, each may leave the interaction with very different beliefs about what was communicated. In Suchman's terms [63], both user and system must provide, in some way, for the ongoing identification and repair of those troubles.

(In computational research in this area, disparities between the beliefs of user and system have been treated as *misconceptions* on the user's part. That is, the system's view of the world is assumed to be correct. Of course, in actual fact, the user may hold the correct belief, but the important thing is that the disparity be pointed out so that the interaction can be correctly interpreted. In what follows, we shall also refer to these disparities as "misconceptions" to be consistent with the literature.)

Heretofore most of the computational work on recognizing and responding to user misconceptions has been done in the context of NL front-ends to database systems. Notice that in these relatively simple systems, where it is only facts and requests for them that are exchanged, users may be able to recognize a disparity between the system's view of the world and their own: for example, they may be able to tell that the system's answer to their question is wrong or strange. Nevertheless, because the user may not recognize the disparity at all or not recognize it quickly enough to avoid confusion, people have felt the need to put as much responsibility on the system as they have been able to. The problem is much greater in user interactions with expert systems, where users are likely to have less domain knowledge and expertise than the system and thus are less likely to recognize a disparity. The result is confusion, or worse, misinterpretation. In what follows, we discuss work on recognizing and responding to user misconceptions, first in the context of database question/answering and then in the context of interactions with expert systems.

There are at least two kinds of beliefs that are easy to recognize. One is a belief that a particular description has a referent in the world of interest; i.e., that there is an individual or set which meets the description. Such a belief is revealed when a noun phrase is used in particular contexts. For example, for Q to ask the question "Which French majors failed CIS531 last term?" Q would have to believe that: 1) CIS531 was given last term, 2) there are French majors, and 3) there are French majors who took CIS531 last term. If R does not hold these beliefs as well, it would be uncooperative for R to give a direct answer, since any direct answer, including "None," implicitly confirms (incorrectly) that R holds these beliefs. The other kind of belief concerns what can serve as an argument to a particular relation. Such beliefs are revealed linguistically in terms of the subject and/or object of a verb. For example, the question "Which graduate students have taught CIS531?" shows that Q believes that graduate students can teach CIS531. Again, if R does not hold this belief, it would be uncooperative to give a direct answer, since any answer would implicitly confirm this belief, including the answer "None." In both examples, a direct answer is insufficient, and must be augmented in a response which addresses the disparity.

Computational work on providing such responses has been done by several researchers. Kaplan [33], in his CO-OP system, addresses the misconception that a description of an entity or set has a referent in the system's database, when according to the system's information, it does not. CO-OP looks for a possible misconception if its response to the user's query comes back empty or with a "no" answer. In these cases, CO-OP checks whether all the entities mentioned in the query exist in the database (i.e., correspond to nonempty sets). If one or more does not, CO-OP assumes a misconception on the user's part and responds accordingly; e.g.,

User: Which French majors failed CIS531 last term?
System: I do not know of any French majors.

Mercer [46] addresses both the "extensional" misconceptions (Kaplan's term) that Kaplan considers, and certain misconceptions inferrable from the use of particular lexical items, such as

User: Has John stopped taking CIS531?
System: No, he has not stopped since he hasn't started.

Mercer's system depends on its ability to determine the presuppositions[11] of the answer it plans to give. Where Mercer's and Kaplan's approaches differ is that Kaplan's system essentially infers what the user would have to believe about the world in order to have asked the particular question. If any of those beliefs conflict with the beliefs of the system, corrective information is included in addition to or instead of an answer. Mercer's system, on the other hand, computes the presuppositions of its planned answer because according to Gazdar's interpretation [21] of Grice's Maxim of Quality [24], a speaker should not say anything which has a nontrue presupposition. The difference between the systems would be apparent if CO-OP did not make the "Closed World Assumption." Then CO-OP could not necessarily conclude there were no individuals satisfying a particular description if it could not find them. For example,

User: Do any professors that teach CSE101 teach CSE114?

If no such individuals could be found nor could individuals satisfying the embedded description "professors that teach CSE101," CO-OP would reply

System: I do not know of professors that teach CSE101.

On the other hand, if Mercer's system can prove that there is no professor who teaches CSE114 but cannot prove that its presupposition is false—that there is no professor who teaches CSE101—it will respond

System: No. Moreover I don't even know if any professor teaches CSE101.

Mays [40] and, more recently, Gal [20] address misconceptions that assume that a relationship can hold between two objects or classes, when according to the system's information it cannot. For example

User: Which graduate students have taught CIS531?
System: CIS531 is a graduate course. Only faculty can teach graduate courses.

Mays' system uses a rich data model containing three types of information against which to validate the system's interpretation of the user's query:

- taxonomic information such as a woman is a person, a teacher is a person, etc.;
- mutual exclusion information such as an undergraduate cannot be a graduate student (and vice versa);
- selectional constraints such as the subject of (one sense of) the relation 'teach' must be interpretable as the concept 'teacher' and its object must be interpretable as the concept 'course.'

Every query interpretation is checked against this data model for possible selectional constraint violations. By virtue of the taxonomic and mutual exclusion information, stating the selectional constraints at the highest place in the taxonomy at which they hold captures all and only those concepts that could serve as arguments to a relation. So even though selectional constraints on 'teach' may only be stated in terms of 'teacher' and 'course,' the system can recognize that 'woman' can be the subject of 'teach' (i.e., it is compatible with the concept 'teacher') while 'undergraduate' cannot be (i.e., it is incompatible with the concept 'teacher'). Any violated constraints are taken as evidence of a misconception on the part of the user.

[11]In simple terms, a presupposition of a sentence S is a statement that must be true for either S or its negation to make sense.

Gal's system uses the database system's integrity constraints to achieve a capability which overlaps that of May's system. (Integrity constraints constrain the values that can appear in fields in the database, and are used primarily to maintain consistency when the database is updated.)

By augmenting the system's analysis of the user's query with integrity constraints on the predicates that appear in the query, Gal's system can recognize user misconceptions at two points: 1) before evaluating the query against the database, if an integrity constraint is simply violated by the query, and 2) during query evaluation, if a fact found in the database causes the integrity constraint to be violated. For example, given an integrity constraint like "Professor S teaches courses only in department A," Gal's system would recognize the user's misconception in the query

> User: How many courses does professor S teach in department B?

before query evaluation. On the other hand, it would recognize the user's misconception in the query

> User: Does Professor S teach 120?

during query evaluation, if it found that course 120 was not in department A.

Gal's system, written in Prolog, uses the same parser to analyze the user's NL queries and to generate corrective NL responses. Additional heuristics are used to decide what to say (i.e., to pass to the parser/generator) in case a corrective response is called for. For example, all violated integrity constraints are always displayed, but if an integrity constraint is only found to be violated during evaluation, the database facts signaling the violation are presented first, followed by the violated constraint, for example:

> User: Is John instructed by Professor Smith?
> System: No; because Smith teaches only in the English-for-foreigner department; John is American and no American student can be enrolled in the English-for-foreigner department.

Work on recognizing and responding to user misconceptions in the context of expert systems or knowledge based systems is even more preliminary. We note here two such efforts. Work on misconceptions involving confusions about objects and their properties has been done by McCoy [72], [73]. These misconceptions may be revealed through a user's utterance either describing an object in terms of a class it does not belong to, incorrectly attributing some property to it, or ascribing an incorrect value to one of its properties. For example,

> System: Do you have any liquid assets?
> User: I have a $5000 money market certificate.
> System: A money market certificate isn't a liquid asset. Your money is tied up for several years in a money market certificate. Do you mean a money market account?

> User: What's the interest rate on this stock?
> System: Stocks don't have an interest rate. They may pay a dividend periodically.

One problem with these is whether, in the current context, they actually constitute misconceptions. That is, McCoy shows that the current context/perspective can affect whether two things are judged to be similar or different. Contrasting them in the former situation would reveal a misconception, as would equating them in the latter. For example,

> System: I'd like to see you hold that $10 000 coming due in a money market fund until mid-March and then get into a longer term money market certificate.
> User: Well, I was just wondering, what about a treasury bill instead?
> System: That's not a bad idea but it doesn't replace your money market certificate in any way—it's an exact duplicate. They're almost identical types of instruments, so one, as far as I'm concerned, is about the same as another.

> User: I'm interested in buying some US Government securities. I was thinking about Money Market Certificates, since they are the same as Treasury Bills.

> System: But they're not: they're two very different things. A Treasury Bill is backed by the government. You have to get it from the Federal Reserve. A money market certificate, on the other hand, is backed by the bank that issues it. So one is a government security while the other isn't.

In the first example, both objects are viewed through the perspective of savings instruments, thereby highlighting such attributes as interest rate and maturity date, which treasury bills and money market certificates share. But in the second example, the objects are viewed in terms of their issuer, thereby highlighting such attributes as issuing company and purchase place, which attributes treasury bills and money market certificates do not share. McCoy's approach through perspective is able to identify that contrasting the two in the first example reveals a misconception, as does equating them in the second example.

McCoy presents a variety of correction strategies whose selection in a particular case depends on what the system believes may have led the user to the misconception. For example, in one case, the system will attempt to include in its response

1) a denial of the incorrect classification
2) a statement of the correct classification
3) a concession of similarity by mentioning their common attributes
4) an override to the conceded information by noting features that define the object as belonging to its correct superordinate class.

> User: I want to buy some corporate bonds. Should I get preferred?
> System: Preferred are not bonds, they are stocks. A preferred stock is like a bond in that both have regular payments of a fixed amount. However, unlike bond payments, the company is under no legal obligation to pay preferred stock payments. In addition, preferred stock gives part ownership in the company.

What is missing from McCoy's system is the machinery

necessary for recognizing when the user has an object-related misconception. It will not be as simple a task as that of recognizing an existential misconception, although we hope that a wide range of cases will be possible. For example, the above case requires the system to interpret the user's second sentence "Should I get preferred?" as a more specific request for information about corporate bonds. The user's goal is to buy corporate bonds. To achieve this goal, s/he needs information or wants advice about corporate bonds. In particular, s/he wants information about a specific type of corporate bonds—preferred. By this chain of interpretation (relating the user's goals and his/her information seeking behavior), the system hypothesizes this misconception on the user's part. This is not a simple type of reasoning, and more work is needed on isolating both easy to recognize object-related misconceptions and the knowledge and reasoning abilities needed to deal with more subtle cases.

On the subject of user plans and goals, this is another area rich in misconceptions that may be more visible in user interactions with expert systems and knowledge-based systems. That is, it is frequently the case that someone has a goal they want to achieve, but that the plan they come up with, which motivates their interaction with someone else, will not achieve that goal. Recognizing this, a human respondent may give information that facilitates the questioner achieving his/her goal, as well as (directly or indirectly) answering the given plan-related question. This kind of behavior is illustrated in the following examples [54]:

Q: Is there a way to tell mail that I'd like to deliver a message after some particular time and/or date?
R: No. I can't really see any easy way to do this either (and guarantee it). If you really wanted to do that, you could submit a batch job to run after a certain time and have it send the message.

Q: What's the combination of the printer room lock?
R: The door's open.

Questioners presume experts know more about a domain than they do, and they expect the above sort of cooperative behavior. If it is not forthcoming, expert systems are likely to mislead their users [31]. If they mislead their users, they are worthless.

This behavior depends on the ability to infer when and how inappropriate plans underlie a user's queries, an ability addressed in [54]. In this work, Pollack develops a model of responding to questions that does not rest on what she has called *the appropriate query assumption*. Abandoning this assumption requires the development of a model of plan inference that is significantly different from the models discussed by Allen [2], Cohen [13], and others. Where their inference procedures are essentially heuristically guided graph searches through the space of valid plans, Pollack's procedure involves R's 1) reasoning about likely differences between R's own beliefs and Q's beliefs and 2) searching through the space of plans that could be produced by such differences. To facilitate this reasoning, Pollack has had to re-analyze the notion of plans, giving a careful account of the mental attitudes entailed by having a plan. Her account derives from the literature in the philosophy of action, especially [23] and [8].

One belief that is claimed to be necessary for a questioner to have a plan is that all the actions in the plan are related either by **generation** [23] or by **enablement**.[12] Consider the first example above. The response can be explained by proposing that the respondent infers that the questioner's plan consists of the following actions:

- telling mail to deliver a message after some particular date and/or time;
- having mail deliver a message after some particular date and/or time;
- having a message delivered after some particular date and/or time.

What the respondent believes is that the questioner believes that the first action *generates* the second, and the second *generates* the third. In the response, the relation that holds between the mentioned action (submitting a batch job) and the inferred goal (having a message delivered after some particular date/time) is also generation, but here, it is the respondent who believes the relation holds. While Pollack does not address the actual information that should be included in a response when the system detects an inappropriate plan underlying a query, it seems clear that an ability to infer such plans is requisite to responding appropriately.

Pollack's analysis can also be applied to handling other types of requests that reflect inappropriate plans. As discussed above, the phenomenon she addresses requires reasoning about the relationship of one action to another, and, more significantly, people's knowledge or beliefs about such relationships: the actions themselves can be treated as opaque. In order to handle the second example above, Pollack's analysis would have to be extended to allow reasoning about people's knowledge or beliefs about objects and their properties and the role that particular objects and their properties play in actions. Reasoning about the relationship between knowledge (primarily of objects) and action has been discussed in [2], [47], [3].

In summary, there are many types of belief disparities that can interfere with the successful transmission of information between user and system. One of the most subtle for expert systems involves the user's belief in the relevance of some concept to his/her question, where for the system, it is irrelevant. Lack of attention to such misconceptions can make it very difficult for the user to understand the system's advice. Examples of this can be found in Cassell's transcripts of doctor's taking medical histories [10]. It is something that the developers of advisory systems will eventually have to deal with, if they are to be used by the lay public.

V. Information Intended to Prevent Misconstrual

In Section IV, we noted that many of the troubles that arise in an interaction come from the fact that participants hold different views about the world and either do not realize it or fail to do anything about it. These different views may lead to utterances being misconstrued. But even when the participants appear to hold the same view of the

[12]Generation corresponds to what is roughly captured as the *by* locution in English: the questioner believes one can have a message delivered *by* telling mail to deliver a message. Enablement corresponds to the relationship captured in the STRIPS model of actions: finding out the combination to the printer room enables one to open the door. Generation and enablement are not the same: telling mail to deliver a message after some particular date and/or time does not alter the current world state in such a way that one can then have mail deliver the message.

world, there are still circumstances where utterances may be misconstrued. To avoid such misunderstandings, responsible human speakers know to augment their utterances with additional information. It is this latter information that we discuss in this section, and what expert systems and knowledge-based systems will require to avoid such misunderstandings as well.

In answering questions, there are at least three bases for a respondent anticipating that the questioner may draw a false conclusion from his/her answer A: *deductive reasoning* involving A and a false belief that the questioner holds; *default reasoning*, where the respondent knows the default does not hold; and *implicature*, where A may suggest, by virtue of its ordinary use, something that is not the case here. The first case is similar to the situation discussed in Section IV, in which the respondent adds information to correct the questioner's misconception. If the respondent does not point out, or better, correct this false belief, the questioner will draw an additional false conclusion from the respondent's answer. The additional information s/he provides in response may point out the false belief or correct it as well [32], [31].

The second case is illustrated by the following hypothetical exchange between user and system:

User: What's the combination of the printer room lock?
System: It's 8-9-8, but Ira changes it every week.

One might call this an "inertial default"—typically, the referent of a definite descriptor stays the same. If the system just answers the user's question ("It's 8-9-8"), s/he may falsely conclude (by default reasoning) that the referent of the definite descriptor "the combination of the printer room lock" will stay 8-9-8. The system provides the additional information ("but Ira changes it every week") to prevent this misconstrual.

The third case, misconstrual based on implicature, is illustrated by the following example:

User: Which ships have a doctor on board?
System: The JFK, the Admiral Nimitz,...

From the system's answer, the user may conclude that the ships named there have a doctor on board and other ships not named there do not. That is, the user may be led to believe that there is a *non-empty set of ships which do not have a doctor on board*. However, if the named ships are the only ships (and the user does not know it), the system's answer—while correct and useful—is nevertheless misleading: the system should have answered "all of them," which would not have mislead the user to believe in a doctor-less subset of ships.

A fourth possible basis for anticipating a false conclusion from a correct answer is the respondent's belief that people typically believe X (i.e., as opposed to his/her belief that people believe that typically X). Consider a user interacting with an employment database, checking out a job that looks interesting. S/he asks

User: What's the salary?
System: There isn't any.

Because of this, the user rejects the job and moves on to consider others. The system's answer may be factually correct (i.e., there is no salary entry for the job), but the user's conclusion may be false: that is, the job may not pay a salary, but remuneration may come in other ways; e.g., tips, commission, stock options, etc.

To anticipate and avoid such a false conclusion, the system could reason about what people typically believe, and if that does not match the current situation, it could inform the user. That is, suppose it believes that people typically will not take nonpaying jobs. It also believes that people typically believe that if a job does not pay a salary, it is a nonpaying job. (Here, SB stands for "System believes" and UB, for "User believes".)

(SB (*typically* U not take a nonpaying job))
(SB (*typically* UB (no salary ⇒ nonpaying job)).

The system, on the other hand, believes that all jobs either pay a salary or have tips or offer commission on sales or provide stock options or other free benefits or are nonpaying. Moreover, the system knows that this job offers a 20-percent commission on sales.

(SB∀x : job. x pays salary or x has tips or x offers commission ... or x is nonpaying job)
(SB JOB617 not pay salary)
(SB JOB617 offers commission)

Given the system's own beliefs and beliefs about user's typical beliefs, the system should be able to anticipate that the user will not take this job (JOB617) because it is nonpaying. However, it knows that while the job does not pay a salary, it does pay commission. Thus the system should also describe what reimbursement it provides, to prevent any misconstrual.

System: There is no salary, but there is a 20-percent commission on every squash racquet you sell.

The principles that guide this behavior are Grice's Maxim of Quantity [24]:

> Make your contribution as informative as is required for the current purposes of the exchange. Do not make your contribution more informative than is required for the current purposes of the exchange.

and Joshi's modification to Grice's Maxim of Quality [30]:

> If you, the speaker, plan to say anything which may imply for the hearer something that you believe to be false, then provide further information to block it.

What this modified Quality Maxim provides is a criterion for the level of informativeness needed to satisfy the Quantity Maxim: enough must be said so that the hearer does not draw a false conclusion.

Hirschberg has investigated a class of indirect and modified direct responses to yes–no questions in her work on scalar implicatures [28]. It is possible to view a large class of the responses she considers as attempts by cooperative speakers to block potential false inferences which hearers might otherwise (wrongly) infer to be implicatures arising from a direct response—while also providing information from which direct response can be derived [28]. We can illustrate this by the following example.

A: Has Mary had her medication?
B: She has taken the Excedrin.

A simple direct response "No" by B, although correct, would be misleading because it licenses A to draw the false inference that Mary has not taken any of her medication.

B's response above blocks this inference. Note also B indirectly conveys to A the fact that Mary has not taken all of her medication [28].

The major assumption of this work in general is that one can limit the amount of reasoning that R is responsible for doing, to detect a possible reason for misconstrual. This is an important assumption, and further work is necessary to show that such constraining can be done in some principled manner.

VI. Adapting the Interaction to Individual Users

As noted earlier, system-generated justifications and explanations can eliminate potential troubles in user–system interactions before they start. So too can adapting the interactions to the abilities and interactional style of the user. Although this adaptive behavior is not unique to systems employing an NL interface, it is strongly evident in normal human interaction and will be expected by users of systems with sophisticated NL interfaces. The issues we will discuss are 1) deciding what questions to ask and not to ask of the user; 2) interpreting a "don't know" answer; and 3) enlarging the range of acceptable answers.

A. Deciding What to Ask

Existing interactive expert systems employ a number of techniques to guide their search for appropriate conclusions. For the most part, they employ strategies which take into account such preferences as:

- prefer rules which can yield definite conclusions over indefinite ones (e.g., MYCIN's *unitypath* heuristic) [59];
- prefer rules whose conclusions can have the greatest impact on the ultimate goals of the system (e.g., the rule selection function in PROSPECTOR [17]);
- prefer rules which do not require asking the user any questions over those which do (e.g., as in MYCIN and ARBY [41]).

Most expert systems fail, however, to take into account any factors which depend on the individual they are interacting with—for example, the user's ability to understand a question or to give a reliable answer. This section discusses some of the ways that an expert system can improve its interaction with a user by incorporating an explicit model of the user's knowledge and beliefs.

Interactive expert systems vary as to when they ask for information and when they rely on their own deductions. However, in this decision, they ignore the user's ability to understand and respond reliably. For example, in PROSPECTOR [17], goals are simply marked as being either "askable" or "unaskable" (never both). In MYCIN [59], the user is only asked for information if either the system's attempt to deduce a subgoal fails; i.e., if no rules were applicable or if the applicable rules were too weak or offset each other; or the user's answer would be conclusive (e.g., lab results). In KNOBS [18], a system for assisting users in mission planning, the user is only asked for preferences, not facts. If the user prefers not to answer at any point, s/he can turn over control to the system and let it compute an appropriate value.

Any attempt to customize a system's interaction to the current user must allow for wrong guesses. The user may not be able to answer its questions or will answer them incorrectly or will find them annoying. Thus customization has several aspects—1) *deciding* what question to ask next; 2) *recovering* from a wrong decision (i.e., from having asked a "bad") question; and 3) *modifying* subsequent decisions about what sort of questions to ask.

One approach might be to evaluate strategies according to how much work is required of the user to provide the information requested of him/her. This can be factored into:

1) How much work is required to understand the question?
2) How much work is required to acquire the information?
3) How much work is required to communicate the information to the system?

A strategy that required no further information from the user would receive an "easy" rating. Somewhat harder would be a strategy that required the user to make an observation, and harder still would be a strategy that required a test. Much harder would be one that needed to access the user's prior knowledge for the information. For example, if a piece of equipment failed to work, an obvious question to ask is "Is it plugged in?" This is simple to ascertain by observation and has a high payoff. Somewhat harder to answer because it involves a test, but again something with a high payoff, is the question "Is the battery working?"[13]

Of course there might be several alternative procedures the user could employ in acquiring the information the system wants, each of different difficulty for him/her, each requiring somewhat different resources. While the system's evaluation of a strategy might be based on the assumption that the user can and will use the easiest of these procedures, more refined evaluations might take into account the *resources available* to the particular user as well. (This information about alternative procedures—their level of difficulty and resource requirements would also be useful for certain cases where a user cannot answer the system's question, as will be discussed in the next subsection.)

A strategy evaluation based solely on how much work is required of the user would not be sufficient however. Another factor in the system's choice of reasoning strategy must be its *a priori* beliefs about the reliability of the user's information. The system should prefer a line of reasoning which depends on facts it believes the user can supply reliably over one which it believes the user can supply with less reliability.

For example, consider a system taking a patient's medical history. If it needs to know whether the patient has a drinking problem it can either ask the question directly (e.g. "Are you an alcoholic?") or ask a set of questions from

[13] This resembles somewhat the strategies embodied in INTERNIST [55]: when simply trying to eliminate a hypothesis from a large number of possible ones, INTERNIST limits its questions to information obtainable via history or physical exam. Later, when trying to discriminate between a small number of contending hypotheses, INTERNIST may request information which comes from more "costly" procedures. Finally, when trying to confirm a particular hypothesis, there are no constraints on the data it may request: biopsies may be required, etc.

which it can conclude the answer (e.g. "Do you drink socially?" "Do you drink at home?" "Would you say you drink as much as a bottle a day?"[14], etc). Often such an indirect approach can provide more reliable data.

B. When the User Cannot Answer

When the system does choose to go to the user for information, the user may still only be able to answer "I don't know." A flexible system should be able to take this response, hypothesize the underlying reasons for it, and act to help the user out of his/her predicament. This section describes some of the underlying reasons why a person cannot (or will not) answer a question, each of which calls for a slightly different response on the system's part. Note that no matter what the reason, the user may still respond with a simple "*I don't know*."

The user *may not understand* the question; e.g., it may contain unfamiliar terms or concepts. Here the system should have one of two possible responses: either it should be able to supply the user with definitions of the terms or concepts s/he is unfamiliar with (cf. Section II) or it should be able to rephrase the question so as not to use them.

The user *may not be sure* that the question means the same to the system as it does to him/her. In this case, the system should be able to generate a paraphrase of the question that clarifies it or allow the user to verify his/her understanding (cf. Section III).

The user may understand the question but *not know how* to go about determining the answer. In this case, the system should be able to suggest one or more procedures for doing so.

The user may understand the question but *may not remember* the information needed for an answer. In this case, the system should be able to ask the user for information that provides strongly suggestive clues to the information it needs.

The user may understand the question but *may not have at hand* the information needed for determining an answer (e.g., lab results). The system should be able to figure out whether it might be able to perform some preliminary reasoning without the missing information and finish things off when it is provided.

The user *may not know why* the system wants the information and will not divulge it until s/he does. In this case, the question needs to be expanded to include the system's reason for asking it.

The user *may not believe* the requested information is *relevant* to solving the problem and is trying to force the expert to adopt another line of reasoning. In this case, the system should be able to identify and pursue an alternative strategy, if one exists. If not, it should be able to explain to the user why it cannot continue. The same holds if it is the case that the user does not wish to divulge the information.

Since its strategy for possible success differs in each case, a system must be able to 1) determine why the user did not answer and 2) offer an appropriate response to help the user formulate an answer. There are several techniques it can use in achieving the first goal:

- It can provide the user with a way to express the problem directly (e.g., through a menu or a set of predefined key words like PARAPHRASE).
- It can use a model of the individual user to determine the most likely problem (e.g., some terms are very likely to be unknown by certain classes of users).
- It can interpret a user's NL response against the set of possibilities

Once the problem has been found, there are also several possible response techniques. If the user does not understand terms in the question, then the techniques discussed in Section II are applicable. If the user is uncertain of the overall meaning of the question, then the paraphrasing techniques of Section III are in order. Some other useful techniques to implement include:

1) Procedures for the user's acquiring the information, annotated by the level of resources, time and sophistication needed to carry them out, as in these exchanges:

System:	Does eating certain foods cause a problem for you?
User:	HOW_KNOW?
System:	Do you get a rash when you eat them, or become short of breath or vomit?

System:	Is it an ASCII file?
User:	HOW_KNOW?
System:	Is it readable when printed on your terminal?

2) "Script"-like information relating events of interest with events that are perhaps more easily remembered by the user [35], as is suggested by this interaction:

System:	Have you had a tetanus shot in the last two years?
User:	DON'T_RECALL
System:	Were you bitten by a dog, cat, or some other animal in the last two years?
User:	Yes.
System:	Were you treated for the bite?
User:	Yes.
System:	Were you given a shot as part of the treatment?
User:	Yes.
System:	Then that was probably a tetanus shot.

3) Teleological information concerning the reasons for making the request, as is shown in this:

System:	How tall are you?
User:	WHY?
System:	We would like to know whether your weight is right for your height.

4) Ability to do case analysis reasoning; i.e., considering the effect on the outcome, under the alternative assumption of each possible value or class of values. (It is possible that it will not make a difference which value is assumed.); e.g.:

System:	What is your mother's Rh factor?

[14]This last question is interesting in its own right, as Cassell [10] notes that one is more likely to get a reliable answer to a question that lets the patient off the hook. "No, not that much, only 5 or 6 shots."

User: DON'T_KNOW
System: Could you ask her? We'll continue now without it.

System: Has anyone in your family committed suicide?
User: SKIP_IT

Note that most of these examples show a system providing another question rather than a definition for the unknown or unshared term or a procedure for determining the answer. The reason for this is that it constrains the subsequent interaction, so that the user knows what is expected of him/her, thereby avoiding some problems of interpretation.

C. Increased Leeway in User's Responses

We have been talking all along about enlarging the *range* of interactions that responsible users need to be able to have with expert and knowledge based systems to ensure that the advice the user receives is appropriate and is understood correctly. As the range widens, the entire interaction becomes more natural. As a result, users may become frustrated if their normal interactive behavior is curtailed. One place that it might be is in responding to the system's requests for information.

People vary in the amount and kind of information they are used to providing in response to questions. They are frustrated if they must act differently. For example, van Katwijk *et al.* [68] report experiments comparing three sources of information on local train schedules—a person, a taped recording, and a simulated computerized information system. What upset people most about the latter was its refusal to process any information that was not explicitly requested. Obviously, the person volunteered it because s/he felt the system would need it. Frame-based systems such as KNOBS [18] and GUS [6], an interactive travel assistant developed at XEROX, permit such behavior, and other systems must do so as well. Here we indicate several other ways (besides *volunteering additional information*) in which an expert system should allow a user more flexibility in answering a question.

Offering Facts from which an Answer can be Deduced: Often the user provides an indirect answer, in the belief that the system can and will deduce a direct answer from it. We have identified four situation in which this occurs.

1) The user is *unable* to determine an answer to the question but has what s/he believes to be information from which the system can deduce an answer, as is shown in the following examples.

System: What is your employee classification: A-1, A-2, or A-3?
User: I'm an assistant professor in Oriental Studies.
System: All faculty members are A-1 employees, thank you.

System: Are you a senior citizen?
User: I'm 62 years old.

Of course the user can be wrong, and the information s/he offers either inadequate or irrelevant, as in this exchange.

Systems: What is your employee classification: A-1, A-2, or A-3?
User: I've been here for over 5 years.
System: Sorry, could you tell me either your job title or position?

2) The user is *able but unwilling* to perform the computations necessary for answering the question, as in:

System: What is your yearly salary?
User: $1840 per month.

3) The user goes beyond a direct literal answer to provide a more precise and hence possibly more informative answer, as in the second of the following two responses:

System: Are you 65 or older?
User 1: Yes.
User 2: I'm 72.

4) The user answers a slightly different question because a direct answer to the given question would be *logically correct but misleading*. This is in the same behavior we discussed in Section V in terms of the system modifying its correct but potentially misleading direct answers. Users must be free to do the same, for example:

System: Did you delete all the KLONE files?
User 1: Yes.
User 2: I deleted the .LISP files.

System: Did you delete all the KLONE files?
User 1: No.
User 2: I deleted the .LISP files.

If the user were confined to answering just yes or no, the system would be misled: if the user just answered "no," the system might think that no files were deleted (which is incorrect, in the second case). If it just answered "yes," the system might think that just those files had been deleted (which is incorrect, in the first case).

Hedging: Often a user may wish to hedge on his answer, say by attaching a certainty measure, offering a range of values, or providing a disjunctive answer. Many current expert systems provide for the first case. On the other hand, the possibility of hedging by specifying a disjunctive answer or a range of values has not been explored to any extent. There are several problems with allowing a user these possibilities. One involves discriminating between an *imprecise single value* and a *precise range of values*. Contrast the following two interactions:

System: How tall is your wife?
User: Between 5'2" and 5'4".

System: How much does a Volvo cost?
User: Between 12K and 18K.

In the first the user is clearly giving a range in which the single answer lies. In the second, s/he is more likely to be indicating that **a Volvo** specifies a set of objects, the costs of which fall in the range given. Determining which of these two cases the user intends requires a model of how the user views the object and attribute.

Deferring an Answer: The user may wish to defer answering a question. This differs from declining to answer (i.e., the SKIP_IT response discussed earlier) in signaling that the user wants to supply the information later, but not now. In an expert system that responds to user preferences (e.g., any kind of planning assistant), one reason for deferring is that for the user the answer is *conditional* on something that has not been resolved yet. A deferral gives the user a method to control the reasoning process of the expert system by forcing it to pursue lines of reasoning which do not require the requested information. For example, consider an expert system which recommends restaurants. The system may ask about cost before cuisine, whereas for the user what s/he's prepared to spend depends on the type of cuisine; e.g.:

System: How much do you want to spend?
User: DEFER
System Do you prefer a restaurant within walking distance?
User: Yes.
System: What kinds of cuisine do you prefer?

In order to handle the kinds of responses to direct questions that we have shown here, a system needs to determine 1) the *relationship* that holds between question and response; 2) the *answer* to the given question; and 3) the *additional information*, if any, that the response provides.

VII. Conclusion

Artificial Experts, often called Knowledge Based Systems, Expert Systems, or Advisory Systems, involve more than simple question answering. NL interfaces for such systems do more than identify or retrieve facts. They must support a variety of cooperative behaviors for the interactions to be successful. The user and the system have to come to terms with each other, they must understand each other's terminology, problems, analyses, and recommendations. The user cannot be expected to come to the system with an already well-formed problem and show an accurate understanding of the system's capabilities and terminology, or even a good understanding of the domain. In this paper we have discussed specific NL capabilities that support aspects of such cooperative behaviors. We have focussed exclusively on such interactions (and not on the fact retrieval aspect of question-answering) because we believe that NL interfaces, of very rich functionality, are critical to the effective use of artificial experts. These aspects of interaction are precisely those where NL interfaces really pay off and will do so even more in the future.

Acknowledgment

The authors like to thank their referee for his or her excellent comments. They would also like to thank those people whose work they have reported on here: Kathy McKeown, R. Rubinoff, Eva Mueckstein, J. Kaplan, B. Mercer, E. Mays, Annie Gal, Kathy McCoy, Martha Pollack, and Julia Hirschberg. Finally, they would like to thank MaryAngela Papalaskaris for suggesting this title.

References

[1] J. Allen and C. R. Perrault, "Analyzing intention in utterances," *Artificial Intell.*, vol. 15, pp. 143–178, 1980.

[2] J. Allen, "Recognizing intentions from natural language utterances," in M. Brady, Ed., *Computational Models of Discourse*. Cambridge, MA: MIT Press, 1982.

[3] D. Appelt, *Planning English Sentences*. Cambridge, England: Cambridge University Press, 1985.

[4] M. Bates and R. J. Bobrow, "Natural language interfaces: What's here, what's coming, and who needs it," in W. Reitman, Ed. *Artificial Applications for Business*. Norwood, NJ: Ablex Publishing, 1984.

[5] ____, "A transportable natural language interface," in *Proc. 6th Ann. Int. SIGIR Conf. on Research and Development in Information Retrieval*, ACM, 1983.

[6] D. Bobrow, R. Kaplan, D. Norman, H. Thompson, and T. Winograd, "GUS, A frame driven dialog system," *Artificial Intell.*, vol. 8, 1977.

[7] B. K. Boguraev and K. Sparck-Jones, "A natural language front end to databases with evaluative feedback," in G. Gardarin and E. Gelenbe, Eds., *New Applications of Data Bases*. London, England: Academic Press, 1984, pp. 159–182.

[8] M. Bratman, "Taking plans seriously," *Social Theory and Practice*, vol. 9, pp. 271–287, 1983.

[9] S. Carberry, "A pragmatics-based approach to understanding intersentential ellipses," in *Proc. 23rd Ann. Meet.* (Association for Computational Linguistics, University of Chicago, Chicago IL, July, 1985), pp. 188–197.

[10] E. J. Cassell, *Talking with Patients. Volume I: The Theory of Doctor Patient Communication*. Cambridge, MA: MIT Press, 1985.

[11] W. Clancey and R. Letsinger, "NEOMYCIN: Reconfiguring a rule-based expert system for application to teaching," in W. Clancey and E. Shortliffe, Eds., *Readings in Medical Artificial Intelligence: The First Decade*. Reading, MA: Addison-Wesley, 1984, pp. 361–381.

[12] E. F. Codd, "Seven steps to rendezvous with the casual user," in J. W. Klimbie and K. L. Koffeman, Eds., *Data Base Management*. Amsterdam, The Netherlands: North-Holland, 1974, pp. 179–199.

[13] P. Cohen, "On knowing what to say: Planning speech acts," Dep. of Computer Science, Univ. of Toronto, Toronto, Ont. Canada, Tech. Rep. 118, Jan. 1978.

[14] R. Creagh-Osborne, *This is Sailing*. Boston, MA: SAIL Publ., 1972.

[15] F. J. Damerau, "Operating statistics for the transformational question answering system, *AJCL*, vol. 7, no. 1, pp. 30–42, 1981.

[16] R. Davis, "Interactive transfer of expertise," *Artificial Intell.*, vol. 12, no. 2, pp. 121–157, 1979.

[17] R. Duda, J. Gaschnig, and P. Hart, "Model design in the PROSPECTOR consultant system for mineral exploration," in D. Michie, Ed., *Expert Systems in the Micro-electronic Age*. Edinburgh, UK: Edinburgh Univ. Press, 1979.

[18] C. Engleman, E. Scarl, and C. Berg, "Interactive frame instantiation," in *Proc. 1st Nat. Conf. on Artificial Intelligence (AAAI)*(Stanford CA, 1980).

[19] S. Feiner, "APEX: An experiment in the automated creation of pictorial explanations," *IEEE Comput. Graphics and Applications*, vol. 5, pp. 29–37, Nov. 1985.

[20] A. Gal and J. Minker, "A natural language database interface that provides cooperative answers," in *Proc. 2nd IEEE Ann. Conf. on Artificial Intelligence Applications* (Miami, FL, Dec., 1985), pp. 352–357.

[21] G. Gazdar, "A solution to the projection problem," in C.-K Oh and D. Dinneen, Eds., *Syntax and Semantics*. New York: Academic Press, 1979, pp. 57–90.

[22] J. Ginsparg, "A robust portable natural language data base interface," in *Proc. Conf. on Applied Natural Language Processing*, pp. 25–30, ACL, 1983.

[23] A. Goldman, *A Theory of Human Action*. Englewood Cliffs, NJ: Prentice-Hall, Englewood Cliffs, NJ: Prentice-Hall, 1970.

[24] H. P. Grice, "Logic and conversation," in P. Cole and J. L. Morgan, Eds., *Syntax and Semantics*. New York: Academic Press, 1975.

[25] L. Harris, "The advantages of natural language programming," in Sime and Coombs, Eds., *Designing for Human-Computer Communication*. London, England: Academic Press, 1983, pp. 73–85.

[26] D. Hasling, W. Clancey, and G. Rennels, "Strategic explana-

tions for a diagnostic consultation system," *Int. J. Man-Machine Studies* vol. 20, pp. 3–20, Jan. 1984.

[27] G. Hendrix, E. Sacerdoti, D. Sagalowicz, and J. Slocum, "Developing a natural language interface to complex data," *ACM Trans. Database Syst.*, vol. 3, no. 2, pp. 105–147, June 1978.

[28] J. Hirschberg, "The theory of scalar implicature," Tech. Rep. MS-CIS-85-56, Dep. Computer and Information Science, Univ. of Pennsylvania, Philadelphia, Dec. 1985.

[29] J. Hollan, E. Hutchins, and L. Weitzman, "STEAMER: An interactive inspectable simulation-based training system," *AI Mag.* vol. 5, no. 2, pp. 15–28, Summer 1984.

[30] A. K. Joshi, "Mutual beliefs in question answering systems," in N. Smith, Ed., *Mutual Belief*. New York: Academic Press, 1982.

[31] A. K. Joshi, B. Webber, and R. Weischedel, "Living up to expectations: Computing expert responses," in Proc. *AAAI-84* (Austin TX, Aug. 1984).

[32] ——, "Preventing false inferences," in *Proc. COLING-84 (Stanford, CA, July*, 1984).

[33] S. J. Kaplan, "Cooperative responses from a portable natural language database query system," in M. Brady, Ed., *Computational Models of Discourse*. Cambridge, MA: MIT Press, 1982.

[34] M. Kay, "Functional grammar," in *Proc. 5th Ann. Meet. of the BLS* (Berkeley Linguistics Society, Berkeley, CA, 1979).

[35] J. K. Kolodner, "Organization and retrieval in a conceptual memory for events or CONS54, where are you?" in *Proc. IJCAI-81*, pp. 227–233. August. 1981.

[36] D. Lanam, R. Letsinger, S. Rosenberg, P. Huyun, and M. Lemon, "Guide to the heuristic programming and representation language. Part 1: Frames," Tech. Rep. AT-MEMO-83-3, Application and Technology Lab., Computer Res. Cen., Hewlett-Packard, Jan. 1984.

[37] W. G. Lehnert and S. P. Shwartz, "EXPLORER: A natural language processing system for oil exploration," in Proc. Conf. on Applied Natural Language Processing, *ACL*, 1983.

[38] D. Litman and J. Allen, "A plan recognition model for clarification subdialogues," in *Proc. 10th Int. Conf. on Computational Linguistics, COLING-84* (Stanford, CA, July, 1984), pp. 302–311.

[39] P. Martin, D. Appelt, and F. Pereira, "Transportability and generality in a natural language interface system," in *Proc. 8th Int. Conf. on Artificial Intelligence* (Karlsruhe, West Germany, Aug. 1983), pp. 573–578.

[40] E. Mays, "Failures in natural language system: Application to data base query systems," in *Proc. 1st Nat. Conf. on Artificial Intelligence (AAAI)* (Stanford, CA, Aug. 1980).

[41] D. McDermott and R. Brooks, "ARBY: Diagnosis with shallow causal model," in *Proc. Nat. Conf. on Artificial Intelligence* (Carnegie-Mellon Univ., Pittsburgh, PA, Aug. 1982), pp. 314–318.

[42] D. McDonald, "Dependency directed control: Its implications for natural language generation," in N. Cercone, Ed., *Computational Linguistics*. New York: Pergamon, 1983, pp. 111–130.

[43] K. McKeown, "Paraphrasing using given and new information in a question-answer system," in *Proc. 17th Ann. Meet. of the ACL* (Association for Computational Linguistics, Aug. 1979), pp. 67–72. Also in *Amer. J. Comp. Ling.*, vol. 9, no. 1, pp. 1–11, Jan.–Mar. 1983.

[44] K. McKeown, M. Wish, and K. Matthews, "Tailoring explanations for the user," in *Proc. 1985 Conf.* (Int. Joint Conf. on Artificial Intelligence, Los Angeles, CA, Aug. 1985).

[45] K. McKeown, *Text Generation: Using Discourse Strategies and Focus Constraints to Generate Natural Language Text*. Cambridge, England, Cambridge Univ. Press, 1985.

[46] R. Mercer and R. Rosenberg, "Generating corrective answers by computing presuppositions of answers, not of questions," in *Proc. 1984 Conf.* (Canadian Society for Computational Studies of Intelligence, University of Western Ontario, London, Ont., May, 1984).

[47] R. C. Moore, "A formal theory of knowledge and action," in R. C. Moore and J. Hobbs, Eds., *Formal Theories of the Commonsense World*. Norwood, NJ: Ablex Publishing, 1984.

[48] M. Moser, "Domain dependent semantic acquisition," in *Proc. 1st Conf. on Artificial Intelligence Applications*, R. M. Haralick, Chairman (Denver, CO, Dec. 1984); published by the IEEE Computer Soc. Press.

[49] E.-M. Mueckstein, "Q-Trans: Query translation into English, in *Proc. 8th Int. Joint Conf. on Artificial Intelligence* (IJCAI, Karlsruhe, West Germany, Aug. 1983), pp. 660–662.

[50] R. Neches, W. Swartout, and J. Moore, "Explainable (and maintainable) expert systems," in *Proc. 1985 Conf.* (Int. Joint Conf. on Artificial Intelligence, Los Angeles, CA, Aug. 1985).

[51] W. Ogden and S. Brooks, "Query languages for the casual user," in *CHI '83 Conf. Proc.* (ACM Special Interest Group on Computer and Human Interaction, Boston, MA, Dec. 1983).

[52] C. R. Perrault and B. J. Grosz, "Natural language interfaces," *Annu. Rev. in Comput. Sci.* (Annual Reviews Inc., Palo Alto, CA), vol. 1, pp. 47–82, 1986.

[53] S. J. Petrick, "Transformational analysis," in R. Rustin, Ed., *Natural Language Processing*. New York: Algorithmics Press, 1973, pp. 27–41.

[54] M. Pollack, "Information sought and information provided," in *Proc. CHI '85* (Assoc. for Computing Machinery (ACM), San Francisco, CA, Apr. 1985), pp. 155–160.

[55] H. E. Pople, Jr., "Heuristic methods for imposing structure on ill-structured problems," in P. Szolovits, Ed., *Artificial Intelligence in Medicine*. Boulder, CO: Westview Press, 1982, ch. 5, pp. 119–190.

[56] R. Rubinoff, "Explaining concepts in expert systems: The CLEAR system," in *Proc. 2nd Conf. on Artificial Intelligence Applications* (IEEE, Miami, FL., Dec. 1985).

[57] ——, "Adapting MUMBLE," in *Proc. 24th Ann. Meet.*, Tech. Rep. MS-C1S-86-22, Dep. Comput. and Inform. Sci., Univ. of Pennsylvania, Philadelphia, PA, Apr. 1986.

[58] M. Sergot, "A query-the-user facility for logic programming," in P. Degano and E. Sandewall, Eds., *Integrated Interactive Computing Systems*. Amsterdam, The Netherlands: North-Holland, 1983, pp. 27–41.

[59] E. Shortliffe, *Computer-based Medical Consultations: MYCIN*. New York: Elsevier, 1976.

[60] ——, "Details of the consultation system," in B. Buchanan and E. H. Shortliffe, Eds., *Rule-Based Expert Systems*. Reading, MA: Addison-Wesley, 1984, pp. 78–132.

[61] C. Sidner, "What the speaker means: The recognition of speakers' plans in discourse," *Int. J. Comput. Math.*, vol. 9, pp. 71–82, 1983.

[62] P. Slattery, "PATHEX: A diagnostic expert system for medical pathology," Master's thesis, University of Pennsylvania, Philadelphia, Aug. 1985.

[63] L. Suchman, "Common sense in interface design," in *Conf. on Work and New Technology*, NAWW, Oct. 1982.

[64] W. Swartout, "XPLAIN: A system for creating and explaining expert consulting systems," *Artificial Intell.*, vol. 21, no. 3, pp. 285–325, Sept. 1983.

[65] ——, "The GIST behavior explainer," in *Proc. Nat. Conf. on Artificial Intelligence* (AAAI, Aug. 1983), pp. 402–407.

[66] H. R. Tennant, K. M. Ross, R. M. Saenz, C. W. Thompson, and J. R. Miller, "Menu-based natural language understanding," in *Proc. 21st Ann. Meet.* (ACL, Cambridge, MA, 1983), pp. 151–158.

[67] F. B. Thompson and B. H. Thompson, "Practical natural language processing: The REL system Prototype," in M. Rubinoff and M. C. Yovits, Eds., *Advances in Computers*, New York: Academic Press 1975, pp. 109–168.

[68] A. van Katwijk, F. van Nes, H. Bunt, H. Muller, F. Leopold, "Naive subjects interacting with a conversing information system," Tech. Rep. IPO Ann. Progress Rep., Eindhoven, The Netherlands, 1979.

[69] J. Wallis and E. Shortliffe, "Customized explanations using causal knowledge," in B. Buchanan and E. H. Shortliffe, Eds., *Rule-Based Expert Systems*. Reading, MA: Addison-Wesley, 1984.

[70] D. L. Waltz, "An English language question answering system for a large relational database," *CACM* vol. 21, no. 7, pp. 526–539, July 1978.

[71] W. Woods, "Semantics and quantification in natural language question answering," in M. Yovits, Ed., *Advances in Computers, Vol. 17*. New York: Academic Press, 1978, pp. 2–87.

[72] K. McCoy, "Correcting misconceptions: What to say," in *Proc. Conf. on Human Factors in Computing Systems—CHI '83* (Cambridge, MA, Dec. 1983).

[73] ——, "Correcting object-related misconceptions, "Dept. of Computer and Information Science, Univ. of Pennsylvania, Philadelphia, Tech. Rep. MS-CIS-85-57, 1985.

Interactive Classification: A Technique for Acquiring and Maintaining Knowledge Bases

TIMOTHY W. FININ, MEMBER, IEEE

The practical application of knowledge-based systems, such as in expert systems, often requires the maintenance of large amounts of declarative knowledge. As a knowledge base (KB) grows in size and complexity, it becomes more difficult to maintain and extend. Even someone who is familiar with the knowledge domain, how it is represented in the KB, and the actual contents of the current KB may have severe difficulties in updating it. Even if the difficulties can be tolerated, there is a very real danger that inconsistencies and errors may be introduced into the KB through the modification. This paper describes an approach to this problem based on a tool called an ***interactive classifier****. An interactive classifier uses the contents of the existing KB and knowledge about its representation to help the maintainer describe new KB objects. The interactive classifier will identify the appropriate taxonomic location for the newly described object and add it to the KB. The new object is allowed to be a generalization of existing KB objects, enabling the system to learn more about existing objects.*

I. Introduction

The practical application of knowledge-based systems, such as in expert systems, requires the maintenance of large amounts of declarative knowledge. As a knowledge base (KB) grows in size and complexity, it becomes more difficult to maintain and extend. Even someone who is familiar with the domain, how it is being represented, and the current KB contents may introduce inconsistencies and errors whenever an addition or modification is made.

One approach to this maintenance problem is to provide a special **KB editor**. Schoen and Smith, for example, describe a display-oriented editor for the representation language STROBE [28]. Freeman *et al.*, have implemented an editor/browser for the KNET language [14], [15]. Lipkis and Stallard are developing an editor for the KL-ONE representation language (personal communication). There are several problems inherent in the editor paradigm, for example:

- The system must take care that constraints in the KB, such as those defined via subsumption, are maintained.
- The system must distinguish at least two different kinds of reference to a KB object: reference *by name* and reference *by meaning*. A reference *by name* to an object should not be effected if the underlying definition of the object is changed by the editor. If one refers to an object *by meaning*, however, and later edits the object referred to, then the reference should still refer to the original description.
- The system must keep track of the origin of the subsumption relationship to distinguish between those explicitly sanctioned by the KB designer and those inferred by the system (e.g., by a classifier).
- Editors tend to be complex formal systems requiring familiarity with the editor and with the structure and content of the KB being modified.

Manuscript received June 20, 1985; revised April 6, 1986.
The author is with the Department of Information Science, University of Pennsylvania, Philadelphia, PA 19104-6389, USA.

This paper describes another approach to the KB maintenance problem based on a tool called an interactive classifier. This kind of tool is not as general or powerful as a full KB editor but avoids many of the problems described above. The interactive classifier can only be used to make monotonic changes to the KB. New objects can be added to the taxonomy and additional attributes can be added to objects already in the KB. It does not allow objects to be deleted or their existing attributes changed or overridden.

Although this may sound like a severe restriction, we believe that there are many situations where this is just the kind of KB update that is to be allowed. Consider, for example, the *computer configuration* problem which has been the domain of several recent expert system projects [21], [14], [23]. Such a system needs to have an extensive KB describing a large number of computer components and their attributes, including their decomposition and interconnection constraints. An important feature of this domain is that new components are constantly being introduced as the underlying technology advances. Older components still need to be represented in the KB since there are many installations in the field which may still need them. We may, however, want to predicate additional attributes of these older components to distinguish them from newer ones. For example, at some point in time we may add a new *laser printer* to the line of hardcopy devices. At a later time, we may want to add a new model, a *high-speed laser printer.* This might involve adding two new objects: one to represent a generic laser printer with an attribute *printing speed* and another to represent the new high-speed laser printer. The original *laser printer* object would be seen as a specialization of the newly created generic laser printer.

Knowledge-based systems often represent declarative knowledge using a set of nodes, corresponding to discrete

Reprinted from *Proceedings of the IEEE,* Volume 74, Number 10, October 1986, pages 1414-1421.

"concepts" or descriptions, which are partially ordered by a subsumption, or inheritance relation. One concept subsumes another if everything that is true about the first is also true about the second. Whenever a new node is added to the knowledge base, either during its initial construction or later maintenance, it must be placed in the appropriate position within the ordering—i.e., all subsumption relationships between the new node and existing nodes must be established. This is called *classification* because a subsuming node can be considered as a representation of a more abstract category than its subsumees. The notions of subsumption and automatic classification are very useful and have been offered as central features of several recent knowledge representation languages (see [5], [26], [9], and [2], for example).

Current classifiers require a complete description of the node to be added before they begin. (See [32] for a description of classification, and [20] or [27] for examples of a classifier for the representation language KL-ONE [4].) When the classifier is used directly by a user to add a new node, the user must know the descriptive terms in use in the existing KB and something of its structure in order to create a description which will be accurately classified. If the classifier places the new node in the wrong place, or if the description of the node contains errors or omissions, the user must repeatedly modify the node and redo classification until he is satisfied. The process of adding a node is much more efficient if done interactively, so that immediate feedback based on the contents of the KB is available to the user as each piece of information about the new node is entered.

The rest of this paper describes an interactive classification algorithm, which has been implemented in Prolog. Together with a simple knowledge representation language, this implementation forms a system called KuBIC, for Knowledge Base Interactive Classifier. The system takes a user's initial description of a new node and a (possibly empty) KB and either classifies the node immediately, if enough information has been specified, or determines relevant questions for the user that will help classify it. Thus a user who is familiar with the knowledge base may completely avoid the question/answer interaction with KuBIC, and use it only as a classifier, while someone who has never seen the knowledge base before may use the interaction to be presented with just those portions of the KB which are relevant to the classification of the new concept. The algorithm could be applied, for example, to knowledge representation systems or environments for building expert systems which contain classifiable knowledge bases, such as KEE [18], HPRL [19], SRL [13], or LOOPS [3].

II. The Representation Language

In order to explore the underlying ideas of interactive classification, a simple knowledge representation language was chosen. The KB is constrained to be a tree structure, so each node has at most one parent. Nodes have single-valued attributes which represent components or characteristics that apply to the object or concept described. Values of attributes can be numbers, intervals, symbols, or sets of symbols. The meaning of a set or range with multiple values is *disjunctive*; children of a node with an attribute with multiple values can have any subset or subrange (including single values) of the parent's value. Each node inherits all the attributes of its parent node, but its values can be restrictions of the parent attribute's values. Finally, no procedural attachment is allowed.

The Subsumption Relation: The tree structure of the knowledge base is formed by the partial ordering of its nodes with respect to the subsumption relation. The intended meaning of "*X subsumes Y*" is that whatever is represented by description *Y*, is also represented by the more general description *X*. All of *X*'s characteristics are inherited by *Y*, perhaps with some restriction. Since the subsumption relation is transitive, *Y* also inherits the characteristics of *X*'s subsumers (i.e., all its ancestors in the tree). In KuBIC, subsumption information is used to achieve *economy of description* and to *localize distinguishing information.*

Economy of description is a direct consequence of the inheritance of attributes and attribute values. Each description is considered to be a *virtual* description whose attributes are either local to the real description, or inherited from an ancestor. Only the most restricted value of an attribute appears in the attribute of the virtual description, even if the value occurs in an attribute of more than one ancestor description.

Classification is aided by the structure of the knowledge base. In such a taxonomic database, distinguishing information is localized. Once a new description has been determined to be subsumed by node *X*, only *X*'s subsumees are possible candidates for a more specific subsumer of the new description. The information stored at *X*'s immediate subsumees allows the classifier to select questions which will determine which node is this more specific subsumer.

III. Interactive Classification

The interactive classification process is divided into three phases: acquiring the initial description of the new concept, finding the appropriate parent concept in the existing taxonomy (the most specific subsumer), and finding the appropriate immediate descendants in the existing taxonomy (the most general subsumees).

A. Acquiring the Initial Description

To make the interaction more efficient and minimize the number of questions the user has to answer, the user is allowed to specify an initial description of the new node. Attributes of the new node can be given, and a subsumer can be stated directly if known. Note that the user can say only that a node subsumes the new node, not that it is the **most specific** subsumer. If enough information is given, it is possible to classify the new node immediately without any further interaction. If not, KuBIC must determine what attributes to ask about so that classification can be completed.

If the initial description includes an attribute which is not currently in the KB, then the user is asked to supply certain information about the new attribute. In the simplified representation language used in KuBIC, this information is just the general constraint on possible values that the attribute can take on and a *question form* that the system can use to ask for a value for this attribute.

B. Establishing the Most Specific Subsumer

Because the characteristics of a node are shared by all its descendants, it is most efficient to search the tree for the new node's most specific subsumer (MSS) in a top-down manner, starting with the root. Two strategies are used to speed the search for the most specific subsumer: classification by *attribute profile* and classification by *exclusion.* The first strategy is used to take the partial description of the new KB object the user initially presents and to identify a likely ancestor as low in the taxonomy as possible. The second strategy, classification by exclusion, is used to *push* the new KB concept lower in the taxonomy, eliciting new information from the user as needed. This second strategy is more basic to the interactive classifier and will be described in detail first.

Classifying Using Exclusion: Classifying by exclusion makes use of the fact that every node (except the root) has exactly one immediate subsumer, or parent. At all times during classification, there is one node which has been verified to be a subsumer of the new node, and is the most specific such node (the current most specific subsumer, or MSS). Only subsumees of this node need be considered as more-specific subsumers. Moreover, at most <u>one</u> of the immediate subsumees of this node may be a more-specific subsumer.

Exclusion therefore proceeds by looking for inconsistencies between the current description of the new node and the immediate subsumees of the current MSS. If no subsumees are consistent, the current MSS remains the actual MSS, and classification continues with the search for the new node's subsumees. If only one node is consistent, it must be verified to be a subsumer of the new node. This is done by asking the user, if necessary. If two or more nodes are consistent, attributes must be found to ask the user about which will help exclude as many of them as possible, until less than two nodes remain consistent.

The word "consistent" is a bit inadequate—what is actually meant by *"node S is consistent with the new node N"* is *"node S subsumes the* ***current*** *description of the new node."* (Note that this consistency relation is <u>not</u> symmetric.) Because the new node's description changes during its interactive classification as the user adds new information, it is possible for *S* to become inconsistent with it. Thus the meaning of the term *consistent* that we are using is similar to that used in discussions of nonmonotonic logic [22].

Verifying Subsumption of a Consistent Node: Because the new description is entered interactively, one attribute at a time, it is incomplete during classification. Suppose that there is only one candidate node in the set of consistent children of the current MSS. This is not enough to ensure that the candidate is a more specific subsumer of the new node since the candidate may have additional attributes that the new node does not have. We are assuming, of course, that the user is giving us a *partial description* of the new KB object. If the candidate has additional attributes, we must verify that it indeed subsumes the new node. For each such attribute, the user is presented with its value in the candidate node and is asked to confirm, deny, or restrict the value as appropriate for the new node (see Fig. 2). This is done to ensure that the values of the new node's attributes are restrictions of the values of the candidate's values. Note that the new node may have attributes which the candidate does not have; this does not affect subsumption. If the user does **not** verify that the single candidate node is a subsumer of the new node, then the current MSS of the new node is established as the final MSS.

An example showing a KB fragment which requires such verification is shown in Fig. 1, and the verification interaction is shown in Fig. 2. For each node in Fig. 1, only the attributes which are either defined locally or locally restricted are displayed at that node. The new description,

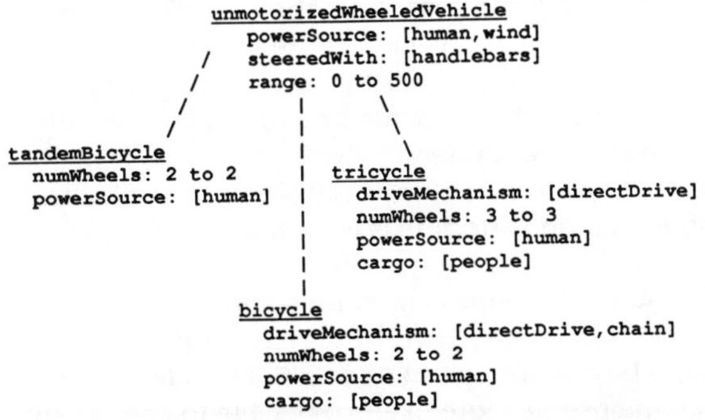

Fig. 1. KB fragment just before verification.

There is evidence that the new description is a *bicycle*. I will now question you on each unverified aspect of *bicycle*. Please confirm, deny, or restrict the value, for each attribute.

What is the cargo?
cargo = [people]
Enter yes, no, or a restriction of the answer: ***yes.***

What is the drive mechanism?
driveMechanism = [directDrive,chain]
Enter yes, no, or a restriction of the answer: ***[chain].***

I've verified that *bicycle* subsumes *tandemBicycle*

Is this acceptable?: ***yes.***

Subsumer changed from *unmotorizedWheeledVehicle* to *bicycle*.

Fig. 2. Interaction during verification (user's response in ***bold italics***).

tandemBicycle, has been determined to be subsumed by *unmotorizedWheeledVehicle*. Since one attribute of *tandemBicycle* is that it has two wheels, only *bicycle* is a consistent candidate for a more specific subsumer of *tandemBicycle*. Before asserting that *tandemBicycle* is subsumed by *bicycle*, the user is asked to verify that *tandemBicycle* also has a *cargo* attribute whose value is *people* and has a *driveMechanism* whose value is a subset of {*directDrive, chain*}. If the user does not agree, *tandemBicycle*'s MSS remains *unmotorizedWheeledVehicle*.

Determining the Next Question: If there are two or more candidate nodes in the set of consistent children of the current MSS, more information about the new node is required to exclude some of them. This is done by selecting an attribute to ask about, getting the answer from the user, and repeating until the set of consistent children has been reduced to zero or one node, or there are no more attributes which will help reduce the set. Two strategies are used to select an attribute to ask about from the set of attributes which apply to the set of consistent children: *explicit attribute ranking* and *maximal restriction*.

In our simple representation language, one can attach to a concept a list of some of the concept's attributes which are ranked with respect to their importance in classifying by exclusion. If such a ranking has been defined, then the attributes are selected in the given order. This strategy supercedes the next one, because the ranking contains external information which is not otherwise available to the system. The ranking could be based on numerical weights, but here it is a non-numerical ordering.

If there are no more attributes in the ranked list, the attribute selected to ask about is the one which *maximally restricts* the set of consistent children, in the worst case. In other words, no matter what answer is given as the value of this attribute, the minimum number of consistent children which are excluded by the answer is greater than or equal to the same minimum for any other relevant attribute. If more than one attribute is best, one is selected without regard to other considerations.

The above strategies could be augmented by using information about the particular user. Since not all questions need to be asked to perform one classification, questions which the user is more likely to be able to answer should be asked first. The user's ability to answer can be decomposed into his or her ability to understand the question, determine an appropriate response, and communicate the response to the system. The user model could be created initially by asking the user several questions intended to establish a stereotype of the user, and refined later as the user answers (or does not answer) questions. (See [25] and [10] for examples of this use of stereotypes.)

Classifying Using Attribute Profiles: The second classification strategy is a heuristic for searching the tree more quickly. Given the operations of determining consistency and asking the user to verify subsumption described above, if a guess could be made about possible subsumers of the new node, it would be a simple matter to verify the subsumption. A good guess is necessary, however, because the user must get involved in the verification.

The particular heuristic used in KuBIC examines the set of attributes specified by the user in the initial description to try to restrict the possible subsumers of the new node. The heuristic could also be used whenever volunteered information is allowed. It works by picking an attribute of the initial description, finding the common ancestor of all nodes in the KB which have the attribute, and using this common ancestor as a guess. The guess must be a subsumee (immediate or not) of the current MSS of the new node.

If the user verifies the guess, then it becomes the current MSS and the process continues. The user has been spared from having to answer questions about attributes of concepts which lie between the original MSS and the guess. The deeper the guess in the tree, the more questions avoided. If the user does not verify the guess, perhaps because the attribute has more than one meaning in the current KB, all is not wasted. Questions asked during verification can contribute information to the new node, or, if the attribute in question is not an attribute of the new node, KuBIC knows not to ask the question again. The system can keep guessing, whether a previous guess succeeded or not, until it runs out of attributes, or until the user becomes weary of incorrect guesses.

C. Establishing the Most General Subsumees

The task of classification is half completed once the most specific subsumer of the new node has been established. Finding the most general subsumees (MGSs) is the other half. Fortunately, this half is much less work because of the constraint that the KB form a tree structure.

The only possible candidates for most general subsumees are children of the MSS of the new node—i.e., siblings of the new node. (This assumes that the KB is well-constructed, so that the immediate subsumer of each node is its MSS, and then immediate subsumees are its MGSs.) Thus to find all the MGSs, it is only necessary to check whether the new node is "consistent" with each sibling in turn, and to ask the user to verify that there is no missing information about either node which misled the classifier. Note that by establishing a node as the MGS of the new node, the interactive classifier can implicitly change the descriptions of the MGS and all its subsumees—nodes which were already in the KB—because they inherit new attributes from the new node.

If the subsumption relationship is allowed to define a lattice rather than a tree, then determining the MGSs is more difficult. A newly entered node may not subsume its siblings, but could subsume some of its sibling's descendants. For example, consider a taxonomy for living things which includes a concept *livingThing* with two immediate children: *animal* and *plant*. We could use the interactive classifier to enter a new node *genderedLivingThing* to represent the concept of a *livingThing* with an attribute gender whose values come from the set {male,female}. This concept would initially be an immediate descendant of the concept *livingThing*. Neither *animal* nor *plant*, however, is a descendant of this new concept, since there are genderless animals and genderless plants. Many of their descendants, however, are subsumed by the new concept *genderedLivingThing*.

IV. Interactive Classification in More Expressive Languages

The KuBIC system is limited by the extremely simple nature of the knowledge representation language we have used. Using this simplified language was a conscious research strategy choice. It allowed us to focus on the notion of an interactive classifier in a simple surrounding. The two major shortcomings in KuBIC's representation language are that nodes are organized in a tree rather than a lattice, and that values of attributes must be explicit sets of values or value intervals in the case of totally ordered domains such as integers. Neither of these limitations is a serious obstacle to extending the idea of interactive classification to more general representation languages. This section briefly describes some additional work on two experimental interactive classifiers, CHPRL and KLASSIC, as well as a proposal for an extension to a KL-ONE-like language to better support interactive classification.

A. The CHPRL Classifier

Our first experiment was to build an interactive classifier for the frame-based representation language HPRL [19].[1] HPRL is a frame-based representation language which relaxes the restrictions found in the simple KuBIC language.

[1]HPRL was developed at the Hewlett Packard Laboratories. The version used in this research ran in Portable Standard Lisp on an HP9836 workstation

In particular, in HPRL concepts can have any number of immediate ancestors and attributes have a much richer structure. A concept in HPRL is called a **frame** and consists of a name, a set of immediate ancestors, and a set of **slots** which correspond to our notion of attributes. In HPRL, a slot has a number of pieces, or facets, of information attached to it. The facets important to classification include a restriction on the type of values a slot can take on, a restriction of the number of values it can have, a description of a default value for the slot, and (possibly) a set of actual values for the slot. We extended the HPRL language slightly to allow the type information to be expressed in one of two ways—either as a procedure (as was normally the case) or as a reference to another HPRL frame.

The CHPRL classifier [24] takes a newly defined HPRL frame and interactively classifies it. One interesting aspect of this system is its ability to handle *exceptions* and *default values*. An *exception* is a specification of a value or a restriction of a value for a slot which is inconsistent with an inherited value or value restriction. A *default value* is an annotation on a slot which specifies a value to be used for the slot if one is needed and there are no local or inherited values for the slot. We have addressed the problems of classification of descriptions involving exceptions and defaults because they are an integral part of the HPRL language, unlike languages in the KL-ONE family. The CHPRL classifier handles defaults by treating them as "virtual values." Thus a slot S_1 with a set of default values D can be subsumed by a slot S_2 only if S_2 has no values, or has a value restriction which subsumes all the defaults in D and either has no defaults of its own or has a set of defaults which is a superset of D.

It is generally held that exceptions introduce many serious problems in a knowledge representation system, particularly with those that include automatic classifiers [6]. It is also widely believed that the general notion of an exception is a useful one, especially in representational systems which attempt to model people's representations. Our attempt to combine the notion of automatic classification and exceptions in CHPRL is based on the following ideas:

- **The classifier never introduces an exception.** All exceptions must by explicitly sanctioned by the user. The classifier is not allowed to hypothesize an exception to enable one concept to be subsumed by another. If a user asserts a subsumption relation between two concepts, any inconsistencies that are detected are marked as exceptions.
- **Exceptions are efficiently indexed.** All exceptional facts are linked to the general facts that they violate. This allows the classifier to efficiently compare the concept being classified to any "exceptional" concepts when appropriate, without searching the entire knowledge base.
- **Exceptions are exceptional.** It is assumed that the number of exceptions in a knowledge base is "small" relative to the size of the knowledge base. If this is not the case, then it is likely that the KB needs to be redesigned. This assumption ensures that the classification process will not bog down in exception checking.

B. The KLASSIC Representation Language

Our second experiment involved building a new representation language with an integral interactive classifier "on top" of HPRL. This language, KLASSIC [17], is very similar to KL-ONE in that it has a more formally defined semantics. One frame C1 subsumes another C2 if everything which is true of C1 is necessarily true of C2. In addition, KLASSIC implements some constraints expressed through *role value maps* [7]. Our approach to building KLASSIC was to define the significant components, or units, out of which a description is built (e.g., concept, role, and role value map) as frames in the underlying HPRL representation language. This made it very easy to modify and extend the KLASSIC language. All three classes of units are organized into abstraction hierarchies as well. Thus we can represent the fact that the role **address** specifies a relation between people and places and has two immediate subsumees, **homeAddress** and **officeAddress** as well as one immediate subsumer, **location**.

In building an interactive classifier for KLASSIC, we had to address the additional problems of *primitive concepts* and *role subsumption*. A *primitive concept* is a concept which is only partially defined in the KB. Typically, a primitive concept has some necessary attributes specified but lacks a specification of all of the sufficient attributes. Realistic KBs typically contain many primitive concepts (see, for example, [16]). When the KLASSIC interactive classifier is trying to determine whether or not a new concept C_{new} is subsumed by a primitive concept C_p, it first tries to verify that C_{new} is consistent with C_p by showing that all of C_{new}'s attributes are true of C_p as well. If this is the case, then KLASSIC asks the user to verify that C_{new} is indeed subsumed by C_p.

There are many primitive concepts, however; that should never be considered as potential subsumers of any new user-defined concept. For example, many applications require an object to represent the concept of an integer. Such an object typically functions as a "name" for the concept and is not elaborated in any way (i.e., has no attributes or constraints). It is not anticipated that the user will want to introduce new descriptions which are to be classified as specializations. To prevent the classifier from pestering the user with questions about such basic concepts, we have introduced a new type of primitive concept, a *primordial* concept. A primordial concept is one whose descendants are fixed.[2] No user is allowed to introduce new descendants through classification. The interactive classifier will never consider a primordial concept as a potential subsumer of a newly entered concept unless it is explicitly mentioned in the description.

In KLASSIC, as in most languages in the KL-ONE family, a concept's roles are organized into an abstraction hierarchy just as the concepts themselves are. Since a description of a new concept to be added to the KB is an expression containing both concepts and roles, the classifier must be

[2]More specifically, the descendants are fixed with respect to the classifier. There may be other ways for new descendants of a primordial concept to be introduced, such as the syntactic recognition of individual integers.

able to compute role subsumption as well as object subsumption. The basic idea is that, given a new description, its roles must be classified before classifying the concept itself.

C. Classification and Primitive Concepts

The basic idea behind classification, whether interactive or not, is that given any two *concept definitions* it is possible to determine if one subsumes the other. However, it is often the case that many concepts we would like to represent do not seem to have precise definitions. To represent such concepts in a KL-ONE knowledge base requires that they be specified as **primitive concepts** [7]. Primitive concepts may have some information specified for them, but they do not have complete definitions.

Primitive concepts hinder classification since the user must explicitly specify the relationship of new concepts to any primitive concepts in the knowledge base. In real applications, the number of primitive concepts may comprise over half of the concepts in the knowledge base [16]. A user wishing to enter a concept must manually classify the concept with respect to all known primitive concepts to ensure the concept is placed correctly in the knowledge base. For large knowledge bases this can be both difficult and error-prone.

We are exploring an extension to languages like KL-ONE which reduces the burden on the user when adding new concepts to a knowledge base while maintaining the soundness of the knowledge representation language. This extension consists of adding an explicit definitional component to concepts in the knowledge base. Within this component the strictness of concept definitions is itself relaxed. The benefits of this modification are threefold:

- The relaxed form of definitions will reduce the number of primitive concepts in a knowledge base.
- The explicit definitional component can be used by the classifier with concepts that do not have complete definitions.
- The definitional component improves the utility of an interactive classification.

Providing an Explicit Definitional Component: We are designing a representation language which can be seen as an extension to a KL-ONE-like language which permits an explicit **definitional** component for each concept. This definitional component has the form

$$def(X) = N_1 \wedge N_2 \wedge \ldots \wedge N_k \wedge D_1 \wedge D_2 \wedge \ldots \wedge D_l.$$

The N_i are necessary conditions for something being an X. The D_i-terms[3] represent disjunctions of sets of contingent conditions (i.e., non-necessary conditions) and have the form

$$D_i = S_{i1} \vee S_{i2} \vee \ldots \vee S_{in}.$$

The S-terms consist of conjunctions of contingent conditions or C-terms. An N-term or C-term may be either a simple term or a reference to another concept's definition.[4] A simple term is akin to a role and its value restriction in KL-ONE. Our notation also allows for a negated simple term, whether it is a necessary or contingent attribute. Finally, a D-term can be a covering disjunction for the concept it defines. Additional details are given in [16].

Consider the following hypothetical example of the definition of a concept X:

$$def(X) = N_1 \wedge N_2 \wedge N_3 \wedge ((C_1 \wedge C_2) \vee (\neg C_1 \wedge C_3)).$$

In this case N_1, N_2, and N_3 are necessary conditions for X. In addition to these conditions C_1, C_2, and C_3 play a role in the definition of X, but are not in themselves necessary. In fact (together with N_1, N_2, and N_3), the clauses $(C_1 \wedge C_2)$ and $(\neg C_1 \wedge C_3)$ form two sets of sufficient conditions for being an X.

Ramifications for Interactive Classification: The proposed extension makes it much easier for the classifier to determine subsumption on its own. Assuming the creators of the knowledge base take full advantage of the extended definitional capability for concepts, the classifier should be able, in many cases, to find a sufficiency set which fits the new description. Even if a perfect fit cannot be found, the classifier can look for the best matching set.

In the case of primitive concepts, the gain is even greater. Whereas the interactive classifier previously had to check *every* primitive concept with the user, it can now autonomously decide about subsumption in many of the cases. The user will be called upon only in cases where the new description could be a *new exception*. If the user sanctions the exception, it will be reflected as a change to the contingent features of the definition of the existing subsuming concept.

V. Providing a Model of the User

If an interactive classifier is to live up to its promise as a tool for building and maintaining large and complex knowledge bases, it must be good at interacting with its users. We are currently working on the incorporation of a more sophisticated model of the user to support this interaction. Such a model can be used to select attributes to ask about next and also to provide the user with appropriate help and guidance in answering questions. This is related to work in the context of interfaces to expert systems (see [30], [31], [11], for example).

There has been some previous research on how expert systems get information from their users. For example, Fox [12] considered integrating reasoning with knowledge acquisition from a resource management perspective. Aikins [1] addressed the seemingly random question-asking behavior of systems which pursued lines of reasoning opportunistically, jumping around to whatever line looked

[3]Note that a definition may have several D-terms, each representing a range of possible attributes. For example:

def(employee)=
preson∧. . .∧
(EmplStatus-FullTime ∨ EmplStatus-PartTime)∧
(pay-salaried ∨ pay-hourly ∨ pay-commissioned).

[4]Referencing another definition is simply a matter of convenience. The fully expanded form could be substituted for the reference.

most promising and asking for whatever information they needed at that point. This randomness annoyed and confused users. Aikins suggested an organization for reasoning that would result in related questions being asked together. Brooks [8] considered the amount of information systems may end up requesting from their users and found that a large number (30 or more) of requests is generally considered unacceptable. He suggested ways of cutting down on the amount of information requested, by enriching systems' models of their domains. These same sorts of considerations can be employed in the context of interactive classification.

For an interactive classifier, the system's goal should be to classify a new concept while burdening the user as little as possible. There may be several outcomes to the system asking a question of the user:

- The user may be unable or unwilling to answer the question.
- The user may need to invoke a subdialogue with the system in order to get additional information to enable him to answer the question. This additional information may be provided in the form of definitions of terms, question paraphrases, or other kinds of help.
- The user may provide an uncertain answer. In general, we might assume that any answer can be qualified or hedged by associating a *degree of belief* with it.

Thus we can define the *effectiveness* of a query as a function of the amount of information returned in the answer and the amount of "user interaction" required. A sensible heuristic for choosing the next attribute to ask about is to choose the most effective query.

Our estimate of the amount of interaction required to get an answer from the user and of the certainty of the answer we will obtain will have to depend on our model of the user. We have developed some general domain-independent tools for building user models [10] that we will use for this purpose.

VI. Summary

This paper presented the design and implementation of an interactive, incremental classifier which is used to add nodes to a hierarchical frame-oriented knowledge base. A knowledge representation language was defined, complex enough to resemble in certain aspects representations of current knowledge-based systems, yet simple enough to allow focusing on interactive classification (for more detail and the Prolog implementation of KuBIC, see [29]). The problem of classification was described as determining most specific and most general subsumption relationships between the new node and nodes already in the knowledge base. Two components to the classification strategy were presented. Classification using exclusion uses a special "consistency" relation and asks questions to exclude whole portions of the KB at a time. Classification using attributes uses a heuristic based on what attributes the user says the new node has in order to take shortcuts in the search. Both of these serve to establish the most specific subsumer; the most general subsumees are then relatively simple to find.

We have built several additional experimental interactive classifiers in representation languages of differing points of view. In one, CHPRL, we looked at some of the issues involved in a language with a system of defaults and in which exceptions are allowed. In another, KLASSIC, we examined the problems of interactive classification in a language in which the abstraction hierarchy forms a lattice rather than a tree and includes definitions of primitive concepts. Current work is focused on extending the concept of an interactive classifier to a more powerful representation language that includes explicit mechanisms for specifying definitions and incorporating a more sophisticated user model.

Bibliography

[1] J. Aikins, "Prototypes and production rules: A knowledge representation for computer consultations," Heuristic Programming Project HPP-80-17, Stanford University, Stanford, CA, Aug. 1980.

[2] H. Ait-Kaci, "Type subsumption as a model of computation," in *Expert Database Systems*, L. Kerschberg, Ed. Menlo Park, CA: Benjamin/Cummings Publ., 1985.

[3] D. G. Bobrow and M. Stefik, "The loops manual," Xerox PARC Tech. Rep. KB-VLSI-81-13, 1981.

[4] R. Brachman, "A structural paradigm for representing knowledge," Bolt Beranek and Newman Inc., Tech. Rep. 3605, May, 1978.

[5] R. Brachman, R. Fikes, and H. Levesque, "KRYPTON: A functional approach to knowledge representation," Tech. Rep. 16, Fairchild Laboratory for AI Research, 1983.

[6] R. Brachman, "I lied about the trees," *AI Mag.*, vol. 6, p. 3, 1985.

[7] R. Brachman and J. G. Schmolze, "An overview of the KL-ONE knowledge representation system," *Cogn. Sci.*, vol. 9, pp. 171–216, 1985.

[8] R. Brooks and J. Heiser, "Controlling question asking in a medical expert system," in *Proc. Int. Joint Conf. on Artificial Intelligence* (Tokyo, Japan, 1979), pp. 102–104.

[9] F. Corella, "Semantic retrieval and levels of abstraction," in *Expert Database Systems*, L. Kerschberg, Ed. Menlo Park, CA: Benjamin/Cummings Publ., 1985.

[10] T. Finin and D. Drager, "$GUMS_1$: A general user modeling system," in *Proc. 1986 Conf. of the Canadian Society for Computational Studies of Intelligence* (Montreal, Que., Canada, CSCSI, May 1986).

[11] T. Finin, A. Joshi, and B. Webber, "Natural language interactions with artificial experts," *Proc. IEEE*, vol. 74, no. 7, pp. 921–938, July 1986.

[12] M. S. Fox, "Reasoning with incomplete knowledge in a resource limited environment: Integrating reasoning with knowledge acquisition," in *Proc. 7th Int. Joint Conf. on Artificial Intelligence* (University of British Columbia, Vancouver, B.C., Canada, Aug. 1981).

[13] M. S. Fox, J. Wright, and D. Adam, "Experiences with SRL: An analysis of a frame-based knowledge representation," in L. Kerschberg, Eds., *Expert Database Systems*. Menlo Park, CA: Benjamin/Cummings Publ., 1985.

[14] M. Freeman, L. Hirschman, and D. McKay, "A logic based configurator," SDC, A Burroughs Co., Tech. Memo LBS 9, May 1983.

[15] —, "KNET–A logic based associative network framework for expert systems," SDC, A. Burroughs Co., Tech. Memo LBS 12, Sept. 1983.

[16] R. Kass, R. Katriel, and T. Finin, "Breaking the primitive concept barrier," CIS, Univ. of Pennsylvania, Tech. Rep. MS-CIS-86-36 (LINC LAB 11), May 1986.

[17] R. Katriel, "KLASSIC," unpublished report.

[18] T. P. Kahler and G. D. Clemenson, "An application development system for expert systems," *Systems & Software*, Jan. 1984.

[19] D. Lanam, R. Letsinger, S. Rosenberg, P. Huyun, and M. Lemon, "Guide to heuristic programming and representation language. Part 1: Frames," Application and Technology

Lab., Computer Res. Ctr., Hewlett-Packard Corp., AT-MEMO-83-3, Jan. 1984.

[20] T. A. Lipkis, "A KL-ONE clasifier," USC/Inform. Sci. Inst., Consul Note 5, Oct. 1981.

[21] J. McDermott, "R1: A rule-based configurer of computer systems," Carnegie-Mellon Univ., Pittsburgh, PA, 1980.

[22] D. McDermott and J. Doyle, "Non-monotonic logic I," *Artificial Intell.*, vol. 13, pp. 1–2, 1980.

[23] J. McDermott, "XSEL: A computer salesperson's assistant," in *Machine Intelligence 10*. Chichester, UK: Horwood, Ltd., 1982, pp. 325–337.

[24] R. Petterson, "The CHPRL classifier for HPRL," unpublished report, 1984.

[25] E. Rich, "User modeling via stereotypes," *Cogn. Sci.*, vol. 3, pp. 329–354, 1979.

[26] J. G. Schmolze and D. Israel, "KL-ONE: Semantics and classification," Bolt Beranek and Newman Inc., Cambridge, MA, Rep. 5421, 1983.

[27] J. G. Schmolze and T. A. Lipkis, "Classification in the KL-ONE knowledge representation system," in *Proc. Int. Joint Conf. on Artificial Intelligence* (Karlsruhe, West Germany, 1983).

[28] E. Schoen and R. Smith, "IMPULSE: A display oriented editor for STROBE," in *Proc. Nat. Conf. on Artificial Intelligence* (AAAI, Washington, DC, Aug. 1983), pp. 356–358.

[29] D. L. Silverman, "An interactive, incremental classifier," Univ. of Pennsylvania, Philadelphia, Tech. Rep. MS-CIS-84-10, Apr. 1984.

[30] B. Webber and T. Finin, "In response: Next steps in natural language interaction," in *Artificial Intelligence Applications for Business*, W. Reitman, Ed. Norwood, NJ: Ablex, 1984.

[31] ——, "Expert questions—Adapting to user's needs," Computer and Information Sci., Univ. of Pennsylvania, Philadelphia, Tech. Rep. MS-CIS-84-19, 1984.

[32] W. Woods, "Theoretical studies in natural language understanding: Annual report," Bolt Beranekand and Newman Inc., Cambridge, MA, Tech. Rep. 4332, 1979.

A Case Study: Acquiring Strategic Knowledge for Expert System Development

Duane Sharman and E.J.M. Kendall
The University of Calgary

When human expertise is available, knowledge acquisition can displace traditional analysis and design. This article presents a case study illustrating the development of an expert system for diagnosing operational faults in a cellular mobile telephone system. We will focus on methodological issues that establish knowledge engineering as an alternative to structured analysis and design, concentrating on the acquisition of strategic knowledge as an implementation structure determinant.

We will analyze benefits derived from applying expert systems in cellular telecommunications systems, discuss related AI work, define structural characteristics of knowledge-based systems, examine a verbal problem-solving protocol from an expert, describe a prototypical knowledge-based program, and review methods and results.

A perspective on expert systems

Since this article discusses the development of expert systems, as distinguished from other software types by their development methodology, we view knowledge acquisition as an alternative to "structured" development. Our case study—an expert system for cellular telecommunications—shows how expert systems can be partitioned.

Structured development divides system development into four phases: analysis, design, implementation, and testing. Analysis defines what the system has to do in detail, and has the important side effect of defining problem structure. Design matches the solution structure to the problem structure as defined by analysis to produce an abstract description of the solution. Implementation is the

EH0303-8/90/0000/0305$01.00 © 1988 IEEE

design's realization using available tools and components. Testing verifies the implementation's fidelity against the specification produced by analysis, and validates the implementation as a satisfactory solution to the original problem statement.

Applying this methodology to software development often fails, mainly because incomplete analysis has left structural characteristics of the problem and its solution to the designer's imagination. Structural flaws arising from poor analysis inhibit system evolution through the design phase to the desired goal.

Knowledge engineering is an alternative analysis-and-design method of particular value when problems are ill-structured (that is, when solution procedure differs on every problem) but are solved routinely by human experts. Knowledge engineering derives structure from human expertise that reflects implicit understanding of problem structure. We can then add knowledge base details incrementally; that is, the system evolves.

It takes three entities to develop an expert system (although not necessarily three separate people)—a human expert, a representative user, and a knowledge engineer. Each assumes a distinct role in the development process.[1] Human experts contribute their compiled experience. Since knowledge engineers must adopt methods from cognitive psychology to analyze the expert's cognitive processes, and since they must also draw on traditional and structured software development techniques when the expert system shell needs enhancement or when subproblems can be solved using algorithms, they should be part psychologist and part computer scientist. In exposing an expert's cognitive structure, knowledge engineers determine fundamental constraints on program function and establish a structure for transferring expert knowledge to the program. User involvement in expert system development does not differ much from other development processes; user concerns relate to program competence, performance, and human factors of the man-machine interface. When employing incremental development methods, users must be deeply involved from an early stage.

In developing knowledge-based systems, development focuses not so much on the problem to be solved but on strategies and methods used by experts in solving the problem. Expert system technology is most suitable when either (1) finding any solution is difficult, or (2) the criteria for an optimal solution can be stated but no algorithm exists that can provide an exact solution within reasonable resource constraints. Expert systems are developed by evolutionary development within a psychologically valid structure, offering the following practical results:

- **Users** can understand program results and explanations at all phases of development;
- **Human experts** can understand how their knowledge, which was previously not consciously considered, relates to rules and declarations assembled into a program's knowledge-base; and
- **Knowledge engineers** can correlate program anomalies with knowledge that is directly accessible within the program.

Researchers in cognitive psychology and computer science have established some general principles about human use and development of knowledge. Their consensus is that we can achieve expert performance levels by carefully trading off generality and power. Problem-solving strategies with the widest scope tend to be weaker (that is, they provide little control over the problem solver's activities). Conversely, we can attain better results within a narrow domain. Expert performance is possible only when a balance is found. The expert system described in this article helps diagnose operational faults in cellular telecommunications networks based on data given in problem reports from customers and maintenance personnel. Although the specific results reported are valid only within a narrowly defined domain, the methodology and techniques used apply in other domains.

Automating maintenance activities

Before addressing various theoretical and practical issues confronting this project's expert system developers, let's consider problems that maintenance personnel face in a cellular telephone operation, and review factors motivating the use of expert system technology. First, let's examine switching-system functions in a cellular telephone system. In contrast to early mobile telephone systems, cellular mobile telephones are highly automated and provide a full range of services compatible with standard telephone exchanges. This automation has brought greater efficiency, enabling reuse of limited bandwidth allotments within smaller geographic areas than can be achieved by competing technologies. However, these benefits have necessitated greater system complexity—a consequence most visible to maintenance personnel.

Cellular telecommunications systems synthesize existing technologies. Thus, although operating companies have difficulty recruiting enough cellular telecommunications specialists, they can recruit technical personnel with experience relevant to specific subsystems within a cellular system. The major subsystems are

(1) Radio-space base-station equipment, providing a medium for voice and data communications between cellular systems and mobile terminals;

(2) Telephone voice-switching systems, linking cellular radio-space base-station equipment to the international telephone network; and

(3) Data communications networks, for remote control of base-station and switching equipment.

The situation confronting maintenance personnel in cellular operations is one of overspecialization: Most radio frequency (RF) technicians understand little about telephone switch equipment. To respond promptly to customer complaints, one must understand all three subsystems. Expert system technology can help interpret customer reports by drawing on automatically collected data such as alarm logs and billing records.

The fierce competition among cellular operating companies underscores the importance of responding to customers promptly. Throughout North America, legislation provides for two competing services in each geographical area. One license is issued to the telephone company operating in the area, and the second license to a competing company. Standardization is enforced, so that consumers are not locked into one company. Operating companies can differentiate their service from the competition's by offering either better voice quality, lower price, greater reliability, or more features. Responding promptly to customers contributes to better voice quality and greater reliability. Both are highly visible system characteristics.

To summarize, an expert system for interpreting customer problem reports in a cellular telephone system would improve an operating company's competitive edge (1) by facilitating prompt response to customer complaints, and (2) by improving voice transmission quality and service reliability. Such systems can help technical personnel understand processes in the system that are outside their field of expertise. To achieve this goal, expert systems must make extensive use of data available on-line through automated switching-system facilities.

Our references cite related work, including research on problem structures and solution strategies covering generate-and-test, means-end analysis, planning, taxonomic specialization, and memory recognition and matching.[2-10] We are planning a paper that will extend one of these cognitive strategies to synthesize deep-knowledge explanations and means-end simulation.

A case study

To understand expert problem-solving methods, we used protocol analysis of a retrospective report to gain insight into a human expert's behavior. The scenario we presented to the expert was based on a hypothetical cellular system. Cellular telephone systems provide telephone service to mobile customers, using radio links to provide paths for voice and data transmission between a radio base station (transceiver) and a mobile telephone (terminal). When a mobile customer moves too far from his current transceiver, or the path between the transceiver and the mobile is broken, the cellular telephone system tries to switch the call from that transceiver to one closer to him.

The problem chosen for our test case involves a possible failure mode for a handoff. Since data communications must control the transceiver from the telephone system, a communications link failure could cause the call to be lost unexpectedly. We gave design descriptions, a hypothetical customer report, various maintenance and billing summaries that would be available, and the following problem report to the expert:

(1) Party A is on Route 1 at 6:00 a.m.
(2) Party A is moving (that is, he is driving).
(3) Using his mobile telephone, party A calls party B.
(4) Parties A and B talk for a while.
(5) The connection fades suddenly.
(6) The call disconnects.

We asked the expert to explain event (6); that is, to determine the event sequence internal to the system that caused the call to disconnect prematurely. The following protocol summarizes the expert's reasoning:

(7) The telephone system successfully set up the call. Event (3) shows that the system should have tried to set up the call. Event (4) shows that the set-up was successful.

(8) Event (6) is not a normal occurrence; the system should have maintained the call.

(9) Since party A is a mobile subscriber, successfully setting up the call (7) involved assigning a radio channel to party A.

(10) Because party B is outside the cellular network, the public switched-telephone network (PSTN) was involved in setting up the call (7).

(11) From the location given for party A (1), we know that Cell 1 was responsible for originally establishing the link to party A.

(12) The fading noticed by party A (5) suggests that party A was on the outskirts of good radio reception (coverage) for Cell 1.

(13) Knowing that a radio channel from Cell 1 was successfully assigned to party A (from (9) and (11)), and that party A was on the outskirts of Cell 1's coverage area (12), we can assume that the cellular telephone system tried to hand off party A from Cell 1 to another cell—a reasonable assumption, since he was moving at the time (2).

(14) To hand off a call (13), the system controller would query neighboring cells (for instance, Cell 2 and Cell 3).

(15) On receiving this query (14), the neighboring cells would measure the received signal's strength and report it to the system controller.

(16) From these measurements (15), the system controller would select either Cell 2 or Cell 3 as the new radio link for the call.

(17) The cellular telephone system would try to hand off party A from Cell 1 to the selected target cell, which requires establishing a new voice path for party A. Communications between the system controller, the target cell, and party A's mobile terminal would route the PSTN link to the new cell, and direct the mobile terminal to the new link.

(18) Since it is early in the morning (2), the system is not overloaded. The rerouting probably succeeded.

(19) The complete sequence did not succeed: Failure of events (14) or (15) would have blocked the handoff. But, after checking the list of alarms for that time, the expert finds no reports of internal communication failures.

(20) Thus, the most plausible explanation for losing the call is that the mobile terminal did not receive the redirect command.

Knowledge acquisition, using human expertise, can displace traditional design methods.

A cognitive strategy analysis

Our goal in studying the preceding report is to identify problem-solving strategies used by the expert. Analysis proceeds by breaking the protocol into episodes, where each episode's utterances relate problem-solving behavior directed towards the same high-level goal. The strategy dominating each episode is classified according to categories discussed previously.

We can identify the following five episodes showing different strategies:

• **Initial description, (1) through (6)**—An initial description of the call sequence is collected. Nothing in the protocol suggests that the expert's reasoning is sensitive to the problem context. This phase uses memory-based recognition to recall expected actions and to create information-gathering goals. Stereotypical information is collected to characterize the known actors (party A and party B and their equipment) and the basic events within a call (setup, holding, and disconnection).

• **Identifying spatial and temporal boundaries, (7) through (12)**—This sequence is concerned primarily with identifying the call's spatial and temporal extent. The expert infers spatial boundaries by consideration of the interconnection of the equipment in the call. Telephone industry jargon recognizes two (possibly identical) paths through the telephone network: a path for voice transmission, and a path for "signaling" control information between pieces of equipment providing the voice path for a call. Point (5) suggests that voice path quality was degraded; this probably motivated the tracing of equipment along the voice path. It would also be reasonable to assume that this is a normal activity in interpreting a call scenario. Memory-based recognition (triggered by the knowledge that voice quality is related to the equipment involved) can account for this goal.

• **Hypothesis: Handoff failure (13)**—This point raises the hypothesis that voice path degradation resulted from a failed handoff sequence. The handoff's purpose is to maintain acceptable voice communication quality. Teleological reasoning based on means-end-analysis strategy offers a good explanation for this hypothesis.

• **Simulation, (14) through (17)**—These points show the expert tracing events involved in a correct handoff sequence to check the hypothesis that the handoff failed, which allows the expert to use deep knowledge for inferring causal links. The phrase "try to" (17) indicates goals about the reasons behind system actions. These goals are not part of the system's state, of course; its designers built them into the system. Failed events and nonevents can be analyzed by teleological reasoning to find a related goal. Forward simulation is a form of planning here, resulting in the construction of an action sequence leading to a hypothesized state.

• **Consideration of candidate solutions, (18) through (20)**—This sequence shows the application of means-end analysis to the generation and evaluation of fault possibilities. The expert tests different possibilities sequentially. The goal of modifying the event sequence envisioned above (to account for the hypothesized failure) is explicit in the protocol. It is also possible that the expert used generate and test here, proposing as a candidate each fallible event in the normal handoff sequence and testing against each individually to find causal or teleological relationships between the candidate event and the disconnected state.

Deep-knowledge strategies

In the above analysis, the simulation uses a different problem space to justify heuristically determined conclusions—an interesting point. Whereas preceding inferences developed a temporally disjoint picture of the problem, facts are reported during simulation in chronological order. The problem-state component explicitly representing equipment actions and states also differs. Operators that are applied manipulate a composite picture of the system state. Simulation enables the expert to introduce many constraints, both temporal and stative, that cannot be managed heuristically.

Following simulation, the proposed sequence of events is available. Cognitive activities before and after simulation permit (1) consideration of constraints on states, and (2) more flexible reasoning about temporal relationships. For instance, the approximate time of day enters into the picture in step (18). We know that all relevant events occurred during what is typically a time of low load on the telephone system. Knowing only the endpoints, reasoning processes may consider certain steps without considering intermediate steps. For example, assuming the failed handoff hypothesis and knowing that the call reached a disconnected state, disconnection sequences are generated for consideration (19).

Although simulation is restrictive regarding temporal inference, it can be a powerful strategy for handling complex, nondeterministic relationships between device states. In general, it is not possible to determine the outcome when multiple functions use the same tool. The cellular telephone system above requires a voice channel to complete the handoff operation. It's difficult to account for other demands on the available voice channel pool without simulating the handoff sequence in some detail.

Knowledge states during simulation are complex: The expert must maintain concurrently a conjunction of state information relevant to various system components. Other reasoning processes are required to identify the equipment for simulation and to direct the search for alternatives.

Knowledge states in other phases permit users to manipulate conjunctions of state information associated by relationships other than temporal concurrency. In effect, simulation is "at 90 degrees" to other strategies; concurrent relationships between components are strongly bound, whereas nonsimulation strategies deal with simpler intercomponent relationships and more varied temporal relationships. Simulation provides a good opportunity for analyzing causal relationships, which depend on information flow between components. In this way, we can apply domain knowledge to confirm and elaborate conclusions reached earlier in heuristic phases.

A preliminary computer implementation

A program was written in OPS4 and Lisp to mimic the strategies discussed above.[10] The accompanying sidebar provides an example of the program's execution. A program design description and an explanation of its operation follow.

Program design. According to our cognition model, systems have three parts: working memory, the representation system, and cognitive processes. Encoding these in OPS4 and Lisp is straightforward: The system's working memory is encoded in OPS4 working memory, the representation system is realized as a simple frames-based representation system written in Lisp, and inference processes are implemented as OPS4 rules. The OPS4 production memory initially contains basic man-machine interface production rules, in addition to those rules needed to retrieve from the representation system other rules that apply to a specific problem class. Thus, all rules are not simultaneously loaded in production memory, but are added as needed.

The preceding program design description should give some insight into representation system and working memory structure, showing how the actual implementation maps onto knowledge structures discussed earlier. The program's OPS rules are divided into groups, according to the cognitive strategies evident in the previously analyzed report.

The representation system. The program's representation system is organized around conceptual frames in the task environment. Frames are composed of roles, each role describing a relation between frames or providing rules relevant within that frame's context. Much of the scenario's reasoning was associated with RF properties, requiring representation of geographic concepts including regions and cells. An aggregate relationship holds between cells and regions; part of a "cell" concept's definition describes regions covered by it. This relation is one-to-many, and is encoded in the representation system as a role whose filler is a list of values.

As the following example illustrates, cell-1 has a list of place values denoting geographic regions (downtown, north hill, and properties) stored under its coverage slot:

```
;
;Cells—zones of radio coverage
;
(frame cell-1
    (coverage (places
        (downtown)
        (north-hill)(properties))))
```

An annotated execution trace

Numbered lines are actual program transcripts (we have omitted the introductory 13 lines). User input is prefixed with an asterisk. The call is initially characterized by collecting a standard set of facts.

.
.
.

```
14 Type of calling telephone?(Enter mobile or land)
15*mobile
16 Type of telephone called?(Enter mobile or land)
17*land
18 Was the call setup successfully?
19*yes
.
.
.
```

Additional case-dependent information is then required to develop a more detailed problem description. The following sequence determines that the problem relates to RF propagation.

```
.
.
.
28 What is the location of partyA?
29 (Enter downtown, hwy-1, mcleod-tr, deerfoot-tr)
30*hwy-1
31 Did either party hang-up intentionally?
32*no
.
.
.
37 How was the initial connection quality?(Enter good or bad)
38*good
.
.
.
43 Was the connection quality degraded immediately prior to disconnection?
44*yes
.
.
.
```

Before setting up the simulation, the equipment involved must be determined and initial conditions for the simulation selected. This information enables more precise specification of the hypothesis.

```
.
.
.
49 Which cell provides radio coverage for hwy-1 ?
(Enter 1 or 2)
50*1
.
.
.
68 Was partyA in the fringe area for cell 1?
69*yes
70 Was partyA moving at the time the call failed?
71*yes
.
.
.
```

The hypothesis is that a handoff failed during its transfer sequence, losing the RF path to the mobile. The initial conditions for simulation are: Cell 1 is serving partyA; partyA is in conversation state; and partyB is now in conversation state.

```
.
.
.
116 partyA moves into a region of poor RF coverage . . .
.
117 The system locates a better cell for partyA . . .
.
.
.
```

Information gathered during simulation is then analyzed further to specialize the hypothesis according to the problem's cause.

```
.
.
.
123 Were more than 90% of the circuits in cell 1 in use?
124*yes
.
126 Is there a hole in the RF coverage for cell 1 at or near hwy-1?
127*yes
.
129 Have there been problems tuning the handoff parameters for
130 cell 1?
131*no
.
.
.
```

The resulting hypotheses are then reported to the user.

```
.
.
.
137 It is possible that a locate request for partyA was received
138 from cell 1, but no better location was found.
.
140 A hole in the RF coverage in cell 1 could have resulted in
141 no better cell being selected for partyA.
.
143 Heavy usage of voice circuits in cell 1 could have blocked a handoff attempt for partyA.
```

The inverse of the coverage role is the serving-cell role, defined as a role of the "region" concept. Serving-cell roles can be differentiated into primary and secondary serving cells, an approximation of the base station coverage that permits overlapping regions to be described simply. For example, the downtown location has cell-1 listed as its primary serving cell and cell-2 as its secondary serving cell. Inference rules associated with defined frames must handle role differentiation currently, although work is underway to declaratively encode such knowledge.

Working-memory structure. Little structure is imposed on elements in working memory. Instead, representations in working memory are based on first-order predicate logic: We represent zero place predicates as atoms, and describe n-ary predicates by lists of length n+1, where the first element of the list is the predicate symbol (as shown below).

(terminal-type partyA mobile)

Working-memory elements serve three purposes. One set explicitly identifies progress of the currently active strategy, a second set realizes various programming control mechanisms, and a third set encodes a problem presentation. Most elements are of the third type. For example, we should interpret the assertion (terminal-type =P mobile) to mean "the type of terminal used by party =P was a mobile telephone" where the symbol "=P" is an OPS4 variable that can be replaced by either "party A" or "party B."

Inference processes. The program implements the protocol's strategies as distinct OPS rule sets. Strategic control elements in working memory are turned on and off (by rules associated with each strategy) to control progress through the phases. Once all processing within a phase has been exhausted, termination rules take effect to end the current phase and start the next phase.
Below, we exemplify an OPS4 rule activating the program's representation system. The program uses this rule when identifying spatial boundaries.

```
Check-serving-cell  ;Rule name
(Identify-boundaries    ;Rule antecedent
 (setup-succeeded yes)
 (terminal-type =P mobile)
 (location =P =L)
 - (=L serving-cell primary =)
—)
( frrecall =L serving-cell primary)) ; Rule Consequence.
```

If the system determines that the call was set up successfully, and a party involved in the call used a mobile telephone terminal, then the program retrieves the primary serving cell for the party's geographic location at call setup from the representation system. OPS4 does not permit use of functional access forms to working memory on the right-hand side of rules; thus, we need an extra left-hand-side term to retrieve the concerned party's location from working memory. The right-hand-side function "frrecall" transfers information stored under a specified frame, role, and facet from long-term memory to short-term memory.

Execution trace phases. The first phase (lines 8 through 44 in the sidebar) acquires information about the problem in a nondescript manner. The first action taken requests information about basic call roles from the user. No inferential reasoning occurs.

The second action in the first phase classifies the problem from initial information. The program groups (according to failure type) problems that it can diagnose. To specialize the problem-related reasoning that follows, this section acquires additional call-related information. Possible problem-type frames that could be used include setup-failed, call-not-maintained, intermittent-loss, and handoff-failed.

The second phase (lines 47 through 55) identifies equipment used in connecting the call. Here, the program assumes that few options exist within the switch (it ignores the possibility of a blocking-switch matrix) so that only the connection points to the switch are needed. The program uses spatial knowledge to determine RF equipment used.

The next phase selects the handoff-failure hypothesis (lines 62 through 110) and chooses a script for analyzing this hypothesis. In using this script, one or both parties must be connected via RF channels. To confirm the feasibility of this hypothesis, additional facts must be assembled. In addition, these facts will establish an initial state for the next phase.

The phase that follows (lines 113 through 119), performs a step-by-step simulation of the telephone system according to (1) constraints imposed by the handoff script and (2) known facts. The program assumes the handoff script allows only one path; thus, no backtracking or other control mechanisms are needed. If an action in the script cannot be applied, the program abandons the script. The model is based on a Petri-net formalism for expressing procedure. Transitions between successive states simultaneously change the states of all system components playing a role in the action.

For instance, a message sent from one component to another simultaneously changes the states of both components in the model. Temporal characteristics of real-world communication actions can be introduced into the model by expanding transitions, although this is not done currently. As simulation progresses, a rule set monitors the simulation process, saving information about states and events that appears relevant to the hypothesis raised above.

Following simulation, the program evaluates monitor rule output in considering alternative solutions. It first evaluates the handoff hypothesis (lines 121 through 131). When insufficient inferences can be made from available data to establish hypothesis validity, the program gathers more information from the user. It evaluates alternative hypotheses heuristically, with no benefit from knowledge explicit in simulation rules. This phase results in a set of consistent fault hypotheses. Among the hypotheses considered are hole-in-RF-coverage, parameter-tuning-error, communication-error (at various points in the simulation), and all-trunks-busy.

Finally, the program generates an explanation by adding "canned" causal links from the hypothesis frame (lines 134 through 145). This would not be reasonable if the hypothesis set available and relationships to inferred system activity became large. Special-purpose rules dominate the inference process throughout this section.

Further developments

Research underway to further develop the telecommunications expert system includes efforts to

(1) **Organize** procedures for knowledge acquisition that adopt a stronger knowledge acquisition model. This effort could also develop automated tools to help encode large volumes of protocol data;

(2) **Collect and analyze** talk-aloud protocols for diagnostic problems. This effort requires development of an inference engine compatible with knowledge acquisition tools so that acquired knowledge can be verified;

(3) **Validate** the expert system against a realistic base of problems; and

(4) **Embed** the inference engine in an operational telecommunications environment.

Methods and results—a review

We believe that this case study permits us to examine several points relevant to further related research. To reach this stage, we made many decisions; by reevaluating some, we hope to present a valuable perspective.

Knowledge acquisition in a rich task environment. Analyzing operational problems is a cognitive activity that exhibits a rich task environment. The expert's strategies aim at reducing cognitive strain by considering a selected subset of global knowledge at any given time. Accordingly, assessing the expert's state of knowledge is difficult. Expert systems seek to mimic the strategies of human experts in domain-specific problem solving. Thus, we treat problem solving as a dynamic activity. Traditional rule-based expert systems that do not employ pattern matching (MYCIN and Prospector, for example) have mimicked only the final outcome of an expert's reasoning process, ignoring the "how" of the expert's success.

In less complex task environments, such as classification problems, the state of a subject's knowledge can be summarized simply: Relationships between concepts do not excessively complicate the problem state. This is seen in "flat" problem spaces characteristic of structured-selection expert system shells. All that we need is a hypothesis set with associated certainty factors. In structured-selection problems, relationships between domain entities are not heavily problem dependent; inferred relationships do not figure prominently in the reasoning. Relationships between beliefs derived by the program are maintained only for purposes of explanation. In these cases, knowledge acquisition can focus on the expert's conceptual framework rather than on his reasoning processes.

When interpreting problem reports in a cellular telephone system and similar tasks, the structure of evolving knowledge structures is important. Typically, only a small part of this structure is evident in final results. To achieve performance at expert standards, knowledge engineers must be attentive to expert cognitive processes, although this doesn't diminish the importance of understanding the final conceptual structure. We do not believe that adequate analysis is obtained by analyzing static domain concepts.

Deep knowledge. Our example demonstrated the use of deep knowledge for simulation, along with shallow heuristic strategies. Questions about the relationship between deep and shallow knowledge remain open; the information processing system model of cognition does not provide a suitable handle for resolving many of these questions.

Choosing a suitable expert. When developing an expert system in domains such as cellular telephony, few "old hands" exist. But, while experts in our study had no experience diagnosing field problems, they did have extensive design experience. This would bias results toward procedural models—prominent during the design process. In future work, we will use other subjects to avoid strong biases in the encoding effort. However, these subjects should have a design background so that they will possess the knowledge required (although it will not be applied for diagnostic purposes). Such subjects are more likely to draw on deep reasoning strategies than are subjects with less theoretical understanding of systems.

Evaluating protocol analysis. Protocol analysis is laborious. We can make few generalities about problem-solving behavior, however, without intensive analysis. Unfortunately, it's difficult to amass enough data to confirm or refute hypotheses in rich task environments, and manual protocol analysis alone is inadequate. Verbal probes may be advantageous, although we need a stronger reference model for knowledge acquisition. Some automation would be desirable, particularly if the task of producing and validating a programmed realization for observed cognitive processes was addressed.[11]

Attention to cognitive processes and strategies is essential to knowledge acquisition. Protocol analysis addresses process explicitly, and helps when determining strategic and structural characteristics.

Knowledge acquisition can displace traditional analysis and design when human expertise is available. We have presented a case study in the development of an expert system for telecommunications system maintenance, concentrating on the acquisition of strategic knowledge as a determinant of implementation structure. We have also discussed the role of deep knowledge in maintenance tasks, presented the resulting system design, and examined the relevant development methodology.

References

1. F. Hayes-Roth, D.A.Waterman, D.B. Lenat, *Building Expert Systems,* Addison-Wesley, Reading, Mass., 1983, pp. 129-132.
2. W.J. van Melle, *System Aids in Constructing Consultation Programs,* UMI Research Press, Ann Arbor, Mich., 1981.
3. A. Newell and H.A. Simon, *Human Problem Solving,* Prentice-Hall, Englewood Cliffs, N.J., 1972.
4. N.J. Nilsson, *Principles of Artificial Intelligence,* Tioga, Palo Alto, Calif., 1980.
5. S. Vere, "Temporal Scope of Assertions and Window Cutoff," *Proc. IJCAI,* Morgan Kaufmann, Los Altos, Calif., 1985, pp. 1055-1059.
6. D. McDermott, "A Temporal Logic for Reasoning About Processes and Plans," *Cognitive Science,* June 1982, pp. 101-155.
7. B.C. Moszkowski, *Reasoning About Digital Circuits,* PhD dissertation, Stanford University, Stanford, Calif., 1983.
8. D.G. Bobrow, ed., *Qualitative Reasoning about Physical Systems,* MIT Press, Cambridge, Mass., 1985.
9. D. Gentner and A.L. Stevens, eds., *Mental Models,* Lawrence Erlbaum, Hillsdale, N.J., 1983.
10. C. Forgy and J. McDermott, "OPS4: A Domain-Independent Production System Language," *Proc. IJCAI,* Morgan Kaufmann, Los Altos, Calif., 1977.
11. H.A. Simon and K.A. Ericsson, *Protocol Analysis: Verbal Reports as Data,* MIT Press, Cambridge, Mass., 1984.

Duane Sharman received his MSc in computer science from the University of Calgary in 1987, concentrating on expert system telecommunication applications. He received his BSc in electrical engineering from the University of Calgary in 1977, and worked from 1977 to 1984 as a design engineer, developing real-time software for pipeline and petrochemical control applications. He has supervised development teams at Sentrol Systems Ltd., NovAtel Communications Ltd., and the Nova-Husky Research Corporation. His present interests include automatic test generation for software systems and provably correct software design.

John Kendall is professor and computer science department chair at the University of Calgary. He has worked in semiconductor and device physics and engineering and electronic instrumentation in both academy and industry. He returned to the University of Calgary in 1984 to work in VLSI design and expert system development. He received his BSc and MSc in physics, and his PhD in electrical engineering, from the University of Birmingham (UK). He is a Fellow of the Royal Institute of Physics (UK), and a Professional Engineer (Alberta).

The authors can be reached at the Computer Science Dept., University of Calgary, 2500 University Dr. NW, Calgary, Alberta, Canada T2N 1N4.

CHOOSING KNOWLEDGE ACQUISITION STRATEGIES FOR APPLICATION TASKS

Catherine M. Kitto and John H. Boose

Knowledge Systems Laboratory, Boeing Advanced Technology Center
7L-64, Boeing Computer Services, P.O. Box 24346, Seattle, Wa., 98124

ABSTRACT

In constructing a knowledge-based system, one of the most difficult and time-consuming activities is the elicitation and modeling of knowledge from the human expert about the problem domain. A major difficulty is that little guidance is available to the domain expert or knowledge engineer to assist with 1) classifying the problem task, 2) choosing the problem-solving method appropriate for the task, and 3) given a task and method, selecting knowledge acquisition strategies and tools to be applied in developing the knowledge base.

We describe a method for providing automated assistance to the knowledge engineer or domain expert in analyzing the problem domain, classifying the problem tasks and sub-tasks, identifying problem-solving methods, suggesting knowledge acquisition tools, and recommending specific strategies for knowledge acquisition within those tools. We have implemented this approach in a subsystem called the Dialog Manager, which was developed to provide process guidance to users of the AQUINAS knowledge acquisition workbench.

INTRODUCTION

The process of acquiring knowledge from an expert is still recognized as one of the most troublesome steps in the development of an expert system. One difficulty encountered frequently is determining the right problem-solving paradigm for the knowledge-based system being constructed. Attempts to use inappropriate problem-solving methods often fail. For example, a tool intended for solving classification problems such as diagnosis of engine malfunctions, selection of photographic equipment, identification of an organism, etc. would produce poor results if used for a computer configuration problem. Even when suitable problem-solving methods are identified for the task, successful knowledge acquisition strategies vary depending on the particular application task. Specific pieces of information need to be elicited for a medical diagnostic task that are irrelevant in the context of an identification task such as selecting a programming language. It is critical to determine the correct knowledge acquisition tools, techniques, and specific problem-solving method required for an application task in developing an expert system.

The selection of knowledge acquisition strategies for specific application tasks has been addressed in the implementation of an automated Dialog Manager. The Dialog Manager (Kitto, Boose)[1] provides guidance and assistance in constructing a knowledge base using the AQUINAS knowledge acquisition workbench. AQUINAS provides an integrated environment for a variety of knowledge acquisition tools and techniques (Boose, Bradshaw)[2], including facilities for elicitation of knowledge from multiple knowledge sources, tools to hierarchically structure knowledge, techniques to combine and propagate knowledge, and a reasoning engine for incremental testing of the knowledge using different methods of reasoning.

Automated assistance from the AQUINAS Dialog Manager is provided at three stages in the knowledge acquisition process: 1) in the initial dialog with the domain expert or knowledge engineer, the Dialog Manager helps to identify the type of application task and the appropriate problem-solving methods; 2) the Dialog Manager suggests appropriate tools for knowledge acquisition; and 3) when AQUINAS is the recommended knowledge acquisition tool for an expert system application, the Dialog Manger continues to provide guidance during the use of AQUINAS, recommending specific knowledge elicitation techniques and suggesting appropriate knowledge acquisition strategies for elicitation and analysis of expert knowledge for the application.

The AQUINAS Dialog Manager has been used to guide knowledge acquisition for a number of knowledge-bases. We show examples from a diagnostic problem for jet engine diagnosis.

APPLICATION TASKS AND PROBLEM-SOLVING METHODS

Several categorizations of knowledge engineering applications have been suggested (Stefik et. al.[3], Hayes-Roth et. al.,[4] Clancey[5], Waterman[6]). Frequently mentioned are application task categories that include interpretation, prediction, diagnosis, design,

Reprinted from *Proceedings: WESTEX-87—Western Conference on Expert Systems,* 1987, pages 96-103.

EH0303-8/90/0000/0314$01.00 © 1987 IEEE

planning, monitoring, debugging, repair, instruction, and control (Hayes-Roth[4]). The generic applications could be divided into two broad groups: those associated with *analysis* (interpretive) tasks, and those involved with *synthesis* (constructive) activities. These tasks can be further organized into taxonomies or classification frameworks of generic tasks and sub-tasks. Figure 1 depicts one such hierarchy of generic application tasks (Clancey[5]).

Our thesis is that given any reasonable classification scheme for application tasks, we can identify problem-solving methods and suggest knowledge acquisition tools and strategies where they exist. We have based our initial work on a hierarchy of problem tasks proposed by Clancey, but the procedure could be applied using any recognized taxonomy of application tasks.

Problem-solving Methods

Methods have also been identified for solving particular classes of problems (Clancey[5]). His two basic problem-solving methods are *heuristic classification* and *heuristic construction*. Heuristic classification is a common problem-solving method in which concepts in different classification hierarchies are heuristically related. Heuristic classification involves three operations: data abstraction, heuristic matching of data abstractions to solution abstractions, and solution refinement. The heuristic classification method is generally appropriate for problems in diagnosis and repair, catalog selection, and skeletal planning.

In heuristic construction, the problem-solver may either generate complete solutions or assemble solutions from components while meeting constraints and requirements. This problem-solving method is best suited for synthesis application tasks (design, configuration, planning, scheduling, etc.). Knowledge of how pieces of a solution fit together is essential, and solutions are assembled incrementally. This involves a technique of proposing a solution and reasoning about it, much like the Propose-and-Revise strategy used in the VT expert system for elevator configuration (Marcus, McDermott, and Wang[7]). Heuristic classification can be viewed as a sub-method of heuristic construction. Heuristic classification can be the problem-solving method used for selecting the components to be configured in a configuration application task. For a planning problem using heuristic construction for problem-solving, heuristic classification may be used as a step in selecting the activities to be included in the plan.

Another common problem-solving method is simple classification, in which a solution is selected from a pre-enumerated set of possible solutions by matching observations with the features of the solution classes which may be organized in hierarchies. An object is identified as a member of a class of objects or events.

Chandrasekaran (1985)[8] identifies a set of six generic tasks in knowledge-based reasoning which ressemble the problem-solving methods proposed by Clancey. These generic reasoning tasks include classification, state abstraction, knowledge-directed retrieval, object synthesis by plan selection and refinement, hypothesis matching, and assembly of compound hypotheses for abduction. Like Clancey, Chandrasekaran views these generic "tasks" as problem-solving methods which can be combined to perform an application task in an expert system.

A complex application task can be decomposed in terms of generic tasks and problem-solving methods applied. For example, in diagnosis, reasoning methods may include classification, knowledge-directed data retrieval, and abductive assembly of multiple diagnostic hypotheses.

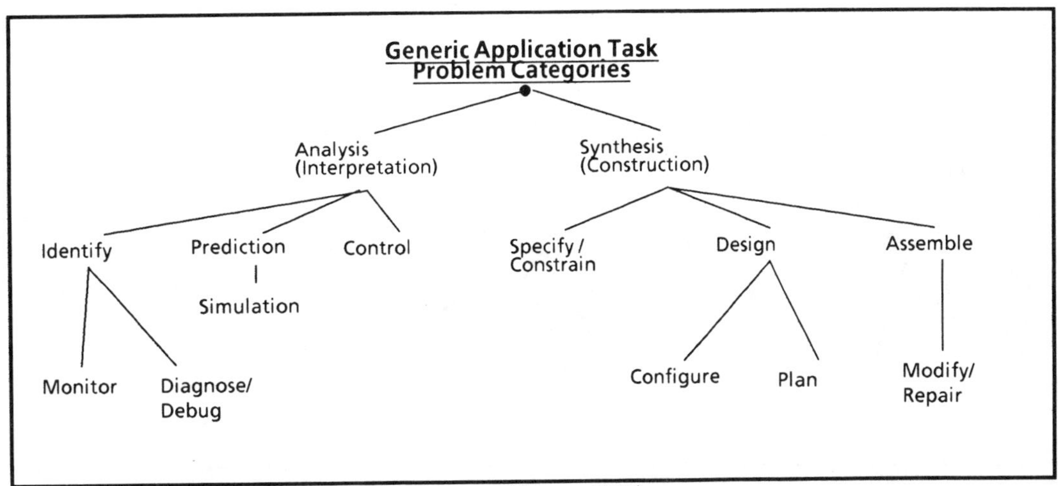

Figure 1. Hierarchical framework for application tasks (Clancey, 1985).

KNOWLEDGE ACQUISITION STRATEGIES AND TOOLS

A set of knowledge acquisition strategies has been proposed (Boose and Bradshaw, 1987)[9], each of which links certain sub-tasks with appropriate problem-solving methods. Knowledge acquisition strategies include establishing distinctions between alternatives, problem decomposition, combining and propagating information, testing of knowledge, combining multiple sources of knowledge, incremental expansion of knowledge, and providing process guidance. Similarly, Bylander and Chandrasekaran[10] suggest that generic problem-solving tasks be associated with a specific knowledge acquisition methodology, using considerations for classification problem-solving as an illustration.

Knowledge acquisition tools make use of these strategies to link an application task and the appropriate problem-solving method. In Figure 2, a hierarchy of application task categories is shown below with the problem-solving methods above. The broken lines represent examples of links supported by existing knowledge acquisition systems. Several knowledge acquisition systems enable diagnostic applications to use the heuristic classification problem-solving method, including MDIS (Antonelli[11]), MORE (Kahn, Nowlan and McDermott[12]), MOLE (Eshelman, Ehret and McDermott[13]), TEIRESIAS (Davis[14]), and TKAW (Kahn, Breaux, Joseph, and DeKlerk[15]). These knowledge acquisition systems are relatively specialized for diagnostic applications, but have considerable domain-related power. Likewise, ETS (Boose [16]), KITTEN (Shaw and Gaines[17]), and STUDENT (Gale[18]) link identification application tasks with heuristic classification capability.

In contrast, AQUINAS (Boose and Bradshaw[2]), which also supports heuristic classification problem-solving, has more generalized knowledge acquisition capablities (diagnosis, interpretation, debugging, identification) with some loss of depth in the resulting domain knowledge.

For synthesis applications, two knowledge acquisition systems are illustrated: SALT (Marcus

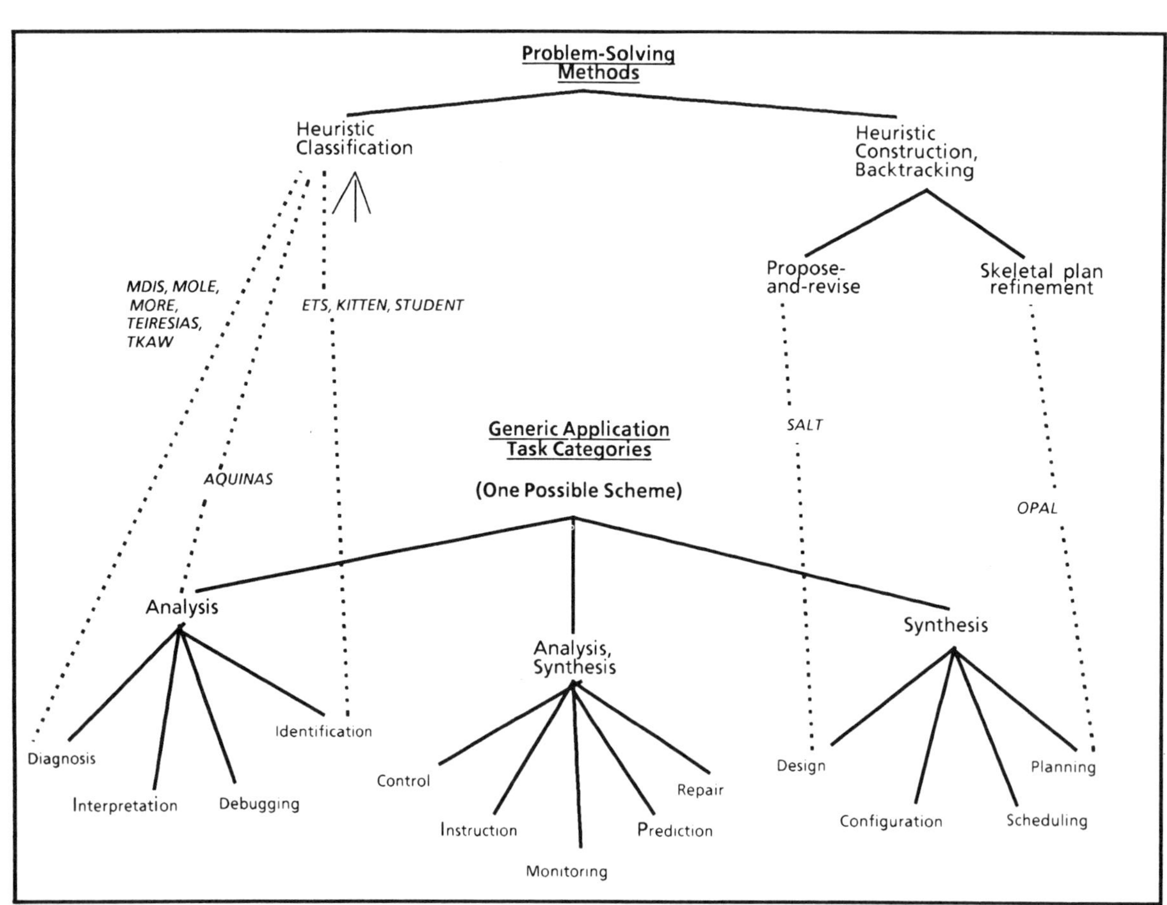

Figure 2. Some examples of knowledge acquistion tools linking application tasks and problem-solving methods.

7,19,20) linking design tasks with a propose-and-revise problem-solving approach, and OPAL (Musen, Fagan, Combs and Shortliffe[21]), a planning application. Not shown is KNACK, a knowledge acquisition system for report generation and evaluation (Klinker, Bentolila, Genetet, Grimes and McDermott[22]) which has characteristics of a synthesis task (assembly of report fragments and evaluation in respect to constraints).

A knowledge acquisition system must be able to elicit information to meet the unique problem-solving requirements of an application task category. For example, in electronic or electromechanical diagnosis, component failure rates, test times, test costs, test complexity, test utility or benefit, and test dependencies are useful to the human expert. According to Simpson and Dowling[23], a fault isolation strategy for troubleshooting systems must be able to incorporate and use information on test complexity, test cost, length of tests, and frequency of component failure. They describe WRAPLE, a system which uses procedures based on information theory to determine fault paths for fault isolation. WRAPLE initially analyzes symptoms, known test results, test equipment information, and skill level availability. It then performs fault isolation considering weighted measures of test time, test cost, field data on mean time between failure (MTBFs), minimum test cost and minimum test time to determine the smallest number of tests used to isolate the most frequently failing components.

Another example of specific application task knowledge requirements is described by Hand[24] for the domain of medical diagnosis. According to Hand, experts in medical diagnosis typically use the following problem-solving strategies:

1). Generate a list of multiple competing hypotheses compatible with the symptoms.
2). Examine the most frequently occurring candidate diseases first.
3). Consider the "utility" of a diagnosis. If there is a high cost associated with failure to diagnose a problem (for example, the patient dies) , then focus on that hypothesis.
4). Consider diagnoses for which treatments are available first.
5) Do not discriminate between alternative diagnoses where treatment is the same.
6). Avoid redundant or superfluous questions or tests. Each piece of information requested should play a role in diagnosis.
7). Evaluate tradeoffs between the benefit of obtaining information against its cost and risk.

New knowledge acquisition strategies are needed to capture the medical diagnosis problem-solving knowledge. Strategies should permit elicitation of probabilities of diseases, utility of a correct diagnosis, and the relative cost and risk of obtaining information. Experts should be able to specify the ordering of questions as well as initial data to be collected at the start of each consultation (for example, patient name, age, sex, complaint, and observations of the physician).

AUTOMATED ASSISTANCE IN KNOWLEDGE ACQUISITON DECISIONS

Dialog Management In AQUINAS

The selection of knowledge acquisition strategies for particular application tasks has been addressed in the implementation of the *Dialog Manager* (Kitto and Boose[25]), part of the AQUINAS system (Figure 3). The Dialog Manager provides guidance in constructing a knowledge base with the AQUINAS system. The methodology in AQUINAS is based on personal construct theory (Kelly[26]). A major component of AQUINAS is ETS (Boose, 1985[27]) which elicits, analyzes, and refines knowledge represented in repertory grids.

The Dialog Manager is an expert system itself, whose domain knowledge concerns the use of

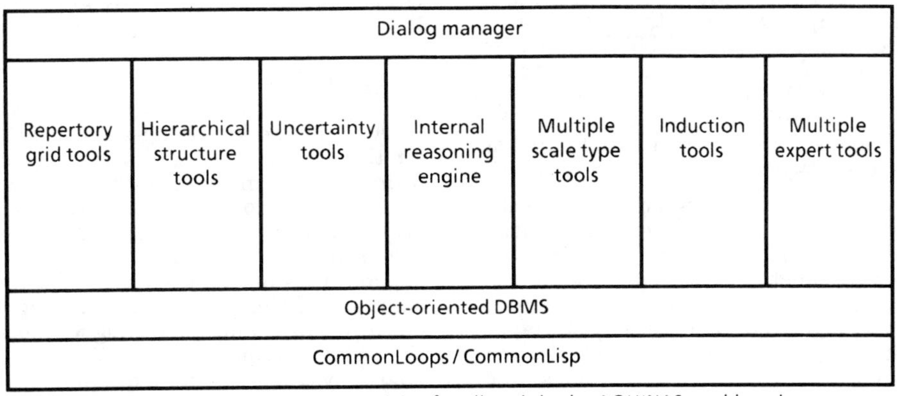

Figure 3. The Dialog Manager provides advice for all tools in the AQUINAS workbench.

AQUINAS as well as the characteristics of other knowledge acquisition tools. The Dialog Manager consists of heuristics and strategies for knowledge acquisition encoded in the form of rules and procedures. It offers a mixed initiative environment with options to set the levels of control, guidance, and prompting. The Dialog Manager examines its set of rules to select the most appropriate knowledge acquisition technique to be applied based upon heuristics which consider the application task, problem-solving method, current state of the knowledge base, user experience, temporal considerations, and user preferences.

The Dialog Manager offers advice in choosing knowledge acquisition strategies for application tasks at three stages: First, in an initial interview with the domain expert, the Dialog Manger queries the expert about the nature of the application task. Based on the task's characteristics and similarities to types of application tasks with which the Dialog Manager is familiar, the Dialog Manager classifies the application task and determines the appropriate problem-solving methods for the task. Second, the Dialog Manager will recommend a specific knowledge acquisition tool to the user based on task characteristics and requirements for the problem-solving method. Finally, for those application tasks for which AQUINAS is the suggested knowledge acquisition system, the Dialog Manager will then guide the expert or knowledge engineer in acquiring the necessary problem-solving knowledge. The Dialog Manager considers the type of application and problem-solving methods to select appropriate knowledge acquisition strategies within AQUINAS.

Determining Problem-Solving Methods for an Application Task

The Dialog Manager assists in classifying a problem into one or more sub-tasks; each sub-task implies the use of a problem-solving method. This step is essential since it will determine the type of knowledge acquisition tools to be applied. For example, if the task is identification, useful knowledge acquisition techniques include those embodied in tools such as ETS, and heuristic classification would be the method employed. However, for a design or planning task, other methods should be applied such as the Propose-and-Revise method used in the SALT knowledge acquisition tool for constructive systems (Marcus[7,19,20]). The Dialog Manager determines the knowledge acquisition strategies to be employed and suggests the use of a specific tool or technique associated with AQUINAS.

The Dialog Manager conducts a dialog with the domain expert or knowledge engineer in the initial stages of expert system development. Like the ROGET system (Bennett[28]), the expert can select similar tasks and characteristics from menus of tasks already categorized. The Dialog Manager determines the nature of the application tasks and sub-tasks, specifies problem-solving methods, and recommends appropriate knowledge acquisition tools. For those domains for which AQUINAS is suggested, the Dialog Manager continues to guide and assist the user throughout the knowledge acquisition process, suggesting specific knowledge acquisition strategies for the problem-solving sub-task.

-- AQUINAS SYSTEM --
Would you like to work on an OLD knowledge base, or create a NEW one?
AQU****NEW**
Do you already know the type of problem task or sub-tasks that you are addressing in this knowledge base?
AQU****NO**
--**[DM]** Analyzing Problem Tasks and Sub-Tasks--
Give a description of the general class of problem you are trying to solve:
AQU****JET ENGINE DIAGNOSTICS**
Please give a short description of the problem:
AQU****DIAGNOSES JET ENGINE FAULTS BASED ON EVENTS AND SYMPTOMS**
Do you already have a list of pre-enumerated solutions to the problem you are trying to solve?
AQU****YES**
Approximately how many possible conclusions or solutions are there?
AQU****100**
Can these solutions be organized into a hierarchy of solution classes and solution children?
AQU****YES**
What would you say is the maximum number of "solution children " that would occur in any solution class?
AQU****10**
Can you select a subset of the most significant solutions and limit your problem domain to this sub-set?
AQU****YES**
Approximately how many possible conclusions or solutions are there in this sub-set?
AQU****15**
--**[DM]** Your application task should be treated as a **DIAGNOSIS** sub-task.
The appropriate problem-solving method is **HEURISTIC CLASSIFICATION**.

Figure 4. AQUINAS interviews an expert to determine application task and problem-solving method.

Example

A prototype knowledge-based system was built for the diagnosis of faults in a jet engine. The following example illustrates how the Dialog Manager employs its heuristics for knowledge acquisition to guide an expert in developing the *"Jet Engine Diagnostics System"* and provides explanations of its strategies. Figure 4 shows an example of the preliminary dialog between the domain expert and the Dialog Manager concerning the task to diagnose malfunctions in the generic jet engine system. Recommendations from the Dialog

Manager are highlighted and preceded by **[DM]**, while AQUINAS user entry is in underlined bold type. Based on information in this interview, the Dialog Manager determines that the application is a diagnostic task and that heuristic classification is the appropriate problem-solving method.

```
--[DM]--Knowledge acquisition tools for electromechanical
DIAGNOSTIC problems include:   MOLE, TKAW, and
AQUINAS.

--[DM]--For your application, I recommend using the
AQUINAS system since it is available at your site.

Do you wish to continue using AQUINAS?
AQU**YES
```

Figure 5. The Dialog Manager recommends knowledge acquisition tools based on application task and problem-solving method.

The Dialog Manager then identifies several knowledge acquisition tools for diagnostic applications, including MOLE, TKAW, and AQUINAS. Figure 5 shows the recommendations of the Dialog Manager.

Dialog Manager Assistance in Choosing Knowledge Acquisition Strategies

The AQUINAS system and the Dialog Manager were used to develop the knowledge base for the generic jet engine diagnostic system. Diagnostic problems have specific requirements for problem-solving. The AQUINAS Dialog Manager will suggest different knowledge acquisition strategies for a diagnostic task than for a more general task of identification. For example, in electronic diagnosis, an expert will assess the relative probabilities of occurrence of various malfunctions in determining how to focus the diagnostic process. This estimate of frequency may take the form of "mean time between failure" data, or it may be a subjective judgement based upon the expert's past experience. Likewise, in medical diagnosis, the physician will consider the frequency of various disorders in selecting hypotheses for consideration, seeking evidence for the most frequently occurring diseases first (Hand).

In order to replicate this approach to problem-solving, a knowledge acquisition system must capture knowledge about frequencies of possible problem solutions for diagnostic problems. In our example, the Dialog Manager applied a heuristic based upon domain characteristics of diagnostic problems and recommended that the expert enter the relative frequencies of malfunctions. These estimates will be used as prior probabilities in the reasoning process in the completed expert system. (Figure 6).

Another knowledge acquisition strategy which is appropriate for diagnostic applications is to elicit information about the relative "costs" of obtaining information. Costs may be a subjective estimate

```
---[DM]--- I recommend that you
     EDIT FILL.IN. SOLUTION WEIGHTS
since the application task is diagnosis and since you have
already entered and rated a set of solution candidates,
and since component failure rates (prior probabilities) are
important in diagnostic problem-solving.

AQU**EDIT FILL.IN.SOLUTION WEIGHTS  (expert agrees)

Please estimate the relative frequencies (weights) of each
of the following solution classes, entering values from 0 to
1. For electromechanical diagnosis, weights may be
estimates of component failure rate or MTBF's.

AQU**FUEL.SYSTEM.MALFUNCTION ** .35
AQU**BLEED.VALVE.CONTROL.UNIT.MALFUNCTION**
.45
AQU**COMPRESSOR.STALL ** .05
     :
```

Figure 6. The Dialog Manager suggests that solution weights (relative frequencies) be entered for a diagnostic problem.

and may include such aspects as the dollar cost of a test, length of test, level of skill of testing personnel, and risk of the test. This information is often used in diagnostic problem-solving to determine whether to schedule a specific diagnostic test. Figure 7 depicts the recommendation of the Dialog Manager.

```
---[DM]--- I recommend that you
   TEST CONFIGURE DIAGNOSTIC.PROBLEMS TRAIT.COSTS
since the application task is diagnosis and since cost of
obtaining information is important in diagnostic problem-
solving. "Cost" may include estimates of risk, cost in
dollars, cost in time, cost in resources, etc.

AQU** TEST CONFIGURE DIAGNOSTIC.PROBLEMS
         TRAIT.COSTS                    (expert agrees)

Please estimate the relative costs of obtaining information
on each problem-solving trait on a scale of 1 to 5 where 1 is
inexpensive and 5 is costly. "Cost" should represent
dollars, time, and risk. Enter CHANGE to specify a different
scale.

AQU** PART.POWER.TRIM.TEST **3
AQU** ACCELERATION TO SOME POWER SETTING --
         IN.FLIGHT ** 5
AQU** ACCLERATION TO SOME POWER SETTING --
         GROUND TEST ** 4
AQU** IDLE.TRIM.TEST ** 2
AQU** CIRCUIT CONTINUITY TEST ** 1
         :
```

Figure 7. The Dialog Manager recommends that trait costs (cost in time, money, risk to obtain information) be entered for a diagnostic problem.

Diagnostic applications also suggest the use of a knowledge acquisition strategy to recommend an appropriate method of reasoning for testing the knowledge base and for the finished expert system.

For a diagnostic problem-solving task, the Dialog Manager may suggest that the expert test the knowledge base using a Bayesian probability-based method of reasoning to employ the prior probabilities elicited from the expert above. Another reasoning technique that might be suggested for a diagnostic problem is one based on entropy which minimizes the number of pieces of information requested in the diagnostic process, asking only for the most important tests or data. This could be very critical in the domain of medical diagnosis if the patient were under stress and the physician wished to minimize the number of tests performed to reduce risk to the patient.

A final example (Figure 8) illustrates a knowledge acquisition strategy tailored for diagnostic applications. The expert is asked to identify any problem-solving traits which should be requested initially in a consultation. This might be necessary background data, (for example, the patient's name, age, and insurance in a medical consultation), or it might be information critical to the focus of the consultation such as primary symptoms or values of key parameters. The expert is also asked to designate a specific order of inquiry to assure that the consultation does not violate rules of good conversation.

---[DM]--- I recommend that you
TEST CONFIGURE DIAGNOSTIC.PROBLEMS INITIAL.QUESTIONS
since the application task is diagnosis and each consultation may require specific background data or information to focus the consultation (major symptoms, etc.).

AQU** **TEST CONFIGURE DIAGNOSTIC.PROBLEMS INITIAL.QUESTIONS** *(expert agrees)*

Please enter any traits that should always be asked in a consultation. (For example patient's name and age in medical diagnosis). You may select traits from the menu. Enter them in order.

AQU** **PRIMARY SYMPTOMS**

AQU ****ABNORMAL ENVIRONMENTAL.CONDITIONS**

AQU ****ENGINE.PARAMETERS**

⋮

Figure 8. The Dialog Manager recommends that initial consultation trait questions be identified for a diagnostic problem.

RESULTS

This approach has been used successfully to guide a number of knowledge-based system development efforts using AQUINAS. While still in its infancy, we feel that this approach demonstrates that the knowledge-based system development process can be facilitated by providing automated assistance with problem task categorization, selection of a problem-solving method for a given task, and recommendations of appropriate knowledge acquisition tools and strategies.

Our experience with the Dialog Manager has demonstrated that it is possible to define a knowledge acquisition environment and to formalize knowledge about the transfer and modeling of expertise in a rule-based representation.

ISSUES AND FUTURE WORK

The implementation of the Dialog Manager and the experience of experts and knowledge engineers who have used AQUINAS and the Dialog Manager to develop prototype expert systems have raised a number of issues. We are only beginning to understand the techniques currently used by experts or knowledge engineers using an automated system such as AQUINAS, their underlying heuristics, and whether there are more effective strategies for knowledge acquisition.

We expect to learn more about heuristics for effective transfer of expertise as more experts and knowledge engineers have the opportunity to use AQUINAS with the Dialog Manager. We will observe and interview AQUINAS users of varying backgrounds and experience in the development of expert systems.

We also anticipate that the requirements for guidance from the Dialog Manager will increase as the capability is added to AQUINAS to acquire knowledge for problem domains which do not fit the identification task--heuristic classification paradigm. We hope to extend AQUINAS to handle knowledge acquisition for constructive problem-solving such as in the SALT system and to represent domain models of qualitative causal relations in acquiring diagnostic knowledge as in the MORE or MOLE systems.

ACKNOWLEDGEMENTS

The authors wish to thank Miroslav Benda, Jeff Bradshaw, Jackson Brown, Ted Kitzmiller, Art Nagai, Dave Shema, Lisle Tinglof-Boose, and Bruce Wilson for their contributions and support. AQUINAS and the Dialog Manager were developed at the Knowledge Systems Laboratory at the Boeing Advanced Technology Center of Boeing Computer Services in Seattle, Washington.

REFERENCES

[1] Kitto, C.M., and Boose, J.H., "An Implementation of Knowledge Acquisition Heuristics", *Boeing Computer Services Technical Report* BCS-G-2010-[36], 1986

[2] Boose, J.H. and Bradshaw, J.M., "Expertise Transfer and Complex Problems: Using AQUINAS as a Knowledge Acquisition Workbench for Expert Systems", *Special Issue of the International Journal of Man-Machine Studies on the AAAI Knowledge Acquisition for Knowledge-Based Systems Workshop*, in press, 1987.

[3] Stefik, M., Aikins, J., Balzer, R., Benoit, J., Birnbaum, L., Hayes-Roth, F., and Sacerdoti, E. "The Organization of Expert Systems. A Tutorial.", *Artificial Intelligence* **18(2)**, pp. 135-173, 1982.

[4] Hayes-Roth, F., Waterman, D., and Lenat, D.B., *Building Expert Systems*, Reading, Mass: Addison-Wesley, 1983.

[5] Clancey, W.J., "Heuristic Classification", in *Knowledge Based Problem Solving*, Kowalik, J.S. (ed.), pp. 1-67, Edgewood Cliffs, N.J.: Prentice Hall, 1986.

[6] Waterman, D., *A Guide to Expert Systems*, Reading, Mass: Addison-Wesley, 1986.

[7] Marcus, S., McDermott, J., and Wang, T. "Knowledge Acquisition for Constructive Systems", in *Proceedings of the Ninth Joint Conference on Artificial Intelligence*, Los Angeles, California, August, 1985.

[8] Chandrasekaran, B., "Generic Tasks in Knowledge Based Reasoning: Characterizing and Debugging Expert Systems at the 'Right' Level of Abstraction", *Proceedings of the Second Conference on Artificial Intelligence Applications, Miami Beach Florida*, pp. 294-300, New York: IEEE Computer Society, 1983.

[9] Boose, J., and Bradshaw, J. "Missing Links: Bridging Problem-Solving Tasks and Methods with Knowledge Acquisition Strategies and Tools", *Boeing Computer Services Technical Report*, to be published, 1987.

[10]Bylander, T., and Chandrasekaran, B., "Generic Tasks for Knowledge Based Reasoning: the 'Right' Level of Abstraction for Knowledge Acquisition", *Special Issue of the International Journal of Man-Machine Studies on the AAAI Knowledge Acquisition for Knowledge-Based Systems Workshop*, in press, 1987.

[11]Antonelli, D., "The Application of Artificial Intelligence to a Maintenance and Diagnostic Information System (MDIS)", *Proceedings of the Joint Services Workshop on Artificial Intelligence in Maintenance*, Boulder, Co., 1983.

[12]Kahn, G., Nowlan, S., and McDermott, J., "MORE: An Intelligent Knowledge Acquisition Tool", in the *Proceedings of the Ninth Joint Conference on Artificial Intelligence*, Los Angeles, California, August, 1985.

[13]Eshelman, L., Ehret, D., McDermott, J., and Tan, M., "MOLE: A Tenacious Knowledge Acquisition Tool", *Special Issue of the International Journal of Man-Machine Studies on the AAAI Knowledge Acquisition for Knowledge-Based Systems Workshop*, in press, 1987.

[14]Davis, R., "Interactive Transfer of Expertise", in Buchanan, B. and Shortliffe, E. (eds.) *Rule-Based Expert Systems: The MYCIN Experiments of the Stanford Heuristic Programming Project*, Reading, Mass: Addison-Wesley, 1985.

[15]Kahn, G.S., Breaux, E.H., Joseph, R.L., and DeKlerk, P., "An Intelligent Mixed Initiative Workbench for Knowledge Acquisition", *Special Issue of the International Journal of Man-Machine Studies on the AAAI Knowledge Acquisition for Knowledge-Based Systems Workshop*, in press, 1987.

[16]Boose, J.H., *Expertise Transfer for Expert System Design*, New York: Elsevier, 1986.

[17]Shaw, M.L.G. and Gaines, B.R., "Techniques for Knowledge Acquisition and Transfer", *Special Issue of the International Journal of Man-Machine Studies on the AAAI Knowledge Acquisition for Knowledge-Based Systems Workshop*, in press, 1987.

[18]Gale, W.A., "Knowledge Based Knowledge Acquisition for a Statistical Consulting System", *Special Issue of the International Journal of Man-Machine Studies on the AAAI Knowledge Acquisition for Knowledge-Based Systems Workshop*, in press, 1987.

[19]Marcus, S. and McDermott, J., "SALT: A Knowledge Acquisition Tool for Propose-and-Revise Systems", *Carnegie-Mellon University Department of Computer Science Technical Report*, 1987.

[20]Marcus, S., "Taking Backtracking with a Grain of SALT", *Special Issue of the International Journal of Man-Machine Studies on the AAAI Knowledge Acquisition for Knowledge-Based Systems Workshop*, in press, 1987.

[21]Musen, M.A., Fagan, L.M., Combs, D.M. and Shortliffe, E.H., "Using a Domain Model to Drive an Interactive Knowledge Building Tool", *Special Issue of the International Journal of Man-Machine Studies on the AAAI Knowledge Acquisition for Knowledge-Based Systems Workshop*, in press , 1987.

[22]Klinker, G., Bentolila, J., Genetet, S., Grimes, M., and McDermott, J., "KNACK - Report-Driven Knowledge Acquisition", *Special Issue of the International Journal of Man-Machine Studies on the AAAI Knowledge Acquisition for Knowledge-Based Systems Workshop*, in press, 1987.

[23]Simpson, W.R. and Dowling, C.S., "WRAPLE: The Weighted Repair Assistance Program Learning Extension", *IEEE Design and Test*, pp 66-73, April 1986.

[24]Hand, D. J., *Artificial Intelligence and Psychiatry*, Cambridge: Cambridge University Press, 1985.

[25]Kitto, C.M., and Boose, J.H., "Heuristics for Expertise Transfer: An Implementation of a Dialog Manager for Knowledge Acquisition", *Special Issue of the International Journal of Man-Machine Studies on the AAAI Knowledge Acquisition for Knowledge-Based Systems Workshop*, in press, 1987.

[26]Kelly, G.A., *The Psychology of Personal Constructs*, New York: Norton, 1955.

[27]Boose, J. H., "A Knowledge Acquisition Program for Expert Systems Based On Personal Construct Psychology", *International Journal of Man-Machine Studies*, **23**,(1985).

[28]Bennett, J.S., "ROGET: A Knowledge-based Consultant for Acquiring the Conceptual Structure of an Expert System", *Report No. HPP-83-24, Computer Sciences Department, Stanford University*, 1983.

Chapter 6: Reasoning Methods

The means of representing and acquiring knowledge must be coupled with the power of an inferencing or reasoning method. Several reasoning techniques are described in this chapter's seven papers. Forgy leads with an exposure of Rete, a method of quickly comparing patterns to objects and determining what matches. The next paper, by Pelavin and Allen, outlines a temporal planning logic that accommodates concurrent actions and assertions about future possibilities in addition to the present inference state. Eliot introduces analogies and their application to reasoning. Bic shows how semantic networks can be processed on other than the classical Von Neumann computer architecture. Korf then considers various search algorithms and settles on a combination of two of them. Reasoning under conditions of uncertainty is important in real-world applications. The report by Groothuizen overviews the major methods. Zadeh, an important author in this field, finishes this chapter by detailing fuzzy logic, a widely used method of dealing with uncertainty.

EH0303-8/90/0000/0323$01.00 © 1990 IEEE

Rete: A Fast Algorithm for the Many Pattern/Many Object Pattern Match Problem*

Charles L. Forgy

Department of Computer Science, Carnegie-Mellon University Pittsburgh, PA 15213, U.S.A.

Recommended by Harry Barrow

ABSTRACT

The Rete Match Algorithm is an efficient method for comparing a large collection of patterns to a large collection of objects. It finds all the objects that match each pattern. The algorithm was developed for use in production system interpreters, and it has been used for systems containing from a few hundred to more than a thousand patterns and objects. This article presents the algorithm in detail. It explains the basic concepts of the algorithm, it describes pattern and object representations that are appropriate for the algorithm, and it describes the operations performed by the pattern matcher.

1. Introduction

In many pattern/many object pattern matching, a collection of patterns is compared to a collection of objects, and all the matches are determined. That is, the pattern matcher finds every object that matches each pattern. This kind of pattern matching is used extensively in Artificial Intelligence programs today. For instance, it is a basic component of production system interpreters. The interpreters use it to determine which productions have satisfied condition parts. Unfortunately, it can be slow when large numbers of patterns or objects are involved. Some systems have been observed to spend more than nine-tenths of their total run time performing this kind of pattern matching [5]. This article describes an algorithm that was designed to make many pattern/many object pattern matching less expensive. The algorithm was developed for use in production system interpreters, but since it should be useful for other languages and systems as well, it is presented in detail.

This article attends to two complementary aspects of efficiency: (1) designing an algorithm for the task and (2) implementing the algorithm on the computer. The rest of Section 1 provides some background information. Section 2 presents the basic concepts of the algorithm. Section 3 explains how the objects and patterns should be represented to allow the most efficient implementations. Section 4 describes in detail a very fast implementation of the algorithm. Finally, Section 5 presents some of the results of the analyses of the algorithm.

*This research was sponsored by the Defense Advanced Research Projects Agency (DOD), ARPA Order No. 3597, monitored by the Air Force Avionics Laboratory under Contract F33615-78-C-1551.

The views and conclusions contained in this document are those of the author and should not be interpreted as representing the official policies, either expressed or implied, of the Defense Advanced Research Projects Agency or the US Government.

1.1 OPS5

The methods described in this article were developed for production system interpreters, and they will be illustrated with examples drawn from production systems. This section provides a brief introduction to the language used in the examples, OPS5. For a more complete description of OPS5, see [6].

A production system program consists of an unordered collection of If-Then statements called *productions*. The data operated on by the productions is held in a global data base called *working memory*. By convention, the If part of a production is called its *LHS* (left-hand side), and its Then part is called its *RHS* (right-hand side). The interpreter executes a production system by performing the following operations.

(1) *Match*. Evaluate the LHSs of the productions to determine which are satisfied given the current contents of working memory.

(2) *Conflict resolution*. Select one production with a satisfied LHS: if no productions have satisfied LHSs, halt the interpreter.

(3) *Act*. Perform the actions in the RHS of the selected production.

(4) Goto 1.

OPS5 working memories typically contain several hundred objects, and each object typically has between ten and one hundred associated attribute-value pairs. An object together with its attribute-value pairs is called a *working memory element*. The following is a typical, though very small, OPS5 working memory element; it indicates that the object of class Expression which is named Expr17 has 2 as its first argument, * as its operator, and X as its second argument.

(Expression ↑Name Expr17 ↑Arg1 2 ↑Op * ↑Arg2 X)

The ↑ is the OPS5 operator that distinguishes attributes from values.

The LHS of a production consists of a sequence of patterns; that is, a sequence of partial descriptions of working memory elements. When a pattern P describes an element E, P is said to *match* E. In some productions, some of the patterns are preceded by the negation symbol, −. An LHS is satisfied when

(1) Every pattern that is not preceded by − matches a working memory element, and

(2) No pattern that is preceded by − matches a working memory element.

The simplest patterns contain only constant symbols and numbers. A pattern containing only constants matches a working memory element if every constant in the pattern occurs in the corresponding position in the working memory element. (Since patterns are partial descriptions, it is not necessary for every constant in the working memory element to occur in the pattern.) Thus the pattern

(Expression ↑Op * ↑Arg2 0)

would match the element

(Expression ↑Name Expr86 ↑Arg1 X ↑Op * ↑Arg2 0)

Many non-constant symbols are available in OPS5 for defining patterns, but the two most important are variables and predicates. A variable is a symbol that begins with the character '⟨' and ends with the character '⟩'—for example ⟨X⟩. A variable in a pattern will match any value in a working memory element, but if a variable occurs more than once in a production's LHS, all occurrences must match the same value. Thus the pattern

```
(Expression ↑Arg1 ⟨VAL⟩ ↑Arg2⟨VAL⟩)
```

would match either of the following

```
(Expression ↑Name Expr9 ↑Arg1 Expr23 ↑Op * ↑Arg2
  Expr23)
(Expression ↑Name Expr5 ↑Arg1 0 ↑Op− ↑Arg2 0)
```

but it would not match

```
(Expression ↑Name Expr8 ↑Arg1 0 ↑Op * ↑Arg2 Expr23)
```

The predicates in OPS5 include = (equal), <>(not equal), < (less than), > (greater than), <= (less than or equal), and >= (greater than or equal). A predicate is placed between an attribute and a value to indicate that the value matched must be related in that way to the value in the pattern. For instance,

```
(Expression ↑Op <>*)
```

will match any expression whose operand is not *. Predicates can be used with variables as well as with constant values. For example, the following pattern

```
(Expression ↑Arg1 ⟨LEFT⟩ ↑Arg2 <>⟨LEFT⟩)
```

will match any expression in which the first argument differs from the second argument.

The RHS of a production consists of an unconditional sequence of actions. The only actions that need to be described here are the ones that change working memory. MAKE builds a new element and adds it to working memory. The argument to MAKE is a pattern like the patterns in LHSs. For example,

```
(MAKE Expression ↑Name Expr1 ↑Arg1 1)
```

will build an expression whose name is Expr1, whose first argument is 1, and whose other attributes all have the value NIL (the default value in OPS5). MODIFY changes one or more values of an existing element. This action takes as arguments a pattern designator and a list of attribute-value pairs. The following action, for example

```
(MODIFY 2 ↑Op NIL ↑Arg2 NIL)
```

would take the expression matching the second pattern and change its operator and second argument to NIL. The action REMOVE deletes elements from working memory. It takes pattern designators as arguments. For example

```
(REMOVE 1 2 3)
```

would delete the elements matching the first three patterns in a production.

An OPS5 production consists of (1) the symbol P, (2) the name of the production, (3) the LHS, (4) the symbol −−>, and (5) the RHS, with everything enclosed in parentheses. The following is a typical production.

```
(P Time 0x
     (Goal ↑Type Simplify ↑Object ⟨X⟩)
     (Expression ↑Name ⟨X⟩ ↑Arg1 0 ↑Op *)
 -->
     (MODIFY 2 ↑Op NIL ↑Arg2 NIL))
```

1.2. Work on production system efficiency

Since execution speed has always been a major issue for production systems, several researchers have worked on the problem of efficiency. The most common approach has been to combine a process called *indexing* with direct interpretation of the LHSs. In the simplest form of indexing, the interpreter begins the match process by extracting one or more features from each working memory element, and uses those features to hash into the collection of productions. This produces a set of productions that might have satisfied LHSs. The interpreter examines each LHS in this set individually to determine whether it is in fact satisfied. A more efficient form of indexing adds memory to the process. A typical scheme involves storing a count with each pattern. The counts are all zero when execution of the system begins. When an element enters working memory, the indexing function is executed with the new element as its only input, and all the patterns that are reached have their counts increased by one. When an element leaves working memory, the index is again executed, and the patterns that are reached have their counts decreased by one. The interpreter performs the direct interpretation step only on those LHSs that have non-zero counts for all their patterns. Interpreters using this scheme—in some cases combined with other efficiency measures— have been described by McCracken [8], McDermott, Newell, and Moore [9], and Rychener [10].

The algorithm that will be presented here, the Rete Match Algorithm, can be described as an indexing scheme that does not require the interpretive step. The indexing function is represented as a network of simple feature recognizers. This representation is related to the graph representations for so-called structured patterns. (See for example [2] and [7]). The Rete algorithm was first described in 1974 [3]. A 1977 paper [4] described some rather complex interpreters for the networks of feature recognizers, including parallel interpreters and interpreters which delayed evaluation of patterns as long as possible. (Delaying evaluation is useful because it makes it less likely that patterns will be evaluated unnecessarily.) A 1979 paper [5] discussed simple but very fast interpreters for the networks. This article is based in large part on the 1979 paper.

2. The Rete Match Algorithm—Basic Concepts

In a production system interpreter, the output of the match process and the input to conflict resolution is a set called the *conflict set*. The conflict set is a collection of ordered pairs of the form

⟨Production. List of elements matched by its LHS⟩

The ordered pairs are called *instantiations*. The Rete Match Algorithm is an algorithm for computing the conflict set. That is, it is an algorithm, to compare a set of LHSs to a set of elements in order to discover all the instantiations. The algorithm can efficiently process large sets because it does not iterate over the sets.

2.1. How to avoid iterating over working memory

A pattern matcher can avoid iterating over the elements in working memory by storing information between cycles. The step that can require iteration is determining whether a given pattern matches any of the working memory elements. The simplest interpreters determine this by comparing the pattern to the elements one by one. The iteration can be avoided by storing, with each pattern, a list of the elements that it matches. The lists are updated when working memory changes. When an element enters working memory, the interpreter finds all the patterns that match it and adds it to their lists. When an element leaves working memory, the interpreter again finds all the patterns that match it and deletes it from their lists.

Since pattern matchers using the Rete algorithm save this kind of information, they never have to examine working memory. The pattern matcher can be viewed as a black box with one input and one output.

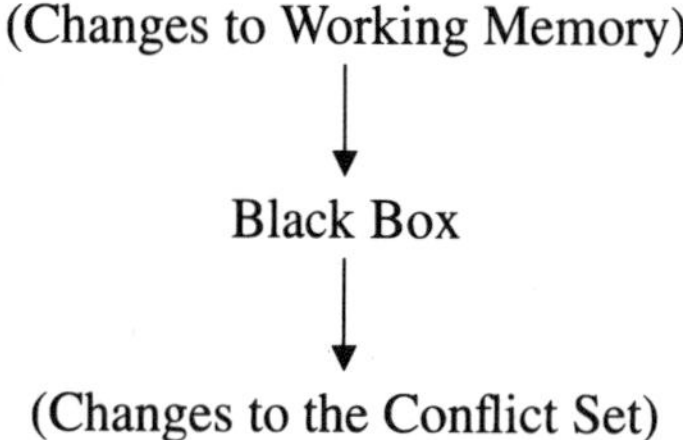

The box receives information about the changes that are made to working memory, and it determines the changes that must be made in the conflict set to keep it consistent. For example, the black box might be told that the element

(Goal ↑Type Simplify ↑Object Expr19)

has been added to working memory, and it might respond that production TimexN has just become instantiated.

2.1.1. *Tokens*

The descriptions of working memory changes that are passed into the black box are called *tokens*. A token is an ordered pair of a *tag* and a list of data elements. In the simplest implementations of the Rete Match Algorithm, only two tags are needed, + and −. The tag + indicates that something has been added to working memory. The tag − indicates that something has been deleted from working memory. When an element is modified, two tokens are sent to the black box; one token indicates that the old form of the element has been deleted from working memory, and the other that the new form of the element has been added. For example, if

(Expression ↑Name Expr41 ↑Arg1 Y ↑Op + ↑Arg2 Y)

was changed to

(Expression ↑Name Expr41 ↑Arg1 2 ↑Op * ↑Arg2 Y)

the following two tokens would be processed.

⟨−(Expression ↑Name Expr41 ↑Arg1 Y ↑Op + ↑Arg2 Y)⟩
⟨+(Expression ↑Name Expr41 ↑Arg1 2 ↑Op * ↑Arg2 Y)⟩

2.2. How to avoid iterating over production memory

The Rete algorithm avoids iterating over the set of productions by using a tree-structured sorting network or index for the productions. The network, which is compiled from the patterns, is the principal component of the black box. The following sections explain how patterns are compiled into networks and how the networks perform the functions of the black box.

2.2.1. *Compiling the patterns*

When a pattern matcher processes a working memory element, it tests many features of the element. The features can be divided into two classes. The first class, which could be called the intra-element features, are the ones that involve only one working memory element. For an example of these features, consider the following pattern.

(Expression ↑Name ⟨N⟩ ↑Arg1 0 ↑Op + ↑Arg2 ⟨X⟩)

When the pattern matcher processes this pattern, it tries to find working memory elements having the following intra-element features.

– The class of the element must be Expression.
– The value of the Arg1 attribute must be the number 0.
– The value of the Op attribute must be the atom +.

The other class of features, the inter-element features, results from having a variable occur in more than one pattern. Consider Plus0x's LHS.

```
(P Plus0x
     (Goal  ↑Type Simplify  ↑Object ⟨N⟩)
     (Expression  ↑Name ⟨N⟩  ↑Arg1 0  ↑Op +  ↑Arg2 ⟨X⟩)
-->. . .)
```

The intra-element features for the second pattern are listed above. A similar list can be constructed for the first pattern. But in addition to those two lists, the following inter-element feature is necessary because the variable ⟨N⟩ occurs twice.

– The value of the Object attribute of the goal must be equal to the value of the Name attribute of the expression.

The pattern compiler builds a network by linking together nodes which test elements for these features. When the compiler processes an LHS, it begins with the intra-element features. It determines the intra-element features that each pattern requires and builds a linear sequence of nodes for the pattern. Each node tests for the presence of one feature. After the compiler finishes with the intra-element features, it builds nodes to test for the inter-element features. Each of the nodes has two inputs so that it can join two paths in the network into one. The first of the two-input nodes joins the linear sequences for the first two patterns, the second two-input nodes joins the output of the first with the sequence for the third pattern, and so on. The two-input nodes test every inter-element feature that applies to the elements they process. Finally, after the two-input nodes, the compiler builds a special terminal node to represent the production. This node is attached to the last of the two-input nodes. Fig. 1 shows the network for Plus0x and the similar production Time0x. Note that when two LHSs require identical nodes, the compiler shares parts of the network rather than building duplicate nodes.

2.2.2. *Processing in the network*

The root node of the network (at the top in Fig. 1) is the input to the black box. This node receives the tokens that are sent to the black box and passes copies of the tokens to all its successors. The successors of the top node, the nodes to perform the intra-element tests, have one input and one or more outputs. Each node tests one feature and sends the tokens that pass the test to its successors. The two-input nodes compare tokens from different paths and join them into bigger tokens if they satisfy the inter-element constraints of the LHS. Because of the tests performed by the other nodes, a terminal node will receive only tokens that instantiate the LHS. The terminal node sends out of the black box the informaton that the conflict set must be changed.

For an example of the operation of the nodes, consider what happens in the network in Fig. 1 when the following two elements are put into an empty working memory.

```
(Goal  ↑Type Simplify  ↑Object Expr17)
(Expression  ↑Name Expr17  ↑Arg1 0  ↑Op *  ↑Arg2 X)
```

First the token

```
⟨+(Goal  ↑Type Simplify  ↑Object Expr17)⟩
```

is created and sent to the root of the network. This node sends the token to its successors. One of the successors (on the right in Fig. 1) tests it and rejects it because its class is not Expression. This node does not pass the token to its successor. The other successor of the top node accepts the token (because its class is Goal) and so sends it to its successor. That node also accepts the token (since its type is Simplify), and it sends the token to its successors, the two-input nodes. Since no other tokens have arrived at the two-input nodes, they can perform no tests; they must just store the token and wait.

When the token

⟨+(Expression ↑Name Expr17 ↑Arg1 0 ↑Op * ↑Arg2 X)⟩

is processed, it is tested by the one-input nodes and passed down to the right input of Time0x's two-input node. This node compares the new token to the earlier one, and finding that they allow the variable to be bound consistently, it creates and sends out the token

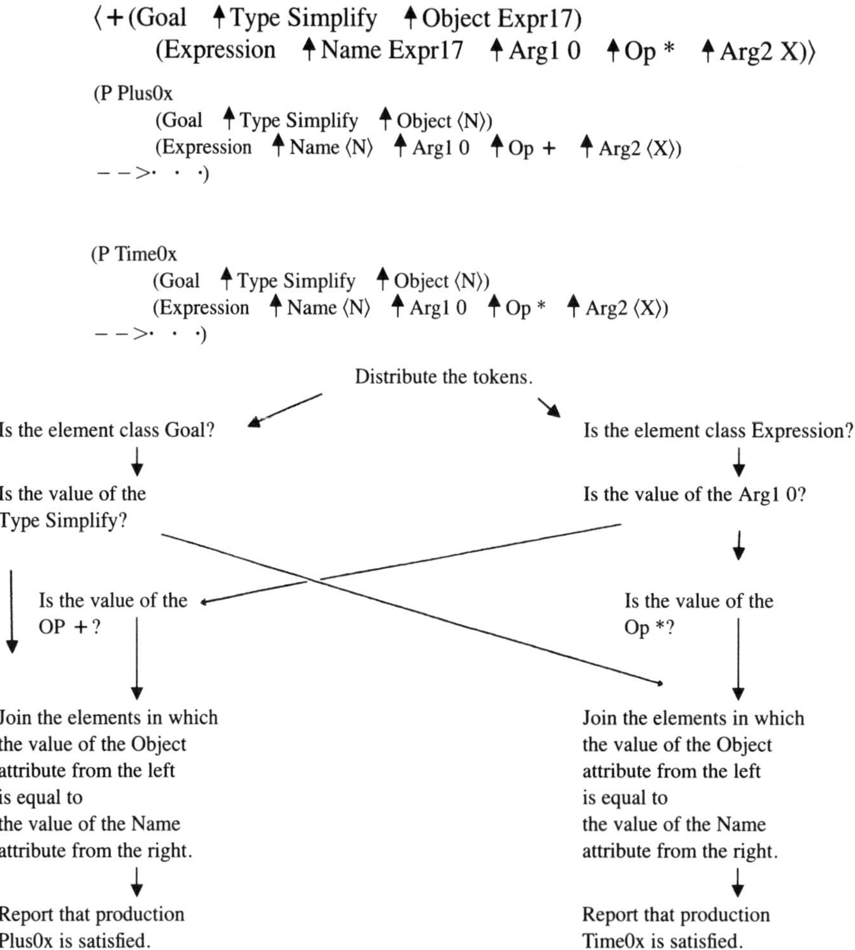

FIG. 1. The network for Plus0x and Time0x.

When its successor, the terminal node for Time0x, receives this token, it adds the instantiation of Time0x to the conflict set.

2.2.3. *Saving information in the network*

As explained above, the black box must maintain state information because it must know what is in working memory. In simple Rete networks all such state is stored by the two-input nodes. Each two-input node contains two lists called its left and right memories. The left memory holds copies of the tokens that arrived at its left input, and the right memory holds copies of the tokens that arrived at its right input. The tokens are stored as long as they are useful. The next section explains how the nodes determine when the tokens are no longer useful.

2.2.4 *Using the tags*

The tag in a token indicates how the state information is to be changed when the token is processed. The + and − tokens are processed identically except:
- The terminal nodes use the tags to determine whether to add an instantiation to the conflict set or to remove an existing instantiation. When a + token is processed, an instantiation is added; when a − token is processed, an instantiation is removed.
- The two-input nodes use the tags to determine how to modify their internal memories. When a + token is processed, it is stored in the internal memory; when a − token is processed, a token with an identical data part is deleted.
- The two-input nodes use the tags to determine the appropriate tags for the tokens they build. When a new output is created, it is given the tag of the token that just arrived at the two-input node.

2.3. Completing the set of node types

The network in Fig. 1 contained four kinds of nodes: the root node, the terminal nodes, the one-input nodes, and the two-input nodes. Certainly one could define many more kinds of nodes, but only a few more are necessary to have a complete and useful set. In fact, only two more kinds of nodes are necessary to interpret OPS5.

A second kind of two-input node is needed for negated patterns (that is, patterns preceded by −). The new two-input node stores a count with each token in its left memory. The count indicates the number of tokens in the right memory that allow consistent variable bindings. The tokens in its right memory contain the elements that match the negated pattern—or, more precisely, the tokens contain the elements that have the intra-element features that the negated pattern requires. The node allows the tokens with a count of zero to pass.

The last node type that needs to be defined is a variant of the one-input nodes described earlier. Those nodes tested working memory elements for constant features (testing, for example, whether a value was equal to a given atomic symbol). The new one-input nodes compare two values from a working memory element. These nodes are used to process patterns that contain two or more occurrences of a variable. The following, for example, would require one of these nodes because ⟨X⟩ occurs twice.

(Expression ↑Arg1⟨X⟩ ↑Op + ↑Arg2⟨X⟩)

3. Representing the Network and the Tokens

This section describes representations for tokens and nodes that allow very fast interpreters to be written.

3.1 Working memory elements

The representation chosen for the working memory elements should have two properties.
- The representation should make it easy to extract values from elements because every test involves extracting one or more values.

– The representation should make it easy to perform the tests once the values are available.

To make extracting the values easy, each element should be stored in a contiguous block in memory, and each attribute should have a designated index in the block. For example, if elements of class Ck had seventeen attributes, A1 through A17, they should be stored as blocks of eighteen values. The first value would be the class name (Ck). The second value would be the value of attribute A1. The third would be the value of attribute A2, and so on. The particular assignment of indices to attributes is unimportant; it is important only that each attribute have a fixed index, and that the indices be assigned at compile time. This allows the compiler to build the indices into the nodes. Thus instead of a node like the following:

Is the value of the Status attribute Pending?

the compiler could build the node

Is the value at location 8 Pending?

With this representation, each value can be accessed in one memory reference, regardless of the number of attributes possessed by an element.

To make the tests inexpensive, the representation should have explicit type bits. One obvious way to represent a value is to use one word for the type and one or more words for the value proper. But more space-efficient representations are also possible. For example, consider a production system language that supports three data types, integers, floating point numbers, and atoms. A representation like the following might be used: One word would be allocated to each value. For integers and atoms, the low order sixteen (say) bits would hold the datum and the seventeenth bit would be a type bit. For floating point numbers, the entire word would be used to store a normalized floating point number. A floating point number would be recognized by having at least one non-zero in the high order bits.

3.2. The network

This section explains how to represent nodes in a form similar to von Neumann machine instructions. This representation was chosen because it allows the network interpreter to be organized like the interpreters for conventional von Neumann architectures.

3.2.1. *An assembly language notation*

To make it easier to discuss the representation for the nodes, an assembly language notaton is used below. A one-input node like

Is the value of locating 8 Pending?

becomes

TEQA 8, Pending

The T, which stands for test, indicates that this is a one-input node. The EQ indicates that it is a test for equality. (It is also necessary to have NE for not equals, LT for less than, etc.). The A indicates the node tests data of type atom. (There is also a type N for integer values, a type F for floating point, and a type S for comparing two values in the same working memory element). Two-input nodes are indicated by lines like the following.

L001 AND (2) = (1)

L001 is a label. AND indicates that this is a two-input node for non-negated patterns. The sequence (2) = (1) indicates that the node compares the second value of elements from the left and the first value of elements from the right: the = indicates that it performs a test for equality. The terminal nodes contain the type TERM and the name of the production. For example

TERM Plus0x

As will be explained below, the ROOT node is not needed in this representation.

3.2.2. *Linearizing the network*

To make the nodes like the instructions for a von Neumann machine, it is necessary to eliminate the explicit links between nodes. Many of the explicit links can be eliminated simply by linearizing the network, placing a node and its successor in contiguous memory locations. However, since some nodes have more than one successor, and others (the two-input nodes) have more than one predecessor, linearizing is not sufficient in itself: two new node types must be defined to replace some of the links. The first of the new nodes, the FORK, is used to indicate that a node has more than one successor. The FORK node contains the address of one of the successors. The other successor is placed immediately after the FORK. For example, the FORK in the following indicates that the node L003 has two successors.

```
L003  TEQA 0, Expression
      FORK L004
      TEQA 3, +
      . . .
L004  TEQA 3, *
```

The other new node type, the MERGE, is used where the network has to grow back together—that is, before two-input nodes. The two-input node is placed after one of its predecessors (say its left predecessor) and the MERGE is placed after the other. The MERGE, which contains the address of the two-input node, functions much like an unconditional jump. Fig. 2 shows the effect of the linearization process; it contains the productions from Fig. 1 and the linearized network for their LHSs.

3.2.3. *Representing the nodes in memory*

This section shows how the nodes could be represented on a computer which has a thirty-two bit word length. The thirty-two bit word length was chosen because it is typical of today's computers; the precise word length is not critical, however. Since the network can be rooted at a FORK (see the example in Fig. 2) it is not necessary to have an explicit root node for the network. Hence only seven classes of nodes are needed; FORKs, MERGEs, the two kinds of one-input nodes, the two kinds of two-input nodes, and the terminal nodes.

FORKs and MERGEs could be represented as single words. Six bits could be used for a type field (that is, a field to indicate what the word represents) and the remaining twenty-six bits could be used for the address of the node pointed to. FORKs and MERGEs would thus be represented:

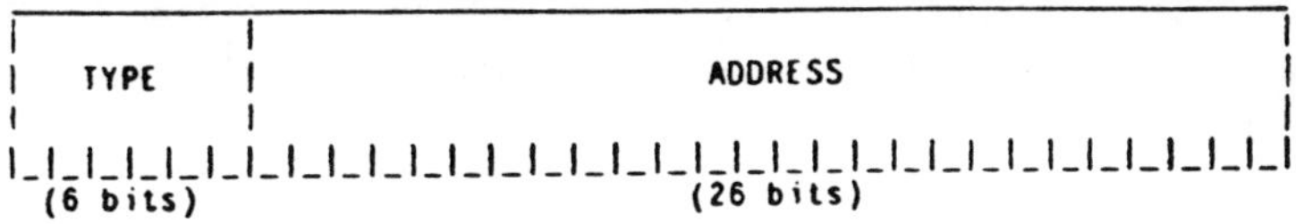

Both kinds of one-input nodes could be represented as single words that are divided into three fields. The first field would hold the type of the node. The second

field would hold the index of the value to test. The third field would hold either a constant or a second index. The bits in a word could be allocated as follows.

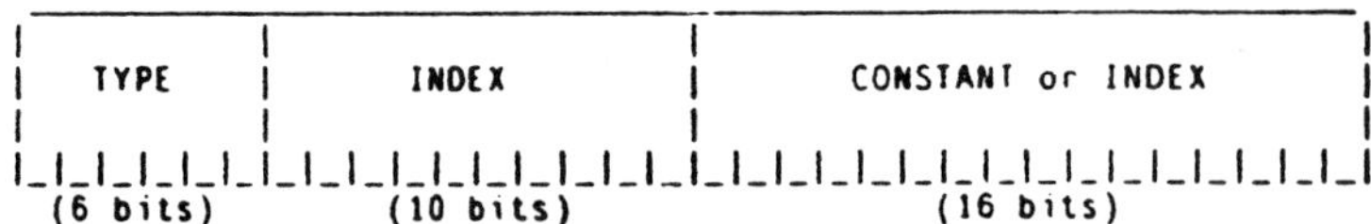

A sixteen-bit field is required to represent an integer or an atom using the format of Section 3.1. Since a floating point number cannot be represented in sixteen bits, in nodes that test floating point numbers, this field would hold not the number, but the address of the number.

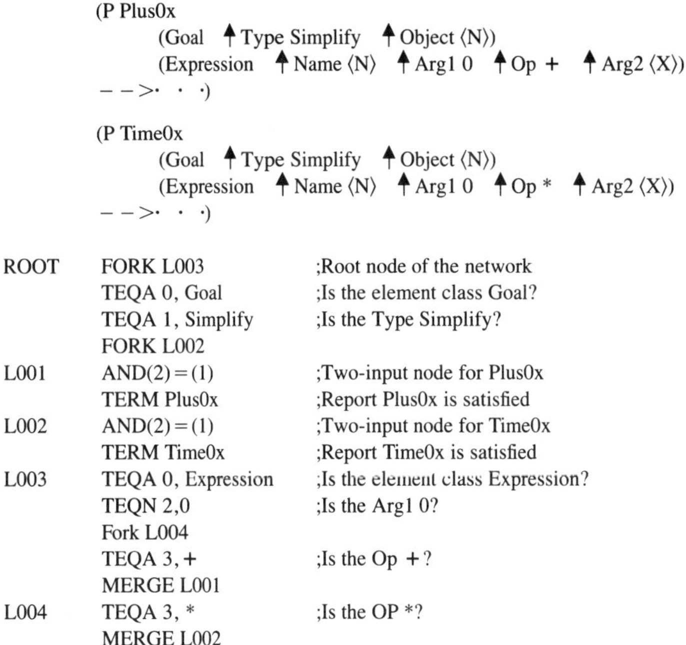

```
(P PlusOx
      (Goal  ↑Type Simplify  ↑Object ⟨N⟩)
      (Expression  ↑Name ⟨N⟩  ↑Arg1 0  ↑Op +  ↑Arg2 ⟨X⟩)
-->· · ·)

(P TimeOx
      (Goal  ↑Type Simplify  ↑Object ⟨N⟩)
      (Expression  ↑Name ⟨N⟩  ↑Arg1 0  ↑Op *  ↑Arg2 ⟨X⟩)
-->· · ·)

ROOT   FORK L003              ;Root node of the network
       TEQA 0, Goal           ;Is the element class Goal?
       TEQA 1, Simplify       ;Is the Type Simplify?
       FORK L002
L001   AND(2)=(1)             ;Two-input node for PlusOx
       TERM PlusOx            ;Report PlusOx is satisfied
L002   AND(2)=(1)             ;Two-input node for TimeOx
       TERM TimeOx            ;Report TimeOx is satisfied
L003   TEQA 0, Expression     ;Is the element class Expression?
       TEQN 2,0               ;Is the Arg1 0?
       Fork L004
       TEQA 3, +              ;Is the Op +?
       MERGE L001
L004   TEQA 3, *              ;Is the OP *?
       MERGE L002
```

FIG. 2. A compiled network.

The terminal nodes could also be stored in single words. These nodes contain two fields, the usual type field plus a longer field for the index or address of the production that the node represents.

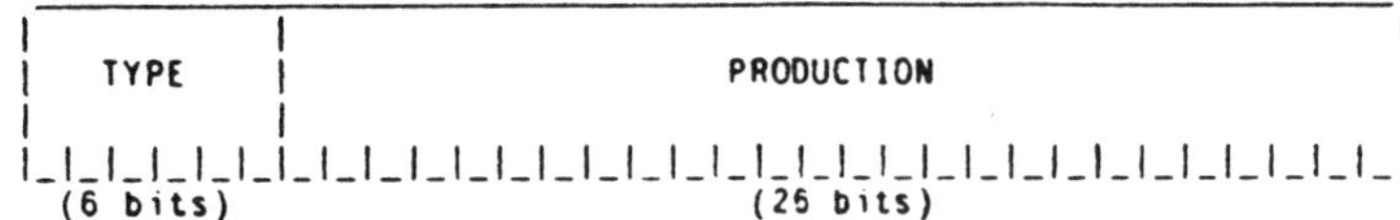

The length of a two-input node would depend on the number of value pairs tested by the node. Each node could have one word of basic information plus one word for each value pair. The first word would contain a type field, a pointer to the memory for the left input, a pointer to the memory for the right input, and a field indicating how many tests are performed by the node. The bits in the word could be allocated as follows.

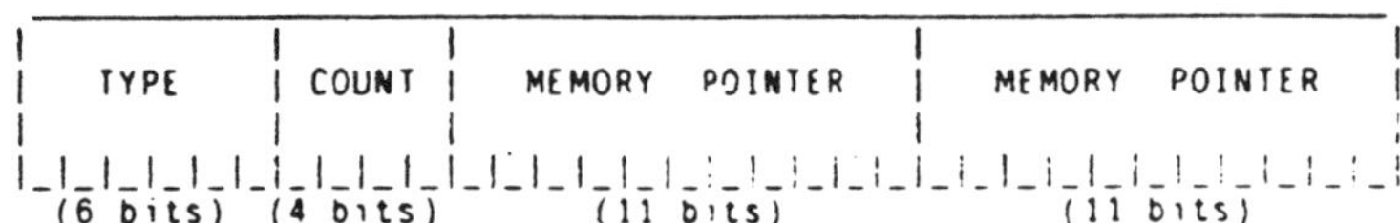

The word for each test would contain three fields. Two fields would hold the indices of the two elements to test. The remaining field would indicate the test to perform; that is, it would indicate whether the node is to test for equality of the two elements, for inequality, or for something else. The bits in the word might be allocated as follows.

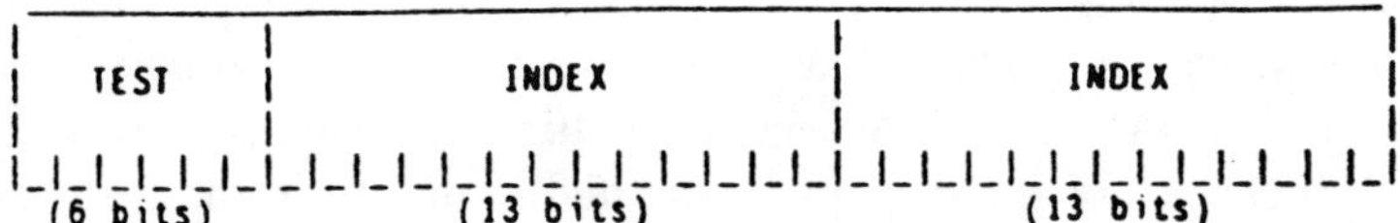

Note that the index fields here are longer than the index fields in the one-input nodes. This is necessary because the indices in the two-input nodes must designate elements in the tokens as well as values in the elements.

3.3. The tokens

This section describes a space-efficient representation for tokens. This representation is not suitable for all interpreters; it requires the interpreter to process only one working memory change at a time, and it requires that certain parts of the network be traversed depth first. Fortunately, these are not serious restrictions. The simplest way to perform the match is process one token at a time, traversing the entire network depth first. Section 4 describes an interpreter that operates in this manner.

If the interpreter operates this way, then it can use a stack to represent its tokens. When a token has to be built, first the tag for the token is pushed onto the stack, and then the working memory elements are pushed onto the stack in order. When tokens have to be extended (a very common operation—see the code in Section 4) the additional working memory elements are just pushed onto the stack.

The one-input nodes will be more efficient if they do not use this stack. Since all the one-input nodes will process the same working memory element—the element that was just added to or deleted from working memory—the element should be made easily available. The element could be copied into a dedicated location in memory, or the address of the element could be loaded into a dedicated base register. Either of these would make it possible for the one-input nodes to access the element without going through the stack.

3.4. The interpreter's state

In addition to the stack for tokens, the interpreter must maintain another stack for its state information. One reason for the stack is to allow the interpreter to find its way about in the network. When the interpreter passes a FORK, it pushes the pointer it does not follow onto the stack. Then when it reaches the end of a path it pops a pointer from the stack and follows it. Another reason for the stack is to provide a place for the two-input nodes to keep their local information. As will be seen in the next section, the two-input nodes sometimes have to suspend themselves while their successors are processed. The stack holds the information that is needed to resume processing the two-input nodes.

4. The Network Interpreter

This section provides a concrete description of the operations performed by the network interpreter. One node from each class has been selected, and the code to interpret the nodes has been written. It might be noted that since the code sequences are short and simple, they could easily be written in microcode.

The code is written in a PASCAL-like language which has literal labels and field extraction. Field extraction is indicated by putting two numbers within angle brackets; the first number is the index of the high order bit in the field, and the second number is the index of the low order bit. The assumption will be made that the bits are numbered from right to left, with the low order bit being bit zero. Thus the expression SELF⟨31:26⟩ indicates that the high order six bits of the value of the variable SELF are to be extracted and right justified.

The main loop of the interpreter is very simple: the interpreter fetches the next node from memory and dispatches on its type field. Let the segment of memory that holds the nodes be called NODE_MEMORY and let the pointer to the current node be called NC. The main loop is then:

```
MAIN:  SELF: = NODE_MEMORY[NC]:
       CASE SELF⟨31:26⟩ OF          !Type field is high order 6 bits
         0: GOTO FORK;
         1: GOTO MERGE;
         2: GOTO TERM;
         3: GOTO TEQA;
         . . .
         END;
```

The node is copied into the variable SELF so that the node programs can examine it. The assignment of numbers to the various node types is arbitrary. Goto's are used instead of procedure calls because these examples make all the state of the interpreter explicit, and not hidden in PASCAL's stack.

TEQA is typical of the one-input nodes for testing constants. If the segment of memory that holds the working memory element being processed is called CURRENT, then TEQA is as follows.

```
TEQA: TEMP: = CURRENT[SELF⟨25:16⟩]:      !Get the word pointed to
                                         !by the index field
      IF (TEMP⟨31:16⟩ = 0) AND           !Test type bits
         (TEMP⟨15:0⟩ = SELF⟨15:0⟩)       !Test value
      THEN GOTO SUCC
      ELSE GOTO FAIL:
```

Either SUCC or FAIL is executed after each one-input node. SUCC is executed when the test succeeds, and FAIL is executed when the test fails. SUCC increments the node counter to point to the next node.

```
SUCC:  NC: = NC + 1;
       GOTO MAIN;
```

FAIL tries to get a node from the stack of unprocessed nodes; if it cannot, it halts the match. Assuming the stack is named NS and the pointer to the top of the stack is called NSTOP, the code is:

```
FAIL:   IF NSTOP<0 THEN GOTO EXIT_MATCH;
        NC: = NS[NSTOP];
        NSTOP: = NSTOP - 1;
        GOTO MAIN;
```

The one-input nodes for comparing pairs of values are similar to the other one-input nodes. TEQS is typical of these nodes.

```
TEQS:   IF CURRENT[SELF⟨25:16⟩] = CURRENT[SELF⟨9:0⟩]
        THEN GOTO SUCC
        ELSE GOTO FAIL;
```

FORK pushes an address onto NS and then passes control to the following node.

```
FORK:   NSTOP: = NSTOP + 1;
        NS[NSTOP]: = SELF⟨25:0⟩;
        GOTO SUCC;
```

A two-input node must be able to determine whether it was reached over its left input or its right input. This can be indicated to the node by a global variable which usually has the value LEFT, but which is temporarily set to RIGHT when a MERGE passes control to a two-input node. If this global variable is called DIRECTION, the code for the MERGE is

```
MERGE:  DIRECTION: = RIGHT;
        NC: = SELF⟨25:0⟩;
        GOTO MAIN;
```

The two kinds of two-input nodes are very similar, so only AND is shown here. In order not to obscure the more important information, some details of the program are omitted. The code does not show how the variables are tested, nor does it show how tokens are added to and removed from the node's memories. Assuming the token stack is called TS and the pointer to the top element is called TSTOP, the program is as follows.

```
!   Control can reach this point many times during the processing
!   of a token. The node needs to update its state and put
!   information on NS only once, however.
AND:IF NS [NSTOP]<>NC                    ! If the state is not in NS
        THEN                             ! Then put it there
          BEGIN
          NSTOP: = NSTOP + 4;
          NS[NSTOP]: = NC;
          NS[NSTOP - 1]: = DIRECTION;
          NS[NSTOP - 2]: = MEMORY__CONTENTS
                              (OPPOSITE(DIRECTION));
          NS[NSTOP - 3]: = TSTOP;
        MODIFY_MEMORY(DIRECTION);        ! Store the token
        DIRECTION: = LEFT;               ! Reset to the default
        END;
```

```
!
!   Go process the tokens
!
        IF NS[NSTOP] - 1] = RIGHT THEN GOTO RLOOP
          ELSE GOTO LLOOP;
!   Compare the token to the elements in the right memory
!
LLOOP:  REPEAT
          TEMP: = NEXT_POSITION(NS[NSTOP - 2]);
          IF TEMP = NIL                    ! If right memory is empty
          THEN                             ! Then clean up and exit
            BEGIN
          TSTOP: = NS[NSTOP - 3];
          NSTOP: = NSTOP - 4;
          GOTO FAIL;
          END
        UNTIL PERFORM_AND_TEST(TEMP.LEFT);
!
!   Fall out of the loop when the test succeeds so that
!   the successors of this node can be activated
!
        ! Extend the token
        TSTOP: = NS[NSTOP - 3] + 1; TS[NSTOP]: = TEMP;
        ! Prepare NS so that control will return to this node
        NSTOP: = NSTOP + 1;NS[NSTOP]: = NC;
        ! Pass control to the successors of this node
        NC: = NC + SELF⟨25:22⟩ + 1;GOTO MAIN;
!
!   Compare the token the elements in the left memory
!
RLOOP:
```

This is similar to LLOOP.

The only remaining node type is the TERM node. Since updating the conflict set is a language-dependent operation, that detail of the TERM node cannot be shown. The rest of the processing of the node is as follows.

```
TERM:   UPDATE_CONFLICT_SET(SELF⟨25 :0⟩);
        GOTO FAIL;
```

5. Performance of the Algorithm

Extensive studies have been made of the efficiency of the Rete Match Algorithm. Both analytical studies (which determined the time and space complexity of the algorithm) and empirical studies have been made. This section presents some of the results of the analytical studies. Because of space constraints, it was not possible to present the empirical results or the proofs of

TABLE 1. Space and time complexity

Complexity measure	Best case	Worst case
Effect of working memory size on number of tokens	O(1)	$O(W^C)$
Effect of production memory size on number of nodes	$O(P)$	$O(P)$
Effect of production memory size on number of tokens	O(1)	$O(P)$
Effect of working memory size on time for one firing	O(1)	$O(W^{2C-1})$
Effect of production memory size on time for one firing	$O(\log_2 P)$	$O(P)$

C is the number of patterns in a production.
P is the number of productions in production memory.
W is the number of elements in working memory.

the analytical results. The proofs and detailed results of some empirical studies can be found in [5].

Table 1 summarizes the results of the analytical studies of the algorithm. The usual notation for asymptotic complexity is used in this table [1]. Writing that a cost is $O(f(x))$ indicates that the cost varies as $f(x)$ plus perhaps some smaller terms in x. The smaller terms are ignored because the $f(x)$ term will dominate when x is large. Writing that a cost is O(1) indicates that the cost is unaffected by the factor being considered. It should be noted that all the complexity results in Table 1 sharp; production systems achieving the bounds are described in [5].

6. Conclusions

The Rete Match Algorithm is a method for comparing a set of patterns to a set of objects in order to determine all the possible matches. It was described in detail in this article because enough evidence has been accumulated since its development in 1974 to make it clear that it is an efficient algorithm which has many possible applications.

The algorithm is efficient even when it processes large sets of patterns and objects, because it does not iterate over the sets. In this algorithm, the patterns are compiled into a program to perform the match process. The program does not have to iterate over the patterns because it contains a tree-structured sorting network or index for the patterns. It does not have to iterate over the data because it maintains state information: the program computes the matches and partial matches for each object when it enters the data memory, and it stores the information as long as the object remains in the memory.

Although the Rete algorithm was developed for use in production system interpreters, it can be used for other purposes as well. If there is anything unusual about the pattern matching of production systems, it is only that the pattern matching takes place on an unusually large scale. Production systems contain rather ordinary patterns and data objects, but they contain large numbers of them, and invocations of the pattern matcher occur very frequently during execution. If programs of other kinds begin to use pattern matching more heavily, they could have the same efficiency problems as production systems, and it could be necessary to use methods like the Rete Match Algorithm in their interpreters as well. Certainly the algorithm should not be used for all match problems; its use is indicated only if the following three conditions are satisfied.

- The patterns must be compilable. It must be possible to examine them and determine a list of features like the lists in Section 2.2.1.
- The object must be constant. They cannot contain variables or other non-constants as patterns can.
- The set of objects must change relatively slowly. Since the algorithm maintains state between cycles, it is inefficient in situations where most of the data changes on each cycle.

ACKNOWLEDGMENT

The author would like to thank Allen Newell and Robert Sproull for many useful discussions concerning this work, and Allen Newell, John McDermott, and Michael Rychener for their valuable comments on earlier versions of this article.

REFERENCES

1. Aho, A.V., Hopcroft, J.E., and Ullman, J.D., *The Design and Analysis of Computer Algorithms* (Addison-Wesley, Reading, MA, 1974).
2. Cohen, B.L., A powerful and efficient structural pattern recognition system. *Artificial Intelligence* 9 (1977) 223–255.
3. Forgy, C.L., A network match routine for production systems. Working Paper, 1974.
4. Forgy, C.L., A production system monitor for parallel computers. Department of Computer Science, Carnegie-Mellon University, 1977.
5. Forgy, C.L., On the efficient implementation of production systems. Ph.D. Thesis, Carnegie-Mellon University, 1979.
6. Forgy, C.L., OPS5 user's manual. Department of Computer Science, Carnegie-Mellon University, 1981.
7. Hayes-Roth, F. and Mostow, D.J., An automatically compliable recognition network for structured patterns, *Proc. Fourth Internat. Joint Conference on Artificial Intelligence* (1975) 246-251.
8. McCracken, D., A production system version of the Hearsay-II speech understanding system, Ph.D. Thesis, Carnegie-Mellon University, 1978.
9. McDermott, J., Newell, A., and Moore, J., The efficiency of certain production system implementations, in: Waterman, D.A. and Hayes-Roth, F. (Eds.). *Pattern-Directed Inference Systems* (Academic Press, New York, 1978) 155–176.
10. Rychener, M.D., Production systems as a programming language for Artificial Intelligence applications. Ph.D. Thesis, Carnegie-Mellon University, 1976.

Received May 1980: revised version received April 1981

A Formal Logic of Plans in Temporally Rich Domains

RICHARD PELAVIN AND JAMES F. ALLEN

This paper outlines a temporal logic extended with two modalities that can be used to support planning in temporally rich domains. In particular, the logic can represent planning environments that have assertions about future possibilities in addition to the present state, and plans that contain concurrent actions. The logic is particularly expressive in the ways that concurrent actions can interact with each other and allows situations where either one of the actions can be executed, but both cannot, as well as situations where neither action can be executed alone, but they can be done together. Two modalities are introduced and given a formal semantics: INEV expresses simple temporal possibility, and IFTRIED expresses counterfactual-like statements about actions.

I. Introduction

This paper presents a formal logic that provides a foundation for a theory of plans in temporally rich domains. Such domains may include actions that take time, concurrent actions, and the simultaneous occurrence of many actions at once. This also includes domains with external events, i.e., actions by other agents and natural forces, that the planner may need to interact with in order to prevent some event, insure the successful completion of some event, or to perform some action enabled by the event.

In general, a planning problem can be specified by giving a description of desired future conditions (the goal), a partial description of the scenario in which the goal is to be achieved (which we will call ***the planning environment***), and for each action that the planning agent can execute, a specification of its effects and the conditions under which it can be executed (which we will call ***the action specifications***). The planner must find a collection of temporally related actions (the plan) that can be executed and if executed would achieve the goal in any scenario described by the planning environment. For example, consider a goal to go to the bank before it closes at 3:00 without getting wet even though it is going to rain. The planning environment might be the following: at 2:30, the time of planning, the agent is at home, an umbrella is in the house, it will start to rain before 3:00, and the agent will get wet if it is outside without an umbrella while it is raining. The actions that the agent can perform might include "walking from home to the bank" which takes 10 min and "taking an umbrella" which can be done as long as, prior to execution, there is an umbrella at the same location as the agent.

Our formal language must be able to express sentences describing the planning environment, the goal, and the action specifications. We will see that this logic diverges from traditional approaches because we are considering planning problems where the world may be affected by events other than the agent's actions. Such a logic must allow us to represent statements about external events that will occur during plan execution and statements describing the interaction between a plan and the external world in which it is being executed. In addition, we are considering plans with concurrent actions and, therefore, our logic must be able to represent concurrent actions and their interactions, such as resource conflicts. By treating concurrency, we will be able to reason about a robot agent that has multiple effector devices that can be operating simultaneously. Furthermore, we can treat plans to be jointly executed by a group of cooperating agents that are simultaneously performing their own part of the plan.

Situation Calculus and the State-Based Planning Paradigm

One of the most successful approaches to representing events and their effects in Artificial Intelligence has been situation calculus [12]. In this logic an event is represented by a function that takes a situation, i.e., an instantaneous snapshot of the world, and returns the situation that would result from applying the event to its argument. Events can be combined to form sequences. The result function for the sequence $e_1; e_2; \cdots ; e_n$ is recursively defined as the result of executing $e_2; \cdots ; e_n$ in the situation that results from applying e_1 to the initial situation.

Situation calculus has given rise to the state-based planning paradigm which has the following form: Given a set of sentences describing conditions that are initially true and a set of sentences describing goal conditions to be achieved, a sequence of actions must be found that when applied to any situation where the initial conditions hold yields a situation where the goal conditions hold. This approach is limited since the description of the planning environment con-

Manuscript received May 1, 1985; revised April 1, 1986. This research was supported in part by the Rome Air Development Center under Grant SU 353-9023-6 and in part by the National Science Foundation under Grant DCR-8502481.

The authors are with the Department of Computer Science, The University of Rochester, Rochester, NY 14627, USA.

Reprinted from *Proceedings of the IEEE*, Volume 74, Number 10, October 1986, pages 1364-1382.

sists only of statements about the initial situation, i.e., the situation that holds just prior to plan execution. As a result, this framework is adequate only for planning problems where all changes in the world result from the planning agent's actions. Conditions that are true in the initial situation will remain true until the agent performs an action that negates it, and thus the future is uniquely determined by the initial situation and the sequence of actions performed starting from this situation.

This simple type of planning environment is not suitable for planning problems where the world may be affected by external events, as well as by the planning agent's actions. Examples of conditions that we want to handle that cannot be represented in the simple state-based model include:

- The bank is going to open at 9:00 and will close at 3:00.
- It will possibly start raining any time between 3:00 and 4:00.
- If the agent is outside without an umbrella while it is raining, the agent will get wet.
- It is possible that the can of paint sitting in the doorway will be knocked over if the agent does nothing about it.

The type of goals considered in state-based planning are also very limited. Goals are just conditions that must hold at the completion of plan execution. In this research, goals may be any temporally qualified propositions describing conditions in the future of planning time. Thus goals might involve avoiding some condition while performing some task, preventing an undesirable condition that possibly will happen, or the achievement of a collection of goals to be done in some specified order. Examples of these types of goals are:

- Do not damage the tape heads while repairing the tape deck.
- Preventing Tom from entering the room.
- Get to the gas station on the way to driving to school (ordered goals).

More precisely, goals are temporally qualified propositions that partition the set of possible futures into the set of possible futures where the goal holds versus the set of possible futures where the goal does not hold. This rules out goals such as "finding the best way to achieve . . ." which presupposes some utility measure or precedence ordering relating the possible futures. Our notion of goals are just black and white. Either a future is good, i.e., the goal holds in it, or it is bad, i.e., the goal does not hold in it.

Finally, in the state-based approach, the type of plans that can be handled are limited to sequences of actions to be executed in the initial situation. Therefore, plans containing concurrent actions and plans that have an execution time that starts later than the initial situation are not treated. These restrictions come about because it is impossible to represent concurrent actions in situation calculus. Secondly, in situation calculus planning problems, there is no need to treat plans with execution times that will start later than the initial situation. Since in this paradigm all changes in the world are due to the planning agent, it makes no difference as to whether a plan is immediately executed or whether it is executed at a later time; the world will not change until the plan is started.

Since we will be handling plans with concurrent actions, our logic must represent the interactions between concurrent actions, some examples being:

- There is only one burner working and only one pan can be placed on it at one time. Thus only one dish can be cooked at a time.
- The agent can always carry one grocery bag to the car, but can only carry two when it is not icy out.
- The agent can move forward at any time, move backward at anytime, but cannot do both simultaneously.

Automated Planning Systems

Most domain-independent planners are essentially state-based planners, or limited extensions of the state-based approach. All these systems model actions as functions that transform one instantaneous state into another. The ancestor of all these systems is the STRIPS planner [6].

The type of planing problems handled by STRIPS is exactly what we have described as the state-based planning paradigm. The major contribution made by STRIPS is the so called "STRIPS assumption" to handle the frame problem, that is, the problem of representing that an action only effects a small part of the world. Thus if situation s_1 is the result of applying action a_1 in situation s_0, then s_0 and s_1 will be very much alike. In STRIPS, an action is defined as ordered triplet consisting of a precondition list, add list, and delete list, each member of these lists being atomic formulas. An action may be applied in any state s where all its preconditions hold yielding a new state that is computed from s, by adding only the formulas on the add list, and deleting all the formulas on the delete list. Implicit in this treatment is that any formula that is not explicitly asserted in a state is taken to be untrue. This allows one to avoid explicitly specifying negated atomic formulas.

Also in this class are the nonlinear hierarchical planners descending from NOAH [16]; the fact that they are nonlinear and hierarchical does not make them any more expressive than STRIPS since these terms refer to the control strategy, rather than the representation of plans. While one might argue that the notion of a procedural network gives us added representational power since it embodies a notion of hierarchical plans that are partially ordered, formally they are just descriptions of the set of linear sequences formed by atomic actions that can be decomposed from the hierarchical descriptions and meet the partial ordering.

There have been a number of domain-independent planning systems that have handled a larger class of planning problems than STRIPS. Wilkins' SIPE [20] and Vere's DEVISER [19] are two of the most sophisticated types. SIPE can handle plans with concurrent actions. The collection of actions making up a plan are only partially ordered. Any two actions where one is not ordered before the other are considered to be in parallel branches. Wilkins introduced the notion of *resources* to reason about the interaction of parallel actions. A resource is defined as an object that an action uses during its execution. If two actions share the same resource, they cannot be executed in parallel. Thus ordering constraints are imposed (if possible) to insure that any two actions that share a resource are not in parallel branches.

Wilkins points out that although resources used in his system are very useful, they are still quite limited in that they do not allow one to treat things such as money or computational power as resources. They also cannot be used to encode actions that conditionally interact, such as the example we posed earlier where the agent can carry two bags only if it is not icy out.

Vere's DEVISER system can represent a planning environment where there are assertions about future events. In this system, a time line is modeled by the nonnegative real numbers, zero being the time of planning and positive real numbers being future times. The system can represent that an event not under the control of the planning agent will occur starting at some specified time and ending at another time. One can specify that some goal condition must hold between two time points. Vere calls the time points in which a goal is to be achieved the goal's (time) window. A goal can be the conjunction of two or more conditions to be achieved simultaneously within one window, or conditions to be achieved at different times, thus each condition has its own specified window.

Vere's treatment of time is in terms of an absolute scale and does not provide a general representation allowing relative temporal orderings. For example, it does not handle goals such as: get to the gas station before getting to school. Furthermore, there is no way to represent that two actions must be nonoverlapping. If two actions, a_1 and a_2 in a plan being constructed must be nonoverlapping, one of the actions must be constrained to end before the other begins. In many cases, the choice as to whether a_1 is constrained to be earlier than a_2, or a_2 is constrained to be earlier than a_1, must be made arbitrarily. If a bad choice is made, the system would have to backtrack. The treatment of external events is also limited. One cannot represent that some external event will start any time between 2:00 and 2:15, the exact start and finish time must be given.

Both Vere's system and Wilkins' system can be viewed as *ad hoc* extensions to the simple state-space planning paradigm. They use a conception of actions that arises directly from situation calculus; namely, that actions transform one state into another. This is manifested in that both systems specify actions by giving what is essentially precondition–add–delete lists and using a STRIPS-like assumption to handle the frame problem. It will be argued that the STRIPS assumption is inappropriate for any planning system that treats either external events or plans with concurrent actions. By explicitly designing a logic that represents concurrent events, we have arrived at a conceptually different way of looking at the frame problem and specifying frame assumptions. This will be described later after the logic is introduced.

An Overview of the Approach

Recently, Allen [3] and McDermott [13] have presented logics of events and time where events are not simply treated as functions from state to state. In both these treatments, a global notion of time is developed that is independent of the agent's actions. In Allen's logic, there are objects that denote temporal intervals which are chunks of time in a global time line. An Event is equated with the set of temporal intervals over which the change associated with the event takes place. Thus there is a notion of what is happening while an event is occurring. Secondly, there may be a number of events that are occurring over the same interval. Therefore, one can treat concurrent actions by asserting that two actions occur over intervals that overlap in time.

McDermott also equates an event with the set of intervals over which the event is occurring, but his notion of intervals is slightly different. While Allen's logic can only express statements about what is actually true, McDermott's logic is based on a world model where there may be many possible futures realizable from a particular world-state (an instantaneous snapshot of the world). A world model consists of a collection of world-states and a "future relation" that arranges the world-states into a tree-like structure that branches into the future. Each branch, which McDermott calls a chronicle, represents a possible complete history of the world extending infinitely in time. A global time line is associated with the real number line and each world-state is mapped to its time of occurrence. Because world-states in different chronicles may map to the same time, one cannot equate the intervals over which events occur with temporal intervals in the global time line. Specifying that an event occurs over a temporal interval does not tell which chronicles it occurs in. Instead, intervals are associated with totally ordered convex sets of world-states (with respect to the future relation). In other words, intervals are contiguous blocks of world-states that lie along some chronicle.

Both logics can be used to describe the outside world, but without extension neither can be used to represent what can and cannot be done by the planning agent. This is essential for a planning system where one must reason about the effects of the various actions that can be executed.

Our formal logic of plans is based on Allen's temporal logic, extended with a modality to express what the agent can and cannot do and a modality that represents future possibility. We introduce a new type of object called a plan instance that refers to an action at a particular time done in a particular way. These objects are discussed in the next section. Following this we give a brief description of Allen's logic and show why it must be extended with these two modalities. We first extend Allen's logic with the inevitability operator arriving at a logic that is similar to McDermott's [13] and Haas's [9] and show why this is insufficient to represent what can and cannot be done. We then introduce a modality called IFTRIED that represents what can and cannot be done. IFTRIED expresses counterfactual-like statements of the form "if the agent were to attempt to execute plan instance *pi* then *P* would be true. Examples are presented illustrating how different types of goals may be represented. This is followed by a section that describes the conditions under which two plan instances, concurrent or not, can be jointly executed and how plan instance interactions, such as resource conflicts, may be represented. We then present a simple planning example which leads into a discussion on how the frame problem manifests itself in our logic. Finally, we give the semantic model and interpretation for our two modalities.

The following notational conventions will be used throughout the rest of the paper. Sentences and terms in our formal language, which is a quantified modal language, will be given in LISP-type notation. For example, the formula (*P a*) refers to the unary predicate *P* with argument *a*. All variable terms will be prefixed with a "?." We will assume that all free variables in a statement are implicitly bound by a universal quantifier. The logical connectives will be specified by: *AND* for conjunction, *OR* for conjunction, *IF* for material implication, *IFF* for equivalence, ∀ for the universal quantifier, and ∃ for the existential quantifier. The formula $(= t_1\, t_2)$ will be used to mean that t_1 and t_2 denote the same object and $(\neq t_1\, t_2)$ will be used to mean that t_1 and t_2 denote distinct objects. When we discuss the semantics for this logic in Section V, the functions and relations in the semantic model will be given in the conventional notation in logic (e.g., "$f(x)$," "$R(a, b)$," etc.).

II. Preliminaries

Plan Instances

In our theory we introduce objects called ***plan instances*** that refer to an action at a particular execution time done in a particular way. A plan instance's execution time is a temporal interval specifying the time over which the plan instance would occur if executed, not just a time point specifying the time that execution would begin. If we wanted to formally introduce plans, they would be functions from intervals to plan instances.

In situation calculus, one could get away with being vague as to whether an action refers to a behavior done in a particular way or whether it refers to a whole class of behaviors. This is because there is no notion of what is happening during the time of execution; as long as two particular behaviors have the same preconditions and effects they are indistinguishable in situation calculus. In a more general model, however, this distinction becomes conspicuous since one can represent what is happening while a plan instance is being executed. Consider a simple scenario where there are two paths of equal length going from location *A* to location *B*. Let us refer to these paths as P_1 and P_2. When talking about the effects of "go from *A* to *B* during interval *I*," one must be clear as to whether it refers to a particular way of going from *A* to *B* during *I* (i.e., whether it refers to going down path P_1, or going down path P_2), or whether it refers to taking either path during *I*. If it refers to a particular behavior, one can simply talk about its effects. If on the other hand, "go from *A* to *B* during *I*" refers to a class of behaviors, one must distinguish between saying "no matter how it is done, EFF will be true" and saying "there is a way that it is done such that EFF is true." In this example, it is correct to say that there is a way for an agent to do "go from *A* to *B* during *I*" such that this agent is on path P_1 during *I*, but incorrect to say that no matter how an agent does "go from *A* to *B* during *I*," this agent is on path P_1 during *I*. Plan instances are defined as "particular ways of doing something" since this is more primitive; later on, plan types referring to "classes of behaviors" can be introduced, defined in terms of plan instances.

Any two plan instances can be composed together to form a more complex plan instance. A composite plan instance occurs iff both its component parts occur. By composing plan instances, plan instances containing concurrent actions can be formed, as well as plan instances that are essentially sequences of actions, and plan instances that contain two components with gaps separating their execution times. In particular, plan instances that have concurrent actions can be formed by composing two plan instances that have execution times that overlap in time.

It is important to point out that a plan instance is not a complete specification of the agent's behavior over its time of execution. If this were so, a plan instance that was the composition of two distinct plan instances with overlapping times could never occur under any circumstances. For example, consider the plan instances "the agent is grasping object1 in its right hand during interval *I*" and "the agent is grasping object2 in its left hand during interval *I*." If plan instances are taken to be complete specifications then we could equivalently describe these plan instances as "during interval *I*, the only action the agent performs is grasping object1 in its right hand" and "during interval *I*, the only action the agent performs is grasping object2 in its left hand." Clearly, these two plan instances could never occur together and therefore their composition could never occur.

The STRIPS Assumption

Implicit in the STRIPS assumption is that actions are complete specifications of the agent's behavior over their time of execution (Georgeff [8] makes a similar point). This leads to problems in planning problems with concurrent actions and external events. We can paraphrase the STRIPS assumption as saying: if condition *C* holds at a time just prior to action a_1's execution, and a_1's effects do not negate *C*, then *C* will be true at a time immediately following a_1's execution. This presents a problem if we have two concurrent actions and we apply the STRIPS assumption to them separately. Suppose at a particular time t_0, P_1 is true. Consider two actions, a_1 and a_2, having the same duration and both having the preconditions that P_1 is true. Let the effects of a_1 be that P_1 is negated and the effects of a_2 be that P_2 is made true. Using the STRIPS assumption to compute the effects of executing a_1 starting at t_0, we get that P_1 is negated at a time immediately following a_1's execution which we will call time t_2. Similarly, using the STRIPS assumption to compute the effects of executing a_2 starting at t_0, we get that both P_1 and P_2 are true at t_2, the time immediately following a_2's execution. This, of course, is a contradiction, P_1 and its negation cannot both hold at the same time (i.e., at t_2).

Problems also arise if the STRIPS assumption is applied in a planning environment where there are external events. For example, suppose that at time t_0 it is asserted that it is raining outside. Consider action a_1 that can be applied at time t_0 and if it is executed will complete at time t_2. If we assume that it is out of the agent's control as to whether or not it is raining, the effects of a_1 will not negate the condition "it is raining." Thus by the STRIPS assumption, if a_1 is executed starting at time t_0, the result is that it is raining out at time t_2. This is unacceptable since, independent of the agent's actions, it might stop raining sometime before t_2.

Automated planning systems that use the STRIPS assumption avoid the above problems by restricting the type of planning problems that can be handled. Clearly, the "concurrent action problem" does not arise if plans are linear sequences of actions. If plans are treated as sets of partially ordered actions, the concurrent action problem can be avoided by making the assumption that the plan will be linearized before execution. Wilkins' system [20] does not make this assumption; if two actions are not ordered then it is possible that they will be executed in parallel. He, therefore, had to introduce the "resource mechanism" to handle conflicts between concurrent actions. As previously mentioned, this mechanism only handles interactions of the form: action a_1 and action a_2 share the same resource. There is no general notion of parallel action interactions.

Clearly, the "external event problem" does not arise if we are planning in a world where all changes are caused by the planning agent. On the other hand, in a system such as Vere's [19] that represents external events, one must be careful in specifying the planning environment. All external events that affect the value of any external property (i.e., a condition out of the agent's control) that is specified in the initial situation must also be included in the planning environment description. For example, if it is asserted that "the bank is open" holds in the initial situation, it is necessary to include the external event that negates this property at the time when the bank is closing. If this is not done, we would get the spurious result that the precondition that the bank is open is always satisfied.

To trace the root of the problem with the STRIPS assumption, let us look at a state-based representation, such as situation calculus, for which the STRIPS assumption was originally intended. Implicit in any state-based representation is that each action is a complete specification of what goes on from one situation to the next. One could re-interpret s_1 = RESULT (a_1, s_0) (i.e., s_1 is the result of applying action a_1 in s_0) as saying that the result of doing only a_1 in s_0 results in situation s_1. The fact that most things do not change in going from s_0 to s_1 is not really due to the fact that a_1 is done, but instead to the fact that all actions other than a_1 are not done. Most things stay the same because of these nonoccurrences. That is, given any property p holding in situation s_0, there is a set of actions $\{a_{p1}, \cdots, a_{p_r}\}$ (possibly empty) that make this property false upon execution in situation s_0. If a_1 does not belong to this set then p will also hold in the situation RESULT (a_1, s_0) due to the fact that none of the actions in $\{a_{p1}, \cdots, a_{p_r}\}$ are executed.

Thus the problem seems to be that the conclusion made by the STRIPS assumption is attributed to the execution of the action, not to the nonoccurrence of any action that can negate p. In effect, the STRIPS assumption is hiding the real reason why a property remains true from situation to situation. We will explicitly treat plan instances that refer to nonoccurrences and do not have to resort to a STRIPS assumption. In a later section, we discuss how the frame problem manifests in our logic and show the use of treating nonoccurrences. Georgeff [8] also presents an approach for handling the frame problem without appealing to the STRIPS assumption. His approach is similar to ours only in the fact that he also treats actions as partial specifications over their time of occurrence (although he does not describe it in these words).

Temporal Intervals, Properties, and Event Instances

The starting point for our logic is the treatment of action and time described in [3]. This logic is cast as a sorted first-order logic with terms denoting temporal intervals, events, properties (static conditions that hold or do not hold over intervals), and objects in the domain. We will slightly modify this theory by considering other event instances which refer to an event at a particular time. Thus we start with a sorted first-order language with terms denoting temporal intervals, event instances, plan instances (which also belong to the event instance sort), properties, and objects in the domain. A small number of predicates are introduced to specify the temporal relation between intervals, to specify that a property holds over an interval, and to specify that an event instance occurs. We also introduce symbols to denote the function that specifies an event instance's time of occurrence and the function that composes two plan instances. We now give the syntax the fragment which we will refer to as interval logic.

There are thirteen different ways in which two intervals can be temporally related. These are described in detail in [1] and so will not be repeated here. In this paper we will just introduce interval relation predicates as needed as they come up in the examples. Two interval relations that will be used frequently are: (PRIOR i_1 i_2) which means the interval denoted by i_1 is before or immediately precedes (meets) the interval denoted by i_2 and (ENDS-BEFORE i_1 i_2) which means that the interval denoted by i_1 ends before or at the same time as the interval denoted by i_2. The HOLDS predicate is used to assert that a property holds over some interval. The formula (HOLDS p i) means that the property denoted by p holds over the interval denoted by i. It is a theorem that if a property holds over an interval, then that property holds over any interval contained in this interval. The OCC predicate is used to specify that an event instance occurs. The formula (OCC *ei*) means that the event instance denoted by *ei* occurs. There is no need to say that it occurs over some interval since there is a time of occurrence associated with each event instance. We will use terms of the form "*e@i*" to refer to an event instance or plan instance whose time of occurrence is denoted by i. The function term (TIME-OF *ei*) denotes the time of occurrence associated with the event instance denoted by *ei*, thus we have (= i(TIME-OF e@i)). Finally, the function term (COMP pi_1 pi_2) denotes the plan instance composed of the plan instances denoted by pi_1 and pi_2. This is the plan instance that occurs iff its component parts both occur together. The time of occurrence of a composite plan instance is the smallest interval that contains both its components' times of occurrence. Throughout the rest of this paper we will be more cavalier in treating the use-mention distinction and, for example, will use "plan instance *pi*" in place of "the plan instance denoted by *pi*."

In laying out this theory of time, Allen did not distinguish an actual time and as a consequence there is no formal notion of the present, past, or future in this theory. For simplicity, we will assume that the specification of the planning environment includes the designation of the interval that represents the time of planning. Throughout the rest of this paper we will use "future plan instance" to refer to a plan instance with an execution time later than the time of planning, use "past condition" to refer to conditions that hold earlier than the time of planning, use "current time" to mean "the time of planning," etc.

Allen's treatment of time can be characterized as a linear time logic. Temporal assertions are only about what is actually true; there is no notion of what will possibly happen or what possibly happened. When planning, the agent must be able to reason about how its actions can bring about different future possibilities. Thus a logic to be used for planning must be able to represent some form of possibility. We will extend interval logic to achieve this end.

III. A Logic for Planning

We extend interval logic by introducing a temporal modality, INEV, referred to as the inevitability operator. The modal statement (INEV *i P*) means that at interval *i*, statement *P* is inevitable, or equivalently, regardless of what possible events occur after *i*, *P* is true. If intervals i_1 and i_2 finish at the same time, then (INEV i_1 *P*) is true iff (INEV i_2 *P*) is true. This is because the same events that are in the future of i_1 are in the future of i_2. The possibility operator POS is the dual of INEV for a fixed time. Thus it is defined in terms of INEV which is given by:

(POS *i P*) $=_{\text{def}}$(NOT (INEV *i* (NOT *P*))).

Theorems involving INEV are given in Fig. 1. INEV1 can be roughly restated as saying if it is possible that property

```
INEV1)
    (IF (ENDS-BEFORE ?ih ?i)
        (IFF (POS ?i (HOLDS ?p ?ih))
             (INEV ?i (HOLDS ?p?ih))))
INEV2)
    (IF (ENDS-BEFORE (TIME-OF?ei) ?i)
        (IFF (POS ?i (OCC ?ei))
             (INEV ?i (OCC ?ei))))
INEV3)
    (IF (PRIOR ?i1 ?i2)
        (IF (INEV ?i1 P)
            (INEV ?i2 P)))
INEV4)
    (IF (INEV ?i (IF P Q))
        (IF (INEV ?i P) (INEV ?i Q)))
INEV5)
    (IF (INEV ?i P) P)
INEV6)
    (IF (INEV ?i P)
      (INEV ?i (INEV ?i P)))
INEV7)
    (IF (POS ?i P)
        (INEV ?i (POS ?i P)))

where P and Q are any sentences in the logic with no free occurrences of ?i
```

Fig. 1. The axiomatization of INEV.

p held in the past or holds in the present, then it is inevitable that property *p* held in the past or holds in the present. INEV2 makes a similar claim about event instances with past or present times of occurrence. INEV3 captures the fact that if a statement is inevitable at time *i*, then it is inevitable at all later times. INEV4 says that material implication distributes out of INEV. INEV5 says that what is inevitable at any time is actually true. INEV6 and INEV7 stem from the fact that INEV is an *S*5 modal operator for a fixed time point. We also have the rule of inference: if *P* is a theorem then (INEV I_p *P*) is a theorem, for all interval terms I_p.

An important property of the INEV operator is: if interval logic statement IL_0 is entailed by the interval logic statements $IL_1 \cdots IL_n$, then (INEV *i* IL_0) is entailed by (INEV *i* IL_1) $\cdots$ (INEV *i* IL_n) for any interval term *i*. This is very useful when doing proofs in our planning examples. Many times, we will have interval logic statements nested within an inevitability operator and must be able to derive other statements of the form: (INEV *i IL*) where *IL* is an interval logic statement. Haas [9] who has an operator similar to INEV elaborates on this argument and claims that most of the modal reasoning needed to do planning (in his system) requires only first-order theorems applied within some modal context. The same argument can be used in our case.

Interval logic extended with the INEV operator can be viewed as a variant of a future branching time logic. These are temporal logics in which one can make statements about future possibility. Typically, these are modal languages with semantic models that consist of a collection of world states (i.e., instantaneous snapshots of the world) arranged in a tree-like structure that "branches into the future." Each branch is a possible history of the world, that is, a totally ordered set of states extending infinitely in time. The set of branches passing through some world-state are all the possible futures realizable from this world-state.

The logic consisting of interval logic and the INEV operator, however, does not have instantaneous world-states. Instead, the semantic model contains world-histories which are complete histories of the world extending throughout time and in the object language, there are temporal interval terms that refer to particular times in a world-history. All statements are interpreted with respect to a world-history. The truth value of a nonmodal statement (i.e., an interval logic statement) at world-history h_0 is only dependent on the event instances that occur and properties that hold in h_0. Thus interval statements are about what is actually true. The truth value of a modal sentence at h_0 depends on world-histories that are accessible from the h_0. This will be discussed in detail in the last section of this paper. Suffice to say that the accessibility relation in terms of which we interpret INEV relates world-histories that have a common past.

Interval logic augmented with the INEV modality can represent statements describing what is inevitable at time *i*, what is possible at time *i*, and also, what is actually true. This poses a slight problem. What does it mean to say that some course of events will possibly happen while also asserting that another course of events will actually happen? One answer to this is that possibilities are what could have happened. If this is the case, why should a planner worry about something that is possibly true if it is asserted that it is actually false?

Thomason [8] describes a formal technique for avoiding the above problem (although his motivation for developing the machinery is different from ours). We will not present this method here, but will do something that for our purposes is equivalent. We will assume that the description of the planning environment consists entirely of modal statements. Thus there will be no assertions about something that is actually false but possibly true. In solving a planning problem we will be looking for a plan instance *pi* such that at time *i*, it is inevitable that *pi* achieves the goal, not simply possible that *pi* achieves the goal.

Before going on, we should note that a term denotes the same object in all possible worlds, thus if two terms are actually equal (unequal) then at all times it is inevitable that they are equal (unequal). Secondly, if a temporal relation between two intervals is actually true, then at all times it is inevitably true. This is because intervals refer to chunks of time in a global time line. Thus their temporal relationship is invariant over different possible worlds. If intervals did not refer to a global time line, we could not talk about different ways the world could have been at some particular time. As a consequence, it is not necessary to embed equality statements, i.e., formulas of the form ($= t_1\ t_2$) and ($\neq t_1\ t_2$), and statements that temporally relate two intervals, i.e., formulas such as (PRIOR $i_1\ i_2$) and (ENDS-BEFORE $i_1\ i_2$), within the POS or INEV modality.

Achieving a Goal

Consider a planning problem where at planning time, which we will denote by I_p, we want to achieve goal G, where G is an interval logic statement describing desired future conditions. A future plan instance must be found that achieves the goal under all possible future conditions as described by the planning environment. Thus we are looking for a future plan instance *pi* such that in any possible future where *pi* occurs, G is also true. This is equivalent to saying that it is inevitable (at I_p) that if *pi* occurs then G is true:

(INEV I_p (IF (OCC *pi*) G)).

One might argue that it is impossible to find a plan instance that works under all possible circumstances, but this is not our objective. We are just looking for a plan instance that works assuming that the planner's view of the world (i.e., the planning environment and the action specifications) is correct. Whether the planning environment describes a great number of possibilities, or just a few of the very likely ones is not of immediate concern here.

If at time I_p, it is inevitable that *pi* does not occur (which would be the case if *pi* was an "impossible plan instance," i.e., one that never can be executed under any circumstance), the above statement would vacuously hold. Therefore, any plan instance *pi* under consideration must also meet the condition that there is a possible future where it occurs. This is simply stated as:

(POS I_p (OCC *pi*)).

This condition, however, does not insure that *pi* can be executed regardless of possible circumstances out of the agent's control or guarantee that *pi* contains all the steps needed for execution.

Consider the following planning problem. The goal is to get into a particular room sometime during the interval I_G. The plan instance WALK-IN-ROOM@I_w refers to the action of walking through the doorway into the room during interval I_w, where we are assuming that I_w ends at a time during I_G. In order to perform this plan instance the door must be unlocked at a time just prior to I_w. Let us suppose that it is possible the door is locked at this time and possible that it is unlocked at this time. Also assume that it is impossible for the robot to perform an action to unlock the door (or to get someone else to unlock the door), if it happens to be locked.

In this scenario, it is possible that the plan instance WALK-IN-ROOM@I_w occurs, thereby achieving the goal, since it is possible that the door is unlocked. We would not, however, be satisfied with such a plan instance, since whether or not it can be done is dependent on a condition out of the agent's control; namely, whether or not the door happens to be locked. What is needed is a plan instance that can be executed under any possible circumstances (as described by the planning environment) out of the agent's control. Thus in the above example, we would be looking for a plan instance that can be executed regardless of whether the door is locked or not.

A logic that represents the above example must be able to represent statements such as: "it is out of the agent's control as to whether or not the door is locked" and to express conditions such as: "under all possible external conditions, plan instance *pi* can be executed." Interval logic augmented with the INEV operator is insufficient in itself to represent these statements. The INEV operator may be used to represent that some future condition is possible, but cannot attribute the cause of the possibility to external factors, the agent's actions, or a combination of both factors.

Finding a plan instance that can be executed under all possible external circumstances, however, is still not sufficient. An even stronger condition that must be satisfied is that the plan instance contains all steps needed for execution (with respect to what is possible at planning time). This can be clarified by the following example. During planning time, which we will denote by I_p, the agent is standing by a locked safe and by a table on which the safe's key is resting. The goal is to open the safe (at some time in the near future). The agent can perform the plan instance OPEN-SAFE@I_o as long as it has the key grasped in its hand just prior to execution time I_o. The agent can also perform GRASP-KEY@I_g which corresponds to grasping the safe's key at a time immediately after planning time and results in the key being grasped just prior to I_o. This plan instance can be performed as long as the agent is by the table on which the key is resting just prior to execution time (I_g). In this example, we assume this condition happens to hold. Thus it is inevitable at planning time that this condition holds since the present is inevitable.

In this case, it is under the agent's control to enable the conditions under which OPEN-SAFE@I_o can be executed. By executing GRASP-KEY@I_g, the agent can bring about the conditions under which OPEN-SAFE@I_o can be executed. Thus it is possible that OPEN-SAFE@I_o occurs. We would not, however, want the planner to simply return that OPEN-SAFE@I_o achieves the goal, leaving out that it must be done in conjunction with some other plan instance. Instead, we would want the planner to return a plan instance such as (COMP GRASP-KEY@I_g OPEN-SAFE@I_o). At planning time I_p, this composite plan instance contains all the steps needed for execution since the conditions under which GRASP-KEY@I_g can be executed are inevitably true (at I_p) and it is inevitably true that OPEN-SAFE@I_o can be executed if GRASP-KEY@I_g is also executed. If we had a different scenario where the condition "the agent is grasping the key just prior to interval I_o" was inevitable, we would have concluded that OPEN-SAFE@I_o could simply be executed alone.

In describing the plan instances in both examples, we gave conditions under which each of them can be executed. If it is possible at planning time I_p that the conditions under which plan instance *pi* can be executed will not hold, and it is not in the agent's control to bring about these conditions, then we do not want to conclude that at time I_p, *pi* can be executed. Even if these conditions can be made true by executing some other plan instance, we consider a plan instance containing *pi* to be under specified unless it also contains a plan instance that enables these conditions. Thus in planning to achieve a goal *G* at time I_p, we are looking for a plan instance *pi* such that i) it is inevitable at I_p, if *pi* occurs then *G* is true and ii) it is inevitable at I_p, the conditions under which *pi* can be executed hold.

To capture the notion of "conditions needed for execution," our theory will make a distinction between plan instance attempts and plan instance occurrences. The conditions under which *pi* can be executed will be equated with the conditions under which attempting *pi* leads to *pi* occurring. As an example, attempting OPEN-SAFE@I_0 might associated with moving one's arm during interval I_0 in such a way that the key being grasped is twisted in the safe's lock (resulting in the safe being opened). This arm twisting movement could be done regardless of whether or not the agent was grasping the key, but only when it is done while the agent is grasping a key, would we say that the agent is opening the safe with the key.

The IFTRIED Modality

We extend our language to include the modal operator IFTRIED. The sentence (IFTRIED *pi P*) is taken to mean that if plan instance *pi* were to be attempted then *P* would be true, where *P* is any sentence in our extended language. This is a counterfactual modality that is used to make assertions about what would result if plan instance *pi* were to be executed, not about what is actually true. Thus it is consistent to assert that (IFTRIED *pi P*) and (NOT *P*) are both true. We must also point out that both arguments to IFTRIED are temporally qualified. The first argument being a plan instance has an associated time of occurrence. The second argument is a sentence in the logic and thus is either an interval logic statement and hence is temporally qualified or is a modal statement containing (temporally qualified) interval logic statements.

The IFTRIED operator is related to INEV by the following axiom schema, which states that if a property *P* is inevitable at some time *t*, then it will remain true no matter what plan is attempted after *t*

IFTRIED1)
(IF (PRIOR ?*t* (TIME-OF ?*pi*))
(IF (INEV ?*t P*) (IFTRIED ?*pi P*))).

From this axiom schema and the two axioms, INEV1 and INEV2, stating that the past and present are inevitable, we get the desired result that attempting a plan instance has no effect on earlier properties and events.

The sentence (IFTRIED *pi* (OCC *pi*)) means that if plan instance *pi* were to be attempted then it would occur, or what we equivalently say: "*pi* is executable." For convenience, we will define the predicate EXECUTABLE in our object language by

(EXECUTABLE *pi*) $=_{def}$(IFTRIED *pi* (OCC *pi*)).

At time I_p, a plan instance *pi* contains all the steps needed for execution iff in all possible futures, if *pi* were to be attempted it would occur. This is expressed by

(INEV I_p (EXECUTABLE *pi*)).

We will also introduce the notion of "choosibility" which can be expressed in terms of IFTRIED. We say that a condition *P* is choosible at time I_p iff there exists a plan instance with execution time after I_p which if attempted would result in *p* being true. The definition is given by:

(CHOOSIBLE I_p *P*) $=_{def}$
(∃ ?*pi*(AND (PRIOR I_p (TIME-OF ?*pi*))
(IFTRIED ?*pi P*)).

Saying that *P* is choosible at time I_p means that there is something the agent could have done (starting after I_p) to make *P* true. By using CHOOSIBLE, we can succinctly state that some condition is out of the agent's control and state that a condition can be made true regardless of external conditions. To express that at time I_p, the agent cannot affect whether or not proposition *P* is true, we state that it is inevitable at I_p that if *P* happens to come out true then it is not choosible at I_p that *P* is untrue, and similarly, it is inevitable at I_p, if *P* comes out to be untrue, then it is not choosible at I_p that *P* is true. For example, the following states that at all times, the agent has no control as to whether it is raining out:

(INEV ?*ie*
(AND (IF (HOLDS (raining) ?i_r)
(NOT(CHOOSIBLE I_p
(NOT (HOLD (raining) ?i_r))))
(IF (NOT (HOLDS (raining) ?i_r))
(NOT (CHOOSIBLE I_p
(HOLDS (raining) ?i_r))))))).

To express "regardless of possible external conditions after I_p, *P* can be made true," we state that regardless of what future possibilities arise, it is in the agent's control to make *P* true. This is expressed by

(INEV I_p (CHOOSIBLE I_p *P*)).

It is illustrative to compare the IFTRIED modality with Moore's RES operator [14] which is a modal operator that captures the result function in situation calculus. Moore equated possible worlds with situations allowing him to integrate a theory of action based on situation calculus with a theory of belief based on possible world semantics. For the comparison here, we will only talk about the RES modality.

Statements in Moore's language are interpreted with respect to a world at a particular time, not with respect to an entire world history as in our logic. Thus statements are about what is currently true and are not temporally qualified. The sentence (RES a_1 *P*) means that currently, action a_1 can be executed and if executed then *P* will be true where *P* is either a sentence in first-order logic, or contains modal operators. Now, it is important to note that neither argument to RES is temporally qualified. Secondly, RES is a "tense shift" operator. (RES a_1 *P*) is making an assertion about what will be true in a situation in the future of the current time. This contrasts to (IFTRIED *pi P*) where both its arguments are temporally qualified and they may have any temporal relation; *P* need not be about a time in the future of *pi*'s occurrence.

In [15] we show how the RES operator can be viewed as a special case of the IFTRIED modality. Very roughly, a situation calculus view of the world is modeled in interval logic by discrete intervals where intervals associated with situations are interleaved with intervals associated with action occurrences. If we assume that interval I_0 denotes the current time, we can translate (RES a_1 P), where P is a nonmodal, to (IFTRIED $a_1@i_1$ (AND (OCC $a_1@I_1$) (HOLDS P I_2)) where we are assuming that i_2 immediately follows i_1 which immediately follows I_0.

IV. Examples

Representing Different Types of Goals

This formalism can easily express goals such as avoiding some condition while performing some task, achieving a collection of goals to be done in some specified order, and preventing an undesirable condition that possibly will happen. In this section, we will examine an example of each of these goals. The simplest example concerns a simple sequence of goals. Suppose at planning time I_p, the goal is to be at school at some time $I_{at\text{-}s}$, while stopping at the gas station on the way. A plan instance *pi* (in the future of planning time I_p) must be found such that the following holds:

(INEV I_p
  (AND (EXECUTABLE *pi*)
    (IFTRIED *pi*
      (∃?*i*(AND (HOLDS (at *agt* school) $I_{at\text{-}s}$)
        (PRIOR ?*i* $I_{at\text{-}s}$)
        (HOLDS (at *agt* gas-station) ?*i*)))))).

The above statements says: under all circumstances possible at time I_p, both *pi* is executable and if *pi* were to be attempted (and thus would occur since it is executable), the agent would be at school during $I_{at\text{-}s}$, and at the gas station prior to this. Similarly, avoiding some condition while achieving another is easily represented by specifying that some condition does not hold during the execution of the plan.

Prevention problems pose a number of problems as other authors [3], [9], [13] have noted. It does not make sense to say that an occurrence is prevented, if it is not possible in the first place. Furthermore, the reason why the occurrence is possible must not be due to the agent's actions alone. This last qualification has been overlooked by these authors. Consider a simple scenario where if an open paint can is sitting in the doorway and an agent happens to walk through the doorway, the paint can will be knocked over spilling the paint on the floor. For simplicity, we will only talk about a particular spilling occurrence at time I_{sp} which will occur iff it is immediately preceded by an occurrence of an agent walking through the doorway while the can is sitting in the doorway:

(INEV ?*ie*
  (IFF (OCC paint-spills@i_{sp})
    (∃ ?i_{wtd} ?*agt*
      (AND
      (MEETS ?i_{wtd} i_{sp})
      (OCC (walks-through-door ?*agt*)@?i_{wtd})
      (HOLDS (position paint-can doorway) ?i_{wtd}))))).

The above statement says: at all times it is inevitable that paint-spills@i_{sp} occurs iff immediately prior to i_{sp} there is an agent that walks through the door while a can of paint is in the doorway.

Now, at planning time I_p, we want to find a plan instance *pi* that under all possible future conditions is executable and if it is attempted then paint-spills@i_{sp} will not occur:

(INEV I_p
  (AND (EXECUTABLE *pi*)
    (IFTRIED *pi* (NOT (OCC paint-spills@i_{sp}))))).

Even if the above statement is true, however, we cannot yet claim that the planning agent can prevent the occurrence of (spills paint)@i_{sp}. To begin with, it might be the case that is inevitable that the paint can is not in the doorway immediately prior to i_{sp} or inevitable that no one will walk through the doorway at a time immediately prior to i_{sp}. If either of the above holds, then it is inevitable that the paint can will not spill. Thus in order to conclude that we prevented the paint can from spilling, it must be the case that it possibly spills, that is, the following must be true:

(POS I_p (OCC paint-spills@i_{sp})).

Even this is not sufficient to conclude that we prevented this occurrence. It might be the case that the only way that the paint can will be in the doorway is if the planning agent puts it there, or the only agent that possibly can spill the paint is the planning agent. To represent that paint-spills@i_{sp} possibly occurs because of external forces, we must introduce a special plan instance that corresponds to the planning agent being inactive over a period of time, which we will denote by do-nothing@i_{dn}. We then check if it is possible that paint-spills@i_{sp} occurs even if the agent does nothing up until the time when paint-spills@i_{sp} would complete:

(IF (AND (MEETS I_p ?i_{dn})(ENDS-BEFORE i_{sp} ?i_{dn}))
  (POS I_p (IFTRIED do-nothing@?i_{dn}
    (OCC paint-spills@i_{sp})))).

Further discussion of these issues can be found in [15].

Composing Two Plan Instances and Plan Instance Interactions

One of the primary goals in developing this logic was to represent plan instance interactions and to provide a formal basis for determining which plan instances can be executed together. In state-based planners, the system determines which linear combinations of actions achieve the goal and which sequence of actions can be executed together. In these planners, the interactions of interest involve one action enabling another by bringing about the other's preconditions, or one action's effects interfering with another's preconditions. All these interactions, however, concern actions that are linearly ordered. In this section, we will consider the interactions between concurrent plan instances and show how to determine whether they can be executed together and if so under what conditions.

Let us first consider the relation between the conditions under which a composite plan instance is executable and the conditions under which each of its components are executable. What it means to say that a composite plan instance is executable is that if both components were attempted together then both components would occur. There are a number of cases to consider. It might be the case that both components are executable when taken alone, but they cannot be executed together since they interfere with each other. Such is the case if two plan instances share the same resource or if two plan instances are alternative choices, one of which can be performed at one time. Thus it is incorrect to assert that if both pi_1 and pi_2 are executable, then (COMP pi_1 pi_2) is executable. The converse of this statement does not hold either. It might be the case that (COMP pi_1 pi_2) is executable while pi_1 is not because the occurrence of pi_2 brings about the conditions under which pi_1 is executable. It might also be the case that (COMP pi_1 pi_2) is executable, but neither pi_1 or pi_2 is executable alone. Such is the case if pi_1 and pi_2 are "truly parallel actions," ones that must be executed together. An example of this is where an object is lifted by applying pressure to two ends of the object, one hand at each end. If pressure were applied to only one end, the result would be a pushing action, not part of a lifting action.

A general theorem will be given that relates (COMP pi_1 pi_2) to its component parts, regardless of their temporal relations. Before presenting this general theorem, however, it will be clearer to first look at special cases concerning the relation between pi_1 and pi_2.

We begin by considering the case where pi_1 and pi_2 do not have overlapping execution times. Without loss of generality assume that pi_1 is before pi_2. Clearly, whether or not a plan instance occurs is not affected by the attempt of another plan instance with a later execution time. This is captured by the following theorem:

```
INTERACTION1)
  (IF (PRIOR (TIME-OF ?pi1) (TIME-OF ?pi2))
     (IFF (OCC ?pi1)
          (IFTRIED ?pi2 (OCC ?pi1))))
```

Since pi_1 is before pi_2, attempting pi_2 has no effect on whether or not pi_1 occurs. Therefore, if (COMP pi_1 pi_2) is executable then pi_1 must also be executable. It need not be the case, however, that pi_2 is executable. The execution of pi_1 may bring about the conditions under which pi_2 is executable. Alternatively, pi_2 may be executable, but it would not be if pi_1 were to occur. This provides justification for the following theorem:

```
INTERACTION2)
  (IF (PRIOR (TIME-OF ?pi1) (TIME-OF ?pi2))
     (IFF (EXECUTABLE (COMP ?pi1 ?pi2))
      (AND (EXECUTABLE ?pi1)
           (IFTRIED ?pi1 (EXECUTABLE ?pi2)))))
```

This theorem says that for any nonoverlapping plan instances pi_1 and pi_2, pi_1 being the earlier one, the composite plan instance (COMP pi_1 pi_2) is executable iff pi_1 is executable and if pi_1 were to be attempted (and thus would occur because it is executable), then pi_2 would be executable.

If pi_1 is earlier than pi_2, but the two plan instances overlap in time, the consequent in the above theorem might not hold while the antecedent does; there are cases where the statements "pi_1 is executable" and "if pi_1 were to be attempted then pi_2 would be executable," are both true but their composition is not executable, since pi_2 interferes with the successful completion of pi_1. For example, consider the function term (walk home store)@I_w which denotes the plan instance where the agent walks from home to the store during interval I_w. This plan instance is executable as long as the agent is at home just prior to execution (i.e., at some time that immediately precedes I_w). The effects of this plan instance are, that the agent is outside during execution and is at the store at a time following execution. Consider also the plan instance (stay-at home)@I_s which refers to the action of staying at home during interval I_s. This plan instance is executable as long as the agent is at home just prior to execution and its effects are that the agent is at home for the duration of its execution. If we assume: i) (stay-at home)@I_s is executable, ii) I_s is earlier than but overlaps I_w, and iii) it is impossible to be at two places at once, then it follows that the earlier plan instance (stay-at home)@I_s is executable, and if the plan instance (walk home store)@I_w were attempted, it would be executable, but their composition is not executable since it is impossible to be outside and at home at the same time.

In general, if plan instance pi_1 is executable, but the conditions prohibit both pi_1 and pi_2 from occurring together, then we have the following:

```
(IFTRIED pi1 (OCC pi1))
(IFTRIED pi1 (IFTRIED pi2 (NOT (OCC pi1))))
```

The first statement is simply the definition of (EXECUTABLE pi_1). The second statement says: If pi_1 were to be executed (and thus would occur), we would get to a scenario where if pi_2 were to be attempted, it would preclude pi_1 from occurring. In other words, pi_2 interferes with pi_1.

If, on the other hand, i) pi_1 is executable, ii) the conditions under which pi_1 and pi_2 can occur together hold, and iii) if pi_2 were to be attempted, pi_2 would be executable, then we would have the following:

```
(IFTRIED pi1 (OCC pi1))
(IFTRIED pi1 (IFTRIED pi2 (AND (OCC pi1) (OCC pi2))))
```

This can be read as saying that if pi_1 were to be attempted (and thus would occur since it is executable), we get to a scenario where attempting pi_2 would result in both pi_2 and pi_1 occurring. In this case, both plan instances can be performed together, and we want to conclude that the composite plan instance is executable.

In the general case, we have the following theorem:

```
INTERACTION3)
  (IF (IFTRIED ?pi1
         (IFTRIED ?pi2 (AND (OCC ?pi1) (OCC ?pi2))))
      (IFTRIED (COMP ?pi1 ?pi2)
         (AND (OCC ?pi1) (OCC ?pi2))))
```

This theorem says that if attempting plan instance pi_1 would result in a scenario where attempting pi_2 would result in both plan instances occurring, then their composition is executable (i.e., if it were attempted, both of its components would occur). Notice that there are no temporal restrictions relating pi_1 and pi_2. Secondly, this general theorem provides for the case where neither pi_1 nor pi_2 are executable alone, but they are executable together. This would be the case if attempting pi_1 alone would not result in pi_1 occurring, but the attempt would make it so that pi_2 is executable, which in turn would answer the conditions for pi_1's successful completion. There is, in fact, an even more general theorem about interaction that is not necessary for the purposes of this paper, which is discussed in [15].

Concurrent Plan Instance Interactions

In this section, a few examples are presented to illustrate how concurrent plan instance interactions may be represented and how these statements lead to conclusions about whether or not the composition of these plan instances are executable. We start with a simple resource conflict example. Consider a very simple case where there is a stove on which only one pan can be placed at a time. Let the function term (heating *pn*)@I_H denote the plan instance where the pan *pn* is being heated on the stove during the interval I_H. The fact that only one pan can be heated at a time is captured by:

HEAT1)
(INEV ?*ie*
(IF (AND (OCC (heating ?pn_x)@?i_{Hx})
(OCC (heating ?pn_y)@?i_{Hy}))
(OR (DISJOINT ?i_{Hx} ?i_{Hy})
(AND (=? pn_x?pn_y)(=?I_{Hx} ?I_{Hy})).

This says that under all possible circumstances (i.e., it is inevitable at all times), if there are two occurrences of a pan being heated on a burner then either their times of occurrence are not overlapping or they refer to the same plan instance (two function terms are equal if all their argument terms are equal). This relationship between plan instances is what Lansky [10] calls a behavioral constraint; it directly relates two plan instances (actions) instead of implicitly relating two plan instances by use of action precondition–effect lists.

Treating such an example with precondition–effect lists would be awkward and inefficient. We would have to introduce a property associated with "burner being used." The action, "heat pan *pn*" could not simply be modeled by a simple action specified by a precondition–effect list, instead it would have to be modeled by two simple actions that must be performed consecutively (note: Vere's system [19] has such a facility). The reason for this is that the effect of "heat pan *pn*" is that the burner is in use during execution, not before or after execution. We would then model this action by two consecutive actions a_1 and a_2, where a_1's effect is that the burner is in use and a_2's effects are that the burner is free. As previously noted, Wilkins' SIPE [20] has a special mechanism for treating a limited class of resource conflicts. This mechanism could be used to solve the above example without recourse to the "in use" properties.

In order for the behavioral constraint HEAT1 to be useful, it must lead to the deduction that a composite plan instance consisting of two overlapping plan instances using the same burner but different pans is not executable. This is easily shown. Let us consider plan instances (heating *pn*1)@I_{H1} and (heating *pn*2)@I_{H2} in the case where *pn*1 and *pn*2 denote different objects, and the intervals, I_{H1} and I_{H2}, overlap in time. We want to prove that the composite is not executable, i.e.,

PRV1)
(NOT (IFTRIED (COMP (heating *pn*1)@I_{H1}
(heating *pn*2)@I_{H2})
(AND (OCC (heating *pn*1)@I_{H1})
(OCC (heating *pn*2)@I_{H2}))))).

From axiom HEAT1, the fact that I_{H1} and I_{H2} overlap in time, and that *pn*1 and *pn*2 are unequal, it logically follows that PRV1 is true. A sketch of this proof is given in Fig. 2.

Sketch of proof for PRV1:

Substituting pn1 for ?pn_x, pn2 for ?pn_y, I_{H1} for ?I_{Hx}, and I_{H2} for ?I_{Hy} in HEAT1:

S1) (INEV ?ie
(IF (AND (OCC (heating pn1)@I_{H1})
(OCC (heating pn2)@I_{H2}))
(OR (DISJOINT I_{H1} I_{H2}) (AND (= pn1 pn2) (= I_{H1} I_{H2})))))

From the fact that pn1 and pn2 are unequal (and thus inevitably unequal), the fact that I_{h1} and I_{H2} are not disjoint since they overlap in time (and thus it is inevitable they are not disjoint), and the theorem that conjunction distributes into INEV, we get:

S2) (INEV ?ie
(NOT (OR (DISJOINT I_{H1} I_{H2}) (AND (= pn1 pn2) (= I_{H1} I_{H2})))))

S1 has form: (INEV ?ie (IF P Q)), and S2 has form: (INEV ?ie (NOT Q)). Moving the conjunction into the INEV operator, and applying Modus Tollens, we get:

S3) (INEV ?ie
(NOT (AND (OCC (heating pn1)@I_{H1})
(OCC (heating pn2)@I_{H2}))))

From S3 and IFTRIED1 (if it is inevitable at time i that P is true then for all pi's with execution times later than i (IFTRIED pi P) is true):

S4) (IFTRIED ?pi
(NOT (AND (OCC (heating pn1)@I_{H1})
(OCC (heating pn2)@I_{H2}))))

And finally, we substitute the composite plan instance for ?pi and get:

S5) (IFTRIED (COMP (heating pn1)@I_{H1} (heating pn2)@I_{H2})
(NOT (AND (OCC (heating pn1)@I_{H1})
(OCC (heating pn2)@I_{H2}))))

Finally, by applying a theorem that negation distributes out of the IFTRIED modality, we arrive at PRV1.

Fig. 2.

Our next example concerns two plan instances that cannot occur together if certain external conditions hold. Suppose that the agent cannot carry two grocery bags from the supermarket to the car (without slipping) if it is icy out. Consider the scenario where at planning time I_p, it is possible that it is going to be icy out during I_{ref}, and also possible that it is not icy out during I_{ref}. This is represented by

ICY-ST1)
(AND (POS I_p (HOLDS (icy) I_{ref})
(POS I_p (HOLDS (not (icy)) I_{ref}))).

We also state that it is impossible for the agent to affect whether or not it is icy out:

ICY-ST2)
(INEV ?*ie*
(AND (IF (HOLDS (icy) ?*ii*)
(NOT (CHOOSIBLE ?*ie*
(NOT (HOLDS (icy) ?*ii*)))))
(IF (HOLDS (not (icy)) ?*ini*)
(NOT (CHOOSIBLE ?*ie*
(NOT (HOLDS (not (icy)) ?*ini*))))))).

Consider the two function terms, (carry *bag*1)@I_c and (carry *bag*2)@I_c, which refer to the plan instances where the agent takes each bag from the shopping cart and carries it to the car during interval I_c which we assume is during I_{ref}. Also assume that *bag*1 and *bag*2 denote distinct objects. Finally, assume that under all circumstances, if it is icy out during I_{ref}, then it is impossible that both plan instances occur together. This is represented by:

ICY-ST3)
(INEV ?*ie* (IF (HOLDS (icy) I_{ref})
(NOT (AND (OCC (carry *bag*1)@I_c)
(OCC (carry *bag*2)@I_c))))).

From ICY-ST3 above, the fact that it is possible that it is icy out (ICY-ST1), and the fact that the agent cannot perform any action that would prevent it from being icy (ICY-ST2), we can prove that it is possible at I_p that there is nothing the agent can do to make (carry *bag*1)@I_c and (carry *bag*2)@I_c occur together.

PRV2)
(POS I_p
(NOT (CHOOSIBLE I_p (AND (OCC (carry *bag*1)@I_c)
(OCC (carry *bag*2)@I_c)))).

The sketch of the proof of PRV2 is given in Fig. 3.

Sketch of the proof of PRV2:

PRV2) (POS I_p
(NOT (CHOOSIBLE I_p (AND (OCC (carry bag1)@Ic)
(OCC (carry bag2)@Ic)))))

We use the two following theorems which we give without proof

CH1: (IF (INEV ?i (IF P Q))
(INEV ?i (IF (NOT (CHOOSIBLE ?i (NOT P)))
(NOT (CHOOSIBLE ?i (NOT Q))))))

POS1: (IF (AND (INEV ?i (IF P Q)) (POS ?i P))
(POS ?i Q))

We first prove that it is possible (at I_p) that it is icy out (during I_{ref}) and in this case the agent cannot choose to make it not icy out. This is derived using the first conjunct of ICY-ST1:

S1) (POS I_p (HOLDS (icy) I_{ref}))

using ICY-ST2 substituting I_{ref} for ?ii and I_p for ?ie, and the theorem that conjunction distributes out of INEV:

S2) (INEV I_p (IF (HOLDS (icy) I_{ref}) (NOT (CHOOSIBLE I_p (NOT (HOLDS (icy) I_{ref}))))))

then applying POS1 substituting I_p for ?i, (HOLDS (icy) I_{ref}) for P and (NOT (CHOOSIBLE I_p (NOT (HOLDS (icy) I_{ref})))) for Q, giving us the desired result:

S3) (POS I_p (NOT (CHOOSIBLE I_p (NOT (HOLDS (icy) I_{ref})))))

Next, we prove that it is inevitable (at I_p) that if it is not choosible that it is not icy out, then it it is not choosible that we can carry both bags together. This is derived using ICY-ST3 substituting I_p for ?ie

S4) (INEV I_p (IF (HOLDS (icy) I_{ref})
(NOT (AND (OCC (carry bag1)@Ic) (OCC (carry bag2)@Ic)))))

and then using CH1 substituting I_p for ?i, (HOLDS (icy) I_{ref}) for P, and (NOT (AND (OCC (carry bag1)@Ic) (OCC (carry bag2)@Ic))) for Q, giving us the desired result:

S5) (INEV I_p
(IF (NOT (CHOOSIBLE (NOT (HOLDS (icy) I_{ref}))))
(NOT (CHOOSIBLE (AND (OCC (carry bag1)@Ic) (OCC (carry bag2)@Ic))))))

Finally, from S3, S5 and POS1 substituting I_p for ?i and making obvious substitutions for P and Q, we prove prv2:

Fig. 3.

This example could not be handled by existing nonlinear planners, because only two types of interactions can be handled: 1) If action a_1 interferes with action a_2's preconditions then either a_1 is ordered after a_2, or another action is inserted between a_1 and a_2 in order to restore a_2's preconditions, and 2) If two actions have effects that contradict each other, then one of the actions must be ordered before the other to avoid the possibility of trying to execute them together and thus failing. Now, in the example above, the plan instances, (carry *bag*1)@I_c and (carry *bag*2)@I_c, do not really contradict each other; the fact that both plan instances occur is consistent with the material implication stating that if it is icy out, then the two plans do not both occur. Thus a naive implementation would allow these two plan instances to simultaneously occur. From the fact that (carry *bag*1)@I_c and (carry *bag*2)@I_c both occur, and this material implication, we would conclude that it is not icy out during interval I_{ref}. Clearly, this is not desired. Instead, we would want the planning system to conclude that the two plan instances can be executed together only if it happens not to be icy out during I_{ref}.

This problem manifests itself differently in the case where the agent can bring about the conditions under which two plan instances can be executed together. Let us suppose that action a_1 and a_2 share the same type of resource and it is under the agent's control how many of these resources are present. Now, it is not a contradiction that a_1 and a_2 occur simultaneously. A naive implementation might return a plan where a_1 and a_2 are executed simultaneously, while failing to include a plan step that guarantees that two resources are present while a_1 and a_2 are being executed. Thus we get a plan that does not have all the steps needed for execution.

These problems can be avoided in a state-based system by explicitly introducing properties associated with resources in use, but as we have previously mentioned this becomes quite cumbersome and would lead to an inefficient search space. Properties and precondition lists are useful, however, for specifying the conditions under which actions taken alone can be executed. Thus an adequate representation needs to handle both behavioral constraints and properties that serve as preconditions. We have just discussed how behavioral constraints can be expressed. Saying that if property *pr* holds over interval *i*, then plan instances *pi*'s preconditions hold (i.e., *pi* is executable) is simply expressed by:

(INEV *?ie* (IF (HOLDS *pr i*) (EXECUTABLE *pi*)).

Most theories can express only precondition properties. Lansky [10] describes a theory that allows behavioral constraints, but it is difficult to treat properties. To define a property one must know all the possible actions and then precompute the patterns of execution that result in the property being true and untrue.

A Simple Planning Problem

In this section, we present a simple planning example that leads into a discussion on how the frame problem manifests itself in our logic. Consider the following example. Suppose at planning time I_p, the planning agent, which we denote by *agt*, is at home and its goal is to get to the bank sometimes during I_G which is an interval that immediately follows I_p. The set of sentences that describe the planning environment are given by:

PE1) (MEETS I_p I_G)
PE2) (INEV I_p (HOLDS (at *agt* home) I_p)).

The interval logic statement describing the goal of getting to the bank at a time during interval I_G is given by

(∃ $?i_b$ (AND (DURING $?i_b$ I_G) (HOLDS (at *agt* bank) $?i_b$))).

In the rest of this example we will use *G* to refer to this interval logic statement describing the goal conditions.

For our simple example, we introduce a class of plan instances that refer to walking from one building location to another. We will let the function term (walk *bldg*1 *bldg*2)@I_{walk} denote the action of walking from building *bldg*1 to building *bldg*2 during interval I_{walk}. The term will denote a plan instance for all arguments, *bldg*1 and *bldg*2, that denote building locations and all interval terms I_{walk}, but unless the duration of I_{walk} is greater than or equal to the minimal time it takes the agent to walk from *bldg*1 to *bldg*2, it denotes an impossible plan instance. These are plan instances that are never executable under any circumstances. For simplicity, we ignore the case where *bldg*1 and *bldg*2 denote the same building or are so far apart that it is impossible to walk from one to the other. If we have two plan instances, (walk *bldg*1 *bldg*2)@I_{w1} and (walk *bldg*1 *bldg*2)@I_{w2}, where I_{w2} and I_{w1} start at the same time, but I_{w1} has a shorter duration, they differ in the rate that the agent walks from *bldg*1 to *bldg*2.

If (walk *bldg*1 *bldg*2)@I_{walk} is not an impossible plan instance, then it is executable as long as the agent is at *bldg*1 just prior to execution time (i.e., I_{walk}). Furthermore, this connection between (walk *bldg*1 *bldg*2)@I_{walk} and the conditions under which it is executable is inevitable at all times. Thus we have the following:

```
EXC-WALK)
  (IF (< = (minimal-time ?bldg1 ?bldg2)
           (DURATION ?i_walk))
     (INEV ?ie
  (IF (AND (HOLDS (at agt ?bldg1) ?i_at-bldg1)
           (MEETS ?i_at-bldg1 ?i_walk))
     (EXECUTABLE
        (walk ?bldg1 ?bldg2)@?i_walk))))
```

where the function term (minimal-time *bldg*1 *bldg*2) denotes the minimal time it takes for the agent to walk from *bldg*1 to *bldg*2.

The effects of (walk *bldg*1 *bldg*2)@I_{walk} are that the agent is outside during the time of execution and will be at *bldg*2 iately after execution. The connection between a plan instance and its effects is inevitable at all times. Thus we have the following:

```
EFF-WALK)
  (INEV ?ie
    (IF (OCC (goto ?bldg1 ?bldg2)@?i_walk)
      (∃ ?i_at-bldg2
        (AND (HOLDS (at agt ?bldg2) ?i_at-bldg2)
             (MEETS ?i_walk ?i_at-bldg2))
             (HOLDS (at agt outside) ?i_walk)))).
```

Given the sentences describing the planning environment and the sentences describing the action specifications, we are looking for a future plan instance such that at the time of planning, it is inevitable that it is executable and if it occurs the goal conditions will obtain. Thus we want to find a plan instance *pi* that has execution time that is later then I_p and it logically follows from the sentence describing the planning environment and the action specifications that the following is true:

```
GOAL)
  (INEV I_p (AND (EXECUTABLE pi) (IFTRIED pi G)))
```

where

```
G = def (∃ ?i_b (AND (DURING ?i_b I_G)
                     (HOLDS (at agt bank) ?i_b)).
```

In order for it to be possible to achieve the goal by executing a plan instance of the form (walk *bldg*1 *bldg*2)@I_{walk}, the minimal time it takes to get from home to the bank must be less than the duration of I_G (since the plan instance must be performed during this interval). We will assume that this condition holds and therefore the planning environment is described by PE1, PE2 along with:

PE3) (< (minimal-time home bank) (DURATION I_G)).

It can be shown that GOAL can be satisfied by any plan instance *pi* belonging to the set:

```
{(walk home bank)@?i_w
   |  (MEETS I_p ?i_w)
      and (< = (minimal-time home bank)
      (DURATION ?i_w))
      and (DURING ?i_w I_G)}.
```

That such an interval exists is guaranteed by PE3 along with an axiom in interval logic concerning the existence of intervals.

These constraints on $?i_w$ result from intersecting the conditions needed to guarantee that it is inevitable (at I_p) that if (walk home bank)@$?i_w$ occurs the goal condition will hold and the conditions needed to guarantee that it is inevitable that the plan instance is executable. The constraint that $?i_w$ finishes before I_G insures that the agent does not arrive at the bank at a time later than I_G, while the conditions $?i_w$ immediately follows I_p and has a duration greater or equal to (minimal-time home bank) insure that the plan instance is executable.

These constraints may be derived by using a control strategy similar to the backward chaining strategy employed to solve a single conjunct in nonlinear planning such as described in Allen and Koomen [2]. We find a set of plan instances each of which would achieve the goal if executed. Call this set S_0. We then see if there is any plan instance (or class of plan instances) belonging to S_0 which are executable. If this is the case we are done. Otherwise, we must try to compose a plan instance *pi* (or class of plan instances) belonging to S_0 with another that achieves the conditions needed for *pi* to be executable. If this composite plan instance is executable, we are done, else we repeat the process looking for a plan instance that achieves the conditions needed for the composite plan instance to be executable, etc.

Now, as we mentioned, by constraining $?i_w$ so that it immediately follows I_p and has duration greater than or equal to the minimal time it takes to go from the home to the bank, we find a class of executable plan instances that achieve the goal. For the sake of illustration, suppose that we must constrain $?i_w$ so that it is strictly after I_p instead. In this case, it does not follow that it is inevitable at planning time that (walk home bank)@i_w is executable. The reason is that (walk home bank)@i_w is executable only if the agent is at home immediately prior to i_w. Although it is inevitable that the agent is at home during I_p, we cannot prove that it is inevitable that the agent is at home (or any other location for that matter) at any time later than I_p.

If we want to execute (walk home bank)@i_w for i_w strictly after I_p, then we must compose this plan instance with another plan instance whose effect is that the agent is at home just prior to i_w. For this purpose, we introduce a class of plan instances that refer to staying at the same location for any period of time. We will use the function term (stay-at bldg)@i_{sa} to denote the plan instance where the agents stays in the building *bldg* during the interval i_{sa}. The conditions under which it is executable are simply that the agent is in the building just prior to execution.

EXC-STAY)
(INEV ?*ie* (IF (AND (HOLDS (at *agt* ?*bldg*) ?$i_{at\text{-}bldg}$)
(MEETS ?$i_{at\text{-}bldg}$?i_{stay}))
(EXECUTABLE (stay-at ?*bldg*)@?i_{stay}))).

The effects of (stay-at *bldg*)@i_{sa} is that the agent is at *bldg* during the time of execution:

EFF-STAY)
(INEV ?*ie* (IF (OCC (stay-at ?*bldg*)@?i_{stay})
(HOLDS (at *agt* ?*bldg*) ?i_{stay}))).

By executing (stay-at home)@?i_{st} where ?i_{st} immediately precedes ?i_w, we can achieve the conditions under which (walk home bank)@?i_w is executable. If ?i_{st} immediately follows I_p, then (stay-at home)@?i_{st} is executable. From this we get

(IF (AND (MEETS I_p ?i_{st})
(MEETS ?i_{st} ?i_w)
(< = (minimal-time home bank)
(DURATION ?i_w)))
(INEV I_p
(EXECUTABLE (COMP (stay-at home)@?i_{st})
(walk home bank)@?i_w)))).

This is derived using the theorem INTERACTION1 which states: if pi_1 is prior to pi_2, pi_1 is executable, and if pi_1 were executed, then pi_2 would be executable, then (COMP pi_1 pi_2) is executable. By also adding the constraint that ?i_w completes before I_G, we can insure that the composite plan instance (COMP (stay-at home)@?i_{sa}) (walk home bank)@?i_w) achieves the goal of being at the bank sometime during I_G.

The Frame Problem

In the example above, we saw that if (walk home bank)@i_w has an execution time that does not immediately follow planning time, we have to introduce another plan instance whose effect is that (at *agt* home) is true immediately prior to i_w. A plan instance of the form (stay-at home)@i_s served this purpose. The property (at *agt* home) holds at planning time and the execution of (stay-at home)@i_s simply maintains this property up to the time when (walk home bank)@i_w is to be executed.

The fact that we have plan instances that maintain properties contrasts with the traditional approach in nonlinear planning. In the nonlinear planning paradigm, if we introduce an action a_1 into the plan and this action has preconditions that are satisfied by conditions that hold in the initial situation, a "ghost node" is created, indicating that it is not necessary (at least at this stage) to explicitly introduce another action to achieve a_1's preconditions [16]. If the system finds another action a_2 in the plan whose effects negate a_1's preconditions, the system would try to order this action to follow a_1. If this ordering is not possible, the system would have to introduce a third action following a_2 and before a_1 that restores a_1's preconditions, or remove a_1 or a_2 from the plan. This strategy can be seen as an implementation of the STRIPS assumption.

In effect, these ghost nodes correspond to our maintenance plan instances, although they are not given the same status as actions and are treated differently by the nonlinear planner. For example, one does not introduce a ghost node into a plan, just like one introduces actions. They result as a side effect of linking an action to an earlier state where one of its preconditions holds. So one cannot explicitly choose between maintaining a condition to achieve a precondition as opposed to introducing an action that explicitly makes the precondition true, in the case where the precondition holds at an earlier state. The first alternative is automatically tried first (which turns out to be a good heuristic).

As we previously described, the use of the STRIPS assumption (in the guise of ghost nodes) leads to problems if we assert that some property, which the agent cannot affect, is true in the initial situation (and in the case of Vere's system, does not include all scheduled external events that affect this property). Since there will be no actions whose effects violate an external property *p*, an action whose precondition is *p* can be ordered anywhere in the plan.

This problem arises because a ghost node can be created for all properties. In our formalism, we do not have the same problem because there will only be maintenance plan instances for properties completely in the agent's control. Thus we might have a maintenance plan instance for "the agent stays in the same location," but clearly would not have a maintenance plan instance for a property such as "the gym is locked." Thus even though the property "the gym is locked" holds during planning time and there are no assertions about whether "the gym is locked" holds in the future, there is no plan instance that can be executed that makes this condition true at any time in the future. The agent is at the mercy of its environment.

Now, the fact that we have maintenance plan instances does not give us a simple solution to the frame problem. The frame problem manifests itself in a different form in our logic. What we do get though is a uniform treatment of the frame problem. In a system such as Wilkins' that uses the STRIPS assumption but allows concurrent actions, he has one mechanism for finding conflicts caused by one action's effects interfering with another's preconditions and another for finding conflicts between concurrent actions, this being the resource mechanism. In our logic, both conflicts can be seen to be of the same form and can be treated by the same mechanism, as follows.

We have already shown that if we want to execute pi_1 and pi_2 together, then we must prove that the following is true:

(INEV I_p (IFTRIED (COMP pi_1 pi_2)
(AND (OCC pi_1) (OCC pi_2)))).

To prove that a plan instance, say pi_1, does not interfere with a later one's preconditions, say pi_2, we show that pi_1 can be executed together with the plan instance(s) that bring about pi_2's preconditions. Thus the problem of determining whether a plan instance interferes with another one's preconditions reduces to the problem of determining whether two plan instances are executable when taken together. The following "blocks-world" example will help to clarify. Consider the plan instance (grasp *blk*1)@i_1 which refers to grasping block *blk*1 during interval i_1. This plan instance is executable as long as the property (clear *blk*1) holds just prior to i_1. Also suppose that property (clear *blk*1) is asserted to be true at planning time and (keep-clear $blk1$)@i_{kc} is a plan instance that maintains property (clear $blk1$) from the time of planning time up to the beginning of i_1. Thus the plan instance (COMP (keep-clear $blk1$)@i_{kc} (grasp $blk1$)@i_1) is executable. Now, consider some plan instance a_0@i_0 that is prior to i_1 and it is inevitable (at planning time) that it is executable. If we want to execute a_0 along with the composite plan instance above, we must prove that (COMP a_0@i_0 (keep-clear $blk1$)@i_{kc} (grasp $blk1$)@i_1)) is executable. Note that, since COMP is associative, without loss of generality we can let COMP take any number of arguments.

Since (keep-clear $blk1$)@i_{kc} guarantees the executability of (grasp $blk1$)@i_1, and (grasp $blk1$)@i_1 cannot interfere with a_0@i_0 since it occurs after a_0@i_0, the above is executable only if the plan instance (COMP a_0@i_0 (keep-clear $blk1$) @i_{kc})) is executable. As a result, reasoning about how (grasp $blk1$)@i_1 and a_0@i_0 interact is reducible to reasoning about how the two overlapping instances (keep-clear $blk1$)@i_{kc} and a_0@i_0 interact.

Thus both forms of interactions involve determining whether it is inevitable at planning time that two overlapping plan instances are executable together. As we discussed earlier, two plan instances that overlap in time are executable together only if they do not interfere with each other, i.e.,

(IF (OCC pi_2) (IFTRIED pi_1 (OCC pi_2))).

It is at this step that we need "frame axioms" or some nondeductive method for solving the frame problem. Typically, when describing an action, one only mentions its effects, not what it does not affect, but to reason about the interaction of overlapping plan instances, we must be able to determine what a plan instance does not affect. Thus we could explicitly encode frame axioms of the form

(INEV I_p (IF C_t (IFTRIED pi C_t)))

to specify the temporally qualified conditions C_t that are not affected by pi's execution.

From a pragmatic standpoint, however, specifying a large number of frame axioms might lead to an inefficient implementation. Such a point has been made by Wilkins [20]. An alternative approach is to restrict the form of sentences allowed in specifying the planning environment and action specifications and use a default-like assumption to compute what an action does not affect. Such is done by adopting the STRIPS assumption where the effects of an action are limited to a list specifying the properties that are negated and a list specifying the properties that are made true. This precludes the use of disjunctive effects among other things. Dean [5] handles the frame problem in a temporally rich domain by appealing to a "persistence assumption." This is a rule of the form: once a property is made true, it remains true until some other property makes it false. To implement this mechanism, one must be able to efficiently compute when a set of properties are inconsistent when taken together. This is done by restricting the way properties can be logically related. We are currently studying several such default reasoning techniques and attempting to characterize the range of problems that each technique can handle.

V. The Semantics

The formal semantics for this logic can be characterized as "possible-world semantics." The basic components of a model structure are a set of world-histories, which are complete histories of the world (our variety of possible worlds), and two accessibility relations. Each sentence is interpreted with respect to a world-history within some model structure. The truth value of a nonmodal sentence (i.e., interval logic sentence) at world-history h_0 is only dependent on the properties that hold and event instances that occur in h_0. The truth value of the sentence (INEV i P) at h_0 depends on the world-histories that are possible with respect to h_0 and share a common past up until the end of interval i. The truth value of the sentence (IFTRIED pi P) at h_0 depends on the "closest" world-histories to h_0 where the plan instance denoted by pi is attempted. The interpretation of counterfactual statements in terms of a "closeness" accessibility relation derives from Lewis' [11] and Stalnaker's [17] work on conditionals. To get at a more concrete notion of what a plan instance attempt is, we appeal to Goldman's theory of actions [7]. This is described in the next section. Following this, we present the model structure and give the interpretation for the two modal statements (INEV i P) and (IFTRIED pi P). We must also note that the interpretation of a term is constant over world-histories, that is, terms are treated as rigid designators. The interpretation for the rest of the language (i.e., atomic formulas, and sentences related by the first-order connectives) is omitted since this is straightforward. A detailed presentation of the semantics is given in [15].

Goldman's Theory of Actions and Basic Generators

In Goldman's theory of action, he defines what it means to say that one act token (an action at a particular time performed by a particular agent) ***generates*** another, under specified conditions. Roughly put, the statement "act token a_1 generates act token a_2" holds whenever it is appropriate to say that a_1 can be done by doing a_2. Associated with each generation relation, "a_1 generates a_2" is a set of conditions C^* such that C^* are necessary and sufficient conditions under which if a_1 occurs then a_2 occurs.

Goldman also introduces the concept of a ***basic*** action token. A basic action token is a primitive action token in the sense that every nonbasic action token is generated by some basic action token (or some set of basic action tokens executed together) and there is no action token, more primitive, that generates a basic action token. They can be thought of as the fundamental building blocks of which all action tokens are composed. In Goldman's work, basic action tokens are associated with particular body movements that can be done "at will" as long as certain "standard conditions" hold. As an example, moving one's arm counts as a basic action since it can be done at will as long as no one is holding the arm down, the agent is not paralyzed, etc.

In our theory, we equate "plan instance pi is attempted" with "pi's basic generator occurs." In each model, a subset of the plan instances are designated as being basic and a basic generator function is specified that associates every plan instance with the basic plan instance that generates it. If the plan instance is basic, then it generates itself. Each plan instance has a unique generator since we are defining a plan instance as an action at a particular time done in a particular way. It is not necessary to provide for a plan instance that is generated by a set of basic plan instances, not just a single one, because the set of basic plan instances are closed under composition. Our conception of "basic" differs from Goldman's treatment in that we allow basic plan instances that are more abstract than collections of body movements at specified times. For example, in modeling a game of chess, we might take our basic plan instances to be simple chess moves such as moving the queen from Q_1 to Q_3 at a particular time, etc. The standard conditions would be that the move is legal by the rules of chess. There is no benefit in looking more closely at a chess move and saying that it is generated by the arm movement that physically moves the piece. Even though each plan instance has a basic generator associated with it, one can make assertions about a plan instance in the object language without knowing its generator. This is analogous to making assertions about physical objects in some first-order language while not knowing all its exact features as captured by the semantic model.

The Model Structure

Formally, a model is a 12-tuple of the form:

$<H, D, OBJ, INT, MTS, PROP, EI, PI, BPI, BGEN, R, F_{c\ell}>$

The constituents of this tuple are described as follows:

H a nonempty set of world-histories.

D a nonempty set of domain individuals. This is partitioned into four disjoint subsets *OBJ*, *INT*, *PROP*, and *EI* which are given as follows:

OBJ a nonempty set of objects that existed at any time in any world-history.

INT a nonempty set of temporal intervals.

The relation ***MTS*** is defined over the set of intervals. $MTS(i_1, i_2)$ means that interval i_1 meets interval i_2 to the left (i_1 immediately precedes i_2). Allen and Hayes [4] present an axiomatization of the MTS relation; these axioms will be adopted here. All other interval relations (such as overlaps, is before, is contained in, etc.) can be defined in terms of MTS as long as we assume that for each interval, there exists another that meets it to the left, and one that is met by it to the right. We also must stipulate that the intersection of any overlapping intervals belongs to the set of intervals and that the concatenation of any two intervals belongs to the set of intervals. The relation $IN\text{-}OR\text{-}EQ(i_1, i_2)$ will be defined as being true when interval i_1 is contained in i_2 or when the intervals are equal. We also define the function $COVER(i_1, i_2)$ which yields the smallest interval that contains both i_1 and i_2

PROP is a nonempty set of properties. Each element of *PROP* is a set containing elements of the form: $<i, h>$ where *i* belongs to *INT* and *h* belongs to *H*.

Property *pr* holds during interval *i* in world-history *h* iff $<i, h>$ belongs to *pr*. For convenience, we define the function HOLDS(*pr*, *i*) which yields the set of world-histories where property *pr* holds over interval *i*:

$$HOLDS(pr, i) =_{def} \{h \mid <i, h> \in pr\}.$$

To capture the constraint: if a property holds over an interval, it holds over any properly contained interval, we have the following:

HOLD1)

For all properties (*pr*) and intervals (i_1 and i_2),
If $IN\text{-}OR\text{-}EQ(i_2, i_1)$
then $HOLDS(pr, i_1) \subseteq HOLDS(pr, i_2)$.

EI is a nonempty set of event instances. Each element of *EI* is an ordered pair of the form: $<i, h\text{-}set>$ where *i* belongs to *INT* and *h-set* is a subset of *H*. The time of occurrence of event instance $<i, h\text{-}set>$ is *i*, and $<i, h\text{-}set>$ occurs only in world-histories belonging to *h-set*.

For convenience, we define the function TIME-OF(*ei*) which yields event instance *ei*'s time of occurrence.

For all event instances ($<i, h\text{-}set>$),

$TIME\text{-}OF(<i, h\text{-}set>) =_{def} i$.

Similarly, we define the function OCC(*ei*) which yields the set of world-histories where event instance *ei* occurs.

For all event instances ($<i, h\text{-}set>$),
$OCC(<i, h\text{-}set>) =_{def} h\text{-}set$.

PI is a nonempty set of plan instances. *PI* is a subset of *EI*.

The set *PI* is closed under plan instance composition. The composition of two plan instances occurs iff both its components occur, and its time of occurrence is the smallest interval that contains both of its components' times of occurrence. For convenience, we define the composition function *CMP* by:

for all pl. inst. ($<i_1, h\text{-}set_1>$ and $<i_2, h\text{-}set_2>$),
$CMP(<i_1, h\text{-}set_1>, <i_2, h\text{-}set_2>) =_{def}$
$<COVER(i_1, i_2), h\text{-}set_1 \cap h\text{-}set_2>$

To state that *PI* is closed under *CMP* we have:

PI1)

For all plan instances (pi_1 and pi_2),
$CMP(pi_1, pi_2) \in PI$.

BPI is nonempty set of basic plan instances. *BPI* is a subset of *PI*. *BPI* is also closed under composition, thus we have:

BPI1)
For all basic plan instances (bpi_1 and bpi_2),
$CMP(bpi_1, bpi_2) \in BPI$.

BGEN is a one-place function with domain *PI* and range *BPI*. For every plan instance *pi*, *BGEN(pi)* is its basic generator. If *pi* is a basic plan instance then *pi* = *BGEN(pi)*.

In all world-histories, if a plan instance occurs then its basic generator also occurs. Thus we have the constraint:

BGEN1)
For all plan instances (*pi*),
$OCC(pi) \subseteq OCC(BGEN(pi))$.

A plan instance and its generator have the same time of occurrence giving us the constraint:

BGEN2)
For all plan instances (*pi*),
$TIME\text{-}OF(pi) = TIME\text{-}OF(BGEN(pi))$.

Finally, we have the constraint that the generator of a composite plan instance is equal to the composition of the generators of the plan instance's components:

BGEN3)
For all plan instances (pi_1 and pi_2),
$BGEN(CMP(pi_1, pi_2)) = CMP(BGEN(pi_1), BGEN(pi_2))$.

The R Accessibility Relation and Interpretation of INEV

The truth value of the sentence (INEV *i P*) directly depends on the *R* accessibility relation, a three-place relation taking an interval and two world-histories as arguments. $R(i, h_0, h_1)$ can be read as: h_1 is an alternate way the future might have unfolded with respect to h_0 at time *i*. The interpretation of (INEV *i P*) is given by

For every world-history (h_0), sentence (*S*),
and interval term (*i*) (INEV *i S*) is true at h_0 iff
for every world-history (h_1)
if $R(V(i), h_0, h_1)$ then *S* is true at h_1

where *V(i)* is the interval (member of *INT*) that the term *i* denotes.

The constraints we place on *R* are as follows:

If $R(i_l, h_0, h_1)$ is true then h_0 and h_1 share a common past up until the end of interval i_ℓ. We therefore impose the following constraints:

R1)
For all world-histories (h_0 and h_1),
properties (*p*) and intervals (i_l and i_e),
if $R(i_l, h_0, h_1)$ and ENDS-BEFORE(i_e, i_l) then
$h_0 \in HOLDS(p, i_e)$ iff $h_1 \in HOLDS(p, i_e)$.

R2)
For all world-histories (h_0 and h_1),
event instances (*ei*) and interval (i_l),
if $R(i_l, h_0, h_1)$ and ENDS-BEFORE(TIME-OF(*ei*), i_l)
then $h_0 \in OCC(ei)$ iff $h_1 \in OCC(ei)$

where ENDS-BEFORE(i_1, i_2) means that i_1 ends before i_2 or the intervals end at the same time.

An alternative world-history at time i_l is an alternative at all earlier times.

R3)
For all world-histories (h_0 and h_1),
properties (*p*) and intervals (i_l and i_e),
if $R(i_l, h_0, h_1)$ and ENDS-BEFORE(i_e, i_l) then $R(i_e, h_0, h_1)$

R is an equivalence relation for a fixed time.

R4) (reflexive)
For all world-histories (h_0) and intervals (*i*),
$R(i, h_0, h_0)$ is true.

R5) (symmetric)
For all world-histories (h_0 and h_1) and intervals (*i*),
if $R(i, h_0, h_1)$ then $R(i, h_1, h_0)$.

R6) (transitive)
For all world-histories (h_0, h_1, and h_2) and intervals (*i*),
if $R(i, h_0, h_1)$ and $R(i, h_1, h_2)$ then $R(i, h_0, h_2)$.

The Selection Function $F_{c\ell}$

The truth value of the sentence (IFTRIED *pi P*) directly depends on the selection function $F_{c\ell}$. This function has domain $BPI \times H$ and range 2^H. If basic plan instance *bpi*'s standard conditions hold in world-history *h*, then, $F_{c\ell}(bpi, h)$ is the set of "closest" world-histories to *h* where *bpi* occurs. If the standard conditions do not hold, $F_{c\ell}(bpi, h)$ is set equal to $\{h\}$. The approach of giving semantics to counterfactuals in terms of a "closeness" accessibility relation follows from the work of Stalnaker [17] and Lewis [11]. Very roughly (using Stalnaker's formalization), the counterfactual "IF *A* then *C*" is true at world w_0, if *C* is true in the closest world to w_0 where antecedent *A* is true (if such a world exists). The reason for having a "closeness" measure on possible worlds seems to stem from a pragmatic principle on how one evaluates counterfactuals. This is best captured by the following test proposed by Frank Ramsy:

> Suppose that you want to evaluate the counterfactual "If *A* then *C*." First you hypothetically add the antecedent *A* to your stock of beliefs and make the minimal revision required to make *A* consistent. You then consider the counterfactual to be true iff the consequent *C* follows from this revised stock of beliefs.

What one can say about a general notion of "closeness" is quite limited. Most matters of "closeness" are decided by pragmatics, not semantics. In both Stalnaker's and Lewis' approaches, only a few, mostly obvious constraints (such as if *A* is true at w_0 then the closest world to w_0 where *A* is true is w_0 itself) are placed on the "closeness" relation. In our theory, we are only treating counterfactuals of a particular form: "If *pi* were to be attempted *P* would be true." This allows us to impose additional constraints on our closeness relation $F_{c\ell}$ that arise from the specific nature of these counterfactuals and their intended use: to reason about which actions can be physically executed together. Our intuitive picture of "closeness" is as follows. Roughly, if h_1 is a closest world-history to h_0 where basic plan instance *bpi* occurs (and not equal to h_0), then h_1 differs solely on the account of executing *bpi*, or any basic action that physically cannot be done in conjunction with *bpi*, or any basic plan instance whose standard conditions are violated by *bpi*. There are alternative conceptions of "closeness," but we argue in [15], that this conception is the most appropriate for reasoning about what actions possibly can be done together.

The interpretation of (IFTRIED *pi S*) is given by:

For every world-history (h_0), sentence (S)
and plan instance term (pi),
(IFTRIED *pi S*) is true at h_0 iff
for every world-history (h_1)
if $h_1 \in F_{c\ell}(BGEN(V(pi)), h_0)$ then S is true at h_1

where $V(pi)$ is the plan instance (member of PI) denoted by the term pi.

This can be read as saying that (IFTRIED *pi P*) is true at world-history h_0 iff in all closest world-histories to h_0 where pi's basic generator occurs, P is true

In this paper, we only present the most general properties that $F_{c\ell}$ must have. In [15] we discuss additional constraints that may be imposed on $F_{c\ell}$ and discuss under what conditions they are appropriate. For convenience, we define $F'_{c\ell}$ by extending $F_{c\ell}$ to take world-history sets as its second argument

$$F'_{c\ell}(bpi, hS) =_{\text{def}} \bigcup_{h \in hS} F_{c\ell}(bpi, h).$$

The constraints we impose are as follows:

There will always be at least one closest world-history

FCL1)
For every world-history (h)
and basic plan instance (bpi), $F_{c\ell}(bpi, h) \neq \emptyset$.

If a basic plan instance *bpi* occurs in a world-history *h* then there is only one closest world-history where *bpi* occurs and it is *h*. In both Stalnaker's and Lewis' models, we have the analogous constraint that if proposition *A* holds in world-history w_0, then the closest world-history to w_0 where *A* holds is w_0 itself.

FCL2)
For every world-history (h) and basic plan instance (bpi),
if $h \in OCC(bpi)$ then $F_{c\ell}(bpi, h) = \{h\}$.

The following constraint relates a composite basic plan instance to its component parts:

FCL3)
For every world-history (h)
and basic plan instances (bpi_1 and bpi_2),
if $F'_{c\ell}(bpi_2, F_{c\ell}(bpi_1, h)) \subseteq OCC(CMP(bpi_1, bpi_2))$
then $F_{c\ell}(CMP(bpi_1, bpi_2), h) = F'_{c\ell}(bpi_2, F_{c\ell}(bpi_1, h))$.

This can be explained as follows. Let *S* be the set of world-histories that are reached by first going to a closest world-history where bpi_1 occurs and then going to a closest world-history where bpi_2 occurs. If both bpi_1 and bpi_2 occur in every world-history belonging to *S* (or equivalently, if $CMP(bpi_1, bpi_2)$ occurs in every world-history belonging to *S*), then the closest world-histories to *h* where the composition $CMP(bpi_1, bpi_2)$ occurs is exactly *S*.

The next constraint says: for all world-histories *h* and times *i*, there is always a basic plan instance *bpi*, later than *i*, such that the closest world-history where *bpi* occurs is *h* itself.

FCL4)
For all world-histories (h), and intervals (i),
there exists a $bpi \in BPI$ such that $PRIOR(i, TIME\text{-}OF(bpi))$
and $F_{c\ell}(bpi, h) = \{h\}$

where $PRIOR(i_1, i_2)$ is defined as: interval i_1 immediately precedes or is before i_2.

The following constraint captures the fact that if a set of world-histories *S* is reachable by first going to the closest world-history where bpi_1 occurs and then going to the closest world-history where bpi_2 occurs, then there is a third world-history bpi_3 that reaches exactly the world-histories in *S* and has a time of occurrence equal to the smallest interval that contains both bpi_1's and bpi_2's times of occurrence:

FCL5)
For every world-history (h)
and plan instances (bpi_1 and bpi_2),
there exists a basic plan instance (bpi_3)
such that $TIME\text{-}OF(bpi_3)) = TIME\text{-}OF(CMP(bpi_1, bpi_2))$
and $F_{c\ell}(bpi_3, h) = F'_{c\ell}(bpi_2, F_{c\ell}(bpi_1, h))$.

$F_{c\ell}$ and R are related by the following constraints:

R-CL-1)
For all world-histories (h_0 and h_1)
basic plan instances (bpi) and intervals (i),
if $h_1 \in F_{c\ell}(bpi, h_0)$ and $PRIOR(i, TIME\text{-}OF(bpi))$
then $R(i, h_0, h_1)$.

R-CL-1 can be read as saying that if h_1 is a closest world-history to h_0 where *bpi* occurs, then h_1 is possible with respect to h_0 at all times up until the beginning of *bpi*'s execution time. Thus h_0 and h_1 share a common past up until the beginning of *bpi*'s execution time.

The following constraint leads to the fact that if h_1 and h_0 share a common past up until the end of *bpi*'s execution time, then any closest world-history to h_0 can be matched with a closest world-history to *h*, that it share a common past with until the end of *bpi*.

R-CL-2)
For all world-histories (h_0, h_1, $h_{c\ell 0}$, $h_{c\ell 1}$)
and basic plan instances (bpi)
if $R(TIME\text{-}OF(bpi), h_0, h_1)$ and $h_{c\ell 0} \in F_{c\ell}(bpi, h_0)$
then there exists a world-history ($H_{c\ell 1}$) such that
$h_{c\ell 1} \in F_{c\ell}(bpi, h_1)$ and $R(TIME\text{-}OF(bpi), h_{c\ell 0}, h_{c\ell 1})$.

VI. Conclusion

The logic presented in this paper extended Allen's linear time logic with the INEV modality which expresses temporal possibility and IFTRIED which is a counterfactual-like modality that can be used to represent sentences describing how an agent can affect the world. This extends the class of planning problems typically handled by the situation calculus. In particular, a planning environment can be represented that has assertions about the past, present, and future, not just assertions about the current state. Plan instances can contain concurrent actions and have any execution time; they are not restricted to be sequences of actions to be executed in the current situation. Finally, any temporal statement can be a goal statement; we are not restricted to goals that describe conditions that must hold just after plan execution.

The IFTRIED modality can be used to specify which propositions can and cannot be affected by the robot's actions. Without making extensions, neither McDermott's or Allen's logic could encode this type of statement. We have shown how to use IFTRIED to represent that some condition cannot be affected by the agent's actions, such as whether or not it is raining, and to represent that some condition could always be made true regardless of the external circumstances.

By nesting the IFTRIED operator, we can represent how plan instances interact with each other. We presented sufficient conditions that guarantee that two plan instances can be executed together. Other forms of interaction can also be represented. For example, the logic can represent that two plan instances can be separately executed, but cannot be executed together. This might be the case if the two plan instances had concurrent execution times and shared the same resource.

In the last section, we presented the model structure which consists of a set of possible world-histories related by the *R* relation which is used to interpret INEV and the $F_{c\ell}$ function which is used to interpret IFTRIED. $F_{c\ell}$ embodies the notion of "closeness." The approach of interpreting a counterfactual-like modality in terms of "closeness" accessibility relation derives from Lewis' and Stalnaker's semantic theories of conditionals.

References

[1] J. F. Allen, "Maintaining knowledge about temporal intervals," *Commun. ACM*, vol. 26, no. 11, pp. 832–843, Nov. 1983.

[2] J. F. Allen and J. A. Koomen, "Planning using a temporal world model," in *Proc. 8th. Int. Joint Conf. on Artificial Intelligence* (Karlsruhe, W. Germany, Aug. 1983), pp. 711–714.

[3] J. F. Allen, "Towards a general theory of action and time," *Artificial Intell.*, vol. 23, no. 2, pp. 123–154, 1984.

[4] J. F. Allen and P. J. Hayes, "A common-sense theory of time," in *Proc. 9th Int. Joint Conf. on Artificial Intelligence* (Los Angeles, CA, Aug. 1985).

[5] T. Dean, "Planning and temporal reasoning under uncertainty," in *Proc. IEEE Workshop on Knowledge-Based Systems* (Denver, CO, 1984).

[6] R. E. Fikes and N. J. Nilsson, "STRIPS: A new approach to the application of theorem proving to problem solving," *Artificial Intell.*, vol. 2, pp. 189–205, 1971.

[7] A. I. Goldman, *A Theory of Human Action.* Englewood Cliffs, NJ: Prentice-Hall, 1970.

[8] M. P. Georgeff, "A theory process," SRI Int., unpublished manuscript, 1985.

[9] A. Haas, "Possible events, actual events, and robots," *Comput. Intell.*, vol. 1, no. 2, pp. 59–70, 1985.

[10] A. L. Lansky, "Behavioral specifications and planning for multiagent domains," Tech. Rep. 360, SRI Int., 1985.

[11] D. K. Lewis, *Counterfactuals.* Cambridge, MA: Harvard Univ. Press, 1973.

[12] J. McCarthy and P. J. Hayes, "Some philosophical problems from the standpoint of artificial intelligence," in B. Meltzer and D. Miche, Eds., *Machine Intelligence, Vol. 4.* New York, NY: Elsevier, pp. 463–502, 1969.

[13] D. McDermott, "A temporal logic for reasoning about processes and plans," *Cogn. Sci.*, vol. 6, no. 2, pp. 101–155, 1982.

[14] R. C. Moore, "Reasoning about knowledge and action," Tech. Rep. 191, SRI Int., 1980.

[15] R. N. Pelavin, "A formal logic that supports planning with external events and concurrent actions," Ph.D. dissertation, Computer Sci. Dep., U. Rochester, expected in 1986.

[16] E. D. Sacerdoti, *A Structure for Plans and Behavior.* New York, NY: Elsevier, 1977.

[17] R. Stalnaker, "A theory of conditionals," in W. L. Harper, R. Stalnaker, and G. Pearce, Eds., *IFS.* Dordrecht, The Netherlands: Reidel, 1981, pp. 41–55.

[18] R. H. Thomason, "Indeterminist time and truth value gaps," *Theoria*, vol. 36, pp. 264–281, 1970.

[19] S. A. Vere, "Planning in time: Windows and durations for activities and goals," Res. Rep., Jet Propulsion Lab., Pasadena, CA, Nov. 1981.

[20] D. Wilkins, "Domain independent planning: Representation and plan generation," Tech. Note 226, SRI Int., May 1983.

Reprinted from *IEEE Expert*, Summer 1986, pages 17-28.

Analogical Problem-Solving and Expert Systems

Lance B. Eliot
University of Southern California

Analogies and analogical thinking are aspects of problem solving that may be found in a variety of domains and can be used by both novices and experts.

What do the terms "analogical thinking" and "analogy" mean? Let us consider the following example that illustrates an intuitive sense of the application of an analogy. The domain in the example falls within software development, though illustrative analogies are identifiable in everyday domains as well.

Use of analogy

John Taylor, an experienced and highly regarded software engineer, is informed that a complex software package has produced erroneous results. Preliminary investigation by other, less experienced software specialists reveals that the problem is not a simple one; instead, it appears to be rooted deep within a maze of code. John studies the code and is intrigued by the novelty of the problem.

After extensive testing of the questioned code, John begins to recognize similarities to a previous problem solved years ago. Using his prior experience, he selects a specific portion of code for additional scrutiny. Though the two problems are not exactly alike, John attacks the current problem with a process that succeeded in determining a solution to the earlier problem. Finally, after additional testing and mental concentration, he arrives at the exact code problem and applies a correcting solution.

Use of prior experiences. What does this example suggest regarding problem solving? In solving the software problem, John made use of a particular problem-solving process involving the identification of prior experiences that could be applied to solving a current problem. Studies of human decision making have formulated theories that account for this kind of problem solving, described here as analogical thinking.

Researchers in artificial intelligence have also studied the analogical problem-solving process. Rather than focus on improving human decision making, though, they tend to seek a machine-based form of analogy solving. Development of automated analogical problem-solving systems is a relatively new subject area, and only a few small-scale systems have been built.

Defining analogy. The previous discussion made use of an example to suggest that the application of prior experiences accounts for the analogy process. However, many other forms of problem solving obviously make use of prior experiences. How then is analogy distinguished from other cognitive acts of problem solving?

First, a survey of the literature (see below) indicates little agreement on an exact definition of analogy. Research in many fields, including machine learning, cognitive psychology, and linguistics, does not make a clear distinction between the psychological phenomenon known as analogy and other types of problem-solving processes.

Second, the study of analogy is still in its infancy. Though studies of analogy and reports of analogy usage are found in a vast body of literature, there has been little focused study on the psychological process. Many prior studies indicate that analogies are critical to other forms of problem solving, such as learning by example and learning by being told, but these same studies generally do not then specifically account for the analogy process.

EH0303-8/90/0000/0362$01.00 © 1986 IEEE

Third, analogy and analogical thinking are complex phenomena and have resisted explanation by a single macroscopic theory. Researchers have concentrated on subsets of analogy and have provided results that are frequently disjoint from other studies. To study a subportion of analogy, researchers tend to create new definitions of the larger notion of analogy (or fail to define it at all, relying instead on an intuitive sense of analogy). These definitions provide a weak and exceedingly simplified framework in which only microscopic examination can take place, and promote further study that is subsequently limited to a particular research concern.

The definition used here. The scope of this article is constrained to an examination of the existing and potential future intersection between expert systems and analogy. A useful classification for examining these two topics is provided later in the article. To proceed with an examination of analogy, we employ a definition encompassing many of the issues inspected by those researchers concentrating on the analogy field. This definition serves as a reasonable vehicle for studying the current research base, and leads to several suggestions for future research on both the nature of analogy and the relationship between expert systems and analogy.

Analogy is an overloaded term, representing both an artifact and a process. Clarification of the overloading can be handled by specifying "analogy" as a reference to the artifact (a static product arising out of the analogy process), and using "analogical thinking" as a reference to the process (the manner in which the analogy was produced). Analogy and analogical thinking are viewed as interdependent and consisting of three dominant features:

(1) Representations for attribute knowledge vital to the analogy and analogical thinking,

(2) Representations for the relationship knowledge vital to the analogy and analogical thinking, and

(3) Operations working upon attribute and relationship knowledge to perform activities that produce the analogy.

A common, specific instantiation of these features is to describe analogy as a set of attributes, measures of distance (or similarity) used in conjunction with the attributes, and a type of pattern matching that provides an application of the attributes and distance measures.

The distinction between analogy and analogical thinking will not be maintained in this article. The term "analogy" will be used to denote both, except in specific contexts where the distinction is of value to the discussion.

Earlier, the use of prior experience was suggested as a critical element in describing the nature of analogy. As part of the definition of analogy used here, storage of the representations identified as dominant features is a critical aspect of analogy and analogical thinking. A knowledge base must exist from which prior analogs may be obtained or derived, and the operations then serve to act on the pre-existing knowledge base. Notice that it is not a *requirement* that the pre-existing knowledge base be in a representational form containing attributes and relationships (though this is a possibility). Instead, it merely must be possible to derive such representations.

Finally, as part of the core definition, there must be identifiable analog pairs consisting of a problem and a corresponding solution. In the case of pre-existing knowledge, the knowledge base must contain potential problem-solution pairs. (Again, it is not necessary that the knowledge base actually contain them, merely that they can be derived). An outcome of analogical thinking is a determination that for a presented problem there is either a successful analog fulfillment, or an unsuccessful analog fulfillment. Analog fulfillment is achieved when some prior analog pair is matched with a new problem and its developed solution.

As part of this definition, the requirement that analogy deals with problem/solution pairs, and matching between pairs, is an aspect that distinguishes it from other forms of similar problem solving (such as the use of metaphors). Also, the definition relies on definitions for many other terms—what is a problem, what is a solution, what is a match, what is proper matching, and so on. These terms and debate over their meaning constitute much of the analogy research focus. Studies of analogy often concentrate on a subset of the full set of terms used in analogy definition and lead to disjoint research results.

The preceding definition for analogy offers a position that is generally open ended, available for modification as our understanding of analogy increases.

Expert systems and analogy. Overall, my definition of analogy is sufficient for the sections that follow. Other key aspects of analogy—such as the need for plans in analogy processing and a delineation of the steps in analogical thinking—have been studied by researchers but are not considered part of this necessary core definition. The definition is conceptually suggestive of what the term "analogy" means, providing a sense of the complexity involved in studying analogy and attempting to automate it.

How do expert systems relate to analogy and analogical thinking? Expert systems, if made to closely approximate the actions of human experts, may require analogical problem-solving features that will allow performance comparable to human problem solvers. As will be shown later, analogy is a pervasive part of human cognition, and is used by both novices and experts to solve problems in widely disparate domains.

This article provides an overview of analogy as it pertains to expert systems concepts, and refers to theories of human analogical thinking, along with related automated systems, as reported in the literature. Studying analogy has several implications for the design and implementation of future expert systems.

Selected expert systems issues

A review of expert systems by Nau indicates that there are several limiting factors to shifting research-oriented expert systems into commercial applications: "For the continued success of this transition from research laboratories to the real world, more progress is needed in resolving several problems."[1]

The three specific problems identified by Nau and studied by others[2] are development issues that commonly appear throughout much of the work on expert systems:

(1) Excessive effort required to build expert systems,

(2) Difficulties in articulating knowledge used by an expert and encoding a knowledge base, and

(3) Proper system interaction with the end user.

An analogical problem-solving (APS) component added into various phases of an expert system could assist in reducing these three problems. The type of APS component proposed would not necessarily be just a single-system construct; instead, several components employing analogy-like capabilities could be used at different developmental stages. A case study description of one expert system, CRIB,[3] illustrates how various aids for development purposes may change during an expert system life cycle. (For example, the type of aid needed during the initial development phase might be quite different from the types needed in later phases, including the final fine-tuning phase.)

Before we consider each of the three problems and offer APS component solutions, a review of the literature on what constitutes analogy will be given. Some familiarity with the analogical thinking process and how automated systems have attempted to recreate this important human decision making capability is needed if we are to understand why APS components may be of use in the expert system life cycle.

Analogy research

Research on analogy can be organized into a simple hierarchy as shown in Figure 1. Studies of analogy tend to concentrate on one of two approaches:

(1) Human decision making—an empirically oriented approach focusing on human performance. This approach often involves conducting experiments with human subjects to discover cognitive mechanisms used in analogical thinking.

(2) Automated decision making—a machine-based approach proposing and possibly constructing a computer system that performs analogical thinking. Resulting systems often have an extremely narrow range of analogy performance capability.

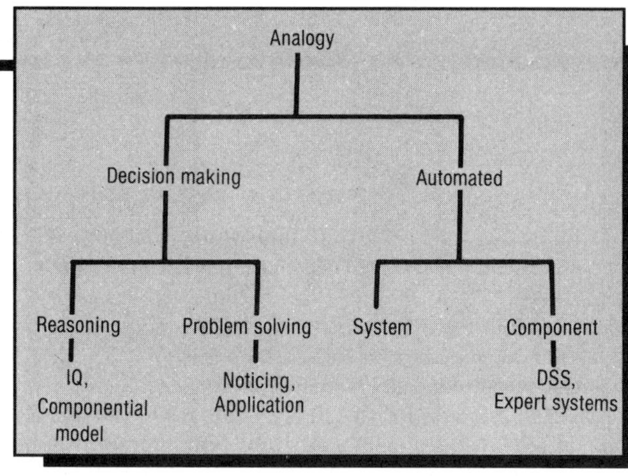

Figure 1. Simple hierarchy of analogical thinking research and topics considered in this article.

A review of the literature indicates that, for the most part, research efforts in one category have not made use of the other; automated systems are often based on researcher intuition about what constitutes analogical thinking, not on behavioral results. Similarly, analogy studies involving human decision making rarely make use of results from automated systems. A recent study of approaches to human reasoning argues that a convergence of alternative approaches (e.g., psychometric, Gestalt, computer-based) will best serve advancement of our understanding of reasoning.[4]

Human decision making. The human decision making category indicated in Figure 1 can be divided into two major research avenues concerning analogical thinking. First, analogical reasoning is an analogy of the proportional form A is to B as C is to D, (A:B::C:D), as commonly used in various intelligence tests. The concept of intelligence quotient, or IQ, has its origins in work by Spearman,[5] and the ability to solve A:B::C:D analogies was considered a central element of general intelligence.[6]

Analogical reasoning. Most studies of analogy in human decision making have concentrated on the proportional A:B::C:?, where the missing D is to be solved. The terms used in analogical reasoning, A, B, C, and D, often consist of word association problems or arithmetic problems. A comprehensive review of analogical reasoning studies has been provided by Sternberg.[7]

Going beyond his comprehensive review, Sternberg has also studied human decision making and analogies by performing a lengthy series of experiments to isolate the analogical reasoning processes step by step. As a result of these studies, Sternberg proposed an information-processing framework for explaining analogy. His framework identified mental operations involved in analogy inspection (encoding), analogy solving (operations consisting of inference, mapping, and application), and analogy answer giving (response and justification).

Research by Sternberg and others (including Whitely and Barnes,[8] and Gardner,[9]) has developed impressive models to explain analogical reasoning. These models, known as componential models, make predictions regarding human performance on A:B::C:D tests. Robust componential models

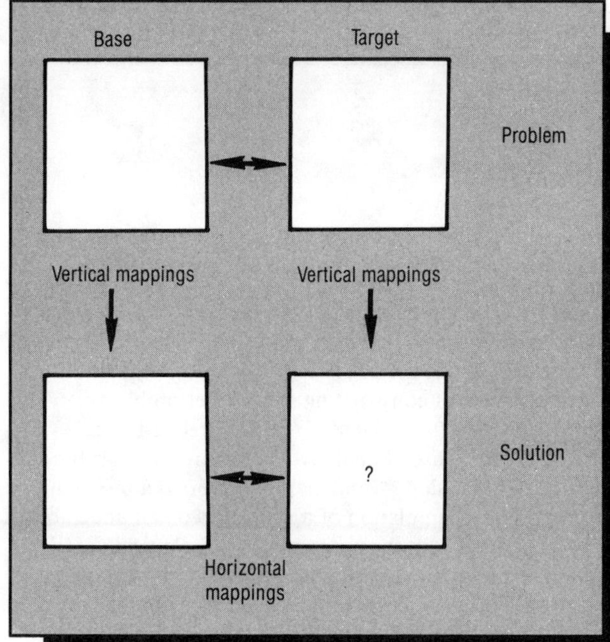

Figure 2. Elements of analogy in problem solving: base and target analogs with their associated mappings.

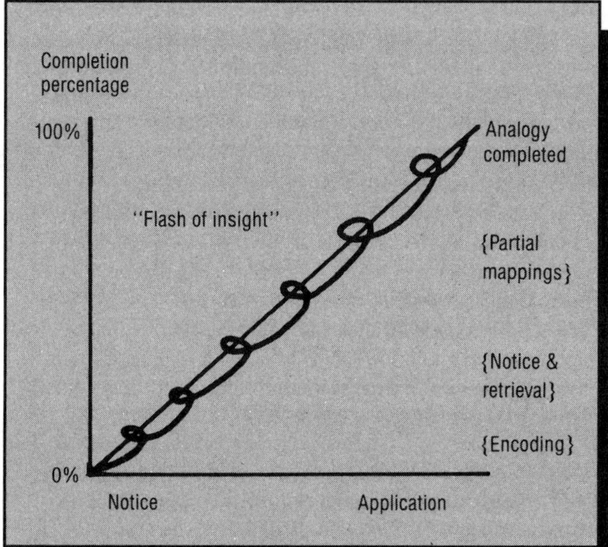

Figure 3. Steps in analogical problem solving suggest APS is a process developed gradually and possibly involving repeated inspections throughout.

attempt to isolate many factors critical to analogical reasoning, and to suggest individual differences (i.e., across age levels of test takers) as part of their prediction of human performance.

Unfortunately, tests requiring simple A:B::C:D proportional problem solving may not be representative of complete real-world problem solving.[10] A test of the A:B::C:D variety may be considered well-defined in its scope—the problem solver knows what kind of problem is being presented. In real-world situations, however, problem solving is often ill-defined—the problem itself is often not even identifiable, even after in-depth inspection.

Other questions confound the relevancy of applying real-world problem solving to A:B::C:D tests. Do these tasks share representations, processes, and strategies used in everyday problem solving? Do these tasks require common-sense knowledge or actual domain knowledge? Do these tasks force us to overlook important discovery acts by suggesting that an analogy exists in the first place?

Analogical problem solving. Another research perspective on analogy, called analogical problem solving, has recently emerged. APS studies human problem solving that involves complex problems in which analogies can be used as solution-deriving alternatives. Representative examples of studies on APS include Reed et al.,[11] Rumelhart,[12] Rumelhart and Abrahamson,[13] and Rumelhart and Norman.[14]

Gick and Holyoak[15,16] have proposed an extensive framework for the APS process. They defined an analogy as consisting of a base analog (prior experience) and a target analog (current problem). Figure 2 indicates the elements of an analogy. The base has a problem-solution pair, while the target has a problem and an unknown solution. APS is a process involving four broad steps:

(1) Formation of a mental representation for the base and the target,

(2) Noticing some aspect of the target (a cue), and then retrieving the base,

(3) Initial partial mapping of base and target, and

(4) Extended mapping to complete the analogy.

Analogy example revisited. The example of analogy usage demonstrated by John Taylor in the hypothetical debugging problem described earlier is a representative instance of Holyoak's framework. John Taylor had prior experiences of software debugging and had encoded these experiences in his mental knowledge base. Upon encountering the new software problem with its associated unknown and perplexing solution, he noticed some part of the newly encoded target that allowed retrieval of his prior experiences. Further study of the relationships between his current problem and prior problem-solution pairs led to a pair selection and an initial mapping of the base and target.

After playing mentally with the code, John completed the mapping to develop a new solution to the problem. His new solution then was tested to ensure that it correctly matched the actual problem being considered. Outside observers—and possibly even John Taylor—may not have realized that an APS process occurred. Quite frequently, analogy usage is reported as a flash of insight (see Leatherdale[17] for a comprehensive history of analogy). Studies of debugging partially recognize the role of analogies, such as a suggestion by Beizer[18] that debugging involves intuitive leaps, experimentation, and conjectures.

Formulation of analogies. Figure 3 illustrates the APS process as a series of interacting steps. This figure is an illustrative, but somewhat oversimplified, summary of empirical results on APS step processing. Analogies, particularly if they are complex, tend to be formulated gradually during problem solving (represented in Figure 3 by lines that gradually spiral upward to analogy completion). Flashes of analogy reported by problem solvers usually come about after the initial formulation. Observers may not be aware of the formulation process at all, assuming that the analogy just "happened" in a sudden instant.

The pervasiveness of analogies is sometimes known but not taken into account by decision makers. Silverman[19] has conducted one of the few studies on analogy usage in software engineering and reports a severe lack of attention to analogy usage by software developers. Under direct questioning, the software developers realized the value of analogies in their work, yet few if any took advantage of this awareness by incorporating formal methods into their work. (Such methods might include improving one's mental ability to use analogies, or making use of automation to assist with analogical thinking.)

Not taking advantage of analogies is unfortunate enough, but falling victim to their disadvantages is worse. Silverman reported a paradox in which software developers used suboptimal analogical knowledge and inferior decision rules without a loss of confidence in their decisions. APS involves many cognitive complications (for studics on misapplications of analogies, see the work by Gilovich[20]), so a cost/benefit analysis should be performed.

Automated analogy. Figure 1 divided analogy into human aspects of analogical thinking and automated aspects of analogy. Automated aspects of analogy can be subdivided into systems and components, a subdivision related to the intent of the automation developers. A developer may choose to propose or build an analogical system that is entirely self-reliant and is not considered part of another automation package. An analogy component refers to circumstances where the analogy automation is intended to work as part of a larger package.

At present most automation of analogy work can be placed in the component category; the few systems reported are only relevant to narrow, experimental projects. In Figure 1, the component category is further subdivided into two major groupings: decision support systems and expert systems. Additional subdivisions of component exist, but these two groupings are directly related to the topics of concern here.

One project involving an analogy decision support system relates back to the software developers studied by Silverman, who is undertaking a project to develop an automated intelligent assistant for software developers.[21] Software requirements would be translated into a building block architecture as directed by analogical procedures. The decision support system envisioned would contain this analogy component along with other software development aids.

The remainder of this section cites selected examples of progress made to date in automated analogy, and provides simplified descriptions of the projects. Additional details concerning these efforts can be found in the referenced sources.

Early analogy automation. Artificial intelligence research has frequently made use of analogy-like features as an incidental aspect of constructing some other problem-solving capability (see Boden[22] for a comprehensive review). One notable early use of analogy-like features was a subroutine called Findanalog.[23] Findanalog supported a program that explored the simulation of a neurotic woman undergoing psychoanalysis. In the program, a psychological conflict might give rise to searches which, if pursued on a random basis, would occur exhaustively. Findanalog computed commonalities between items stored in an information matrix of noun codes and communicated these shared attribute ratings to other parts of the program, thereby helping the other subroutines reduce their searches.

Maps, models, and diagrams are considered analogical representations by some authors (e.g., Barr and Feigenbaum[24]). As a result, programs such as the Geometry Theorem Prover,[25] the General Space Planner,[26] and Whisper,[27] may be classified as making use of analogy. These types of analogical representations are not reviewed here, since the conceptual use of analogy as employed in these studies generally involves direct representations of world objects rather than the ability of a program itself to "reason" in an analogical fashion.

An artificial intelligence program may employ a limited linguistic analogy feature for semantic processing of text. For example, Wilks[28] developed a program that obtained semantic knowledge of words within text chunks by matching problematic words with other portions of the text. The meaning of a problematic word could possibly be determined by selecting similar terms in a relevant text passage. Measures of semantic density were then used to select semantic preferences.

Analogy as a system. One of the first artificial intelligence programs to concentrate on analogy as a *system* (in which analogy is used as the entire basis of machine processing for the tasks being automated) was a program by Evans called Analogy.[29] Evans developed a program to deal with geometric problems of the A:B::C:D variety as encountered on traditional intelligence tests. His program has served as a fundamental starting point for much of the automated analogy work that has followed.

The Evans program was constructed with an analogy-like computational ability to seek out patterns in the geometrical problems presented, and could perform at the human level of older children for A:B::C:D proportional solving (see Stelzer[30] for an interesting axiomatic approach to analogy). Improvements on this computational focus, embodying a more robust sense of analogy, were built into Winston's analogy program.[31-33] Winston attempted to include analogy discovery between conceptual structures and made use of elaborate network-matching procedures.

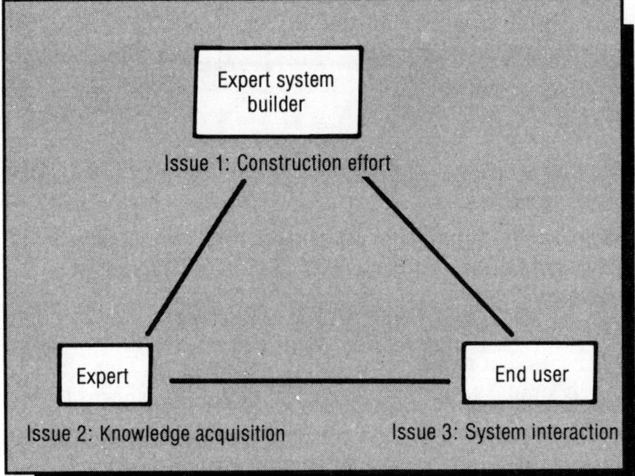

Figure 4. Selected expert system issues and their primary participants.

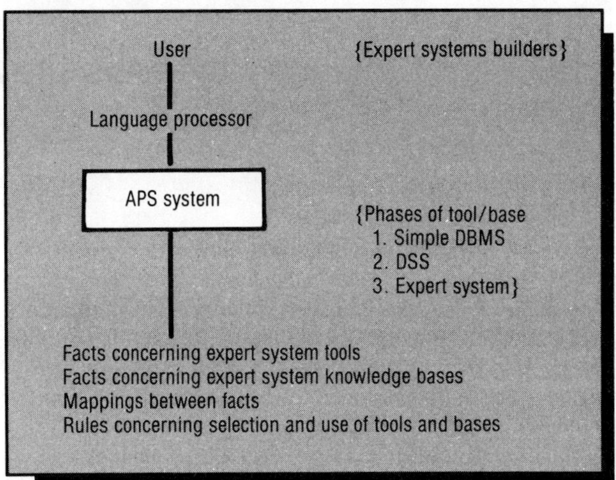

Figure 5. Conceptual anatomy of tool/base package using APS for selection of expert systems tools and knowledge bases.

Recent trends in automated analogy. The focus of automated analogy has gradually shifted from analogical reasoning tasks (A:B::C:D) to analogical problem-solving tasks, in a manner similar to behavioral studies discussed earlier. Newell and Simon[34] identified the importance of APS in their examination of problem solvers playing chess, performing cryptarithmetic work, and doing logical substitutions. The work by previous analogy researchers (including Newell and Simon,[34] Winston,[31,32] and Evans,[29]) led Carbonell to conduct several studies on automated APS.[35-37]

Carbonell extended means/end analysis (MEA as given in Newell and Simon[34]) to include an analogy transform problem space. He then proposed an analogical inference engine that could be incorporated into artificial intelligence models. A general problem-solving process would link into the engine on an "as needed" basis, seeking out possible analogies and then using the results from the engine to pursue other opportunities. Rather than considering analogy as a disjoint activity used for particular classes of problems (e.g., geometric problems), Carbonell argued for analogy to be integrated into larger APS systems.

The program called Politics[35] embodied many of Carbonell's early concepts regarding the development of automated APS. His Politics system planned and counterplanned against adversaries in the realm of international politics. There are many possible analogies for a given situation, and a good analogy developed in one context may be irrelevant in another. (This raises various issues of analogy structure mapping; see Gentner,[38] Gentner and Toupin,[39] and Burstein.[40]) Thus, purpose-directed analogy development is critical to the use of analogies.

Winston's analogy project required guided learning from a human teacher before causal relations between the analogy base and target could be constructed. A new prototype analogy system written in Prolog makes use of purpose-directed analogy.[41] In purpose-directed analogy automated analogy development is supposed to be self-directed, with the system providing its own guidance during the generation of a causal network of relations. This guidance is dependent on the context of a particular analogy.

In the narrow examples chosen for the purpose-directed analogy development prototype,[41] purpose refers to the purpose for which an artifact is to be used. In the prototype analogy settings, artifacts included a ceramic mug (whose purpose concerns the containment of hot liquids) and a vehicle (an object usable for driving and subject to legal restrictions). New research is needed to enlarge the notion of purpose in analogy formation to include such concepts as purpose of explanation and purpose of learning.

Julip[42,43] was implemented to discover analogies used for argumentation. An analogy might be formed for the purpose of arguing a position, such as a debate over the impact of computerization on people's jobs. Parties to such a debate might argue that historical events surrounding horse-drawn carriages and the introduction of automobiles is a fitting analogy. Julip attempts to recognize the presence of just such an analogy when it is presented in a specially prepared editorial statement. Julip follows the work of IPP,[44] a system that compared wire service stories for similar events stored in memory.

Zorba,[45,46] developed in the early 1970s, generated simple analogies between theorem pairs that were stated in a form of predicate calculus. To direct analogy development, Zorba successively constrained the environment of the analogy. Subproblems were drawn out of a larger problem and then attacked by the use of an analogical reasoner, thus decreasing the problem-solving effort by shrinking the scope being examined.

McDermott constructed a production-based system in OPS to illustrate the use of analogies between task-completion methods.[47,48] His system, ANA, began with methods for solving simple problems (i.e., the task of painting a chair) and built upon these methods to form a collection of analogy bases. If ANA was given a new task to perform, it would use analogy to find solution methods. The result was added to the collection of analogy bases so it could be used to detect the same task in future problems, thus avoiding repetitive analogy formation.

Analogies might be used not only for argumentation but also for conflict resolution. Mediator,[49,50] a system for resolving disputes, has been used on a case study of the Sinai conflict. Mediator examined two prior analogs, a Panama Canal dispute and a Korean conflict, in an effort to resolve the Sinai case.

Mediator, Julip, and other relatively new systems have made extensive use of research in other areas of artificial intelligence, including memory organization (e.g., Schank and Abelson[51]) and natural language processing. Analogy studies often must cross disciplinary boundaries and examine issues of automation, experimentation, planning and learning behavior, and other related topics. (For planning and learning see Munyer[52] and Carroll and Mack;[53] a general review of analogy topics is provided by Silverman.[54,55]) Commercial systems with prominent features making use of analogy-like techniques are beginning to emerge from research results. An example is TIMM, an expert system development program.[56,57]

Concepts for expert systems

How does the advancement of analogical systems and our understanding of analogy aid in the development of expert systems? We will now examine the three expert systems issues identified earlier, and offer analogical problem solving as a means of resolving several current difficulties. Currently, APS is neither a cure-all nor a ready "off-the-shelf" tool for expert systems. However, APS does present potential solutions to hard problems. Ongoing research projects on analogy, such as those by the Microelectronics and Computer Technology Consortium (MCC),[58-60] may lead to working systems representing major contributions.

Figure 4 summarizes the three expert systems issues we have selected, grouping each issue according to its impact on the participants in expert system development. First, the expert system builder (knowledge engineer) is concerned with the effort required to construct an expert system. Second, the expert supplying the knowledge for the knowledge base is concerned with knowledge acquisition processes and encoding. Third, the end user is concerned with the interaction that must take place when the expert system is used. Of course, each participant may be concerned with all three issues, and the triangular shape depicted in Figure 4 illustrates the interdependencies. For example, while end user is the primary participant in the issue of system interaction, it is an issue the expert system builder has to be concerned with as well.

Issue one: effort in expert systems construction. Expert system development is contingent on tools for expert system construction and knowledge base collection. New development projects often start without a predetermined tool and without an existing encoded knowledge base. After development of several expert systems, an organization will have selected a large set of tools to be used in future projects, and will also have amassed a substantial number of knowledge bases. In both cases—initial expert system projects and later ones—builders are required to select some expert system tool. In the latter case, builders may wish to reuse existing knowledge bases.

One example of an automated analogical problem solver used to reduce the effort involved in expert systems construction would be a tool/base selection package. Builders of an expert system would be able to consider selecting a tool that relates the current expert system project to previous projects. Similarly, builders could consider previously built knowledge bases that might be reused in new projects. This idea of reuse is similar to the software engineering reuse urged by Silverman.[19]

Tool/base example. Figure 5 illustrates an APS tool/base package serving as an expert system about expert systems. In a more primitive state, lacking an APS capability, the tool/base package merely maintains a database containing historical data on previous expert system projects. Builders of new systems can tap previous successes by making simple queries into the database.

In a second stage of evolution, the tool/base package serves as a significant decision-support aid. The database is now augmented by a model of expert systems projects that specifically relate details of tools with tools, tools with bases, and bases with bases. A builder would work with the package in a more sophisticated manner, allowing input of various new project requirements and inspecting output selections provided by the model.

In its fullest state, the tool/base package uses expert knowledge about expert system selection to analogically relate a new project to prior projects. Mappings between the current project and prior projects find similarities and differences that lead to automated selection of relevant options for reusing previous efforts. Recommendations from the tool/base package would also provide guidelines for transforming any prior matches to the current project.

A tool/base package has been depicted here as an APS component for the initial phase of expert system development. Many current developers might consider the package of little use in light of the current state of expert system construction. (For example, with few tools easily available and few knowledge bases, and with tools and bases existing only on an experimental basis, many developers must already be "experts" on expert system development.) Predictions of future expert system development[61] suggest that there will be increasing demands for expert system construction and wider availabilty of tools and bases. This will open the door for less skilled developers, which would in turn increase the utility of a tool/base package.

Developing a tool/base package requires a deeper understanding of expert system developers and the relationships between expert system tools and knowledge bases. A research project by Eliot[62] is investigating these topics. A tool/base has applications in phases beyond the initial development of an expert system. For example, a library of expert system subroutines could be used by builders to piece together an expert system, with the selection guided by APS. Also, an APS subroutine contained in a library might be combined with other routines to produce a system that requires an embedded analogy function.

Issue two: knowledge acquisition. Assembling a knowledge base can be a very slow, difficult, error-prone process.[2] Already, several systems have emerged to improve knowledge acquisition via automatic interviewing mechanisms applied to domain experts. (Examples include Teiresias[63] and Meta-Dendral.[64]) Analogy can be used to assist in knowledge acquisition and encoding.

Manual procedures automated. Manual procedures have been used to test for expert knowledge usage by applying psychometric methods such as multidimensional scaling.[65] A psychometric method used in this context requires experts to evaluate multiple case problems within their domain. The prescribed intent of the method is to uncover hidden cognitive assumptions made by the experts, and it is frequently used in conjunction with other methods such as verbal protocols. Opinions rendered by experts can then be examined for similarities, and generalizations leading to rules and knowledge elements are developed. A manual procedure employing such a method could be enhanced by an automated APS component detecting these generalizations (a process similar to automating verbal protocol analysis[66]).

Direct automated acquisition. A more powerful form of knowledge acquisition using APS would require an automated system to interview experts directly, incrementally building a knowledge base. There would be several options available to the experts questioned by the APS component:

(1) The experts enter cases, and the system encodes the cases and uses analogy to build relations among them;

(2) The system supplies cases and experts comment on case relationships, allowing APS to build analogies and detect the use of expert rules; or

(3) Experts and the system interact, mutually deriving expert rules and capturing knowledge.

Similar automated acquisition. Programs for learning rules and acquiring factual knowledge have previously used inductive reasoning techniques. For example, AQ11[67] formed rules to diagnose plant diseases by inferring classifications from examples provided by experts. MCC is using analogies between common-sense knowledge topics (e.g., automobile diagnosis and medical diagnosis) to assemble a large "real-world" knowledge base.[58]

Knowledge acquisition is a form of machine learning, as suggested by Carbonell.[37] He has argued that different learning strategies require different levels of inference, and possibly different types of inference mechanisms. Generally, learning by analogy requires a greater amount of inference on the part of the problem solver than does learning by being told (sometimes called learning from instruction).

Learning from example is a special case of inductive learning which tends to require greater inference than analogy learning. However, varying contexts can dramatically sway the amount of inference required by either learning strategy. (Note that in learning from example there are generally no pre-existing knowledge chunks directly usable for coping with the new problem, while pre-existing knowledge chunks are essential to analogy learning.)

Thus, research focused specifically on other automated learning strategies is helpful in developing automated analogy, but has not provided exacting solutions to complete analogical processing.

Robustness automated. Hartley[3] indicated that CRIB, an expert system for computer fault diagnosis, was fragile when faced with an unknown fault: "It is difficult to see how any expert system can be made to function effectively in such a situation; a doctor faced with a new strain of virus in a sick patient is in similar difficulty until an expert microbiologist comes to the rescue."

One viable alternative would be to have an APS component that not only aided knowledge acquisition, but that also could revise an existing knowledge base while in active use. Using the example posed by Hartley, a doctor faced with a new strain of virus is likely to consider analogies from her/his existing knowledge base prior to seeking out other expert advice (similar in some respects to the John Taylor case given earlier). Of course, adding an APS component as an afterthought, after implementation, will make assimilation of the APS feature more difficult; while incremental addition would serve to reduce the effort required and allow mixing in of APS on an as-needed basis.

Issue three: end user interaction. Success of an expert system, like other computing projects, is greatly dependent on ease of use.[2] However, interaction with the end user can be even more problematic with expert systems than with traditional software. Expert systems and AI software in general may be held to higher quality requirements than other systems. (A series of reasons for this claim is provided by Bundy and Clutterback.[68])

Tailoring explanations. One way to improve the quality of interaction is to tailor explanations to the user.[69] Research on explanation tailoring recommends that user beliefs and goals should be determined, either on an a priori basis or during discourse, and then used to explain system performance. Similarly, behavioral studies of analogy suggest that the act of explaining frequently involves having a teacher relate the student's knowledge to the topic of discourse.

Combining these two topics, tailoring and teaching, suggests that a powerful interface can be constructed by relating an expert system to its users via APS.

An APS component built into an expert system during its construction, and actually used in the final phase of development (implementation, maintenance, and refinement), would provide a more accommodating user interaction. Instead of receiving a list of inferences given in system language, end users requesting explanations of system performance would have the explanation provided by a narrative that described the system from the perspective of the user (and could be further enhanced by providing a tutoring capability[70]).

Four classes of explanation tailoring can occur:

(1) System domain to user domain,
(2) System domain to user world knowledge,
(3) World knowledge to user domain, and
(4) World knowledge to world knowledge.

Domains refer to the type of problem/solution pairs being selected from the knowledge base. For example, suppose a boat engine mechanic wanted to consult an expert system designed for car engine repairing. The system domain of car engine repair would have to include the capability to tailor an explanation for a user domain of boat engine repair. The system domain and user domain could range from the same domains to rather disparate ones. To avoid making incorrect analogies, the APS component would need to be able to assess the relative applicability across the particular domains. (These would have to be domains in which analogies would hold; otherwise, the system would not render an analogy and alternative explanation techniques would be used.)

World knowledge refers to common-sense knowledge, and the remaining three classes represent increasingly difficult forms of explanation. Various questions continue to perplex research on common-sense knowledge. How is common-sense knowledge acquired, for example, and how big a role does common sense play in expert judgment?

Tailoring example. Our hypothetical example involving John Taylor can be expanded to include an expert system developed for software debugging. John Taylor would input characteristics of the debugging problem to an APS component of the expert system and receive an answer ("Fix a particular line of code in the manner prescribed as follows."). If John didn't find this answer acceptable, he could invoke an explanatory mechanism that described the answer by matching the current code to analogically related code.

Analogically related code could be transformations of previously stored cases, or a new case based on John's background. For example, suppose John is a Pascal programmer who has just learned the language Ada. During an Ada programming session he needs debugging assistance, but the Ada explanations aren't clear to him. The APS component could relate the problem and solution to Pascal, and help John to master Ada while producing the proper programming code. Further, rather than a domain-to-domain explanation, APS might be invoked to describe the solution by using an analogy to a common-sense world knowledge sequence (domain-to-world explanation).

Other interface issues. Relating explanations to users analogically (primarily a form of output) is only one alternative for the use of APS as an interface aid. User input could be in the form of narrative cases, and APS would work with a natural language component to select user requests and data needed by the system. Input requests and output from the expert system do not have to be in text alone; they may be combined with graphics as well. In this interface category APS components are envisioned as a front end or back end to the expert system, providing a service equivalent to that found in recent revisions of popular database management systems.

Future research directions

The preceding sections have considered our current understanding of analogy and the potential role of analogy in expert systems. From this discussion we can derive a number of research directions that would expand our knowledge of analogy and expert systems. While not exhaustive, the following list does indicate research directions with a strong bearing on progress in these two emerging fields.

(1) Develop a macroscopic theory of analogy that includes human decision making and automated analogical processing.
(2) Increase understanding of cognitive processes involved in analogy discovery, formulation, and usage.
(3) Further distinguish direct analogy work from indirect work, while noting contributions provided by indirect research.
(4) Investigate and build usable APS into existing technology.
(5) Investigate and prototype APS that will be usable in future technology (e.g., taking advantage of parallel processing).
(6) Expand expert systems capabilities by refining expert system technology so that it encompasses automated analogy.
(7) Provide specialized languages or adopt expert systems languages that incorporate analogical structures, relationships, and processing.
(8) Evaluate the cost effectiveness of APS usage across various domains, assessing the increased loads exhibited on expert systems.
(9) Examine interrelationships between expert system processing and APS (such as trade-offs in coordination, intersection points, and subdivision of activities).
(10) Provide measures for matching the expert system life cycle to automated analogy tools, which may assist development (service) or become part of the artifact (product).

These research directions are overlapping and serve simply as broad indicators for possible research projects. Many subtasks are identifiable within each of the listed statements, and can be used to further investigate approaches to the broader statements.

Analogies are found in everyday conversation, appearing to span everything from routine thoughts to complicated mental reasoning. They can be used in both common-sense knowledge and expert-knowledge problem solving. Analogical thinking is often a neglected form of cognition, one that problem solvers are usually unaware of using. Analogical thinking is frequently mistaken for a "flash of insight," and thought by some to be a creative form of problem solving that is not reducible to formal steps of logic.

Behavioral studies indicate that the APS process does appear to have a mental framework that can be formalized. Recent artificial intelligence projects suggest that analogy-like features can be used in the development of artificial intelligence software. Currently, analogy systems and analogy components perform well below human levels of APS. Still, encouraging results have been demonstrated, and analogy is a rich area for further research.

Expert systems research and development can also benefit from the study of analogy. Three issues of immediate concern are (1) the effort required to construct expert systems, (2) knowledge acquisition, and (3) end user interaction. At this stage in our understanding of both expert systems and analogical thinking, concepts for useful means of combining the two fields are important. Suggestions for applying APS to each of the three issues have been outlined in this article, providing several concepts and some impetus for the construction of future expert systems. E

Acknowledgments

I acknowledge the helpful comments provided by Dr. Kamran Parsaye and *IEEE Expert* referees.

References

1. D. S. Nau, "Expert Computer Systems," *Computer*, Vol. 16, No. 2, Feb. 1983, pp. 83.
2. F. Hayes-Roth, D. A. Waterman, and D.B. Lenat, eds., *Building Expert Systems*, Addison-Wesley, Reading, Mass., 1983.
3. R. T. Hartley, "CRIB: Computer Fault-finding Through Knowledge Engineering," *Computer*, Vol. 17, No. 3, Mar. 1984, pp. 76-83.
4. R. J. Sternberg and M. I. Lasaga, "Approaches to Human Reasoning: An Analytic Framework," in A. Elithorn and R. Banerji, eds., *Artificial and Human Intelligence*, Elsevier Science Publishers, The Netherlands, 1984.
5. C. Spearman, *The Nature of "Intelligence" and the Principles of Cognition*, Macmillan, London, 1923.
6. G. Polya, *How to Solve It*, Princeton University Press, Princeton, N.J., 1957.
7. R. J. Sternberg, *Intelligence, Information Processing, and Analogical Reasoning: The Componential Analysis of Human Abilities*, Erlbaum, Hillsdale, N.J., 1977.
8. S. E. Whitely, "Relationships in Analogies: A Semantic Component of Psychometric Task," *Educational and Psychological Measurement*, Vol. 37, 1977, pp. 725-739.
9. M. K. Gardner, *Some Remaining Puzzles Concerning Analogical Reasoning and Human Abilities*, doctoral dissertation, Yale University, 1982.
10. H. L. Dreyfus, *What Computers Can't Do*, Harper and Row, New York, 1979.
11. S. K. Reed, G. W. Ernst, and R. Banerji, "The Role of Analogy in Transfer Between Similar Problem States," *Cognitive Psychology*, Vol. 6, 1974, pp. 436-450.
12. D. E. Rumelhart, "Notes on a Schema for Stories," in D. G. Bobrow and A. Collins, eds., *Representation and Understanding: Studies in Cognitive Science*, Academic Press, New York, 1975.
13. D. E. Rumelhart and A. A. Abrahamson, "A Model for Analogical Reasoning," *Cognitive Psychology*, Vol. 5, 1973, pp. 1-28.
14. D. E. Rumelhart and D. A. Norman, "Analogical Processes in Learning," in J. R. Anderson, ed., *Cognitive Skills and Their Acquisition*, Erlbaum, Hillsdale, N.J., 1981.
15. M. L. Gick and K. J. Holyoak, "Analogical Problem Solving," *Cognitive Psychology*, Vol. 12, 1980, pp. 306-355.
16. M. L. Gick and K. J. Holyoak, "Schema Induction and Analogical Transfer," *Cognitive Psychology*, Vol. 15, 1983, pp. 1-15.
17. W. H. Leatherdale, *The Role of Analogy, Model, and Metaphor in Science*, North-Holland, Amsterdam, 1974.
18. B. Beizer, *Software System Testing and Quality Assurance*, Van Nostrand Reinhold, 1984.
19. B. G. Silverman, "Software Cost and Productivity Improvements: An Analogical View," *Computer*, Vol. 18, No. 2, May 1985, pp. 86-95.
20. T. Gilovich, "Seeing the Past in the Present: The Effect of Associations to Familiar Events on Judgments and Decisions," *Journal of Personality and Social Psychology*, Vol. 40, No. 1, Jan. 1981, pp. 797-808.
21. B. G. Silverman, "Toward an Integrated Cognitive Model of the Inventor/Engineer," Institute for Artificial Intelligence Technical Report GWU/EAD, Washington, D.C., 1984.
22. M. Boden, *Artificial Intelligence and Natural Man*, Basic Books, New York, 1977.
23. K. M. Colby, *Artificial Paranoia*, Pergamon, New York, 1975.
24. A. Barr and E. A. Feigenbaum, eds., *The Handbook of Artificial Intelligence*, Vol. I, William Kaufmann, Los Altos, Calif., 1981.
25. H. Gelernter, "Realization of a Geometry-Theorem Proving Machine," in E. A. Feigenbaum and J. Feldman, eds., *Computers and Thought*, McGraw-Hill, New York, 1963.
26. C. M. Eastman, "Automated Space Planning," *Artificial Intelligence*, Vol. 4, 1973, pp. 41-64.
27. B. V. Funt, "WHISPER: A problem-solving system utilizing diagrams and aparallel processing retina," *IJCAI*, 1977, pp. 459-464.
28. Y. A. Wilks, *Grammar, Meaning, and the Machine Analysis of Natural Language*, Routledge and Kegan Paul, London, 1972.
29. T. G. Evans, "A Program for the Solution of Geometric Analogy Intelligence Test Questions," in M. Minsky, ed., *Semantic Information Processing*, MIT Press, Cambridge, Mass., 1968.
30. J. Stelzer, "Analogy and Axiomatics," *International Journal of Man-Machine Studies*, Vol. 18, 1983, pp. 161-173.
31. P. H. Winston, "Learning and Reasoning by Analogy," *Comm. ACM*, Vol. 23, No. 12, Dec. 1980, pp. 689-703.
32. P. H. Winston, *Learning Structural Descriptions from Examples*, PhD dissertation, MIT AI Lab Technical Report AI-TR-231, Cambridge, Mass., 1970.
33. P. H. Winston, "Learning by Creating and Justifying Transfer Frames," *Artificial Intelligence*, Vol. 10, No. 2, 1978, pp. 147-172.
34. A. Newell and H. A. Simon, *Human Problem Solving*, Prentice-Hall, Englewood Cliffs, N.J., 1972.
35. J. G. Carbonell, "Counterplanning: A Strategy Based Model of Adversary Planning in Real-World Situations," *Artificial Intelligence*, Vol. 16, 1981, pp. 295-329.
36. J. G. Carbonell, "Metaphor: An Inescapable Phenomenon in Natural Language Comprehension," in W. G. Lehnert and M.

H. Ringle, *Strategies for Natural Language Processing*, Erlbaum, Hillsdale, N.J., 1982.

37. J. G. Carbonell, "Learning by Analogy: Formulating and Generalizing Plans from Past Experience," in R. S. Michalski, J. G. Carbonell, and T.M. Mitchell. eds., *Machine Learning: An Artificial Intelligence Approach*, Tioga, Palo Alto, Calif., 1983.
38. D. Gentner, "Structure-Mapping. A Theoretical Framework for Analogy," *Cognitive Science*, Vol. 7, 1983 pp.155-170.
39. D. Gentner and C. Toupin, "Cross-Mapped Analogies: Pitting Systematicity Against Spurious Similarity," *Proc. Seventh Annual Cognitive Science Society Conf.*, Irvine, Calif., Aug. 1985.
40. M. H. Burstein, "A Model of Learning by Incremental Analogical Reasoning and Debugging," *Proc. AAAI*, Washington, D.C., Aug. 1983.
41. K. Smadar, "Purpose Directed Analogy," *Proc. Seventh Annual Cognitive Science Society Conf.*, Irvine, Calif., Aug. 1985.
42. S. E. August and M. G. Dyer, "Analogy Recognition and Comprehension in Editorials," *Proc. Seventh Annual Cognitive Science Society Conf.*, Irvine, Calif., Aug. 1985.
43. S. E. August and M. G. Dyer, "Analogy Recognition and Comprehension in Editorials," UCLA AI Lab Technical Report UCLA-AI-85-7, Mar. 1985.
44. M. Lebowitz, "Generalization and Memory in an Integrated Understanding System," Yale University Computer Science Technical Report No. 186, Nov. 1980.
45. R. E. Kling, "A Paradigm for Reasoning by Analogy," *Artificial Intelligence*, Vol. 2, 1971, pp. 147-178.
46. R. E. Kling, "An Information Processing Approach to Reasoning by Analogy," Stanford Research Institute AI Group TN10, Menlo Park, Calif,, June 1969.
47. J. McDermott, "Learning to Use Analogies," *IJCAI*, Aug. 1979, pp. 568-576.
48. J. McDermott, "ANA: An Assimilating and Accommodating Production System," Carnegie-Mellon University Computer Science technical report, 1978.
49. J. L. Kolodner, R. L. Simpson, and K. Sycara-Cyranski, "A Process Model of Case-Based Reasoning in Problem Solving," *IJCAI*, Aug. 1985, pp. 284-290.
50. R. L. Simpson, "A Computer Model of Case Based Reasoning in Problem Solving," PhD thesis, School of ICS Technical Report GIT-ICS-85/18, Georgia Institute of Technology, Atlanta, Ga., 1985.
51. R. Schank and R. P. Abelson, *Scripts, Plans, Goals, and Understanding: An Inquiry into Human Knowledge Structures*, Erlbaum, Hillsdale, N.J., 1977.
52. J. C. Munyer, *Analogy as a Means of Discovery in Problem Solving and Learning,* doctoral dissertation, University of California, Santa Cruz, Mar. 1981.
53. J. M. Carroll and R. L. Mack, "Metaphor, Computing Systems, and Active Learning," *International Journal of Man-Machine Studies*, Vol. 22, 1985, pp. 39-57.
54. B. G. Silverman, "Expert Intuition and Ill-Structured Problem Solving," *IEEE Trans. Engineering Management*, EM-32, No. 1, Feb. 1985, pp. 29-33.
55. B. G. Silverman, "Analogy in Systems Management: A Theoretical Inquiry," *IEEE Trans. Systems, Man, and Cybernetics*, SMC-13, No. 6, Nov./Dec. 1983, pp. 1049-1075.
56. J. Kornell, "Embedded Knowledge Acquisition to Simplify Expert Systems Development," *Applied Artificial Intelligence Reporter*, Aug./Sept. 1984, pp. 28-30.
57. S. D. Stewart and G. Watson, "Applications of Artificial Intelligence," *Simulation*, June 1985, pp. 306-310.
58. K. A. Frenkel, "Report on the Microelectronics and Computer Technology Corporation Conference," *Comm. ACM*, Vol. 28, Aug. 1985, pp. 808-813.
59. W. Raych-Hindin, "AI: Grading the Hardware and Software Options," *Systems and Software*, Aug. 1985, pp. 38-60.
60. L. Belady and C. Richter, "The MCC Software Technology Program," *ACM Sigsoft Software Engineering Notes*, Vol. 10, July 1985, pp. 33-36.
61. D. Waltz, "Artificial Intelligence: An Assessment of the State-of-the-Art and Recommendation for Future Directions," *The AI Magazine*, Fall 1983, pp. 55-67.
62. L. B. Eliot, "Managing Expert Systems Developers: An Examination of Expert System Development and Software Project Management," USC Systems Science Expert Systems Lab technical report, forthcoming.
63. R. Davis and D. Lenat, *Knowledge-Based Systems in Artificial Intelligence*, McGraw-Hill, New York, 1980.
64. B. G. Buchanan and E. A. Feigenbaum, "DENDRAL and META-DENDRAL: Their Applications Dimension," *Artificial Intelligence*, Vol. 11, 1978, pp. 5-24.
65. S. S. Schiffman, M.L. Reynolds, and F. W. Young, *Introduction to Multidimensional Scaling*, Academic Press, New York, 1981.
66. D. A. Waterman and A. Newell, "Protocol Analysis as a Task for Artificial Intelligence," *Artificial Intelligence*, Vol. 2, 1971, pp. 285-318.
67. R. S. Michalski, "Pattern Recognition as Rule-Guided Inductive Inference," *IEEE Trans. Pattern Analysis and Machine Intelligence*, Vol. 2, 1980, pp. 349-361.
68. A. Bundy and Clutterback, R., "Raising the Standard of AI Products," *IJCAI*, Aug. 1985, pp. 1289-1294.
69. K. R. McKeown, M. Wish, and K. Matthews, "Tailoring Explanations for the User," *IJCAI*, Aug. 1985, pp. 794-798.
70. M. Coombs and J. Alty, "Expert Systems: An Alternative Paradigm," *International Journal of Man-Machine Studies*, Vol. 20, 1984, pp. 21-43.

Lance B. Eliot is an assistant professor of system science and director of the Expert Systems Laboratory at the University of Southern California. He holds CDP, CCP, and CSP certifications. Eliot's current research interests include expert systems, analogical thinking, and cognitive aspects of systems analysis. He is also part of research efforts to develop AI software packages and to use Ada for business-related applications.

Eliot is the author of articles and book reviews, and coauthor of a book. He serves on advisory groups, is a member of IEEE, ACM, and AAAI, and appears in Who's Who guides. He has been a visiting faculty member in the UCLA Graduate School of Management.

Eliot can be contacted at the Expert Systems Laboratory, Systems Science Department, ISSM 102, University of Southern California, Los Angeles, California, 90089-0021.

Reprinted with permission from *Artificial Intelligence*, Volume 19, 1982, pages 17-37.

Processing of Semantic Nets on Dataflow Architectures

Lubomir Bic
Department of Information and Computer Science, University of California, Irvine, CA 92717, U.S.A.

Recommended by Pat Hayes

ABSTRACT

Extracting knowledge from a semantic network may be viewed as a process of finding given patterns in the network. On a von Neumann computer architecture the semantic net is a passive data structure stored in memory and manipulated by a program. This paper demonstrates that by adopting a data-driven model of computation the necessary pattern-matching process may be carried out on a highly-parallel dataflow architecture. The model is based on the idea of representing the semantic network as a dataflow graph in which each node is an active element capable of accepting, processing, and emitting data tokens traveling asynchronously along the network arcs. These tokens are used to perform a parallel search for the given patterns. Since no centralized control is required to guide and supervise the token flow, the model is capable of exploiting a computer architecture consisting of large numbers of independent processing elements.

1. Introduction

Most AI researchers would probably argue that there are no 'architectural' solutions to 'AI problems'. While sympathetic with this point of view, finding architectures which would significantly speed up the execution of a given application seems an effort worth pursuing.

The importance of parallelism has been recognized and a number of AI applications have been implemented on multiprocessor architectures, e.g. ZMOB [5]. Unfortunately, such architectures are based on the von Neumann model of computation and hence the inherent difficulty of dividing computation into independent subtasks and the supervision and synchronization of their execution is the main limitation to massive parallelism.

In this paper we focus on a particular domain—the processing of semantic networks—and show that by adopting a data-driven view of processing, many problems faced by conventional (von Neumann model based) systems are eliminated. As a result, computer architectures consisting of large numbers of processing elements, in particular those designed to execute dataflow programs, may usefully be exploited.

2. Viewing Semantic Nets as Dataflow Graphs

At the architectural level, a semantic net is a collection of nodes interconnected via directed labeled arcs. When implemented on a conventional von Neumann architecture, this graph is a *passive* data structure maintained in primary memory (real or virtual) and manipulated by an outside agent—a program. In our model we adopt a different point of view: The semantic network, rather than being a passive representation of knowledge, is a *dataflow graph* [2, 6]. That is, each node is an active element capable of accepting, processing, and emitting value tokens (messages) traveling asynchronously

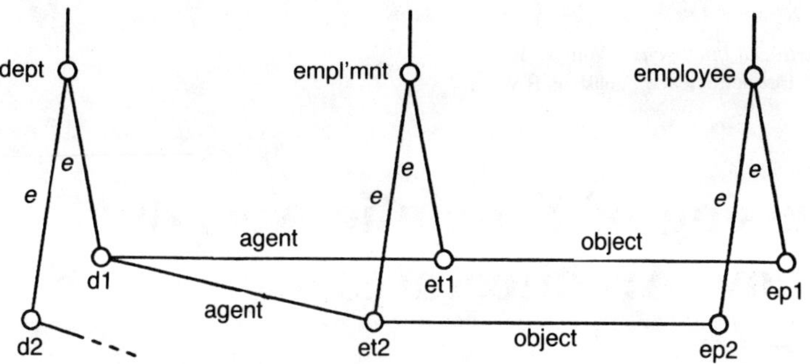

FIG. 1.

– A query is a graph consisting of edges of the form

$$T_1 \quad p \quad T_2$$

along the network arcs. Thus *the computational power is distributed throughout the network*. As will be discussed below, this principle has important ramifications for processing of such 'active' networks.

It has been suggested by several semantic-net based systems, in particular SNIFFER [4], that extracting information from the net may be viewed as a special process of pattern matching: each 'query' is interpreted as a graph template for which the system tries to find matching structures in the semantic net. In a dataflow graph the finding of a given pattern may be carried out by propagation of tokens; the pattern is placed on one or more tokens and injected into selected nodes of the graph. Each token is replicated concurrently from its injection point into many directions along the network arcs in the search of possible matches for the given pattern. (The above principle of extracting information from the net shows some similarity to the marker-propagation scheme proposed in NETL [3]. The main distinction, however, is the need for centralized control in the latter, which imposes significantly different constraints and requirements on the underlying computer architecture.)

Before specifying the procedures for injecting and guiding tokens through the network, we describe the underlying pattern-matching problem more formally:

– The semantic net is a dataflow graph consisting of edges

$$t_1 \quad p \quad t_2$$

where t_1 and t_2 are *constants*, representing nodes, and p is the label of the arc connecting these two nodes. Each node t_1 is capable of receiving, processing, and emitting tokens traveling along arcs. Fig. 1 shows a small portion of a semantic network representing the relationship 'employment' between elements of the two sets of 'department's and 'employee's. All nodes in this network represent constant values, as opposed to variables.

where T_1 and T_2 could be *constants* or *variables*, and p is the corresponding arc label. To answer a given query, the system will try to bind a constant to each variable of the query such that the query matches some portion of the semantic net. In this sense, the query graph is interpreted as a *graph template* to be 'fitted' into the semantic net. For example, Fig. 2 shows a graph template

where the nodes X and Y represent variables.[1] It can be interpreted as finding employees (together with the corresponding instances of the 'employment' relationship) of the department d1: By instantiating X and Y to the pairs et1, ep1, or et2, ep2, respectively, the query matches different portions of the semantic network of Fig. 1.

We will make the following assumptions about the query template:

(1) It forms a connected graph. While it is possible to envision queries consisting of several disconnected components, each of these may be treated as a separate graph template and fitted into the semantic net independently. Hence this assumption can be made without any loss of generality.

(2) At least one of the nodes of the template corresponds to a constant value. This assumption is justifiable on pragmatic grounds since queries containing *only* variables are too general to be of any practical use; they correspond to finding sets or set elements related via unspecified relationships to other unspecified sets.[2]

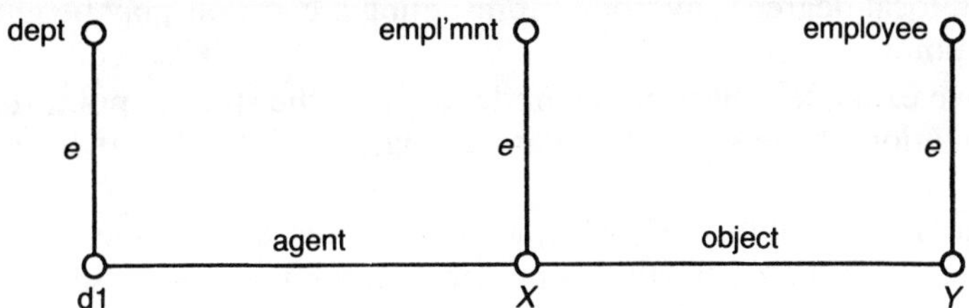

FIG. 2.

3. Transformation of Templates to Simplify Fitting Procedures

As mentioned earlier, the fitting of a graph template into the semantic net is to be performed by tokens propagating asynchronously along the edges of the dataflow graph representing the semantic net. Due to the asynchronous nature of dataflow systems, the detection of cycles is difficult. The problem is greatly simplified if the given template is first transformed into a non-cyclic form. Two approaches may be taken to accomplish this—the template may be transformed into a *tree structure* or into a *linear chain*. The latter is defined as

- a connected chain of edges without cycles or branches, where
- the leftmost node T_1 is a constant while the remaining nodes T_2 through T_n may be constants or variables.

Graphically, we can depict a linear chain as follows:

While the approach of transforming a given template into a general tree structure yields more potential parallelism, we chose to describe only the simpler case of linear chains to illustrate the basic principles. The transformation in this case is based on the idea of finding an *Euler path* through the given graph template, i.e. a path which traverses all arcs exactly once. The following steps illustrate how this is accomplished:

(1) Transform each cluster into an Euler graph, that is:

(i) Examine each node and determine the number of arcs emanating from that node. This number is termed the *local degree* of a node.

[1] Throughout this paper, lower case letters are used to denote constants while capitals are used to denote variables.

[2] The model can easily be extended to cope with queries containing only variables; since such an extension does not increase the usefulness of the model in any significant way, it will not be discussed further.

(ii) If there is a node with an odd local degree, replicate one or more of the edges connected to that node to make the degree even. (Nodes with an odd local degree always occur in pairs and hence it is possible to eliminate them by replicating the paths connecting each pair.) If there are several choices of paths that could be replicated, favor those with the smallest number of free variables in order to minimize the number of possible bindings. Repeat this step until at most two nodes (one of which must be a constant) have odd local degree.

To illustrate this step, assume that the graph template of Fig. 2 is to be fitted into the semantic net, a portion of which is shown in Fig. 1. By replicating the edge connecting the nodes 'employment' and '*X*', as shown by the dotted line in Fig. 3, the number of odd-degree nodes has been reduced to two; these are the nodes 'dept' and 'employee'.

(2) Starting with one of the odd-degree nodes representing a constant, find an Euler path through the graph template. Simple algorithms for finding such a path are discussed in most books on graph theory. (In the case where all nodes have an even local degree, any node representing a constant may be chosen as the starting point.)

In the above example, selecting the node 'dept' as the starting point results in the transformation of the graph template of Fig. 3 into the linear form shown in Fig. 4.

(3) Due to the possibility of multiple occurrences of variables in a linear graph template a final modification is performed: the template is scanned from left to right and all repeating occurrences of variables are linked together via pointer chains. These will be used during execution to bind all occurrences of the same variable to the same constant. In Fig. 4 the variable *X* occurs twice; the two instances are linked via pointers, resulting in the final form of the linear template shown in Fig. 5.

It can be shown that the above algorithm is guaranteed to produce a linear graph template comprising all edges and nodes of the original nonlinear template, some of which may be in duplicate. We now devote our attention to describing how patterns corresponding to such a linear template are found in the underlying semantic networks using tokens.

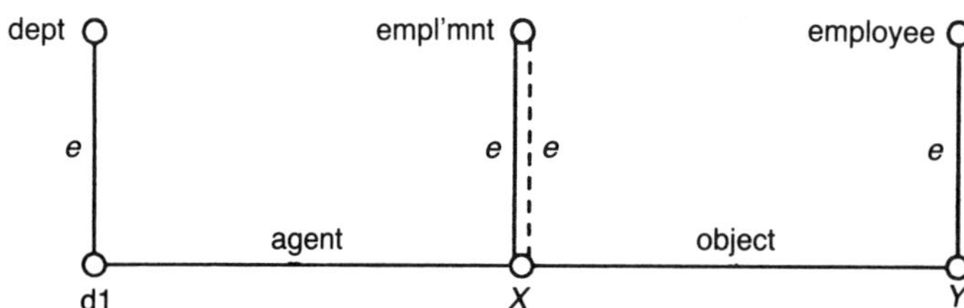

Fig. 3

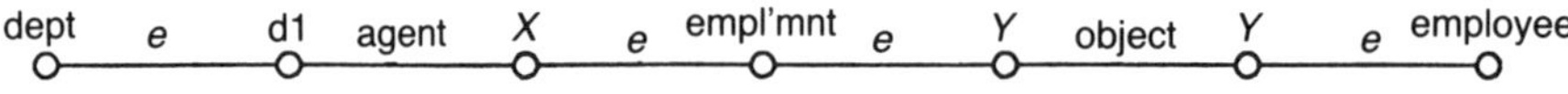

Fig. 4.

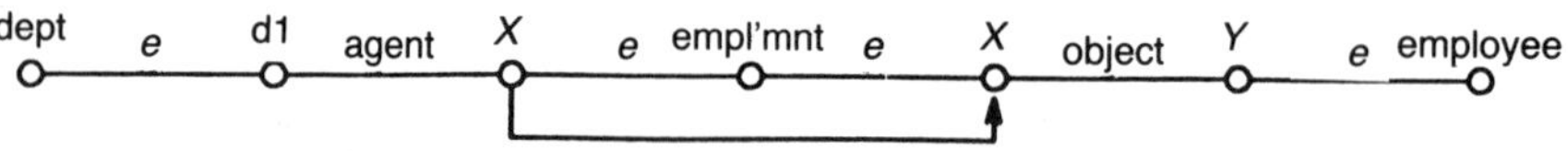

Fig. 5.

4. Graph Fitting Using Token Propagation

The processing of a linear graph template

$$\overset{T_1}{\circ}\!\!\xrightarrow{\;p_1\;}\!\!\overset{T_2}{\circ}\!\!\xrightarrow{\;p_2\;}\cdots\cdots\xrightarrow{\;p_{n-1}\;}\!\!\overset{T_n}{\circ}$$

as defined in Section 3, takes place as follows: The template is placed on a token a copy of which is injected into all nodes of the semantic net that match the leftmost node T_1 of the template. From each of these nodes the token is replicated along all edges that match the first template edge p_1. Nodes receiving this token will compare their own contents to the second template node T_2 and, if the match is successful, perform the same operation; that is, replicate the token along all edges matching the next template edge p_2, etc. By this stepwise expansion of the template into many directions of the network, all possible matches are found without the need for any centralized control.

The above actions are formalized as the following procedure which specifies the individual steps performed by a node t_i upon receiving a token carrying the following linear graph template

$$\overset{T_i}{\circ}\!\!\xrightarrow{\;p_i\;}\!\!\overset{T_{i+1}}{\circ}\!\!\xrightarrow{\;p_{i+1}\;}\cdots\cdots\xrightarrow{\;p_{n-1}\;}\!\!\overset{T_n}{\circ}$$

(1) The node t_i examines the first node T_i of the cluster carried by the received token. If T_i is a constant different from t_i then the node returns a token to the sender, indicating a failure. If, on the other hand, T_i is a constant that matches t_i, or if T_i is a variable, then the node detaches T_i (including the first arc p_i) from the graph template, and replicates the modified token

$$\overset{T_{i+1}}{\circ}\!\!\xrightarrow{\;p_{i+1}\;}\cdots\cdots\xrightarrow{\;p_{n-1}\;}\!\!\overset{T_n}{\circ}$$

along all edges labeled p_i. In the case where T_i is a variable, the node must also bind its own value t_i to all other occurrences of T_i within the template (using the pointers created during step (3) of the token construction procedure of Section 3) prior to replicating the token.

(2) The node t_i then awaits response tokens from all directions into which a token has been sent. Each response token carries the bindings of constants to variables made during the forward propagation of the corresponding token; it represents one successful match, or indicates failure. All tokens received by the node t_i are returned to its sender. Thus answers propagate backward along the same paths taken by tokens during their forward propagation. Computation in each node terminates when responses to all tokens emitted previously have been received.

To illustrate the above procedure, consider again the linear template of Fig. 5. It is placed on a token and injected into the node 'dept' of the semantic network of Fig. 1. From there copies of the token are replicated along all edges labeled '*e*', thus arriving at the nodes 'd1' and 'd2'. Since the node 'd2' of the network does not match the corresponding node 'd1' of the template, 'd2' discards the token and instead returns a failure token to its sender—the node 'dept'. The match in the other network node, 'd1', is successful and hence copies of the token are propagated along all edges labeled 'agent'. These are received by the two nodes 'et1' and 'et2'. The corresponding node in the template is the variable '*X*', which implies a successful match in both cases. However, since there is another occurrence of the same variable '*X*' in the template, (indicated by the pointer), it must be bound to the same constant as the first occurrence. Hence the node 'et1' binds (replaces) the variable '*X*' with the constant 'et1' before forwarding the token along the next edge '*e*' to the node 'employment'. Similarly, the node 'et2' binds the variable to 'et2'. Using the same basic procedure, both tokens continue their journey, traversing once more the nodes 'et1' and 'et2' respectively, until the nodes 'ep1' and 'ep2' are reached. Since the corresponding template node is a variable, '*Y*', the match will succeed in both cases. This indicates that two independent solutions for fitting the original template into the network have been found.

5. Dataflow Architectures

The main objective of the approach described in this paper was to develop a model which would be suitable for processing on a highly-parallel computer architecture. We have therefore rejected the von Neumann model of computation in which a semantic network is a stored data structure manipulated by a program. Instead, the network is viewed as a dataflow graph, where procedures for token manipulation and communication are part of each node. It should be emphasized that each node of the network contains the same set of procedures, which are triggered solely by the arrival of tokens at that node. Hence *computation is strictly data-driven—there is no need for any centralized control to synchronize the operation of individual nodes*. Each token, once injected, contains sufficient information to be guided through the network independently of other tokens. This is in contrast to NETL, where sets of marker bits are pushed through the network in lock step under the supervision of a central controller; the technological difficulties resulting from such an approach have been pointed out in [3]. The architectural requirements of the data-driven approach, on the other hand, may be satisfied by a variety of different architectures; these requirements are as follows:

(1) A *mapping function f* must be provided which assigns each node of the semantic network to one of the PEs. This PE then receives and processes tokens for all nodes mapped onto that PE. In the simplest case, a hashing function may be used as in the data-driven database machine described in [1].

(2) A *communication network* must be provided which permits nodes residing in different PEs to communicate with one another along the (logical) arcs. This implies that, in general, each PE must be able to communicate—directly or via other PEs—with any other PE. A number of possible interconnection schemes satisfying this requirement exist; the choice is largely a cost/performance tradeoff.

(3) It must be possible to *inject tokens* into any node of the semantic net. At the architecture level, this implies finding the address of the PE holding the receiving node. This is very similar to the second requirement above, which is the ability to provide communication channels (logical arcs) between nodes mapped onto different PEs. Hence the same mechanisms may be used: the sender, be it an internal node or an external input source, can apply the same mapping function *f* (used initially to map the net onto the architecture) to find the receiver's PE. The PE number is carried by the token as its destination address when it is being routed through the architecture. In a similar manner, results extracted from the semantic net may be routed to PEs connected to an output device.

From the above discussion it follows that a number of existing architectures, in particular those designed to execute dataflow programs, could easily be adapted to effectively support the processing of semantic nets according to the proposed model. In closing, let us contrast the two sources of parallelism available through the proposed approach:

Intra-request parallelism. As described in Section 4, a token injected into a node of the dataflow graph is replicated into many directions in the search of a match. Each node receiving a copy of the token may process and forward it concurrently with other nodes. The theoretical time complexity for finding all matches for a given graph template is proportional to the number of edges constituting that template.

Inter-request parallelism. Assuming a significant number of PEs constitute the underlying computer architecture, only a small subset of these will, in general, be busy processing a given query. This implies that the unused computational power of idle PEs may be exploited by executing more than one query at a time. A simple token coloring scheme is sufficient to distinguish tokens belonging to different graph templates and thus to permit their coexistence in the system.

REFERENCES

1. Bic, L. and Hartmann, R., Hither hundreds of processors in a database machine, in: *Proceedings Fourth International Workshop on Database Machines* (Springer, Berlin, 1985).
2. *Computer* **15** (2) (1982) Special Issue on Dataflow Systems.
3. Fahlman, S.E., *NETL: A System for Representing and Using Real-World Knowledge* (MIT Press, Cambridge, MA, 1979).
4. Fikes, R.E. and Hendrix, G.G., A network-based knowledge representation and its natural deduction system, in: *Proceedings Fifth International Joint Conference on Artificial Intelligence*, Cambridge, MA, 1977.
5. Rieger, C., Trigg, R. and Bane, B., ZMOB: A new computing engine for AI, in: *Proceedings Seventh International Joint Conference on Artificial Intelligence*, Vancouver, BC, 1981.
6. Treleaven, P.C., Brownbridge, T.R. and Hopkins, R.C., Data-driven and demand-driven computer architecture, *ACM Computing Surveys* **14** (1) (1982).

Reprinted with permission from *Artificial Intelligence*, Volume 27, 1985, pages 97-109.

Depth-First Iterative-Deepening: An Optimal Admissible Tree Search*

Richard E. Korf**

Department of Computer Science, Columbia University, New York, NY 10027, U.S.A.

ABSTRACT

The complexities of various search algorithms are considered in terms of time, space, and cost of solution path. It is known that breadth-first search requires too much space and depth-first search can use too much time and doesn't always find a cheapest path. A depth-first iterative-deepening algorithm is shown to be asymptotically optimal along all three dimensions for exponential tree searches. The algorithm has been used successfully in chess programs, has been effectively combined with bi-directional search, and has been applied to best-first heuristic search as well. This heuristic depth-first iterative-deepening algorithm is the only known algorithm that is capable of finding optimal solutions to randomly generated instances of the Fifteen Puzzle within practical resource limits.

1. Introduction

Search is ubiquitous in artificial intelligence. The performance of most AI systems is dominated by the complexity of a search algorithm in their inner loops. The standard algorithms, breadth-first and depth-first search, both have serious limitations, which are overcome by an algorithm called depth-first iterative-deepening. Unfortunately, current AI texts either fail to mention this algorithm [10, 11, 14], or refer to it only in the context of two-person game searches [1, 16]. The iterative-deepening algorithm, however, is completely general and can also be applied to uni-directional search, bi-directional search, and heuristic searches such as A*. The purposes of this article are to demonstrate the generality of depth-first iterative-deepening, to prove its optimality for exponential tree searches, and to remind practitioners in the field that it is the search technique of choice for many applications.

Depth-first iterative-deepening has no doubt been rediscovered many times independently. The first use of the algorithm that is documented in the literature is in Slate and Atkin's Chess 4.5 program [15]. Berliner [2] has observed that breadth-first search is inferior to the iterative-deepening algorithm. Winston [16] shows that for two-person game searches where only terminal-node static evaluations are counted in the cost, the extra computation required by iterative-deepening is insignificant. Pearl [12] initially suggested the iterative-deepening extension of A*, and Berliner and Goetsch [3] have implemented such an algorithm concurrently with this work.

We will analyze several search algorithms along three dimensions: the amount of time they take, the amount of space they use, and the cost of the solution paths they find. The standard breadth-first and depth-first algorithms will be shown to be inferior to the depth-first iterative-deepening algorithm.

* This research was supported in part by the Defense Advanced Research Projects Agency under contract N00039-82-C-0427, and by the National Science Foundation Division of Information Science and Technology grant IST-84-18879.

** Present address: Department of Computer Science, University of California, Los Angeles, CA 90024, U.S.A.

We will prove that this algorithm is asymptotically optimal along all three dimensions for exponential tree searches. Since almost all heuristic tree searches have exponential complexity, this is a fairly general result.

We begin with the problem-space model of Newell and Simon [9]. A *problem space* consists of a set of states and a set of operators that are partial functions that map states into states. A *problem* is a problem space together with a particular initial state and a set of goal states. The task is to find a sequence of operators that will map the initial state to a goal state.

The complexity of a problem will be expressed in terms of two parameters: the branching factor of the problem space, and the depth of solution of the problem. The *node branching factor* (b) of a problem is defined as the number of new states that are generated by the application of a single operator to a given state, averaged over all states in the problem space. We will assume that the branching factor is constant throughout the problem space. The *depth* (d) of solution of a problem is the length of the shortest sequence of operators that map the initial state into a goal state. The time cost of a search algorithm in this model of computation is simply the number of states that are expanded. The reason for this choice is that we are interested in asymptotic complexity and we assume that the amount of time is proportional to the number of states expanded. Similarly, since we assume that the amount of space required is proportional to the number of states that are stored, the asymptotic space cost of an algorithm in this model will be the number of states that must be stored.

This work is focused on searches which produce optimal solutions. We recognize that for most applications, optimal solutions are not required and that their price is often prohibitive. There are occasions, however, when optimal solutions are needed. For example, in assessing the quality of non-optimal solutions, it is often enlightening to compare them to optimal solutions for the same problem instances.

2. Breadth-First Search

We begin our discussion with one of the simplest search algorithms, breadth-first search. Breadth-first search expands all the states one step (or operator application) away from the initial state, then expands all states two steps from the initial state, then three steps, etc., until a goal state is reached. Since it always expands all nodes at a given depth before expanding any nodes at a greater depth, the first solution path found by breadth-first search will be one of shortest length. In the worst case, breadth-first search must generate all nodes up to depth d, or $b + b^2 + b^3 + \cdots + b^d$ which is $O(b^d)$. Note that on the average, half of the nodes at depth d must be examined, and therefore the average-case time complexity is also $O(b^d)$.

Since all the nodes at a given depth are stored in order to generate the nodes at the next depth, the minimum number of nodes that must be stored to search to depth d is b^{d-1}, which is $O(b^d)$. As with time, the average-case space complexity is roughly one-half of this, which is also $O(b^d)$. This space requirement of breadth-first search is its most critical drawback. As a practical matter, a breadth-first search of most problem spaces will exhaust the available memory long before an appreciable amount of time is used. The reason for this is that the typical ratio of memory to speed in modern computers is a million words of memory for each million instructions per second (MIPS) of processor speed. For example, if we can generate a million states per minute and require a word to store each state, memory will be exhausted in one minute.

3. Depth-First Search

Depth-first search avoids this memory limitation. It works by always generating a descendant of the most recently expanded node, until some depth cutoff is reached, and then backtracking to the next most recently expanded node and generating one of its descendants. Therefore, only the path of nodes from the initial node to the current node must be stored in order to execute the algorithm. If the depth cutoff is d, the space required by depth-first search is only $O(d)$.

Since depth-first search only stores the current path at any given point, it is bound to search all paths down to the cutoff depth. In order to analyze its time complexity, we must define a new parameter, called the *edge branching factor* (e), which is the average number of different operators which are applicable to a given state. For trees, the edge and node branching factors are equal, but for graphs in general the edge branching factor may exceed the node branching factor. For example, the graph in Fig. 1 has an edge branching factor of two, while its node branching factor is only one. Note that a breadth-first search of this graph takes only linear time while a depth-first search requires exponential time. In general, the time complexity of a depth-first search to depth d is $O(e^d)$. Since the space used by depth-first search grows only as the log of the time required, the algorithm is time-bound rather than space-bound in practice.

Another drawback, however, to depth-first search is the requirement for an arbitrary cutoff depth. If branches are not cut off and duplicates are not checked for, the algorithm may not terminate. In general, the depth at which the first goal state appears is not known in advance and must be estimated. If the estimate is too low, the algorithm terminates without finding a solution. If the depth estimate is too high, then a large price in running time is paid relative to an optimal search, and the first solution found may not be an optimal one.

FIG. 1. Graph with linear number of nodes but exponential number of paths.

4. Depth-First Iterative-Deepening

A search algorithm which suffers neither the drawbacks of breadth-first nor depth-first search on trees is depth-first iterative-deepening (DFID). The algorithm works as follows: First, perform a depth-first search to depth one. Then, discarding the nodes generated in the first search, start over and do a depth-first search to level two. Next, start over again and do a depth-first search to depth three, etc., continuing this process until a goal state is reached.

Since DFID expands all nodes at a given depth before expanding any nodes at a greater depth, it is guaranteed to find a shortest-length solution. Also, since at any given time it is performing a depth-first search, and never searches deeper than depth d, the space it uses is $O(d)$.

The disadvantage of DFID is that it performs wasted computation prior to reaching the goal depth. In fact, at first glance it seems very inefficient. Below, however, we present an analysis of the running time of DFID that shows that this wasted computation does not affect the asymptotic growth of the run time for exponential tree searches. The intuitive reason is that almost all the work is done at the deepest level of the search. Unfortunately, DFID suffers the same drawback as depth-first search on arbitrary graphs, namely that it must explore all possible paths to a given depth.

Definition 4.1. *A brute-force search* is a search algorithm that uses no information other than the initial state, the operators of the space, and a test for a solution.

Theorem 4.2. *Depth-first iterative-deepening is asymptotically optimal among brute-force tree searches in terms of time, space, and length of solution.*

Proof. As mentioned above, since DFID generates all nodes at a given depth before expanding any nodes at a greater depth, it always finds a shortest path to the goal, or any other state for that matter. Hence, it is optimal in terms of solution length.

Next, we examine the running time of DFID on a tree. The nodes at depth d are generated once during the final iteration of the search. The nodes at depth $d-1$ are generated twice, once during the final iteration at depth d, and once during the penultimate iteration at depth $d-1$. Similarly, the nodes at depth $d-2$ are generated three times, during iterations at depths d, $d-1$, and $d-2$, etc. Thus the total number of nodes generated in a depth-first iterative-deepening search to depth d is

$$b^d + 2b^{d-1} + 3b^{d-2} + \cdots + db\,.$$

Factoring out b^d gives

$$b^d(1 + 2b^{-1} + 3b^{-2} + \cdots + db^{1-d})\,.$$

Letting $x = 1/b$ yields

$$b^d(1 + 2x^1 + 3x^2 + \cdots + dx^{d-1})\,.$$

This is less than the infinite series

$$b^d(1 + 2x^1 + 3x^2 + 4x^3 + \cdots)\,,$$

which converges to

$$b^d(1-x)^{-2} \quad \text{for } \mathrm{abs}(x) < 1\,.$$

Since $(1-x)^{-2}$, or $(1-1/b)^{-2}$, is a constant that is independent of d, if $b > 1$ then the running time of depth-first iterative-deepening is $O(b^d)$.

To see that this is optimal, we present a simple adversary argument. The number of nodes at depth d is b^d. Assume that there exists an algorithm that examines less than b^d nodes. Then, there must exist at least one node at depth d which is not examined by this algorithm. Since we have no additional information, an adversary could place the only solution at this node and hence the proposed algorithm would fail. Hence, any brute-force algorithm must take at least cb^d time, for some constant c.

Finally, we consider the space used by DFID. Since DFID at any point is engaged in a depth-first search, it need only store a stack of nodes which represents the branch of the tree it is expanding. Since it finds a solution of optimal length, the maximum depth of this stack is d, and hence the maximum amount of space is $O(d)$.

To show that this is optimal, we note that any algorithm which uses $f(n)$ time must use at least $k \log f(n)$ space for some constant k [7]. The reason is that the algorithm must proceed through $f(n)$ distinct states before looping or terminating, and hence must be able to store that many distinct states. Since storing $f(n)$ states requires $\log f(n)$ bits, and $\log b^d$ is $d \log b$, any brute-force algorithm must use kd space, for some constant k. □

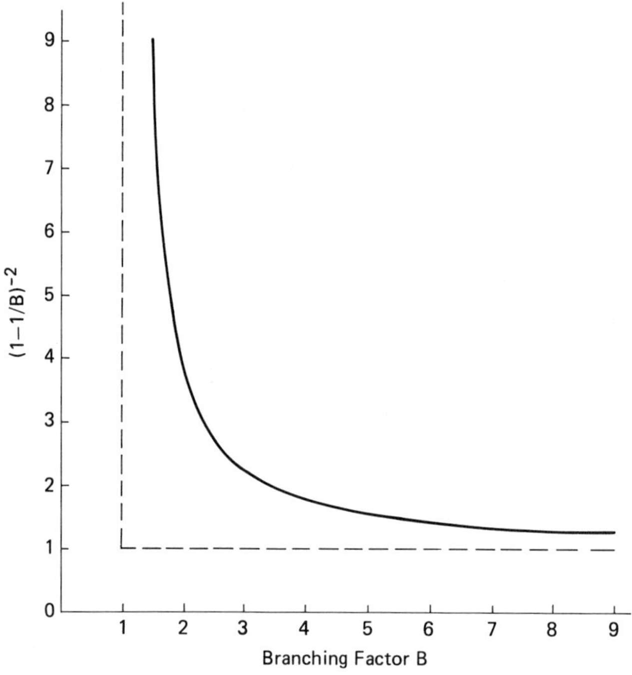

FIG. 2. Graph of branching factor vs. constant coefficient as search depth goes to infinity.

The value of the constant $(1 - 1/b)^{-2}$ gives an upper bound on how much computation is wasted in the lower levels of the search, since it is the limit of the constant coefficient as the search depth goes to infinity. Fig. 2 shows a graph of this constant versus the branching factor. As the branching factor increases, the constant quickly approaches one. For branching factors close to one, however, the value of the constant coefficient approaches infinity as the depth goes to infinity.

5. Bi-Directional Search

For those problems with a single goal state that is given explicitly and for which the operators have inverses, such as the Fifteen Puzzle, bi-directional search [13] can be used. Bi-directional search trades space for time by searching forward from the initial state and backward from the goal state simultaneously, storing the states generated, until a common state is found on both search frontiers. Depth-first iterative-deepening can be applied to bi-directional search as follows: A single iteration consists of a depth-first search from one direction to depth k, storing only the states at depth k, and two depth-first searches from the other direction, one to depth k and one to depth $k + 1$, not storing states but simply matching against the stored states from the other direction. The search to depth $k + 1$ is necessary to find odd-length solutions. This is repeated for k from zero (to find solutions of length one) to $d/2$. Assuming that a hashing scheme is used to perform the matching in constant time per node, this algorithm will find an optimal solution of length d in time $O(b^{d/2})$ and space $O(b^{d/2})$. In experiments involving Rubik's Cube [8], which has an effective branching factor of 13.5, this algorithm was used to find solutions up to 11 moves long on a DEC VAX 11/780.

6. Heuristic Search

Depth-first iterative-deepening can also be combined with a best-first heuristic search such as A* [6]. The idea is that successive iterations correspond not to increasing depth of search, but rather to increasing values of the total cost of a path. For A*, this total cost is composed of the cost so far in reaching the node (g) plus the estimated cost of the path from the node to a goal state (h). Iterative-deepening-A* (IDA*) works as follows: At each iteration, perform a depth-first search, cutting off a branch when its total cost ($g+h$) exceeds a given threshold. This threshold starts at the estimate of the cost of the initial state, and increases for each iteration of the algorithm. At each iteration, the threshold used for the next iteration is the minimum cost of all values that exceeded the current threshold.

A well-known property of A* is that it always finds a cheapest solution path if the heuristic is *admissible*, or in other words never overestimates the actual cost to the goal [6]. This property also holds for iterative-deepening-A*. Furthermore, IDA* expands the same number of nodes, asymptotically, as A* in an exponential tree search.

The proofs of these results are much simpler and more intuitive if we restrict our attention to cost functions which are monotonically non-decreasing along any path in the problem space. Such a heuristic is called *monotone or consistent* [11]. Formally,

Definition 6.1. A cost function $f(n)$ is *monotone* if for all nodes n and $s(n)$, where $s(n)$ is a successor of n, $f(n) \leq f(s(n))$.

This restriction is not essential, and slightly more complex proofs will establish the same results without it. As a practical matter, however, almost all reasonable cost functions are monotone [11]. In fact, using an idea proposed by Mérõ [17], we can formally make this assumption without loss of generality, as shown in the following lemma.

Lemma 6.2. *For any admissible cost function f, we can construct a monotone admissible function f' which is at least as informed as f.*

Proof. We construct f' recursively from f as follows: if n is the initial state, then $f'(n)=f(n)$; otherwise, $f'(s(n)) = \max[f(s(n)), f'(n)]$. Clearly, f' is monotone since $f'(n) \leq f'(s(n))$. In order to show that f' is admissible, note that $f'(n)$ is equal to the maximum value of f applied to all the predecessors of n along the path back to the initial state. Since f is admissible, the maximum value of f along a path is a lower bound on the cost of that path, and hence a lower bound on the cost of n. Thus, f' does not violate admissibility. Furthermore, f' is at least as informed as f since for all n, $f'(n) \geq f(n)$ and hence $f'(n)$ is at least as accurate an estimate as $f(n)$. □

Note that this lemma provides a simple and intuitive proof of the admissibility of A*. If we restrict our attention to cost functions which are monotone non-decreasing, and A* always expands the open node of least cost, it is clear that the first solution it finds will be one of least cost. Similarly, the result below follows just as easily.

Lemma 6.3. *Given an admissible monotone cost function, iterative-deepening-A* will find a solution of least cost if one exists.*

Proof. Since the initial cost cutoff of IDA* is the heuristic estimate of the cost of the initial state, and the heuristic never overestimates cost, the length of the shortest solution cannot be less than the initial cost cutoff. Furthermore, since the cost cutoff for each succeeding iteration is the minimum value which exceeded the previous cutoff, no paths can have a cost which lies in a gap between two successive cutoffs. Therefore, since IDA* always expands all nodes at a given cost before expanding any nodes at a greater cost, the first solution it finds will be a solution of least cost. □

Not only does IDA* find a cheapest path to a solution and use far less space than A*, but it expands approximately the same number of nodes as A* in a tree search. Combining this fact with several recent results on the complexity and optimality of A* allows us to state and prove the following general result:

Theorem 6.4. *Given an admissible monotone heuristic with constant relative error, then iterative-deepening-A* is optimal in terms of solution cost, time, and space, over the class of admissible best-first searches on a tree.*

Proof. From Lemma 6.3, we know that IDA* produces a solution of optimal cost.

To determine the time used by IDA*, consider the final iteration, in other words the one which finds a solution. It must expand all descendents of the initial state with values greater than or equal to the initial cost estimate and less than the optimal solution cost, plus some number of nodes whose cost equals the optimal solution cost. If A* employs the tie-breaking rule of 'most recently generated', it must also expand these same nodes. Thus, the final iteration of IDA* expands the same set of nodes as A* under this tie-breaking rule. Furthermore, if the graph is a tree, each of these nodes will be expanded exactly once. IDA* must also expand nodes during the previous iterations as well. However, Pearl has shown that if the heuristic used by A* exhibits constant relative error, then the number of nodes generated by the algorithm increases exponentially with depth [11]. Thus, we can use an argument similar to the proof of Theorem 4.2 to show that the previous iterations of IDA* do not affect the asymptotic order of the total number of nodes [18]. Thus, IDA* expands the same number of nodes, asymptotically, as A*. Furthermore, a recent result of Dechter and Pearl [5] shows that A* is optimal, in terms of number of nodes expanded, over the class of admissible best-first searches with monotone heuristics. Therefore, IDA* is asymptotically optimal in terms of time for tree searches.

Since the number of nodes grows exponentially, we can again appeal to the argument in the proof of Theorem 4.2 to show that the space used by IDA* is also asymptotically optimal. □

Is the assumption of constant relative error, i.e. that the error in the estimate grows at the same rate as the magnitude of the actual cost, valid for heuristics? Pearl observes that heuristics with better accuracy almost never occur in practice. For example, most physical measurements are subject to constant relative error [11]. Thus, we can conclude that heuristic depth-first iterative-deepening is asymptotically optimal for most best-first tree searches which occur in practice.

An additional benefit of IDA* over A* is that it is simpler to implement since there are no open or closed lists to be managed. A simple recursion performs the depth-first search inside an outer loop to handle the iterations.

As an empirical test of the practicality of the algorithm, both IDA* and A* were implemented for the Fifteen Puzzle. The implementations were in PASCAL and were run on a DEC 2060. The heuristic function used for both was the Manhattan distance heuristic: for each movable tile, the number of grid units between the current position of the tile and its goal position are computed, and these values are summed for all tiles. The two algorithms were tested against 100 randomly generated, solvable initial states. IDA* solved all instances with a median time of 30 CPU minutes, generating over 1.5 million nodes per minute. The average solution length was 53 moves and the maximum was 66 moves. A* solved none of the instances since it ran out of space after about 30 000 nodes were stored. An additional observation is that even though IDA* generated more nodes than A*, it actually ran faster than A* on the same problem instances, due to less overhead per node. The data from this experiment are summarized in Table 1. These are the first published optimal solution lengths to randomly generated instances of the Fifteen Puzzle. Although the Fifteen Puzzle graph is not strictly a tree, the edge branching factor is only slightly greater than the node branching factor, and hence the iterative-deepening algorithm is still effective.

7. Two-Person Games

In the discussion so far, we have assumed a single-agent search to find a solution to a problem, and have been concerned with minimizing time and space subject to a fixed solution depth and branching factor. However, a two-person game such as chess with static evaluation and mini-max search is a somewhat different situation. In this case, we assume that accuracy of the static evaluation increases with increasing search depth, and hence we want to maximize search depth subject to fixed time and space constraints. Since depth-first iterative-deepening minimizes, at least asymptotically, time and space for any given search depth, it follows that it maximizes the depth of search possible for any fixed time and space restrictions as well.

Another reason that DFID is used in game programs is that the amount of time required to search the next deeper level in the tree is not known when the ply begins, and the search ply may have to be aborted due to time constraints. In this case, the complete search at the next shallower depth can be used to make the move.

TABLE 1. Optimal solution lengths for 100 randomly generated Fifteen Puzzle instances using iterative-deepening-A* with Manhattan distance heuristic function

NUMBER	INITIAL STATE	ESTIMATE	ACTUAL	TOTAL NODES
1	14 13 15 7 11 12 9 5 6 0 2 1 4 8 10 3	41	57	276,361,933
2	13 5 4 10 9 12 8 14 2 3 7 1 0 15 11 6	43	55	15,300,442
3	14 7 8 2 13 11 10 4 9 12 5 0 3 6 1 15	41	59	565,994,203
4	5 12 10 7 15 11 14 0 8 2 1 13 3 4 9 6	42	56	62,643,179
5	4 7 14 13 10 3 9 12 11 5 6 15 1 2 8 0	42	56	11,020,325
6	14 7 1 9 12 3 6 15 8 11 2 5 10 0 4 13	36	52	32,201,660
7	2 11 15 5 13 4 6 7 12 8 10 1 9 3 14 0	30	52	387,138,094
8	12 11 15 3 8 0 4 2 6 13 9 5 14 1 10 7	32	50	39,118,937
9	3 14 9 11 5 4 8 2 13 12 6 7 10 1 15 0	32	46	1,650,696
10	13 11 8 9 0 15 7 10 4 3 6 14 5 12 2 1	43	59	198,758,703
11	5 9 13 14 6 3 7 12 10 8 4 0 15 2 11 1	43	57	150,346,072
12	14 1 9 6 4 8 12 5 7 2 3 0 10 11 13 15	35	45	546,344
13	3 6 5 2 10 0 15 14 1 4 13 12 9 8 11 7	36	46	11,861,705
14	7 6 8 1 11 5 14 10 3 4 9 13 15 2 0 12	41	59	1,369,596,778
15	13 11 4 12 1 8 9 15 6 5 14 2 7 3 10 0	44	62	543,598,067
16	1 3 2 5 10 9 15 6 8 14 13 11 12 4 7 0	24	42	17,984,051
17	15 14 0 4 11 1 6 13 7 5 8 9 3 2 10 12	46	66	607,399,560
18	6 0 14 12 1 15 9 10 11 4 7 2 8 3 5 13	43	55	23,711,067
19	7 11 8 3 14 0 6 15 1 4 13 9 5 12 2 10	36	46	1,280,495
20	6 12 11 3 13 7 9 15 2 14 8 10 4 1 5 0	36	52	17,954,870
21	12 8 14 6 11 4 7 0 5 1 10 15 3 13 9 2	34	54	257,064,810
22	14 3 9 1 15 8 4 5 11 7 10 13 0 2 12 6	41	59	750,746,755
23	10 9 3 11 0 13 2 14 5 6 4 7 8 15 1 12	33	49	15,971,319
24	7 3 14 13 4 1 10 8 5 12 9 11 2 15 6 0	34	54	42,693,209
25	11 4 2 7 1 0 10 15 6 9 14 8 3 13 5 12	32	52	100,734,844
26	5 7 3 12 15 13 14 8 0 10 9 6 1 4 2 11	40	58	226,668,645
27	14 1 8 15 2 6 0 3 9 12 10 13 4 7 5 11	33	53	306,123,421
28	13 14 6 12 4 5 1 0 9 3 10 2 15 11 8 7	36	52	5,934,442
29	9 8 0 2 15 1 4 14 3 10 7 5 11 13 6 12	38	54	117,076,111
30	12 15 2 6 1 14 4 8 5 3 7 0 10 13 9 11	35	47	2,196,593
31	12 8 15 13 1 0 5 4 6 3 2 11 9 7 14 10	38	50	2,351,811
32	14 10 9 4 13 6 5 8 2 12 7 0 1 3 11 15	43	59	661,041,936
33	14 3 5 15 11 6 13 9 0 10 2 12 4 1 7 8	42	60	480,637,867
34	6 11 7 8 13 2 5 4 1 10 3 9 14 0 12 15	36	52	20,671,552
35	1 6 12 14 3 2 15 8 4 5 13 9 0 7 11 10	39	55	47,506,056
36	12 6 0 4 7 3 15 1 13 9 8 11 2 14 5 10	36	52	59,802,602
37	8 1 7 12 11 0 10 5 9 15 6 13 14 2 3 4	40	58	280,078,791
38	7 15 8 2 13 6 3 12 11 0 4 10 9 5 1 14	41	53	24,492,852
39	9 0 4 10 1 14 15 3 12 6 5 7 11 13 8 2	35	49	19,355,806
40	11 5 1 14 4 12 10 0 2 7 13 3 9 15 6 8	36	54	63,276,188
41	8 13 10 9 11 3 15 6 0 1 2 14 12 5 4 7	36	54	51,501,544
42	4 5 7 2 9 14 12 13 0 3 6 11 8 1 15 10	30	42	877,823
43	11 15 14 13 1 9 10 4 3 6 2 12 7 5 8 0	48	64	41,124,767
44	12 9 0 6 8 3 5 14 2 4 11 7 10 1 15 13	32	50	95,733,125
45	3 14 9 7 12 15 0 4 1 8 5 6 11 10 2 13	39	51	6,158,733
46	8 4 6 1 14 12 2 15 13 10 9 5 3 7 0 11	35	49	22,119,320
47	6 10 1 14 15 8 3 5 13 0 2 7 4 9 11 12	35	47	1,411,294
48	8 11 4 6 7 3 10 9 2 12 15 13 0 1 5 14	39	49	1,905,023
49	10 0 2 4 5 1 6 12 11 13 9 7 15 3 14 8	33	59	1,809,933,698
50	12 5 13 11 2 10 0 9 7 8 4 3 14 6 15 1	39	53	63,036,422
51	10 2 8 4 15 0 1 14 11 13 3 6 9 7 5 12	44	56	26,622,863

TABLE 1. *Continued*

NUMBER	INITIAL STATE	ESTIMATE	ACTUAL	TOTAL NODES
52	10 8 0 12 3 7 6 2 1 14 4 11 15 13 9 5	38	56	377,141,881
53	14 9 12 13 15 4 8 10 0 2 1 7 3 11 5 6	50	64	465,225,698
54	12 11 0 8 10 2 13 15 5 4 7 3 6 9 14 1	40	56	220,374,385
55	13 8 14 3 9 1 0 7 15 5 4 10 12 2 6 11	29	41	927,212
56	3 15 2 5 11 6 4 7 12 9 1 0 13 14 10 8	29	55	1,199,487,996
57	5 11 6 9 4 13 12 0 8 2 15 10 1 7 3 14	36	50	8,841,527
58	5 0 15 8 4 6 1 14 10 11 3 9 7 12 2 13	37	51	12,955,404
59	15 14 6 7 10 1 0 11 12 8 4 9 2 5 13 3	35	57	1,207,520,464
60	11 14 13 1 2 3 12 4 15 7 9 5 10 6 8 0	48	66	3,337,690,331
61	6 13 3 2 11 9 5 10 1 7 12 14 8 4 0 15	31	45	7,096,850
62	4 6 12 0 14 2 9 13 11 8 3 15 7 10 1 5	43	57	23,540,413
63	8 10 9 11 14 1 7 15 13 4 0 12 6 2 5 3	40	56	995,472,712
64	5 2 14 0 7 8 6 3 11 12 13 15 4 10 9 1	31	51	260,054,152
65	7 8 3 2 10 12 4 6 11 13 5 15 0 1 9 14	31	47	18,997,681
66	11 6 14 12 3 5 1 15 8 0 10 13 9 7 4 2	41	61	1,957,191,378
67	7 1 2 4 8 3 6 11 10 15 0 5 14 12 13 9	28	50	252,783,878
68	7 3 1 13 12 10 5 2 8 0 6 11 14 15 4 9	31	51	64,367,799
69	6 0 5 15 1 14 4 9 2 13 8 10 11 12 7 3	37	53	109,562,359
70	15 1 3 12 4 0 6 5 2 8 14 9 13 10 7 11	30	52	151,042,571
71	5 7 0 11 12 1 9 10 15 6 2 3 8 4 13 14	30	44	8,885,972
72	12 15 11 10 4 5 14 0 13 7 1 2 9 8 3 6	38	56	1,031,641,140
73	6 14 10 5 15 8 7 1 3 4 2 0 12 9 11 13	37	49	3.222,276
74	14 13 4 11 15 8 6 9 0 7 3 1 2 10 12 5	46	56	1,897,728
75	14 4 0 10 6 5 1 3 9 2 13 15 12 7 8 11	30	48	42,772,589
76	15 10 8 3 0 6 9 5 1 14 13 11 7 2 12 4	41	57	126,638,417
77	0 13 2 4 12 14 6 9 15 1 10 3 11 5 8 7	34	54	18,918,269
78	3 14 13 6 4 15 8 9 5 12 10 0 2 7 1 11	41	53	10,907,150
79	0 1 9 7 11 13 5 3 14 12 4 2 8 6 10 15	28	42	540,860
80	11 0 15 8 13 12 3 5 10 1 4 6 14 9 7 2	43	57	132,945,856
81	13 0 9 12 11 6 3 5 15 8 1 10 4 14 2 7	39	53	9,982,569
82	14 10 2 1 13 9 8 11 7 3 6 12 15 5 4 0	40	62	5,506,801,123
83	12 3 9 1 4 5 10 2 6 11 15 0 14 7 13 8	31	49	65,533,432
84	15 8 10 7 0 12 14 1 5 9 6 3 13 11 4 2	37	55	106,074,303
85	4 7 13 10 1 2 9 6 12 8 14 5 3 0 11 15	32	44	2,725,456
86	6 0 5 10 11 12 9 2 1 7 4 3 14 8 13 15	35	45	2,304,426
87	9 5 11 10 13 0 2 1 8 6 14 12 4 7 3 15	34	52	64,926,494
88	15 2 12 11 14 13 9 5 1 3 8 7 0 10 6 4	43	65	6,009,130,748
89	11 1 7 4 10 13 3 8 9 14 0 15 6 5 2 12	36	54	166,571,097
90	5 4 7 1 11 12 14 15 10 13 8 6 2 0 9 3	36	50	7,171,137
91	9 7 5 2 14 15 12 10 11 3 6 1 8 13 0 4	41	57	602,886,858
92	3 2 7 9 0 15 12 4 6 11 5 14 8 13 10 1	37	57	1,101,072,541
93	13 9 14 6 12 8 1 2 3 4 0 7 5 10 11 15	34	46	1,599,909
94	5 7 11 8 0 14 9 13 10 12 3 15 6 1 4 2	45	53	1,337,340
95	4 3 6 13 7 15 9 0 10 5 8 11 2 12 1 14	34	50	7,115,967
96	1 7 15 14 2 6 4 9 12 11 13 3 0 8 5 10	35	49	12,808,564
97	9 14 5 7 8 15 1 2 10 4 13 6 12 0 11 3	32	44	1,002,927
98	0 11 3 12 5 2 1 9 8 10 14 15 7 4 13 6	34	54	183,526,883
99	7 15 4 0 10 9 2 5 12 11 13 6 1 3 14 8	39	57	83,477,694
100	11 4 0 8 6 10 5 13 12 7 14 3 1 2 9 15	38	54	67,880,056

LEGEND

GOAL STATE

```
 0  1  2  3
 4  5  6  7
 8  9 10 11
12 13 14 15
```

0 1 2 3 4 5 6 7 8 9 10 11 12 13 14 15

ESTIMATE	Initial heuristic estimate
ACTUAL	Length of optimal solution
TOTAL NODES	Total number of states generated

Finally, the information from previous iterations of a DFID search can be used to order the nodes in the search tree so that alpha-beta cutoff is more efficient. In fact, the best move at a given iteration has been shown experimentally to terminate the next iteration in about 70% of cases. This improvement in ordering, which is critical to alpha-beta efficiency, is only possible with the use of iterative-deepening [4].

8. Conclusions

The standard algorithms for brute-force search have serious drawbacks. Breadth-first search uses too much space, and depth-first search in general uses too much time and is not guaranteed to find a shortest path to a solution. The depth-first iterative-deepening algorithm, however, is asymptotically optimal in terms of cost of solution, running time, and space required for brute-force tree searches. DFID can also be applied to bi-directional search, heuristic best-first search, and two-person game searches. Since almost all heuristic searches have exponential complexity, iterative-deepening-A* is an optimal admissible tree search in practice. For example, IDA* is the only known algorithm that can find optimal paths for randomly generated instances of the Fifteen Puzzle within practical time and space constraints.

ACKNOWLEDGEMENT

Judea Pearl originally suggested the application of iterative-deepening to A*. Hans Berliner pointed out the use of iterative-deepening for ordering nodes to maximize alpha-beta cutoffs. Michael Lebowitz, Andy Mayer, and Mike Townsend read earlier drafts of this paper and suggested many improvements. Andy Mayer implemented the A* algorithm that was compared with IDA*. An anonymous referee suggested the shortcomings of depth-first search on a graph with cycles. Finally, Jodith Fried drew the figures.

REFERENCES

1. Barr, A. and Feigenbaum, E.A. (Eds.), *Handbook of Artificial Intelligence* (Kaufmann, Los Altos, CA, 1981).
2. Berliner, H., Search, Artificial Intelligence Syllabus, Department of Computer Science, Carnegie-Mellon University, Pittsburgh, PA, 1983.
3. Berliner, H. and Goetsch, G., A quantitative study of search methods and the effect of constraint satisfaction, Tech. Rept. CMU-CS-84-147, Department of Computer Science, Carnegie-Mellon University, Pittsburgh, PA, 1984.
4. Berliner, H., Personal communication, 1984.
5. Dechter, R. and Pearl, J., The optimality of A* revisited, in: *Proceedings of the National Conference on Artificial Intelligence,* Washington, DC (August, 1983) 95–99.
6. Hart, P.E., Nilsson, N.J. and Raphael, B., A formal basis for the heuristic determination of minimum cost paths, *IEEE Trans. Systems Sci. Cybernet.* **4**(2) (1968) 100–107.
7. Hopcroft, J.E. and Ullman, J.D., *Introduction to Automata Theory, Languages, and Computation* (Addison-Wesley, Reading, MA, 1979).
8. Korf, R.E., *Learning to Solve Problems by Searching for Macro-Operators* (Pittman, London, 1985).
9. Newell, A. and Simon, H.A., *Human Problem Solving* (Prentice-Hall, Englewood Cliffs, NJ, 1972).
10. Nilsson, N.J., *Principles of Artificial Intelligence* (Tioga, Palo Alto, CA, 1980).
11. Pearl, J., *Heuristics* (Addison-Wesley, Reading, MA, 1984).
12. Pearl, J., Personal communication, 1984.
13. Pohl, I., Bi-directional search, in: B. Meltzer and D. Michie (Eds.), *Machine Intelligence* **6** (American Elsevier, New York, 1971) 127–140.
14. Rich, E., *Artificial Intelligence* (McGraw-Hill, New York, 1983).
15. Slate, D.J. and Atkin, L.R., *CHESS* 4.5 – *The Northwestern University Chess Program* (Springer-Verlag, New York, 1977).
16. Winston, P.H., *Artificial Intelligence* (Addison-Wesley, Reading, MA, 1984).
17. Mérõ, L., A heuristic search algorithm with modifiable estimate, *Artificial Intelligence* **23** (1984) 13–27.
18. Korf, R.E., Iterative-deepening-A*: an optimal admissible tree search, in: *Proceedings Ninth International Joint Conference on Artificial Intelligence,* Los Angeles, CA, 1985.

Received December 1984; revised version received March 1985

INEXACT REASONING IN EXPERT SYSTEMS: AN INTEGRATING OVERVIEW

by

R.J.P. Groothuizen

1 INTRODUCTION

In this report several methods for modeling the uncertainty in expert systems are discussed. An overview is given of methods used in various existing expert systems and an attempt is made to put these methods into their proper perspective.

Modeling of uncertainty is important because often the information expert systems get is uncertain and often the rules they use are not well-defined in a logical sense. Nevertheless conclusions have to be drawn, based on such information and such rules, since alternatives may not be available. But then some measure of reliability of these conclusions has to be defined in order to evaluate them. Moreover, it is possible that contradictory conclusions are derived. In that case, the system should notice this and adequate actions should be taken.

2 UNCERTAINTY IN AN EXPERT SYSTEM

The model of an expert system we consider is a set $\mathcal{P}$ of propositions (usually finite) satisfying:

1. if $p \in \mathcal{P}$ then $\neg p \in \mathcal{P}$. (negation)
2. if $p \in \mathcal{P}$ and $q \in \mathcal{P}$ then $p \wedge q \in \mathcal{P}$ (conjunction).

A proposition can represent a (measurable) fact (also called evidence) such as: 'John has fever', a hypothesis such as: 'John has influenza', or a logical combination of (not necessarily similar) propositions such as: 'If patient has fever then the patient has influenza' (rules, etc.). The problem of uncertainty is twofold. First there is uncertainty about the propositions themselves, second there is the problem how to reason under uncertainty.

2.1 Uncertainty of propositions

A proposition can be uncertain either because its truth (or falsity) cannot be established definitely, or because a proposition states the value of a variable in a not sufficiently determined way. The latter case is also called imprecision.

The former can be caused by:

a. imprecision of the language in which the proposition is expressed.

b. incomplete information.

c. inconsistencies, for example coming from incorrect or contradictory information or inexact reasoning.

The uncertainty or imprecision has also as consequence that a proposition need not be either true or false. So more 'truth-values' should be assignable.

2.2 Reasoning under uncertainty

In logical systems reasoning is performed using modus ponens and modus tollens. Considering two propositions p and q and assuming the implication $p \rightarrow q$ to be true modus ponens allows to derive the truth of q from the truth of p, while modus tollens allows to derive the falsity of p from the falsity of q. That is, propositions are true or false and so is $p \rightarrow q$. This exactness is lost when dealing with uncertain propositions. Given a proposition p and/or an implication $p \rightarrow q$ whose truth cannot be established definitely, or with truth values different from 0 and 1, what can be said about q?

In this connection the terms approximate and plausible reasoning are used. Approximate reasoning is deductive inference with uncertain or imprecise premises, plausible reasoning can give uncertain conclusions even when the premises are certain.

Note that the problem we described is also (or in fact) an update problem. It concerns the problem to determine the consequences of the insertion of new evidence into the system.

2.3 Modeling uncertainty

In the following sections several ways to model uncertainty will be discussed. Starting from the oldest mathematical theory of uncertainty, that is probability theory, this will lead to a discussion of the methods based on the relatively new theory of fuzzy sets and measures.

Some well-known expert systems will be referred to. The survey is not intended to be exhaustive but most methods applied at the moment will be discussed while some possibilities (!) of recent methods will be touched upon.

3 METHODS ORIGINATING FROM PROBABILITY THEORY

Application of probability theory requires the prescription or the determination of some probability distribution on a set of objects. This implies that to every object i a probability p_i has to be assigned so that $\Sigma_i \, p_i = 1$. We consider first the case that we have two sets of objects: a set of pieces of evidence and a set of propositions saying conclusion i is true. The question is then how to relate the two probability distributions, that is we are interested in $P(c_i|e_j)$: the probability that conclusion c_i is true given evidence e_j. This leads directly to a Bayesian attack of the problem:

$$P(c_i|e_j) = \frac{P(c_i)\, P(e_j|c_i)}{P(e_j)} \quad .$$

Note that not necessarily all probabilities $P(e_j)$ have to be specified since $P(e_j) = \Sigma_k \, P(e_j|c_k)\, P(c_k)$. So an underlying probability distribution can be assumed.

Several problems related to the Bayesian approach can be mentioned here.

a. assignment of two probability distributions requires two sets of exhaustive and mutually exclusive pieces of evidence and conclusions.

b. a large amount of statistical data is needed in the form of all $P(e_j|c_i)$.

c. incompleteness is not captured.

d. ignorance cannot be represented: if some piece of evidence e gives no reason to consider one conclusion more likely than another, this can only be described by assigning equal probabilities $p = P(c_j|e)$ for all j in question. However, in that case any combination becomes more likely.

e. in this approach: $P(\text{not } c_i|e_j) = 1 - P\,(c_i|e_j)$, that is evidence supports both c_i and (not c_i), although not to the same degree. This is intuitively very peculiar.

f. in relation to point a.: it is difficult to combine evidence if it is not independent. Even then intuitively it is not necessarily true that more evidence always gives more belief (whatever that is: see later) in a conclusion as probability theory tells us.

We will now first discuss for two expert systems using derivations of this approach the way they tackle some of the problems mentioned above.

3.1 PROSPECTOR ([2])

This expert system uses a modified odds-likelihood form of Bayes' rule in order to obtain the updating formula

$$O(c|e_1, \ldots, e_n) = \{\prod_{i=1}^{n} \lambda_i\} O(c).$$

Here $O(c) = \frac{P(c)}{1-P(c)}$, $O(c/e) = \frac{P(c|e)}{1-P(c|e)}$ and $\lambda_i = \frac{P(e_i|c)}{P(e_i|\text{not } c)}$ $i=1, \ldots, n.$

Two assumptions are made:

1. $P(e_1, \ldots, e_n|c) = \prod_{i=1}^{n} P(e_i|c)$,
   conditional independence under conclusion.
2. $P(e_1, \ldots, e_n|\text{not } c) = \prod_{i=1}^{n} P(e_i|\text{not } c)$
   conditional independence under complement of conclusion.

Unfortunately, it was shown ([6]) that in case there are m conclusions c_i such that $m > 2$ and $\sum_{i=1}^{m} P(c_i) = 1$, then all λ_i are equal to one, that is no updating takes place, evidence is irrelevant! Later it was shown ([5]) that the second assumption is superfluous. However, still the first assumption has to hold, requiring a careful design of rules. Finally we note that in this system none of the problems mentioned above are solved.

3.2 MYCIN ([1])

During the design of MYCIN it was recognized that application of the Bayesian method required too huge an amount of statistical data, in particular in case of new evidence:

$$P(c/e) = \frac{P(e_1|c \text{ and } e_o)\ P(c|e_1)}{\sum_j P(e_1|c_j \text{ and } e_o)\ P(c_j|e_1)}$$

Here e_o is all evidence to date, e_1 some new evidence, $e = e_o$ and e_1 and c is one of n disjoint conclusions. This is closely related to problem f above. Therefore approximation methods were sought for computing $P(c|e)$ in terms of $P(c|s_j)$ where s_j are elementary pieces of evidence. Such methods will in general not be exact, due to dependencies. Moreover it was recognized that probability cannot always express the belief an expert has in a conclusion given some pieces of evidence (in particular see problems e. and f.). Therefore to rules two numbers were assigned not to be interpreted as probabilities but as possibilities (beliefs), indicating the strength of a rule. Relations were sought with confirmation theory, considering the strength to be of relative importance only for singling out conclusions that are comparatively more likely.

Two measure MB(c,e) and MD(c,e) are defined. MB(c,e) = x (MD(c,e) = x) means that the measure of increased belief (disbelief) in c based on e is x. The evidence e need not be an observed event, but may be some hypothesis (which itself is subject to confirmation). Relation with probability theory is expressed in the formulas:

$$MB(c,e) = \begin{cases} 1 & \text{if } P(c) = 1 \\ \dfrac{\max[P(c|e),\ P(c)] - P(c)}{\max[1,0] - P(c)} & \text{otherwise} \end{cases}$$

$$MD(c,e) = \begin{cases} 1 & \text{if } P(c) = 0 \\ \dfrac{\min[P(c|e),\ P(c)] - P(c)}{\min[1,0] - P(c)} & \text{otherwise} \end{cases}$$

Moreover, a third measure is defined, the certainty factor CF(c,e) as $CF(c,e) = MB(c,e) - MD(c,e)$.
The formulas hold if probabilities are available, otherwise the values of MB and MD have to be estimated (of course the question here is how?).

Using the definitions above it is not possible to express for two pieces of evidence e_1 and e_2 $CF(c,\ e_1 \text{ and } e_2)$ in terms of $CF(c,\ e_1)$ and $CF(c,\ e_2)$. Still some way of handling new evidence had to be developed. The following combining functions were introduced:

$$MB(c,\ e_1 \text{ and } e_2) = \begin{cases} 0 & \text{if } MD(c,\ e_1 \text{ and } e_2) = 1 \\ MB(c,e_1) + MB(c,e_2) & \text{otherwise} \\ \quad - MB(c,e_1)\ MB(c,e_2) & \end{cases}$$

$$MB(c_1 \text{ and } c_2, e) = \min (MB(c_1,e), MB(c_2,e))$$

$$MB(c_1 \text{ or } c_2, e) = \max (MB(c_1,e), MB(c_2,e))$$

$$MB(c, e_2) = MB'(c,e_2) \times \max(0, CF(e_2,e_1))$$

Here $MB'(c,e_1)$ is the degree of belief in c when e_1 is known to be true with certainty.
For MD there are similar rules.
These definitions of combining functions are rather arbitrary. They are so that in some particular cases results match with an expert's intuitive feeling of (propagation of) belief. In general however, the definitions may lead to very unexpected and unwanted evaluations. Note that the first combining function assumes independence of e_1 and e_2. Whether or not the definitions give a good approximation of reality will depend on the set of rules. In MYCIN, with its short chains of reasoning and simple conclusions, they gave good results.

The idea of interpreting probability as a measure of belief has been developed further. In section 3.3 we will discuss the Dempster-Shafer theory. In the next chapter other results will be exposed.

3.3 Dempster-Shafer theory

It is questionable whether belief can be expressed in a single number. This is recognized in MYCIN by means of MB and MD. However, introduction of CF returns a one-valued measure. Moreover, abandoning a well founded theory, as has been done in MYCIN, may lead to unsatisfactory results without knowing exactly why.
These points are addressed to in the theory of evidence developed by Dempster and Shafer ([1]). This theory also addresses the question how to deal with ignorance.

For every piece of evidence a basic probability function (bpf) $m: \mathcal{P}(\Theta) \rightarrow [0,1]$ is defined such that $m[\phi] = 0$ and $\Sigma\, m[A] = 1$. Here Θ is the set of all conclusions (hypotheses), $\mathcal{P}(\Theta)$ is the power set of Θ and summation is over all subsets A of Θ. Note that m is not a probability distribution on Θ. m[A] measures that portion of the total belief committed precisely to A. Note also that $m[A^c] \neq 1-m[A]$ in general. The total belief in A is measured by the belief function Bel:

$$Bel[A] = \sum_{B \subset A} m[B].$$

For the combination of two pieces of evidence with bpf m_1 and m_2 a combined bpf $m_1 \oplus m_2$ is defined by:

$$(m_1 \oplus m_2)[A] := \sum_{X \cap Y = A} m_1[X]\, m_2\, [Y].$$

with an associated belief function.
The belief interval is then defined as $[Bel[A], 1-Bel[A^c]]$. The larger the belief that A and A^c are clearly separated, the less mass will be divided between A and A^c, so the smaller the interval. In case of complete ignorance the interval is [0,1].
The quantity $1-Bel[A^c]$ expresses the extent to which the evidence allows one to fail to doubt A (the plausibility of A).

Although this theory seems to be quite satisfactory, several problems arise. In the first place, computation of combined bpf's is very time-consuming; second the definition of these combined bpf's indicates some kind of assumed independence and finally there remains the problem of contradictory evidence. The last two problems are addressed to in the next expert system to be dealt with, in which the idea of an interval-valued measure of belief is further developed.

3.4 INFERNO ([8])

All methods discussed until now did not tackle the problem of inconsistent information: how to discover it and how to deal with it. Moreover, these methods were primarily developed for systems in which the flow of inference is in one direction only. Also they were based on assumptions about probability distributions and were not able to use explicitly known relations, for instance known independencies. INFERNO was designed to give a solution for these problems.

Given a set of propositions, to every proposition A two values t(A) and f(A) are assigned such that $P(A) \geq t(A)$ and $P(\neg A) \geq f(A)$ which is to be interpreted as t(A) being a lower bound on the probability of A derived from the evidence for A and f(A) is a lower bound on the probability of $\neg A$ derived from the evidence against A. Evidence is for A if it allows the inference $P(A) \geq X$ and against A if it gives $P(A) \leq X$. Here X is a value determining the strength of the evidence.

The information about A is defined to be consistent if $t(A) + f(A) \leq 1$. The relations between propositions that can be used are listed in table 1 together with their interpretation.
Initially each proposition A has the trivial bounds $t(A) = f(A) = 0$. New information causes these values to be changed according to the propagation constraints given in Table 2.

That is as soon as some $t(A)$ or $f(A)$ changes all other values have to be changed so that the constraints are still satisfied. If some information proves to be inconsistent, the input value of the information has to be changed or the value of a strength value X.
Problems related to this approach are:
Should the strengths be determined a priori; what is their significance (Bayesian multipliers?)? Can adaptation of values really be done efficiently and consistently? Note also that as t and f are not exact, inconsistencies are not always noticed. Finally, there is the general decision problem between alternatives measured in terms of intervals.

3.5 Probability measures and fuzzy measures

The previous sections have described some methods for modeling uncertainty. These methods emerged from probability theory and evolved into measures of belief. This evolution can be given a theoretical background (see also [7]).
Consider again a set of propositions $\mathcal{P}$ with properties as given in section 2. Let $\mathbb{0}$ be the ever-false and $\mathbb{1}$ be the ever-true proposition. A probability measures P on $\mathcal{P}$ is then a function from $\mathcal{P}$ to [0,1] such that $P(\mathbb{0}) = 0$, $P(\mathbb{1}) = 1$ and if $p \wedge q = \mathbb{0}$ then $P(p \vee q) = P(p) + P(q)$. P can be considered to be derived from a probability distribution. The distribution of probability among the propositions will change as new (external) information is obtained. If after some amount of information for some proposition p we have $P(p) = 1$ then p has turned into the ever-true proposition, that is p is true (with probability 1).
Note that for P we have: 1) $\forall p$: $P(p) + P(\neg p) = 1$
2) if $(p \rightarrow q = \mathbb{1})$ then $P(q) \geq P(p)$.
For expert systems the first property is not desirable as we have stated before. Therefore other measures were considered generalizing the concept of probability measure.
Let g be a function from $\mathcal{P}$ to [0,1] such that $g(\mathbb{0}) = 0$, $g(\mathbb{1}) = 1$ and if $(p \rightarrow q = \mathbb{1})$ then $g(q) \geq g(p)$.

Such a function is called a fuzzy measure. The set of fuzzy measures is actually too large. In order to obtain fuzzy measures for practical use more requirements have to be added.

Note that the belief functions of Dempster and Shafer can also be considered as fuzzy measures. In analogy we obtain a restricted class of fuzzy measures as follows. Let m be a function from $\wp$ to [0,1] such that $m(\emptyset) = 0$ and $\sum_{p\in\wp} m(p) = 1$. Then the credibility (belief) function based on m is defined as $Cr(q) = \sum_{(p\to q=\mathbb{1})} m(p)$.

By duality a plausibility function Pl is defined by $Pl(p) = 1-Cr(\neg p)$. Always $Cr(p) \leq Pl(p)$.

The use of such fuzzy measures and the use of fuzzy sets for the description of uncertainty and imprecision will be the subject of discussion in the next chapter.

4 FUZZY REASONING

The previous chapter has indicated that probability theory is not adequate enough to describe uncertainty. In this chapter we will show how the relatively new theory of fuzzy sets and fuzzy measures can be applied to describe uncertainty and imprecision. In particular the notions of necessity and possibility are introduced. These notions are commonly used to model the intuitive feeling people have for a minimal amount of evidence needed to support a proposition and for all propositions supported by some piece of evidence. As they are related to intuition, they are modelled as fuzzy measures. See also [7] or [9].

4.1 Fuzzy measures

Fuzzy measures on propositions were introduced in the previous chapter. More generally fuzzy measures are used to evaluate the certainty that some object belongs to a well-defined set. For instance it can be an assessment of the answer to the question: 'Do you consider this piece of furniture to be more than 200 years old?'. Returning to the set of propositions $\wp$ defined in the previous chapters, the objects are propositions, the well-defined set is the set of all true propositions.

Consider a credibility function Cr defined by a function m. Propositions p with $m(p) > 0$ are called focal. If the focal propositions can be ordered in $p_n, p_{n-1}, \ldots, p_1$ so that $\forall i \leq n$: $(p_i \rightarrow p_{i-1} = \mathbb{1})$ then it can be shown that $\forall p$: 1) $Cr\ (p \wedge q) = \min\ (Cr(p), Cr(q))$

2) $Pl\ (p \vee q) = \max\ (Pl(p), Pl(q))$.

Any credibility function satisfying 1) is called a necessity measure. Any plausibility function satisfying 2) is called a possibility measure. Necessity and possibility measures have intuitively appealing properties with respect to the representation of the uncertainty of a proposition ([7]), for example (with N and Π denoting necessity and possibility measures, respectively): $\forall p \in \mathcal{P}$:

$$\min\ (N(p), N(\neg p)) = 0 \quad , \quad \max\ (\Pi(p), \Pi(\neg p)) = 1$$

$$\Pi(p) < 1 \rightarrow N(p) = 0 \quad , \quad N(p) > 0 \rightarrow \Pi(p) = 1$$

$$N(p) = 1 - \Pi(\neg p)$$

The uncertainty of propositions is then represented by a pair of numbers $(N(p), \Pi(p))$.

The functions MB(c,e) and MD(c,e) defined in MYCIN can be viewed as a necessity measure and as the complement to 1 of a possibility measure. In this set up, if logical independence of conclusions is not the case, in MB(c_1 and c_2, e) = min (MB(c,e), MB(c_1,e)) the sign '=' should be replaced by '≦'.

4.2 Fuzzy sets

Fuzzy sets are introduced in order to represent the fact that the location of a given set is not known exactly. In classical set theory a subset A of a set X is represented by its characteristic function χ: $\chi(x) = \{^1_o$ if $x \,{}^{\in A}_{\notin A}$. $\chi(x)$ is the probability that $x \in A$. For a fuzzy set A not every x has to have a probability equal to zero or one. It is characterized by its characteristic function μ_A: $X \rightarrow [0,1]$. Again $\mu_A(x)$ can be interpreted as a probability. The statement "John is tall" can be represented by: length (John) $\in A$, where length is known exactly and A is a fuzzy set, e.g.

$$\mu_A(x) = \begin{cases} 0 & \text{if } x < 170 \\ 1 & \text{if } x > 190 \\ \frac{x-170}{20} & \text{otherwise} \end{cases} .$$

Union, intersection and complement of fuzzy sets are defined by:

$$\mu_{A\cup B}(x) = \max(\mu_A(x), \mu_B(x))$$

$$\mu_{A\cap B}(x) = \min(\mu_A(x), \mu_B(x))$$

$$\mu_{A^c}(x) = 1-\mu_A(x)$$

However, other definitions are possible (see for instance [3]). Inclusion is defined by $A \subset B \leftrightarrow \forall x \in X: \mu_A(x) \leq \mu_B(x)$.

Given a variable u with values in X the range of u can be some fuzzy subset A of X. The possibility $\Pi(u=x)$ that u assumes the value x is then defined as

$$\Pi(u=x) = \Pi_u(x) = \mu_A(x).$$

Note that nothing is said about a probability that u assumes the value x. Π_u is called the possibility distribution of u.

Imprecision is now dealt with as follows.
Consider a proposition of the form 'u is A', which restricts the possible values of a variable u. Imprecision is expressed by allowing A to be a fuzzy subset of the space X of all acceptable values. $\mu_A(x)$ is interpreted as the possibility that the proposition p_x: 'u takes the value x' is true, knowing that 'u is A'. Thus the proposition 'u is A' is translated into $\forall x : \Pi_u(x) = \mu_A(x)$. Here Π_u is supposed to be normalized, that is for some x $\Pi_u(x) = 1$: at least one value must be assumed by u with certainty. Conjunction and disjunction of propositions 'u is A' and 'v is B' can be expressed in terms of μ_A and μ_B. So can 'u is not A'.
Knowing that 'u = A' (that is $\Pi_u = \mu_A$), for a non fuzzy subset F of X the possibility that the proposition q = 'u is F' is true can be calculated as $\Pi_u(F|A) = \max_{x\in F} \mu_A(x)$.
The necessity is calculated as $N_u(F|A) = \min_{x\notin F} (1-\mu_A(x))$.
This has an extension to fuzzy subsets F:

$$\Pi_u(F|A) = \sup_x \min(\mu_F(x), \Pi_u(x))$$

$$N_u(F|A) = \inf_x \max(\mu_F(x), 1-\Pi_u(x)).$$

These measures indicate uncertainty of the proposition 'u = F'. It is important to note that these values are purely indicative since it is difficult to give the most relevant definition of μ_A (and μ_F).

4.3 Inexact reasoning

The classical modus ponens given by the truth table

$$\frac{\begin{array}{l} p \rightarrow q \\ p \end{array}}{q}$$

can be generalized to the case of the uncertain premises as

$$\frac{\begin{array}{ll} \text{Nec } (p \rightarrow q) \geq a & , \text{Pos } (p \rightarrow q) \geq A \ (\geq a) \\ \text{Nec } (p) \geq b & , \text{Pos } (p) \geq B \ (\geq b) \end{array}}{\text{Nec } (q) \geq \min (a,b), \ \text{Pos } (q) \geq \max \left(\begin{cases} 0 & \text{if } A+b \leq 1 \\ A & \text{if } A+b > 1 \end{cases}, \begin{cases} 0 & \text{if } a+B \leq 1 \\ B & \text{if } a+B > 1 \end{cases} \right)}$$

$$\frac{\begin{array}{ll} \text{Prob } (p \rightarrow q) & \geq a \\ \text{Prob } (p) & \geq b \end{array}}{\text{Prob } (q) \quad \geq \max (o, a + b - 1).}$$

In MYCIN the product ab is used instead of min (a,b).

Results can also be derived for more complicated cases of inexact reasoning. We mention:

1. Multivalued logics: each proposition has a degree of truth in [0,1].
2. Use of fuzzy valued measures, fuzzy valued truth values.
3. Rules as 'If u is F then v = G', F and G fuzzy sets, can be expressed by means of fuzzy relations relating μ_G to μ_F. However, in general several relations are possibile ([4]).

   These relations allow one to consider the generalized modus ponens:

$$\frac{\begin{array}{l} \text{if u is F then v is G} \\ \quad \text{u is F'} \end{array}}{\text{v is G'}.}$$

Let us finally note that in fuzzy set theory union and intersection can be defined in more than one way. So can cartesian products. This implies that many relations and formulas can get other interpretations than given by min/max.

5 CONCLUDING REMARKS

Traditionally probability theory has been the theory to handle uncertainties. This theory has been built on a solid mathematical foundation and is rich in results and applications, such as in statistics and information theory. These probability based theories seemed to do very well in representation and management of uncertainty in information. However, the rise of expert systems and the need to model reasoning of experts appeared to lead to such serious difficulties in the application of probabilistic methods that other theories were developed which are now competing with probability theory. Some of these theories have been highlighted in this report. We mention theories of fuzzy sets, belief functions, possibility distributions, approximate and plausible reasoning, multi-valued logics etc.
In order to evaluate the competetiveness of these alternative theories the following can be noted.
Whether or not the arguments against probability mentioned in section 3 really apply is at the moment a point of intense discussion, in which the last word has not been said yet. Alternative theories pretend to be able to represent the more intuitive way of reasoning as experts do. The theories themselves are well founded but as to their use there are still questions. First, the theories are recent and the number of applications is still limited so that an evaluation is restricted. Second, to what extent do they really model intuition and if so, whose intuition? Application of these theories may be difficult because it may be hard to specify the fuzzy operations and it may be hard to keep the results meaningful and interpretable after longer chains of reasoning.

Given this state of the art it is recommendable not to abandon probability theory at once in favor of the alternative theories. First it should be investigated to what extent probability theory can be applied to model uncertainty. Should this lead to unsatisfactory results or should the required effort be too large, then alternatives can be considered. Note that alternative methods can always be applied to enrich the language that can be used to describe the problem. Therefore investigation of all theories is advisable to optimize the usability of systems dealing with uncertainties.

REFERENCES

[1] Buchanan, B.G., Shortliffe, E.H. — Rule-based Expert Systems. Addison-Wesley 1985.

[2] Duda, R.O., Hart, P.E., Nilsson, N.J. — Semantic Network Representation in Rule-based inference systems. Pattern-directed inference systems, Academic Press Inc., 1978, pp. 203-221.

[3] Dubois, D., Prade, H. — Outline of fuzzy set theory: an introduction. Advances in fuzzy set theory and applications. North Holland 1979.

[4] Mizumoto, M., Fukami, S., Tanaka, K. — Some methods of fuzzy reasoning. Advances in fuzzy set theory and applications. North Holland 1979.

[5] Pearl, J. — Reverend Bayes on inference engines: a distributed hierarchical approach. Proceedings of Nat. Conf. on AI, pp. 133-136, 1982.

[6] Pednault, E.P.D., Zucker, S.W., Muresan, L.V. — On the independence assumption underlying subjective Bayesian updating. Artificial Intelligence, 16, 213-222, 1981.

[7] Prade, H. — A computational approach to approximate and plausible reasoning with applications to expert systems. IEEE transactions on pattern analysis and machine intelligence, Vol. Pami-7, no. 3, May 1985.

[8] Quinlan, J.R. Inferno: a cautious approach to uncertain inference. The computer journal, vol. 26, no. 3, 1983.

[9] Zadeh, L.A. A theory of approximate reasoning. Machine Intelligence, Vol. 9, Elsevier, New York, 1979, pp. 149-194.

TABLE 1

INFERNO relations and their interpretation

Relation	Interpretation
A enables S with strength X	$P(S\mid A) \geqq X$
A inhibits S with strength X	$P(\neg S\mid A) \geqq X$
A requires S with strength X	$P(\neg A\mid \neg S) \geqq X$
A unless S with strength X	$P(A\mid \neg S) \geqq X$
A negates S	$A \equiv \neg S$
A conjoins $\{S_1,S_2,\ldots,S_n\}$	$A \equiv \bigwedge_i S_i$
A conjoins-independent $\{S_1,S_2,\ldots,S_n\}$	$A \equiv \bigwedge_i S_i$; and for all $i \neq j$, $P(S_i \wedge S_j) = P(S_i) \times P(S_j)$
A disjoins $\{S_1,S_2,\ldots,S_n\}$	$A \equiv \bigvee_i S_i$
A disjoins-independent $\{S_1,S_2,\ldots,S_n\}$	$A \equiv \bigvee_i S_i$; and for all $i \neq j$, $P(S_i \wedge S_j) = P(S_i) \times P(S_j)$
A disjoins-exclusive $\{S_1,S_2,\ldots,S_n\}$	$A \equiv \bigvee_i S_i$; and for all $i \neq j$, $P(S_i \wedge S_j) = 0$
$\{S_1,S_2,\ldots,S_n\}$ mutually exclusive	for all $i \neq j$, $P(S_i \wedge S_j) = 0$

TABLE 2

INFERNO propagation constraints

A enables S with strength X:

$t(S) \geq t(A) \times X$

$f(A) \geq 1 - (1 - f(S))/X$

A negates S:

$t(A) = f(S)$

$f(A) = t(S)$

A conjoins $\{S_1, S_2, \ldots S_n\}$:

$t(A) \geq 1 - \Sigma_i \, (1 - t(S_i))$

$f(A) \geq f(S_i)$

$t(S_i) \geq t(A)$

$f(S_i) \geq f(A) - \Sigma_{j \neq i} \, (1 - t(S_j))$

A conjoins-independent $\{S_1, S_2, \ldots S_n\}$:

$t(A) \geq \Pi_i \, t(S_i)$

$f(A) \geq 1 - \Pi_i \, (1 - f(S_i))$

$t(S_i) \geq t(A)/\Pi_{j \neq i} \, (1 - f(S_j))$

$f(S_i) \geq 1 - (1 - f(A))/\Pi_{j \neq i} \, t(S_j)$

A disjoins $\{S_1, S_2, \ldots S_n\}$:

$t(A) \geq t(S_i)$

$f(A) \geq 1 - \Sigma_i \, (1 - f(S_i))$

$t(S_i) \geq t(A) - \Sigma_{j \neq i} \, (1 - f(S_j))$

$f(S_i) \geq f(A)$

A disjoins-independent $\{S_1, S_2, \ldots S_n\}$:

$t(A) \geq 1 - \Pi_i \, (1 - t(S_i))$

$f(A) \geq \Pi_i \, f(S_i)$

$t(S_i) \geq 1 - (1 - t(A))/\Pi_{j \neq i} \, f(S_j)$

$f(S_i) \geq f(A)/\Pi_{j \neq i} \, (1 - t(S_j))$

A disjoins-exclusive $\{S_1, S_2, \ldots S_n\}$:

$t(A) \geq \Sigma_i \, t(S_i)$

$f(A) \geq 1 - \Sigma_i \, (1 - f(S_i))$

$t(S_i) \geq t(A) - \Sigma_{j \neq i} \, (1 - f(S_j))$

$f(S_i) \geq f(A) + \Sigma_{j \neq i} \, t(S_j)$

$\{S_1, S_2, \ldots S_n\}$ mutually exclusive:

$f(S_i) \geq \Sigma_{j \neq i} \, t(S_j)$

Fuzzy Logic

Lofti A. Zadeh
University of California, Berkeley

Reprinted from *Computer*, April 1988, pages 83-93.

Logic, according to Webster's dictionary, is the science of the normative formal principles of reasoning. In this sense, fuzzy logic is concerned with the formal principles of approximate reasoning, with precise reasoning viewed as a limiting case.

In more specific terms, what is central about fuzzy logic is that, unlike classical logical systems, it aims at modeling the imprecise modes of reasoning that play an essential role in the remarkable human ability to make rational decisions in an environment of uncertainty and imprecision. This ability depends, in turn, on our ability to infer an approximate answer to a question based on a store of knowledge that is inexact, incomplete, or not totally reliable. For example:

(1) Usually it takes about an hour to drive from Berkeley to Stanford and about half an hour to drive from Stanford to San Jose. How long would it take to drive from Berkeley to San Jose via Stanford?

(2) Most of those who live in Belvedere have high incomes. It is probable that Mary lives in Belvedere. What can be said about Mary's income?

(3) Slimness is attractive. Carol is slim. Is Carol attractive?

(4) Brian is much taller than most of his close friends. How tall is Brian?

There are two main reasons why classical logical systems cannot cope with problems of this type. First, they do not provide a system for representing the meaning of propositions expressed in a natural language when the meaning is imprecise; and second, in those cases in which the meaning can be represented symbolically in a meaning representation language, for example, a semantic network or a conceptual-dependency graph, there is no mechanism for inference.

Fuzzy logic — the logic underlying approximate, rather than exact, modes of reasoning — is finding applications that range from process control to medical diagnosis.

As will be seen, fuzzy logic addresses these problems in the following ways. First, the meaning of a lexically imprecise proposition is represented as an elastic constraint on a variable; and second, the answer to a query is deduced through a propagation of elastic constraints.

During the past several years, fuzzy logic has found numerous applications in fields ranging from finance to earthquake engineering. But what is striking is that its most important and visible application today is in a realm not anticipated when fuzzy logic was conceived, namely, the realm of fuzzy-logic-based process control. The basic idea underlying fuzzy logic control was suggested in notes published in 1968 and 1972[1,2] and described in greater detail in 1973.[3] The first implementation was pioneered by Mamdani and Assilian in 1974[4] in connection with the regulation of a steam engine. In the ensuing years, once the basic idea underlying fuzzy logic control became well understood, many applications followed. In Japan, in particular, the use of fuzzy logic in control processes is being pursued in many application areas, among them automatic train operation (Hitachi),[5] vehicle control (Sugeno Laboratory at Tokyo Institute of Technology),[5] robot control (Hirota Laboratory at Hosei University),[5] speech recognition (Ricoh),[5] universal controller (Fuji),[5] and stabilization control (Yamakawa Laboratory at Kumamoto University).[5] More about

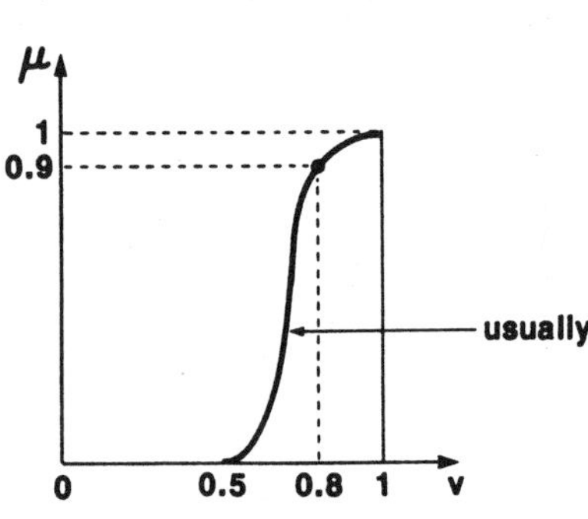

Figure 1. Representation of "usually" as a fuzzy proportion.

some of these projects will be said in the section dealing with applications.

In most of the current applications of fuzzy logic, software is employed as a medium for the implementation of fuzzy algorithms and control rules. What is clear, however, is that it would be cheaper and more effective to use fuzzy logic chips and, eventually, fuzzy computers. The first logic chip was developed by Togai and Watanabe at Bell Telephone Laboratories in 1985, and it is likely to become available for commercial use in 1988 or 1989. On the heels of this important development came the announcement of a fuzzy computer designed by Yamakawa at Kumamoto University. These developments on the hardware front may lead to an expanded use of fuzzy logic not only in industrial applications but, more generally, in knowledge-based systems in which the deduction of an answer to a query requires the inference machinery of fuzzy logic.

One important branch of fuzzy logic may be called *dispositional logic*. This logic, as its name implies, deals with *dispositions*, that is, propositions that are preponderantly but not necessarily always true. For example, "snow is white" is a disposition, as are the propositions "Swedes are blond" and "high quality is expensive." A disposition may be viewed as a usuality-qualified proposition in which the qualifying quantifier "usually" is implicit rather than explicit. In this sense, the disposition "snow is white" may be viewed as the result of suppressing the fuzzy quantifier "usually" in the usuality-qualified proposition

usually (snow is white)

In this proposition, "usually" plays the role of a fuzzy proportion of the form shown in Figure 1.

The importance of dispositional logic stems from the fact that most of what is usually referred to as common sense knowledge may be viewed as a collection of dispositions. Thus, the main concern of dispositional logic lies in the development of rules of inference from common sense knowledge.

In what follows, I present a condensed exposition of some basic ideas underlying fuzzy logic and describe some representative applications. More detailed information regarding fuzzy logic and its applications may be found in the cited literature.

Basic principles

Fuzzy logic may be viewed as an extension of multivalued logic. Its uses and objectives, however, are quite different. Thus, the fact that fuzzy logic deals with approximate rather than precise modes of reasoning implies that, in general, the chains of reasoning in fuzzy logic are short in length, and rigor does not play as important a role as it does in classical logical systems. In a nutshell, in fuzzy logic everything, including truth, is a matter of degree.

The greater expressive power of fuzzy logic derives from the fact that it contains as special cases not only the classical two-valued and multivalued logical systems but also probability theory and probabilistic logic. The main features of fuzzy logic that differentiate it from traditional logical systems are the following:

(1) In two-valued logical systems, a proposition p is either true or false. In multivalued logical systems, a proposition may be true or false or have an intermediate truth value, which may be an element of a finite or infinite truth value set T. In fuzzy logic, the truth values are allowed to range over the fuzzy subsets of T. For example, if T is the unit interval, then a truth value in fuzzy logic, for example, "very true," may be interpreted as a fuzzy subset of the unit interval. In this sense, a fuzzy truth value may be viewed as an imprecise characterization of a numerical truth value.

(2) The predicates in two-valued logic are constrained to be crisp in the sense that the denotation of a predicate must be a nonfuzzy subset of the universe of discourse. In fuzzy logic, the predicates may be crisp—for example, "mortal," "even," and "father of"—or, more generally, fuzzy—for example, "ill," "tired," "large," "tall," "much heavier," and "friend of."

(3) Two-valued as well as multivalued logics allow only two quantifiers: "all" and "some." By contrast, fuzzy logic allows, in addition, the use of fuzzy quantifiers exemplified by "most," "many," "several," "few," "much of," "frequently," "occasionally," "about ten," and so on. Such quantifiers may be interpreted as fuzzy numbers that provide an imprecise characterization of the cardinality of one or more fuzzy or nonfuzzy sets. In this perspective, a fuzzy quantifier may be viewed as a second-order fuzzy predicate. Based on this view, fuzzy quantifiers may be used to represent the meaning of propositions containing fuzzy probabilities and thereby make it possible to manipulate probabilities within fuzzy logic.

(4) Fuzzy logic provides a method for representing the meaning of both nonfuzzy and fuzzy predicate-modifiers exemplified by "not," "very," "more or less," "extremely," "slightly," "much," "a little," and so on. This, in turn, leads to a system for computing with *linguistic variables*,[3] that is, variables whose values are words or sentences in a natural or synthetic language. For example, "Age" is a linguistic variable when its values are assumed to be "young," "old," "very young," "not very old," and so forth. More about linguistic variables will be said at a later point.

(5) In two-valued logical systems, a proposition p may be qualified, principally by associating with p a truth value, "true" or "false"; a modal operator such as "possible" or "necessary"; and an intensional operator such as "know" or "believe." Fuzzy logic has three principal modes of qualification:

- *truth-qualification*, as in

  (Mary is young) is not quite true,

  in which the qualified proposition is (Mary is young) and the qualifying truth value is "not quite true";
- *probability-qualification*, as in

  (Mary is young) is unlikely,

  in which the qualifying fuzzy probability is "unlikely"; and
- *possibility-qualification*, as in

  (Mary is young) is almost impossible,

  in which the qualifying fuzzy possibility is "almost impossible."

An important issue in fuzzy logic relates to inference from qualified propositions, especially from probability-qualified propositions. This issue is of central importance in the management of uncertainty in expert systems and in the formalization of common sense reasoning. In the latter, it's important to note the close connection between probability-qualification and usuality-qualification and the role played by fuzzy quantifiers. For example, the disposition

Swedes are blond

may be interpreted as

most Swedes are blond;

or, equivalently, as

(Swede is blond) is likely,

where "likely" is a fuzzy probability that is numerically equal to the fuzzy quantifier "most"; or, equivalently, as

usually (a Swede is blond),

where "usually" qualifies the proposition "a Swede is blond."

As alluded earlier, inference from propositions of this type is a main concern of dispositional logic. More about this logic will be said at a later point.

Meaning representation and inference

A basic idea serving as a point of departure in fuzzy logic is that a proposition p in a natural or synthetic language may be viewed as a collection of elastic constraints, $C_1, \ldots, C_k$, which restrict the values of a collection of variables $X = (X_1, \ldots, X_n)$.[6] In general, the constraints as well as the variables they constrain are implicit rather than explicit in p. Viewed in this perspective, representation of the meaning of p is, in essence, a process by which the implicit constraints and variables in p are made explicit. In fuzzy logic, this is accomplished by representing p in the so-called *canonical form*

$$p \rightarrow X \text{ is } A$$

in which A is a fuzzy predicate or, equivalently, an n-ary fuzzy relation in U, where $U = U_1 \times U_2 \times \ldots \times U_n$, and U_i, $i = 1, \ldots, n$, is the domain of X_i. Representation of p in its canonical form requires, in general, the construction of an explanatory database and a test procedure that tests and aggregates the test scores associated with the elastic constraints $C_1, \ldots, C_k$.[6]

In more concrete terms, the canonical form of p implies that the possibility distribution[6] of X is equal to A—that is,

$$\Pi_X = A \tag{1}$$

which in turn implies that

$$\text{Poss}\{X = u\} = \mu_A(u),\ u \in U$$

where μ_A is the membership function of A and $\text{Poss}\{X = u\}$ is the possibility that X may take u as its value. Thus, when the meaning of p is represented in the form of Equation 1, it signifies that p induces a possibility distribution Π_X that is equal to A, with A playing the role of an elastic constraint on a variable X that is implicit in p. In effect, the possibility distribution of X, Π_X, is the set of possible values of X, with the understanding that possibility is a matter of degree. Viewed in this perspective, a proposition p constrains the possible values that X can take and thus defines its possibility distribution. This implies that the meaning of p is defined by (1) identifying the variable that is constrained and (2) characterizing the constraint to which the variable is subjected through its possibility distribution. Note that Equation 1 asserts that the possibility that X can take u as its value is numerically equal to the grade of membership, $\mu_A(u)$, of u in A.

As an illustration, consider the proposition

$$p \triangleq \text{John is tall}$$

in which the symbol $\triangleq$ should be read as "denotes" or "is equal to by definition." In this case, X = Height(John), A = TALL, and the canonical form of p reads

Height(John) is TALL

where the fuzzy relation TALL is in uppercase letters to underscore that it plays the role of a constraint in the canonical form. From the canonical form, it follows that

$$\text{Poss}\{\text{Height (John)} = u\} = \mu_{\text{TALL}}(u)$$

where μ_{TALL} is the membership function of TALL and $\mu_{\text{TALL}}(u)$ is the grade of membership of u in TALL or, equivalently, the degree to which a numerical height u satisfies the constraint induced by the relation TALL.

When p is a conditional proposition, its canonical form may be expressed as "Y is B if X is A," implying that p induces a conditional possibility distribution of Y given X, written as $\Pi_{(Y|X)}$. In fuzzy logic, $\Pi_{(Y|X)}$ may be defined in a variety of ways,[7] among which is a definition consistent with the definition of implication in Lakasiewicz's L_{Aleph_0} logic. In this case, the conditional possibility distribution function, $\pi_{(Y|X)}$, which defines $\Pi_{(Y|X)}$, may be expressed as

$$\pi_{(Y|X)}(u, v) = 1 \wedge (1 - \mu_A(u) + \mu_B(v)),\quad u \in U,\ v \in V, \tag{2}$$

where

$$\pi_{(Y|X)}(u, v) \triangleq \text{Poss}\{X = u,\ Y = v\}$$

μ_A and μ_B denote the membership functions of A and B, respectively; and $\wedge$ denotes the operator min.

When p is a quantified proposition of the form

$$p \triangleq Q\ A\text{'s are } B\text{'s}$$

for example,

$$p \triangleq \text{most tall men are not very fat}$$

where Q is a fuzzy quantifier and A and B are fuzzy predicates, the constrained variable, X, is the proportion of B's in A's, with Q representing an elastic constraint on X. More specifically, if U is a finite set $\{u_1, \ldots, u_m\}$, the proportion of B's in A's is defined as the *relative sigma-count*

$$\Sigma\text{Count}(B/A) = \frac{\sum_j \mu_A(u_j) \wedge \mu_B(u_j)}{\sum_j \mu_A(u_j)} \qquad j = 1, \ldots, m \tag{3}$$

where $\mu_A(u_j)$ and $\mu_B(u_j)$ denote the grades of membership of u_j in A and B, respectively. Thus, expressed in its canonical form, Equation 3 may be written as

$$\Sigma\mathit{Count}(B/A) \text{ is } Q$$

which places in evidence the constrained variable, X, in p and the elastic constraint, Q, to which X is subjected. Note that X is the relative sigma-count of B in A.

The concept of a canonical form pro-

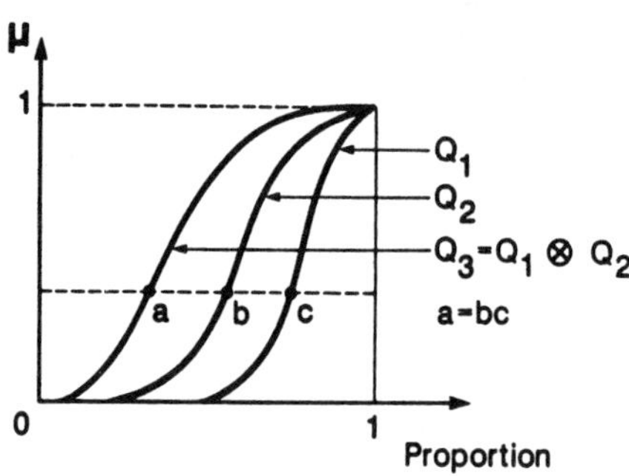

Figure 2. Representation of fuzzy quantifiers in the intersection/product syllogism.

vides an effective framework for formulating the problem of inference in expert systems. Specifically, consider a knowledge base, *KB*, which consists of a collection of propositions $\{p_1, \ldots, p_N\}$. Typically, a constituent proposition, p_i, $i = 1, \ldots, N$, may be (1) a fact that may be expressed in a canonical form as "*X* is *A*" or (2) a rule that may be expressed in a canonical form as "*Y* is B_i" if "*X* is A_i." More generally, both facts and rules may be probability-qualified or, equivalently, expressed as quantified propositions. For example, a rule of the general form "*Q A*'s are *B*'s" may be interpreted as the probability-qualified proposition (*X* is *B* if *X* is *A*) is λ, where λ is a fuzzy probability whose denotation as a fuzzy subset of the unit interval is the same as that of the fuzzy quantifier *Q* and *X* is chosen at random in *U*.

Now if p_i induces a possibility distribution $\Pi^i_{(X_1,\ldots,X_n)}$, where $X_1, \ldots, X_n$ are the variables constrained by p_i, then the possibility distribution $\Pi_{(X_1,\ldots,X_n)}$, which is induced by the totality of propositions in *KB* is given by the intersection[3] of the $\Pi^i_{(X_1,\ldots,X_n)}$. That is,

$$\Pi_{(X_1,\ldots,X_n)} = \Pi^1_{(X_1,\ldots,X_n)} \cap \ldots \cap \Pi^N_{(X_1,\ldots,X_n)}$$

or, equivalently,

$$\pi_{(X_1,\ldots,X_n)} = \pi^1_{(X_1,\ldots,X_n)} \wedge \ldots \wedge \pi^N_{(X_1,\ldots,X_n)}$$

$\pi_{(X_1,\ldots,X_n)}$ is the possibility distribution function of $\Pi_{(X_1,\ldots,X_n)}$. Note that there is no loss of generality in assuming that the constrained variables $X_1, \ldots, X_n$ are the same for all propositions in *KB* since the set $\{X_1, \ldots, X_n\}$ may be taken to be the union of the constrained variables for each proposition.

Now suppose that we are interested in inferring the value of a specified function $f(X_1, \ldots, X_n)$, $f: U \rightarrow V$, of the variables constrained by the knowledge base. Because of the incompleteness and imprecision of the information resident in *KB*, what we can deduce, in general, is not the value of $f(X_1, \ldots, X_n)$ but its possibility distribution, Π_f. By employing the extension principle,[8] it can be shown that the possibility distribution function of *f* is given by the solution of the nonlinear program

$$\pi_f(v) = \max_{u_1,\ldots,u_n} [\pi^1_{(X_1,\ldots,X_n)}(u_1, \ldots, u_n) \wedge \ldots \wedge \pi^N_{(X_1,\ldots,X_n)}(u_1, \ldots, u_n)] \quad (4)$$

subject to the constraint

$$v = f(u_1, \ldots, u_n)$$

where $u_j \in U_j$, $i = 1, \ldots, n$, and $v \in V$. The reduction to the solution of a nonlinear program constitutes the principal tool for inference in fuzzy logic.

Fuzzy syllogisms. A basic fuzzy syllogism in fuzzy logic that is of considerable relevance to the rules of combination of evidence in expert systems is the *intersection/product syllogism*—a syllogism that serves as a rule of inference for quantified propositions.[9] This syllogism may be expressed as the inference rule

Q_1 *A*'s are *B*'s (5)
Q_2 (*A* and *B*)'s are *C*'s
―――――――――――
$(Q_1 \otimes Q_2)$ *A*'s are (*B* and *C*)'s

in which Q_1 and Q_2 are fuzzy quantifiers, *A*, *B*, and *C* are fuzzy predicates, and $Q_1 \otimes Q_2$ is the product of the fuzzy numbers Q_1 and Q_2 in fuzzy arithmetic.[10] (See Figure 2). For example, as a special case of Equation 5, we may write

most students are single
a little more than a half of single students are male
―――――――――――
(most ⊗ a little more than a half) of students are single and male

Since the intersection of *B* and *C* is contained in *C*, the following corollary of Equation 5 is its immediate consequence.

Q_1 *A*'s are *B*'s (6)
Q_2 (*A* and *B*)'s are *C*'s
―――――――――――
$\geq (Q_1 \otimes Q_2)$ *A*'s are *C*'s

where the fuzzy number $\geq (Q_1 \otimes Q_2)$ should be read as "at least $(Q_1 \otimes Q_2)$." In particular, if the fuzzy quantifiers Q_1 and Q_2 are monotone increasing (for example, when "$Q_1 = Q_2 \triangleq$ most"), then

$$\geq (Q_1 \otimes Q_2) = Q_1 \otimes Q_2$$

and Equation 6 becomes

Q_1 *A*'s are *B*'s (7)
Q_2 (*A* and *B*)'s are *C*'s
―――――――――――
$(Q_1 \otimes Q_2)$ *A*'s are *C*'s

Furthermore, if *B* is a subset of *A*, then *A* and $B = B$, and Equation 7 reduces to the *chaining rule*

Q_1 *A*'s are *B*'s (8)
Q_2 *B*'s are C's
―――――――――――
$(Q_1 \otimes Q_2)$ *A*'s are *C*'s

For example,

most students are undergraduates
most undergraduates are young
―――――――――――
most^2 students are young

where "most^2" represents the product of the fuzzy number "most" with itself (see Figure 3).

What is important to observe is that the chaining rule expressed by Equation 8 serves the same purpose as the chaining rules in Mycin, Prospector, and other probability-based expert systems. However, Equation 8 is formulated in terms of fuzzy quantifiers rather than numerical probabilities or certainty factors, and it is a logical consequence of the concept of a relative sigma-count in fuzzy logic. Furthermore, the chaining rule (Equation 8) is robust in the sense that if Q_1 and Q_2 are close to unity, so is their product $Q_1 \otimes Q_2$. More specifically, if Q_1 and Q_2 are expressed as

$$Q_1 = 1 \ominus \varepsilon_1$$
$$Q_2 = 1 \ominus \varepsilon_2$$

where ε_1 and ε_2 are small fuzzy numbers, then, to a first approximation, Q may be expressed as

$$Q = 1 \ominus \varepsilon_1 \ominus \varepsilon_2$$

An important issue concerns the general properties Q_1, Q_2, A, B, and C must have to ensure robustness. As shown above, the containment of B in A and the monotonicity of Q_1 and Q_2 are conditions for robustness in the case of the intersection/product syllogism.

Another basic syllogism is the *consequent conjunction syllogism*

$$\frac{\begin{array}{l} Q_1\ A\text{'s are } B\text{'s} \\ Q_2\ A\text{'s are } C\text{'s} \end{array}}{Q\ A\text{'s are } (B \text{ and } C)\text{'s}} \qquad (9)$$

where

$$0 \oslash (Q_1 \oplus Q_2 \ominus 1) \leq Q \leq Q_1 \oslash Q_2$$

in which the operators $\oslash$, $\oslash$, $\oplus$, $\ominus$, and the inequality $\leq$ are the extensions of $\wedge$, $\vee$, $+$, $-$, and $\leq$, respectively, to fuzzy numbers.

The consequent conjunction syllogism plays the same role in fuzzy logic as the rule of combination of evidence for conjunctive hypotheses does in Mycin and Prospector.[11] However, whereas in Mycin and Prospector the qualifying probabilities and certainty factors are real numbers, in the consequent conjunction syllogism the fuzzy quantifiers are fuzzy numbers. As can be seen from the result expressed by Equation 9, the conclusion yielded by the

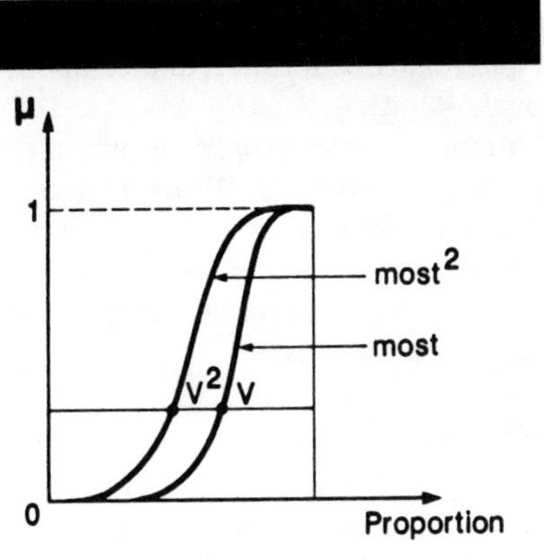

Figure 3. Representation of "most" and "most²."

Inference with fuzzy probabilities

An example of an important problem to which the reduction to a nonlinear program may be applied is the following. Assume that from a knowledge base $KB = \{p_1, \ldots, p_N\}$ in which the constituent propositions are true with probability one, we can infer a proposition q which, like the premises, is true with probability one. Now suppose that each p_i in KB is replaced with a probability-qualified proposition "$p_i^* \triangleq p_i$ is λ_i," in which λ_i is a fuzzy probability. For example

$p_i \triangleq X$ is small

and

$p_i^* \triangleq X$ is small is very likely

As a result of the qualification of the p_i, the conclusion, q, will also be a probability-qualified proposition that may be expressed as

$q^* = q$ is λ

in which λ is a fuzzy probability. The problem is to determine λ as a function of the λ_i, if such a function exists. A special case of this problem, which is of particular relevance to the management of uncertainty in expert systems, is one in which the fuzzy probabilities λ_i are close to unity. We shall say that the inference process is *compositional* if λ can be expressed as a function of the λ_i; it is *robust* if whenever the λ_i are close to unity, so is λ.

By reducing the determination of λ to the solution of a nonlinear program, it can be shown that, in general, the inference process is not compositional if the λ_i and λ are numerical probabilities. This result calls into question the validity of the rules of combination of evidence in those expert systems in which the certainty factor of the conclusion is expressed as a function of the certainty factors of the premises. However, compositionality does hold, in general, if the λ_i and λ are assumed to be fuzzy probabilities, for this allows the probability of q to be interval-valued when the λ_i are numerical probabilities, which is consistent with known results in inductive logic.

Another important conclusion relating to the robustness of the inference process is that, in general, robustness does not hold without some restrictive assumptions on the premises. For example, the brittleness of the transitivity of implication is an instance of the lack of robustness when no assumptions are made regarding the fuzzy predicates A, B, and C. On the other hand, if in the inference schema

$$\frac{\begin{array}{l} X \text{ is } A \\ Y \text{ is } B \text{ if } X \text{ is } A \end{array}}{Y \text{ is } B}$$

the major premise is replaced by "X is A is probable," where "probable" is a fuzzy probability close to unity, then it can be shown that, under mildly restrictive assumptions on A, the resulting conclusion may be expressed as "Y is B is $\geq$ probable," where "$\geq$ probable" is a fuzzy probability that, as a fuzzy number, is greater than or equal to the fuzzy number "probable." In this case, then, robustness does hold, for if "probable" is close to unity, so is "$\geq$ probable."

application of fuzzy logic to the premises in question is both robust and compositional.

A more complex problem is presented by what in Mycin and Prospector corresponds to the conjunctive combination of evidence. Stated in terms of quantified premises, the inference rule in question may be expressed as

$$\frac{\begin{array}{l} Q_1\ A\text{'s are } C\text{'s} \\ Q_2\ B\text{'s are } C\text{'s} \end{array}}{Q\ (A \text{ and } B)\text{'s are } C\text{'s}} \tag{10}$$

where the value of Q is to be determined. To place in evidence the symmetry between Equation 9 and Equation 10, we shall refer to the rule in question as the *antecedent conjunction syllogism*.

It can readily be shown that, without any restrictive assumptions on Q_1, Q_2, A, B, and C, there is nothing that can be said about Q, which is equivalent to saying that "Q = none to all." A basic assumption in Mycin, Prospector, and related systems is that the items of evidence are conditionally independent, given the hypothesis (and its complement). That is,

$$P(E_1, E_2 \mid H) = P(E_1 \mid H)\, P(E_2 \mid H)$$

where $P(E_1, E_2 \mid H)$ is the joint probability of E_1 and E_2, given the hypothesis H; and $P(E_1 \mid H)$ and $P(E_2 \mid H)$ are the conditional probabilities of E_1 given H and E_2 given H, respectively. Expressed in terms of the relative sigma-counts, this assumption may be written as

$$\Sigma\text{Count}\ (A \cap B\ /\ C) = \Sigma\text{Count}\ (A\ /\ C)\ \Sigma\text{Count}\ (B\ /\ C) \tag{11}$$

where $\cap$ denotes the intersection of fuzzy sets.[3]

To determine the value of Q in Equation 10 we have to compute the relative sigma-count of C in $A \cap B$. It can be verified that, under the assumption (Equation 11), the sigma-count in question is given by

$$\Sigma\text{Count}\ (C\ /\ A \cap B) = \Sigma\text{Count}\ (C\ /\ A)\ \Sigma\text{Count}\ (C\ /\ B)\ \delta$$

where the factor δ is expressed by

$$\delta = \frac{\Sigma\text{Count}\ (A)\ \Sigma\text{Count}\ (B)}{\Sigma\text{Count}\ (A \cap B)\ \Sigma\text{Count}\ (C)} \tag{12}$$

Inspection of Equation 12 shows that the assumption expressed by Eqaution 11 does not ensure the compositionality of Q. However, it can be shown that compositionality can be achieved through the use of the concept of a relative *ϱsigma-count*, which is defined as

$$\varrho\Sigma\text{Count}\ (B/A) = \frac{\Sigma\text{Count}(B/A)}{\Sigma\text{Count}\ (\neg B/A)}$$

where $\neg B$ denotes the negation of B. The use of ϱsigma-counts in place of sigma-counts is analogous to the use of odds instead of probabilities in Prospector, and it serves the same purpose.

Interpolation

An important problem that arises in the operation of any rule-based system is the following. Suppose the user supplies a fact that, in its canonical form, may be expressed as "X is A," where A is a fuzzy or nonfuzzy predicate. Furthermore, suppose that there is no conditional rule in *KB* whose antecedent matches A exactly. The question arises: Which rules should be executed and how should their results be combined?

An approach to this problem, sketched in Reference 8, involves the use of an interpolation technique in fuzzy logic which requires a computation of the degree of partial match between the user-supplied fact and the rows of a decision table. More specifically, suppose that upon translation into their canonical forms, a group of propositions in *KB* may be expressed as a fuzzy relation of the form

R	X_1	X_2	.	X_n	X_{n+1}
	R_{11}	R_{12}	.	R_{1n}	Z_1
	.	.	.	.	.
	R_{m1}	R_{m2}	.	R_{mn}	Z_m

in which the entries are fuzzy sets; the input variables are $X_1, \ldots, X_n$, with domains $U_1, \ldots, U_n$; and the output variable is X_{n+1}, with domain U_{n+1}. The problem is: Given an input n-tuple $(R_1, \ldots, R_n)$, in which R_j, $j = 1, \ldots, n$, is a fuzzy subset of U_j, what is the value of X_{n+1} expressed as a fuzzy subset of U_{n+1}?

A possible approach to the problem is to compute for each pair (R_{ij}, R_j) the degree of consistency of the input R_j with the R_{ij} element of R, $i = 1, \ldots, m$, $j = 1, \ldots, n$. The degree of consistency, γ_{ij}, is defined as

$$\gamma_{ij} \triangleq \sup\,(R_{ij} \cap R_j) = \sup_{u_j}\,(\mu_{R_{ij}}(u_j) \wedge \mu_{R_j}(u_j))$$

in which $\mu_{R_{ij}}$ and μ_{R_j} are the membership functions of R_{ij} and R_j, respectively; u_j is a generic element of U_j; and the supremum is taken over u_j.

Next, we compute the overall degree of consistency, γ_i, of the input n-tuple $(R_1, \ldots, R_n)$ with the ith row of R, $i = 1, \ldots, m$, by employing $\wedge$ (min) as the aggregation operator. Thus,

$$\gamma_i = \gamma_{i1} \wedge \gamma_{i2} \wedge \ldots \wedge \gamma_{in}$$

which implies that γ_i may be interpreted as a conservative measure of agreement between the input n-tuple $(R_1, \ldots, R_n)$ and the ith-row n-tuple $(R_{i1}, \ldots, R_{in})$. Then, employing γ_i as a weighting coefficient, the desired expression for X_{n+1} may be written as a "linear" combination

$$X_{n+1} = \gamma_1 \wedge Z_1 + \ldots + \gamma_m \wedge Z_m$$

in which $+$ denotes the union, and $\gamma_i \wedge z_i$ is a fuzzy set defined by

$$\mu_{\gamma_i \wedge z_i}(u_{i+1}) = \gamma_i \wedge \mu_{z_i}(u_{i+1}),\ i = 1, \ldots, m$$

The above approach ceases to be effective, however, when R is a sparse relation in the sense that no row of R has a high degree of consistency with the input n-tuple. For such cases, a more general interpolation technique has to be employed.

Basic rules of inference

One distinguishing characteristic of fuzzy logic is that premises and conclusions in an inference rule are generally expressed in canonical form. This representation places in evidence the fact that each premise is a constraint on a variable and that the conclusion is an induced constraint computed through a process of constraint propagation — a process that, in general, reduces to the solution of a nonlinear program. The following briefly presents — without derivation — some of the basic inference rules in fuzzy logic. Most of these rules can be deduced from the basic inference rule expressed by Equation 4.

The rules of inference in fuzzy logic may be classified in a variety of ways. One basic class is *categorical rules*, that is, rules that do not contain fuzzy quantifiers. A more general class is *dispositional rules*, rules in which one or more premises may contain, explicitly or implicitly, the fuzzy quantifier "usually." For example, the inference rule known as the *entailment principle*:

$$\frac{X \text{ is } A \qquad A \subset B}{X \text{ is } B} \tag{13}$$

where X is a variable taking values in a universe of discourse U, and A and B are fuzzy subsets of U, is a categorical rule. On the other hand, the *dispositional entailment principle* is an inference rule of the form

$$\frac{\text{usually } (X \text{ is } A) \qquad A \subset B}{\text{usually } (X \text{ is } B)} \tag{14}$$

In the limiting case where "usually" becomes "always," Equation 14 reduces to Equation 13.

In essence, the *entailment principle* asserts that from the proposition "X is A" we can always infer a less specific proposition "X is B." For example, from the proposition "Mary is young," which in its canonical form reads

Age(Mary) is YOUNG

where YOUNG is interpreted as a fuzzy set or, equivalently, as a fuzzy predicate, we can infer "Mary is not old," provided YOUNG is a subset of the complement of OLD. That is

$$\mu_{\text{YOUNG}}(u) \subset 1 - \mu_{\text{OLD}}(u),\ u \in [0, 100]$$

where μ_{YOUNG} and μ_{OLD} are, respectively, the membership functions of YOUNG and OLD, and the universe of discourse is the interval [0, 100].

Viewed in a different perspective, the entailment principle in fuzzy logic may be regarded as a generalization to fuzzy sets of the inheritance principle widely used in knowledge representation systems. More specifically, if the proposition "X is A" is interpreted as "X has property A," then the conclusion "X is B" may be interpreted as "X has property B," where B is any superset of A. In other words, X inherits property B if B is a superset of A.

Among other categorical rules that play a basic role in fuzzy logic are the following. In all of these rules, $X, Y, Z, \ldots$ are variables ranging over specified universes of discourse, and $A, B, C, \ldots$ are fuzzy predicates or, equivalently, fuzzy relations.

Conjunctive rule.

$$\frac{X \text{ is } A \qquad X \text{ is } B}{X \text{ is } A \cap B}$$

where $A \cap B$ is the intersection of A and B defined by

$$\mu_{A \cap B}(u) = \mu_A(u) \wedge \mu_B(u),\quad u \in U$$

Cartesian product.

$$\frac{X \text{ is } A \qquad Y \text{ is } B}{(X, Y) \text{ is } A \times B}$$

where (X, Y) is a binary variable and $A \times B$ is defined by

$$\mu_{A \times B}(u, v) = \mu_A(u) \wedge \mu_B(v),\quad u \in U, v \in V$$

Projection rule.

$$\frac{(X, Y) \text{ is } R}{X \text{ is } {}_XR}$$

where ${}_XR$, the projection of the binary relation R on the domain of X, is defined by

$$\mu_{{}_XR}(u) = \sup_v \mu_R(u,v),\quad u \in U, v \in V$$

where $\mu_R(u,v)$ is the membership function of R and the supremum is taken over $v \in V$.

Compositional rule.

$$\frac{X \text{ is } A \qquad (X, Y) \text{ is } R}{Y \text{ is } A \circ R}$$

where $A \circ R$, the composition of the unary relation A with the binary relation R, is defined by

$$\mu_{A \circ R}(v) = \sup_u (\mu_A(u) \wedge \mu_R(u,v))$$

The compositional rule of inference may be viewed as a combination of the conjunctive and projection rules.

Generalized modus ponens.

$$\frac{X \text{ is } A \qquad Y \text{ is } C \text{ if } X \text{ is } B}{Y \text{ is } A \circ (\neg B \oplus C)}$$

where $\neg B$ denotes the negation of B and the bounded sum is defined by

$$\mu_{\neg B \oplus C}(u,v) = 1 \wedge (1 - \mu_B(u) + \mu_C(v))$$

An important feature of the generalized modus ponens, which is not possessed by the modus ponens in binary logical systems, is that the antecedent "X is B" need not be identical with the premise "X is A." It should be noted that the generalized modus ponens is related to the interpolation rule which was described earlier. An additional point that should be noted is that the generalized modus ponens may be regarded as an instance of the compositional rule of inference.

Dispositional modus ponens. In many applications involving common sense reasoning, the premises in the generalized modus ponens are usuality-qualified. In such cases, one may employ a dispositional version of the modus ponens. It may be expressed as

$$\frac{\text{usually } (X \text{ is } A) \qquad \text{usually } (Y \text{ is } B \text{ if } X \text{ is } A)}{\text{usually}^2 \ (Y \text{ is } B)}$$

where "usually^2" is the square of "usually" (see Figure 4). For simplicity, it's assumed that the premise "X is A" matches the antecedent in the conditional proposition; also, the conditional proposition is interpreted as the statement, "The

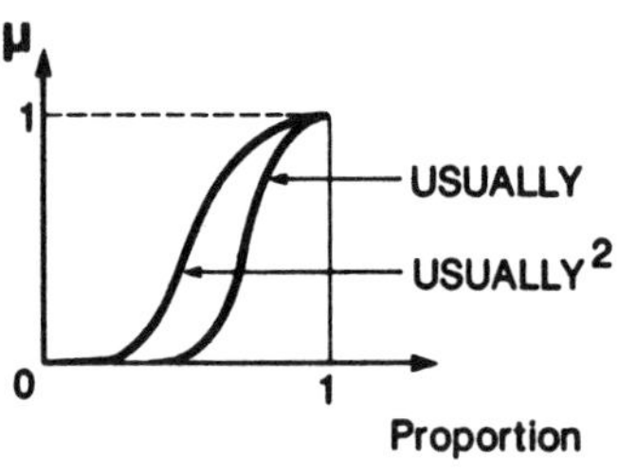

Figure 4. Representation of "usually" and "usually²."

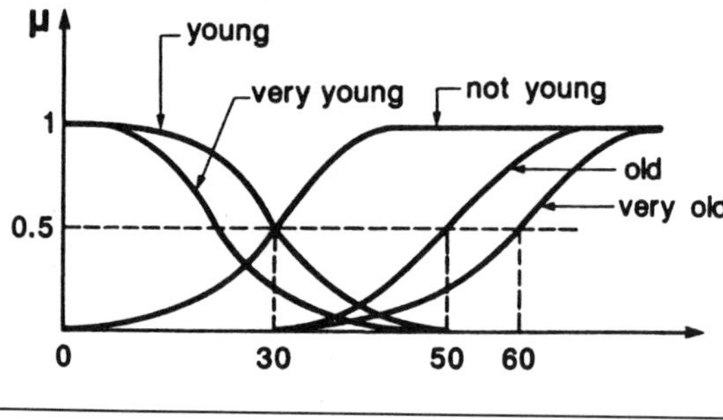

Figure 5. The linguistic values of "Age."

value of the fuzzy conditional probability of B given A is the fuzzy number USUALLY."

Extension principle. The extension principle plays an important role in fuzzy logic by providing a mechanism for computing induced constraints. More specifically, assume that a variable X taking values in a universe of discourse U is constrained by the proposition "X is A." Furthermore, assume that f is a mapping from U to V so that X is mapped into $f(X)$. The question is: What is the constraint on $f(X)$ which is induced by the constraint on X?

The answer provided by the extension principle may be expressed as the inference rule

$$\frac{X \text{ is } A}{f(X) \text{ is } f(A)}$$

where the membership function of $f(A)$ is defined by

$$\mu_{f(A)}(v) = \sup_u \mu_A(u) \qquad (15)$$

subject to the condition

$$v = f(u),\ u \in U,\ v \in V$$

In particular, if the function f is 1:1, then Equation 15 simplifies to

$$\mu_{f(A)}(v) = \mu_A(v^{-1}),\ v \in V$$

where v^{-1} is the inverse of v. For example,

$$\frac{X \text{ is small}}{X^2 \text{ is small}^2}$$

and

$$\mu_{\text{SMALL}^2}(v) = \mu_{\text{SMALL}}(\sqrt{v})$$

As in the case of the entailment rule, the dispositional version of the extension principle has the simple form

$$\frac{\text{usually } (X \text{ is } A)}{\text{usually } (f(X) \text{ is } f(A))}$$

The dispositional extension principle plays an important role in inference from common sense knowledge. In particular, it is one of the inference rules that play an essential role in answering the questions posed in the introduction.

The linguistic variable and its application to fuzzy control

A basic concept in fuzzy logic that plays a key role in many of its applications, especially in the realm of fuzzy control and fuzzy expert systems, is a *linguistic variable.*

A linguistic variable, as its name suggests, is a variable whose values are words or sentences in a natural or synthetic language. For example, "Age" is a linguistic variable if its values are "young," "not young," "very young," "old," "not old," "very old," and so on.

In general, the values of a linguistic variable can be generated from a *primary term* (for example, "young") its antonym ("old"), a collection of modifiers ("not," "very," "more or less," "quite," "not very," etc.), and the connectives "and" and "or." For example, one value of "Age" may be "not very young and not very old." Such values can be generated by a context-free grammar. Furthermore, each value of a linguistic variable represents a possibility distribution, as shown in Figure 5 for the variable "Age." These possibility distributions may be computed from the given possibility distributions of the primary term and its antonym through the use of attributed grammar techniques.

An interesting application of the linguistic variable is embodied in the fuzzy car conceived and designed by Sugeno of the Tokyo Institute of Technology.[5] The car's fuzzy-logic-based control system lets it move autonomously along a track with rectangular turns and park in a designated space (see Figure 6). An important feature is the car's ability to learn from examples.

The basic idea behind the Sugeno fuzzy car is the following. The controlled variable Y, which is the steering angle, is assumed to be a function of the state variables $X_1, X_2, X_3, \ldots, X_n$, which represent the distances of the car from the boundaries of the track at a corner (see Figure 7). These values are treated as linguistic variables, with the primary terms represented as triangular possibility distributions (see Figure 8).

The control policy is represented as a finite collection of rules of the form

$$R^i\text{: if } (X_1 \text{ is } A^i_1) \text{ and} \ldots (X_n \text{ is } A^i_n), \text{ then } Y^i = a^i_0 + a^i_1 X_1 + \ldots + a^i_n X_n$$

where R^i is the ith rule; A^i_j is a linguistic value of X_j in R^i; Y^i is the value of the control variable suggested by R^i; and $a^i_0, \ldots, a^i_n$ are adjustable parameters, which define Y^i as a linear combination of the state variables.

In a given state $(X_1, \ldots, X_n)$, the truth value of the antecedent of R^i may be expressed as

$$W^i = A^i_1(X_1) \wedge \ldots \wedge A^i_n(X_n)$$

where $A^i_j(X_j)$ is the grade of membership of X_j in A^i_j. The aggregated value of the controlled variable Y is computed as the normalized linear combination

$$Y = \frac{W_1 Y^1 + \ldots + W_n Y^n}{W_1 + \ldots + W_n} \quad (16)$$

Thus, Equation 16 may be interpreted as the result of a weighted vote in which the value suggested by R^i is given the weight $W_i/(W_1 + \ldots + W_n)$.

The values of the coefficients $a^i_1, \ldots, a^i_n$ are determined through training. Training consists of an operator guiding a model car along the track a few times until an identification algorithm converges on parameter values consistent with the control rules. By its nature, the training process cannot guarantee that the identification algorithm will always converge on the correct values of the coefficients. The justification is pragmatic: the system works in most cases.

Variations on this idea are embodied in most of the fuzzy-logic-based control systems developed so far. Many of these systems have proven to be highly reliable and superior in performance to conventional systems.[5]

Since most rules in expert systems have fuzzy antecedents and consequents, expert systems provide potentially important applications for fuzzy logic. For example[11]:

IF the search "space" is moderately small
THEN exhaustive search is feasible

IF a piece of code is called frequently
THEN it is worth optimizing

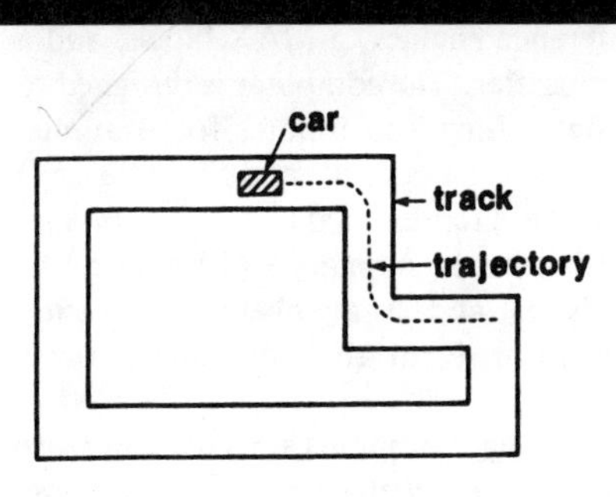

Figure 6. The Sugeno fuzzy car.

Figure 7. The state variables in Sugeno's car.

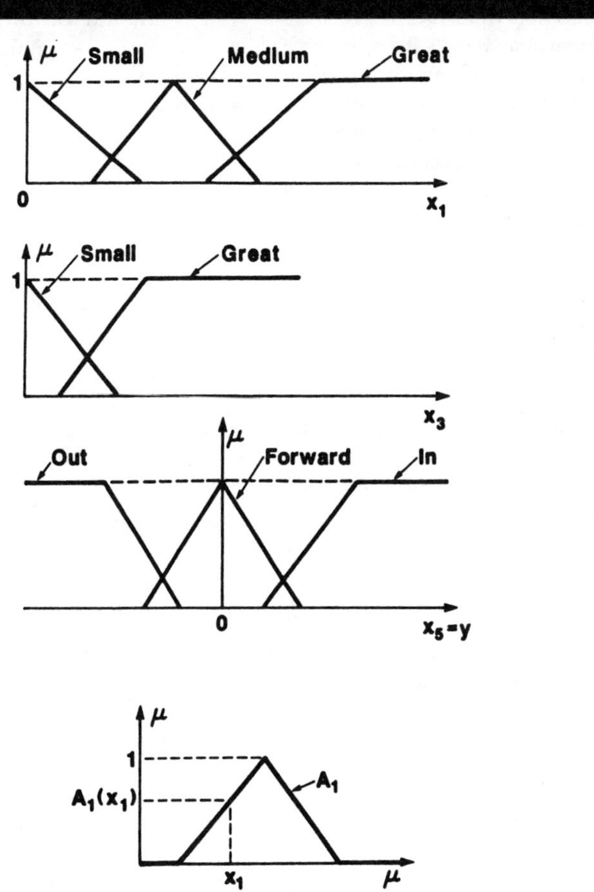

Figure 8. The linguistic values of state variables.

IF large oil spill or strong acid spill
THEN emergency is strongly suggested

The fuzziness of such rules is a consequence of the fact that a rule is a summary, and summaries, in general, are fuzzy. However, in the context of expert systems and fuzzy logic control, fuzziness has the positive effect of reducing the number of rules needed to approximately characterize a functional dependence between two or more variables.

Fuzzy hardware. Several expert system shells based on fuzzy logic are now commercially available, among them Reveal and Flops.[5] The seminal work of Togai and Watanabe at Bell Telephone Laboratories, which resulted in the development of a fuzzy logic chip, set the stage for using such chips in fuzzy-logic-based expert systems and, more generally, in rule-based systems not requiring a high degree of precision.[12] More recently, the fuzzy computer developed by Yamakawa of Kumamoto University has shown great promise as a general-purpose tool for processing linguistic data at high speed and with remarkable robustness.[5]

Togai and Watanabe's fuzzy inference chip consists of four major components: a rule set memory, an inference processor, a controller, and I/O circuitry. In a recent implementation, a rule set memory is realized by a random-access memory. In the inference processor, there are 16 data paths; one data path is laid out for each rule. All 16 rules on the chip are executed in parallel. The chip requires 64 clock cycles to produce an output. This translates to an execution speed of approximately 250,000 fuzzy logical inferences per second (FLIPS) at 16 megahertz clock. A fuzzy inference accelerator, which is a coprocessor board for a designated computer, is currently being designed. This board accommodates the new chips.

In the current implementation, the control variables are assumed to range over a finite set having no more than 31 elements. The membership function is quantized at 16 levels, with 15 representing full membership. Once the Togai/Watanabe chip becomes available commercially in 1988 or 1989, it should find many uses in both fuzzy-logic-based intelligent controllers and expert systems.

Yamakawa's fuzzy computer, whose hardware was built by OMRON Tateise Electronics Corporation, is capable of performing fuzzy inference at the very high speed of 10 megaFLIPS. Yamakawa's computer employs a parallel architecture. Basically, it has a fuzzy memory, a set of inference engines, a MAX block, and a defuzzifier. The computer is designed to process linguistic inputs, for example, "more or less small" and "very large," which are represented by analog voltages on data buses. A binary RAM, an array of registers and a membership function generator form the computer's fuzzy memory.

The linguistic inputs are fed to inference engines in parallel, with each rule yielding an output. The outputs are aggregated in the MAX block, yielding an overall fuzzy output that appears in the output data bus as a set of distributed analog voltages. In intelligent fuzzy control and other applications requiring nonfuzzy commands, the fuzzy output is fed to a defuzzifier for transformation into crisp output.

Yamakawa's fuzzy computer may be an important step toward a sixth-generation computer capable of processing common sense knowledge. This capability is a prerequisite to solving many AI problems — for example, handwritten text recognition, speech recognition, machine translation, summarization, and image understanding — that do not lend themselves to cost-effective solution within the bounds of conventional technology. □

Acknowledgment

The research reported in this article has been supported in part by NASA Grant NCC-2-275 and NSF Grant DCR-8513139. The article was written while the author was a visiting scholar at the Center for the Study of Language and Information at Stanford University. It is dedicated to the Japanese scientists and engineers who have contributed so importantly to the development of fuzzy logic and its applications.

References

1. L.A. Zadeh, "Fuzzy Algorithms," *Information and Control*, Vol. 12, 1968, pp. 94-102.
2. L.A. Zadeh, "A Rationale for Fuzzy Control," *J.Dynamic Systems, Measurement and Control*, Vol. 94, Series G, 1972, pp. 3-4.
3. L.A. Zadeh, "Outline of a New Approach to the Analysis of Complex Systems and Decision Processes," *IEEE Trans. Systems, Man, and Cybernetics*, Vol. SMC-3, 1973, pp. 28-44.
4. E.H. Mamdani and S. Assilian, "A Case Study on the Application of Fuzzy Set Theory to Automatic Control," *Proc. IFAC Stochastic Control Symp.*, Budapest, 1974.
5. *Preprints of the Second Congress of the International Fuzzy Systems Association*, Tokyo, Japan, 1987.
6. L.A. Zadeh, "Test-Score Semantics as a Basic for a Computational Approach to the Representation of Meaning," *Literary and Linguistic Computing*, Vol. 1, 1986, pp. 24-35.
7. D. Dubois and H. Prade, *Theorie des Possibilites*, Masson, Paris, 1985.
8. L.A. Zadeh, "The Role of Fuzzy Logic in the Management of Uncertainty in Expert Systems," *Fuzzy Sets and Systems*, Vol. 11, 1983, pp. 199-227.
9. L.A. Zadeh, "Syllogistic Reasoning in Fuzzy Logic and its Application to Usuality and Reasoning with Dispositions," *IEEE Trans. Systems, Man, and Cybernetics*, Vol. SMC-15, 1985, pp. 754-763.
10. A. Kaufmann and M. M. Gupta, *Introduction to Fuzzy Arithmetic*, Van Nostrand, New York, 1985.
11. B. Buchanan and E. K. Shortliffe, *Rule-Based Expert Systems*, Addison-Wesley, Reading, Mass., 1984.
12. M. Togai and H. Watanabe, "Expert Systems on a Chip: An Engine for Real-Time Approximate Reasoning," *IEEE Expert*, Vol. 1, 1986, pp. 55-62.

Related Publications

Adams, E.W., and H.F. Levine, "On the Uncertainties Transmitted from Premises to Conclusions in Deductive Inferences," *Synthese*, Vol. 30, 1975, pp. 429-460.

Baldwin, J.F., and S.Q. Zhou, "A New Approach to Approximate Reasoning Using a Fuzzy Logic," *Fuzzy Sets and Systems*, Vol. 2, 1979, pp. 302-325.

Bellman, R.E., and L.A. Zadeh, "Local and Fuzzy Logics," in G. Epstein, ed., *Modern Uses of Multiple-Valued Logic*, Reidel, Dordrecht, 1977, pp. 103-165.

Dubois, D., and H. Prade, *Fuzzy Sets and Systems: Theory and Applications*, Academic Press, New York, 1980.

Goguen, J.A., "The Logic of Inexact Concepts," *Synthese*, Vol 19, 1969, pp. 325-373.

Klir, G.J., and T.A. Folger, *Fuzzy Sets, Uncertainty, and Information*, Prentice Hall, Englewood Cliffs, N.J., 1988.

Moisil, G.C., *Lectures on the Logic of Fuzzy Reasoning*, Scientific Editions, Bucharest, 1975.

Prade, H., and C.V. Negoita, *Fuzzy Logic in Knowledge Engineering*, Verlag TÜV Rheinland, Köln, 1986.

Sugeno, M., ed., *Industrial Applications of Fuzzy Control*, Elsevier Science Publishers BV, The Netherlands, 1985.

Watanabe, H., and W. Dettloff, "Fuzzy Logic Inference Processor for Real Time Control: A Second Generation Full Custom Design," *Proc. 21st Asilomar Conference on Signals, Systems, and Computers*, Asilomar, Calif., 1987.

Lofti A. Zadeh joined the Department of Electrical Engineering at the University of California, Berkeley, in 1959, serving as its chairman from 1963 to 1968. Earlier he was a member of the electrical engineering faculty at Columbia University. He has held a number of visiting appointments, among them a visiting scientist appointment at the IBM Research Laboratory, San Jose, and a visiting scholar appointment at the AI Center, SRI International.

Zadeh's work centered on system theory and decision analysis until 1965 when his interests shifted to the theory of fuzzy sets and its applications.

An alumnus of the University of Teheran in Iran, MIT, and Columbia University, Zadeh is a fellow of the IEEE and AAAS and a member of the National Academy of Engineering. He holds a doctorate honoris causa from Paul Sabatier University, France, in recognition of his development of the theory of fuzzy sets.

Zadeh's address is Computer Science Division, Dept. of EECS, University of California, Berkeley, CA 94720.

Chapter 7: Issues and Commentary

Many issues should be considered by anyone contemplating using or building an expert system. The seven papers in this chapter summarize some of these issues. Denning looks at the history of expert systems, as well as some of their limitations. Rolandi describes some of the practical aspects of expert system development. Matthews addresses development issues that impact the maintenance of such systems. Marcot then describes testing knowledge bases relative to their intended behavior. Green and Keyes discuss the validation/ verification process, that process of testing a system against its specifications. Departing from this series of articles, Zeide and Liebowitz talk about the growing legal dilemma being brought about by expert systems. Bahill and Ferrell close by telling of their experiences in teaching this subject.

Towards a Science of Expert Systems

Peter J. Denning, NASA Ames Research Center

We humans have held high opinions of ourselves throughout the ages. We define ourselves, *Homo sapiens*, as the only earthly creatures capable of rational thought. Perhaps as expressions of a racial urge to be godlike, we aspire not only to control our environment, but to create artificial beings. In *Machines Who Think*, Pamela McCorduck gives a fascinating account of how this aspiration has shown itself down through history.[1] She sees the urge to create artificial beings expressed in ancient idols; in medieval legends of homunculi or golems; in the calculating machines, mechanical statues, and chess automatons of later centuries; and in the current quest for thinking machines. I might add that the urge also expresses itself in genetic engineering.

Our creative urge is coupled with a tremendous reverence for logic. The idea that the ability to reason logically—to be rational—is closely tied with intelligence was clear in the writings of Plato. The search for greater understanding of human intelligence led to the development of mathematical logic, the study of methods of proving the truth of statements by manipulating the symbols in which they are written without regard to the meanings of those symbols. By the nineteenth century a search was underway for a universal logic system, one capable of proving anything provable in any other system.

This search came to an end in the 1930's with Kurt Gödel's incompleteness theorem and Alan Turing's incomputability theorem. Gödel showed that, given any sufficiently powerful, consistent system of logic, one can construct a true proposition that cannot be proved within that system. Turing described a universal computer and showed that there are well-defined problems that cannot be solved by any computer program, even in principle. As an example, he showed that there is no program that can determine whether any other program will enter an endless loop when executed. A consequence of these results is that there are always theorems beyond logic and tasks beyond machines. Some philosophers argue that human intelligence is higher: there will always be human accomplishments that defy logic or mechanical simulation.

Despite the fact that logical deductions can be carried out by machines as pure symbol manipulations without regard to content, we continue to associate logic with intelligence. Indeed, the fact that this intelligent activity is well-defined, reliable, and mechanizable appeals strongly to the urge to create a thinking machine. One cannot fully appreciate modern research in machine intelligence, its frequent psychological allusions and occasional mysticism, without understanding the human urges to create life and to explain intelligence with logic. Knowing this, one can peer past the aura to the reality.

The term artificial intelligence (AI) was first used in 1956 by John McCarthy, now at Stanford University. It refers to the subfield of computer science that studies how machines might behave like people. Several technologies popularly associated with AI have reached the marketplace—notably image-pattern recognition, speech recognition, speech generation, and robotics. However, the commercially successful versions of these technologies do not exploit symbolic manipulations, and they were built by engineers who do not profess special skill in AI.

Expert systems

One technology in particular has captured many fancies: expert systems. An expert system is a computer system designed to simulate the problem-solving behavior of a human who is expert in a narrow domain. Examples include prescribing antibiotics, classifying chemical compounds,

This article is reprinted with permission from American Scientist *and is taken from their column, "The Science of Computing," Vol. 74, No. 1, Jan-Feb. 1986, pp. 18-20.*

configuring and pricing computer systems, scheduling experiments in real time for a telescope, classifying patterns in images, and diagnosing equipment failures. Like books, expert systems are a way to make the knowledge of a few available to the many.

In mathematical logic, a system consists of (finite) sets of axioms and rules of inference. A proof is a sequence of strings of symbols, say

$S_1,...,S_p$

such that each string S_i either is an axiom or is derivable from some subset of $S_1,...,S_{p-1}$ by a rule of inference. An expert system is a restricted form of a system of logic. It attempts to compute a sequence of strings representing the steps in the solution of a problem. The sequence serves as a proof of the solution. The rules of inference are of a simple form that represents a pattern in problem-solving: "if *conditions* then *consequences.*"

An expert system has three main parts: a database for storing axioms and rules of inference, an algorithm for constructing proofs, and a user interface. The interface, which often includes powerful interactive graphics, provides a language for the user to express queries and to provide information to the system. In expert system jargon, axioms are called "facts," rules of inference "rules," the database the "knowledge base," the database programmer the "knowledge engineer," the proof-constructing algorithm the "inference engine," proof-construction "reasoning," and a proof of a solution an "explanation."

A number of companies now market expert system shells, expert systems that come with empty knowledge databases. Where does the data come from? A database is built by a trial-and-error process called knowledge engineering. Through extensive interviews with the expert, the knowledge engineer attempts to elicit verbalizations of rules and highly detailed facts, and to program this information into the database. The process includes continual testing of the system and expansion of its database to cover cases previously left out.

Deep vs. shallow systems

Although prototypes of systems based on a few hundred rules can often be brought into operation within two months, it is not unusual for the process of building and testing an expert system to require many months and to produce several thousand rules. The set of rules may be incomplete and inconsistent; the exact behavior of the system may have no precise specification and may not be reliably predictable by its designer. For this reason, the technique of programming an expert system is sometimes called heuristic programming.

This class of systems using heuristic programming and rule evaluation does not contain all systems that simulate human expertise. John Shore points out that autopilots—control systems that solve the equations of motion of the aircraft—fly planes very well. They are expert fliers, but not expert systems.[2]

AI researchers often distinguish between shallow and deep expert systems. Shallow systems are designed for speed. In their limited databases they store more facts than rules. The proofs of their conclusions are usually short, and most of the conclusions strike observers as being straightforward consequences of information in their databases. They often give poor results when applied to problems other than those for which they have been tested. Examples include a system to generate production schedules using bin-packing heuristics and a system to answer electronic inquiries for technical information.

Deep systems derive their conclusions from models of phenomena in their domains, using first principles embedded in the model. The proofs of their conclusions are usually long, and some conclusions may strike some observers as anything but obvious. An example is Dendral, a system that, when given spectral data, generates diagrams of chemical compounds using ball-and-stick atomic models. Other prominent examples include qualitative physics programs, which determine states of behavior of physical systems.[3] These programs can proceed backwards from observations and generate hypotheses about initial conditions; they can proceed forward from initial conditions and explain discrepancies between predicted and observed final states by generating hypotheses about which components may be broken.

The distinction between shallow and deep systems is not sharp. It may not even be important. A system that derives conclusions from first principles may generate long proofs, but how long is a deep proof? Every expert system is capable of generating conclusions that are not obvious to some observers, but what fraction of observers will be surprised by a deep system? A model is a small set of rules that explains a large amount of data, but how much compaction of data into rules is deep?

A system need not be deep to be useful. One of the best-known expert systems is Mycin, whose rules associate diseases with symptoms. By answering questions about symptoms, Mycin leads the user to a diagnosis and recommended antibiotic treatment. The processes by which Mycin reaches its conclusions resemble those used by a doctor in interviewing a patient and analyzing lab data. Mycin is not a deep system, because its knowledge is empirical and is not derived internally from models of diseases, symptoms, or cures.

Representing knowledge

Every expert system, deep or shallow, is likely to use one of three ways of representing knowledge. The three resulting types of system are nicely described in three articles edited by Peter Friedland of Stanford University.[4] I will summarize these systems briefly here.

The first type is the logic programming system. It embodies explicitly the model of proof in mathematical logic. In the language Prolog, for example, rules are represented in the form

q :- $p_1, \ldots, p_n$

interpreted as: "If the predicates $p_1, \ldots, p_n$ are all true, then the predicate q follows." A predicate is a formula that is true for some values of its variables; for example, PARENT(x, y) is true exactly when person x is the parent of person y. A fact is represented by an empty right side; for example, if Alice is the parent of Bob, we write "PARENT(Alice,Bob) :- ". The principle of transitivity allows us to replace a pair of rules,

Q :- $p_1, \ldots, p_n$
s :- $Q, r_1, \ldots, r_m$

with the new rule

s :- $p_1, \ldots, p_n, r_1, \ldots, r_m$

This pattern of reduction is the basis of the Prolog inference algorithm, called resolution, for finding the ultimate consequences of given conditions. The resolution principle is actually much more powerful than described above. It is not necessary that the Q on the left of the first rule be the same as the Q on the right of the second rule; it is only necessary that they can be made the same by substituting for some of the variables. For example, an allowable Prolog resolution takes

PARENT(x,Bob) :- MOTHER(x,Bob)
GRANDP.(Alice, z) :- PARENT(Alice,y), PARENT(y,z)

to

GRANDP.(Alice, z) :- MOTHER(Alice,Bob), PARENT(Bob,z)

after assigning x=Alice and y=Bob. The process of forcing two predicates to be the same by finding matches among components is called unification. With each resolution, Prolog keeps track of the possible values that make the predicates true. Not only can it answer queries with a yes or no, it can report which values variables must have for yes to be the answer.

The second type is the rule-based system. Its database is structured as a hierarchy of rule sets. Each rule set comprises a series of condition-consequence rules as defined earlier. Two strategies are common for evaluating the consequences of the initial conditions given by the user. Under forward chaining, evaluation proceeds by tracing rules from left side to right, until the ultimate consequences of the given conditions have been found. Under backward chaining, evaluation proceeds by tracing rules from right side to left, until the ultimate antecedents of a hypothesis about the given conditions have been found; hypotheses that conflict with the given conditions are discarded along the way. Which method is faster depends on the problem. If much data is available, it is often faster to chain backwards from a few likely hypotheses. If there are many possible hypotheses (for example, in medical diagnosis), it is usually better to chain forward from the limited initial data. The search for applicable rules can be restricted by allowing consequents of rules to name rulesets in which further applicable rules can be found.

The third type is the frame-based system. A frame-based system is centered on a hierarchy of descriptions of objects referred to in the rules. The description of a class of objects or an individual object is called a frame. The relation "is an instance of" organizes the hierarchy of frames. Thus in a computer failure diagnosis system, a frame called Microprocessor No. 3 might be immediately subordinate to a frame called Microcomputers, which in turn might be subordinate to a frame called Computers. Rules are used to describe functional relationships among frames. Any property of an object that can be derived by locating it in the hierarchy of frames need not be explicitly represented in the rule sets. This gains relative simplicity of rule sets in exchange for complexity of the catalogs of frames and of the inference algorithm (which must do more computation).

Limitations of Expert Systems

What are the limitations of expert systems? The most important concern their reliability. Expert systems are limited by the information in their databases and by the nature of the process for putting that information in. They cannot report conclusions that are not already implicit in their databases. The trial-and-error process by which knowledge is elicited, programmed, and tested is likely to produce inconsistent and incomplete databases; hence, an expert system may exhibit important gaps in knowledge at unexpected times. Moreover, expert systems are unlikely to have complete, clear functional specifications, and their designers may be unable to predict reliably their behavior in situations not tested. If not carefully structured as modules, large databases are likely to be difficult to modify and maintain.

This aspect of expert systems has been sharply criticized. David Parnas argues that software developed heuristically is inherently less reliable than software developed from precise specifications.[5] He asserts that to get a reliable program to solve a problem one must study the problem, not the way people say they solve it. For example, to distinguish among objects in a picture, one studies the characteristics of the objects and of the photographic process; if one asks people for the rules, one is unlikely to find a reliable program. The expert may not be consciously aware why his methods work, and the knowledge engineer who interviews him is like an investigative reporter who may not always ask the right questions.

John Shore reminds us that testing is an inherently unreliable way to find errors in programs.[2] Because it can only reveal the presence of bugs, never their absence, testing is inadequate for most software. The situation is worse for expert systems, where the correct behavior in a particular case is often a matter of opinion. In Shore's view, programming by trial and error will produce expert systems that are unreliable and cannot be trusted as autonomous systems. He does believe that expert systems can be useful, with human supervision, as intelligent assistants.

Some AI researchers respond that complete testing is meaningless for expert systems. The best these systems can do is imitate human experts, who are themselves imperfect. Even as one does not expect a doctor to diagnose correctly every illness, one cannot expect an expert system to do better. They respond further that mathematical proof is meaningless for large programs because one cannot tell whether the specifications are complete or accurate. While I find much to agree with in these statements, I do not support their implied conclusion. I believe that we can and often do build machines that overcome human limitations. I believe further that in producing machines on whose decisions lives may depend we should use every available method of establishing that the machines will perform their functions reliably. Although precise specifications may be imperfect, they are more likely to produce a reliable machine than a trial-and-error process.

Insistence by their developers that expert systems cannot be made more reliable than humans will be met by public insistence that software pass tests similar to those for certifying human experts. This may significantly extend the time required to bring an expert system to market. Other tests may be imposed too. For example, the Food and Drug Administration is considering a requirement that medical diagnosis systems be subjected to stringent field tests like other medical apparatus before being allowed on the general market.

Although expert systems developed by trial and error are likely to be unreliable, there is no reason in principle that, with a different programming methodology, expert systems cannot be as reliable as other software. Expert systems are, after all, nothing more than computer programs. The powerful languages of logic, rules, and frames permit deductive processes to be programmed quite rapidly. Like other software, there is every reason to believe that rules can be constructed as sets of modules, that modules can be organized into hierarchies, that calculi for precise specifications can be developed, and that methods for proving that rules meet specifications can be devised.

The notion of precise specifications for human processes has its quixotic side, for how does one specify precisely what an expert does? And yet there are grounds for hope. The most successful expert systems to date are the ones used for diagnosis. They are based on a well-understood model of diagnostic processes. The rules can be organized into a hierarchy of modules corresponding to decision points in the diagnostic process; their completeness and consistency can be evaluated with respect to the model. The hope is that models of other human problem-solving processes can be defined and used similarly for their expert systems.

There is nothing magic about expert systems. Turing's thesis tells us that any mechanical procedure can be programmed on a computer. If we know of a procedure to find a procedure, we can program that. But if we know nothing at all, we are stuck. We cannot expect an expert system to help if we do not know how something is done. Alan Perlis of Yale has said: "Good work in AI concerns the automation of things we know how to do, not the automation of things we would like to know how to do." AI cannot replace RI. ■

Acknowledgments

Work reported herein was supported in part by contract NAS2-11530 from NASA to the Universities Space Research Association (USRA).

References

1. P. McCorduck, *Machines Who Think*, W. H. Freeman & Co., San Francisco, 1979.
2. J. E. Shore, *The Sachertorte Algorithm*, Viking, New York, 1985.
3. *Qualitative Reasoning About Physical Systems*, D. Bobrow, ed., MIT Press, Cambridge, Mass., 1985.
4. P. Friedland, "Knowledge-based Architectures," *Comm. ACM*, Vol. , No. 9, Sept. 1985.
5. D. L. Parnas, "Software Aspects of Strategic Defense Systems," *American Scientist*, Vol. 73, No. 5, Sept.-Oct. 1985.

Peter J. Denning is director of the Research Institute for Advanced Computer Science at the NASA Ames Research Center, where he studies computer systems architecture, operating systems, and performance modeling. Previously, he was head of the Computer Sciences Department at Purdue University and assistant professor of electrical engineering at Princeton University.

Denning received a PhD in electrical engineering from MIT. He was president of the ACM and editor-in-chief of ACM's *Computing Surveys* and has published over 145 technical papers and articles. He is the editor-in-chief of *ACM Communications.*

Denning's address is Research Institute for Advanced Computer Science, NASA Ames Research Center, MS 230-5, Moffet Field, CA 94035.

BY WALTER G. ROLANDI

Knowledge Engineering in Practice

The job of a knowledge engineer is to determine what an expert does during decision-making

As with most anything new, a good bit of mystery and confusion surround the emerging field of software development called knowledge engineering. Enthusiastic AI types hail it as a revolutionary process—a sort of "mind automation." Traditional data processing types dismiss it as a pretentious new name for an already well-understood process—systems analysis. As is also often the case with two-sided arguments, there is some truth to both positions.

The word "practical" is an important component in this paper. This article is intended for software developers who are first groping with the concepts of knowledge engineering and the issues and problems associated with getting a first knowledge engineering project off the ground. As such, references to AI and particularly to expert systems are made in a purely applied sense; no attempt is made to address the theoretical problems associated with the automation of language or thinking. For practical purposes, an expert system is defined as any system that can process data and deliver or administer the advice or expertise a human expert would give under the same circumstances.

The unbridled, enthusiastic mind automation perspective is not the best place for the software developer to start looking for practical advice. This is not to say that research into the nature of language and thinking with regard to the computer analogy are unimportant. AI has made some progress in its first 30 years and is destined to enjoy major breakthroughs once machine learning comes of age. These areas are truly fascinating but they are largely unrelated to project deadlines and production schedules.

Traditional data processing professionals, although perhaps understandably skeptical, are wrong to dismiss knowledge engineering as "nothing new." Although knowledge engineering is not a revolutionary methodology for the mechanization of thought, neither is it

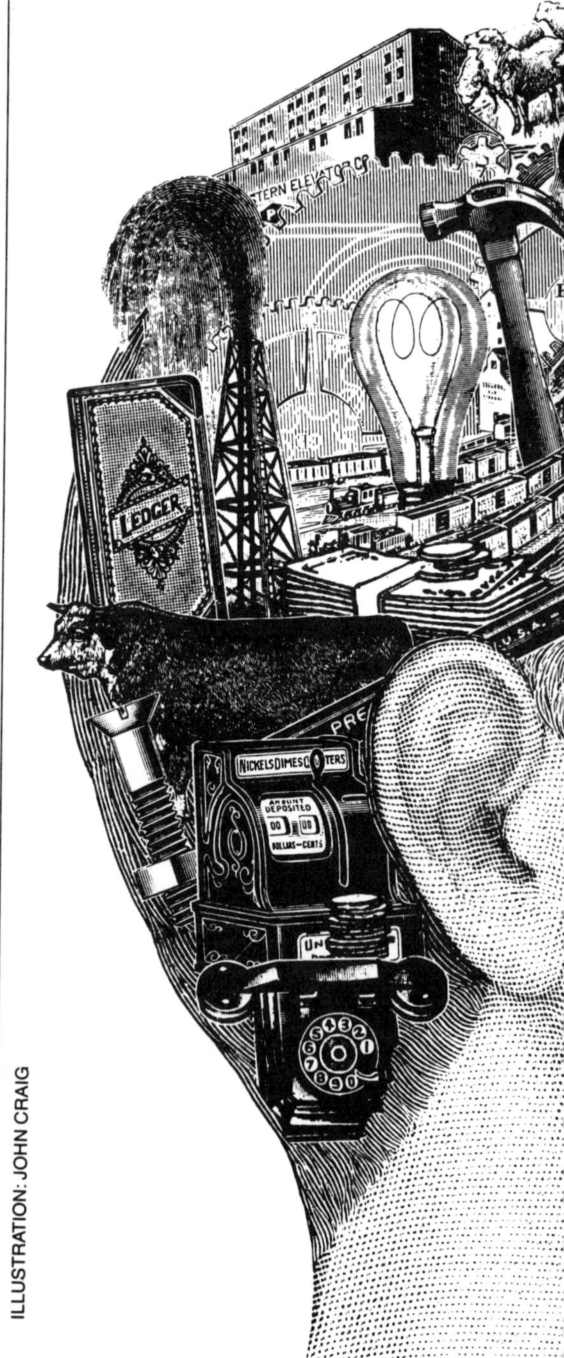

ILLUSTRATION: JOHN CRAIG

simply traditional systems analysis. However, knowledge engineering is more systems analysis than it is not. In other words, the tools and methods of the systems analyst are largely the same as those of the knowledge engineer.

The important difference is the subject of analysis. In traditional systems analysis, the analyst studies the flow of information through (usually) a clerical process. In knowledge engineering, the knowledge engineer studies the flow of information through the decision-making processes of a human expert. This distinction may be the source of some suggestions that knowledge engineering somehow entails the more clinical skills of a psychologist than those of the systems analyst. There may well be cases where such skills provide valuable insights into a knowledge-engineering project, but on the whole they are irrelevant.

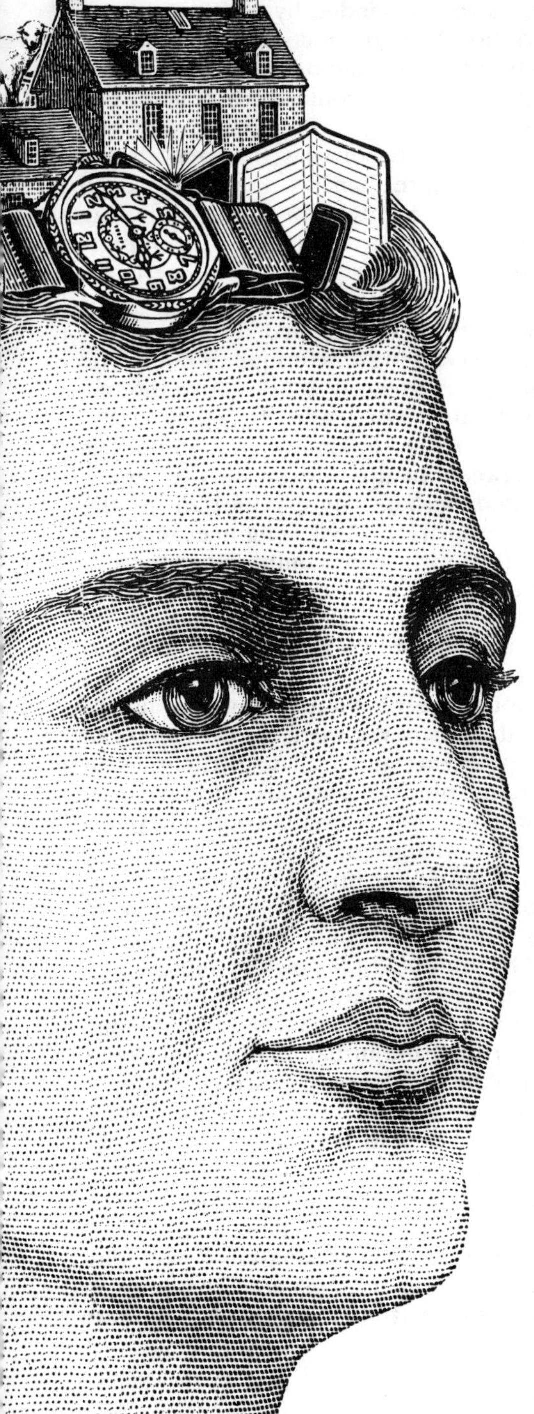

The real job of a knowledge engineer is to determine what an expert does when he or she goes through the decision-making process. Given the intellectual nature of the system to be analyzed, this task may initially appear to involve intangible and somehow inaccessible processes. Even so, the decision-making processes of an expert always involve quite observable behavior. In fact, the most effective tools of analysis are those of applied behavior analysis.

BEHAVIOR ANALYSIS

Behavior analysis usually refers to the quantitative, functional, and causal analysis of behavior. Initially, a behavioral analyst seeks to understand a behavior in terms of the frequency of its occurrence. Simple, central tendency descriptive statistics are compiled to obtain an understanding of the numbers associated with the performance parameters of a given behavior. For example, let's say a knowledge engineer is assigned the task of producing an expert system that emulates the behavior of an expert in an obscure field involving selective classification. Where might the knowledge engineer begin?

The knowledge engineer previously trained in the methods of behavior analysis might first ask the expert to act out or in some way publicly demonstrate the performance of this selective classification process for 10 typical cases. This knowledge engineer turned behavior analyst notes (counts) the instances of each type of classification outcome. Ten cases are certainly insufficient for a thorough understanding of any complex decision-making process, but this can be an encouraging start, especially if a particular outcome appears more likely than the others.

If subsequent repetitions of this experimental procedure produce the same percentage of outcome breakdowns, the knowledge engineer has not only isolated the predominant outcome areas of the process but has also ranked the outcome types in terms of their priority for completion and incorporation into a production system.

Behavior analysis can provide a much more comprehensive understanding than numerical description. Developed as a logical extension of the experimental analysis of behavior, behavior analysis is asserted as a means to obtain predictive control over observable behavior by behaviorists and practitioners of behavior modification.[1-4]

The underlying assumption of behavior analysis is that all behavior can be understood as a function of its consequences: to understand a particular behavior in a causal or predictive sense, the behavior analyst must first examine not only the environmental cir-

Decision-making processes always involve observable behavior

cumstances under which the behavior typically occurs, but also the environmental consequences the behavior is instrumental in affecting. Although perhaps not intuitively obvious, an inference or conclusion drawn by an expert is an environmental consequence. The behaviors associated with observing *A, B,* and *C* can be understood as a function of the conclusion D.

First, a quantitative description of the behavior and the circumstances associated with its occurrence is obtained. This description serves to illuminate the frequency of occurrence of a response or response set. As in the selective classification example, the ability to understand an expert's decision-making behavior in these terms can be a tremendous asset in a knowledge engineering endeavor.

Sometimes expertise is so sprawling it defies conceptualization: the knowledge engineer does not know what knowledge the expert commands, much less where to start in structurally representing the expert's knowledge. By employing the methods of behavior analysis (obtaining a causal and quantitative description of the behaviors entailed in the expert's decision-making process), the knowledge engineer will be directed to those areas of consideration (thought processes) that are typically called into play when the expert is at work.

Let's pick up the preceding example where we left off: the knowledge engineer has succeeded in determining the relative frequencies of classification outcomes and established attack priorities by noting which outcome types are most prevalent. Now it is time to examine the most prevalent outcome. Again, a quantitative approach is required. The knowledge engineer should obtain 10 cases that all result in that particular outcome and have the expert think aloud as they walk through the classification process.

The knowledge engineer should carefully note where in the process the expert draws his or her conclusion and retain lists of the influential factors the expert deems active on the process up to that point. These factors will become evident as the expert demonstrates more and more cases: they will be the words, terms, or concepts that tend to repeatedly emerge when the expert describes a sample case. The most prevalent will become the pivotal concepts the knowledge engineer will employ in prototyping or system modeling.

At this time in the project life cycle, the knowledge engineer will begin to see that classification *D* only takes place under a certain set of circumstances (the presence of *A,* the absence of *B,* and the possibility of *C.* In this sense, *A, B,* and *C* can be understood as a function of conclusion *B*; *D* is the consequence of *A, B,* and *C.*

It is perhaps ironic that a methodology developed by behaviorists for purposes of behavior modification should find such productive employment in a field that has been referred to as "the practice of cognitive psychology." But, again, the work of a knowledge engineer entails first and foremost the comprehensive understanding of exactly what an expert does when making a decision or in some way exercising expertise. Being able to say that out of every *N* typical cases, the outcome is *D* and the consistently evident components were *A, B,* and *C,* is a big step toward mastering the methods of the expert's practice.

SOME PRACTICAL DISCOVERIES

We have adopted several general but important and useful methods for going about the business of knowledge engineering. Some of the following are essentially endorsements of practices recommended by various sources. Others, however, are practices we have either actively evolved or that became painfully apparent to us in the routine performance of our tasks.

One painful lesson was that one should never have more than two knowledge engineers actively involved in the interviewing process. This can result in a seemingly chaotic questioning scheme if the expert's answers are interpreted differently by various interviewers. We developed a two-member team interviewing technique where one knowledge engineer does the active interaction with the expert and the other is employed to police the quality of communication between the interviewer and the expert. Also (and perhaps more importantly), the quality control knowledge engineer serves to head off or diffuse miscommunications and other unavoidable frustrations of prolonged discourse which might otherwise lead to problems between knowledge engineer and expert.

Although we have found that the use of more than two knowledge engineers can cause problems and lead to difficulties in communication, we have also decided that the use of only one is less than optimal. Sometimes a knowledge engineer will concentrate on some particular line of reasoning and in so doing will exclude the exploration of other concepts concomitantly brought into play. The presence of the second (quality control) engineer can largely prevent this from happening. In fact, in all facets of knowledge engineering; from initial reading and exploration of the problem space through in-depth interviewing, prototyping, and production system development, the two-person team approach has repeatedly and consistently demonstrated distinct advantages over the alternatives already described.

The emphasis on case analysis is essential. Knowledge engineering should be an inductive process, but the knowledge engineer should avoid generalizing from a small number of cases. A very large number of cases should be obtained, and the expert should think aloud through them all. The process should be taped so the session can be reviewed or reenacted if needed. Usually, a pattern will emerge as a function of the number of cases demonstrated. Thus the greater the number of cases, the better. This is what a behavior analyst calls "taking a baseline." Even the most obscure processes will eventually surrender to this method.

The knowledge engineer should look for tactics the expert may employ to chunk or manage large bodies of information for themselves. These are important to note not only as a step toward understanding the decision-making process of the expert, but also as clues as to what type of data-structuring schemes the knowledge engineer should be thinking about.

Contamination is a very real problem that can take many forms. Generally speaking, contamination is a passive and subtle problem and is defined as the introduction of uncontrolled influences on the decision-making process of an expert. For example, if an expert typically calls on external sources of information when making a decision, efforts should be made to experimentally determine exactly where in the decision-making process this happens.

The knowledge engineer should experimentally determine the importance of such sources of information by withholding access and asking the expert to attempt to decide without them. If the expert can make a decision in the absence of the support documents or in ignorance of their contents, they are probably irrelevant to the decision-making process. Just like everyone else, experts are sometimes (if not usually) unaware of the causes of their behavior. It is essential to isolate that which is pertinent to the decision-making process from that which is not.

Although it is the job of the knowledge engineer to give structure to the decision-making behavior of the expert, excessive interpretation of that behavior should be avoided. The tendency to interpret will be inversely related to the size of the sample case set; the smaller the sample case set, the more likely the knowledge engineer will be to interpret. As is the case in psychology, excessively interpretive hypothetical constructs are usually untestable. Whenever possible, the knowledge engineer should seek to quantify the behavior of the expert and stick to the numbers.

Becoming personally involved with a particular interpretation can also cause problems. The knowledge engineer should never impose a particular interpretation of the expert's problem-solving strategy on the expert. When a given hypothesis as to how the expert exercises his or her expertise becomes the conceptual darling of the knowledge engineer, objective exploration may be forfeited in favor of possessive defense of the construct. The exploration of such hypotheses is an essential element of knowledge engineering, but the knowledge engineer should actively avoid interviewing practices inadvertantly designed to prove themselves right in some particular hypothetical analysis.

Emphasis on case analysis is essential

On the contrary, noncontaminated interviewing should take an experimental, scientific approach; it should serve the goals of discovery and exploration—wherever those goals may lead. The knowledge engineer should test their hypotheses by attempting to prove them wrong, not by attempting to prove them right.

Sometimes the knowledge engineer will note a relationship among the expert's deciding factors which, although perhaps obviously related to some particular outcome, has remained less than obvious to the expert. One must never express excessive enthusiasm in making such discoveries. For one thing, the knowledge engineer may be wrong. The expert's behavior may actually be under the control of other influences of which the knowledge engineer is totally ignorant.

It is human to want to be right and to want to be recognized for making insightful discoveries. But if at all possible, it is best to let the expert make all the insightful discoveries (that is, to promote the feeling that the expert has been predominantly responsible for the illumination of his or her own decision-making processes).

Never aggressively debate an interpretation of how the expert performs some task with the expert. This sort of tendency can spoil a previously good relationship. When expert and knowledge engineer argue, no one can win. If the knowledge engineer is later proven correct, the triumph will most likely be Pyrrhic, succeeding in alienating the expert or making the expert feel silly or stupid. If the knowledge engineer turns out to be wrong, he or she may have come across as second-guessing the expert or worse, challenging their expertise. Occasionally the knowledge engineer must resist the temptation to demonstrate a prowess for understanding the workings of the expert in an effort to promote harmonious exchange.

The use of tape recorders serves several good purposes although video recorders are not recommended. As a permanent record, audio tapes of interview sessions often serve as a means to settle subsequent disputes between knowledge engineers as to what a per-

Behavior can be understood as a function of its consequences

son said or did not say. Some experts will be intimidated by the presence of any kind of recording device, and their objection should be respected. In any case, a noiseless and inconspicuous microcassette recorder should be used. As a general rule, it is good idea to attempt to replicate the circumstances under which the expert normally performs expert decision-making. To minimize contamination, avoid introducing artificiality.

KNOWLEDGE ENGINEER QUALITIES

A knowledge engineer should be broadly educated and generally well informed. The computer skills of the knowledge engineer should be complemented by a healthy exposure to the liberal arts. Course work in systems design and analysis, psychology, logic, and linguistics are among many disciplines that have something to offer the practicing knowledge engineer.

A knowledge engineer should not be totally ignorant of the expert's area of expertise and should be comfortable with working with accomplished professionals. But while demonstrating a general competence, the knowledge engineer should never presume to command the expertise of the expert. The inconsistencies of the expert should never be pointed out; it is better to adopt a more nurturing posture toward the expert.

In the end, the most important abilities the knowledge engineer must possess are social in nature. The successful knowledge engineer must be an amiable and effective communicator. In addition to being generally intelligent, a knowledge engineer should be patient and tolerant. A good sense of humor can be an important asset as well.

Diplomacy is invaluable. It is important that the knowledge engineer be sensitive to the feelings, pride, and prestige of the expert in general. Sometimes the way an innocent question is posed can cause unforeseen consequences. Advanced, socially sophisticated verbal skills are called for. The knowledge engineer who can always find the most palatable way to phrase some doggedly persistent line of questioning will always enjoy a happy working relationship between expert and expert analyzer. AI

Special thanks to Kenneth Andrews for his comments and contributions.

Walter G. Rolandi is a knowledge engineer with Blue Cross and Blue Shield of South Carolina and a graduate student in psychology at the University of South Carolina at Columbia.

REFERENCES

1. Skinner, B.F. *Science and Human Behavior.* New York: Macmillan, 1953.

2. Skinner, B.F. *About Behaviorism.* New York: Knopf, 1974.

3. Ferster, C.B., S. Culbertson, M.C.P. Boren. *Behavior Principles.* Englewood Cliffs, N.J.: Prentice-Hall, Inc., 1975.

4. Bandura, A. *Principles of Behavior Modification.* New York: Holt, Rinehart and Winston, Inc., 1969.

MAINTENANCE & Language Choice

BY M. HAYTHAM MATTHEWS

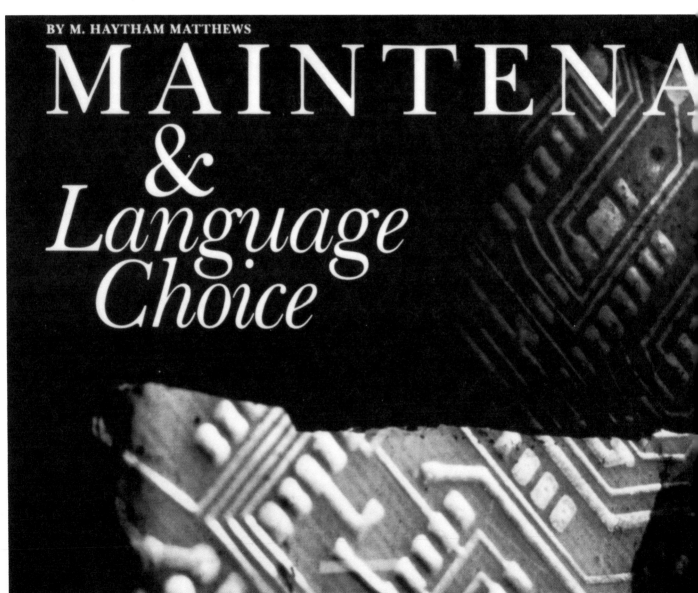

Once you've got a real system working in a real company, you need to keep it that way

Is maintenance, that (ever-present) bugbear of the software world, destined to become the nightmare of the corporate expert systems group? Or will the same software technology that gave birth to expert systems eventually produce a maintenance-free variety? Either way, maintenance of today's expert systems—in particular, those written using fifth generation languages—is a matter to be reckoned with.

Some people argue that an entirely new view of the software life cycle is needed for AI systems[1] because it is impractical to develop complete and accurate advance specifications for them. This idea should come as no real surprise since software engineers confront the non-AI version of this reality on a routine basis.[2] In fact, a small battery of techniques has been developed just for dealing with the specification problem. For example, rapid prototyping, a concept that, in conjunction with successive refinement or iterative enhancement, is becoming a standard approach to expert systems development.[3,4]

FIFTH GENERATION LANGUAGES

Indeed, one of the primary achievements of fifth generation languages has been the extension of prototyping to more effectively model complex, incompletely specified systems. Developers using these newer tools experiment more easily with design alternatives and structure their prototypes to mirror human problem-solving processes.

But what specific advantages do these unusual new languages have to offer? The answer falls roughly into two categories. The first has to do with language features that increase the ease with which real world problems and solutions can be expressed in executable code. These facilities help programmers to concentrate more of their time and attention on design problems and less

BY M. HAYTHAM MATTHEWS

MAINTENA & *Language Choice*

Reprinted with permission from *AI Expert*, September 1987, pages 42-48.

on implementation details. Such features can include, for example:

- Declarative representation of application concepts in rules or object classes.
- Nonprocedural, knowledge-oriented control regimes, such as forward and backward chaining, resolution theorem proving, spreading activation, truth maintenance, and message passing.
- Abstraction mechanisms, such as object-oriented packaging of data with the procedures that operate on it.
- A variety of features that help insulate the programmer from the minutiae of lower-level implementation. These include support for the symbol as a data type, automatic memory management, dynamic binding of data objects and types, automatic pattern matching, inheritance, and operator overloading.

A different set of features available in many fifth generation languages can increase programmer productivity in a different way. A number of tools, particularly those first developed for use on LISP machines, come with an integrated programming environment that can significantly speed up program development. Sometimes these environments are provided by the operating system, with enhancements and customizations supplied by the tool vendors. The programming support one finds in such environments include multitasking, smart editors, elaborate debugging facilities, incremental compilation, windowing, extensive utilities, and the seamless integration of all of these programming features.

Powerful programming environments and elaborate language features are highly effective weapons in the prototyping arsenal. But the long-term cost-effectiveness of an installed system is dependent upon a host of other concerns.

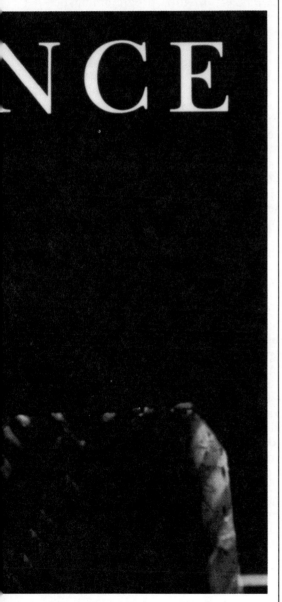

MAINTENANCE ISSUES

Software maintenance encompasses bug fixing, porting of applications onto new or upgraded hardware, and program enhancement. Among the maintenance concerns relevant to each of these categories are portability, new releases, interfaces, performance, and program modifiability. Let's take a look at each of these as it pertains to the use of fifth generation languages and tools.

Will the same technology that gave birth to expert systems eventually produce a maintenance-free variety?

Portability of a software system across different classes and brands of hardware gives an organization the flexibility to upgrade easily to more powerful machines, adapt to the hardware requirements of other (newer) software, and perform program maintenance tasks more easily on less expensive computers. Keeping old or highly specialized computers around just to continue running a nonportable application is costly and increases the maintenance load. Also, downloading corporate data for processing on a workstation or high-end microcomputer may present a problem for data security.

Vendors of fifth generation languages have recently been paying a lot of attention to portability in their aggressive search for expanded markets. Many have implemented their languages in C (for example, ART and M1/S1), which now runs on almost everything, or in Common LISP (for example, KEE), which is also becoming widely available on mainstream hardware.

Looking forward, the advent of distributed computing environments conceivably could relieve some of the concern over portability while also leveraging the performance benefits of specialized hardware. The rapid expansion of the fifth generation tool market has created a continuous push to add new features to entice new clients and keep up with the competition.

PHOTO BY DAVID JOHANSON

New releases of computer languages represent a potential hazard for existing programs

New releases of computer languages represent a potential hazard for existing programs. For example, if the syntax or the semantics of the language has changed, upgrades can effectively incapacitate a previously healthy software system. Even worse, new releases may suffer from new bugs. Hurried updates can incorporate substantial revisions that may not have been subjected to sufficient quality control. Furthermore, the vendor companies usually do not have the staff to provide adequate support for older versions and may be eager to phase them out.

Fifth generation language and tool vendors have started paying more attention to software interfaces—with good reason. Most significant expert systems need to access external data bases or I/O routines. But interface standards do not exist for most fifth generation languages. Also, language interface support is often limited and subject to change with new releases. (Constructs in some of the new languages, for example, can invoke external procedures written in other languages, but cannot themselves be invoked by them.)

The long-term success of a software system is, in part, dependent on how well it scales up to the stress of the production environment. Performance, though not a primary concern during initial prototyping, can mean the difference between an interesting experiment and a viable system. Fifth generation languages tend to make heavy demands on memory and the CPU. The execution speed of programs written in such languages is closely related to the volume of data being processed at any given moment and the number of rules (or the amount of message passing) in the system.

PROGRAM MODIFIABILITY

How to best insure the future modifiability of software programs has been the subject of much study and creative effort. Many of the concepts and methodologies that have emerged over the last 20 years are no less useful in the context of expert systems than in mainstream systems projects. Among these are the project library, chief programmer teams, program walk-throughs, documentation standards, version control systems, the use of naming conventions and *make* utilities, and, of course, rapid prototyping. System design methodologies, such as the Yourdon, Jackson, and Warnier-Orr techniques, though of questionable applicability to rule-based program design, remain important for designing integrated systems that include an expert system component.

The aspect of program development and maintenance most clearly different for projects employing fifth generation languages is programming style. Structured programming, while it has analogies in top-down structured languages like PROLOG, is not easily transferable to programs that make extensive use of nonprocedural control regimes. Nevertheless, the stylistic goals of consistency, efficiency, modularity, and reusability can still be used to distinguish an elegant fifth generation language program from sloppy, unmaintainable code.

Furthermore, as experience with these languages grows, it behooves us to watch carefully for new stylistic conventions that can potentially help improve the structure—and hence the maintainability—of the programs we create. Here are a few examples, representing the observations of a variety of authors and practitioners in the expert systems field.

Object-oriented programming: 1. Avoid accessing the internal instance variables of another object directly; adhere to message-passing conventions.[5] Many languages that support object-oriented programming permit the assignment of values to the instance variables (or slots) of an object without going through message-handling functions. While this sometimes results in slight performance benefits, it can also destroy the distinction between modular objects—whose contents are secured against accidental or arbitrary modification—and unprotected global data structures.

2. Exploit inheritance to express commonality and increase reusability.[5] For example, when two object classes have something in common, consider creating a higher-level object class so that common definitions can be inherited.

3. Make new object classes to highlight important differences instead of expanding the definition of an existing class.

4. Avoid unnecessary object definitions; when an object class has only one child object inheriting from it, merge the child object class into the parent.[6]

Forward-chaining rule-based programming: 1. Modularize/segment the rule base using rule sets or context condition elements.[4] The safe modification of a rule-based program requires the programmer to understand the conditions under which individual rules are enabled or disabled. A standard approach to organizing a rule-based program is to partition rules into groups that represent meaningful subtasks.

The enabling of these subtasks is accomplished either by explicitly invoking the associated rule set or, more commonly, changing a context element in the fact base. In the latter case, the change to the fact base satisfies an antecedent condition of all rules in one subtask while (usually) disabling rules in other subtasks.

2. Exploit hierarchical problem structures; consider making contexts hierarchical

by allowing subcontexts, sub-subcontexts, and so on. A context or subtask can be thought of as being analogous to a procedure in mainstream languages. Organizing a program hierarchically is the rule-based equivalent of top-down design.

3. Refine the rule base by making the antecedents more specific and combining rules that perform identical actions.[4] The rules in early versions of a rule-based system are often broader than those in later versions. As rules are tailored to handle specific cases, they become more specialized and more numerous. Eventually, however, rules that respond to different situations but do the same things may be legitimately combined.

4. Reduce redundancy by writing functions to package sets of actions that occur repeatedly in right-hand sides of rules.[7]

5. Reduce uncertainties about how and when a rule is applied; associate individual rules not only with the subtask or context for which they are applicable but also in accordance with a set of standardized knowledge roles. Examples of such knowledge roles are *propose-operator, reject-operator, evaluate-operator, apply-operator, recognize-success,* and *recognize-failure.*[8]

6. Avoid applying rules when no use is made of the resulting inference(s).[6]

7. Avoid using rules to implement tables or algorithms that would be better handled by calling a function or subroutine.[6]

8. Organize condition elements in individual rules for maximum efficiency. For example, placing more restrictive antecedent conditions before less restrictive ones in the left-hand side of a rule will often improve the efficiency with which that rule is processed. *Programming Expert Systems in OPS5* includes an excellent discussion of ways to improve the efficiency of rule processing.[4]

Other examples: 1. Limit, localize, and document diversions from the predominant control paradigm (forward chaining, backward chaining, message passing, access driven, or procedural).

2. Experiment with alternative paradigms to determine which best suits the problem.

3. In PROLOG, avoid overuse of the *cut* and *or* (! and ;) operators.[9]

Another issue affecting the maintainability of fifth generation language programs is the shortage of programmers competent to modify the developed systems. This can present a special problem to the extent that the tool is relatively unknown, complicated to learn, and lacking in similarity to other, more familiar languages.

LANGUAGE CHOICE

One way to cope with these issues is not to deliver in fifth generation languages at all. We can benefit from them during prototyping and then reimplement the system in a mainstream language we already know how to maintain, such as C. In fact, the potential usefulness of fifth generation prototyping tools in mainstream systems development has scarcely begun to be explored. Language translators (such as LISP to C) can sometimes facilitate this process. However, the generated code itself usually needs customization and reworking to improve its structure and efficiency.

By contrast, if the system involves the modeling of expert problem-solving behavior, implementing it in a mainstream language is less advantageous. The design of an expert system may require changes due to obsolesence, expansion, or shifts in priorities or points of view. Fifth generation languages help by keeping the assumptions and problem-solving logic on the surface of a program, where they can be better understood, supplemented, and revised. Instead of giving up this benefit, the best approach, on balance, may be to deliver in an intermediate language that provides a critical subset of the features found in the more advanced tools while having fewer disadvantages. Ideally, such languages should exhibit:

- Good run-time performance and response to data or transaction volume.
- A reasonable probability of continued existence and future growth.
- Relatively stable syntax and language features, preferably accompanied by a standard textbook devoted to the language.
- A relatively short learning curve.
- Portability across different classes of hardware (micros, minis, and mainframes).
- Good software interface facilities.
- An integrated procedural language for those cases where the most direct approach is a well-crafted procedure.

No fifth generation language scores 100% against these criteria. For example, the future growth of most fifth generation languages cannot be assured. Likewise, the syntax of new features added to any language is subject to some instability at first. But such realizations should motivate us to more carefully identify the maintenance-relevant dimensions along which these languages vary, then employ them effectively in the language selection process.

SOME OPTIONS

Some of the languages that today score reasonably well on these measures are OPS5, OPS83, PROLOG, C++, and Common LISP with object-oriented extensions. Each of these has distinctly different strengths and limitations.

C++ is an object-oriented expansion of the C language sponsored by AT&T. It re-

Adopt stylistic conventions appropriate to the architecture of each delivery language

tains most of C as a subset, runs on a wide variety of hardware, and does not sacrifice speed to achieve its added modularity and abstraction capabilities. C++ supports message passing, hierarchical inheritance, and operator overloading.[5] Unlike the other languages mentioned here, it does not have a symbol data type or automatic memory management. The primary distributors of C++ in the U.S. are AT&T, OASYS, and Lifeboat.

OPS5 is probably the most widely known rule-based language in the U.S. It has a forward-chaining architecture based on the fast Rete pattern-matching algorithm. It features a simple, straightforward syntax and well-understood programming techniques.[4] OPS5 was the language used by Digital Equipment Corp. for its now-famous R1/XCON expert system for VAX configuration.

LISP-based versions of OPS5 are generally slower than other implementations. While an interface to the base language is usually provided, standard OPS5 does not have an integrated procedural language. OPS5 is available from DEC and Computer Thought Corp. YES/OPS, an OPS5 derivative, is also being developed at IBM.

OPS83 is a fast, extended rule-based language expressly designed to provide a high-performance vehicle for the delivery of expert systems.[10] With a rule definition syntax derived from the OPS4 and OPS5 languages, OPS83 adds a Pascal-like procedural language to a forward-chaining architecture. Unlike other rule-based languages, OPS83 gives the programmer complete control over the mechanism that determines the sequence of rule firings. It was developed by the author of the Rete algorithm and is marketed for a range of machines by Production Systems Technologies.

PROLOG is a backward-chaining rule-based language based on Horn logic and resolution theorem proving. This language is one of the more popular fifth generation languages, particularly in Europe and Japan. Most implementations of PROLOG provide the same semantic language kernel[9] but differ significantly in syntax and especially in the unfortunately critical extensions to the basic elements of the language.

One key to PROLOG's future probably lies in how well its proponents are able to standardize some of these critical extensions (or built-in predicates). Another key may be how well PROLOG vendors can forge a link with relational data base technology since PROLOG is by nature a powerful extension of relational languages (such as SQL). Most PROLOG implementations lack an integrated procedural language, but built-in predicates can sometimes substitute for this.

Common LISP[11] has become the standard AI procedural language in the U.S. Full Common LISP is a large, lexically scoped, symbol-manipulation language with automatic memory management and over 700 built-in functions. One of the shortcomings of Common LISP as a maintainable delivery vehicle is also one of its strengths as an AI tool—namely, its ultraflexibility in embodying a large number of semantic ideas and having numerous ways of achieving identical effects. Ironically, straight Common LISP lacks the semantics of a standard rule-based or object-oriented programming facility. The latter, however, is a restriction likely to be removed soon, when the Common LISP standards committee comes to agreement on object-oriented extensions to the langauge.

Despite their differences, these languages all share some common elements. All have compiled versions that run on IBM PCs in addition to versions for a variety of larger machines. Each has a standard commercial textbook that details the basic syntax and language features (OPS83 is an exception, but the standard textbook for OPS5 can be substituted). Each can usually be interfaced to the C language (and others) if needed.[12]

On the other hand, PROLOG, Common LISP, and OPS5 have interpreted or incrementally compiled versions whereas C++ and OPS83 currently do not. The latter two, however, are noteworthy for their ability to deliver particularly fast run-time execution speeds.

SINGLE-PARADIGM LIMITATION

An important restriction characterizing all of these intermediate languages is that, unlike their more elaborate competitors, they do not provide built-in support for multiple programming paradigms (apart from varying degrees of integration with a procedural language). While it is generally possible to achieve multiparadigm effects through custom coding, each language excels in one paradigm only—that is, either in its provisions for (object-oriented) message passing (C++) or in its support for (rule-based) forward chaining (OPS5 and OPS83) or backward-chaining (PROLOG) features. Furthermore, some control paradigms, such as assumption-based truth maintenance, are directly supported only in the larger, multiparadigm languages.

The chief disadvantage of the single-paradigm restriction is clearly that one cannot rely on the same language to be equally well suited to all applications. This makes prototyping in these languages less convenient than in powerful multiparadigm tools such as ART, KEE, KnowledgeCraft, GoldWorks, and Nexpert Object. On the other hand, programs that mix fundamentally different programming paradigms are not likely to be as easily comprehended and modified in the long run as those rooted in a smaller, consistent set of semantic ideas.

IMPLICATIONS

The implications of the preceding discussion can be summarized as follows:

- Consider using intermediate languages as expert system delivery vehicles. (Language selection might best be accomplished in stages, responding to alternate delivery language needs as they arise.)
- Develop in-house expertise and training materials in support of your choices.
- Prototype in your tool of choice.
- Use standard maintenance protocols and systems development techniques wherever possible and appropriate.
- Evaluate and adopt stylistic conventions appropriate to the architecture of each selected delivery language and the control paradigm(s) it supports.

Although much still remains to be discovered, past experience has already provided us with much of the guidance we need to exercise prudent control over expert systems projects. One of the most critical decisions affecting maintainability is the selection of an implementation language. A balanced approach to language selection requires an understanding of the trade-offs involved in each potential choice. Also, language selections must eventually be backed up by appropriate stylistic conventions, training programs, and maintenance protocols for long-term cost-effectiveness to be achieved. AI

REFERENCES

1. Partridge, D., and Y. Wilks. "Does AI Have a Methodology Which is Different from Software Engineering?" *Artificial Intelligence Review* 1(2): 111-120 (1987).
2. Parnas, D.L., and P.C. Clements. "A Rational Design Process: How and Why to Fake It." *IEEE Transactions on Software Engineering* SE-12(2): Feb. 1986.
3. Smith, R.G. "On the Development of Commercial Expert Systems." *AI Magazine* Fall 1984, p. 68.
4. Brownston, L., R.G. Farrell, and E. Kant. *Programming Expert Systems in OPS5.* Reading, Mass.: Addison-Wesley, 1985, pp. 87, 165, 172-173, 241.
5. Stroustrup, B. *The C++ Programming Language.* Reading, Mass.: Addison-Wesley, 1986, pp. 9, 191.
6. Kline, P.J., and S.B. Dolins. *Choosing Architectures for Expert Systems,* CCSC Technical Report 85-01-001. Dallas, Texas: Texas Instruments Inc., 1985, pp. 122-123.
7. Blackwell, S. Personal communication, April 15, 1987.
8. van de Brug, A., J. Bachant, and J. McDermott. "The Taming of R1." *IEEE Expert,* 1(3): 34-35 (Fall 1986).
9. Clocksin, W.F., and C.S. Mellish. *Programming in PROLOG,* 2nd ed. New York, N.Y.: Springer-Verlag, 1984, p. 183.
10. Neiman, D., and J. Martin. "Rule-Based Programming in OPS83." *AI EXPERT* 1(1): 54-65.
11. Steele, G.L., Jr. *Common LISP: The Language.* Billerica, Mass.: Digital Press, 1984.
12. Matthews, M.H. "PROLOG and C Join Forces." *COMPUTER LANGUAGE* 4(7): 34-44 (July 1987).

M. Haytham Matthews is an associate of Information Resource Management Associates, Inc. He is currently serving as a consultant in New York, N.Y., specializing in business applications of AI and fifth generation languages.

AUTHOR'S UPDATE TO "Maintenance and Language Choice"

In the two years since my article on "Maintenance and Language Choice" appeared in **AI Expert**, significant changes have occurred in the AI language market, not to mention the expert systems field in general. My own views on some maintenance-related issues have also changed. The following random notes are included here in the spirit of an update.

The AI Language Market Today

Over the last two years, we have witnessed a tremendous push by AI product vendors to make their tools integrateable, or at least interfaceable, with applications written in mainstream languages. In addition, those general purpose tools that have achieved a degree of market success, such as AION's ADS and Neuron Data's Nexpert Object, are noteworthy for their portability between PC and mainframe. Such portability has been a huge selling point, in that it enables companies to start small by developing applications on the PC and yet have the assurance that their code can then be moved into mainframe production environments.

Vendors have specialized in various ways. The Carnegie Group, for example, has focussed on developing customized shells for specific application domains, such as manufacturing and text processing. Also, mainframe AI tools today tend to specialize in either the IBM or the DEC hardware platforms, rarely supporting both.

Language Choice for Prototyping

I no longer support the idea that the choice of a prototyping tool can safely be considered independently from the selection of the implementation language for the production version. While the notion of giving a free hand to the prototyping process may seem attractive, too much freedom can contribute to a phenomenon that has plagued the expert systems field - the dead end prototype (one that never makes it into production).

Prototyping is probably best done in the delivery language (which makes the choice of the latter that much more important). "Throw-away" prototyping should be avoided in favor of a more deliberate, incremental approach. The difference between the latter and "rapid" or "proof-of-concept" prototyping is that concern for the speed of initial prototype development is moderated by the goal of minimizing the need for future backtracking over design and coding decisions. The best way of achieving this is probably (1) by investing sufficient time up front to acquire a fundamental understanding of the domain and then (2) by building and expanding a prototype in stages, each ending with a short-term deliverable.

Typically, the area where experimentation is most needed is in user interface design. Fortunately, interactive screen design and user interface prototyping tools are plentiful and are now available for a wide spectrum of computer languages and hardware platforms (not to mention their inclusion in many popular AI shells). Unfortunately, good-quality, user-oriented explanation facilities - a vital part of most business-critical expert systems - are often not as easy to prototype as other aspects of the user interface. Not surprisingly, the best explanation facilities are those which were designed hand-in-hand with the knowledge representation scheme.

Reusability of Rule-Based Code

One of the great maintenance-related advantages of object-oriented systems development is the natural way in which code can be written for reuse in multiple modules and future applications. Theoretically, the knowledge expressed in rules should likewise be reusable. However, most rules are still too application-specific for this to be the case. One of the reasons for this is the over-simplicity, and lack of programmability, of the control system in most rule-based languages. The control system is the mechanism by which the next rule instantiation is selected for activation/execution. These systems are typically built into the language or shell and use simple heuristics, such as rule "specificity," or simply the textual ordering of clauses, to select rules or rule instantiations for "firing." The result is that the programmer is forced to contaminate the design of rules meant to express domain knowledge by taking application-dependent measures to ensure the control system will select the right rule at the right time.

Over the past two years, I have been experimenting with an alternative approach in the form of a programmer's toolkit for the OPS83 language. This alternative is to associate a descriptor element with every rule, which describes the rule in some standard way so that a customized, application-specific control system can choose whether, and when, to employ it. Of course, this approach requires that the host language permit the developer to take full charge of the control system - a freedom still relatively rare in the context of today's AI shells.

Reuse of Causal or "Deep" Knowledge

One area that has been gaining ground in the AI research arena is that of model-based reasoning. Here, the approach is to develop a causal model of a particular domain in order to support a form of "deep reasoning" or "reasoning from first principles." Ideally, developers can then write application programs which employ this causal model either as a driving force or as a backup mechanism. Theoretically, one could prototype an application by initially relying heavily on the model. Then, the application could be optimized by replacing some or all of the reliance on the model by speedier "shallow" reasoning components which might need to be application-specific.

A related trend is toward development of deductive and object-oriented database systems. There is not space here to go into these hybrid technologies. However, both could offer tremendous reusability benefits. More than one object-oriented database system is already on the market. Ontologic, for example, has one specifically designed for use with the C++ language.

M. Haytham Matthews,
Director of Knowledge-Based Systems Consulting
MJR Associates
10/15/89

BY BRUCE MARCOT

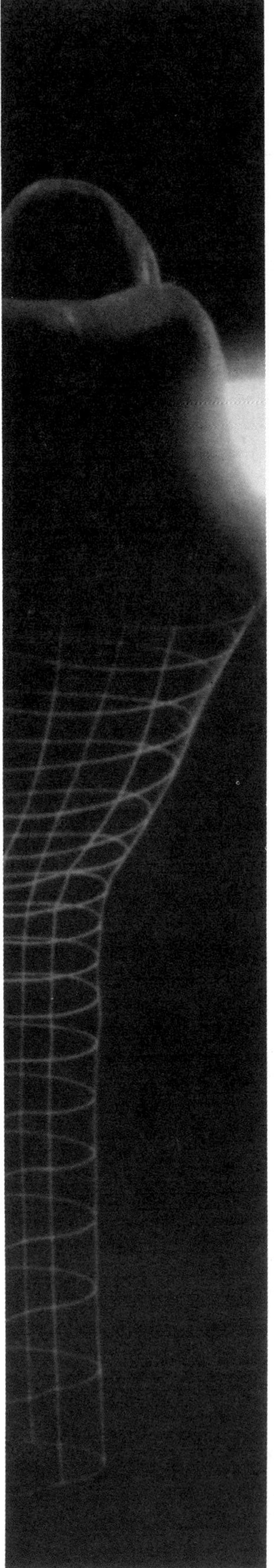

TESTI[NG] your knowledge base

In the development of a knowledge base, eventually the time comes to put its contents to the test. This article describes how knowledge bases can be tested and presents a set of validation criteria. Although knowledge bases take many forms, this article focuses on those founded on production rules. However, the general concepts and criteria discussed here apply to other types of knowledge bases as well.

I draw on my own experience developing and testing knowledge-based systems for Perkin-Elmer Corp., where I helped develop an expert system to monitor respiration conditions of patients.[1] I also worked for Oregon State University, Corvallis, Ore., and the U.S. Dept. of Agriculture Forest Service developing expert systems to assess conditions of wildlife habitat[2] on USDA Forest Service lands.

Testing a knowledge base is not a trivial matter, regardless of the exactitude with which the problem was defined and the production rules (with associated probabilities) were devised. Testing cuts to the heart of why the knowledge base was initially developed and how one expects it to be used.

WHAT REQUIRES TESTING?

At best, testing a knowledge base is more an operant philosophy of programming than a discrete stage in the development of production rules. It is good programming practice to test any program in various ways throughout its development. Testing should be integrated into the development-application cycle.

Generally, there are two levels of testing any computer code: verification and validation. Verification involves insuring that the computer code—whether it is LISP code or a rule base written in some expert system shell—is written without bugs. This level of testing is relatively straightforward and may be pursued on a regular basis as the knowledge base is developed and revised. Many commercially available expert system shells have tracing functions or filters for catching common errors in syntax or rule redundancy. These tools are of immense value when the knowledge base approaches the size of even several dozen rules.

Validation, on the other hand, involves the more deceptively difficult task of insuring that the meaning and content of the rules meet some carefully defined criteria of adequacy. Defining such criteria is the key to successfully conducting a validation procedure and demonstrating the level of acceptability of the knowledge base.

Performance may be tested progressively as a knowledge base is built. For example, S.M. Weiss and C.A. Kulikowski[3] suggest testing the initial model design, knowledge base data, system performance, model refinements, and the effect of model changes on case conclusions.

Development of a knowledge base commonly proceeds through four phases: creating the prototype, developing the first generation rule set, testing and expanding the rule set to the second generation, and testing the second generation rule set. Attention must also be given to how well the system might be marketed and used. Specific criteria and procedures for testing the valid-

Reprinted with permission from *AI Expert,* August 1987, pages 42-47. Copyright © 1987 by Miller Freeman Publications, 500 Howard Street, San Francisco, CA 94105.

NG

Testing cuts to the heart of why a knowledge base was designed and how one expects it to be used

ity of the knowledge base (see the sidebar to this article entitled "Validation Criteria") should be integrated within each of these phases.

CREATING THE PROTOTYPE

Validation should be part of the initial stage of creating the prototype knowledge base. Prototypes that typically consist of only a dozen or so rules demonstrate the application and feasibility of a knowledge engineering approach to the problem at hand. Validating the prototype is integral to defining the original problem and developing the first set of rules to frame the inference structure.

When testing the prototype, it is important to determine the specific domain within which the system should operate. More fundamentally, testing the prototype should involve asking whether the original problem domain was defined narrowly enough. Knowledge base projects have a greater likelihood of succeeding—and, in a sense, of being valid—when they address a narrowly defined problem.

The prototype should also show a fair degree of usefulness to demonstrate the desirability of continuing work on a fuller system. A system is useful when it contains necessary and adequate parameters to solve at least some problems. Today's expert system shells are so user-oriented that bogus rule bases may be readily developed, creating what might be called "amateur systems." Although usefulness may be judged subjectively, it is more than just a trivial or self-evident criterion.

PHOTO BY MICHEL TCHEREVKOFF COURTESY OF BELL ATLANTIC

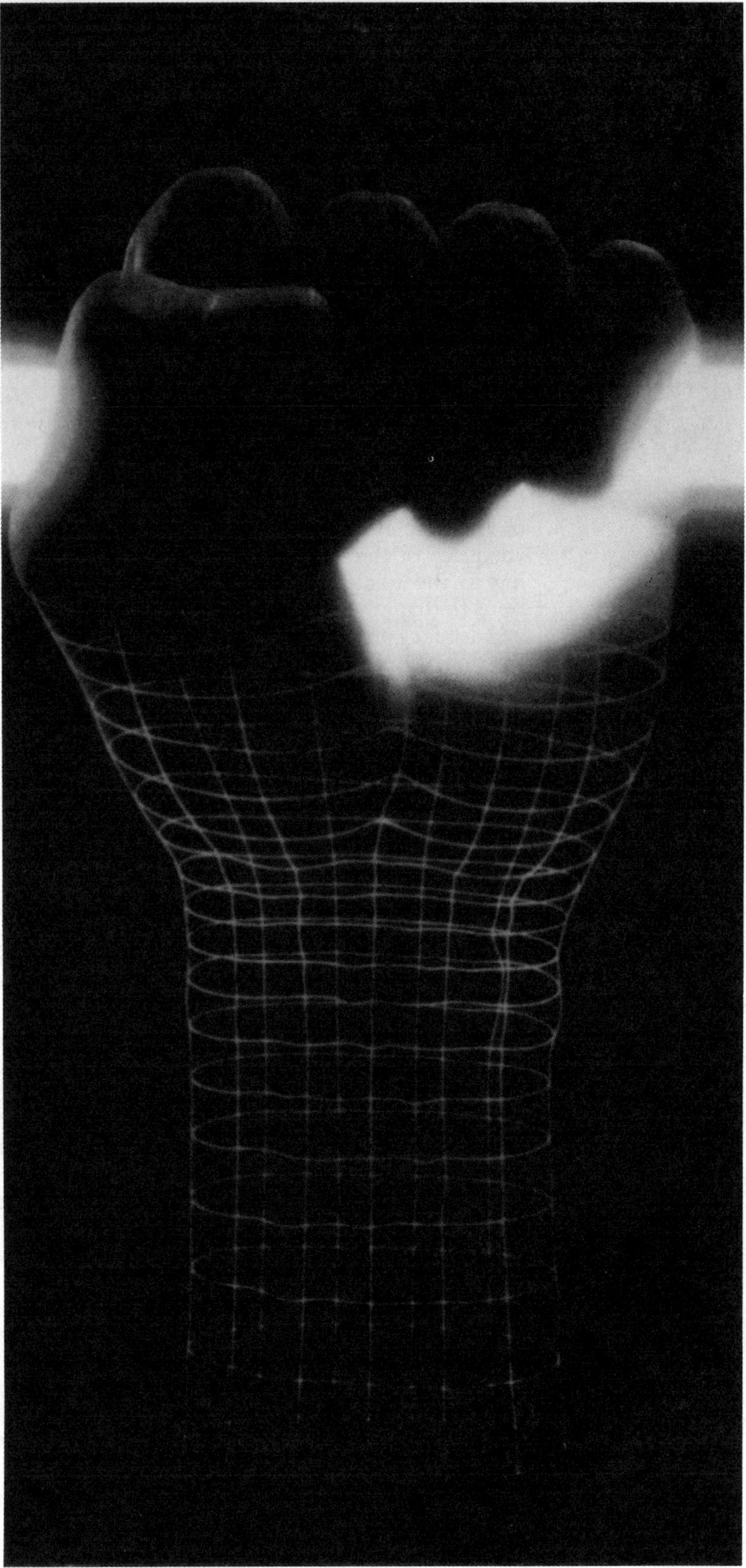

At the alpha testing stage it is important to determine the accuracy and adequacy of the system

THE FIRST GENERATION RULE SET

After the prototyping stage, the first generation rule set typically consists of several dozen to a few hundred rules. The main goals of validation at this stage are to reexamine the original objectives, more precisely determine the problem domain, and establish the degree of detail desired in the system.

At this alpha-testing stage it is important to determine the accuracy and adequacy of the system. Accuracy is measured by comparing the number of correct predictions with known data. It may be assessed statistically with a χ^2 or goodness-of-fit model that compares observed proportions of correct answers (the fraction of predictions that were observed to be empirically correct) with expected proportions (the desired rate of accurate predictions).

Adequacy, on the other hand, is a measurement of the fraction of actual conditions included in the system. For example, the breath gas monitoring system I helped develop diagnosed 12 waveform patterns of CO_2. Each waveform pattern corresponded to a particular physiological condition. The adequacy of the system was judged according to the total desired number of conditions to diagnose.

Adequacy may be expressed as a simple fraction. The breath gas monitoring system was able to diagnose 12 of 27 important physiological conditions, so the system was $100 \times (12/27) = 44\%$ adequate. As an alternative, subjective weights may be added to particular conditions that are more important to recognize (such as cessation of breathing).

Another question at the alpha-testing stage concerns the degree of precision required. Precision may be measured as the capacity of the knowledge base to predict, diagnose, classify, or monitor within a specified statistical confidence interval. Precision is also a measure of the number of significant figures used in calculations and the error of estimate of the parameters. Precision error may be expressed as the standard error or confidence interval of the observed or inferred values of parameters.

The alpha-testing stage also entails statistical tests of the reliability of the inferences, diagnoses, or classifications made by the knowledge base. Evaluating reliability may involve a complex series of statistical tests that apply prior knowledge to empirical evidence, such as in the use of Bayesian statistics.[3,4] A discussion of such tests is beyond the scope of this article, but the reader is encouraged to explore the concepts.

EXPANDING THE RULE SET

Alpha-testing the first generation rule set should result in revision of the objectives for developing the fuller rule base. As rules are added and amended at this stage, performance standards that determine the characteristics of the full-scale system should be carefully described, especially for determining utility or ultimate marketability. These standards include the flexibility or adaptability of the system to future applications. An adaptable system can be readily enhanced and its user interface modified as contexts warrant.

Expanding the rule set to full scale should also identify the number of parameters necessary to address the problem domain (resolution) and the desired complexity (wholeness) of the knowledge base. Expansion typically involves recrafting inference structures, high-level control rules, and lower-level facts and relationships among facts. To this end, determining the robustness and sensitivity of conclusions to rules and variables helps direct the knowledge engineering efforts. Several workers have explored using empirical information for modifying the rule base (as used in SEEK).[5,6]

SECOND GENERATION RULE SET

Testing of the first full-scale knowledge base is often done directly on the site where the system will be used. Determining the accuracy, precision, and reliability of the full-scale rule base is necessary to ultimately determine the specific contexts within which the user should expect the system to work well (technical and operational validity). This beta-testing should also involve determining the usability of the system by the intended audience, including how well the system fits into existing procedural and administrative structures.

Testing the audience, so to speak, was one of the most important limitations in developing and implementing a knowledge engineering approach to solving problems of natural resource management in the USDA Forest Service. The weakest facet of the knowledge base approach for diagnosing and ameliorating the condition of wildlife habitat on National Forest lands was selling the basic idea to management.

Existing administrative processes for evaluating wildlife habitat simply did not allow for this new approach. Thus the knowledge base was, in an important sense, invalid. The solution in this case was to educate veteran supervisors and managers within the agency as to the utility of such a system and revise the general habitat evaluation process.

MARKETABILITY

In this article the term "marketability" refers not only to the specifics of selling a product in the marketplace but also to selling the idea and use of a system to people who may be reluctant to adopt it in the course of their work. Selling a system involves determining the practicability and utility of a system in specific already-functioning work environments. From the commercial perspective, this may also include market feasibility studies to determine desirability and availability. The system must also be adaptable to different work environments, changes in integrated hardware and software, and changing information needs over time.

If a system is to be used by technicians, professionals, administrators, or managers in the course of their work, it should be appealing and credible (face validity). In the development of the breath gas analysis system, we contacted recognized medical specialists in the field of breath gas analysis during the early phases of developing the prototype. The appeal and face validity of the prototype was important for securing research and development funds within the corporation and for helping sell the system to customers once the fuller system became available.

Another facet of these less-tangible criteria of validity concerns how the human user is to be integrated into the knowledge system. Users of knowledge-based monitoring systems will want to remain an integral part of the information-gathering/analysis/interpretation cycle. For example, the breath gas monitoring system will be more readily accepted and used if the clinician has a role in interpreting and at times overriding the system's diagnoses and conclusions.

How specifically can these various validation criteria be tested? The main steps in conducting validation tests under each of the phases of developing the knowledge base include:

- defining the domain and context within which a system is expected to perform well and thus the contexts in which its performance is poor or unknown.
- identifying specific performance criteria for validation.
- conducting the validation tests and analyzing and evaluating the results.

DOMAIN AND CONTEXT

Like any model, the domain and context of a problem determine how well a knowledge base can be expected to perform. Applying the model outside the arena for which it was intended will likely produce unreliable and poor performance.

The creator of the model should specify precisely the conditions under which the model is to be used. The audience must adhere to these conditions for the model to insure predictable levels of performance. This proved especially important in the expert system that evaluated wildlife habitat. Similar habitats in different geographic areas have vastly different species of wildlife associated with them, and the expert system failed to predict reliably outside of the area in which it was developed.

Testing a knowledge base may also involve assessing its generality or breadth. Generality is the range of contexts within which a system can be expected to perform reliably. Breadth is the number of conditions and parameters an expert system contains, proportional to the number of rules and clauses in its knowledge base.

A system should be broad when the range of conditions and contexts within which it should operate reliably is wide. Such testing involves applying the knowledge system to problems outside the specific confines of its domain under controlled circumstances and observing the accuracy or reliability of the outcomes.

Identifying a specific problem domain may also entail describing the number and kinds of variables chosen to represent each component of the knowledge structure (depth). This in turn helps identify the degree of realism of the knowledge model; that is, which relevant variables and relations have been included in the knowledge base. However, it may not be possible to simultaneously maximize realism and generality. Understanding how well a system performs under both criteria is essential for developing realistic expectations about its performance.

Although a system like the breath gas monitoring system must perform reliably, the realism of the production rules is equally important. Because a system like this one will be closely scrutinized by many medical experts, even if it performs reliably, the variables and relations expressed in each specific rule must conform to generally accepted expert understanding.

This is particularly important with knowledge bases such as medical monitoring systems, where legality and culpability are significant issues. For other systems, such as the wildlife habitat expert system, the degree of realism of the rules is secondary to insuring that the system provides reasonable and useful conclusions.

SPECIFIC PERFORMANCE CRITERIA

Performance criteria must be specified for each stage in the development of a knowledge base. For example, the degree of accuracy of the system may be assessed during testing of the second generation rule set.

Acceptable levels of each performance standard should be determined before the tests are conducted. What allowable fraction of predictions (outputs) may be in error? How accurate or precise do diagnoses or classifications have to be for a system to

If human experts incur a 15% error rate, should the system be expected to perform better?

be acceptable? How realistic should the variables and relations be within each rule?

In determining the overall validity of a system, it is instructive to determine how well human experts do in the problem area and to thereby create reasonable expectations of the system's performance. If human experts incur a 15% error rate, should the system be expected to perform better? If so, how much better, and why?

A Turing test of the output of a system may help determine its overall validity.[7] In a Turing test, human experts (evaluators) are given results of running the knowledge base model for a specific problem and results furnished by a human expert for the same problem. Both sets of results are unlabeled.

The degree to which the evaluators can distinguish between results from the model and the human expert is a test of how well the system mimicked human performance. Furthermore, having the evaluators explain exactly how they distinguished model from human performance can help pinpoint which model parameters require further refinement.

Another performance criterion that may be tested is to analyze the type of errors a rule set produces. The conclusions of particular rules—or an entire knowledge base—for a given run may be wrong in two ways. They may fail to accept what is actually a correct conclusion (a type I error) or accept and report what is actually an incorrect conclusion (a type II error).

These two types of error may have vastly different implications, depending on the purpose of the knowledge base. For example, the breath gas monitoring system alerts the clinician when a serious problem, such as increased CO_2 content of breath gas or cessation of breathing (anoxia), has been detected. False alarms (type II errors) would be much more tolerable with such a monitoring system than failures to detect serious problems (type I errors). Thus the rules that detect and interpret anoxic conditions allow for a wide level of tolerance.

On the other hand, false alarms may be much less tolerable with other kinds of systems. For example, the wildlife habitat expert system predicts the effects on the distribution and abundance of bird species from various forest management activities. This knowledge base provides advice on additional and sometimes costly activities that can help protect wildlife species from detrimental impacts the result from the harvesting of timber.

In this case, false alarms (type II errors) are much less tolerable than in the medical monitoring system, because high costs may be incurred. Thus the rules that predict negative impacts on wildlife entail narrow tolerance levels.

Validation Criteria

The following set of criteria may be used to test and evaluate the validity of a knowledge base:

Accuracy: how well a simulation reflects reality. Compare inferences made by rules with historic (known) data, observe correctness of the outcome.

Adaptability: possibilities for future development and application. Keep I/O and control rules general; revise facts and rules when new information is available. Periodically review the desirability of integrating with existing or proposed hardware or software systems. Should the system be self-modifying or context sensitive? Can it be customized for particular user needs?

Adequacy: the fraction of pertinent empirical observations that can be simulated. Establish list of parameters (variables, conditions, and relations) that influence inference outcome, determine which to include in rule set.

Appeal: usability; how well the knowledge base matches our intuition and stimulates thought; practicability. Appeal is a potentially key criterion for marketability; test usability by assessing I/O friendliness relatively early in the development process. Test simulation and practicability on site in beta-development stage.

Availability: existence of other, simpler, validated knowledge bases that solve the same problem(s), important for determining eventual marketability. Will users perceive the need for a new rule-based system if other tools are already available and meet their needs?

Breadth: proportional to the number of rules used in the knowledge base. Determine the number of contexts within which the system should be expected to perform, and thus the number of pertinent parameters to account for in the rule set.

Depth: proportional to the number and kinds of variables chosen to describe each component in the model. Determine the range of conditions the system will address and which parameters are necessary to diagnose, classify, and/or advise for each condition. Depth will in turn determine necessary input data and user interface.

Face validity: model credibility. Have knowledge base, inference structure, and output reviewed by credible human experts during early development of prototype and later expansion of full-scale system. Compile and report results.

Generality: capability of a knowledge base to be used with a broad range of similar problems. Define the general contexts within which the system can be expected to perform at expert levels and provide strong caution that use beyond these contexts may not yield accurate results.

Precision: capability of a model to replicate particular system parameters; also the number of significant figures used in numeric variables and computations. Ensure that all pertinent variations of parameters are represented in the rule base and facts. Express numbers as floating point or real format as necessary; use double precision for calculations, especially those involving matrix or linear algebra calculations.

Realism: accounting for relevant variables and relations.

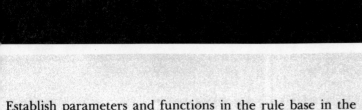

Establish parameters and functions in the rule base in the same terms and with the same conceptual models used by experts or end user audience. Realism is particularly important when developing the full-scale knowledge base, and also involves the logical order with which queries are made.

Reliability: the fraction of model predictions that are empirically correct (actually, part of a complex statistical analysis of the accuracy and correctness of the entire rule base). Reliability includes conditional and posterior probabilities of correct diagnoses and classifications. Ultimately describes statistical utility of the likelihoods in the rules and outputs.

Resolution: the number of parameters of a system the model attempts to mimic. Identify which parameters need to be defined and represented in detail and which can be grouped into more general conditions or ignored.

Robustness: conclusions that are not particularly sensitive to model structure. Determine which input parameters are least and most significant in the form of the interim (diagnosis, classification) and final (advice, alarm) results and output. Be sure the latter are well defined in the rules and functions.

Sensitivity: the degree to which variations of knowledge base parameters induce outputs that match historical data. Specifically determine sensitivity of results to each input parameter by varying that parameter incrementally, holding all other parameters constant and matching model output with historical (known) data.

Technical and operational validity: identification and importance of all divergence in model assumptions from perceived reality. Carefully explicate the contexts, conditions, and assumptions that underlie the rules and relations. Discuss how each assumption limits model results. How do they affect model accuracy, reliability, robustness, and generality?

Turing test: assessing the validity of a knowledge base by having human evaluators distinguish between the model's conclusions on a specific problem and a human expert's conclusions solving the same problem.

Usefulness: validates that the system contains necessary and adequate parameters and relationships for use in problem-solving contexts (if at least some model predictions are empirically correct). Usefulness is trivial for a full-scale system but important for prototyping and adding onto existing rule sets.

Validity: a knowledge base's capability of producing empirically correct predictions. Given the contexts within which the system is expected to operate well, determine how many actual conditions the system can accurately diagnose, classify, and advise. Determine the level of correctness with human experts in the same area and set realistic objectives for correctness of the knowledge base.

Wholeness: the number of processes and interactions reflected in the model. How complex is the rule base? How many factors does it use? Consider wholeness in light of adaptability.

ANALYZING RESULTS

Analysis and evaluation entail drawing conclusions from the results and assessing the implications of each conclusion for further development, marketing, or revision of the system. The analytical steps involved should be clearly determined before each test proceeds, especially where quantitative tests are concerned.

For example, during the testing of the second generation rule set, assessing the accuracy of the system should involve determining the specific problem domain, selecting a representative spectrum of known cases against which to test the system (these should be cases that were not used in the initial development of the knowledge base), operating the system and recording its output (diagnosis, classification, advice), and comparing the output with the known conditions of each case. A statistical comparison between system predictions and known cases may help determine the frequency and types of errors the system made.

Finally, the results of the validation tests should be used to evaluate the development of the system. Does the system meet originally defined objectives and standards? Can it be used in the intended problem domain within an acceptable error rate? Can it be successfully placed in the market? What are the needs for adding to or revising the knowledge base? Are there additional considerations to address, such as legal or regulatory constraints? What can be learned to help in the development of the next system? Every knowledge engineering project is unique in its intent and application, and every project should embrace carefully designed tests of the validity of a knowledge base. AI

REFERENCES

1. Rader, C.D., V.M. Crowe, and B.G. Marcot. "CAPS: A Pattern Recognition Expert System Prototype for Respiratory and Anesthesia Monitoring." Presented at Westex-87 IEEE Expert Systems Conference, June 1987, Anaheim, Calif.
2. Marcot, B.G. "Use of Expert Systems in Wildlife-Habitat Modeling." In *Wildlife 2000: Modeling Habitat Relationships of Terrestrial Vertebrates,* edited by J. Verner, M.L. Morrison, and C.J. Ralph. Madison, Wisc.: Univ. of Wisconsin Press, 1986, pp. 145-150.
3. Weiss, S.M., and C.A. Kulikowski. *A Practical Guide to Designing Expert Systems.* Totowa, N.J.: Rowman & Allanheld, 1984.
4. Rich, E. *Artificial Intelligence.* New York, N.Y.: McGraw-Hill, 1983.
5. Politakis, P., and S.M. Weiss. "Using Empirical Analysis to Refine Expert System Knowledge Bases." *Artificial Intelligence* 22(1984):23-48.
6. Politakis, P. "Empirical Analysis for Expert Systems." In *Research Notes in Artificial Intelligence 6.* Boston, Mass.: Pitman Advanced Publishing Program, 1985.
7. Chandrasekaran, B. "On Evaluating AI Systems for Medical Diagnosis." *AI Magazine* 4(2):34-37.

Bruce G. Marcot, Ph.D., is an AI consultant with Thought Services, Portland, Ore.

Reprinted from *Proceedings: WESTEX-87—Western Conference on Expert Systems*, 1987, pages 38-43.

VERIFICATION AND VALIDATION OF EXPERT SYSTEMS

Christopher J. R. Green
Marlene M. Keyes

Structured Systems & Software, Inc.
23141 Plaza Pointe Drive
Laguna Hills, CA 92653

Abstract. Verification and validation (V&V) are formal methods used to determine whether computer programs will satisfy user requirements. An obstacle to the acceptance of expert systems is the lack of a methodology for V&V of expert systems. V&V of expert systems is hampered by a lack of stable documentation, inadequate methods to evaluate test results, and a vicious circle that hinders development of V&V methods. Accepted methodologies for performing V&V on expert systems would further the acceptance of expert systems technology.

An initial method proposed for V&V of expert systems comprises requirements definition, verification, test case preparation, test execution, and evaluation.

V&V should be conducted on a sample of expert systems in order to refine and develop confidence in the proposed methodology. Candidate expert systems for experimental V&V are those where improved customer confidence in the expert system would be beneficial and problems of logistics in validation and evaluation are tractable.

Verification and Validation

Verification and validation (V&V) comprise formal tests of software conducted to determine whether:

> Each level of specification and the deliverable code are *traceable* to a superior specification; that is, the specification or code fully and exclusively implements the requirements of the superior specification. This process of determining traceability is called *verification.*
>
> The deliverable code correctly implements the original user requirements. The process by which the deliverable code is directly shown to satisfy (or fail to satisfy) user requirements is called *validation.*

In many cases, verification is a "paper" activity in which specifications are read, compared, and cross-referenced, while validation is a "live" activity in which the software is tested in both contrived and operational conditions. Verification and validation are effectively employed together, because each is effective at detecting errors that the other is ill-equipped to detect.[1] The current state of V&V practice is summarized by Deutsch.[2]

As in any product assurance activity, independence from the developing activity is essential in ensuring diligent and unbiased performance of V&V.[1,3] Software development contractors may employ an independent organization within the firm, or may employ an outside firm to perform V&V. The use of an outside firm to perform V&V is known as *Independent Verification and Validation* (IV&V). IV&V is often mandated by procuring agencies for mission-critical computer programs.

The Problem with Expert Systems

V&V as understood by its practitioners today cannot be used to determine whether an expert system computer program is correct. Because of this, many organizations that might otherwise consider applying expert systems to their mission are forced to employ more costly and less capable conventional software for mission-critical functions. Let us examine some of the difficulties in applying V&V to expert systems.

> Expert system software requirements specifications are often nonexistent, imprecise, or rapidly changing.[4] Expert systems are often procured in situations where the user does not fully understand his own needs. Some procurements omit requirements specifications as too constraining or not cost-effective. When expert systems are built by refinement and customer interaction, requirements may change rapidly or go unrecorded.
>
> The success of verification demands that the requirements of the superior specification be at least recognizable in the subordinate specification; if this is not so, then requirements tracing is futile. Expert systems are typically developed from a system specification (Type A) or an informal specification by prototyping and refinement. Intermediate specifications (Type B5 and C5) are either not produced, not precise enough, or too subject to change to serve in verification.
>
> Even if adequate specifications for requirements tracing were available, it is doubtful that conventional verification would yield many answers concerning whether the implemented system indeed satisfied the requirements. The ability of the expert system to produce the desired outputs is often an emergent property of the interaction between its knowledge base and an inference engine.

Conventional validation demands precise test procedures. As long as reasonably precise requirements and design specifications can be obtained, test procedure preparation should be of no greater difficulty than for conventional software. When requirements and design information is unavailable, imprecise, or changing, test procedure design becomes a matter of guesswork.

There is no widely accepted, reliable method for evaluating the results of tests of expert systems. The approach of having human experts in the domain of the expert system evaluate the results has numerous drawbacks.[5] There may be no expert available, or the expert may not be independent when independent evaluation is needed. Human experts may be prejudiced or parochial. The problem for which the expert system was written may be one that no human can solve reliably or efficiently.

The difficulties of applying V&V to expert systems have caused organizations to refrain from requiring V&V in expert system procurements. If V&V were applied to expert systems, the following benefits would ensue:

Expert systems would be fielded with less risk of software failure. This would promote the use of expert systems technology in mission critical systems.

Organizations wary of expert systems because of the lack of V&V would be more inclined to employ this new and desirable technology.

Experimental application of V&V to expert systems would permit the development of effective V&V methodologies for expert systems.

The lack of understanding of how to perform V&V of expert systems creates a vicious circle (Figure 1): V&V of expert systems is not done because nobody requires it. Nobody requires V&V of expert systems because nobody knows how. Nobody knows how to do V&V of expert systems because nobody has done it.

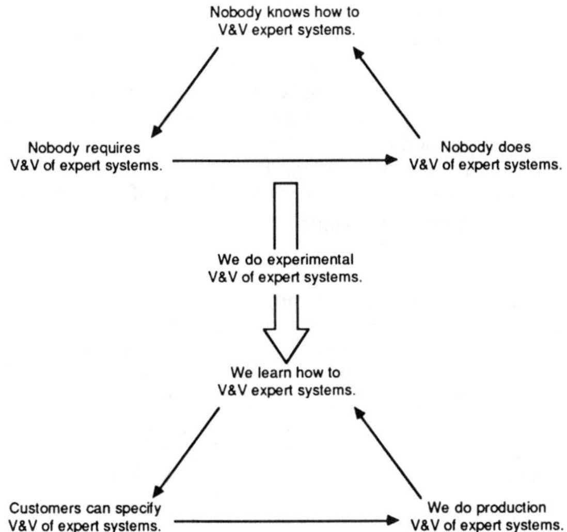

Figure 1. The vicious circle that impedes V&V of expert systems (and how to break it).

What is needed to break the circle and make progress toward an effective V&V methodology for expert systems is experimentation with application of V&V techniques to expert systems (Figure 1). The remainder of this paper deals with the selection of appropriate problems for experimental V&V and methods that may find applicability in the V&V of expert systems.

Efforts to define the problem of testing expert systems and to define test tools have been few, but encouraging. Buchanan and others discuss extensively what testing is appropriate for expert systems.[6,7,8] Scambos[8] and Bliss and others[9] have demonstrated test tools for evaluating expert systems. Scambos[8] presents a tool for driving the expert system FRESH with previously generated scenarios. Bliss and others[9] present a methodology for verifying expert systems based on measures of complexity and correctness. It is unlikely, however, that the answer to the problem of testing expert systems lies in a single tool or single method: even for conventional software, an individual method may only find 30 to 70 percent of the defects in a program.[1] The solution lies rather in a methodology that exploits as many independent methods of defect detection as are practical and suitable.

A Verification and Validation Method for Expert Systems

The shortage of practical and theoretical knowledge on the subject of V&V of expert systems suggests that a "cut-and-try" approach to the selection of appropriate methods will be most fruitful. The first-cut method for V&V of expert systems proposed here comprises five tasks: define requirements, verify the knowledge base and supporting software, prepare test cases, execute the tests, and evaluate the results.

Requirements Definition

Often the development of expert systems proceeds without the formal definition of software requirements. This may happen for a number of reasons:

The user and the developer plan to begin prototyping from a statement of purpose and refine the requirements during prototyping.

The user or the developer is unwilling to commit the resources necessary to prepare and maintain a requirements specification.

The developer believes that imposing detailed, documented requirements would unduly constrain the developer's creativity.

The user does not plan to perform a formal acceptance test.

If a software requirements specification is written and maintained throughout the development of an expert system, the job of V&V is made easier. But if there is no requirements specification, or if the requirements are vague, obsolete, or untestable, then V&V must start with the definition of requirements. There are several useful standards for requirements specifications: MIL standard 483[10] and DOD standard 2167[11] for military projects; ANSI/IEEE standard 830-1984[12] for commercial projects.

When development has proceeded without a usable requirements specification, the V&V practitioner is in the unenviable position of trying to set down requirements for software that has been written. There may be political pressure to conform the requirements to the software, or the user or developer may be uninterested in stating the requirements. The V&V practitioner needs the organizational authority to set the requirements over the developer's objections and then must use this authority to obtain concurrence among the user, the developer, and V&V.

The quality of any requirements specification, whether for expert systems or for conventional software, depends on certain characteristics of the document itself and the requirements stated therein. A good requirements specification is unambiguous, complete, verifiable, consistent, modifiable, traceable, and usable in operations and maintenance. A fuller discussion of the characteristics of a good requirements specification may be found in ANSI/IEEE 830-1984.[12]

The requirements specification appropriate to an expert system may be expected to differ from the requirements specification of a conventional computer program. Some of the differences to be expected are:

> A conventional requirements specification divides the computer program into functions, then describes each function in terms of its inputs, processing, and outputs. This architecture is often satisfactory to describe a system of objects and methods, or when a rule system is controlled by rule classes, frames, or metarules. It may be unsatisfactory when "any rule can fire at any time" or "blackboard" architectures are planned. It should always be possible to define the inputs and outputs of the expert system as a whole, although the inputs and outputs will consist mainly of symbols and text rather than numeric data.
>
> Most expert systems include a knowledge base that may itself be the subject of requirements. If the knowledge base consists of objects with methods, or rules that cause actions or assert facts, then it may be possible to define knowledge base requirements in input/processing/output terms. If the knowledge base consists mainly of regularly structured data, then data base definition models such as the one in MIL-STD-483 may be employed. Many knowledge bases will fit neither of these criteria, in which case the requirements writer will have to be content with defining what knowledge must appear.

Numerous iterations of writing requirements, developer review, and customer review are necessary to get a requirements specification acceptable to all parties. Once a usable requirements specification is in hand, the remaining tasks of verification and validation can begin.

Verification
Verification of conventional software comprises the tracing of requirements through the hierarchy of specifications, design, and code and such engineering analyses as are appropriate to demonstrate the suitability of the design. The requirements trace encompasses the system and software requirements specifications, the design specification, and the source code. Engineering analyses regularly performed cover criticality, sensitivity and stability, efficiency, and maintainability. We consider here how these activities might be performed in the verification of expert systems.

Requirements Tracing
Requirements tracing is the task of determining whether all of the requirements of a superior specification are met in the subordinate specification and whether the subordinate specification introduces any extraneous requirements. Requirements tracing is used to assure that errors, omissions, and "free features" do not intrude in passing from one phase of development to the next.

If a requirements specification is available for the expert system, requirements tracing is just as practical as it is for conventional software. If there is no design document, then the requirements specification may be traced directly to the source code. The well-written expert system is sufficiently self-documenting that the design document is of reduced importance. The requirements tracer should remember that because expert systems programming is declarative rather than procedural the processing done to implement a requirement may be diffuse and troublesome to identify.

Engineering Analyses
Engineering analyses are often conducted as part of verification. These analyses are used to predict whether the system will work as intended before effort is expended in coding and testing. Several types of engineering analyses may be conducted, depending on the needs of the particular project:

> Criticality: What would be the impact of failure of this software on other components or on the mission?
>
> Sensitivity and Stability: How sensitive to minor variations in input are the algorithms and heuristics employed, particularly around singular points and operating limits? Could any such sensitivity result in inaccuracy, instability, or failure? Numerical analysis is often employed in judging the sensitivity of an algorithm.
>
> Efficiency: Can the software do the job intended within the given limits of time and resources?
>
> Maintainability: Is the software written so as to facilitate maintenance over its life cycle?

All of the above analyses are generally applicable to expert systems. The importance of criticality analysis should be obvious. Sensitivity and stability are particularly important in expert systems because of the possibility that an unidentified sensitivity could lead to a catastrophically wrong result.[7] Especially as expert systems become targeted for real-time applications, efficiency analysis will become important.[4,7] Maintainability is of great concern for expert systems since many are intended to be flexible in meeting new objectives or incorporating new knowledge. Froscher and Jacob have made progress in defining the characteristics of expert system software that promote maintainability.[13,14]

Three engineering analyses not employed with conventional software become important in the verification of expert systems. These are interaction analysis, truth analysis, and uncertainty analysis. Interaction analysis deals with the ability of the interacting components of the expert system to produce the desired results. Truth analysis evaluates the truth of the knowledge base contents. Uncertainty analysis is applicable when the expert system makes use of measures of uncertainty. It determines whether the uncertainties assigned and the method of combining uncertainties yield correct results.

The structure commonly employed for expert systems makes interaction analysis important.[15] Expert systems often comprise a number of small and quasi-independent modules (whether rules or objects), and the intended behavior of these systems results from the interaction of these modules under the control of an inference engine. Unlike conventional, procedural computer programs, the sequence and results of processing are not readily apparent. Interaction analysis strives to predict the results of interaction of modules and the inference engine, with particular attention paid to the possibility of unintended or deleterious results arising.

Truth analysis is employed to evaluate the truth of facts, rules, etc. stored in the knowledge base. Since expert systems encode substantially more knowledge than do conventional computer programs, verifying the truth of the knowledge stored is proportionately more important. Truth analysis may be performed by comparing the knowledge base contents to published data or by review of the knowledge base by an independent expert.

Certain expert systems tools support multiple "viewpoints" (ART) or "worlds" (KEE 3.0) for hypothesis testing and "what if" analysis. The intent of such tools is to permit the making of hypotheses that are not subject to objective truth analysis. Of greater concern in these systems is the system of truth maintenance: the ability of the expert system to detect an internally inconsistent world. Truth analysis of expert systems that employ multiple worlds therefore must also consider whether the truth maintenance system employed is correct, appropriate to the application, and sufficiently robust.

Uncertainty analysis is required when the expert system employs measures of uncertainty. It addresses the following questions:

> Is the system of uncertainty consistent with the way uncertainty is expressed in the domain, and are the assumptions made valid for the domain? For instance, Bayes' Theorem is appropriate if the expression of uncertainty is a probability, but not if it is a measure of belief.[16]
>
> Are the rules used to combine uncertainty monotonic, commutative, and associative? When uncertain evidence is combined, is the result of the combination credible? Is the cutoff for drawing a conclusion appropriate?
>
> Are the uncertainties assigned to facts and assertions reasonable, and do they give reasonable results when combined?

Test Case Preparation

Selection of test cases for expert systems should be similar in principle and no more difficult than for conventional computer programs. Myers[1] covers the subject of test case design for conventional software in detail. The criteria for selecting test cases for expert systems may be summarized as follows:

> Test every requirement. For each requirement, there must be a test case capable of showing that the requirement has not been met. (This is why there must be a requirements specification that contains only testable requirements.)
>
> A string of test cases that repeatedly pushes the system against one of its limits or causes error-recovery or housekeeping activities to be invoked is often more effective at revealing errors than a set of independent test cases.
>
> Test every item of code and every possible decision.[1] Equivalence partitioning may be used to keep the number of test cases reasonable. For expert systems, this means that every fact, object, rule, etc. must be invoked, every line of every method must be run, and every possible outcome of each rule must be caused. Particularly in expert systems, where the flow of processing is not always clear, diligent study may be required to determine how to sensitize a particular decision.
>
> In expert systems it is necessary to test both the outputs of the system and the process by which the system arrived at those outputs. The ability of an expert system to justify its conclusions is considered at least as important as the accuracy of the conclusions.[4,7] Most expert systems provide audit trails or "WHY" or "HOW" facilities. Test the products of these facilities.
>
> Test every singular point and boundary condition.[1] Test cases just below, right on, and just above a boundary are needed to characterize behavior around the boundary.
>
> Test a judicious selection of combinations of conditions.[1] (For any nontrivial system, it is impractical to test all combinations of inputs.) Select those believed to be particularly significant in operation or that have the best chance of revealing errors.
>
> For mission-critical software, or when the test budget allows, it is important to test beyond the required or designed limits to identify the points at which the system degrades or fails.[4]

Test Execution

Execution of tests of expert systems is similar to execution of tests of conventional software. An operator conducts the tests from a script prepared in the test case preparation phase, and any deviations from, or clarifications to, the script are recorded. It has proven profitable to conduct tests of conventional software using "script" and "log" files that provide test inputs and record test results automatically. Encouraging results from scenario-based testing of expert systems have been obtained.[8]

Evaluation

Most of the controversy over the verification and validation of expert systems is due to the evaluation phase of testing. Some of the problems with evaluation of expert systems are:

> The objective and exact expected results and evaluation criteria employed with conventional software testing may not be applicable. In problems to which expert systems are applied, there may be no single best answer; there may instead be a set of acceptable answers, or there may be no way to determine *a priori* what is an acceptable answer.[7]
>
> There may be no agreement as to what are acceptable results. Tests of the expert system MYCIN employed a number of distinguished physicians as evaluators.[5] Even the elite evaluators exhibited:

Prejudice: some evaluators refused to consider the possibility that the expert system's answers were as good as theirs. Blind evaluation was used to overcome this problem.

Parochialism: evaluators in different regions of the country applied different criteria and different bodies of knowledge.

Inconsistency: some evaluators rated answers substantially identical to their own as unacceptable.

Especially when the problem domain involves reasoning with uncertainty, the expert system may not give results that are one hundred percent correct. The best expert systems have typically done about as well as the best human experts at solving problems involving uncertainty.[5] When neither the expert system nor the human expert can give perfect answers, what is the acceptable rate of error?

Evaluating an expert system is to evaluating conventional software as grading an essay examination is to grading a true-false examination. Tests of conventional software yield true-false results: either the expected result appears, or it does not. Tests of expert systems yield more complex results. There may be more than one acceptable answer, or there may be more than one way of stating the answer. Whether an answer is acceptable may be a matter of opinion. Often the process by which the expert system arrived at the answer is of interest. Finally it may be impossible for the expert system or the human evaluators to provide a perfect answer every time.

The need for independent evaluation is plain. If experts employed in developing the expert system are employed in its evaluation, then their evaluations are neither statistically nor psychologically unbiased. Independent evaluation may be achieved in a number of ways:

Independent experts may be available within the organization.

Another organization may be able to provide independent experts.

A V&V organization may be able to provide independent expertise through research or by hiring consultants.

The factors considered in evaluation of an expert system are more diverse than those considered for conventional software. The evaluation of an expert system should take into account:

Whether the expected result, an equivalent result, or an equally acceptable result is obtained.

If the expert system is expected to justify its results, whether the justification provided is of the quality and form needed by the customer.

Whether the inference process followed by the expert system is sufficiently complete and robust, and whether the knowledge employed is of sufficient depth, to assure customer confidence in the expert system.

If the expert system is not expected to achieve one hundred percent correct results, what is the impact of incorrect results: are the incorrect results merely suboptimal, bad enough to render the expert system ineffective, or actually damaging to the mission the expert system supports? If incorrect results are to be tolerated, then what fraction of incorrect results is acceptable, and what is the worst acceptable consequence of an incorrect result?

Evaluation of an expert system is expected to be more difficult and more costly than evaluation of conventional software. Fortunately evaluation plans can be worked out in the systems analysis phase of development[15], or at the latest during test planning, so the cost and difficulty of evaluation will be known going in. The products of evaluation are likely to be more complex than a true-false decision and will require careful consideration by project management and the customer before proceeding with a decision to field the expert system.

Experimental Verification and Validation

The development of a methodology for the verification and validation of expert systems will require experimentation. Experimentation will permit the refinement of the techniques employed to verify and test expert systems and will increase confidence in their effectiveness. Some criteria for the selection of expert systems for experimental V&V are:

V&V should enhance the customer's confidence in the completed expert system.

The expert system should address a problem domain in which conventional software is normally subjected to V&V, e.g. domains employing mission-critical computer systems.

The expert system should pose challenges in several aspects of the proposed methodology: V&V should need to address at a minimum requirements tracing, criticality, sensitivity, interaction, and truth analyses, and tests of processing, outputs, and explanation facilities.

Several expert systems developed by different organizations should be subjected to V&V. Lessons learned in V&V of one system should be applied to the next.

The transition from experimental to production V&V of expert systems will occur when the experimental V&V methodology has demonstrated its ability to:

Detect errors in all phases of its application and across a spectrum of expert systems.

Assure the correctness of an expert system that has passed V&V.

Contribute to an overall reduction in life-cycle cost of expert systems software.

The ability to verify and validate expert systems will contribute substantially to the acceptance of expert systems technology.

References

[1] Myers, G. J., *The Art of Software Testing*. John Wiley & Sons, New York, NY. 1979.

[2] Deutsch, M. S., *Software Verification and Validation: Realistic Project Approaches*. Prentice-Hall, Englewood Cliffs, NJ. 1982.

[3] DeMarco, T., *Controlling Software Projects: Management, Measurement, and Estimation*. Yourdon Press, New York, NY. 1982.

[4] Lane, N. E., "Global Issues in Evaluation of Expert Systems". In *Proceedings of the 1986 IEEE International Conference on Systems, Man, and Cybernetics*, pp. 121-125. IEEE, Piscataway, NJ. 1986.

[5] Yu, V. L., L. M. Fagan, S. W. Bennett, W. J. Clancey, A. C. Scott, J. F. Hannigan, B. G. Buchanan, and S. N. Cohen, "An Evaluation of MYCIN's Advice". In *Rule-Based Expert Systems* (Buchanan, B. G. and E. H. Shortliffe, eds.), pp. 589-596. Addison-Wesley, Reading, MA. 1984.

[6] Buchanan, B. G. and E. H. Shortliffe, "The Problem of Evaluation". In *Rule-Based Expert Systems* (Buchanan, B. G. and E. H. Shortliffe, eds.), pp. 571-588. Addison-Wesley, Reading, MA. 1984.

[7] Gaschnig, J., P. Klahr, H. Pople, E. Shortliffe, and A. Terry, "Evaluation of Expert Systems: Issues and Case Studies". In *Building Expert Systems* (Hayes-Roth, F., D. A. Waterman, and D. B. Lenat, eds.). Addison-Wesley, Reading, MA. 1983.

[8] Scambos, E. T., "A Scenario-Based Test Tool for Examining Expert Systems". In *Proceedings of the 1986 IEEE International Conference on Systems, Man, and Cybernetics*, pp. 131-135. IEEE, Piscataway, NJ. 1986.

[9] Bliss, J., P. Feld, and R. E. Hayes, "Decision Aid Measurement and Evaluation (DAME)". In *Proceedings of the 1986 IEEE International Conference on Systems, Man, and Cybernetics*, pp. 126-130. IEEE, Piscataway, NJ. 1986.

[10] MIL-STD-483, *Military Standard: Configuration Management Practices for Systems, Equipment, Munitions, and Computer Programs*. U. S. Government Printing Office, Washington, DC. 1970.

[11] DOD-STD-2167, *Military Standard: Defense System Software Development*. U. S. Government Printing Office, Washington, DC. 1985.

[12] ANSI/IEEE Standard 830-1984, *IEEE Guide to Software Requirements Specifications*. IEEE Standards Board, New York, NY. 1984.

[13] Froscher, J. N. and R. J. K. Jacob, "Designing Expert Systems for Ease of Change". In *IEEE Proceedings of the Expert Systems in Government Symposium*, pp. 246-251.

[14] Jacob, R. J. K. and J. N. Froscher, "A Software Engineering Methodology for Facilitating Ease of Change in Rule-Based Systems". In *Expert Systems: The User Interface* (J. A. Hendler, ed.). Ablex, Norwood, NJ. 1987.

[15] Buchanan, B. G., D. Barstow, R. Bechtal, J. Bennett, W. Clancey, C. Kulikowski, T. Mitchell, and D. A. Waterman, "Constructing an Expert System". In *Building Expert Systems* (Hayes-Roth, F., D. A. Waterman, and D. B. Lenat, eds.). Addison-Wesley, Reading, MA. 1983.

[16] Shortliffe, E. H. and B. G. Buchanan, "A Model of Inexact Reasoning in Medicine". In *Rule-Based Expert Systems* (Buchanan, B. G. and E. H. Shortliffe, eds.), pp. 233-262. Addison-Wesley, Reading, MA. 1984.

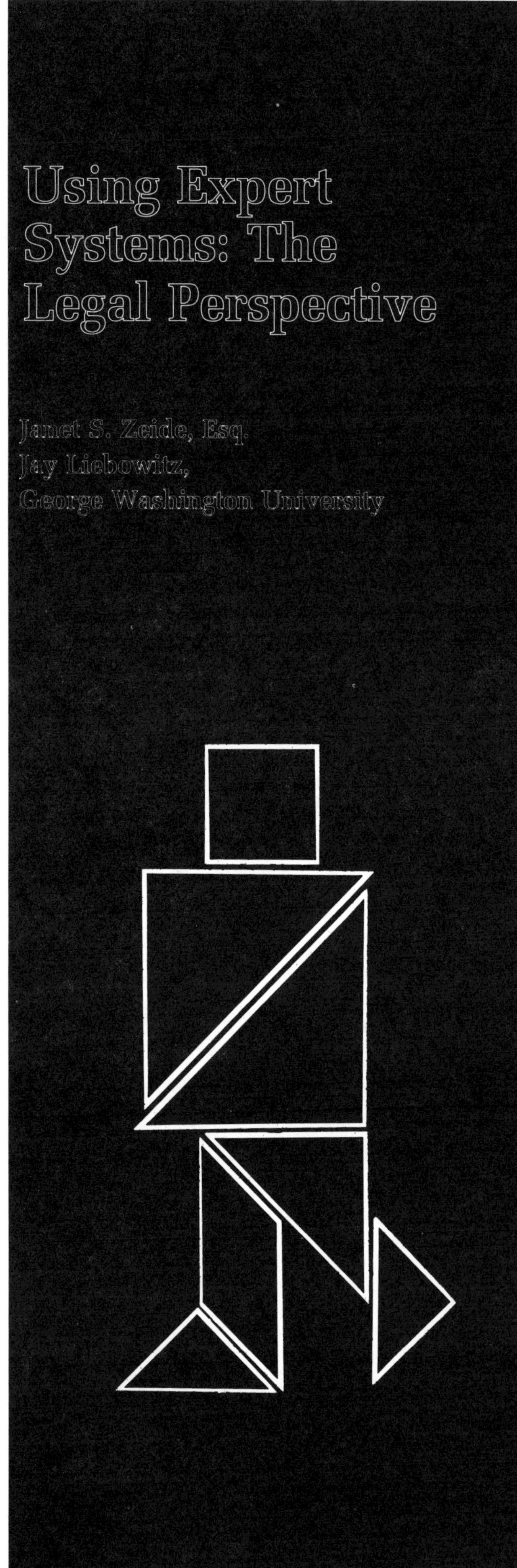

Using Expert Systems: The Legal Perspective

Janet S. Zeide, Esq.
Jay Liebowitz,
George Washington University

Automation has brought many benefits, and concomitant problems. Consequently, our courts now face a dilemma: How should they handle lawsuits arising from computer program errors? As expert systems proliferate, especially in medicine and business, the potential for litigation increases proportionately. Defective computer programs can cause serious injury. Controversy over expert system use, misuse, and nonuse is inevitable.

Computer programs causing personal or economic injury present novel causes of action. The courts can apply no single legal theory. Traditionally, manufacturers are held responsible under the theory of strict liability when their products cause injury. However, when persons offering services cause harm in performing those services, they are charged with negligence.

The debate begins when one considers whether computer programs are products or services; they have qualities of both. The courts treat liability for injury from products differently than they treat liability for injury from services. And public policy considerations factor heavily into this debate.

Negligence

"Negligence" and "products liability" are legal terms that fit under the heading of torts (wrongful acts subject to civil action). Under negligence theory, given the circumstances of each case, the courts consider reasonableness of conduct or care. Injured users of services, suing under theories of negligence, must prove they were owed a duty of care; that is, that performance of the service fell below the standard of care (or that reasonable care could have prevented the injury and was not used), and that damages provide appropriate compensation.

To compensate for loss or injury resulting from computer program error, consumers would have to prove computer programmers were negligent in creating or copying programs. A plaintiff consumer might have to specify the very mistake among thousands of information bits in the program, and then prove that the injury was reasonably foreseeable.[1]

Since proving such negligence could be difficult, it might be hard for plaintiffs to recover for injuries caused by computer program failures. Public policy considerations make strict products liability more desirable because it eliminates the need to prove fault.

Products liability

Strict liability deals with unreasonably dangerous or hazardous products, regardless of conduct or care. The Restatement (Second) of Torts, section 402 (1965), defines strict products liability as follows:

Special liability of seller of product for physical harm to user or consumer

1. One who sells any product in a defective condition unreasonably dangerous to the user or consumer or to

his property is subject to liability for physical harm thereby caused to the ultimate user or consumer, or to his property if: (a) the seller is engaged in the business of selling such a product, and (b) it is expected to and does reach the user or consumer without substantial change in the condition in which it is sold.

2. The rule stated in Subsection 1 applies although (a) the seller has exercised all possible care in the preparation and sale of his product, and (b) the user or consumer has not bought the product from or entered into any contractual relation with the seller.

Under strict products liability, manufacturers of unreasonably dangerous products are liable for injuries caused by their products regardless of reasonableness of care. Policy considerations for holding manufacturers to strict liability are partly based on manufacturers being in a better position than consumers to test for and identify problems, and to absorb or spread losses. The law expects assembly line products to have uniform quality and design over which the user has no control, but over which the manufacturer does. If products are defective and unreasonably dangerous, then the court applies strict liability to manufacturers even if reasonable care was used in production and even if no amount of care in production could have prevented the injury.

Product vs. service

Strict liability applies only to defective products. Products are usually viewed as tangible items, with set monetary values, that can be owned.[2] Computer programs have an intangible nature but are sold in physical (tangible) form.

Computer programs, consisting of electronic impulses, are not tangible; a program's single tangible aspect is the punch card, magnetic tape, or magnetic disk that stores and transports the program.[3] Furthermore, only the underlying ideas translated by a computer have real value.[4] It is interesting to note that ideas, not being concrete forms, cannot be copyrighted. But computer programs can be copyrighted.

Some courts, treating computer programs as intangibles, have scrutinized them under negligence theories (in the few cases that have arisen). Tax courts facing the question of computer program tangibility have arrived at different findings: Several have ruled that software is not taxable as personal property because of the program's intangible nature. Others have held that the tangible output is taxable even though the program itself may not be. One ruling maintained that a computer program should be taxable for its full value even though its nature is intangible, and even though it is recorded on relatively inexpensive material.[5]

Some intangible properties are not normally considered products capable of inflicting injury (licenses, for instance).[4] However, strict liability has been applied to other intangible properties (leases of real and personal property, energy, and certain types of service transactions such as applying a defective hair treatment).[6] Manufacturers, realizing the enormous responsibility they bear for their computer programs, have attached disclaimers to products to lessen their liability. The legality of disclaimers, so far uncertain, centers around the product vs. service question in terms of applying the Uniform Commercial Code.

Expert systems

Expert systems haven't experienced much legal trouble yet. But great litigious potential exists over their use, misuse, and nonuse (potential that will soon be realized).

A lawsuit could arise if expert systems are consulted and fail to perform correctly, or give inaccurate or misleading answers leading to injury. Such problems could be caused by hardware malfunction, knowledge base or program errors, unintended use, misuse, or undue reliance on expert systems.[7]

Someone unfamiliar with a program's subject matter can misuse an expert system, leading to legal action. Conversely, a professional who consults expert systems as standard practice could be liable for the nonuse of an expert system; that is, it may be "just as bad not to use available technology as it is to rely on a system when you should have used your own brain."[8] Such situations, if not current problems, are not far off. As Gemignani asserts, for "certain sensitive, delicate or hazardous tasks [such as aircraft requiring fast and accurate response beyond human capability], it may be unreasonable not to rely upon an expert system."[9]

Expert systems such as MYCIN (medical), Prospector (mineral mining), ACE (telephone cable maintenance), and R1/XCON (computer configuration) are already used daily. These systems have succeeded, in some cases proving more accurate than human specialists. Moreover, they store more data and transmit that data faster than human experts can.[7]

Applications. Nonexperts can misuse expert systems, and systems can be inaccurate in particular situations. For example, a family using a tax preparation program may take a home office deduction for which they are ineligible, and the expert system will not spot the oversight. Later, when the IRS discovers the mistake, the family will have to pay additional taxes, plus interest and penalties. A tax accountant would probably have found the error and alerted the family.[10] Indeed, accountants would be negligent in performing their service if they failed to warn clients of such errors. But who is at fault when an expert system fails?

Assuming that appropriate expert systems are available at an accessible price, consider the following situations.

- **Scenario 1:** Patient B describes her symptoms to Dr. Y. Dr. Y treats Patient B without consulting an expert system although one was available. Dr. Y misdiagnoses and incorrectly treats Patient B. Patient B dies.

- **Scenario 2:** Patient A describes his symptoms to Dr. X. Dr. X consults an expert system and treats Patient A based on the expert system's diagnosis and treatment recommendation. But the expert system has erred, and Patient A dies.

In both scenarios, the doctors can be sued for medical malpractice. But beyond that are questions of who is to blame, and how plaintiffs can recover when computers err. Are expert systems products or services? Which theory of liability applies?

Such questions resist quick answers, because expert systems are both products and services. As with other computer programs, they are tangible since they are sold, owned, and transported on disks and tape. On the other hand, they can be "individually tailored to perform a specific task" like services, rather than mass-produced products.[7]

When expert systems perform like human experts, and cause injury, the courts must trace liability back to a human source, even if no direct or nearly direct human negligence caused the injury. The law on expert systems will be a long time evolving due to the frequent advent of newer and more efficient knowledge-based technology, the growing prevalence of expert systems, and their various applications.

While expert telecommunication systems may cause little physical harm to people, errors can cost millions of dollars. Poorly designed satellites, or improper frequency-spectrum management caused by expert systems, have profound financial and legal implications. Moral and ethical issues will also be raised in cases where expert systems could have been consulted but were not. Conversely, sole reliance on expert systems will evoke ethical considerations if misdiagnosis occurs.

Since teams of experts and programmers produce expert systems, it is difficult to prove negligence by any one person, leaving injured users out in the cold. However, viewing programs as products eliminates the need to prove negligence. In light of public policy considerations and since courts tend to view expert systems as unreasonably dangerous products (especially in fields like medicine, finance, and space), the law will probably hold manufacturers responsible under strict products liability for injuries caused by system defects.

As knowledge-based technology becomes more commonplace, public expectations of expert system capabilities must be put into proper scope. Otherwise, exaggerated expectations will create an even more litigious society.

References

1. "Computer Software and Strict Liability," *San Diego Law Rev.*, Vol. 20, No. 2, Mar. 1983, p. 441.
2. *Corpus Juris Secundum*, Vol. 73, Property Section 1, 1951.
3. R. Ducker, "Liability for Computer Software,"*Business Law*, Vol. 26, 1971, p. 1081.
4. V.M. Brannigan and R.E. Dayhoff, "Liability for Personal Injuries Caused by Defective Medical Computer Programs," *Am. J. Law and Medicine*, Vol. 7, 1981, p. 123.
5. Greyhound Computer Corp. v. State Dept. of Assessments and Taxation, *Atlantic Reporter*, Vol. 320, 1974, p. 52.
6. J. Prince, "Negligence: Liability for Defective Software," *Oklahoma Law Rev.*,Vol. 33, 1980, p. 848.
7. S.H. Nycum et al., "Artificial Intelligence and Certain Resulting Legal Issues," *The Computer Lawyer*, May 1985, p. 1-10.
8. "Artificial Intelligence Goes to Court," *IEEE Expert*, Vol. 1, No. 2, Summer 1986, p. 101.
9. M. Gemignani, "Laying Down the Law to Robots," *San Diego Law Rev.*,Vol. 21, 1984, p. 1045.
10. L. J. Kutten, "Are Expert Systems More Trouble Than They Are Worth?", *Computerworld*, May 20, 1985, p. 72.

Janet S. Zelde, an attorney and a member of the Maryland and District of Columbia bars, also teaches broadcast journalism at Montgomery College in Rockville, Maryland. She has served on the legal staff at the FCC, in the National Association of Broadcasters' legal department, and has worked at WCBS-TV, WGBB radio, and other radio stations and law firms. She received her BA from McGill University, and her JD from the Nova University Law Center in Fort Lauderdale, Florida.

Currently, she is assisting in the development of Evident, an expert system used for determining admissibility of evidence under federal rules. She serves as legislative policy and focus coeditor for the *International Journal of Telematics and Informatics*, and also writes for *Broadcasting and the Law*.

Jay Liebowitz is an assistant professor of management science at George Washington University. He has worked in expert systems technology at the NASA Goddard Space Flight Center, American Management Systems, and currently works at the Navy Center for Applied Research in AI. He received his BBA, MBA, and D.Sc. from George Washington University. His research interests include expert systems for software engineering, law, and telecommunications.

Before joining George Washington University, he worked for the Computer Sciences Corporation; Peat, Marwick, Mitchell, and Company; and the Program Analysis and Evaluation Division at the Pentagon. He is associate editor of the *International Journal of Telematics and Informatics*, and is on the editorial advisory boards of the *International Journal of Robotics and Computer-Integrated Manufacturing*, the *International Journal of Computers and Operations Research*, and the *International Journal of Computers and Industrial Engineering*. He is a member of Sigma Xi, Tau Beta Pi, Omega Rho, Omicron Delta Kappa, the AAAI, the IEEE, the ACM, and WORMSC.

The authors can be reached at the Department of Management Science, George Washington University, Washington, DC 20052.

Special Feature

Teaching an Introductory Course in Expert Systems

A. Terry Bahill and William R. Ferrell
University of Arizona

In the fall of 1985, 25 students and 10 auditors (including three faculty members) attended our new course on expert systems—an introductory engineering course aimed at teaching useful technical skills and applying those skills to real problems. The course had three objectives: to share understanding, to impart techniques, and to apply these in practice. We wanted our students

(1) To understand the nature, limitations, and suitable applications of expert systems;

(2) To effectively use expert system shells (software packages allowing rapid prototyping of expert systems, shells can be considered special-purpose high-level languages designed to help write if-then production rules); and

(3) To produce a small expert system.

In these objectives, the course differed markedly from many other AI and expert system courses. Other courses have attempted, variously, either

(1) To teach the design and programming of inference engines;

(2) To provide an understanding of heuristic programming methods;

(3) To survey AI research;

(4) To explore AI and expert system applications in specific problem domains; or

(5) To teach programming methods or languages especially suited to knowledge-based systems.

The teaching experience convinced us that students can learn, in a one-semester course, to apply effective fundamental skills in expert systems development—skills providing a sound basis for further growth. But we also found, from observing student attempts at applications, that the course needed improvement in imparting technique and in developing the understanding needed for technical application. Consequently, we have revised the course for the fall of 1986.

We shall first present what we did, then discuss the students' projects, and finally describe the changes needed in light of those projects.

The course

We used Randy Davis videotapes[1] for a broad AI overview, Forsyth's edited volume on expert systems,[2] and an expert systems textbook by Harmon and King.[3] We bought the M.1 shell instructor's package from Teknowledge including 10 copies of M.1, overhead transparencies, and lecture notes. The package, while expensive ($5000), was worthwhile.

Building a knowledge-based system using Pascal, Prolog, or other general-purpose, high-level language is too big a task for a one-semester course. And using a big knowledge-engineering tool such as KEE or ART is too complicated for an introductory course. We concluded that the best way to teach expert systems in a one-semester, introductory course would be to have students build an expert system on a PC using an expert system shell.

Reprinted from *IEEE Expert,* Winter 1986, pages 59-63.

This paper has been translated into Chinese and republished in *Microcomputer Application Technology,* Volume 17, Number 1, 1988, pages 38-43.

EH0303-8/90/0000/0454$01.00 © 1986 IEEE

The class discussed several recent papers comparing currently available expert system shells that run on PCs.[3-6] Comparing and contrasting these PC shells, and examining such historically famous expert systems as MYCIN,[7] was an important part of the course.

Shells we considered for course use

Even back in early 1985, when we were planning this course, there were scores of commercially available expert systems shells. We purchased three—two for the VAX, and one for PCs.

The Rand Corporation's ROSIE,[8] though several years old, is not heavily used. It was inexpensive ($200), ran on a VAX, and needed InterLisp (although the latest version runs with portable standard Lisp). We abandoned ROSIE because it was big, cumbersome, and slow.

M.1 from Teknowledge, a backward-chaining shell for making expert systems on an IBM-compatible PC, was easy to learn and easy to produce user-friendly systems with. M.1 has an effective system for dealing with uncertainty, many automatic features (such as formulating its own questions when they are not specified by the developer), good trace capability, and its knowledge base can be written without complicated formats. We did not buy other shells similar in power and features, such as Personal Consultant Plus, because we could not afford two tools from the same class.

OPS5, the expert system shell supported by DEC, is a forward-chaining system suitable for different problems than are the backward-chaining M.1 and ROSIE—forward chainers work *toward* a goal state, while backward chainers work *from* a goal state. DEC has used OPS5 in-house for dozens of projects, although soon DEC will also support S.1 from Teknowledge. There is a textbook explaining OPS5,[9] and the shell is available for PCs and the VAX for between $75 and $3000. We found OPS5 was difficult for students to use and it was difficult for them to make user friendly systems with.

After completing the course, we discovered Micro Expert from McGraw Hill—costing $50.[10] It could serve as the basis for a low-budget expert systems course.

Student projects

We required students, individually or in pairs, to find an expert and (using an expert system shell) to make an expert system. Their experts formed a motley group—professors, physicians, siblings, roommates, parents, teachers, and user consultants. The availability of experts was a prime factor determining expert system topics, although our suggestions on their one-page proposals had some effect.

Most students enjoyed their projects. Several have continued working while enrolled for independent study, which brings up a disappointing aspect of Teknowledge's license agreement: Students cannot take a working version of their expert system with them. They are allowed to use M.1 only while enrolled in a course.

From these class projects, listed in the accompanying box, we gleaned clues about what problems are suitable for expert systems and what problems are best left for conventional computer programs. Some class problems were inappropriate for expert systems because the implementations were simple checklists or questionnaires—some because they involved numerical computation rather than symbolic manipulation. Others were inappropriate because no human expert could solve the problem, or because the solution could have been represented unequivocally as a decision tree in which users were shunted up different branches depending on what answer the system posed. For such cases, a diagram on paper would have been as effective as the computer system—a simple recipe for an answer.

How can one identify a task that is appropriate for an expert system? First, there must be a human expert who performs that task better than most other people. For example, Julia Child is an expert chef; most of us do not cook as well. Designing an expert system to add single digit numbers is silly, because almost everyone does this well. On the other hand, designing an expert system to predict the stock market is doomed to failure because no human expert does this consistently well (if someone does, they are just quietly salting away their millions). Figure 1 shows the type of performance histogram ideal for an expert system.

Second, the task's solution must be explainable in words rather than requiring explanatory pictures—imagine Michelangelo creating an expert system to direct the painting of the Sistine Chapel!

The constraint that experts not draw pictures leads to a third criterion—the telephone test: Can the problem be solved routinely in a 20-minute (or even a one-hour) telephone conversation with the expert? If so, the problem suits a PC-based expert system. If the problem takes a human two days to solve, however, then it's far too complicated for an expert system. And if a human can answer in two seconds, it's too simple.

Fourth, problems inviting one of many possible solutions are ideal candidates for expert systems—problems such as "What disease does the person have?"

Evaluating student projects was difficult. We could not afford to hire experts to do our grading or to test the "expert systems" on various realistic problems and compare their performance with the humans whose knowledge they were intended to capture. Furthermore, it would be unfair for two generalist professors of Systems and Industrial Engineering to judge student expert systems designed for use by specific-domain authorities. Therefore, our evaluations tended to emphasize (1) how much we could "fool" systems by plausible (but nonsensical) inputs, and (2) the difficulty we had using systems or making sense of their queries or output. We found a typical consultation's tree diagram greatly helped us to understand the knowledge base, and more fairly evaluate the expert systems.

It is also helpful having an overseer expert to critique systems. We had four students, working with four different experts, making expert systems on autism. When our primary autism expert, Linda Swisher of our Speech and Hearing Department, ran one of those expert systems she exclaimed "That lady is no expert! I mean, the rules are correct, but I can see what books she got them from." Swisher was right. The student had interviewed a pediatric neurologist who, being too busy, volunteered the services of her resident—a very intelligent professional but not an autism *expert*. Swisher correctly noted such lack of real expertise in two student-generated expert systems.

Our experts enjoyed participating and often told us so. "Working through these production rules," Linda Swisher observed, "and seeing the decision tree of a consultation made me realize what I actually do during a consultation. If nothing else, this experience will help me teach better in the future. What I have been telling my students to do for the last 15 years is not really what I do myself." As has been noted before, expressing a knowledge base can lead experts to clearer understandings of their own expertise.[11]

We were surprised to find the best (or what seemed to us the best) expert systems were written not by our systems engineering students, but rather by psychology, communications, and business majors. Moreover, our brilliant foreign students did not produce the best expert systems; interacting with their programs was far too difficult, leading us to conclude that, for making expert systems, communication skills are at least as important as engineering analysis skills. This conclusion complements advice given by Feigenbaum and Davis in the AI video conference broadcast November 13, 1985: They described ideal knowledge engineers (1) as reductionists who love to wade through myriad rules, heuristics, hunches, and intuitive notions, digesting them into a production-rule nutshell, (2) as aggressive and not afraid to take a risk, and (3) as engineers who enjoy poking their noses into other people's business. For that is what knowledge engineering is about, after all—to go into someone else's domain, learn it, cast this knowledge into an expert system, then a half a year later move off into another person's domain and repeat the process.

If they'd had time to think about it, we suspect Feigenbaum and Davis would also have mentioned the need for communications skills that aid in understanding problems, dealing with experts, and making systems that work.

Planned improvements

We want to improve interviewing techniques. Half of our students' time went into gathering knowledge from human experts and translating that knowledge into if-then production rules. The few students who were their own experts, or who got their knowledge out of textbooks, were deprived of a valuable experience.

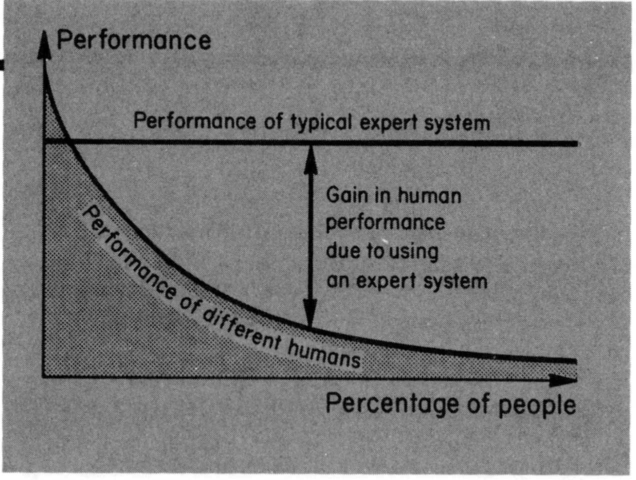

Figure 1. Problems having a performance histogram like this are ideal for expert system technology. An expert system's purpose is not to replace humans, but to increase mass productivity; systems should perform almost as well as human experts, as the figure shows.

We will acquire an induction system such as Expert-Ease, 1st-Class, or KDS because they're different in that, instead of requiring if-then production rules, they accept examples and case studies and then derive production rules. However, we're not sure it will be easier for students to generate adequately general examples than to provide the rules.

Not surprisingly, students were reluctant to face dealing with uncertain data and vague rules—expert system requirements. We are similarly reluctant. But expert systems must incorporate capacities to do so if they are to deal effectively with real problems.

To deal with imprecision and vagueness, M.1 uses certainty factors. Chapters 5-7 of Forsyth's book (rated by our students as the best chapters) helped us broaden the discussion of uncertainty by including Bayesian probability updating and fuzzy sets. In future classes, we'll spend more time dealing with imprecision, uncertainty, and their effects on solutions. Although no dogma leads to salvation (these matters being perennial problems), appreciating them thoroughly is, we think, prerequisite for designing good expert systems.

After studying the 22 student-built expert systems, we realized that students do not intuitively build good human/computer interfaces allowing easy and effective communication. In fact, they are not adequately aware of the problems involved. For example, most systems require users to enter large amounts of data. When requests for data come in long lists of questions, each presented as a separate frame without comment or explanation, users have no sense of control, no idea of where the consultation is going, and no idea of why the information is needed. When questions jump from one context to another, users think the computer is scatterbrained. For example, would you trust a computer that asked you, in order, "What is your height? What is your weight? What is the capital of Kurdistan?"

We can ameliorate these problems by (1) entering data in related blocks with visual prompts, as when filling out a form, (2) ordering questions so that meaningful preliminary conclusions can be drawn and presented to users (for example, "That rules out such and such"), and (3) removing interrogative redundancy (if you ask a child's age, for instance, you should not subsequently ask if the child is past the age of puberty, or if the child is of preschool age). In the future, our expert-system course will deal more fully with such problems. But a companion course in human/machine interaction would be better. Expert systems should be designed so that users almost feel they are conversing with an intelligent human.

Finally, we found our students all too willing to join in uncritical enthusiasm for expert-system technology. We presented problems and criticism, and emphasized negative as well as positive points of view—but we will pursue this approach more systematically next time. We plan to assign and discuss such readings as articles by Parnas[12] and by Dreyfus and Dreyfus.[13]

Specific suggestions about M.1

If you want users to choose menu items, then identify those items with letters instead of numbers because M.1 treats numbers strangely; in addition, many experienced typists enter the lowercase letter "l" instead of the numeral "1."

Your system will be more user friendly if your last menu entry is "none of the above." Although this will create more work for knowledge engineers, it will accommodate inadvertent menu displays.

The consultation's beginning will probably give users instructions that will then disappear forever. It might be nice to provide an optional review of these instructions, perhaps

A list of the systems our students produced.

Program	Commentary
autism	Help a psychiatrist diagnose autistic children.
autism2	Help a neurologist diagnose autistic children.
AUTIS	Help a special education field worker diagnose autistic children.
ESIAC	Help diagnose autistic children based on speech behavior.
stutter	Detect disfluent (stuttering) children and suggest prognosis.
ANES	Aid an anesthesiologist during surgery.
Chromie	Congenital chromosomal defect diagnosis system.
Cogito	Help install 4.2BSD Unix on a VAX computer or add new devices (this project was expanded into a masters thesis and is a useful expert system; if you would like to use the system, please contact the authors).
diagnosis	Discover cause of failure for RS232 terminal/computer interface (written in OPS5).
labdes	Help design a PC laboratory.
invest	Help develop an individual investment portfolio.
solar	Help design solar-energy home-remodeling plans.
STATCON	Aid a biomedical engineer in selecting BMDP statistical-analysis programs.
Rockbolt	Help design rockbolt support systems for coal mines.
plans	Process planner and operations scheduler for a machine shop.
schedule	Find the best scheduling rule for a job shop.
advice	Help SIE graduate students formulate study plans (a routine, but very complicated problem; this expert system used M.1, C, C_to_dBase hooks, and dBaseII).
major	Help incoming freshmen choose a major (unsuitable for M.1 because it was just a question-and-answer session).
Diplomacy	Help a person to play the game of Diplomacy.
backgammon	Offer advice about the best move for backgammon (unsuitable for M.1 because it primarily involved numerical computation, not symbolic manipulation).
EXSYS	Identifies problems where an expert system is appropriate (unsuitable for an expert system because there is no human expert who does this much better than everyone else).
bid	Select correct bid in a bridge game (unsuitable for M.1 because it was a simple table lookup).

by labeling the banner knowledge base entry of your end-user system as follows:

```
instructions:
configuration(banner)=
['Welcome to your M.1 advisor.
Type "list instructions." to get this message.
Type "help." to get M.1"s help message.
Remember to end all your answers with a period and a
return.
Good luck!',nl].
```

When the semester began, we were not sure what we would teach in this experimental course on expert systems; however, the course evolved nicely. We ended up using the MIT video tapes for nine hours, and the Teknowledge M.1 instructor notes for 10 hours. Guest lectures occupied four hours, Forsyth's book nine hours, Harmon and King's book five hours, and other material took eight hours. Although it seems a hodgepodge, everything fit together well and (even if we say so ourselves) it proved satisfactory. Our students seemed to agree—they rated the course as "very good" on the CIEQ course evaluation questionnaires. ■

Acknowledgement

Grants from Bell Communications Research and the AT&T Foundation supported this course and related research.

References

1. R. Davis, "Expert Systems" videotapes, MIT, Cambridge, Mass., 1984.
2. R. Forsyth, ed., *Expert Systems: Principles and Case Studies,* Chapman and Hall, New York, N.Y., 1984.
3. P. Harmon and D. King, *Expert Systems: Artificial Intelligence in Business,* John Wiley and Sons, New York, N.Y., 1985.
4. E. Tello, "Knowledge Systems for the IBM PC," *Computer Language,* July 1985, pp. 71-83, and Aug. 1985, pp. 87-102.
5. J. Goldberg, "Experts on Call," *PC World,* Sept. 1985, pp. 192-201.
6. *Expert Systems,* Vol. 2, 1985, pp. 188-265.
7. B.G. Buchanan and E.H. Shortliffe, *Rule-Based Expert Systems,* Addison-Wesley, Reading, Mass., 1984.
8. F. Hayes-Roth, D.A. Waterman, and D.B. Lenat, *Building Expert Systems,* Addison-Wesley, Reading, Mass., 1983.
9. L. Brownston, R. Farrell, and E. Kant, *Programming Expert Systems in OPS5,* Addison-Wesley, Reading, Mass., 1985.
10. B. Thompson and W. Thompson, *Microexpert,* McGraw-Hill, New York, N.Y., 1985.
11. M.J. Horvath, C.E. Kass, and W.R. Ferrell, "An Example of the Use of Fuzzy-Set Concepts in Modeling Learning Disability," *Am. Educational Research J.,* Vol. 17, No. 3, 1980, pp. 309-324.
12. D.L. Parnas, "Software Aspects of Strategic Defense Systems," *Am. Scientist,* Vol. 73, No. 5, Sept.-Oct. 1985, pp. 432-449.
13. H. Dreyfus and S. Dreyfus, "Why Computers May Never Think Like People," *Technology Rev.,* Vol. 89, 1986, pp. 42-61.

A. Terry Bahill is a professor of systems and industrial engineering at the University of Arizona at Tucson. His research interests include control theory, modeling physiological systems, head and eye coordination of baseball players, expert systems, and computer text and data processing. He is the author of *Bioengineering: Biomedical, Medical, and Clinical Engineering* (Prentice-Hall, 1981). He received his BS from the University of Arizona and his MS from San Jose State University in electrical engineering, and his PhD in electrical engineering and computer science from the University of California at Berkeley.

In addition to being on the *IEEE Expert* editorial board, Bahill is a member of several IEEE societies including Engineering in Medicine and Biology, Automatic Controls, Professional Communications, and Systems, Man, and Cybernetics. He was vice president for publications, and is now vice president for meetings and conferences and an associate editor for the Systems, Man, and Cybernetics Society. He is a member of Tau Beta Pi, Sigma Xi, and Psi Chi.

William R. Ferrell, professor of systems and industrial engineering at the University of Arizona, teaches courses in human factors and in mathematical modeling of human performance. He earned his BA with honors in English literature at Swarthmore College. He studied mechanical engineering at MIT where he earned his SB (also with honors), his SM, ME, and PhD. For four years during that period he worked in machine and product design at Polaroid. From 1962 to 1969, he taught at MIT where he was ultimately associate professor and codirector of the Man-Machine Systems Laboratory in the Mechanical Engineering Department.

His research has covered a wide range of human performance: driver behavior, aids for the handicapped, remote manipulation, robotics, human information processing, subjective judgment, and expert systems. He has more than 30 publications in such areas and, with T.B. Sherman, is author of *Man-Machine Systems: Information, Control, and Decision Models of Human Performance* (MIT Press).

Ferrell is a member of the administrative committee of the IEEE Systems, Man, and Cybernetics Society, a fellow of the Human Factors Society, a founder and currently a director of its Arizona chapter, and a member of Sigma Xi.

The authors' address is the Systems and Industrial Engineering Dept., University of Arizona, Tucson, AZ 85721.

Appendix: Further Reading

Schutzer, D., *Artificial Intelligence: An Applications Oriented Approach,* New York, NY, Van Nostrand Reinhold, 1987. (An introduction for the non-technical manager. Also offers an extensive bibliography.)

Barr, A., Feigenbaum, E.A., and Cohen, P.R., *The Handbook of Artificial Intelligence; Vol. I, II, & III,* Reading, MA, Addison-Wesley, I: 1981, II & III: 1982. (A widely recognized desk reference. Very useful to the person who has some knowledge of computer science. Offers detailed introductions to specific topic areas. Each topic area has its own bibliography.)

Gevarter, W.B., *An Overview of Artificial Intelligence and Robotics,* Franconia, VA, National Technical Information Service, 1983. (These are government reports available at little cost to the public. They are well written and useful to the beginner and the experienced person. Request the following document numbers: N83-24193, PB83-217554, N84-10834, PB84-178037, N84-14805, PB83-217547, and PB83-217562.)

Hayes-Roth, F., Waterman, D.A., and Lenat, D.B., *Building Expert Systems,* Reading, MA, Addison-Wesley, 1983. (Designed to provide a broad introduction to the concepts and methods of expert systems. Contains lengthy examples. For the person new to the field.)

Klahr, P., and Waterman, D. A., [Eds.], *Expert Systems: Techniques, Tools, and Applications,* Reading, MA, Addison-Wesley, 1986. (Discusses current techniques, methods, and uses.)

Hamming, R.W., and Feigenbaum, E.A., *Problem-Solving Methods in Artificial Intelligence,* New York, NY, McGraw-Hill, 1971. (Once the reader has covered introductory material, this book is a good follow-up discussion of the major AI reasoning methods.)

Brown, D., and Chandrasekaran, B., *Design Problem Solving: Knowledge Structures and Control Strategies,* Palo Alto, CA, Morgan Kaufmann, 1986. (Outlines problem-solving using expert systems and a general theory of knowledge-based reasoning. Illustrates an example expert system.)

Naylor, C., *Build Your Own Expert System,* New York, NY, Halstead Press, 1985. (For the beginner, this book starts very simple and goes through all the basics. Brings the reader from a good foundation to the realization of an actual system written in generic Basic.)

Krutch, J., *Experiments in Artificial Intelligence for Small Computers,* Indianapolis, IN, Howard W. Sams, 1981. (For the beginner, this book discusses basic principles and then illustrates them with short programs written in generic Basic. Each chapter covers one principle and gives one program.)

About the Author

Peter G. Raeth is a senior Air Force Captain stationed with the Wright Research and Development Center in Dayton, Ohio. He holds a M.S. in computer engineering from the Air Force Institute of Technology and a B.S. in electrical engineering from the University of South Carolina. Mr. Raeth manages and conducts research in artificial intelligence as applied to signal detection, classification, and identification. He is a member of Tau Beta Pi, Eta Kappa Nu, Omicron Delta Kappa, the IEEE Computer Society, and the Association of Old Crows.

Other IEEE Computer Society Press Texts

Monographs

Integrating Design and Test: Using CAE Tools for ATE Programming:
Written by K.P. Parker
(ISBN 0-8186-8788-6 (case)); 160 pages

JSP and JSD: The Jackson Approach to Software Development (Second Edition)
Written by J.R. Cameron
(ISBN 0-8186-8858-0 (case)); 560 pages

National Computer Policies
Written by Ben G. Matley and Thomas A. McDannold
(ISBN 0-8186-8784-3 (case)); 192 pages

Physical Level Interfaces and Protocols
Written by Uyless Black
(ISBN 0-8186-8824-6 (case)); 240 pages

Protecting Your Proprietary Rights in the Computer and High Technology Industries
Written by Tobey B. Marzouk, Esq.
(ISBN 0-8186-8754-1 (case)); 224 pages

Tutorials

Ada Programming Language
Edited by S.H. Saib and R.E. Fritz
(ISBN 0-8186-0456-5); 548 pages

Advanced Computer Architecture
Edited by D.P. Agrawal
(ISBN 0-8186-0667-3); 400 pages

Advanced Microprocessors and High-Level Language Computer Architectures
Edited by V. Milutinovic
(ISBN 0-8186-0623-1); 608 pages

Communication and Networking Protocols
Edited by S.S. Lam
(ISBN 0-8186-0582-0); 500 pages

Computer Architecture
Edited by D.D. Gajski, V.M. Milutinovic, H.J. Siegel, and B.P. Furht
(ISBN 0-8186-0704-1); 602 pages

Computer Communications: Architectures, Protocols and Standards (Second Edition)
Edited by William Stallings
(ISBN 0-8186-0790-4); 448 pages

Computer Grahics (2nd Edition)
Edited by J.C. Beatty and K.S. Booth
(ISBN 0-8186-0425-5); 576 pages

Computer Graphics Hardware: Image Generation and Display
Edited by H.K. Reghbati and A.Y.C. Lee
(ISBN 0-8186-0753-X); 384 pages

Computer Grahics: Image Synthesis
Edited by Kenneth Joy, Max Nelson, Charles Grant, and Lansing Hatfield
(ISBN 0-8186-8854-8 (case)); 384 pages

Computer and Network Security
Edited by M.D. Abrams and H.J. Podell
(ISBN 0-8186-0756-4); 448 pages

Computer Networks (4th Edition)
Edited by M.D. Abrams and I.W. Cotton
(ISBN 0-8186-0568-5); 512 pages

Computer Text Recognition and Error Correction
Edited by S.N. Srihari
(ISBN 0-8186-0579-0); 364 pages

Computers for Artificial Intelligence Applications
Edited by B. Wah and G.-J. Li
(ISBN 0-8186-0706-8); 656 pages

Database Management
Edited by J.A. Larson
(ISBN 0-8186-0714-9); 448 pages

Digital Image Processing and Analysis: Volume 1: Digital Image Processing
Edited by R. Chellappa and A.A. Sawchuk
(ISBN 0-8186-0665-7); 736 pages

Digital Image Processing and Analysis: Volume 2: Digital Image Analysis
Edited by R. Chellappa and A.A. Sawchuk
(ISBN 0-8186-0666-5); 670 pages

Digital Private Branch Exchanges (PBXs)
Edited by E.R. Coover
(ISBN 0-8186-0829-3); 400 pages

Distributed Control (2nd Edition)
Edited by R.E. Larson, P.L. McEntire, and J.G. O'Reilly
(ISBN 0-8186-0451-4); 382 pages

Distributed Database Management
Edited by J.A. Larson and S. Rahimi
(ISBN 0-8186-0575-8); 580 pages

Distributed-Software Engineering
Edited by S.M. Shatz and J.-P. Wang
(ISBN 0-8186-8856-4 (case)); 304 pages

DSP-Based Testing of Analog and Mixed-Signal Circuits
Edited by M. Mahoney
(ISBN 0-8186-0785-8); 272 pages

End User Facilities in the 1980's
Edited by J.A. Larson
(ISBN 0-8186-0449-2); 526 pages

Fault-Tolerant Computing
Edited by V.P. Nelson and B.D. Carroll
(ISBN 0-8186-0677-0 (paper) 0-8186-8667-4 (case)); 432 pages

Gallium Arsenide Computer Design
Edited by V.M. Milutinovic and D.A. Fura
(ISBN 0-8184-0795-5); 368 pages

Human Factors in Software Development (Second Edition)
Edited by B. Curtis
(ISBN 0-8186-0577-4); 736 pages

Integrated Services Digital Networks (ISDN) (Second Edition)
Edited by W. Stallings
(ISBN 0-8186-0823-4); 404 pages

(Continued on inside back cover)

Interconnection Networks for Parallel and Distributed Processing
Edited by C.-l. Wu and T.-y. Feng
(ISBN 0-8186-0574-X); 500 pages

Local Network Equipment
Edited by H.A. Freeman and K.J. Thurber
(ISBN 0-8186-0605-3); 384 pages

Local Network Technology (3rd Edition)
Edited by W. Stallings
(ISBN 0-8186-0825-0); 512 pages

Microprogramming and Firmware Engineering
Edited by V. Milutinovic
(ISBN 0-8186-0839-0); 416 pages

Modern Design and Analysis of Discrete-Event Computer Simulations
Edited by E.J. Dudewicz and Z. Karian
(ISBN 0-8186-0597-9); 486 pages

New Paradigms for Software Development
Edited by William Agresti
(ISBN 0-8186-0707-6); 304 pages

Object-Oriented Computing—Volume 1: Concepts
Edited by Gerald E. Peterson
(ISBN 0-8186-0821-8); 214 pages

Object-Oriented Computing—Volume 2: Implementations
Edited by Gerald E. Peterson
(ISBN 0-8186-0822-6); 324 pages

Office Automation Systems (Second Edition)
Edited by H.A. Freeman and K.J. Thurber
(ISBN 0-8186-0711-4); 320 pages

Parallel Architectures for Database Systems
Edited by A.R. Hurson, L.L. Miller, and S.H. Pakzad
(ISBN 0-8186-8838-6 (case)); 478 pages

Programming Productivity: Issues for the Eighties (Second Edition)
Edited by C. Jones
(ISBN 0-8186-0681-9); 472 pages

Recent Advances in Distributed Data Base Management
Edited by C. Mohan
(ISBN 0-8186-0571-5); 500 pages

Reduced Instruction Set Computers
Edited by W. Stallings
(ISBN 0-8186-0713-0); 384 pages

Reliable Distributed System Software
Edited by J.A. Stankovic
(ISBN 0-8186-0570-7); 400 pages

Robotics Tutorial (2nd Edition)
Edited by C.S.G. Lee, R.C. Gonzalez, and K.S. Fu
(ISBN 0-8186-0658-4); 630 pages

Software Design Techniques (4th Edition)
Edited by P. Freeman and A.I. Wasserman
(ISBN 0-8186-0514-0); 736 pages

Software Engineering Project Management
Edited by R. Thayer
(ISBN 0-8186-0751-3); 512 pages

Software Maintenance
Edited by G. Parikh and N. Zvegintzov
(ISBN 0-8186-0002-0); 360 pages

Software Management (3rd Edition)
Edited by D.J. Reifer
(ISBN 0-8186-0678-9); 526 pages

Software-Oriented Computer Architecture
Edited by E. Fernandez and T. Lang
(ISBN 0-8186-0708-4); 376 pages

Software Quality Assurance: A Practical Approach
Edited by T.S. Chow
(ISBN 0-8186-0569-3); 506 pages

Software Restructuring
Edited by R.S. Arnold
(ISBN 0-8186-0680-0); 376 pages

Software Reusability
Edited by Peter Freeman
(ISBN 0-8186-0750-5); 304 pages—

Software Reuse: Emerging Technology
Edited by Will Tracz
(ISBN 0-8186-0846-3); 392 pages

Structured Testing
Edited by T.J. McCabe
(ISBN 0-8186-0452-2); 160 pages

Test Generation for VLSI Chips
Edited by V.D. Agrawal and S.C. Seth
(ISBN 0-8186-8786-X (case)); 416 pages

VLSI Technologies: Through the 80s and Beyond
Edited by D.J. McGreivy and K.A. Pickar
(ISBN 0-8186-0424-7); 346 pages

Reprint Collections

Selected Reprints: Dataflow and Reduction Architectures
Edited by S.S. Thakkar
(ISBN 0-8186-0759-9); 460 pages

Selected Reprints on Logic Design for Testability
Edited by C.C. Timoc
(ISBN 0-8186-0573-1); 324 pages

Selected Reprints: Microprocessors and Microcomputers (3rd Edition)
Edited by J.T. Cain
(ISBN 0-8186-0585-5); 386 pages

Selected Reprints in Software (3rd Edition)
Edited by M.V. Zelkowitz
(ISBN 0-8186-0789-0); 400 pages

Selected Reprints on VLSI Technologies and Computer Graphics
Edited by H. Fuchs
(ISBN 0-8186-0491-3); 490 pages

Technology Series

Artificial Neural Networks: Theoretical Concepts
Edited by V. Vemuri
(ISBN 0-8186-0855-2); 160 pages

Computer-Aided Software Engineering (CASE)
Edited by E.J. Chikofsky
(ISBN 0-8186-1917-1); 132 pages